ENVIRONMENTAL SCIENCE
Living Within the System of Nature

ENVIRONMENTAL SCIENCE
Living Within the System of Nature

THIRD EDITION

Charles E. Kupchella

Western Kentucky University

Margaret C. Hyland

PRENTICE HALL
Englewood Cliffs, New Jersey 07632

Library of Congress Cataloging-in-Publication Data

KUPCHELLA, CHARLES E.
    Environmental science : living within the system of nature /
Charles E. Kupchella, Margaret C. Hyland. — 3rd ed.
        p.   cm.
    Includes bibliographical references and index.
    ISBN 0-13-282740-9
    1. Human ecology.   2. Power resources.   3. Pollution.
    4. Conservation of natural resources.   5. Environmental policy.
I. Hyland, Margaret C., (date).   II. Title.
GF41.K86   1992
363.7—dc20   92-15627
                                                        CIP

Acquistion Editor: David Kendrick Brake
Production Editor: Judy Winthrop
Marketing Manager: Kelly Albert
Prepress Buyer: Paula Massenaro
Manufacturing Buyer: Lori Bulwin
Cover Designer: Lisa Dominguez
Cover Art: (Background) Robert Winslow/Viesti Associates;
(Earth) Photo courtesy of NASA
Editorial Assistant: Mary DeLuca

© 1993, 1989, 1986 by Prentice-Hall, Inc.
A Simon & Schuster Company
Englewood Cliffs, New Jersey 07632

Printed in the United States of America

10 9 8 7 6 5 4 3 2 1

The credits section begins on page 566, which is considered an extension of the copyright page.

ISBN   0-13-282740-9

Prentice-Hall International (UK) Limited, *London*
Prentice-Hall of Australia Pty. Limited, *Sydney*
Prentice-Hall Canada Inc., *Toronto*
Prentice-Hall Hispanoamericana, S.A., *Mexico*
Prentice-Hall of India Private Limited, *New Delhi*
Prentice-Hall of Japan, Inc., *Tokyo*
Simon & Schuster Asia Pte. Ltd., *Singapore*
Editora Prentice-Hall do Brasil, Ltda., *Rio de Janeiro*

We dedicate this third edition to Mary Ann, Pat, Jane, Cory and Kyle
and to the ideal of a clean, healthy environment for them
and for all of the Earth's people.

Contents

part one

Basic Principles of Ecology

part two

Homo sapiens in the Scheme of Natural Things 109

7 Mineral and Water Resources _____ 180

8 Population, Food, and Hunger _____ 208

part three

The Impact of Human Activities on Health and the Environment   243

part four

Points of View

527

Preface

In the not too distant future, the environment will be required subject matter for all college and university students. Knowing about the dependence of humans on a healthy environment is at least as important to all college graduates as knowing about history, science, math, language, or the arts.

Environmental issues are already so predominately important throughout the world—so critical to the future of human civilization, that we simply cannot have a world whose leaders are ignorant of the dynamics of the interaction between humans and the environment.

The governor of a western state told a group of us recently that about half the issues he deals with day to day are environmental issues. In Kentucky, the legislature met in special session in 1991 to deal mainly with a crisis in solid waste disposal. Debris washing up on East Coast beaches illustrated the need to find better general ways of dealing with the solid waste problem. Air quality is a worsening problem throughout the world. Ozone in the ozone shield now appears to be decreasing even faster than originally feared. We face the threat of global warming as the levels of carbon dioxide and other "greenhouse" gases build up in the atmosphere. We continue to diminish our survival capital–fossil fuels, groundwater, and the earth's biodiversity. Each year the farmers of the world have to feed nearly 100 million more people on nearly 25 billion tons less topsoil.

Since our first edition was published, world population passed the five billion mark, an ozone hole was discovered, and there was a serious nuclear accident at Chernobyl in the Soviet Union. Tens of thousands of square miles of tropical rain forest continue to be cleared each year and wetlands continue to disappear at an alarming rate. Urban sprawl continues to overrun good farmland in the United States and our groundwater is becoming increasingly contaminated. Trees in Central Europe and in Eastern North America are dying; and lakes throughout the world are being afflicted by acid deposition.

Our prospects for dealing with these kinds of problems effectively are not good. Some of them are global problems whose solutions will require unprecedented international cooperation. All of them are complex problems whose solutions will require people able to grasp the big picture, able to appreciate and deal with complex problems and able to support strategies having long-way off impacts. We don't have many of these kinds of people in the world; our schools must generate more of them. We wrote this book to help make this happen.

We wrote this book because we believe that the seriousness of our global ecological problems demands a serious introductory environmental science textbook. Many environmental science textbooks seem to assume that students of environmental science are uninterested in facts and substance, that they want to be told how to think and what to do, that they wish simply to be moved to indignation by pictures of sewers and smokestacks. Our book takes our readers beyond the superficial.

This book is designed to be used by students headed for leadership positions in business, science, law, government, education, engineering, agriculture, and other fields, students who are taking their first environmental science course. Our book assumes no science background other than good secondary-level courses in biology and general science.

Our approach is human-centered. Our case for concern about the environment is that when the environment is harmed, humans are harmed, either now or in the future—or both. However, this book is based on the idea that the principles of ecology are the foundation upon which environmental science must be based. Our extended coverage of the interactions between humankind and the environment takes place on a solid framework of ecological principles. We also give heavy emphasis to the economic, social, political, ethical, and legal aspects of our environmental problems. All of these are important in environmental decision making.

xvii

There are no lists of simple solutions in this book. We want our readers to appreciate that most environmental problems are too complex for simple solutions. We have made every effort to present balanced, fair treatment of controversial topics, and we encourage our readers to formulate their own conclusions. One of our main objectives is to provide facts that students can use in making their own reasoned decisions. Some other aspects of our strategy and some distinctive features of this book are as follows.

Organization

The book is divided into four parts.

The principles of ecology are considered in Part One. Chapter 1 is a summary or preview chapter. It outlines, in broad brush fashion, how nature is organized and identifies energy and nutrient flow as the foundations of ecological interaction. Chapter 2 is devoted to **energy flow**; it begins with the **laws of thermodynamics**. Chapter 3 outlines the principal **biogeochemical cycles** and introduces the concept of **self-regulation in ecosystems**. Chapter 4 covers **population and community dynamics** in nature and ends with sections on **biomes** and on **ecological succession** designed to illuminate the dynamic nature of self-regulation of ecosystems. Succession, in turn, leads to Chapter 5 which introduces **evolution** as a kind of super-succession. In outlining how the ecosphere came to be, Chapter 5 uses evolution to drive home the point that the interactions between and among ecosystem elements are very powerful forces; they not only serve to regulate ecosystems, but also actually gave them their very form and shape.

Part Two does for human beings what Part One does for animals and plants in general; it explores the human need for energy and nutrients. It goes beyond the needs of individuals to examine the need for energy and mineral resources to support human civilization. Chapter 6 is a thorough review of our **energy resources past and present**. It looks at the environmental advantages and disadvantages and the long term prospects for **coal, oil, nuclear power, solar power** and for every other energy resource—and for **energy conservation**. Because energy use is a root cause of so many of our environmental problems, we cover energy in much more depth than most other environmental science textbooks. Chapter 7 is divided into two parts, one each for mineral and water resources. It presents world trends and imbalances in the use of **mineral resources** and strategies for increasing reserves and reducing consumption. Mining the deep sea and recycling are two of the strategies

that are examined. Supply and demand as it relates to the **water resources** of the world are also considered in this chapter. International water conflicts and concepts relating to water marketing are presented.

Chapter 8 is also divided into two sections, one each for population and food-production/hunger. It presents historical and projected data on **human population growth** and explores the concept of overpopulation and its relationship to lifestyle. The ability of the land to produce sufficient food to feed the growing world population is explored in topics including the **Green Revolution, sustainable agriculture, desertification,** and the promise of **biotechnology**.

Part Three could have been titled simply "Pollution"; it explores the impacts of human activities on air, water, land, and on our biological resources. Chapter 9 begins with an account of the classic **air-pollution disasters** as a means of illustrating the complex nature of air pollution; it then goes into a straightforward description of the **types and sources of air pollutants**. Chapter 10 describes the effects of air pollution on **human health**, on **vegetation**, on **materials** and on **climate**. Chapter 10 has an unusually detailed description of the health effects of air pollutants including a distinctive section on the science of **epidemiology** as well as special sections on **acid rain, the ozone hole,** and on **global warming**. Chapter 11 looks at the **economics of air pollution**—the costs of air pollution control and the cost of uncontrolled air pollution. Chapter 12 is devoted to **water pollution**. It outlines the types and sources of water pollution and pollution control technologies. Some notable "cases" of water pollution including **Lake Erie**, the **Mediterranean Sea**, and the **oceans** in general are considered in detail. **Groundwater** pollution and legal approaches to protecting water resources are also considered in Chapter 12.

Chapter 13 explores the uses and misuses of land. It explains the very basic concepts of **soil formation** and **soil conservation**. The impact of public policy on **land use** in urban and rural areas is discussed. The use of land for the long term storage of **solid, hazardous,** and **radioactive waste** is also examined.

In Chapter 14 we consider the impacts of human activities on biological resources. We discuss concerns about **threatened** and **endangered species** and loss of habitat such as **tropical rainforests** and **wetlands**. International efforts at species and habitat protection are presented. U.S. policy regarding species and habitat protection on public lands is reviewed in depth.

In Chapter 15 we look at **"Noise"** as a special kind of air pollution. In Chapter 16, 17, and 18 we examine the kinds of impacts of human activities that affect all the ecosphere in a general way. In Chapter 16 we examine the impacts of mis-spent **pesticides** and other persistent **hazardous materials**; we look at alternatives to the complete dependence on pesticides such as **integrated pest management** and at strategies for reducing the generation of hazardous materials. Chapter 17's focus is **solid and hazardous waste**. Here we look at the patterns of waste generation and at **strategies for minimizing waste**. Chapter 18, the final chapter of Part Three, examines the **environmental causes of cancer** and explores how we might develop strategies for preventing cancer.

Part Four consists of two editorial chapters. Chapter 19 examines the roles of **government** and other "institutions" in environmental problem-solving and in environmental problem-causing. Chapter 20 is an editorial summary of the entire book; it outlines our environmental problems, points to the hurdles to be overcome in dealing with these problems, and assesses our prospects for dealing with them effectively.

The book is designed to be used flexibly; the chapters can be taken up in any order.

Big-Picture Approach

One of the elements in our philosophy of teaching is that learning and appreciating details come easily if a framework is provided on which to hang them. This is reflected in the organization of the text and in the art program. Many of our illustrations are visual metaphors designed to give our readers working images to help keep the framework in mind.

Global Scope

We describe environmental problems using examples and illustrations from all over the world, and we explain the need for international approaches to global environmental problems. We make it clear that neither problems nor solutions are limited by national boundaries.

Among the international topics we consider in depth: pollution of the Mediterranean Sea (Ch 12); air pollution problems in Eastern Europe (Ch 20); energy use and energy resources throughout the world (Ch 6); Chernobyl (Ch 6); tropical rainforests (Ch 14); global warming and worldwide acid deposition problems (Ch 10); the future of Antarctica (Ch 7); and human population growth in Third-World countries (Ch 8).

Mini-Glossaries

We define key terms associated with particular topics in clusters called mini-glossaries. In these mini-glossaries, terms that might be confusing are clearly defined. All key terms are also given in **boldface** the first time they are defined in context, and the locations of all key terms are indicated by boldface in the **index**. We believe that presenting definitions in this way favors learning over memorization.

Enrichment Boxes

Throughout the text we present special topics that expand on points made in the text or that provide interesting asides. These special features range from an essay on how beavers influence the environment to an expanded explanation of the expression, "micrograms per cubic meter."

Environmental Career Profiles

All signs indicate that the world's attention will be focused ever more increasingly in the years ahead on the restoration and maintenance of environmental quality. In 1992 in the United States alone, well over $100 billion will be spent on environmental protection. Large numbers of well educated, well trained, and motivated college graduates will be needed to do the work of environmental protection in careers that will offer excitement, good pay, and satisfaction. In order to help students see the variety of environmental career possibilities, we have added a feature in this edition called "Environmental Career Profiles." There are a total of 15 people profiled in this edition, one at the end of each of the last 15 chapters. We have included people working for federal and state agencies, people working in environmental protection and environmental education; people working for private agencies such as The Nature Conservancy, The Audubon Society, and the National Wildlife Federation; and people working in private industry. We have included a wide variety of environmental jobs so that students might get some idea of the array of the possibilities.

For Further Thought ... Boxes

Things called "issues" usually have more than one perspective and/or more than one set of values through which perspectives are filtered. One of our principal objectives in writing this book was to help the next generation of leaders develop the ability to recognize, consider, and evaluate the various sides of issues—as a life-long skill. We have built this objective into the way we present various issues throughout this book. In order to carry this objective even further, we have also added a special feature titled, "For Further Thought ..." In each chapter there are boxed inserts in which specific questions are posed about timely environmental issues and topics such as the Gaia debate, the solidarity and stability of the environmental movement throughout the world; land-use policies; the need for energy conservation; the need to do something now about global warming; our responsibility to future generations; the importance of endangered species and the relative value of strategies designed to save endangered species.

Concepts to Remember

In a direct outline format we enumerate the major points at the end of each chapter so that students can easily review and reconsider the most important things in each chapter.

Discussion Questions and Food for Thought

The Instructor's Manual has a chapter by chapter feature called "Discussion Questions and Food for Thought" which is intended to help instructors help students to connect the text to the real world. This material is also intended to serve as a substrate for **critical thinking**.

Food for Thought also contains some tried and tested ideas for class discussions, outside speakers, and field trips.

Revised Features

We updated the entire book in this third edition. Many of our tables, text, and figures were updated and we have included several dozen completely new tables. Among the features revised in this edition:

1. an updated analysis of the nuclear accident at Chernobyl and the status of nuclear power worldwide.
2. new separate chapters on pesticides and hazardous materials, solid and hazardous waste, and

environmental cancer—topics covered in a single, long chapter in the second edition.
3. an analysis of the **Exxon Valdez** oil spill.
4. a presentation on the controversy over Old Growth Forests and the Spotted Owl in Western North America.
5. an expanded discussion of low-input, sustainable agriculture and the impact of biotechnology on agriculture.
6. an expanded discussion of birth control technologies.

We also made the book a bit shorter by condensing some of the material on environmental law, noise, and by eliminating some tables.

References and Further Reading

The articles and books listed at the end of each chapter are readily available in nearly all college libraries. They offer the reader a diversity of points of view and help to identify sources of current information.

Think Metric

Throughout the text, volumes, weights, and areas are often given in *both* English and metric measurements. Where only one system is used, the conversion charts in Appendix A will permit easy interconversion.

How This Book Came to Be

What kind of environmental science book does the world need? Our answer to this question was developed over a period of twenty years as we watched interest in the environment take hold throughout the world. From our unique combination of vantage points we were able to watch the environment develop and mature as a public policy issue and as educational subject matter from kindergarten through the graduate level. We have taught environmental science as a foundation for environmental studies majors, to medical students, and to college students taking it as a general education course. We have helped illuminate environmental issues for legislative groups and for other government officials. We have worked with teachers at the secondary and elementary school levels to make use of the environment as a means of making math, science, and other subjects more interesting, more relevant to their students. We have had countless discussions with colleagues from throughout the world about how best to teach environmental science. Through all these interactions, we have developed what we think is a practical sense of what it is that all college graduates need to know about the environment. This need became the basis of the outline and the approach for this book.

Our mosaic of experiences told us that an environmental science book needs a solid underpinning in the concepts of ecology—the basic rules governing the interactions of all animals and plants. These concepts are a necessary foundation for understanding human ecology. We have also come to believe that it is important that environmental science depict human beings as integral parts of nature rather than as destructive trespassers. It must show human beings as having the same basic needs as all other species. Even when showing the negative impacts of humans on the environment, this should be presented within the context of the tendencies on the part of all species to go further in exploiting their environment than nature will allow. We have been convinced that an environmental science book should not simply give students long lists of things they must believe and do; it should give them an understanding that lets them construct their own belief system.

Our experience persuaded us that environmental science books must help teach critical thinking. They should not present environmental issues as if they were black and white. Environmental issues should be presented as they *really* are—in shades of gray. We have also come to believe that a good environmental science book should have real data in it—so that students can practice interpreting real data, data that is sometimes inconclusive. We saw the need to write a book that demands that students be tested with essay questions rather than true/false questions.

Acknowledgments

We have many people to thank. The support and encouragement of our friends, including the members of our families, were crucial. The time we spent on this project was time we would have spent with them, yet they responded to this theft of time with genuine encouragement. We thank the federal, state and other public agencies that provided us with recent data, photographs, and other kinds of help. We thank in particular the following EPA officials who provided us

with much useful advice and information: Alan Basala, Sam Colon-Velez, Bruce Jordan, John Rasnick, and B.J. Steigerwald. We also thank Jean Almand of Western Kentucky University's Science Library for her considerable help in tracking down references. We acknowledge elsewhere the invaluable expert assistance provided to us by our reviewers. Some of these were colleagues we asked for help; others were identified by our editor. We very much appreciate the extra effort made by our editors in assembling an impressive group of helpful reviewers. Our editors managed to persuade an unusually large number of very busy people to provide us with timely, detailed, constructive criticism. We also acknowledge Keith Clement's help conceptualizing illustrations for the "For Further Thought ... Boxes". Typing the manuscript and otherwise getting it into final form over three editions required the able assistance of many typists and other kinds of help. We especially wish to thank:

Sharon Abney	Debbie Lynn
Donna Alexander	Patty Mann
Krista Button	Donna Marine
Keith Clements	Janice Melton
Marillyn Clements	Vicki Miller
Carol Holton	Angie Mobley
Vera Howerton	Joan Pedigo
Jane Hyland	Gail Raspberry
Mary Helen Hyland	Kaye Summers
Joyce Johnson	Lynn Summers
Susan Johnson	Janet Terry
Helga Keller	Pat Thomas
Michele Kupchella-Adams	Susan Vance
Jean Lynch	

We are also grateful to the very able production team at Prentice Hall. We would especially like to thank Dave Brake, Mary DeLuca, Barbara DeVries, and Judy Winthrop.

After you have been through the book, let us know what you liked or did not like about it. We would very much like to have your suggestions as to how we might make our book better as a teaching and learning tool.

Charles E. Kupchella, Dean
Ogden College of Science, Technology and Health
Western Kentucky University
Bowling Green, Kentucky 42101
Phone: 502-745-4448

Review and Development

No two people could write an environmental science text alone. The subject is far too broad for anyone to be an expert in even a fraction of the areas that constitute it. We needed and received help from many very capable people.

The individuals listed below reviewed various parts of the manuscript in this edition and in earlier editions and made numerous valuable suggestions, corrections, and comments. Some of these individuals were and are co-workers who served quadruple duty as reviewers, sounding boards, sources of inspiration, and providers of elusive key words. Some were selected by the editor because they are outstanding teachers of environmental science courses; they ably assisted in guiding the manuscript toward the intended target. Others were chosen by the editor for their expertise in particular subject areas; they evaluated individual chapters for accuracy, timeliness, and balance. No reviewer, however, had any control after submitting suggestions and criticisms. The responsibility for any errors that may remain is ours.

— **John W. Adams,** University of Texas at San Antonio, Earth and Physical Sciences

— **Robert Anderson,** Environmental Protection Agency

— **Kenneth B. Armitage,** University of Kansas, Systematics and Ecology

— **John Bachman,** Environmental Protection Agency

— **Linda R. Berg,** University of Maryland, Botany

— **Richard J. Borden,** College of the Atlantic, Human Ecology

— **J. Phillip Bromberg, Esq.,** Pittsburgh, Pennsylvania

— **David A. Brown,** Aquatic Research Consultants, Long Beach, California

— **William L. Brown,** National Academy of Sciences, Board of Agriculture

— **Carl A. Carlozzi,** University of Massachusetts, Fish, Game, and Wildlife Management

— **James E. Carrel,** University of Missouri—Columbia, Biological Sciences

— **Ken Carstens,** Murray State University, Anthropology

— **Ann S. Causey,** Auburn University, Biology

— **Bernard L. Clausen,** University of Northern Iowa, Biology

— **John D. Cunningham,** Keene State College, Environmental Studies

— **Edward J. Daniels,** Institute of Gas Technology, Energy Development Center

— **Edward J. DiPolvere,** National Association of Noise Control Officials, Trenton, New Jersey

— **Donald Duncan,** Murray State University, Physics

— **Frank Edwards,** Murray State University, Economics

— **Frederick I. Eilers,** University of South Florida, Biology

— **Robert Etherton,** Murray State University, Physics

— **Samuel Faust,** Rutgers University, Environmental Sciences

— **Franklin B. Flower,** Rutgers University, Environmental Sciences

— **Norma L. Fowler,** University of Texas–Austin, Botany

— **Hugo D. Freudenthal,** Holzmacher, McLendon and Murrell, P.C., Consulting Engineers, Environmental Scientists and Planners, Farmingdale, New York

— **Gene Garfield,** Murray State University, Political Science

— **Lowell L. Getz,** University of Illinois, Ecology

— **Ralph J. Gorton,** Lansing Community College, Natural Science

— **Richard Greenberg,** University of Louisville School of Medicine, Epidemiology

— **John P. Harley,** Eastern Kentucky University, Biological Sciences

— **Dardy Hassell,** Murray State University, Biology

— **Ross Hemphill,** Ohio State University, Agricultural Economics and Rural Sociology

— **Fred Hitzhusen,** Ohio State University, Agricultural Economics and Rural Sociology

— **Donald W. Humphreys,** Temple University, Biological Sciences

— **Charles E. Jenner,** University of North Carolina–Chapel Hill, Biology

— **Edward J. Kormondy,** University of Maine, Biology

— **Louis Krumholz,** University of Louisville, Biology

— **Larry Kupchella,** Indiana University of Pennsylvania, Chemistry

— **David Lincoln,** University of South Carolina, Biology

— **Timothy F. Lyon,** Ball State University, Natural Sciences

— **Gene Maddox,** Murray State University, Physics

— **Dwight Meyer,** Queensborough Community College, Biological Science

— **Richard L. Meyer,** University of Arkansas, Botany and Microbiology

— **W. Craig Meyer,** Los Angeles Pierce College, Earth Science

— **V. Monsour,** McNeese State University, Biological and Environmental Sciences

— **Condict Moore,** University of Louisville School of Medicine, Oncology

— **Conrad Moore,** Western Kentucky University, Physical Geography

— **Mark O. Morgan,** Rutgers University, Biology

— **Michael J. Mueller,** University of Oklahoma, Economics

— **John Mylroie,** Mississippi State University, Geology

— **Steven B. Oppenheimer,** California State University–Northridge, Biology

— **Edward O. Oswald,** University of South Carolina, Environmental Health Sciences

— **Glenn Parsons,** The University of Mississippi, Biology

— **Carol S. Pennenga,** Rutgers University, Environmental Science/Noise Technical Assistance Center

— **David Pimentel,** Cornell University, Entomology/Ecology

— **C. Everett Pitt,** Northern Montana College, Biology

— **Dennis H. Rainear,** Rainear Consulting Company/Environmental Law, Toledo, Ohio

— **Henry A. Raup,** Western Michigan University, Geography

— **Harold Robertson,** Murray State University, Mathematics

— **C. Lee Rockett,** Bowling Green State University, Biological Sciences

— **Elmer A. Rosauer,** Iowa State University, Materials Science and Engineering

— **William J. Rowland,** Indiana University, Biology

— **John P. Russell,** Western Kentucky University, Engineering

— **Mitchell Rycus,** University of Michigan, Agriculture and Urban Planning

— **Alan M. Schwartz,** St. Lawrence University, Environmental Studies

— **Manuel Schwartz,** University of Louisville, Physics

— **Michael R. Smith,** Illinois State University, Environmental Health

— **John Spratt, M.D.,** University of Louisville School of Medicine, Oncology

— **Frederick R. Steiner,** Washington State University, Horticulture and Landscape Architecture

— **B. A. Stout,** Texas A&M University, Agricultural Engineering

— **Donald R. Strong,** Florida State University, Biological Sciences

— **Donald Stucky,** Southern Illinois University—Carbondale, Plant and Soil Science

— **Myron F. Uman,** National Academy of Sciences, Environmental Studies Board

— **Steve White,** Murray State University, Biology

— **Donald R. Whitehead,** Indiana University, Biology

— **Duke Wilder,** Murray State University, Biology

— **Joe Winstead,** Western Kentucky University, Ecology

— **Sylvan H. Wittwer,** Michigan State University, Agriculture Experiment Station

— **Timothy Wood,** Wright State University, Biological Sciences

To the Student

We're glad you're here. Most of you, we hope, have come to this book and to this course because you have already decided to get involved in helping humankind find better ways of living in harmony with the system of nature. The world does need your help.

Half the world's people live in overcrowded, polluted cities. In Jakarta, Indonesia, only a fraction of the houses have piped-in water. Lagos, Nigeria, one of the world's fastest growing cities, essentially has no sewage system. Millions of children throughout the world are diseased and malnourished. In cities throughout the world, including U.S. cities like Denver and Los Angeles, eyes water, people have trouble breathing, and outdoor exercise is often unsafe. Dozens of U.S. cities have ozone levels well above the safe limit. Air quality is bad in cities worldwide.

While urban sprawl continues to overrun some of America's best farmland, many of the rest of the world's people live on land that is parched, overused, and abused. Sub-Saharan countries in Africa—Mauritania, for example—are in danger of being wiped off the face of the earth by human-accelerated desertification. Tens of thousands of square miles of tropical forest are burned off each year, changing the chemistry of our atmosphere and wiping out unknown numbers of species.

Misspent pesticides and hazardous chemicals contaminate our water, land, air, and food. Toxic materials leak from landfills and from illegal dump sites. As we write, a cholera epidemic rages in South America—caused by the contamination of coastal waters and drinking water by human waste. Thousands of lakes and acres of forests throughout North America and Europe are dying, apparently from the effects of acid rain. Precious resources grow scarce. The basic necessities—clean air, adequate food, fresh water and raw materials—are enjoyed by relatively few people. Yet over the next 30 years, the population equivalent of a thousand Clevelands will be added to the world.

The world needs more people in all walks of life who know why and how human life is dependent upon a healthy environment and who can help shape ways of living accordingly. A few of you will become scientists and technicians, working on technical solutions. Some of you will become lawmakers, judges, and government workers who will legislate and enforce solutions. Others will become philosophers, psychologists, artists, and teachers who will help the rest of us learn to live with the changes associated with the new solutions. But most important, all of you are already citizens in positions to demand that environmental problems be effectively addressed.

This book is intended to help you make environmentally-sound decisions in whatever roles you will play in society. One of the purposes of our book is to present the most basic principles of ecology as the foundation upon which all environmental problem-solving must be based. Another purpose is to illuminate the world's most important environmental problems and to describe the pros and cons of possible solutions to these problems. We present multiple sides of every issue, first of all because there are at least two sides to every issue, but also because we want to help you learn to deal more effectively with shades of gray. We all need to see environmental science as a way of finding the truth rather than as a simple list of scientifically-derived great truths that everyone must believe.

Although we have strong opinions on what should be done about our environmental problems, we have tried very hard to keep our preaching to a minimum. Instead, we offer facts and balanced presentations of opposing views—to give you a basis for making your own judgements and to allow you to practice making up your own minds.

How This Book Will Help You Learn

We have tried to make this book easy to study. Each chapter contains one or more mini-glossaries to define important concepts. A list of "concepts to remember" at the end of each chapter will help you review the most important points within the chapter. References and "suggested reading" at the end of each chapter will guide you to additional information about specific issues. We have identified sources that should be available in all college-level libraries. The end-of-chapter references also serve to point to the kinds of places to look for future information. Your instructor will be able to suggest ongoing sources of specific information as well. All of this is to say that we hope your interest in the environment will continue long after this course is completed. We have included a feature in this edition, called "Environmental Career Profiles" designed to help you consider the possibility of a career in environmental protection. A real person, engaged in some aspect of environmental protection, is profiled at the end of each of the last 15 chapters.

The problems that you will study in this course are not theoretical ones that will fade away after you graduate. For the rest of your life, you will be called upon, in your workplace, in your home, and as a citizen, to make up your mind about environmental issues. You are about to begin to build a strong foundation from which you will be able to do just that.

About the Authors

Charles E. Kupchella is the Dean of the Ogden College of Science, Technology and Health at Western Kentucky University in Bowling Green.

Since 1968, Dr. Kupchella has worked as a teacher and researcher in the environmental field. At Bellarmine College in the early 1970s he developed a course called "Man and His Environment" as an offshoot of the first Earth Day observation. During this time he also formulated an environmental education program at the King Center in Nazareth, Kentucky.

From 1973 to 1979, Dr. Kupchella served as Associate Director of the Cancer Center at the University of Louisville School of Medicine. His research at the University of Louisville was in tumor biology as part of a group engaged in a cancer control program associated with the plastics industry. He taught a course on the health effects of environmental contaminants and helped to develop and coordinate a course on the biology of cancer in the School of Medicine.

From 1979 to 1985, Dr. Kupchella was Professor and Chairman of the Department of Biological Sciences at Murray State University, Murray, Kentucky, where he taught a range of courses including Human Ecology. He is a member of Sigma Xi, the American Institute of Biological Sciences, the North American Association for Environmental Education, and the Kentucky Science and Technology Council. He is also chairman of the Basic Sciences Section of the American Association for Cancer Education and past president of the Kentucky Academy of Sciences and the Kentucky Association for Environmental Education. Dr. Kupchella is the author of the book *Dimensions of Cancer*. He received his Ph.D. in Biology from St. Bonaventure University.

Margaret C. Hyland has spent the last fifteen years as a teacher, practitioner, and researcher in the environmental field.

Since 1976, she has been associated with the Kentucky Legislative Research Commission: first as a Legislative Analyst responsible for policy research, fact finding, and drafting of legislation in the areas of agriculture, natural resources, and environmental protection and at present as Assistant Director responsible for long-range issues and public policy development.

In her legislative work, Ms. Hyland has kept up to date on environmental and natural resource issues at the federal, state, and local levels; briefed legislators on pending environmental laws; and helped to develop an ecology workshop for state legislators that explained basic ecological processes and showed how to apply biological concepts to environmental problems.

Prior to her work with the Kentucky State Legislature, Ms. Hyland was the Director of the Environmental Education Program at the King Center in Nazareth, Kentucky, where she developed, implemented, and evaluated environmental education programs for teachers, school groups, and citizen groups.

Ms. Hyland has also served on the Board of Directors of the North American Association for Environmental Education and the Steering Committee of the Environmental Studies Section of that organization. She is a charter member of the Kentucky Association for Environmental Education. She graduated from Ohio State University in 1973 with a Master of Science degree in Natural Resources, specializing in natural resource management and environmental education.

Ecology is the study of how the living and nonliving things in nature relate to one another. Ecologists have discovered many basic truths about these relationships and have forged these into principles that serve as a foundation for modern ecology and for environmental science in general. The basic principles of ecology must be the starting points in understanding any and all environmental problems. Part One of this book is devoted to a consideration of (1) the overall framework of structure and function in living systems, (2) the flow of energy through living systems, (3) the movement of chemicals within living systems, (4) the basic patterns of interaction between members of the same species and between members of different species, and (5) the basic patterns of change in living systems.

part one

Basic Principles of Ecology

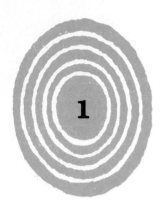

1 The Framework of Ecology

At first glance, nature seems almost hopelessly complex. The webs that connect every living thing to every other living and nonliving element of our world are made of countless finely drawn and far-reaching threads. Everything is connected to everything else. The poet Francis Thompson went so far as to suggest that when one touches a flower, a star is disturbed. He was essentially correct.

Consider the task we would face if, in order to understand nature, we had to be familiar with *all* of its details. To appreciate completely how a rabbit relates to its world, for instance, we would have to take into account how each rabbit relates to other rabbits, and to owls, grass, and humans as well. We would also have to consider how physical and chemical environmental changes affect each rabbit relationship, subtle or dramatic, human-made or otherwise. To appreciate how the entire natural world functions, we would have to do this for the millions of species, subspecies, varieties, and subtypes of living and nonliving things—an impossible task.

Fortunately, there are patterns in nature. Nature functions as one big system made up of countless little systems, all with similar basic parts organized in the same basic patterns of interaction. The key to understanding the natural world is to recognize that the same types of activities go on everywhere in nature. The big picture will be our focus in Chapter 1. Our aim is to depict the living world in its simplest terms, stripped of the differences that distinguish a desert from a jungle, a forest from a prairie, or a rabbit from a mouse.

Levels of Organization in Nature and the Scope of Ecology

Levels of Organization in Nature

Figure 1.1 illustrates the levels of organization in nature. Shown here are the subjects studied by chemists, physicists, geologists, biologists, hydrologists, astronomers, sociologists, and political scientists. Although ecology technically encompasses nearly all of the levels depicted, ecologists concern themselves mostly with those levels above the individual organism. **Ecologists** study how organisms interact with one another and with the nonliving environment.

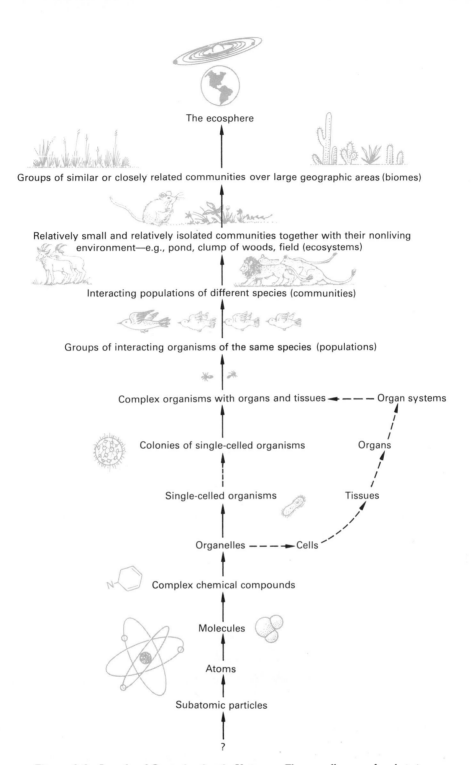

The ecosphere

Groups of similar or closely related communities over large geographic areas (biomes)

Relatively small and relatively isolated communities together with their nonliving environment—e.g., pond, clump of woods, field (ecosystems)

Interacting populations of different species (communities)

Groups of interacting organisms of the same species (populations)

Complex organisms with organs and tissues ◄ — — — Organ systems

Colonies of single-celled organisms   Organs

Single-celled organisms   Tissues

Organelles — — — ► Cells

Complex chemical compounds

Molecules

Atoms

Subatomic particles

?

Figure 1.1 Levels of Organization in Nature. The overall scope of ecology is broad, but its core begins just above the level of the individual organism and extends up through the ecosphere to include all of the plants and animals of the earth and the parts of the nonliving environment with which these interact. Ecology is the study of what goes on in the living systems called ecosystems.

Ecology, the Study of Ecosystems

Ecology is the study of the structure and function of nature. Ecology can also be defined as the study of ecosystems. **Ecosystems** are self-regulating communities of different kinds of living creatures in-teracting with one another and with their nonliving setting. The words *ecology* and *ecosystem* come from the Greek *oikos,* a word meaning "house" or "place of residence." Ecology deals mainly with the roles filled by organisms in nature and how environmental

Beavers

While it is easy to imagine how the environment can influence an individual organism, it may be a little more difficult to appreciate (at least for organisms other than humans) the extent to which an individual organism affects its environment. Beavers offer a particularly good example of how much organisms can influence their surroundings.

Beavers build dams that hold streams in check. These dams provide a steadier water supply (and a diminished threat of drought), reduce erosion, and somewhat stabilize silt and sediment loads. The water behind the dams provides a habitat for fish, turtles, ducks, and many other aquatic organisms. Beaver dams spread out and divert water, creating marshes and other wetlands. This can increase humidity and soil moisture over a large area and have a profound impact on vegetation.

Beavers carry wood into the water. Some kinds of wood are rich in nitrogen and other potentially scarce nutrients. In this and other ways, beaver dams tend to release steady amounts of nutrients to the waters downstream, actually making it possible for certain organisms to live in the water.

Beavers cut down certain trees, making it possible for the sun to shine through the canopy to reach the forest floor. This in turn influences the humidity in the forest, increases air temperature, and increases the average temperature of the forest soil surface. Warming of the soil influences the insect and microbial populations in the soil. The reduction in shade allows sun-loving plants to grow on the forest floor and lets shade-intolerant trees like pine and spruce grow.

---

conditions affect and are affected by these roles (see Enrichment Box 1.1).

The **ecosphere** is the grand system that includes *all* life forms on earth together with the parts of the earth on which, and in which, living things exist: the **atmosphere**—the earth's gases; the **hydrosphere**—the earth's water; and the **lithosphere**—the earth's rock and soil. *Biosphere* is commonly used as a synonym for *ecosphere,* but it makes more sense for *biosphere* to mean all of the plants and animals on earth. The ecosphere could then be defined as the biosphere plus those parts of the hydrosphere, atmosphere, and lithosphere in which, and with which, plants and animals interact (see Figures 1.2 and 1.3). Living things tend to be found at the junctions of these three "great spheres" (Figure 1.3).

Ecosystems are real systems like ponds, woodlots, and fields, but the ecosystem is also an abstraction. Just as the triangle is a generalized concept applicable to many specific triangles, the ecosystem concept is a generalization about natural systems that grew out of observing what goes on in ponds, woodlots, and fields. An ecosystem has no particular size. In a sense, asking how big an ecosystem is would be like asking how big a city government, an engine, or a triangle is. All of these things are abstractions. A pond is an ecosystem and so is a forest.

The ecosphere is the system of the earth's land, air, and water and all of its life forms.

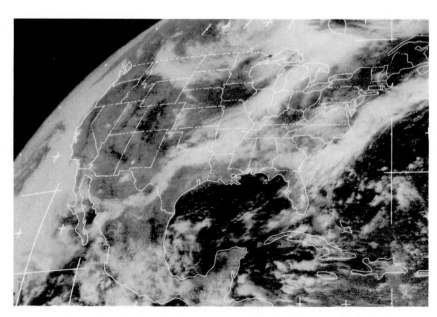

Ecology: the study of the structure and function of ecosystems, dealing mainly with the interaction of organisms with one another and with their nonliving setting.

Ecosphere: all of the living things on earth together with the part of the nonliving world in which, and with which, they interact.

Ecosystem: a self-regulating community of plants and animals interacting with one another and with their nonliving environment.

System: any collection of interrelated parts that form a functional, defined whole.

Mini-glossaries such as this one are scattered throughout the book to highlight and show the relationship between key terms. Words and concepts found in mini-glossaries, along with many other terms indicated in **bold type,** are also defined in context throughout the text. Some of the terms introduced in this chapter are defined more completely in Chapters 2, 3, and 4.

Figure 1.2 The Earth's Great Spheres. The ecosphere comprises all living organisms and the parts of the atmosphere, lithosphere, and hydrosphere that are reached by living organisms. Other parts of each of the latter three spheres are relatively inaccessible to living things.

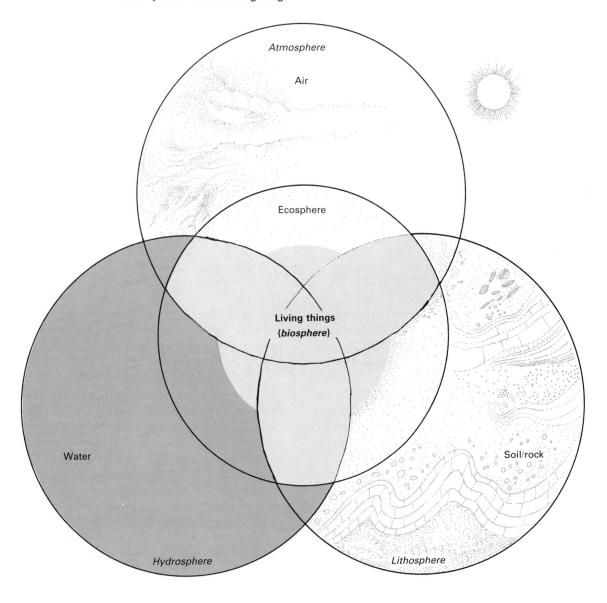

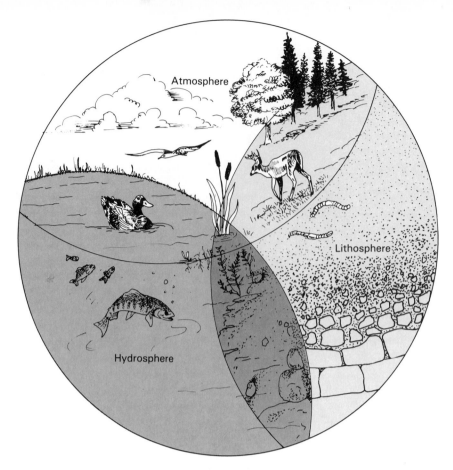

Figure 1.3 The Junctions of the Earth's Great Spheres. Most life on earth is found where land, air, and water meet. This tells us something fundamental about life. Living things are agents that need and use and—in the process—mix, maneuver, and manipulate the elements in air, soil, and water.

No two ecosystems are exactly alike, just as no two city governments are alike.

The Structure of Ecosystems

Consider a pond and an abandoned field (Figures 1.4 and 1.5). The sun is the source of energy for both ecosystems, and both contain plants that provide food. In a pond, microscopic algae carry out most of the food making; in a field, grasses and other plants do precisely the same thing. Both ponds and meadows contain animals that eat plants. In a pond, certain fish and insects, and snails eat plants; in a field, mice, other kinds of insects, and rabbits do the plant eating. Both ponds and meadows contain animals that eat the animals that eat plants. Snapping turtles and bass eat the pond's plant eaters; in a field this same ecological role is filled by hawks and foxes.

There are actually four basic components of all ecosystems:

1. Food makers, or **producers**

2. **Consumers,** or eaters of plants or animals
3. A special class of consumers that gets food from decaying plants and animals, the **decomposers**
4. Nonliving components

The first three of these are **biotic** (living) components; the nonliving things are called **abiotic.** Let's consider the abiotic components first.

Nonliving Components of Ecosystems

The nonliving parts of ecosystems have both physical and chemical features. Physical features include wind, terrain, soil moisture, water current, temperature, soil porosity, and light level. The chemicals of the nonliving environment make up all the materials that are not, for the moment, a part of an animal or plant—water, gases such as oxygen, minerals such as iron and sulfur, compounds such as acids, and a wide variety of other complex chemicals.

A chemical can be part of a living thing at one moment and part of the nonliving environment a moment later. The carbon in a molecule of protein can

6

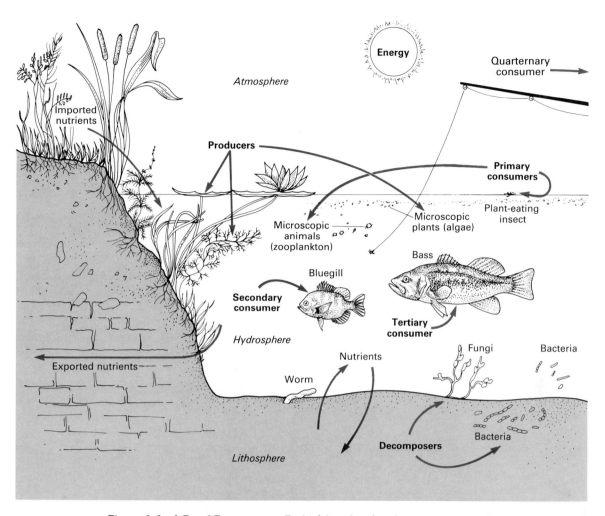

Figure 1.4 A Pond Ecosystem. Each of the roles of producer, consumer, and decomposer is filled by a number of different organisms in a pond ecosystem. For example, additional secondary or tertiary consumers might be water snakes, snapping turtles, and various birds of prey. Although ecosystems are often thought of as closed systems, none of them really are closed. Typically, both living and nonliving things are imported and exported.

Nonliving components such as water, minerals, light, weather, and terrain help to fashion forest and stream ecosystems. The living components in turn affect the physical and chemical features.

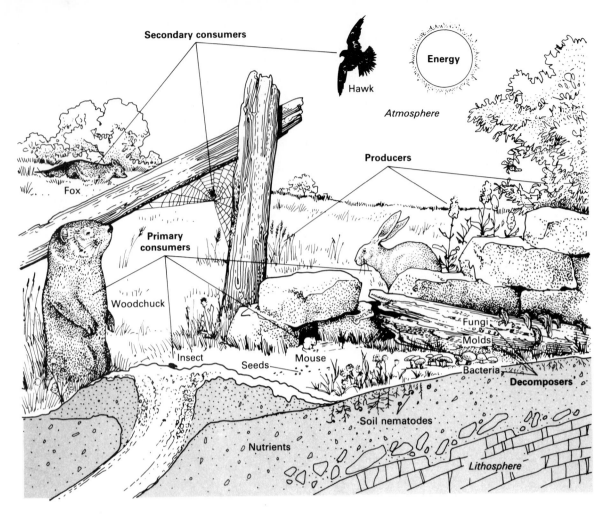

Figure 1.5   A Field Ecosystem.   Each of the roles of producer, consumer, and decomposer is filled by a number of different organisms in a field ecosystem. The same fundamental relationships exist between these classes of organisms in every ecosystem.

be part of a functioning enzyme of some animal, but after the enzyme is broken down and reduced to its component parts, that same carbon atom could be exhaled as a molecule of carbon dioxide and thus become a part of the *gaseous* abiotic environment. Chemicals move into and out of living organisms and are used over and over again. Some of the carbon atoms forming a protein molecule in the muscle of your left arm may have once been part of a chicken liver, the hide of a dinosaur, or even a limestone formation.

The importance of chemical substances to living systems varies with the type, location, and form of the chemical. Certain chemicals—carbon, for example—make up much more of the structures of animals and plants and play a far greater role in their activities than do other chemicals—for example, copper. Some chemicals may lie dormant deep within the earth's crust for millions of years, so living organisms do not have access to them; they may be

too far removed from the junctions of the great spheres (Figure 1.3). Certain other substances, because of their physical or chemical forms, may be inaccessible to living things even when they are in continuous contact with a variety of organisms. Only a small fraction of the earth's chemicals exist in forms usable by living things.

Nitrogen and its compounds offer a particularly good example of the importance of *chemical form.* All plants need nitrogen in order to manufacture proteins. The atmosphere is nearly 80% nitrogen, but most plants cannot use nitrogen in its gaseous form; plants grown in an atmosphere of 80% nitrogen may turn yellow because of a nitrogen deficiency. Most plants require their nitrogen in the form of compounds in which nitrogen is combined with other elements in the form of nitrates, for example, which we sometimes provide for them in fertilizer.

The kinds and amounts of chemicals available in an ecosystem regulate the activities of the plants, and thus the animals, in that system. They may even determine which organisms can or cannot be part of that system. At any particular time there may be too much of one chemical substance or too little of another for a given organism or group of organisms in an ecosystem.

To complicate matters further, the living components of an ecosystem have a great effect on chemical and physical features of the environment. One of the most important concepts of ecology is that while the physical and chemical features of an ecosystem have an impact on animals and plants, plants and animals also have an impact on their physical and chemical surroundings. Trees and grasses help form soil, and hold soil and sand dunes in place. Trees can buffer the wind and make the climate cooler. Plants, as we will see later, are generally responsible for the fact that the oxygen in the atmosphere is in the form that *we* can breathe.

Living Components of Ecosystems: Producers, Consumers, and Decomposers

Living things are made of carbon and other chemicals with a lot of water added. Living things are beautifully organized combinations of nonliving materials—but so are diamonds. Although most living things are readily distinguishable from diamonds by the magnitude of their complexity, a more distinctive quality of life is that life is more or less in constant complex and dynamic action. Living organisms exchange, expel, convert, assemble, disassemble, organize, and otherwise manipulate the constituents of earth, air, and water. The energy-requiring manipulation of earth, air, and water by living things enables individual organisms to grow, repair themselves, and persist.

We have already identified producers, decomposers, and consumers as the basic kinds of *biotic* ecosystem components. As we will now see, the distinctions between these groups are based on their sources of energy and materials.

Producers. All green plants are *producers.* They produce in the sense that they **assimilate** (take in) simple chemicals from the soil and from the air (Figure 1.6) and, with the help of energy from the sun (Figure 1.7), transform them by photosynthesis into more complex energy-rich chemicals that eventually make up the substance of the plant. Obviously, plants

Figure 1.6 Cycling of Ecosystem Materials or Nutrients.   Several of the constituent parts of this diagram could be removed without stopping the cyclic flow of nutrients within a functional ecosystem.

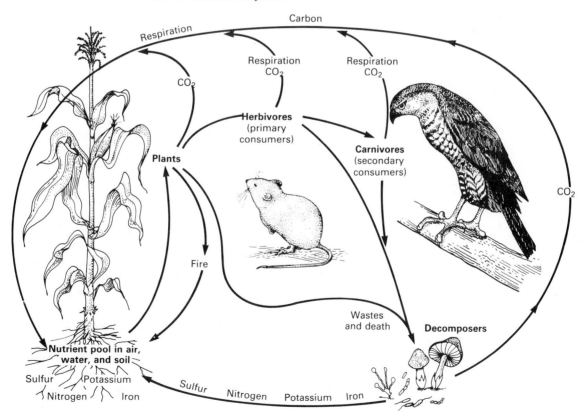

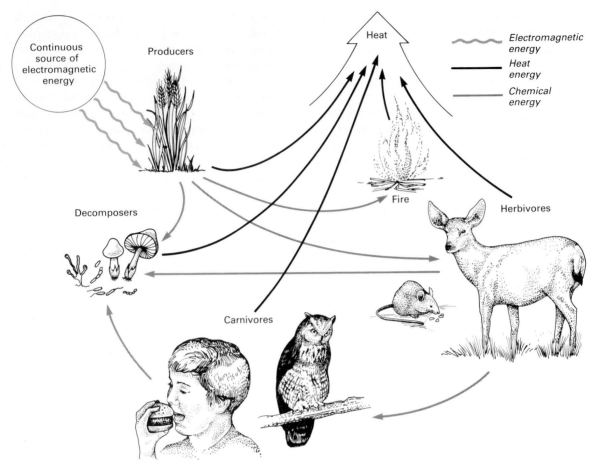

Figure 1.7 Energy Flow through Ecosystems. Three basic kinds of energy are involved in energy exchanges within an ecosystem. These are electromagnetic energy, heat energy, and chemical energy.

do not produce something from nothing. Perhaps a better name for producers would be "converters" or "transformers." The term *producer* will suffice, however, as an indication of the relative role that plants have in all ecosystems. From the perspective of the consumer, producers make food.

In recent years, scientists have found ecosystems based on chemical energy at great ocean depths (more than a kilometer), far below the limits of light penetration. The producers in these systems are bacteria that are able to gain energy from the oxidation of hydrogen sulfide that seeps from volcanic vents in the ocean floor. Since these organisms get their energy from chemical reactions rather than light, they are called **chemotrophs** rather than **phototrophs**—literally, "chemical feeders" rather than "light feeders." The roles of consumers in these ecosystems are filled by bacterium-eating relatives of the consumers that feed on organic matter derived from sunlight elsewhere in the ocean.

Except for a few obscure species such as those just described, all living things other than green plants are *consumers.* They must consume the chemical energy and chemical nutrients derived from other living things.

Consumers. Cattle graze. They consume plant material from which they extract energy and chemical building blocks. Some of the food energy is used as a cow moves, and some is used in assembling chemical subunits into new cells for the cow or into a calf fetus—for growth or reproduction. Similarly, an eater of cows extracts energy from some of the chemicals in beef, and in the same way that a cow breaks down plant chemicals to get energy, a cow eater breaks down cow chemicals to get energy. A cow eater might use some of this energy to square-dance or to swim; some is used to reassemble beef chemicals into human chemicals (if the cow eater happens to be human).

Because cows eat plants, they are called **herbivores.** They could also be called vegetarians. Because they get their food directly from producers, cows and all other strict herbivores are also called

Who Needs Consumers Anyway?

We humans tend to consider ourselves the focal point of the biosphere. However, the principles outlined in this chapter indicate that consumers have a role, but not necessarily an indispensable one, in maintenance of the ecosphere. Together, producers and decomposers could constitute complete ecosystems—without consumers. Consumers do add complexity and diversity to the system, but they aren't essential. This realization leads to a fundamental question. Should we evaluate other species in light of their direct use to humans or in light of their role in maintenance of the ecosystem? Which should be emphasized on endangered species lists, mushrooms or mammals? What do you think?

primary consumers. Organisms that eat cows or that eat any other plant eater are called **secondary consumers,** since their food is one step removed from plants. They are also called **carnivores.** This pattern would predict that an animal that eats a cow eater would be called a **tertiary consumer,** and so on. Obviously, there are organisms in ecosystems that consume plants *and* plant eaters. A person eating steak and potatoes would qualify as a *multilevel consumer,* or an **omnivore** ("eats all").

Decomposers. Decomposers comprise a special class of consumers that get energy and nutrients by digesting waste matter and dead plant or animal material. Decomposers are the organisms—mostly bacteria and fungi—responsible for decay, decomposition, or rotting. Sometimes animals die for reasons other than being killed and eaten by a predator. If their carcasses are not found and picked clean by scavengers, they rot. The energy-rich and mineral-rich treasure that these carcasses hold goes to bac-

Decomposers such as fungi consume dead plant material and return nutrients to the soil.

teria and fungi. Similarly, plants die for reasons other than being picked and eaten by a primary consumer. The chemicals in dead plant material—the enormous volume of leaves that forests give up every fall, for instance—also go to the decomposers.

Decomposers fill a very important role in ecosystems. They are responsible for the completion of ecosystem mineral cycles. (In Chapter 2 this statement will be qualified somewhat in discussions of ecosystem storage and ecosystem export/import.) By breaking down residual organic chemicals and by using up any remaining energy, decomposers figuratively "mop up"; they ensure that nutrients do not remain tied up in nonfunctioning plant or animal mass; they complete cycles.

Fire is like decomposition. Fire can also release chemical energy and nutrients held in plant and animal material. Fire also mops things up. It also completes cycles and closes circles. Fire is much more traumatic than decomposition and is obviously different in other ways. Many kinds of ecosystems are not compatible with regular fires; others depend on them (see Keeley, 1987). Still, the net results of fire and decomposition are the same: Complex organic material is broken down into mineral nutrients, carbon dioxide, and water; energy is released as heat.

Ecosystem Function

Given the proper physical setting—temperature, light level, slope, etc.—living organisms really require only two things from the environment: (1) *chemicals* to make up the substance of the organisms and (2) *energy,* the power necessary to make the substance "go." All of the principles of ecological interaction are concerned in one way or another with the acquisition of chemical substance (matter) and energy. Energy and matter thus form the basis of what ecologists consider to be the cardinal principles of ecology, namely:

1. Energy *flows through* and propels ecosystems (Figure 1.6).
2. Chemical matter *cycles within* and/or *between* ecosystems (Figure 1.7).

Chemicals are used over and over again within ecosystems. Although in most ecosystems there is a certain amount of import and export, *matter* need not be imported from outside an ecosystem in order for the system to continue to function.

Energy cannot be used over and over. The energy that gets into ecosystems is *not* restored to its original form as ecosystems use it. Except in fireflies and negligibly few other luminescent creatures, the *chemical* energy in the herbivore or primary consumer is never restored to *light* energy. Also, energy transfers are inefficient. As energy is converted from one form to another form—from light to chemical, for instance—much of it is necessarily lost as heat. As energy passes from plants through consumers and decomposers, all of it is eventually dissipated as heat (see Figure 1.7). Thus energy does not cycle. Energy comes into the ecosystem, it is used, and most of it is lost to the system forever as it is converted to heat. Ecosystems must all have continuous sources of energy.

For *all* living things an energy crisis is a crisis indeed.

Populations and Communities

Most ecosystems have many different kinds (species) of producers, decomposers, and consumers. This makes ecosystems lively and variably complicated. Individual organisms in an ecosystem relate to other members of their species and to members of other species in ways that range from competition for space and food to mating. Ecologists study these interactions, lumping them into the categories of **intraspecific** (between members of the same species), or *population* interactions, and **interspecific** (between members of different species), or *community* interactions (review Figure 1.1).

Many of the principles of ecology relate to populations and communities. A **population** is a group of interacting individuals of the same kind or species. All the gray squirrels in a clump of woods, all the bass in a pond, or all the deer mice in a field would qualify as a population. The individuals in a population relate to one another in many ways. While some animals—the lynx, for instance—lead an almost solitary existence, insects—ants, for instance—carry in-

teraction among members of the same species to the extreme. As we will see in Chapter 4, a population is considerably more than the sum of its individual members.

A **community,** in the ecological sense, is made up of all of the interacting populations of a number of species in a given area. Examples of communities are all the plants and animals of a desert, all the plants and animals in a pond, all the plants and animals in a meadow, and all the plants and animals in an aquarium. The community is the biotic part of an ecosystem (see Enrichment Box 1-1). Interaction at the community level can be extremely complex. In a pond community a bass population may feed on smaller fish, which in turn feed on minnows, which in turn eat microscopic animals, which in turn subsist on microscopic plants; larger plants provide shelter and food for other animals in the pond food web. In Chapter 4 we will see that a community is also much more than the sum of its parts.

The Molecular Basis of Production and Consumption: Photosynthesis and Respiration

Photosynthesis is the process by which plants convert light energy into chemical energy. **Respiration** is the process by which chemical energy is released to do work in both plants and animals. Photosynthesis and respiration are at the molecular "heart" of the mechanisms by which matter and energy are moved within ecosystems. A simplified summary of the processes of photosynthesis and respiration, and the interrelationships among these processes, is presented in Figures 1.8 through 1.10.

Plants, as illustrated in Figures 1.8 and 1.10, use some of the chemical energy they derive from sunlight to make fruit, seeds, stems, and leaves. Plants synthesize molecules like glucose and starch, and from some of this they extract the energy they need to make new plant tissue through plant respiration (Figure 1.9).

The net effect of respiration is photosynthesis in reverse. Figures 1.8 and 1.9 illustrate that the processes of photosynthesis and respiration exemplify cycling in natural systems. The chemicals used in photosynthesis are regenerated by respiration.

Fortunately for animals, plants do not themselves reverse all photosynthesis through plant respiration. This fact makes it possible for consumers to exist. As shown in Figure 1.10, herbivores and carnivores are completely dependent on plants.

Figure 1.8 shows that three major effects of photosynthesis are significant to our consideration of

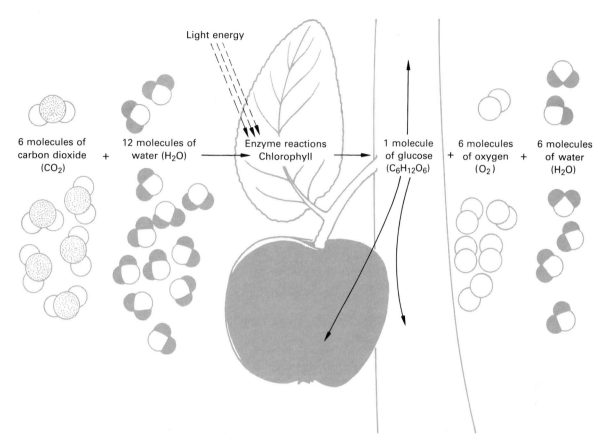

6 molecules of carbon dioxide (CO_2) + 12 molecules of water (H_2O) → Enzyme reactions Chlorophyll → 1 molecule of glucose ($C_6H_{12}O_6$) + 6 molecules of oxygen (O_2) + 6 molecules of water (H_2O)

Light energy

Figure 1.8   Photosynthesis Simplified.   Carbon dioxide and water are "consumed" in the process of photosynthesis, and oxygen, water, and glucose are "produced." Much more is involved than is illustrated in this diagram; only the *net* effect of photosynthesis is shown here.

Figure 1.9   Respiration Simplified.   In respiration, oxygen, water, and glucose are "consumed," energy, water, and carbon dioxide are "produced." The energy is used by the respiring organism; the carbon dioxide and water are by-products. The specific enzyme reactions that take place are not represented in this simplified diagram. Note that the *net* effect of respiration is the reverse of the *net* effect of photosynthesis.

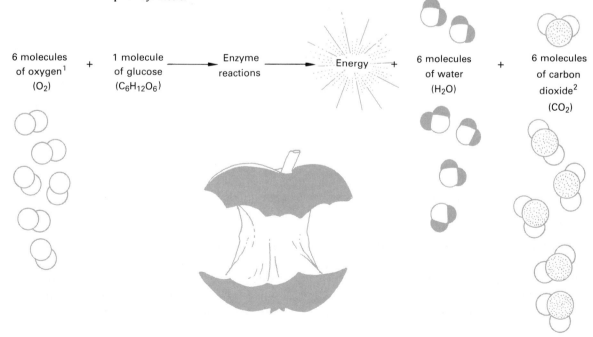

6 molecules of oxygen[1] (O_2) + 1 molecule of glucose ($C_6H_{12}O_6$) → Enzyme reactions → Energy + 6 molecules of water (H_2O) + 6 molecules of carbon dioxide[2] (CO_2)

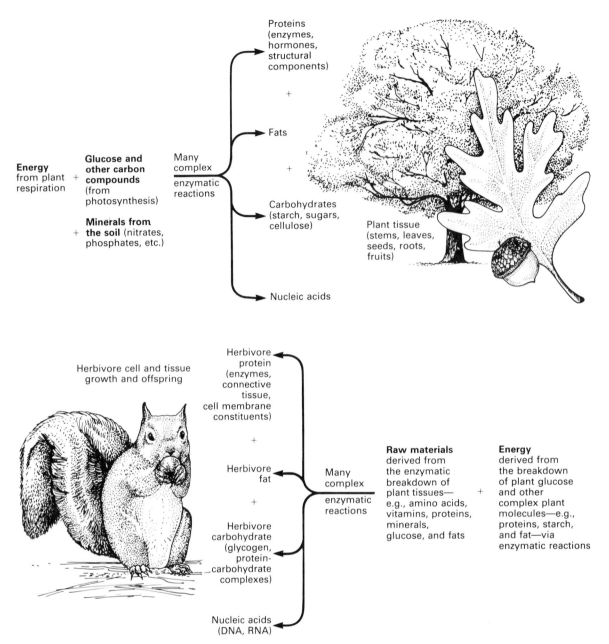

Figure 1.10 Plant and Herbivore Biosynthesis. Plants use some of the chemical energy they produce to convert some of the chemicals they produce into proteins, fats, and carbohydrates. Animals use some of the chemical energy they consume to convert derivatives of some of the chemicals they consume into proteins, fats, and carbohydrates.

ecosystems. First, energy is converted from electromagnetic or light energy into chemical energy. This brings solar energy into another form useful to plants and animals. The second effect is that simple chemicals are converted into more complicated ones. The third is that oxygen is released as a by-product of photosynthesis. All the oxygen in the atmosphere, approximately 20% of the air, is probably a result of photosynthesis that has never been reversed by respiration. Chapter 5 will explain this further.

While plants get their energy from photosynthesis and their nutrients from the soil, all animals ultimately rely on plants for both the energy and nearly all of the nutrients they require. Each species of organism satisfies its need for minerals and energy and finds a good place to live in the unique way that constitutes that species' functional role in the natural scheme. Species compete with one another, and some are dependent on others. Some organisms like it hot; some like it wet. Some do best where it is hot and dry. Some plants flourish on slopes, and some do better in the shade than in the sun. Physical variables exist in numerous combinations over the surface of the earth, and these factors all bear on the ability of organisms to survive. Therefore, we should expect great differences in species composition from one place to another, and there are indeed great differences. The differences are qualitative, however. All ecosystems, all living systems, have their equivalents of producers, consumers, and decomposers.

Balance in Nature

If producers produce and give rise to more producers that also produce, how is it that the Earth is not overrun by producers and production? Why isn't the Earth covered by a thick tangle of brush and vines? If decomposers decompose and give rise to more decomposers, which also cause decomposition, why hasn't everything rotted? To belabor the point: If consumers consume, if they have babies that also consume, why hasn't everything consumable been consumed by now? From time to time, grasshoppers do strip areas of North Dakota clean. Why doesn't this sort of thing happen more often? How is it that photosynthesis, respiration, and the other processes happening very rapidly in most ecosystems do not seem to change things very much? The answer surely lies in some system of checks and balances.

Although nature may not change very much, it needs to be made clear that this constancy is real only in an overall sense and even then only on certain time scales. Nature certainly does have dramatic ups and downs from place to place over short periods of time. There are lots of apples one year and hardly any the next. There are cycles in game animal populations. There are red tides and other algal blooms. Occasionally there are explosions in lemming, cicada, Gipsy moth, and grasshopper populations. We are currently in the midst of a dramatic human population explosion. On local scales there are volcanic

The Gaia Debate

Scientists generally agree that the environment affects living things and that living things in turn interact with the environment and can affect or change it. Feedback mechanisms are crucial to this operation. But a British scientist has taken the concept further to maintain that living things as a totality interact with the environment to keep it relatively stable over time and favorable to life. The implication is that life can effectively manipulate the environment for its own purpose. It is a "biological cybernetic system with homeostatic tendencies." This biological control system for earth is called Gaia, the Greek name for the goddess of the earth or Mother Earth. Gaia was proposed by James Lovelock, a British scientist, and Lynn Margulis, an American microbiologist. They began developing the concept of Gaia in the late 1960s and early 1970s. The first book on the topic published by Lovelock in 1979 was called *Gaia: A New View of Life on Earth.*

The hypothesis has been criticized by scientists because it appears to give "life" the ability to plan or be purposeful in its interactions with the abiotic world. Other scientists believe that Gaia is useful as a framework for examining the global ecosystems of the world—the three great spheres and the biosphere—in an integrated manner. It is this aspect of Gaia coupled with the awareness of global environmental problems that has brought the concept back into the arena of scientific review and debate.

In a more recent book, *The Ages of Gaia,* Lovelock updates his concept in response to criticisms. Initially Gaia seemed to support the idea that the earth could adjust to any major environmental disturbance and, consequently, that concern over pollution was blown out of proportion. Lovelock maintains that Gaia will indeed resist environmental change and attempt to maintain the current conditions. But

he concedes that the regulatory apparatus can be stressed beyond its limits. When this happens, a jump is made to a new stable state, and many species may be eliminated.

Lovelock maintains that his theory does not demand that "life" have some foresight or planning. Regulation is possible without such "purpose." He describes a world whose only form of life is several kinds of daisies that range from light to dark in color. In Daisyworld, homeostasis is maintained by the rise and fall in numbers of light- and dark-colored daisies. The dark daisies absorb the sun's heat and warm the planet. As the temperature starts to rise, it becomes too warm for the dark daisies to survive. As they die off, the lighter daisies, which reflect sunlight, proliferate. Temperature is thus kept relatively stable.

In regard to his theory, which addresses global environmental interactions, Lovelock has described a new field called geophysiology. The control-system that maintains the earth is comparable to the physiological systems that maintain humans and other organisms. According to Lovelock's theory, the earth can be studied as an organism—a single living whole.

eruptions, fires, floods, and droughts that have profound, albeit temporary, impacts on the populations of many species. On long time scales, dinosaurs have come and gone. Most of the species of plants and animals that have ever existed are gone.

Still, overall, on time scales measured in decades or even in hundreds of years, nature does not tend to change very much. But even this "constancy" is dynamic to the extent that some ecologists question the use of the word *balance* to describe what goes on in nature. Some argue that *balance* is really an inaccurate term to describe what goes on in nature because population densities are continually changing. Species numbers tend to oscillate about a mean that is relatively stable, though subject to change over long periods of time. Other ecologists say **"balance in nature"** is a useful concept—if it can be defined as the persistence of ecological systems as a result of their ability to compensate for disruptions. Perhaps the truth about balance in nature can best be illustrated by a short fictionalized tale.

Once upon a time, on an island in a large lake there was a large moose herd. On the same island there was a large wolf pack. The island had lots of succulent vegetation and some tall trees but not much else. For many, many years things stayed pretty much the same on the island. Moose were born to spend a lifetime standing knee-deep in water, browsing on vegetation, eventually dying. Some moose, usually the sick ones and occasionally a young one, were eaten by wolves. Wolves were born to spend a lifetime seeking out and running down old or young moose and eventually dying. A new crop of lush vegetation came up every year, and some of it was eaten by the moose. Not much else happened.

One year there was not much rain on the island. There was even less the following year and still less the year after that. This went on until the island's ponds began to dry up, and the vegetation was not lush anymore. Six years into the drought, things really got bad for the moose. Coincidentally, things really got *good* for the wolves. More moose died than were born. More wolves were born than died. Eight years into the drought, the island's vegetation was even less lush. There were not many moose, but there were lots of sleek, healthy wolves. As the moose became increasingly scarce, things began to go badly for the wolves. Nine years into the drought, more wolves died than were born.

Eleven years after the drought began, the rains came again in abundance. The spring of the following year was glorious; ponds rose, vegetation came back, and the moose that remained experienced no decline in numbers. Because there were not many moose, they were not too hard on the vegetation that year, and lots of it "went to seed," so there was even more

vegetation the next year. The wolves were still relatively few in number. Things really began to look good for the moose. The following year, many more moose were born than died. Because there were not many wolves to harass them, most of the moose that were born lived to reproductive age, to produce new moose.

Soon, with large numbers of young, inexperienced moose about, things got a lot easier for the wolves, and they too eventually began a slow climb in numbers. About 35 or 40 years after the drought began, the number of moose and the number of wolves on the island returned to the levels that existed before the drought. Nature is resilient.

Resilience is unquestionably a better term than *balance* to describe nature. Nature has the ability to cope with disruptions, but this ability is limited. Limitations in the ability of natural systems to snap back from insult and disturbances wrought by humans are at the heart of many of our environmental problems.

In the extreme, nature—in some form, at least—is sufficiently resilient to withstand *any* human-induced disruption. Humanity simply cannot survive the systems on which it depends.

Even though ecosystems are resilient and can—within limits—compensate for disruptions, they also are subject to change. In a later chapter we will consider how natural systems drift. We will examine the factors in natural systems that determine the kinds and degrees of disruptions for which they can compensate, and we will be looking for some ideas about why natural systems sometimes collapse altogether when their elastic limits are reached and exceeded.

Human effects on an ecosystem need not be disruptive. Crown vetch planted along roadways helps to inhibit erosion.

Has the Environmental Movement Taken Hold?

In many countries of the world, regard for the environment has come a long way. Many countries of western Europe now have "green" political parties, and there has been significant growth in the membership of environmental organizations throughout the world, including eastern Europe and the states of the former Soviet Union. Many countries of the world now have fairly strong environmental laws, and these are increasingly being enforced—all in response to an apparently strong public insistence that the environment needs protecting. In the United States, environmental issues have been getting more and more attention in political campaigns at all levels of government. Do you think this is genuine and likely to be long-lasting? Some observers argue that only a small fraction of the earth's people have been behind the environmental movement and that most people really don't care about the environment one way or another. Some people believe that if things *really* get tough economically and remain that way for a while, concern for the environment will fade away. Do you think that concern for the environment is now so thoroughly built in to our way of looking at the world that it is here to stay?

Environmental Problems and the Principles of Ecology

Our species is now having tremendous reproductive success worldwide. Cultural evolution and human inventiveness have, to a large degree, overcome many of the restraints and controls previously imposed by nature on humanity. However, our current reproductive success is a temporary phenomenon. Already our species is beginning to reach new limits. We are approaching the limit of our supplies of some fossil fuels—oil, for example. Some say we have reached the point at which there are too many people, too many pollutants, too much waste, too many poisons, and too little food.

The critical question has become: Will natural controls automatically come down hard on human beings, or will we recognize how seriously we strain and disrupt the systems that support us and find ways through our inventiveness to reduce or minimize this impact? Can the human brain, which made possible the tremendous successes our species has enjoyed, also provide the means by which to avert or diminish the disaster that many say looms on the horizon? It is with these questions in mind that we will consider in more detail the ecological principles introduced in this chapter. We will specifically take a very close look at how these principles apply directly to *Homo sapiens*. Our species will not be presented as a villain or as a defiler of nature. We are an integral part of nature. Conceivably, if we can restrict our activities to the kinds and levels that are compatible with the system of which we are part, our species may persist for a very long time. The question to keep in mind from here on is: What level of human activity and what kinds of human activity can the ecosphere sustain indefinitely with humankind as one of its integral components?

Concepts to Remember

1. Ecology deals with organisms at the level of their interaction with other organisms, both of their own species and of different species, and with their nonliving environment.
2. Different parts of the ecosphere (deserts, lakes, forests, etc.) may look very different, but the same basic structural and functional relationships are found in all of them.
3. The two most basic kinds of relationships in any ecosystem are those having to do with energy flow and those having to do with chemical cycles.
4. Producers, consumers, and decomposers are the basic classes of organisms that comprise ecosystems.
5. Physical and chemical features of the environment (moisture, light level, temperature, soil texture, etc.) influence the organisms that live there, but the reverse is also true: Living things influence their physical and chemical environment.
6. Nature appears to stay the same because ecosystems are self-regulating and resilient, not because populations are static. The size of the population of a particular species may vary within wide limits because of how that

population relates to the other elements in the ecosystem.

7. All human environmental problems have their roots in one or more of the fundamental principles of ecology.

References and Further Reading

References marked with an asterisk are cited in the chapter.

Ehrlich, P. R., and Birch, L. C., 1967. "The 'Balance of Nature' and 'Population Control,'" *American Naturalist* **101**:97–107. This is a classic article that makes the point that the idea of balance with respect to population sizes in nature is at best misleading and at worst false.

Hadley, N. F., and Szarck, S. R., 1981. "Productivity of Desert Ecosystems," *BioScience* **31**:741–753. This easy-to-read article discusses just how water limitation sets the ecological character of the desert.

*Keeley, J. E., 1987. "Role of Fire in Seed Germination of Woody Taxa in California Chaparral," *Ecology* **68**:434–443.

Kormondy, E. J., 1984. *Concepts of Ecology.* 3d ed. Englewood Cliffs, N.J.: Prentice Hall.

Odum, E. P., 1983. *Basic Ecology.* Philadelphia: Saunders. For readers interested in more detail, this is a good, textbook designed for college courses in general ecology.

Romme, W. H., and Despain, D. G., 1989. "The Yellowstone Fires," *Scientific American* **261**(5):37–46.

Seastedt, T. R., and Crossley, D. A., 1984. "The Influence of Arthropods on Ecosystems," *BioScience* **34**:157–161. This article describes the influence of insects on nutrient cycling and other aspects of ecosystems.

Slobodkin, L. B.; Smith, F. E.; and Hairston, N. G., 1967. "Regulation in Terrestrial Ecosystems and the Implied Balance of Nature," *American Naturalist* **101**:109–124. This classic article, a response to the article by Ehrlich and Birch (also cited here), makes the point that use of the term *balance* does in fact have utility in ecology.

2 Energy in Ecosystems

Without energy there would be nothing. Never has a statement made anywhere been meant more literally. Without energy, nothing could walk, fly, prowl, dive, swim, chew, slither, hiss, bark, or grow. Einstein showed that even matter is a form of energy. It should be obvious, then, why energy is central to one of the cardinal principles of ecology presented in Chapter 1. One of the most basic ways in which organisms relate to their environment is through their need for energy.

Energy and Entropy

Energy

Energy is difficult to define. Let's begin with a simple definition and expand on it.

Energy is the ability to do work. Energy is something that, given the right device, can be converted into a push. Energy can be used to move an object against an opposing force over some distance and thus account for work. Energy is stored work.

In the relationship of energy to work, energy must be defined in terms of space, distance, and spatial order. In a strict physical sense, work always involves movement or displacement of one body rela-

tive to another. It takes energy to move a stone away from the earth. The actual amount of **work** done against gravity to move a stone away from the center of the earth equals the force applied to the stone times the distance it is moved. Having been moved "up," the stone has acquired a certain amount of potential energy. Released and allowed to respond to gravity, the stone falls to earth, and its potential energy is converted into **kinetic energy** (the energy of motion) and other forms of energy. In this example it should be fairly obvious that the forms of energy being discussed have meaning only in terms of the spatial relationship between two or more bodies.

With other forms of energy the importance of spatial arrangement may be less obvious, but the concept still holds. Consider chemical energy—that contained in gasoline, for example. The **hydrocarbons** (chemicals that contain carbon and hydrogen in a $1:2$ ratio) in a droplet of gasoline represent a certain amount of chemical energy or chemical potential energy—but only if in the vicinity of this droplet of gasoline there are molecules of oxygen or some other potentially interreactive chemical. Under certain conditions, hydrocarbons and oxygen will combine to form carbon dioxide and water and yield heat energy in the process.

19

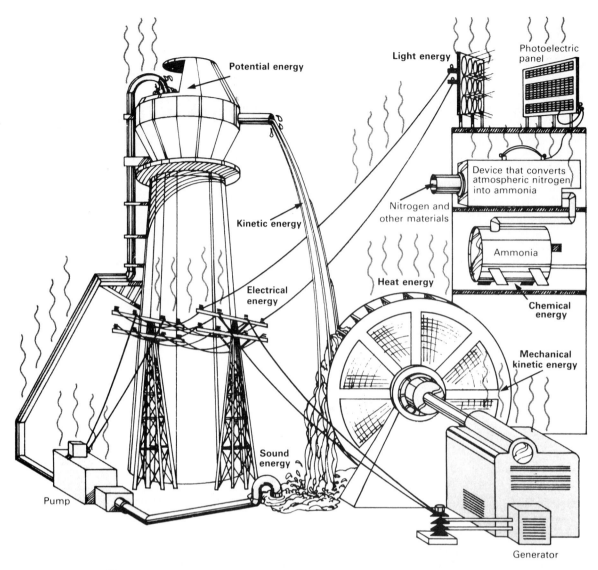

Figure labels: Potential energy, Light energy, Photoelectric panel, Kinetic energy, Nitrogen and other materials, Device that converts atmospheric nitrogen into ammonia, Ammonia, Electrical energy, Heat energy, Chemical energy, Mechanical kinetic energy, Sound energy, Pump, Generator

Figure 2.1   Some of the Forms of Energy and Examples of How They Can Be Converted to Other Forms.   No conversion from one form of energy to another is 100% efficient; some energy is lost as heat. (Actually, since heat is a form of energy, the energy it represents is not lost in the absolute sense.) Heat, which tends to disperse rapidly within a system, is the least useful (harnessable) form of energy. If this were a closed system, all activity in it would quickly come to a halt; all the energy would be converted to heat evenly distributed throughout the system.

There are actually many different kinds of energy; most of them are illustrated in Figure 2.1. All forms of energy can be expressed in terms of capacity for work or in terms of the amount of heat into which the energy could be converted.

Entropy

If a body of water, or anything else, were heated to a high temperature, the energy held by that body by virtue of its heat would not be available to do work unless another unheated body were available nearby to serve as a sink (a place to which the heat could be moved). Steam can move an engine only if there is some place to which the steam can push a piston, somewhere where there is not as much steam. A positive electrical charge by definition bears potential work only because there is a relatively negative charge nearby. To be harnessed, energy must flow; it must have someplace to go *to, from* where it is. Thus the usefulness of energy in doing work is directly related to how it is distributed within a system. The more uniformly heat is distributed within a closed system, the less harnessable the heat is for work. **En-**

Calorie: the unit of energy needed to raise one gram of water one degree Celsius. Any form of energy has an equivalent in calories. Calories used in connection with diets or food are actually "big" calories, equal to 1,000 "little" or *gram calories*. A big calorie will raise the temperature of one kilogram (liter or 1,000 cm²) of water one degree Celsius. Big calories are also called *kilogram calories*.

Closed system: strictly speaking, a system (see Chapter 1) that is totally isolated from everything else. A closed system exchanges neither matter nor energy with its surroundings. Physicists speak of *totally* closed systems, but ecologists sometimes use the phrase *closed system* to refer to systems that do not exchange matter but do exchange energy with their surroundings. The earth is a closed system in the ecological sense.

Energy: ability to do work. In common terms, energy is anything that can be used to change something by moving or heating it.

Entropy: a measure of the energy within a closed system that is unavailable to do work within that system.

Gross primary production: total amount of solar energy converted into chemical energy by producers within a unit of time.

Kinetic energy: the energy an object has by virtue of its motion relative to another object.

Net primary production: total amount of organic matter made available by plants to consumers per unit of time; gross primary production minus plant respiration.

Potential energy: the energy an object has by virtue of its position relative to another object; for example, a stone 2 meters above the ground has a certain amount of potential energy and could be used to accomplish some work.

Solar constant or **solar flux:** the average amount of solar energy the earth gets every day from the sun, 2 gram calories per square centimeter per minute.

Work: strictly speaking, the product of the force applied to an object and the distance that force moves the object against some opposing force. In common terms, *work* is the name given to what is accomplished when an object is moved against an opposing force such as gravity.

---

tropy is the amount of energy in a system *not* available to work within that system.

Consider a bar of soap in a large volume of water. Each time the soap is used, less of it is available for soaping, even though the body of water—the system—has the same total amount of soap. What happens to soap for soaping—or energy for working—as it is used is that it becomes dispersed, unharnessable. The key to understanding entropy or the "usefulness" of energy is to see that it is related to order, dispersion, and randomness.

Heat is a relatively random form of energy because heat tends to become rapidly dispersed. Figure 2.2 depicts a system, room A, heated to uniform temperature. If a colder room were placed next to the uniformly heated room, the two-room system would have a nonuniform distribution of heat that could be harnessed to do work as shown. Note that once the two-room system reached a uniform temperature, the heat energy would become unharnessable within that system—the air would stop moving, the fan blades would stop, and no work could be done—even though

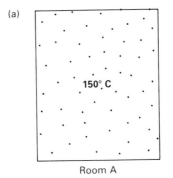
(a)
Room A

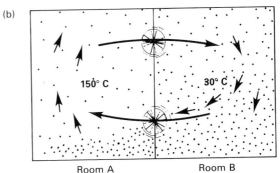
(b)
Room A   Room B

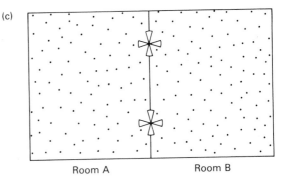
(c)
Room A   Room B

Figure 2.2 A Crude Illustration of Entropy. A room heated uniformly (a) has heat energy, but none can be harnessed to do work in the room. If the room were connected to a cold room and holes were made between the rooms at the top and bottom of the common wall (b), the colder, more dense air from room B would flow through the lower opening into room A. The hot air in room A would tend to rise into room B. If vanes were placed in the openings, the flow of air could be harnessed to make the blades rotate as in a windmill. Once the two rooms became uniformly heated (c), the heat would no longer be harnessable within the two-room system. What might happen if a third room were then hooked up to this system?

The Second Law of Thermodynamics and Lunch

The second law of thermodynamics holds that every time energy is converted from one form to another, some of that energy is lost as heat. A corn plant is a producer that harnesses solar energy and stores it as chemical energy in kernels of corn. That corn can be ground, made into cornbread, and eaten directly by a person. Or that corn could be fed to animals to produce meat for human consumption. If the corn is fed to an animal, however, most of the chemical energy is "lost" or "used up" by the animal; consequently, there is not as much energy available from that corn for humans. With this in mind, it has been proposed that we should eat lower on the food chain. With fewer energy transfers in the food chain, more people could be fed with the same amount of food. Others note that grass-fed and forage-fed animals convert food that is *not* consumable by humans into a usable form. Should we all move toward vegetarianism? What do you think?

the two-room system had the same amount of heat energy that it had at the outset.

It is the tendency toward uniform distribution of energy as heat that physicists speak of when they say that the universe is running down.

The Laws of Thermodynamics

To appreciate the behavior of energy in living systems, we must become acquainted with two universal laws of energy, the first and second laws of thermodynamics. These two laws govern all energy transactions in the universe, including those of animals and plants. The laws of thermodynamics form the basis for the principles that energy flows *through* ecosystems and that ecosystems must have continuous inputs of energy.

The **first law of thermodynamics** specifies that energy can be neither created nor destroyed by any process or event but that energy *can* be changed from one form to another. Stated in other words, the total amount of energy in the universe or in any closed system is always the same, though the proportions of the various forms of energy may vary. In still other words, energy that seems to disappear from a so-called dead flashlight battery is really still around somewhere, in some form.

Consider Figure 2.1 again. Depicted there are a number of forms of energy and some arrangements by which they could be interconverted. The first law tells us that no matter what arrangement exists in a system like the one shown in Figure 2.1, no matter what kind of energy is converted into what other kind, the total amount of energy *in a closed system* remains the same.

The **second law of thermodynamics** says, among other things, that some heat is produced—entropy is increased—each time energy is converted

from one form to another. Regardless of the kind of energy, not all of it can be converted into work; some of it is always "lost" as heat in the process of being used. No energetic conversion is 100% efficient. The second law also specifies that spontaneous processes are those in which entropy tends to increase. Energy transformations (conversions from one form to another) in which entropy could *decrease* are not likely to occur spontaneously. The first law of thermodynamics would not be violated if a lake suddenly cooled down a number of degrees (losing heat energy) and at the same time rose several feet (gaining potential energy). Some of the lake's heat energy would simply have been converted into an equal amount of potential energy. Nothing created, nothing destroyed. The second law says not to expect such a thing because it describes a decrease in entropy, an increase in order—a conversion from a less useful to a more useful energy state. What are some implications of the second law in everyday life?

Life, including human life, is a struggle against the trends described by the second law. After the keys are turned over to the owner of a new house, the owner in effect begins a continuous response to the second law of thermodynamics. A new home is a form of order; a complex combination of chemical and potential energies is represented in its structure. Over time, the assault of weather on shingles and paint, the reaction of air pollutants with paints and fabrics, and the effects of use tend to reduce a house to a scattering of its molecular constituents. A house must be maintained, repaired, and kept up at great expense. It is because of the second law of thermodynamics that homeowners spend Saturday mornings traveling to and from hardware stores.

Think of an abandoned house. Gradually, the roof leaks. Gradually, water gets inside, and weather-

Energy flows through ecosystems—from sunlight to plants to herbivores (like these American bison) to the organisms that consume the herbivores and to decomposers.

ing takes its toll. Rotting begins. Eventually, the house collapses into a heap. If such a house were located in the eastern part of the United States, the lot on which the house stood would eventually become part of a deciduous forest. One kind of order would be replaced by another. While at first this might seem to belie the universal tendency toward disorder, forest order is made possible only by a colossal amount of solar energy. While the forest was replacing the house, the sun was losing order. The system made up of the house, the forest, and the sun actually experienced a great increase in entropy in the process.

Life is characterized by an ability to extract order from the environment, to bring about an increase in order at the expense of a decrease in order somewhere else. This violates neither the first nor the second law of thermodynamics, since these laws govern only total, closed systems. For plants and ecosystems the "somewhere else" is ultimately the sun. Ecosystems extract order from solar energy as it passes by on its way to dilution throughout the universe.

Suppose we filled a large, clean jar with a handful of humus and leaves dug from a forest floor and then sealed the jar and had some way of measuring exactly what went on in the jar from that point forward in terms of energy. If the jar were properly shielded from light or from any other energy input, we would in effect have given to the jar system a certain amount of chemical energy. We would have invariably included organisms of decay with the leaves and humus, and these decomposers would break

down the chemicals in the leaves, extracting the chemical components and the energy for their activities and growth. The second law of thermodynamics says that such a closed system must inevitably wind down because every time one chemical is converted into another, some energy is converted to heat. The jar would give off heat over time. Eventually, all of the original energy would be converted to heat (entropy), and all living activity in the jar would grind to a halt (though this is a simplified description of the process).

The Earth's Energy Budget

The earth is like a jar of rotting leaves in that it has a decreasing amount of chemical energy stored in the form of coal and oil. The earth is unlike a jar full of leaves in that it has a daily energy income—from the sun.

The Source

If our coal and oil reserves are like a candle, which some claim we are burning at both ends, the sun is figuratively a lifetime battery for the ecosphere. Literally, the sun is a thermonuclear fusion reactor. Every second, some 4.2 million tons of the sun's mass is converted into 10^{26} calories of energy. Although the rate of conversion of mass into energy on the sun is enormous, the sun is so large that there is enough of it left to allow this process to go on for billions of years.

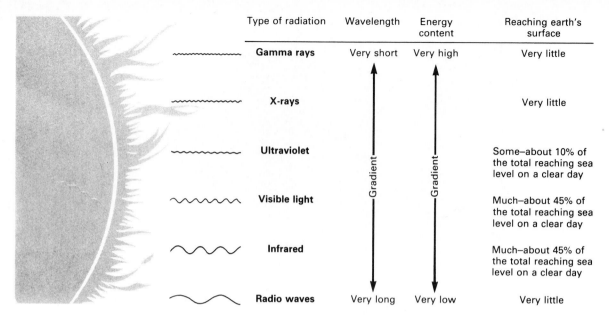

Type of radiation	Wavelength	Energy content	Reaching earth's surface
Gamma rays	Very short	Very high	Very little
X-rays			Very little
Ultraviolet			Some—about 10% of the total reaching sea level on a clear day
Visible light			Much—about 45% of the total reaching sea level on a clear day
Infrared			Much—about 45% of the total reaching sea level on a clear day
Radio waves	Very long	Very low	Very little

Figure 2.3   The Solar Emission Spectrum and the Relative Amounts of Each Type of Radiation Reaching the Earth. The earth's atmosphere is relatively opaque to most high-energy forms of radiation such as X-rays and gamma rays. Most of these get absorbed high in the atmosphere.

The specific energy-releasing reaction taking place on the sun is the fusion of hydrogen into helium. Two molecules of hydrogen are fused to form a molecule of helium, and a helium molecule weighs *almost* as much as two molecules of hydrogen. The difference, or the lost mass, is converted into energy according to the equation $E = mc^2$.

The energy generated from solar mass emerges in the form of electromagnetic radiation (including light), heat, and other forms of energy. Because neither heat nor sound can cross the near vacuum of space, the only kinds of energy that reach us here on earth are the electromagnetic varieties. The electromagnetic spectrum of solar radiation is presented in Figure 2.3. Though a rather wide spectrum of energy is shown in this illustration, about 99% of the radiation that reaches the earth is in the range encompassing *visible* light, near-infrared, and (a little) ultraviolet radiation.

Solar Constant

As electromagnetic energy streams away from the sun in all directions, it becomes less concentrated with distance. The earth, a small target some 150 million kilometers (93 million miles) from the sun, intercepts only one 50-millionth of the sun's energy output. This amounts to about 2 gram calories per square centimeter of the earth's surface per minute. This figure of 2 calories per square centimeter per minute, a constant sometimes referred to as the **solar flux** or **solar constant,** is the average amount of energy the earth gets every day—some to be used for

photosynthesis, some to heat the atmosphere, and some to be expended in other ways important to living things.

Solar Energy, Weather, and Climate

As illustrated in Figure 2.4, of the energy that reaches the outer atmosphere of the earth, on the average about 42% is reflected by clouds and dust in the atmosphere. Another 10% or so is absorbed by molecules of ozone, water vapor, and other gases in the atmosphere, leaving less than half the solar flux to reach the surface of the earth. Although our primary interest in this chapter is in the tiny fraction of solar energy that is converted into chemical energy by plants, we should pause and consider some of the other things that solar energy does for living systems.

Solar energy warms the earth; it causes evaporation of water and brings rain; and it drives the winds. The influences that the earth has on the absorption of solar energy by virtue of its shape, its atmosphere, its orbit around the sun, its rotation, and its topography mean the differences between night and day, the seasons, deserts and tropical rain forests, the poles and the equator, life and no life. Let's consider some of the details.

Weather and climate are the direct and indirect results of solar heating. The atmosphere is the first thing to experience the sun's rays in the morning, but actually only a little of the incident radiation heats the gases of the atmosphere as sunlight comes in. Of the energy that is absorbed by water vapor and other atmospheric gases, some is reemitted as infrared ra-

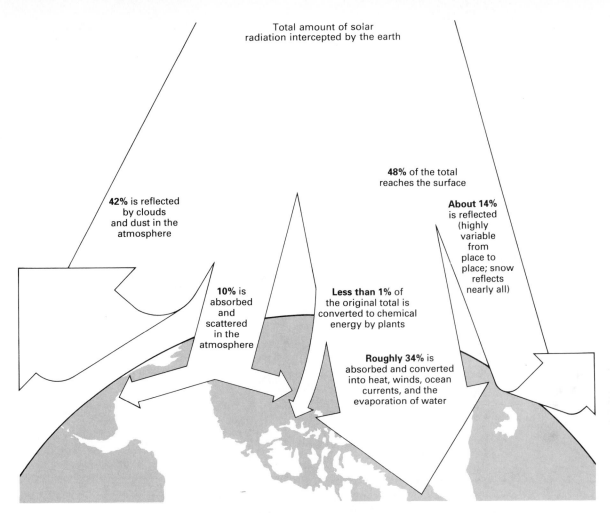

Figure 2.4   The Fate of the Solar Energy Intercepted by the Planet Earth.
About half of the intercepted energy remains in the atmosphere or is reflected into space.

diation, and some of this reaches and warms the ground. Some of the energy absorbed by atmospheric gases is converted directly into molecular kinetic energy (heat). However, most of the heating of the atmosphere occurs from the bottom up (see Figure 2.5).

A highly variable fraction of the sunlight that reaches the earth's surface is reflected all the way into space. Snow or white sand reflects much more than black volcanic rock or vegetation. In any case, some of the radiant energy that reaches the earth's surface (including both direct sunlight and radiation reemitted by clouds and atmospheric gases) is absorbed (see Figure 2.4). Some of this energy heats the ecosphere directly; some is reemitted as infrared radiation. Reemitted infrared radiation would also travel back into space except that some of the same gases that allowed the visible light *in* will not let infrared radiation *out*. The fact that carbon dioxide and other gases will not absorb much visible light but will absorb infrared radiation is the basis of the so-called *greenhouse effect*, which we will discuss in Chapter 10.

The atmosphere can be heated by the earth in three general ways: conduction, radiation, and heat transfer. Heat can be passed from the earth to air by direct conduction (contact). Infrared radiation emitted from the earth's surface can be absorbed by water vapor, carbon dioxide, and other gases, increasing their kinetic energy. Warming of the surface can also cause surface waters to evaporate; water that evaporates at the surface and later condenses in the atmosphere actually transfers heat into the atmosphere.

The atmosphere is heated unevenly. Heating causes air to expand, and uneven heating leads to uneven expansion and differential densities of air. Hot air rises, and cold air falls and moves to fill in the space left by the hot air (Figure 2.5). In other words, uneven heating leads to *wind*. The winds and other air movements can carry water vapor from warm places to cold places, where the vapor condenses and falls back to earth. Thus air movement leads to rain and snow. In Chapter 3 we will consider the role of sun-driven weathering by wind and rain in biogeochemical cycles (see also Figure 2.7).

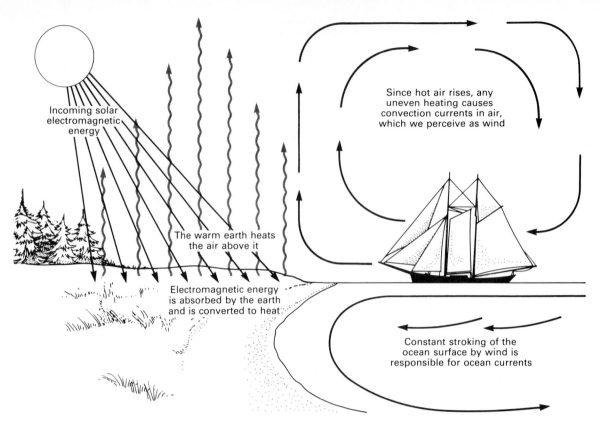

Incoming solar
electromagnetic
energy

Since hot air rises, any
uneven heating causes
convection currents in air,
which we perceive as wind

The warm earth heats
the air above it

Electromagnetic energy
is absorbed by the earth
and is converted to heat

Constant stroking of the
ocean surface by wind is
responsible for ocean currents

Figure 2.5   The Energy That Causes Winds and Ocean Currents Comes from the Sun by Way of the Surface of the Earth.   During most of the year in many places the sun heats up the land during the day, so it is warmer than the ocean. As the air over the land is heated, it rises, and cooler air over the ocean moves in as a sea breeze to take its place. At night, the ocean is warmer than the land, so the breeze shifts from the land out to sea.

Keep in mind that energy budgets for particular parts of the earth vary considerably. The earth rotates such that half of it is in darkness at any given moment. Since the earth is a globe, solar radiation strikes it at a complete spectrum of angles with the surface ranging from perpendicular to parallel; the earth is farther from the sun at some times of the year than at others, with different parts of the earth at different angles (accounting for the seasons). What all of this means is that the actual amount of radiation striking a field, pond, or lake varies considerably from one place to the next, from one day to the next, from one season to the next, and from one hour of the day to the next. It is only *on the average* that 2 calories per square centimeter per minute strikes the outer atmosphere and approximately 1 calorie per square centimeter per minute strikes the surface of the earth.

One final meteorologic note: Radiation equilibrium does not prevail at all latitudes. That is, not all areas of the earth radiate as much into space as they receive from the sun. Some of the solar energy coming into equatorial regions, for instance, is dissipated into other zones via winds and ocean currents.

This meteorologic distribution of heat helps to moderate climate; it blends what would otherwise be much more distinct zones having profound differences in climate.

Ecosystem Energy Budgets

Of the electromagnetic energy striking a typical ecosystem, less than half is absorbed by the plants directly. Only a small fraction is of the right wavelengths to be absorbed by chlorophyll and converted into chemical energy in photosynthesis. On the average, the efficiency of energy fixation by green plants is less than 1%.

As shown in Table 2.1, the efficiencies of energy fixation in various ecosystems actually range from a fraction of 1% up to 7% or so. During optimal growing conditions, corn, one of the most efficient energy capturers, converts about 3.2% of solar radiation into the chemical energy in kernels, cobs, silk, stalks, roots, and leaves and uses another 3.6% (for a total of 6.8%) in its own metabolic activities.

Care must be taken in interpreting *efficiency* as

Table 2.1 Solar Input versus Primary Productivity in Nature and Agriculture. Note that a direct comparison cannot be made between the natural and agricultural efficiencies here. The natural figures are computed on a yearlong basis (including nongrowing season), whereas only the best conditions for productivity were used in calculating the agricultural efficiency. This was done to show that even with this advantage, the increase over natural efficiencies is not particularly large.

	Efficiency of Gross Production[1] %	Efficiency of Net Production[2] %	Net Production as a Percent of Gross Production %
In Nature[3]			
Maximum	5.0	4.0	80
Average favorable condition	1.0	0.5	50
Average for biosphere	0.2	0.1	50
With Human Intervention[4]			
Sugarcane	7.6	4.8	62
Corn	6.8	3.2	47
Sugar beets	7.7	5.4	72

[1] Efficiency of gross production is the percentage of electromagnetic energy reaching a system that is captured by plants and converted into chemical energy.
[2] Efficiency of net production is the percentage of electromagnetic energy reaching the system that is captured by plants and available to plant eaters.
[3] Long-term averages over one entire year or longer.
[4] Short-term productivity (best days) with intense cultivation.
Sources: Odum (1963); Montieth (1965).

it is used here. We have based our figures on *total* solar energy reaching leaves, even though some portions of the visible light spectrum cannot be absorbed by plants. This, you might say, really is not fair. But we think it is, as long as we qualified what we did. Qualifiers are important here. Efficiency would be one thing if computed for corn that is free of disease on a hot sunny day following a soaking rain in a fertilized field and quite another if computed on a per-year basis (counting the winter, in which there might be less solar input but *no* production).

Figures 2.6 and 2.7 summarize what happens to

Figure 2.6 What Happens to Solar Energy on Its Way to and through an Ecosystem. Only one part in 50 million of the sun's energy reaches the earth. Of that amount, less than 1% is converted to chemical energy by green plants.

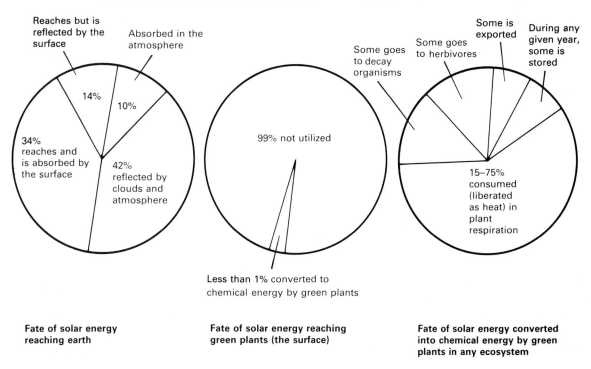

Reaches but is reflected by the surface

Absorbed in the atmosphere

Some is exported

During any given year, some is stored

Some goes to decay organisms

Some goes to herbivores

14%

10%

34% reaches and is absorbed by the surface

42% reflected by clouds and atmosphere

99% not utilized

15–75% consumed (liberated as heat) in plant respiration

Less than 1% converted to chemical energy by green plants

Fate of solar energy reaching earth

Fate of solar energy reaching green plants (the surface)

Fate of solar energy converted into chemical energy by green plants in any ecosystem

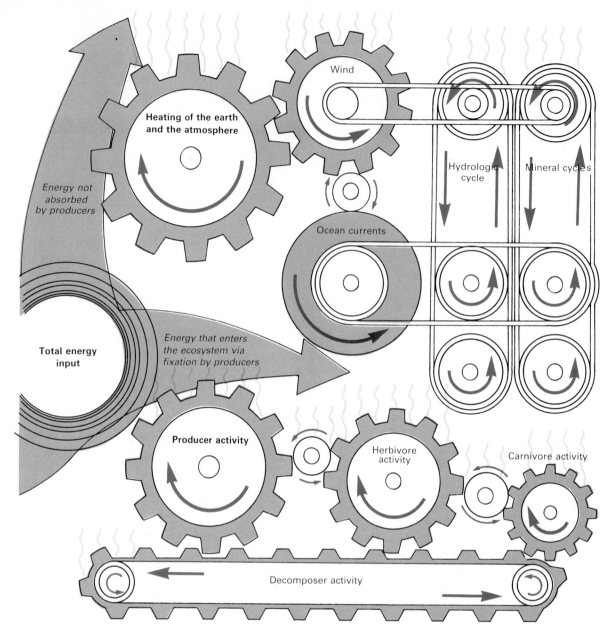

Figure 2.7 Almost without Exception, If It Is Happening, the Sun Is Making It Happen. The earth and its ecosystems can be thought of as rather inefficient interrelated engines, all powered by the sun. Mineral cycles driven by the sun are covered in Chapter 3.

energy from the time it starts out on the sun until it passes through an ecosystem here on earth.

Energy Flow Through Living Systems

About one-tenth of one percent (0.1%) of the energy that the earth gets from the sun is fixed or converted into chemical energy by plants in photosynthesis. One-tenth of one percent of what the earth receives as solar energy is a considerable amount of energy. In

the temperate zone the daily conversion of sunlight into chemical energy is equivalent to 300 to 400 calories per square centimeter (Odum, 1983).

Net Versus Gross Production
You may remember from Chapter 1 that the conversion of solar energy into chemical energy (photosynthesis) constitutes **primary production.** Primary production is the accomplishment of producers. We can also speak of **primary productivity,** the *rate* of production by plants—that is, how much production goes on per hour, per day, per minute, or per

year. Primary production or productivity is the rate at which the photosynthetic equation moves from left to right:

$$6CO_2 + 12H_2O \xrightarrow[\text{energy}]{\text{solar}} C_6H_{12}O_6 + 6H_2O + 6O_2$$

In the absolute sense, productivity means the total amount of this reaction that occurs in a unit of time. Of course, not all of the sugar made in photosynthesis goes on to become plant matter; the plants oxidize (respire) some of the glucose for the energy they themselves need. In effect, in plant respiration, plants reverse some of the photosynthesis they carry out. Because the proportion of photosynthesis that leads to accumulated plant mass varies from plant to plant and with varying conditions of a plant's physical and chemical environment, ecologists have found it useful to differentiate *net production* and *gross production*.

In all the earth's ecosystems, both terrestrial and marine, about 160 to 170 billion metric tons of dry organic matter are produced annually (Woodwell, 1970). This amount of organic matter is presented to the earth by plants each year and constitutes *net primary* production for the ecosphere. As was pointed out, plants use up some chemical energy each year for their own metabolic needs. The chemical energy plants use up plus *net* production equal *gross* production. The **gross primary productivity** of an ecosystem, then, is the total rate of the conversion of solar energy into chemical energy in that system. **Net primary productivity** is the rate at which plants make chemical energy available to consumers and decomposers or for storage or export. Depending on the type of plant and other factors, the amount of organic matter used up in plant respiration ranges from 15% to 75% of the total amount of organic matter produced by the plants (see Table 2.1).

Of the net primary production in an ecosystem, herbivores and decomposers may not consume all of the organic matter produced by plants in any given period. The expression **net community productivity** or *net ecosystem productivity* is sometimes used to describe the rate at which organic matter is accumulated (stored) in an ecosystem. Net community production (or net ecosystem production) is equivalent to net primary production minus the organic matter respired or used by consumers and decomposers. Sometimes the chemical energy that accumulates at the level of the consumer (e.g., kilograms of rabbit per year) is referred to as **secondary production.**

On a year-to-year basis, for the earth as a whole there is virtually *no* net ecosphere production (total annual added plant and animal biomass). Nearly all

that is produced is consumed in the respiration of plants, herbivores, carnivores, omnivores, and decomposers. The result is the near complete annual return to the environment of carbon dioxide and water, effectively reversing the chemical effects of the photosynthesis that went on during that year.

Plant material does not pile up in a mature forest. The thickness of the deposit of leaves on a forest floor stays about the same from year to year. During any year there seems to be about as much death, decomposition, and rotting as there is new plant growth and as there are new offspring.

At certain periods in the early development of an ecosystem, however—starting from bare, burned-over land, for instance—there are periods of considerable annual net ecosystem production. Net production in an ecosystem varies with its age and certain other factors.

Factors That Affect Productivity

The productivity of a particular plant or population of any plant species depends on growing conditions, that is, the factors that limit plant growth: access to water and nutrients, temperature, sunlight, wind, and slope, to name a few. Still other factors include physical characteristics of the soil, humidity, and, in aquatic environments, cloudiness (transparency) of the water. With variations in any or all of these, the same plant species would exhibit different annual productivities at different latitudes (see Enrichment Box 2.1) or even in different locations in the same field. Genetics is another factor. Different kinds of plants have different rates of production under similar circumstances; some are simply more efficient producers than others.

In general, net productivity is affected by the same factors that affect gross productivity. Some of these factors also affect the ratio of net production to gross production. Human beings long ago learned to identify plants with high *net* productivity as particularly useful for food or fiber. They also long ago discovered that some of the factors affecting gross and net productivity can be manipulated. Fertilizer use and irrigation are two examples. Phosphates and nitrates are limited in most ecosystems, and adding them can enhance the rate at which plants can convert solar energy into chemical energy, to human advantage.

Carbon dioxide may be another one of the limiting factors in production. The idea is plausible. Woodwell (1970) has suggested that the increase of about 10% in carbon dioxide in the atmosphere that has occurred since the mid-nineteenth century may have increased net production worldwide by as much as 10%.

The development of "miracle grains" has a spe-

Solar Energy and Climate

Although every place on the earth gets an average of 12 hours of sunlight per day over the course of a year, not all places get the same amount of solar energy or heat. The amount of energy delivered to a square meter of the earth's surface is a function of the angle of the sun's rays. The most energy is delivered per unit of area when the sun is directly overhead.

The plane of the earth's rotation and the plane of its orbit around the sun are such that the place on earth where the sun is directly overhead changes with both the time of day and the season. On any given day the sun will pass directly over all of the points over a certain line of latitude—a line or narrow stripe that goes all the way around the earth from east to west.

Over the course of a year the path of the overhead sun moves north and south from the extreme of the Tropic of Cancer (23.5° N) in the northern hemisphere at summer solstice, gradually to the equator, which it reaches at fall equinox, past the equator to the Tropic of Capricorn (23.5° S) in the southern hemisphere at winter solstice (shortest day of the year in the northern hemisphere). The *torrid zone* between these latitudes obviously gets much more solar energy during the year than other more northerly and southerly places. This accounts directly for the contrast between the richness of the flora and fauna in the tropics and the relative lifelessness of the polar regions.

The amount of energy delivered to each latitude during the course of the year is illustrated to the right. Note that in certain months the northernmost latitudes lose more energy than they gain. The mean temperature of the earth decreases 4°C for every degree of latitude at sea level.

The basic air movements or wind patterns for the globe are illustrated below. On the average, air tends to rise directly over the equator because it is heated most and becomes relatively least dense. Air moves in from either side of the equator to replace the rising air because of the earth's rotation; this "moving in" becomes the northeast and southwest trade winds. The air that rises over the equator falls again over 30 degrees north and south latitude, and this coupled with the air falling over the poles drives the prevailing westerlies.

Actually, because of the north-south movement of the overhead sun band, the latitude above which air rises most strongly also shifts from one side of the equator to the other during the year. The zone where the air rises most strongly is called the **intertropical convergence zone.** Shifting of the intertropical convergence zone causes seasonal rains in the tropics. It rains in the intertropical convergence zone, but to either side north and south there is no rain unless moving air is forced to go up over mountains.

This brings us to another point. Wherever air rises very far, it cools such that its water vapor eventually condenses and falls as rain. This accounts both for intertropical convergence zone rains, which fall practically every afternoon, and for the fact that where air descends, there is little, if any, rainfall. The latter explains why many deserts of the world are located near 30 degrees latitude. Elsewhere on the globe, air is forced to give up its moisture as rainfall when it must go up over mountains or where warm air is forced up over cooler air masses (see Chapter 3).

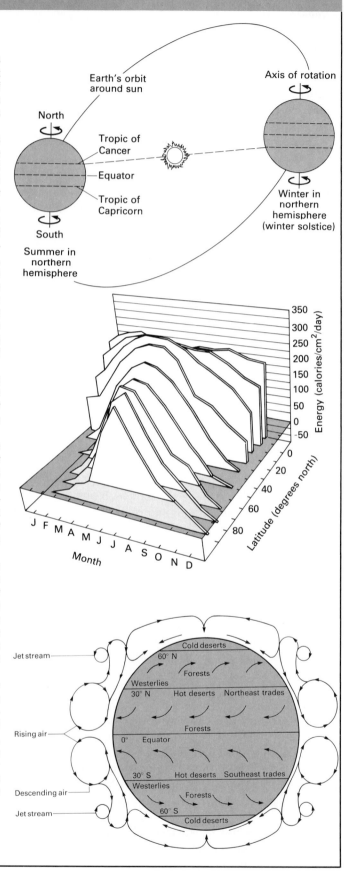

cial relationship to the concept of net production. Plant breeding and artificial selection have been used extensively to maximize net production and also to maximize net *edible* production. Not all of what plants provide as net production can be eaten by any one species. Plant geneticists have developed plant varieties that put maximum amounts of chemical energy into the parts (seeds and fruits, for example) that human beings or domestic animals can eat.

Productivities of the Major Subdivisions of the Ecosphere

Because the ecosystems of the world vary in almost every factor that influences net production, it is not surprising that annual rates of production vary considerably from one subdivision of the ecosphere to another (see Figure 2.8). Relatively high rates of production are characteristic of the tropics, swamps, estuaries, and marshes, places in which conditions are ideal for the growth of plants. Intensive agriculture, with its "artificial" addition of fertilizer and water and with pest and disease control, also favors relatively high plant productivity. Deserts, oceans, and tundras have low rates of primary production, attributable to combinations of limiting factors such as temperature, water, and nutrients. In the oceans, light penetration is a chief limiting factor.

The open oceans are relatively unproductive in comparison with terrestrial ecosystems. Although they cover most of the earth's surface, the oceans

Figure 2.8 Relative Rates of Primary Production in Some Natural and Managed Systems.   We have chosen to express annual production in terms of big calories or kilocalories (1,000 little or gram calories) here, but we could just as easily express it in terms of grams dry weight. An approximate average conversion factor is 4 kilocalories per dry gram.

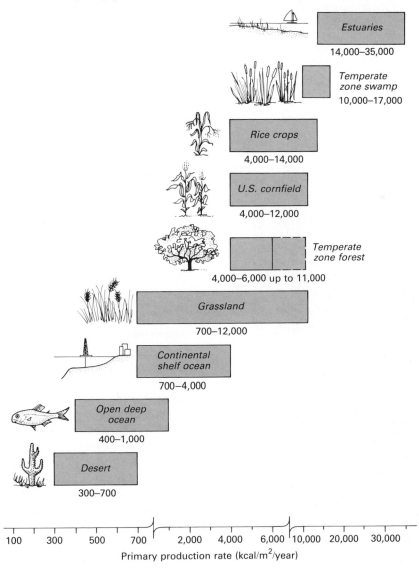

Estuaries
14,000–35,000

Temperate zone swamp
10,000–17,000

Rice crops
4,000–14,000

U.S. cornfield
4,000–12,000

Temperate zone forest
4,000–6,000 up to 11,000

Grassland
700–12,000

Continental shelf ocean
700–4,000

Open deep ocean
400–1,000

Desert
300–700

100   300   500   700   2,000   4,000   6,000 10,000   20,000   30,000

Primary production rate (kcal/m²/year)

Ocean ecosystems have feeding levels like those on land. The man-of-war is a secondary (or higher-order) consumer.

Productive ecosystems like forests may contain a great variety of species. The Indian pipe plant, which lacks chlorophyll, gets its food from decaying organic matter through an association of its roots with fungi.

contribute less than a third of the annual primary production on earth. With a mean productivity of 50 to 55 grams of dry organic matter per square (surface) meter per year (Woodwell, 1970), oceans may not represent the "vast exploitable source of food energy" described by some. Some authors, in fact, have suggested that the oceans are already being exploited for fish at close to the maximum sustainable rate (based on average measurable productivity) and that the continued use of the oceans as dumps for wastes may actually bring this resource to an even lower level of productivity (see Chapter 12).

The oceans are not all the same, obviously; some parts of some oceans are highly productive. In the open sea, relatively high productivity can be found in places where there are favorable combinations of warm temperatures, upwelling of nutrients, and sunlight. Estuaries, where oceans meet land and mineral-laden rivers, are among the most highly productive places on earth.

Forests, in contrast to oceans, are quite productive. Gross production in one forest on Long Island was shown by Woodwell and Whittaker (1968) to be in the neighborhood of 2,600 grams per square meter per year. Net community production in this particular forest system approached 1,200 grams per year, an amount suggesting that this ecosystem was immature; that is, it was still accumulating organic matter on a year-to-year basis. The forest in the Woodwell and Whittaker study had, in terms of solar energy availability, an efficiency of gross primary production approaching 1%. Other forests have been reported to have annual efficiencies of up to 3%.

According to the ecologist Eugene Odum (1983), on the average for all kinds of ecosystems, about 50% of gross production is available to plant eaters and decomposers. It is important to remember, however, that this is highly variable.

How Is Primary Productivity Measured?
The most important activity-determining, life-supporting characteristic of an ecosystem is its primary productivity. One can derive a lot of information about an ecosystem if its primary productivity is known. For otherwise suitable habitats we would know roughly how many herbivores—be they cattle, rabbits, or deer—could be supported in a system of known productivity. Because of its importance, ecologists have devised a number of techniques and methods for measuring primary productivity.

All of the techniques used to measure productivity are based on the fact that the photosynthetic equation is a *balanced equation*. The net effect of the very complex photosynthetic process is that 6 moles of carbon dioxide (CO_2) (264.066 g) and 12 moles of water (H_2O) (216.192 g) are incorporated into every

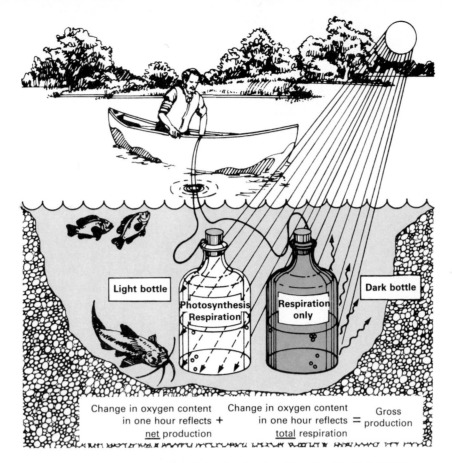

Figure 2.9 **Estimating (Gross) Production in a Pond by the Light Bottle/Dark Bottle Method.** The equation shows how gross productivity is calculated. Net productivity during one daylight hour can be estimated from the change in oxygen content in the light bottle.

mole of glucose ($C_6H_{12}O_6$) (180.162 g) and every 6 moles of oxygen (O_2) (192 g) and 6 moles of water (H_2O) (108.096 g) produced in photosynthesis. (A **mole** is the molecular weight of a compound expressed in grams.)

Some relatively simple ways of estimating *net* productivity include (1) measuring the amount of oxygen "produced" in a system per unit time, (2) measuring how much carbon dioxide is "consumed" per unit time, and (3) harvesting, drying, and weighing the plant material that accumulates in a period of time. (Recall that oxygen and carbon dioxide are not literally produced or consumed but are converted or result from conversions.) Estimates can also be made by determining how much chlorophyll is present in an ecosystem. For each of the methods, the facts that respiration and photosynthesis are coupled in living systems (the raw material for one comes from the other) and that they are going on simultaneously must be taken into account.

Light Bottle, Dark Bottle: A Method for the Separation of the Effects of Respiration and Photosynthesis. Suppose we lowered two

bottles into a pond 1 meter with the bottles weighted somehow so that they could be lowered to that level while empty (Figure 2.9). Suppose we also had a device for opening the bottles at the level of 2 feet and then resealing them. Such a device would allow water and pond organisms to enter the bottles; consumers and producers would be included. Suppose we filled the bottles on a particular day at noon, sealed them, and permitted them to remain in place for two hours.

In both bottles, production and respiration would go on. Carbon dioxide and water would be taken up in *photosynthesis* and released in *respiration.* If we had a way of measuring changes in these substances, in carbon dioxide content, or in oxygen and glucose that were produced, we would be measuring the net effects of these two generalized equations. We would be measuring *net pond production* or *net ecosystem production.*

Suppose we had painted one of the bottles black so that no sunlight could get to the inside. What then? Obviously, photosynthesis could not go on; only respiration could. Oxygen and organic material would be consumed, and carbon dioxide and water

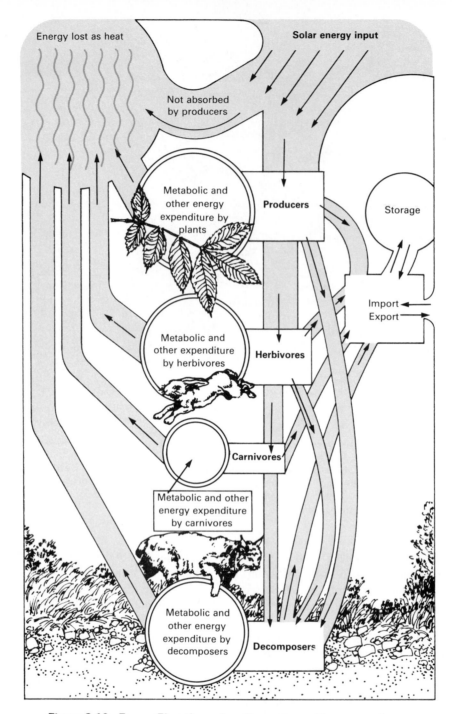

Figure 2.10 Energy Flow through an Ecosystem. The amount of energy traveling by each path may vary greatly from one system to another, but the pathways are basically identical.

would be produced as a by-product; no oxygen or glucose would be *produced*. With both a light and a dark bottle we would have a way of estimating how much *gross production actually takes place.*

By measuring the amount of oxygen in both bottles at the beginning and at the end of a period of time we would know two things: (1) how much oxygen was consumed in respiration (dark bottle oxygen change) and (2) how much oxygen was produced (light bottle oxygen change) in a bottle in which *both*

respiration and photosynthesis were going on at the same time. Assuming that the same amount of respiration went on in both bottles, if we subtracted the change in oxygen in the dark bottle from the change in oxygen in the light bottle, we would have an estimate of the amount of oxygen that was produced as a by-product of photosynthesis per unit time

Net productivity can also be measured in terms of biomass, the weight of (dry) organic material accumulated over a period of time. If we have such data in grams, let's say, all we would need to do is sample some of that biomass in a device in which the caloric content of organic material can be measured, that is, by the amount of heat released when that organic material is combined with oxygen. We would then have a precise measurement of the energy content of that biomass and could express the material that accumulated in an ecosystem over time as energy in calories.

Specialized Ways to State Production.
The U.S. Department of Agriculture periodically publishes estimates of the number of bushels of corn, wheat, and other crops that will be produced in one year in the United States. This is a useful expression of net production. As we stated earlier, humans are consumers of only certain kinds of food. We are not really interested in the weight of wheat stalks produced; we are interested in the weight of the grain. As we will see in Chapter 6, the use of biomass as fuel may change that focus.

What Happens to Energy in Ecosystems Once It Is Produced?
Figure 2.10 summarizes how energy flows into, through, and out of ecosystems. As we have already discussed, plants utilize some of the chemical energy that they produce for their own activities. The same pattern applies to herbivores and to carnivores higher up in the food chain. Herbivores and carnivores take some of the energy they capture and use it to do the things that they do, that is, grow, carry out metabolic activities, and expend energy through various mechanical activities. Some of the energy assimilated or fixed at each level is available to the next level up—to the carnivores in the form of herbivores. In

every case, some of it ends up going to the decomposers; decomposers use up the last bits of "useful" energy that enter an ecosystem, releasing it ultimately as heat.

Note in Figure 2.10 that not all of the energy that comes into an ecosystem comes in as solar energy; some of it is imported. Leaves blown into a pond from a forest are an example of imported chemical energy. A frog that eats an insect that flew out of a forest is importing energy. A forest bird that eats a fish from a pond is carrying energy in the opposite direction, exporting it from the *pond* ecosystem. Similarly, a raccoon that eats a freshwater mussel from a pond is an exporter of energy from the pond ecosystem.

Chemical energy may also be temporarily stored out of the mainline of energy flow in an ecosystem. Coal is an example of a long-term version of such storage. Organic matter remaining as humus over a period of time would qualify as a short-term form of energy storage.

Efficiencies of Energy Transfers Within Ecosystems
We have already mentioned a number of times that the efficiencies of primary production are very low; less than 1% of available energy is captured by plants and ecosystems. Beyond the producers, efficiencies of transfer increase a bit. Here, too, there may be considerable variation from one ecosystem to another. Depending on the ecosystem, and certain physical and chemical factors, up to 50% of the chemical energy present in the form of producers is transferred to the primary consumers. In turn, somewhere between 5% and 30% of the energy present in

---

---

the form of primary consumers makes its way to the secondary consumers or carnivores. *On the average, about 10% of the energy entering a particular feeding level is transferred to the next level.* That is, if 10% of the energy available in the form of plants ends up being incorporated into secondary consumers, then 10% of that 10% is available to the next, or tertiary, level of consumption. The *10% law,* as it is sometimes called, is one of the reasons why if you set out to count the number of animals you see in a forest, you will count more herbivores than carnivores. It is much easier to find plants as you walk through the woods; it is somewhat more difficult to find a primary consumer; and only rarely does one encounter a secondary consumer such as a mountain lion.

Like the cabbage looper, all primary consumers must eat huge amounts of vegetation to satisfy their energy needs.

Pyramids: Why There Are More Plants Than Plant Eaters

The second law of thermodynamics forces the "shape" of ecosystems into so-called ecological pyramids; the number of ecosystem components *decreases* with distance from the primary producer level (see Figure 2.11). A rapidly diminishing amount of energy is available as the number of steps away from the primary producer increases. Because of this, typically only three to five feeding levels will exist in an ecosystem. We have been referring up to this point to feeding levels, but ecologists call them **trophic levels** (*trophic* coming from the Greek word meaning "to eat" or "to feed"). Figure 2.11 illustrates that the higher the trophic level, the fewer organisms there are. Since some organisms are big and some are little, perhaps this relationship should more accurately

be expressed in terms of mass—or even better, in terms of energy. If we dried and weighed all the organisms in a representative area of any ecosystem, we would have to infer that it takes hundreds of thousands of kilograms of (dry) producer biomass to support tens of thousands of kilograms of herbivore biomass to support in turn less than a kilogram of carnivore mass.

By measuring the caloric content of the different types of biomass at each level we could express biomass in terms of caloric (energy) content by assuming an average caloric value of dry organic matter. Kormondy (1984) uses the value of 4 kilocalories per dry gram as an arbitrary average (this, according to Kormondy, is somewhat on the low side). At any rate,

Figure 2.11 The Shape of Ecosystems. Practically any way of looking at producers and consumers—in terms of numbers, weight, or energy content—yields a pyramid shape that results from the inefficiency of energy transfers, the use of some energy by the members of each trophic level, and the fact that some energy goes directly to decomposers and storage from each level.

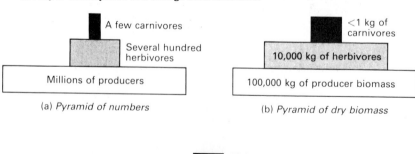

A few carnivores

Several hundred herbivores

Millions of producers

(a) *Pyramid of numbers*

<1 kg of carnivores

10,000 kg of herbivores

100,000 kg of producer biomass

(b) *Pyramid of dry biomass*

10,000 calories in carnivores

1 million calories in herbivores

10 million calories in producers

(c) *Pyramid of energy content*

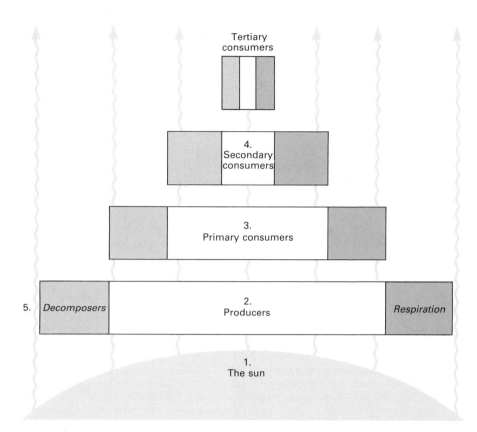

Figure 2.12   The Dynamics of the Ecosystem Pyramid.   Energy dissipates as heat as it passes through an ecosystem. 1. Electromagnetic energy comes to an ecosystem from the sun. 2. A small part of solar radiation is changed into chemical energy by producers. 3. Some plant chemical energy ends up in herbivores or primary consumers. 4. Some herbivore chemicals go to carnivores (secondary consumers). 5. Some of the chemical energy at each level goes to decomposers. The organisms at each level also use up some energy for their own metabolic activities. Because of the inefficiency of each energy transfer step, not much energy is left after four or five transfers.

tens of millions of calories in the form of producers are needed to support millions of calories in the form of herbivores, to support tens of thousands of calories in the form of carnivores, and so on. We depict just three trophic levels in Figure 2.11, but some ecosystems may have a few more. They may all have multi-level consumers. Figure 2.11 also ignores some of the dynamics of energy flow; for instance, no indication is given of the energy lost to respiration and decomposition.

Regardless of what units are used, it should be appreciated that in nature, pyramids can change shape from moment to moment. They can even become *inverted* for short periods. Fill an aquarium bowl with pond water and put it on the windowsill in the sun. Cover the bowl with something to reduce evaporation. If you are lucky, the aquarium will become pea soup green and then clear up from time to time. Clear-ups are caused by the alga eaters' experiencing great population booms. Unfortunately for

them, their populations peak just as the algae are nearly all eaten up. At that point the pyramid in the aquarium is inverted. Can this last? Obviously not.

The dynamics of biological pyramids are perhaps better illustrated in Figure 2.12, which shows that some of the chemical energy at each level is destined to pass through the decomposers. In essence, Figure 2.12 illustrates the energy that comes to an ecosystem from the sun and dissipates as it moves through the trophic levels, eventually to be converted completely into heat.

Biological pyramids such as those illustrated in Figures 2.11 and 2.12, if properly derived for a given ecosystem, can provide valuable information about the nature of that ecosystem. Such data can be used to make general comparisons between ecosystems of various types in various locations. Such data provides a way of expressing the absolute natural value of a particular piece of land in terms of biological production. Such data could allow questions to be answered such as how vital a piece of territory is as a spawning ground or what the natural impact would be of diminishing the habitat for certain kinds of producers or consumers in a particular place—a swamp or an estuary that somebody wanted to fill in, for instance.

Food Webs and Food Chains

The figures mentioned in our discussions are simplified; they do not take into account that each trophic level of an ecosystem is occupied by a wide variety of species. This is illustrated in Figure 2.13. One of the characteristics of an ecosystem is the number and nature of the species that occupy its various trophic levels. We generally find that although many species can occupy each trophic level in any ecosystem (for example, buffalo, grasshoppers, and field mice may all occupy the same trophic level in a prairie ecosystem), certain species may *predominate* at each level. A certain kind of grass may dominate a field ecosystem, a few species of trees may dominate a forest ecosystem, and a certain type of alga may carry on most of the photosynthesis in a stream ecosystem.

Figure 2.13 also shows that there are complex relationships between the constituents of one trophic level and constituents of adjacent trophic levels. Ecologists sometimes refer to the relationship between trophic levels as *food chains;* at other times they are

Figure 2.13   A Food Web.   Many species can occupy the same trophic level in any ecosystem. Certain species may predominate, however, and some species occupy more than one level. This illustration does not capture the essence of nature in the raw. We could not possibly represent something so dynamic and ever-changing on paper. To cite just a few of the things this figure fails to show: Predators change their prey from season to season; some predators and herbivores eat different things when they are young than they do when they grow up; some predators are more flexible than others when their food of choice decreases in supply.

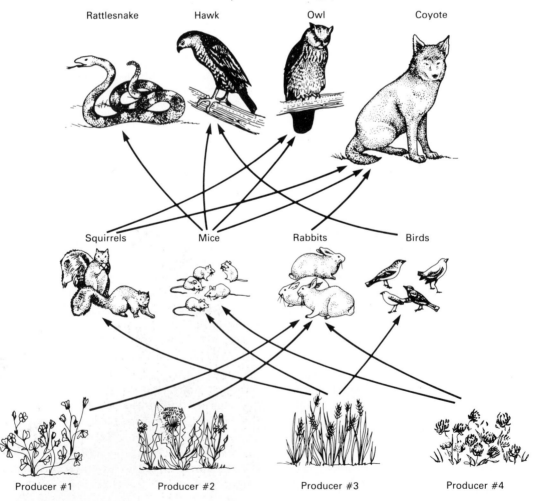

A consumer may feed on many different species. This seven-spotted ladybird beetle has found one of its favorites, a pea aphid.

somewhat more accurately described as *food webs*. The web is a much better analogy to the actual nature of the interactions between trophic level components. As Figure 2.13 shows, the occupants of a primary consumer level may eat (draw energy and material from) a number of producer species. Similarly, carnivores draw their food from a number of individual species at the primary consumer level. Ecosystems vary considerably in the design of their energy-nutrient webs.

Complexity and Stability

Generally speaking, the more complex the combination of trophic level interrelationships in a particular ecosystem, the more stable the system. It would be considered a somewhat unstable condition if an industry depended on a raw material available from only one source. Any disruption at the source or in the line to the source would mean a critical problem for that industry. Similarly, if all of the herbivores in an ecosystem depended on a single species of producer, the system would not be very stable. A disease or anything else that caused the demise of that producer would be immediately felt at the consumer level. Ecosystems that are in some kind of relative equilibrium and in which many different kinds of plants and animals fill similar niches should be better able to adjust to problems with any one of the species. It does not necessarily follow from this that the

more species an ecosystem has, the better it is able to withstand disruption. Many other factors go into determining the ability of an ecosystem to resist alteration or to return to its original condition after being altered. These are discussed in Chapter 4.

We will see in Chapter 4 that the word *stability* has several ecological meanings and that the ability of an ecosystem to persist in the face of disruption is only loosely connected to the number of species in that ecosystem. You should also begin to see more clearly that ecosystems are nearly always in some state of turmoil, responding to or recovering from severe winters, dry summers, spring floods—some sort of environmental change.

Moving On

It should be obvious that the concepts covered in this chapter have much to do with problems that face humankind in its relationship to the natural environment. There are starting points here for improving our system of agriculture, our energy policies, and our prospects for solving the world hunger problem. Arguments could be developed from t[...] sented here to support the develo[...] power. Explanations could be develop[...] terial presented here on the following [...]

— Why the oceans may not be the answer to the problems of feeding the people in this world
— Why people in poorer countries eat very little meat

— Why you cannot eliminate coyotes and expect no environmental repercussions
— Why there is danger in humanity's relying on fewer than a dozen plant species for nearly all of its food

Concepts to Remember

1. There are many forms of energy; what they all have in common is the ability to do work. Work can be thought of as all of the things that plants and animals do—running, growing, fruiting, flowering, growling, squeaking, and so on. All living things must have energy sources in order to do what they do.

2. The second law of thermodynamics says, in essence, that as the energy in a closed system is used, it becomes less useful to do work—it wears out in the functional sense even though the first law of thermodynamics says that energy cannot be destroyed. A consequence of this is that all living systems need continuous supplies of new energy.

3. Practically all the energy that drives ecosystems comes from the sun. Of the total amount of solar energy that strikes the earth's outer atmosphere, less than half reaches the surface. A significant fraction of the solar energy intercepted by the earth determines the earth's weather and climate. Because of the earth's shape, its orbit, its surface topography, and its atmospheric composition, different amounts of incident solar radiation spell the difference between a tropical paradise and an inhospitable polar region.

4. All of the solar energy intercepted by the earth is eventually returned to space.

5. Only a tiny fraction of the solar energy that strikes vegetation is converted into chemical energy—less than 1% on the average.

6. About half of the chemical energy fixed by green plants is available to consumers; plants use the rest themselves.

7. Productivity varies from one kind of ecosystem to another and from one time to another. The availability of water, the amount of minerals, and many other factors (in addition to incident radiation) limit productivity in different ecosystems.

8. Energy dissipates as heat as it flows through ecosystems.

9. On the average, about 10% of the energy entering a particular trophic level is available to the next level in an ecosystem.

10. A web is a better metaphor than a chain to describe the flow of energy through most ecosystems.

References and Further Reading

References marked with an asterisk are cited in the chapter.

During, A. B., 1991. "Fat of the Land," *World Watch*, May-June, 11–17. An argument for eating lower on the food chain.
*Kormondy, E. J., 1984. *Concepts of Ecology*, 3d ed. Englewood Cliffs, N.J.: Prentice Hall.
*Lewin, R., 1986. "In Ecology, Change Brings Stability." *Science* **234**:1071–1073.

Montieth, J. L., 1965. "Light Distribution and Photosynthesis in Field Crops." *Annals of Botany* (new series) **29**:17–37.
*Odum, E. P., 1983. *Basic Ecology*. Philadelphia: Saunders.
*Woodwell, G., 1970. "The Energy Cycle of the Earth," in *The Biosphere*. New York: Freeman.
*Woodwell, G., and Whittaker, R., 1968. "Primary Production in Terrestrial Ecosystems," *American Zoologist* **8**(1):19–30.

See also the references cited at the end of Chapter 1.

3

Material Cycles in Living Systems

The alarm has just gone off. You awaken to face another day. After pulling yourself out of bed, you wash up, relieve yourself of the wastes you accumulated through the night, dress, have a bite to eat, and depart. Already you have had some impact on the chemistry of your environment.

As you breathed, you exchanged carbon dioxide (CO_2) for oxygen (O_2). As you washed, you added chemicals to the water—hydrocarbons and oils from your skin and soap. Your wastes added other chemicals. During breakfast you took in nutrients produced by plants or animals, and later you broke some of these down into materials your system could use. Big deal, you say? But over 5 billion people around the globe are doing the same thing. Surely this *is* a big deal, chemically speaking. It would be a big deal even if modern civilization did not also have a chemically intense technology.

We humans move a lot of chemicals around. It follows that if chemicals regulate ecosystems, our chemical manipulations must be having an impact on living systems. Here are some examples of environmental problems related to the things we humans do with chemical substances.

__ Nutrients in organic wastes that humans flush or drain into aquatic systems accelerate plant growth, resulting in the choking of waterways and, as a result of the ultimate death and decay of this vegetation, the depletion of oxygen.

__ Certain agricultural practices have depleted nutrients from some soils and added harmful salts to others.

__ Certain human-made chemical compounds that have been released in our environment—vinyl chloride, EDB, and benzene, for example—cause cancer.

__ The rapid release of carbon and sulfur from burning coal has increased the amounts of these elements in harmful chemical forms in the atmosphere—for example, as carbon monoxide (a poison), carbon dioxide (which may alter the earth's climate), and sulfur dioxide (a poison). The transport of sulfur dioxide by wind has resulted in acid rain hundreds of miles from the source of the chemical.

__ We have added ozone to the lower reaches of the atmosphere, where it acts as a poison, and we have removed it from the upper reaches where it serves to shield us from harmful ultraviolet radiation.

So that we might understand these and other problems related to the movement of materials in nature, we will review in this chapter some of the cycles particularly important in the regulation of ecosystems.

Chemicals in Motion: Cycles in the Ecosphere

We saw in Chapter 2 that energy flows through ecosystems. The emphasis was on the word *through*. Energy is generated from matter in the sun; as energy passes through living systems, enabling organisms to perform various kinds of work, all the energy is ultimately dissipated as heat. It is gone forever in terms of usefulness to the system. As we have already suggested, *matter* is quite another matter.

Of the more than 100 elements, only 20 to 30 are constituents of living things. Some are needed in relatively large amounts (**macronutrients**); some are needed only in trace amounts (**micronutrients**). The major macronutrients are found in all organisms. The minor macronutrients and the micronutrients are found in varying amounts in different species (see Table 3.1).

The basic chemical elements or nutrients that make up the substance of living systems need not be supplied continuously from the outside because the *substance* of ecosystems is continuously cycled. Since chemical elements are not used up but rather "used

over," one might wonder how they can ever be in short supply. The answer lies in the fact that the availability of an element can be limited by its chemical form, its physical location in relation to living things, and the rate at which it passes from one form or location to another.

Chemical Form

Individual chemical elements can exist in combination with other elements, that is, in different molecular forms or compounds, the smallest units of which are called **molecules.** Some of these compounds cannot be used by individual organisms to gain access to individual essential elements within them. Plants need carbon, for example, but they generally cannot get it from methane (CH_4), sugar ($C_6H_{12}O_6$), carbon monoxide (CO), or any of thousands of other carbon compounds. Carbon must be supplied to plants in the form of carbon dioxide (CO_2). Similarly, plants all need nitrogen, but most of them must have it in the form of soluble nitrate (NO_3^-) or ammonia (NH_3).

Location

A second general factor that limits the availability of essential chemicals for living systems is physical location. We mentioned in Chapter 1 that most organisms are found where air, water, and earth meet. Conditions that favor life diminish with distance from the thin film at the surface of the earth. Deep within the oceans there is not enough light or oxygen to support much life. Deep within the earth

Table 3.1 Relative Amounts of Chemical Elements That Make Up Living Things

Major Macronutrients (>1% dry organic weight)		Relatively Minor Macronutrients (0.2–1% dry organic weight)		Micronutrients (<0.2% dry organic weight)	
Name of Element	Symbol	Name of Element	Symbol	Name of Element	Symbol
Carbon	C	Calcium	Ca	Aluminum	Al
Hydrogen	H	Chlorine	Cl	Boron	B
Nitrogen	N	Copper	Cu	Bromine	Br
Oxygen	O	Iron	Fe	Chromium	Cr
Phosphorus	P	Magnesium	Mg	Cobalt	Co
		Potassium	K	Fluorine	F
		Sodium	Na	Gallium	Ga
		Sulfur	S	Iodine	I
				Manganese	Mn
				Molybdenum	Mo
				Selenium	Se
				Silicon	Si
				Strontium	Sr
				Tin	Sn
				Titanium	Ti
				Vanadium	V
				Zinc	Zn

Source: Data from Kormondy (1984).

Dissociation: the separation of ions from one another in water, e.g., $NaCl \rightleftharpoons Na^+ + Cl^-$.

Element: any of the more than 100 fundamental units of matter that consist of atoms of one kind (some examples: boron, carbon, chlorine, hydrogen, oxygen, nitrogen, sulfur).

Essential element: an element needed by an organism, usually in some particular chemical form from the environment. Phosphorus is an essential element for plants, but the element phosphorus is poisonous. Plants usually require phosphorus in the form of phosphate (PO_4^-) salts. In this text we sometimes use the phrase **mineral nutrient** as a synonym for essential element.

Ion: an atom or group of atoms with a positive or negative charge; the charge is the result of the loss or gain of electrons. Positive and negative ions attract one another, e.g., $Na^+ + Cl^- \rightarrow NaCl$.

Molecule: the smallest unit of either an element or a **compound** (combination of two or more atoms) still having the chemical characteristics and identity of that element or compound. Molecules may contain more than one element. For example, water molecules (H_2O) contain two atoms of the element hydrogen (H) and one atom of the element oxygen (O).

Organic: generally speaking, of living material or material derived from living things. The strict chemical definition pertains to chemicals formerly found only in living things or derived from living things but now including these and other hydrocarbon compounds. **Inorganic:** generally speaking, not of or derived from living things; chemically, compounds that do not contain hydrocarbons.

---

there is too little oxygen, little or no water, and *no* light. High in the atmosphere there is little chemical matter, and it is too cold. The point is that regardless of chemical form, key nutrients can be lost to living systems for short periods or geologically long periods by being lifted high into the atmosphere, by becoming incorporated in deep ocean sediments (and eventually into rock), or by becoming embedded in silt, to give just a few examples.

Reservoirs. There are places in the ecosphere that serve as major abiotic reservoirs for chemical elements. Consider nitrogen. Though the sizes are not drawn to scale, the relationship between the various reservoirs of nitrogen might look as shown in the following diagram:

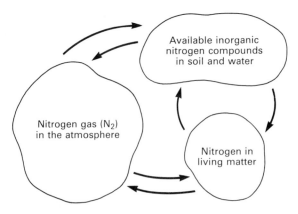

The reservoirs are not uniform. There is far more nitrogen in the form of nitrogen gas than there is in the form of available nitrogen compounds in the soil or in the form of organic compounds.

Chemicals move from one reservoir to another

in cycles. This means that the size of any reservoir and the accessibility of nutrients to living things depend on the patterns of movement and the rates at which nutrients move from one reservoir to another.

Pathways and Rates

The cycling of chemicals may be long, complex loops or short loops. This is a short carbon loop involving plants and animals:

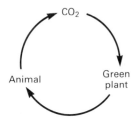

This is a long carbon loop involving plants:

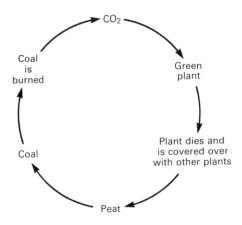

Living things participate in the movement of

Congestion and One-Way Traffic Patterns

Our failure to consider biogeochemical cycles is a root cause of many environmental problems. Sometimes we humans simply dump too much waste too rapidly into ecosystems, and the systems can't move it along fast enough. We are doing this with carbon dioxide, for example. By burning fossil fuels, we are putting carbon dioxide into the air faster than the ecosphere can move it along. The resulting increase in atmospheric carbon dioxide may in turn cause global warming. Sometimes we fail to consider natural cycles as we force materials to move in a one-way flow. Given the ecological reality of biogeochemical cycles, the ways in which we use material and handle waste go against the grain of nature. We extract minerals from mines at one end of a line and at the other end, we pile up mineral-laden materials (which we call ''junk'' or ''garbage'') into landfills. It seems there should be a better way. What do you think?

chemicals into and out of various reservoirs and into and out of various chemical forms, hence the term **biogeochemical cycles.**

The rate at which any particular chemical nutrient makes a complete loop is determined by the number of steps in the cycle and the nature of the specific steps involved. Again using the carbon loops just illustrated, the loop with the coal formation in it would take much longer than the one in which animals are involved. Many plants that long ago converted carbon dioxide into plant matter were themselves later converted into coal. The burning of coal releases carbon that had been tied up for millions of years. Another example of a long-term loop involves calcium. An-

cient crustaceans whose shells were made of calcium carbonate have ended up as carbonate rocks, where the carbon has been held for millions of years. Living things also have important short-term effects on the availability of mineral nutrients. In certain aquatic systems, plant growth can take up the limited available phosphate, tying up the phosphate in the substance of the plants such that any additional plant growth is limited by the lack of phosphate.

The most ecologically important fact about the steps or stages of chemical cycles is that individual steps can be slowed, blocked, or accelerated. Throughout this book there will be examples of how blocking a portion of a cycle by interfering with it in

During some periods of the earth's history, enormous amounts of water have been locked up in glaciers.

some way can result in a harmful accumulation of materials at one point or a harmful depletion of a component in the cycle at some other point.

Cycles and Energy

Biogeochemical cycles are dynamic processes in which atoms are recombined and rearranged and the starting materials are regenerated. Progress through portions of all such cycles requires energy, and the principal energy source driving biogeochemical cycles is the sun. Solar energy powers the processes of evaporation and condensation that move water in cycles. "Water power" in turn becomes the mechanism for weathering and erosion, which moves minerals through long stretches of various geochemical cycles. Solar energy, converted into chemical energy through photosynthesis, powers the portions of the biogeochemical cycles that involve living things. Without the sun, essentially all biogeochemical cycles on the earth's surface would run down and stop. It turns out that water (as a solvent and as an agent of weathering) is the principal medium through which solar energy is applied to the movement of chemicals in the ecosphere. For this reason, though water is not a nutrient in the sense that other chemicals are, we will begin our consideration of mineral cycles with the water cycle.

The Hydrologic Cycle

Water: A Unique Solvent

Water is the universal internal medium of all living things; living things are made of 90% or more water. Water is also the external medium of all aquatic life forms. The special relationship of life and water stems from the fact that water is a relatively universal solvent; almost anything will dissolve in water to some degree.

Water is the most essential and the most unique of all compounds. Water is clear—transparent to light. This means that photosynthesis can occur in water, at least to certain depths. Consequently, producers—green plants, the foundation of the biosphere—can grow, flourish, and serve as the nutrient source for aquatic consumers. Water has a high **heat capacity;** this means that it can absorb much heat energy with a relatively small increase in temperature in comparison with other substances. Because of this high heat capacity, aquatic life forms do not have to be adaptable to a wide range of temperatures or to rapidly changing temperatures.

Water becomes increasingly dense and thus heavier with lower temperature down to 4°C. This is why deeper water feels cooler in a lake. Because the maximum density of water occurs at 4°C, water be-

comes increasingly lighter at 3°C, 2°C, 1°C, and 0°C (freezing). Thus water is heavier as a liquid than as a solid; this is why ice floats in a glass of water. Another important consequence of these properties is that water freezes from the top down. This protects aquatic organisms because ice acts as an insulator to prevent further decreases in temperature in the remaining water and decreases the chances that ponds and lakes will freeze solid.

Water Does Not Always Flow Downhill

Although we do not usually think about it, water is moving around us constantly. It is not always flowing downhill; a good deal of it is being pulled up as individual molecules into the atmosphere by the sun. Solar energy causes water on the surface of the earth to **evaporate** (change from a liquid to a gas) and enter the atmosphere. When water evaporates from the exposed parts of plants, such as leaves, the process is called **transpiration.** (Transpiration is the driving force that raises water and dissolved substances from the roots into the rest of the plant.) Once in the atmosphere, water vapor is carried by the wind and may meet any of several ends, all of

Water movement is cyclical. Surface water evaporates into the atmosphere, where it condenses and returns to the earth as precipitation, which will eventually evaporate once again.

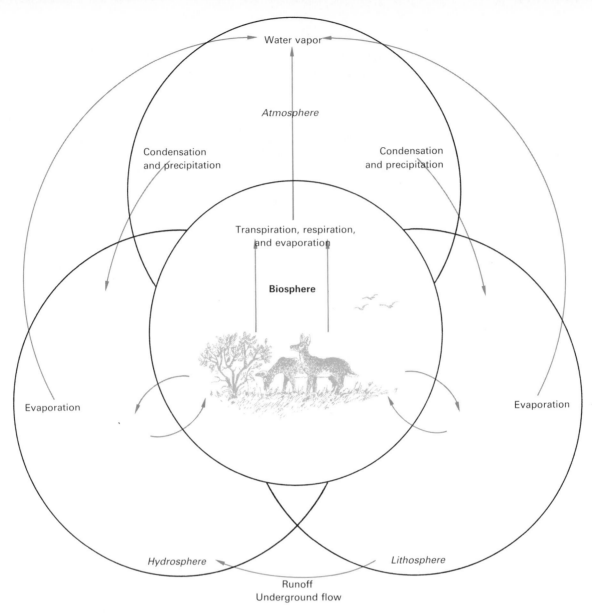

Figure 3.1 The Hydrologic or Water Cycle. This cycle consists primarily of the processes of evaporation, condensation, and precipitation. Water passes through all of the great spheres, carrying with it various mineral substances, some of them important nutrients for plants and animals.

which ultimately involve **condensation** (changing from a gas to a liquid or a solid) as a result of cooling. Following condensation into larger and larger water droplets or ice crystals, water returns to the surface of the earth as precipitation. Thus the complete **hydrologic cycle** includes evaporation, condensation, and precipitation—with intermediate effects and stops throughout (Figure 3.1).

The amount of water in circulation annually is only a small percentage of the total. The oceans constitute a reservoir for over 97% of all the water on earth. A tiny fraction is held in the atmosphere, and the rest exists as ice, ponds, lakes, streams, and subsurface water.

Water that falls as rain or snow on land may seep into the soil, where it becomes groundwater.

Some rock layers are more permeable to water than others, and if groundwater reaches a rock layer that is relatively impermeable, the water tends to collect above the layer and may form an **aquifer**—an underground water storage area (Figure 3.2). The upper surface of such underground water is called the **water table.** Some precipitation ends up as runoff carrying sediment and dissolved minerals to lakes, streams, and rivers, from which some water evaporates and the rest makes its way to the oceans.

Terrestrial organisms influence the hydrologic cycle in many ways. Plant cover diminishes the force of impact of raindrops and reduces erosion. Organic matter in soil acts as a sponge holding water in place between rains. Plants extract water from soil and give off water vapor from leaves via transpiration. Transpi-

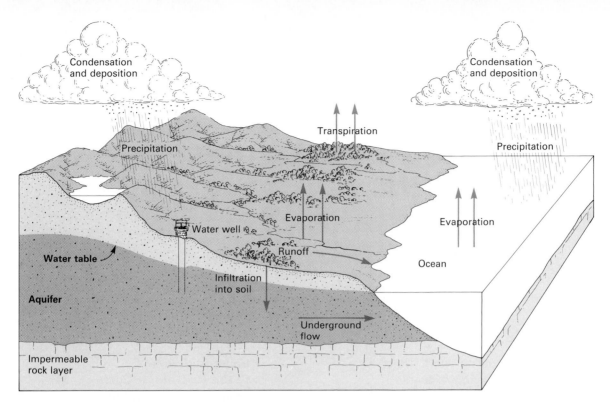

Figure 3.2 The Formation of Aquifers. Water that seeps into the ground may eventually reach relatively impermeable layers of rock. The water then tends to accumulate above this layer, forming aquifers that may be tapped for use via wells. The upper level of the groundwater, called the water table, fluctuates somewhat depending on rainfall, rates of withdrawal by humans, and the rate at which the groundwater flows to still lower levels.

Figure 3.3 Global Wind and Rainfall Patterns. Sunlight warms the atmosphere in some places more than others, causing movement of the air or winds in predictable patterns. The trade winds and westerlies are the world's major wind patterns. Also shown on this map is the mean annual rainfall in various parts of the world.

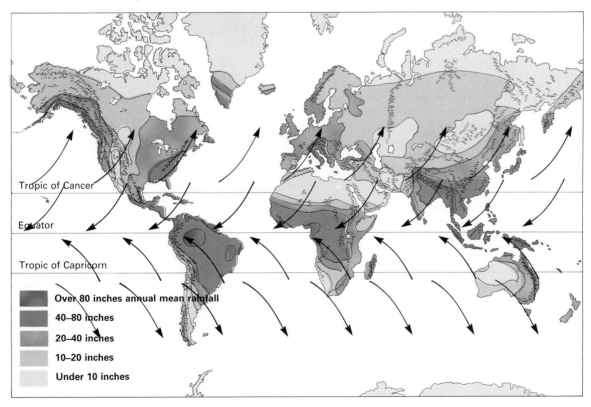

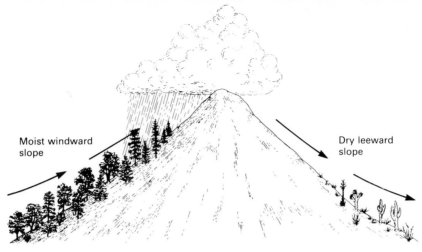

Figure 3.4   The Effect of Air Flow over Mountains on Rainfall Distribution.   As moist air rises and cools, its ability to hold water decreases, and the moisture falls as precipitation. As cool air is forced up and over mountains, it tends to drop its moisture going up on the windward side, creating a dry area (rain shadow) or even a desert on the other side.

ration in turn has a cooling effect, which helps to moderate temperature and thus climate.

Rainfall Patterns: Water, Wind, and Weather

We described the relationship between global rainfall patterns and the patterns by which the sun heats the earth's atmosphere in Chapter 2. Figure 3.3 shows how the major global wind patterns and topography combine to produce wet places. A graphic reminder that rain tends to fall on the windward side of mountains as air is forced upward and cooled is presented in Figure 3.4. Certain land-water junctions have high annual precipitation even where no mountains are involved because of the differential air temperatures over land and water and the resultant more frequent meeting of cold and warm moisture-laden air masses. (See Enrichment Box 3.1)

Human Activities and the Hydrologic Cycle

Humans affect the hydrologic cycle in several ways. For example, if flowing water is pooled by dams, the evaporation rate changes. Standing water over a large surface area can absorb more heat energy from the sun to vaporize water molecules. This in turn may increase precipitation somewhere else. Clear-cutting a forest or building a parking lot reduces water seepage and increases runoff, thus reducing the amount of groundwater and increasing the risk of flooding. In terms of direct human need, if water is withdrawn from a stream or an aquifer faster than it is replenished, a water shortage is sure to follow. If overuse were not bad enough, we humans continue to decrease our access to high-quality water by polluting it. There will be much more on human water problems in Chapters 7 and 12.

Enrichment Box 3.1

El Niño and La Niña

The tropical Pacific Ocean is thought to have a significant impact on short-term climatic changes. You will from time to time hear reference to El Niño and La Niña to explain some abnormal weather occurrence. El Niño refers to a large mass of exceptionally warm water that occurs in the eastern Pacific at the equator near the coast of South America. This strip of warm water shows up periodically and has a counterpart with which it alternates—an exceptionally cold equatorial strip of water referred to as La Niña. These two phenomena alternate every three to six years, resulting in worldwide weather changes. The atmospheric changes that result from this periodic shift in temperature are called the *southern oscillation*. The impetus for the occurrence of El Niño or La Niña is the trade winds. Strong trade winds tend to blow warm surface waters toward the western Pacific, causing cooler waters to prevail in the eastern Pacific and bringing on the cold phase. In the spring, trade winds tend to be weaker, and warmer surface waters prevail. This change in oceanic temperatures causes heavy storms and atmospheric disturbances. These atmospheric disturbances can in turn affect rainfall patterns and temperatures worldwide. El Niño in 1982–1983 caused droughts in India and Australia and heavy rains on the western coast of South America. La Niña on the heels of El Niño in the spring of 1988 resulted in extremely cold temperatures in Alaska and Canada the following winter. Have you heard reference to El Niño or La Niña lately to explain unusual weather occurrences?

Atmospheric Cycles

We will now look at the carbon cycle and the nitrogen cycle, which are good examples of **atmospheric cycles,** cycles in which the primary reservoir or source (as far as living systems are concerned) is the air.

The Carbon Cycle

The carbon cycle intimately involves all living things; a molecular carbon "skeleton" is found—by definition—in every organic chemical. Carbon is thus a key element in the chemistry of all life.

The carbon cycle is categorized as an atmospheric cycle because most of the carbon that passes through the biosphere comes from the air. Approximately 0.03% to 0.04% of the air is carbon dioxide (CO_2), and this minute percentage is the main source of carbon for all living things. It is the reservoir from which plants withdraw carbon in photosynthesis, the key process by which carbon enters the biosphere (see Figure 3.5).

Through photosynthesis, green plants pick up carbon from carbon dioxide in the air and, through complex chemical processes, combine and rearrange these carbon atoms into carbon skeletons of various molecular shapes and sizes. Oxygen, nitrogen, phosphorus, sulfur, and other elements are attached to carbon to produce the substance of living matter—fats, carbohydrates, proteins, and nucleic acids. In turn, this plant matter may be passed on to consumers, who rearrange the elements to suit their own specific chemical needs.

Figure 3.5   The Carbon Cycle.   The chemical skeleton for the substances that compose living things is based on carbon. Illustrated here are the patterns of carbon cycling through the great spheres.

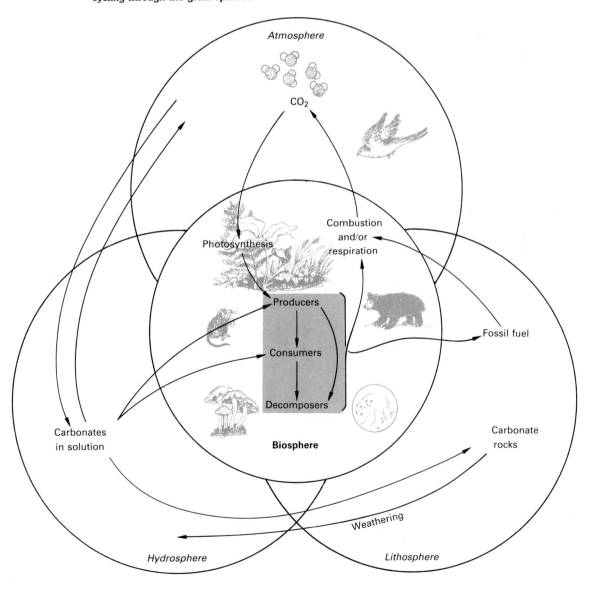

Limestone, a form of calcium carbonate, is relatively insoluble in water. Being sedimentary, it tends to occur in layers. It has been found on all continents, in strata from every geologic age.

Carbon Equilibrium

Oxides of carbon are also found in water and in the form of rock. Carbon dioxide is slightly soluble in water; in solution, carbon dioxide forms carbonic acid (H_2CO_3). Carbonic acid dissociates to form hydrogen ions (H^+) and bicarbonate ions (HCO_3^-). The bicarbonate ions in turn dissociate further to form hydrogen ions and carbonate ions (CO_3^{2-}). Negatively charged ions can combine with positive ions to form various salts—calcium carbonate, for example. Some calcium salts are relatively insoluble in water and accumulate as carbonate sedimentary rock (limestone in the case of calcium carbonate). The reactions involved in carbon dioxide–carbonate equilibrium are as follows:

$$CO_2 + H_2O \rightleftharpoons H_2CO_3 \rightleftharpoons H^+ + HCO_3^- \quad (1)$$

(carbon dioxide)   (water)   (carbonic acid)   (hydrogen ion)   (bicarbonate ion)

$$HCO_3^- \rightleftharpoons H^+ + CO_3^{2-} \quad (2)$$

(bicarbonate ion)   (hydrogen ion)   (carbonate ion)

$$CO_3^{2-} + \text{positive ions} \rightarrow \text{salts} \quad (3)$$

(carbonate ion)   (e.g., Ca^{2+})   (e.g., $CaCO_3$)

Phenomena that would tend to remove carbon dioxide from the air in a given locality would shift all of the reactions to the left, in effect replacing the carbon dioxide. Conversely, if some process or event tended to add to the CO_2 in the atmosphere, the reaction series would shift to the right, again having the effect of tending to keep the CO_2 concentration constant. The reactions are affected by pH, temperature mixing, and other factors, and on a global scale the adjustments are far from instantaneous. It is worth noting that the percentage of carbon dioxide in the atmosphere has actually increased over the past century. Because of the role of CO_2 in holding in heat, this has some potentially serious implications for the earth's climate. This problem will be discussed in detail in Chapters 9 through 11.

The Nitrogen Cycle

A basic difference between the nitrogen cycle and the carbon cycle is that whereas all green plants can extract the "raw material" carbon dioxide directly from the atmosphere, most green plants are dependent on a few species of bacteria and blue-green algae to convert atmospheric nitrogen into a form they can use. The earth's atmosphere contains approximately 79% nitrogen—about four times more nitrogen than oxygen and 2,000 times more nitrogen than carbon dioxide. Yet plants that can utilize the minute amount of carbon dioxide *cannot* utilize the large amount of atmospheric nitrogen.

The nitrogen cycle offers a good illustration of the complexity and fragility of nature. Figure 3.6 shows that atmospheric nitrogen can be changed into ammonia (NH_3) and nitrate (NO_3^-) (the primary forms usable by plants) by at least two processes. Lightning may serve as an energy source to cause atmospheric nitrogen and oxygen to react to form nitric acid (HNO_3), which is then carried to the soil in rainfall. However, the contribution of this reaction to the overall availability of nitrogen to organisms is minimal. Certain nitrogen-fixing bacteria and blue-green algae utilize atmospheric nitrogen in their metabolic processes and produce ammonia as an end product. Most nitrogen that is fixed naturally occurs as a fortuitous by-product of the activities of these bacteria and blue-green algae.

The most important nitrogen-fixing bacteria are those of the genus *Rhizobium,* which grow in the roots of legumes such as clover and alfalfa. *Rhizo-*

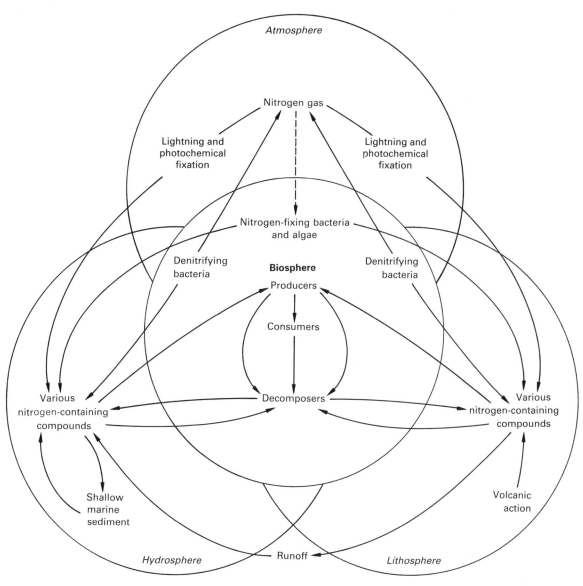

Figure 3.6 The Nitrogen Cycle. Although nitrogen comprises almost 80% of the atmosphere of the earth, most living things cannot use nitrogen as a gas. This diagram shows how nitrogen enters the biosphere primarily through the action of bacteria and algae, which convert nitrogen gas to ammonia and nitrates, which are usable by plants. Many types of bacteria and algae keep the nitrogen cycle moving. Each uses products from one or more of the others as raw material for its own metabolic activities. Note that there are different kinds of decomposers, some of which are bacteria.

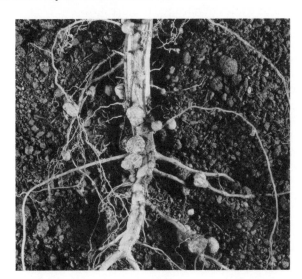

The supply of nitrogen available to plants depends on nitrogen fixers like the *Rhizobium* bacteria found in these soybean root nodules, which in turn depend on the availability of other minerals.

bium resides in nodules on the roots and receives nourishment from the plant while producing "nitrogen fertilizer," usable nitrogen compounds for that plant. When such legumes decay, ammonia is released to serve as nitrogen sources for other plants. This is why a farmer will periodically plant a field with a cover crop of legumes such as beans, alfalfa, or clover.

Anaboena is a blue-green alga capable of fixing nitrogen. It is a free-living alga sometimes found in association with a certain water fern. Farmers in China take advantage of this relationship by growing the fern in their rice fields. Most nitrogen fixation in aquatic systems is accomplished by free-living bacteria and algae. Free-living bacteria include those of the genus *Azotobacter,* which are **aerobic** (oxygen-requiring), and the genus *Clostridium,* which are **anaerobic** (able to live in the absence of oxygen).

Most plants absorb ammonia and nitrates from the soil through their roots and incorporate the nitrogen into protein, nucleic acids, and other biochemicals. Animals acquire these various organic forms of nitrogen by eating plants and rearranging the nitrogenous compounds to suit their own needs. When animal wastes are excreted or when plant or animal tissue dies, certain (ammonifying) decomposers convert organic nitrogen compounds into ammonia gas (NH_3) or ammonium salts (in which nitrogen occurs in the form of an ammonium ion, NH_4^+). These in turn may be acted on by still other species of bacteria. Through a process called **nitrification,** ammonia and ammonium salts are converted to nitrites ($NH_4^+ \rightarrow NO_2^-$), and then other bacteria convert nitrites to nitrates ($NO_2^- \rightarrow NO_3^-$). Bacteria of the genera *Nitrosomonas* and *Nitrobacter* are associated with this process. There are also bacteria that reverse nitrification; they are called **denitrifying bacteria.** Some convert nitrates to nitrites, and others convert nitrites to ammonia; others convert nitrates to free nitrogen or oxides of nitrogen (Figure 3.6).

The remarkable thing about the nitrogen cycle is that a few obscure species of bacteria and algae provide the principal link between the primary reservoir of inorganic nitrogen and living systems. Each participating species in the chain is operating using raw materials that are the waste products of other species, and through it all, life as we know it is made possible. The nitrogen cycle is an important example of the "tightness" and complexity of interrelationships in the biosphere.

Agriculture and Nitrogen
Modern farming practices include the use of commercial, synthetic, nitrogen-containing fertilizers in addition to nitrogen-building cover crops and natural compost or manure to increase crop yields. Al-

though the effect of natural and commercial nitrates is basically identical for the crops, there are some significant problems with commercial fertilizers from an environmental and cultural perspective. For one thing, it takes a lot of energy to fix nitrogen artificially. This is relevant to the energy problems we will discuss at length in Chapter 6. Second, the availability of so much synthetic nitrate aggravates the problem of what to do with manure and other naturally occurring organic wastes. Excess nitrates from over-fertilization or improper disposal of organic wastes can be washed into waterways, where they stimulate excessive plant growth and indirectly deplete oxygen, or can seep into groundwater, contaminating water supplies. Kormondy (1984) indicates that biological fixation of nitrogen produces about 54×10^6 metric tons per year compared to 30×10^6 metric tons annually fixed in the manufacture of synthetic fertilizers. However, he also states that the amount fixed synthetically is increasing at such a rate as to reach 100×10^6 metric tons by the year 2000.

Lithospheric Cycles

Another class of cycles are the **lithospheric cycles** in which the principal and most important nutrient reservoir is the lithosphere. Lithospheric cycles are characterized by the occurrence of **sinks,** or places where elements are tied up, taken out of the cycle, or removed from the juncture of the spheres for extremely long periods of time. Over long time spans,

In lithospheric cycles, minerals may be held out of the cycle for very long periods. These stalactites, formed from a solution of calcium carbonate in groundwater that has seeped into rock, may represent thousands of years of groundwater action on rock.

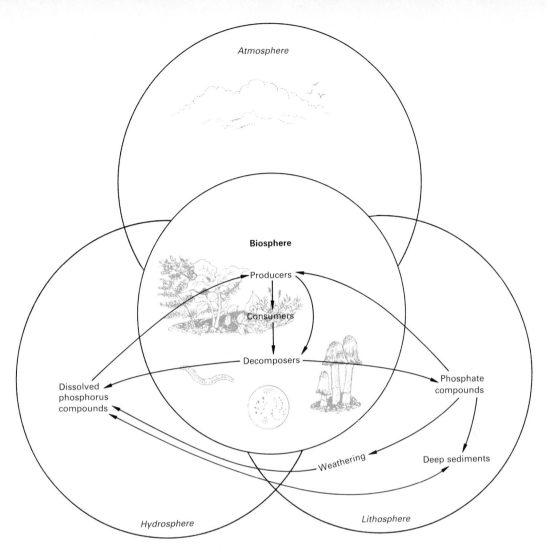

Figure 3.7 The Phosphorus Cycle. Physical and geologic processes are important in the cycling of phosphorus and in all other lithospheric cycles. The rate of cycling of phosphorus is extremely important to growth and activity in living things.

lithospheric cycles depend on geologic uplifts; they are driven by weathering over the short term. The forces of erosion and weathering cause minerals to break off and dissolve in water and to be carried into rivers and streams and eventually into the oceans. These minerals may enter numerous loops or cycles in the biosphere before an ultimate redeposition as ocean sediment.

The Phosphorus Cycle

The role of phosphorus, in the form of phosphate, in reactions that store and release energy for use in cells makes it a very important nutrient. The availability of phosphate through the phosphorus cycle is often a limiting factor in ecosystem productivity.

Through erosion and weathering of rock, inorganic phosphate is made available to plants through uptake from the soil or, in the case of aquatic plants, from the aqueous environment. Once taken into the plant, the phosphate may become part of an energy-rich molecule of ATP (adenosine triphosphate), a nu-

cleic acid, or some other organic compound. The phosphate may be returned to the soil or sediments when the plant dies and decomposes (phosphate is also one of the constituents of the ash left following the burning of a plant or animal). Phosphorus may also be passed on to a consumer.

In the consumer, phosphorus may be incorporated into bones or teeth, in which case the phosphorus is bound for an extended period of time. Some of it is excreted as waste and is immediately accessible to decomposers; it may, by a short loop, be converted back to inorganic phosphate and be reassimilated by plants very quickly. It may become tightly bound as phosphorus salts to iron, calcium, aluminum, and clays in the soil to be washed away or "lost" in sediments. In an aquatic system, released phosphorus may eventually become part of deep sediments, where it is out of reach of organisms at the air-water interface and only geologic uplifts, shifting vertical currents, or the like will get it back into the cycle (Figure 3.7). In coastal marine ecosystems, in which the sea gull may be the ultimate consumer (or

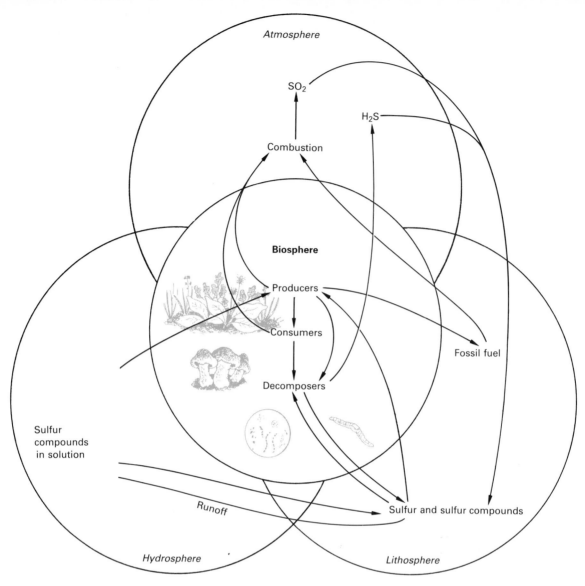

Figure 3.8 The Sulfur Cycle. Although most troublesome in its atmospheric form of sulfur dioxide, sulfur moves in what is predominantly a lithospheric cycle. Bacteria are important factors in the sulfur cycle, changing sulfur into various forms in the soil.

top carnivore), the fate of phosphorus—whether it goes to a long-term sink or is quickly cycled—depends on where gull droppings fall: on the beach, the cliffs, or the ocean. Bird droppings, called guano, are actually high-phosphate reservoirs "mined" as a resource along some coastlines such as the western coast of South America.

Phosphorus is scarce in many—especially aquatic—environments, where it is the key limiting factor for biological activity. The turnover (cycling) rate for phosphorus may actually determine the level of productivity in many aquatic ecosystems. This is one reason why phosphates in fertilizer runoff can disrupt aquatic communities (see Chapter 12). This is also the reason why some states have banned the use of detergents containing phosphates. It is interesting, to say the least, that one kind of environmental prob-

lem is keeping excess phosphates out of living systems while others have to do with the need for phosphates in agricultural production.

The Sulfur Cycle

Sulfur is an important element in proteins and various protein-carbohydrate complexes. Sulfur is found in the soil primarily in the form of elemental sulfur and sulfates (compounds containing the SO_4^- group) (Figure 3.8). Sulfur is taken in by plants in the form of sulfates and is eventually returned to the soil through the decomposition of wastes or dead materials. Organic sulfur is converted into inorganic sulfate by common decomposers (bacteria and fungi). The sulfur cycle involves numerous species of bacteria and fungi that interconvert sulfates, sulfides, and elemental sulfur.

One of the most common sulfides is a gas, hydrogen sulfide (H_2S), well known for its "rotten egg" odor. The bacteria that convert sulfates to hydrogen sulfide are adapted to anaerobic (oxygen-deficient) habitats. The foul smell of hydrogen sulfide is characteristic of decomposition in an oxygen-deficient system.

Sulfur may enter the atmosphere as hydrogen sulfide or as sulfur dioxide (SO_2). Sulfur dioxide is formed in the combustion of organic material—leaves, carcasses, or coal. Sulfur dioxide is toxic to animals and plants, and it reacts in the atmosphere to form sulfuric acid. When this falls to the ground with various forms of precipitation, the result is called *acid precipitation*. Acid precipitation will be considered in detail in Chapter 10.

Nutrient Cycles as Regulators of Ecosystem Activity

We have seen that biogeochemical cycles involve complex interactions between the abiotic environment and living organisms. We have also seen that any one cycle is composed of numerous loops and steps. As we look back over Figures 3.1, 3.5, 3.6, 3.7, and 3.8, we find that the representative cycles all overlap, that they operate simultaneously in the same functioning systems. We have observed that the biosphere both regulates and is regulated by various biogeochemical cycles. The nature and speed of certain chemical interactions among producers, consumers, decomposers, and the abiotic environment determine how well any given ecosystem functions.

The Law of the Minimum

In 1840, Justus von Liebig expressed the idea that organisms and the living systems they comprise are held in check by the scarcest of the things they need. What we refer to today as the **law of the minimum** was derived from Liebig's observation that a plant tends to grow only to the limit of the foodstuff available to it in the most extreme minimum quantity. This would be analogous to saying that an automobile assembly plant can make cars only as fast as it can get tires during a rubber shortage. In the general world of real ecosystems, the applicability of Liebig's law is far from being as simple and straightforward as this analogy might suggest.

For one thing, the law of the minimum is a law only under steady-state conditions. During periods of change, ecosystems tend to be held in check by the availability of many raw materials, the particular organisms involved, and many physical conditions, all at the same time. There are many potential limiting factors in any ecosystem. For example, the rate at which photosynthesis occurs depends on the availability of water; carbon dioxide; nutrients such as nitrogen, phosphorus, iron, zinc, and magnesium; the ambient temperature; and the light intensity—any one of which could be rate-limiting. Even under steady-state conditions these things vary over time and influence one another in complex ways. A given plant might need less rainfall when it is cool. Its need for a particular mineral might depend on the availability of one or more other minerals. A plant might very well need less of a particular mineral if it is growing in the shade.

Because conditions tend to be somewhat more constant in aquatic ecosystems, the law of the minimum tends to be more apparent in such systems. Phosphorus as phosphate is often a limiting nutrient in lakes and rivers; this is why phosphate detergents can cause a disruptive acceleration of plant growth in these aquatic systems. Nitrogen is commonly a limiting or near-limiting nutrient in terrestrial (and aquatic) systems; it is for this reason that nitrogen is such a common ingredient in fertilizers.

Factors Governing the Supply of Mineral Nutrients

There are a number of influences on the supply of nutrients within any living system. For one thing, nutrients affect one another. Phosphorus is often a key regulator of activity in ecosystems because of its relative scarcity in the form of soluble phosphate and because of its key role in storing and releasing energy as part of the ATP molecule. In an aquatic system the supply of phosphorus to living systems may be linked to the presence of sulfur and iron. The binding of sulfur with iron under anaerobic conditions in an aqueous environment creates conditions that convert phosphorus from an insoluble form to a soluble form that can be used by living things. Bonding with sulfur may also tie up minerals such as copper, cadmium, zinc, and cobalt.

The supply of usable nitrogen is dependent primarily on nitrogen fixers, which take atmospheric nitrogen and convert it to ammonia and nitrates. The activity and reproduction in nitrogen fixers may be controlled in turn by the availability of other minerals such as phosphorus, iron, calcium, molybdenum, or cobalt.

Because of variations in geochemical influences, the amount of nitrates, phosphates, water, and other minerals available to living systems varies from one area to another and may even change from one season to another or be influenced by periodic "imports" and "exports" (see also Chapter 2). Runoff from a feedlot or a heavily fertilized cornfield may import phosphorus and nitrogen to a nearby lake and result in a large increase in the availability of these

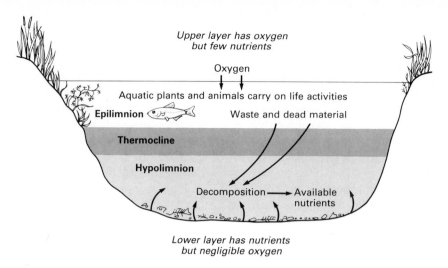

Figure 3.9 **Stratification in Deep Lakes.** The thermocline is a horizontal interface between less dense, warmer, upper layers of water and colder, denser lower layers. This effectively prevents mixing of the upper (epilimnion) and lower (hypolimnion) layers of the stratified lake.

usually limiting nutrients. Erosion may export nutrients in topsoil, resulting in less fertile farmland and more fertile nearby streams.

Lakes: Fall and Spring Overturn

Changes in physical environment can also cause changes in the supply of chemicals within a system. One such example is the effect of seasonal temperature changes on bodies of water. A pattern that occurs twice yearly in certain deep lakes is known as fall and spring **overturn.** This phenomenon is related to the nature of water, which, as was mentioned previously, has its greatest density at 4°C.

Normally, in the summer, temperate zone lakes become stratified into layers having different temperatures (Figures 3.9 and 3.10). Being somewhat insulated, the lower layer (**hypolimnion**) stays cool while the surface layer (**epilimnion**) warms. As the temperature and density differentials increase, a fairly dramatic vertical temperature boundary forms between the upper and lower layers. This boundary is called a **thermocline.** This stratification prevents mixing of the upper and lower layers. Algae grow best at the surface, where the light is strong. As they grow, they take up nutrients such as phosphate and nitrate; and as they (along with consumers) die, they sink to the bottom, carrying their accumulated nutrients with them. This tends to deplete the surface layer of nutrients; the depletion soon limits plant growth. Meanwhile, at the bottom, decomposers break down the organic matter slowly because of the cool temperatures and low oxygen concentration.

During the fall the surface layer gets cooler, and a point is eventually reached at which the surface layer begins to become more dense (heavier) than the lower layer, and a dramatic mixing of the layers, or overturn, occurs (Figure 3.10).

Going into the winter months, surface water temperatures cool to below 4°C. Toward the end of winter and into early spring, surface water temperatures begin to rise. As the surface temperature approaches 4°C, surface water becomes heavier than the lower layers, and mixing occurs again.

Overturns bring nutrients from the lower level to the sunlit upper level, creating ideal conditions for plant growth. This is why algal blooms often occur shortly after overturns. This is not necessarily a good thing, however. The growth of algae may be so prodigious as to form a mat over parts of the lake. Plants below this mat no longer receive sufficient sunlight and begin to die. This provides a huge food source for decomposers and results in a rapid increase in the number of decomposers and an accelerated rate of decomposition. The increased respiration by decomposers may use up oxygen faster than it can dissolve in the water from the atmosphere and faster than the aquatic plants can release it. The lake may become oxygen-deficient, and organisms unable to tolerate this condition will perish.

Note that during this entire sequence, several *different* factors (sunlight, oxygen, nutrients) were limiting for *different* parts of the aquatic community. Overturn illustrates clearly that physical conditions are able to influence living systems through an impact on chemistry.

The Law of Tolerance

We have just seen that there can be problems in nature associated with both too little and too much of a chemical regulator. Figure 3.11 demonstrates this fact graphically; it applies to individual organisms and to communities of living things. Similar generalized graphs could be drawn for features of the envi-

Part One   Basic Principles of Ecology

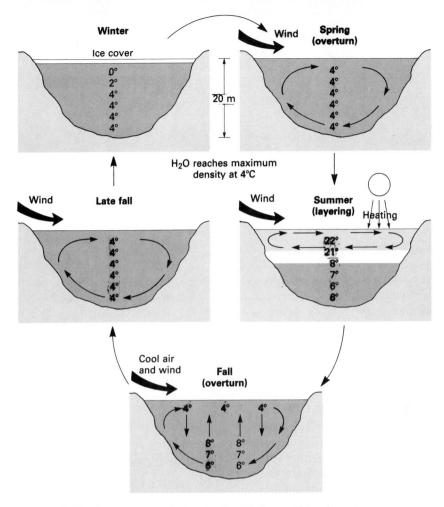

Figure 3.10 Temperature Cycles in Deep Lakes. This schematic demonstrates how the seasonal change in water temperature results in a layering or mixing of water in deep lakes. The movement of the water is the result of the fact that water is heaviest at 4°C.

ronment other than nutrients. Temperature, light level, soil moisture, and slope are among the most important. All of this could be summarized by saying that organisms and, in the aggregate, whole systems of organisms have ranges of physical and chemical conditions within which they can flourish but beyond which they change and much beyond which they cannot exist.

Shelford's law of tolerance states, in essence, that both too much and too little of various environmental chemical and physical factors can serve as limiting factors or regulators in ecosystems.

Figure 3.11 Shelford's Law of Tolerance. Both too much and too little of a chemical nutrient or given environmental factor (heat, rainfall, etc.) can be harmful to living things.

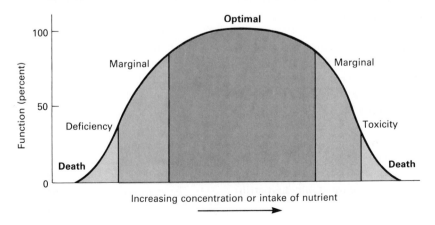

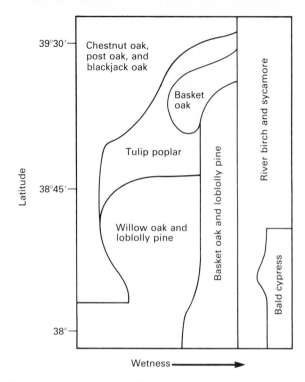

Figure 3.12 Ranges of Tolerance in Some Tree Species. The distribution of tree species can be shown to vary according to their tolerance ranges for moisture and latitude; latitude affects factors like temperature and day length. Ranges of tolerance overlap.

If the lowest annual temperature reached in a particular environment is below the lowest temperature that a particular plant can tolerate, *that* plant is not likely to be found in *that* environment, even if moisture or soil type is sufficient.

As one might expect, some organisms have wider ranges of tolerance than others, and individual organisms have varying ranges of tolerance for different environmental factors (Figure 3.12). Animals and plants having the widest ranges of tolerance overall can be found in many different kinds of environments. Limits of tolerance are often interrelated. Bacteria, for instance, can tolerate higher heat longer when they are dry and when the soil or substrate is not too acid or too alkaline. The concepts of tolerance limits and limiting factors link the disciplines of ecology and physiology and form the basis of **biogeography,** the study of the abundance and distribution of species in nature.

The human species has learned to manipulate limiting factors to some extent. We cultivate plants by adding water and nutrients to the level that should be optimum for growth. We have learned to develop methods of importing nutrients for human consumption into areas that of themselves could not support many people. Humans can live in all types of temperature extremes because of the ability to heat and cool shelters and to put on and take off layers of clothing. The human species is among the most widely distributed of all species.

The Ecological Connection

The ecosphere is in chemical motion. Elements are being taken out of the abiotic environment by plants, moved up the food chain, and excreted. As you read this, chemicals you consumed at breakfast are being incorporated into proteins and other chemicals. Yesterday's proteins are being torn down, and the constituents are being incorporated into other proteins and other chemicals. Calcium is moving into and out of your bones. You may feel stable, but you are in a dynamic chemical state with the ecosystem around you. This is true for the entire global system of which we are part. The cycling of nutrients is complex, and individual cycles are interrelated.

Human activity can alter cycles. Alterations can be for the good or for the bad, but inadvertent alterations are seldom for the good in intricate ecological systems. Many of the environmental and health problems we face today are the result of unanticipated impacts of synthetic substances on the ecosphere, overloading cycles and piling up concentrations of chemicals that exceed tolerance limits, and revving up natural systems by adding nutrients of which small amounts had been limiting.

Cycles and Environmental Problem Solving

In attempting to solve environmental problems, knowledge of biogeochemical cycles can be very useful. It helps us to avoid the practice of simply shifting the problem from one sphere to another. For example, if we require all utilities to scrub air emissions to trap sulfur, we also need a plan for what will be done with all of the sulfur. Otherwise, we would have done nothing more than shift the problem from air to land. If we want to promote desalinization of ocean water to help to meet water needs, we also need a plan for what to do with the salt. If we want to diminish the use of land for waste disposal, we must look at ways to reduce waste and otherwise keep it from piling up. It is important to consider the long-range ramifications of all policy choices. Consider the trade-offs involved in deciding to compost yard waste or

send it to the incinerator or the landfill. Should waste services refuse to accept grass clippings and leaves—in effect requiring lawn owners to recycle these materials on site? What do you think?

Concepts to Remember

1. For the most part, the total amount of each chemical element in the ecosphere remains the same. Chemical elements are used over and over by living things.

2. Chemical elements are combined to form the substance of living and nonliving things.

3. At any point in time a specific chemical element in an ecosystem is distributed in various chemical forms in air, rock (or soil), water, and in some cases living things. Over time the elements move from one sphere to another in patterns called biogeochemical cycles.

4. Chemical elements that are essential to organisms are called **chemical nutrients.** The availability of chemical nutrients regulates growth and other activities in living systems. Plant productivity, for example, can be held in check by the lack of certain compounds of nitrogen or phosphorus.

5. The availability of specific chemical elements to living things is determined by three factors: (1) molecular form—the element must be in a form that is usable by that organism; (2) physical location—the element has to be

accessible to living things; and (3) patterns and rates of cycling—how fast the element moves from one form to another and from one location to another.

6. Individual organisms and communities of organisms have characteristic tolerance ranges for individual chemical elements and their compounds. Too much or too little can be detrimental. Living things also have tolerance ranges for physical conditions such as light and temperature.

7. Human activities have altered biogeochemical cycles in ways that are detrimental to humans and other living things. We have influenced chemical form, physical location, and the rate of cycling of chemical elements. We have discovered and have added new compounds to biogeochemical cycles. Some of these synthetic substances have had serious unanticipated effects on the ecosphere.

8. Human beings are integral components of biogeochemical cycles and share a dynamic chemical commonality with all other things, living and nonliving.

References and Further Reading

References marked with an asterisk are cited in the chapter.

Cook, R. B., 1984. "Man and the Biogeochemical Cycles: Interacting with the Elements," *Environment* **26**(7):10–15ff.

Holdgate, M. W., 1991. "The Environment of Tomorrow," *Environment* **33**(6):14–20ff.

"How Plants Cope: Plant Physiological Ecology Today," *BioScience* **37**:18–67, 1987. (Special issue)

* Kormondy, E. J., 1984. *Concepts of Ecology*, 3d ed. Englewood Cliffs, N.J.: Prentice Hall.

Malone, T. F., and Corell, R., 1989. "Mission to Planet Earth Revisited: An Update on Studies of Global Change," *Environment* **31**(3):7–11ff.

Morris, D., 1991. "As if Materials Mattered," *The Amicus Journal* **13**(4):17–18.

Odum, E. P. 1984. *Fundamentals of Ecology*. Philadelphia: Saunders.

See also the references cited at the end of Chapter 1.

4

Populations and Communities

In Chapter 2 we looked at how organisms depend on the environment for the energy they need; in Chapter 3 we looked at how organisms depend on the environment for their material substance. It should be clear now that energy and minerals are important limits to the organisms in any ecosystem. In this chapter we will examine in more detail both the biological and the abiotic aspects of the interdependences and limits that operate in ecosystems. We will examine relationships between and among organisms of the *same* species, and we will examine relationships between and among organisms of *different* species. We will consider how the constraints imposed by these interrelationships plus those imposed by the abiotic environment result in characteristically different types of ecosystems in different places on earth.

On the way through this chapter we should find the basis for the answers to the following kinds of questions:

— How do particular groups of living things get to be what they are? For example, why are there characteristic kinds of plants and animals in nearly all North Carolina swamps?

— Why do the numbers and kinds of plants and animals in, say, a Pennsylvania forest stay about the same from year to year? Or do they?

— Why do population explosions occur in nature? Why don't they occur more often? Why does the explosive growth of a population go only so far?

— What happens when a new species is introduced into an ecosystem? What happens when an old one is taken out?

— What if there is a fire? Do ecosystems eventually recover—get back to their original condition? If so, why? If so, how? If not, what does happen? How long does it take in either case?

— Assuming that we all know why deserts are where they are, why are there prairies in some places, deciduous forests in some places, and evergreen forests in other places, each a characteristic community of living things?

— What is the ecological significance of territoriality and social organization in some species—packs of wolves, troops of baboons, and colonies of ants or termites?

These are all questions about regulation, and regulation is the theme of Chapter 4. The cardinal question we will be addressing throughout is this: Just how do the organisms and the abiotic features of an ecosystem interact so as to result in distinctive self-regulating communities of living things?

We will start at the population level, the level of interaction between members of the *same* species. This will set us up to look at the more complex community level, the level of interaction of *different* species. It will help to keep in mind that the interdependence of living things is the essence of ecology.

Populations: Some General Characteristics

A **population** is a group of similar organisms or, more specifically, interbreeding members of the same species in a given area. The term *population* can apply to all the members of a species in a field or to those in larger areas such as the Atlantic Ocean. Population is another example of an abstraction with variable applications. Because populations are composed of many individuals, they exhibit certain patterns or traits that individuals cannot.

Patterns of Distribution

Populations can vary in the way their individual members are distributed (Figure 4.1). The individuals in populations of some species tend to be regularly

Figure 4.1 Distribution Patterns of Organisms in Nature. The range of distribution results from interactions with abiotic factors such as climate and soil type and with each other (social factors). Organisms may be distributed randomly, uniformly, or in clumps.

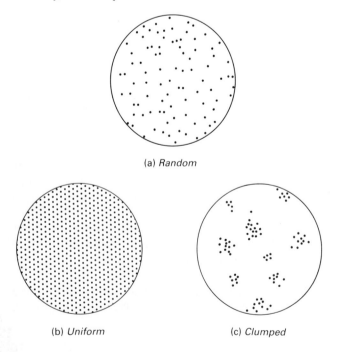

(a) *Random*

(b) *Uniform*

(c) *Clumped*

dispersed with no aggregation or clumping; an apple orchard is a familiar example, though it is not wild. Such *uniform distributions* are not common in nature, but they can occur, usually where there is severe competition for resources. The creosote bush of the American Southwest is thought to produce a substance in its roots that prevents creosote bush seeds from germinating and growing too close. Competition for water and other needs may also play a role. In any case, there are rather uniform distributions of creosote bushes in the desert.

The distributions of the individuals in some species approach complete *randomness*. Plants with airborne seed dispersion mechanisms might follow this pattern. Random patterns are found in environments with relatively uniform conditions throughout and consequently are relatively rare.

In old fields, sassafras, hawthorn, and other shrubs exhibit a third pattern of distribution, called *clumped*. This pattern is the most common. Generally speaking, clumping is a result of short-range seed dispersal or nonuniform environmental conditions (remember the concept of tolerance explained in Chapter 3). Within a large, apparently uniform area, different microenvironments may favor some species over others. Moss usually grows best on the side of a tree that is most moist. Different plant species are often found on the north-facing and south-facing slopes of the same woods. Within a plant community, plants requiring more moisture may be found in shady areas or in slight depressions or low places. Social factors can likewise cause clumping in species distribution. The formation of packs or herds by some kinds of animals serves to help them get food and protect themselves.

Density

Apart from its distribution pattern, a population may exhibit various **densities,** that is, different numbers of individuals (or amounts of biomass) per unit area or volume. If there are 100 mice in a 2-acre field, the density is 50 mice per acre. Sometimes more refined expressions of density must be used when not all of a given area is suitable habitat for particular species under study. If a 1/2-acre pond were located in the middle of the 2-acre field, for example, it would make more sense to speak of 100 mice per 1.5 acres. Density per unit of *habitat* is called **ecological density** and is a more useful expression of density.

Social Organization

Animal populations exhibit varying degrees of social organization. Social organization can take the form of territoriality or hierarchical relationships and extends to the highly specialized divisions of labor that are found among the members of insect colonies.

Part One   Basic Principles of Ecology

Biomes: the major community types of the world, characterized by specific climate conditions and plant types (examples: prairie, tundra, desert, deciduous forest).

Biotic community: a group of populations of different plant and animal species that interact within any prescribed area or physical unit of the ecosphere. A biotic community is all of the living portion of an ecosystem.

Climax community: the end point of ecological succession. A climax community is a community that is in a relatively steady state; the impact of each species tends to be neutralized by the other species. Climax communities do not change much, even over long periods of time.

Diversity: a measure of the number of different species in a biotic community. Diversity is high when there are many different species and low when there are few.

Ecological dominance: a measure of the degree to which one or a few species have the greatest biological impact in an ecosystem and thus characterize that system.

Hierarchy: pecking order, leadership, or dominance patterns among the members of a species.

Home range: area in which a member of a population roams and carries on all of its activities.

Interspecific relationships: interactions between the individuals of different species in a community. Interaction can be directly beneficial to one and harmful to another (predation or parasitism), beneficial to both (mutualism), or beneficial to one and neutral to another (commensalism). Competition is a form of interaction.

Patterns of distribution: the way in which the members of a population are dispersed throughout their habitat. Individuals in a population may be found in clumps or distributed randomly or uniformly.

Population: a group of interbreeding members of the same species in a given area.

Population density: the number of individuals of a given population per unit of area.

Social organization: the characteristic way in which members of animal populations relate to one another.

Species: the basic unit in the classification of organisms. All of the individuals of like kind that are related to one another through relatively recent common ancestry and share the same ecological role. For organisms that reproduce sexually, a species can also be defined as all of the organisms potentially able to interbreed with one another.

Territoriality: the tendency of a member (or members) of a population to establish an area of its own, defending it against intrusion by other members of the same species.

Many species of animals exhibit **territoriality,** wherein individuals vie for the "rights" to units of habitat. The winner gets to use the territory; the loser is driven off. Usually, the defended territory is somewhat smaller than the **home range,** the total area in which an animal lives, eats, and functions. Territories and home ranges vary in size for different species.

Territories may cover several miles in the case of large animals or birds; they may be limited to a single plant in the case of some insects. In confrontations over territory between two members of the same species, physical harm is rarely done even to the loser. Quite often the interaction consists only of scent signals or vocal and behavioral displays. Territoriality

In the hierarchy of breeding colonies, the older male seals take the choice locations; the younger males, in their less desirable spots, can attract fewer females.

serves to diminish destructive competition for resources such as food or habitat by limiting the number of organisms of a species in a given area.

A second form of social organization expresses itself in patterns of **hierarchy** in species in which interaction is more extensive and continuous. Perhaps the most familiar example of this is the **pecking order** among chickens. At the top of the pecking order is a chicken that can peck (dominate) all others and in turn is pecked by none. In the middle are chickens that can peck some chickens but are pecked by others. At the bottom is a chicken that all other chickens can peck but that cannot peck back. These dominant-subordinate relationships become more apparent in some animals during mating season. The result is that the more dominant and usually stronger have first choice of mates. Individuals at the lower end of the hierarchy may or may not have the opportunity to mate.

Extreme social organization is found in the structure of colonies of insects like termites, ants, and bees. Insect colonies may be so highly integrated that an individual may not be able to survive outside of the society. Bees are a prime example. So interdependent are the individuals in a beehive that a hive is more like an organism made up of cells than it is like a population made up of individuals.

Growth of Populations

Factors Affecting Population Growth

To determine the growth rate of a population, one would have to look at the factors that tend to *increase* the number of individuals within that population and those that tend to *decrease* the number of individuals in that population. There are only two ways by which new individuals can enter a population. One is by new *births,* and the other is by *immigration.* The birth rate is determined by the inherent reproductive potential of the species and the number of individuals in the population capable of reproducing. The overall reproductive success of the species is also affected by environmental factors such as the food and habitat available and favorable or unfavorable climate. The immigration rate is determined by species mobility and the extent of the exploitable opportunity. A population loses individuals by *death* and by *emigration* (see Figure 4.2). Death and emigration rates are determined by such things as abundance of food, environmental conditions, disease, predation, and competition for food and habitat.

For a given population in a given place we can think of growth rate as a function of the maximum reproductive rate characteristic of that species, various factors that tend to keep maximum reproductive

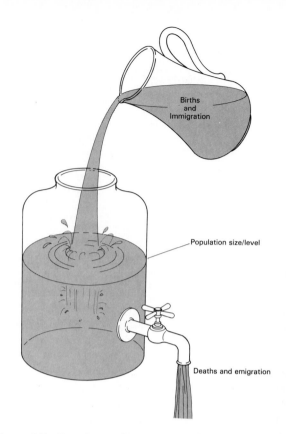

Figure 4.2 Population Change. Population growth results from the net effect of all factors adding to the number of individuals in that population (births and immigration) and those decreasing the number of individuals in that population (deaths and emigration). These factors are in turn the result of species characteristics and environmental conditions.

potential from being realized, and various characteristics of the population itself. Age structure would be an example of the last of these.

Biotic Potential and Environmental Resistance

The **biotic potential** of a species is its maximum possible growth rate; it is a concept based on births per female in the "ideal" absence of any limiting conditions, for example, limitations in habitat or food supply. Biotic potential differs widely from species to species but is a relatively fixed characteristic for any one species. Ideally, one female fly and her offspring could produce 6 trillion flies in one year. In 750 years a pair of elephants and their offspring could theoretically produce 19 million elephants. Likewise, in ten years a pair of sparrows could generate 275 billion offspring. Biotic potentials can be expressed as population doubling times, that is, how long it takes a population to double in size. Doubling times can vary from minutes in the case of bacteria to many years in the case of elephants.

Biotic potentials are never realized for very long; in fact, most species never realize their full biotic potential. Disease, lack of food, predation, lack of

Part One   Basic Principles of Ecology

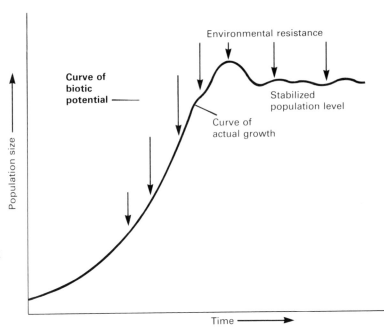

Figure 4.3 Biotic Potential and Environmental Resistance. Seldom does the growth rate of a population equal its biotic potential—its growth rate under ideal, nonlimiting conditions. Environmental factors such as food supply, habitat, and disease limit growth.

space, and other factors, collectively termed **environmental resistance,** keep growth rates well below biotic potentials. Ecologists spend a great deal of time determining which factors or combinations of them are most important in influencing population growth in specific kinds of organisms (Figure 4.3).

There are two basic types of environmental resistance factors in nature. **Density-independent factors** tend to exert the same pressure on a population regardless of the number of individuals present in that population. Often climate and weather are density-independent factors. Seasonal changes and even storms and rapid temperature changes tend to exert the same basic influence on each individual regardless of how many individuals there happen to be. The same is true of chemical pollutants. Under some circumstances, density independence might be somewhat less than complete. Within certain limits, for example, the availability of shelter or social behavior might help to ease the effects of certain physical environmental changes such as climate. In this case the degree of the impact might be somewhat density related and could even be called "density-vague."

Density-dependent factors come into (and out of) play gradually as population size increases (or decreases). They are also called "density-governing." Density-dependent factors include such things as access to food, access to suitable habitat, and the impact of parasites, pathogens, and predators. (Recall the wolves and moose in Chapter 1.) Density-dependent control mechanisms influence population growth by influencing the probability that an individual will live to reproductive maturity or by influencing fertility. High concentrations of geese, for example, can (at the same time) attract more predators, offer conditions in which disease is more likely to spread,

and create conditions in which food is scarce. Each of these effects can in turn influence the average number of viable offspring per nest and the probability that a given goose will live to reproductive maturity.

An often cited example of predation *and* food supply limitations as density-dependent population control mechanisms is based on data for Canadian lynx pelts and the pelts of the snowshoe hare taken over a 90-year period. It was observed that the hare population decreased every nine to ten years and that the lynx followed a similar pattern but a year or two behind the hare. It seems that the hare, being a principal source of food for the lynx, is playing some role in the regulation of the lynx population—and possibly vice versa. Other factors are surely involved, because similar hare population cycles occur even where there are no lynx. However, one plausible partial explanation for this is that when the hare population is large, it provides a large source of food for the lynx, resulting in an increase in the lynx population. An increase in the lynx population, however, means that more hare will be taken for food. This alone or in combination with other elements of resistance to hare reproduction causes the hare population to decrease. A loss of food for the lynx means that the lynx population will have to compete for a smaller amount of food, and its population will decrease. With fewer lynx, more hare survive, their population flourishes again, and the system repeats. The length of the interval between ups and downs would be a function (at least in part) of how long it takes hare to become plentiful and lynx to become plentiful once conditions become favorable for each. In each case, density-dependent factors come into play (predation and food supply) to keep the upward growth of the populations in check.

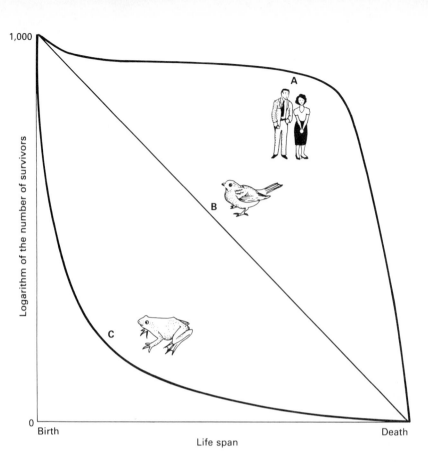

Figure 4.4 Survivorship Curves in Populations. This graph shows three basic patterns for survivorship. Each curve begins with 1,000 live offspring and indicates the decline in numbers of survivors with time as a logarithm of the number of survivors over the normal life span. There are three characteristic survivorship curves: Most members of the species die late (curve A), mortality is relatively constant throughout the life span (curve B), and most members die young (curve C). The exact configuration of each curve varies from species to species. Survivorship is also affected by factors such as food supply, density, and quality of habitat.

Graph labels: vertical axis — Logarithm of the number of survivors (1,000 at top, 0 at bottom); horizontal axis — Life span (Birth at left, Death at right); curves labeled A, B, C.

Certain physiological control mechanisms may also serve as density-dependent growth regulators in some species, especially vertebrates. Christian and Davis (1964) monitored dense populations of rats in the laboratory. They found that crowding resulted in an enlargement of the adrenal gland (an indication of stress) and decreased fertility.

Density-dependent factors tend to keep populations at levels that the environment can support, but the influence of these factors wanes as numbers fall, keeping populations from getting too small. For most species there are *optimum* densities—too few as well as too many individuals may be limiting. If populations fall too low, mates may not be found, or the signals that lead to successful reproduction may be missed. For some species there may even be a threshold level of density below which a population is doomed to extinction.

Density-independent factors are generally more important in population regulation in ecosystems with *low* species diversity and with regular or periodic physical disruptions such as fires or flooding; density-dependent factors tend to be more important in ecosystems with *high* species diversity and a relatively constant physical environment. Density-independent factors are usually more important growth rate regulators when a species is first introduced into a new habitat, and density-dependent factors become increasingly important, by definition, as the population gets larger. For organisms in situa-

tions in which the density-independent factors are *not* limiting, that is, where physical factors are such that the populations are not continuously knocked off balance, density-dependent factors ultimately regulate population size. Generally, density-dependent factors combine to increasingly suppress growth as the population gets large, and a point is reached at which growth stabilizes. A theoretical parameter known as **environmental carrying capacity** is the largest population of a given species that can be sustained indefinitely in a particular ecosystem. This is a theoretical balance point between biotic potential and environmental resistance.

Survivorship

The age-specific pattern in which the individuals in a population survive or die varies from species to species, and such patterns are important determinants of the way populations grow. Figure 4.4 shows three different kinds of **survivorship curves.** In one type of survivorship pattern (curve A), most of the organisms survive to their maximum life spans because there are few factors to cause death among young organisms. Humans and mountain sheep tend to follow this pattern. In species exhibiting an equal probability of death at any point in a typical life span, survivorship is represented as a straight line in a logarithmic graph, as in curve B. This pattern is found in most species of songbirds and in small aquatic organisms such as hydra. In the third pattern, characteris-

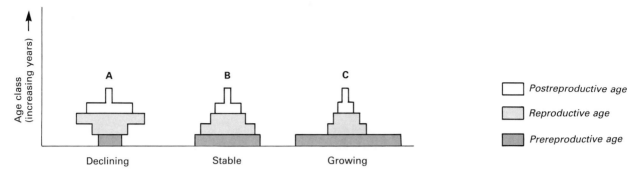

Figure 4.5 The Age Distribution Patterns and Growth Potentials in Populations. The patterns show the relative proportion of a population that is reproducing and the proportion that soon will be reproductive, assuming a certain pattern of survivorship.

tic of organisms like frogs, oysters, and oak trees, large numbers of offspring never reach maturity but die at early ages (curve C). Typically, the reproductive pattern of such organisms amounts to the production of extremely large numbers of offspring, only a small fraction of which then usually survive to reproduce. Organisms following this pattern have the greatest explosive growth potential if, for whatever reason, survival improves for a time.

These curves represent basic patterns. It is unlikely that any species has exactly a straight-line logarithmic pattern. The exact nature of a survivorship curve varies from one population to another for the same species. In humans, for example, higher infant mortality rates tend to occur in segments of the population where poverty is prevalent. Males and females of a species have slightly different patterns of survivorship. Changing population density, food availability, and other such factors can even change the survivorship pattern for a given population for a period of time.

Survivorship curves for individual species can be derived through detailed field studies. One of the first was done on Dall mountain sheep by Edward S. Deevey using data collected by Adolf Murie. Murie collected skulls from a population of Dall sheep in Alaska, and he could tell their age from annual rings on the horns of the skulls. Later he published this data, and Deevey developed a "life table" from them whereby he was able to show and compare the probabilities of death for various age groups. Similar tables have been constructed for humans for use by insurance companies.

Age Structure

The age structure of a population reflects its growth *potential* (see Figure 4.5). **Age structure** is usually represented as the percentage of individuals within incremental age ranges, for example, 0–5 years, 5–10 years, 10–20 years, and so on. In terms of potential population growth, the most useful way of classifying the age structure of populations is according to ability or potential to reproduce. The members of a population that have not yet matured sexually form a *prereproductive* group. The two other categories are the *reproductive* group and the *postreproductive* group. Again, by examining the age structure of a given population at one point in time *in light of current survivorship for that species,* it is possible to tell whether that particular population is growing or shrinking.

If there is a bulge in a population in the actively reproducing group, the prereproductive group will obviously not entirely replace those presently reproducing, and barring any surge in **fertility rate** (births per reproductive-age female), that population can be expected to decline (pattern A).

In many species, including humans, if the prereproductive fraction of a population is comparable in size to the reproductive fraction, if survivorship is high, and if fertility is constant, the population is in a steady state and will stay the same size (pattern B). If the postreproductive group comprises a very large fraction, this is generally an indication that the population is in decline.

If the prereproductive class is much larger than the reproductive class and if survivorship is high, the prereproductive group will eventually more than replace those now in the reproductive state (pattern C). Of course, if survivorship is low and stays low, the configuration may well change very little, if at all.

Any change in survivorship would influence just what will happen for any given population profile. These basic ecological facts have direct application to questions about human population growth (Chapter 8).

Population Growth Patterns

Exponential Growth

To understand population growth patterns we must first examine what is meant by exponential growth. Exponential rates of growth can be interestingly deceptive. For example, if you offered to give a person one penny and to double it each day thereafter, on the seventh day you would have to pay that person 64¢, on the fourteenth day you would owe $81.92, and on the twenty-first day you would have to pay $10,485.76! (see Table 4.1). Just for fun, estimate the number of grains of corn in a kilogram. Then calculate the weight of corn you would have coming if someone promised to give you one grain on the first square of a checkerboard, two on the second, four on the third, eight on the fourth, and so on. (Mathematically, the number of grains on the sixty-fourth square would be 2^{63}, or $9.2234 \times 10^{18} = 9.2$ quintillion grains. How would this compare to the world production of corn for last year?)

When a species is introduced into a new suitable habitat, following a brief lag period its growth rate often soon becomes exponential. The growth rate is invariably something less than its biotic potential, but it may be dramatic nevertheless. Although the *rates* of initial growth may vary considerably from one species to another, the *pattern* of lag followed by exponential growth tends to be the same for all species. What happens after that seems to fall into one of two general patterns. Depending on the species and to some degree on circumstance, overall growth patterns assume either a J-shaped pattern or a sigmoid (S-shaped) pattern.

The J-shaped Curve

The **J-shaped growth curve** (Figure 4.6) describes populations in which growth is exponential to a point at which a "crash" or a dying off of most of

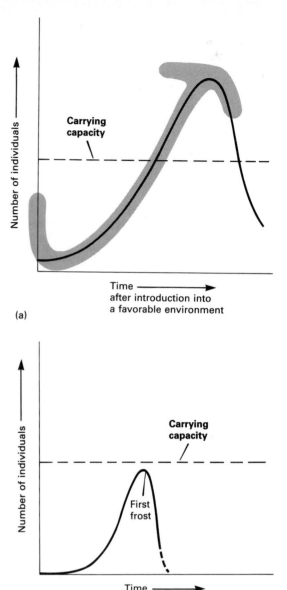

Figure 4.6 Exponential Growth Rate in Populations (J-shaped Curve). Species with this type of growth pattern are "opportunists." (a) Some are environmental opportunists that grow rapidly in response to favorable conditions and then literally eat and live themselves out of house, home, and local existence. (b) Some are seasonal or cyclic opportunists whose numbers go up during favorable conditions and fall sharply when conditions change (for example, annual plants, some insects).

Table 4.1   Exponential Growth Rates

First Week		Second Week		Third Week	
Day	Amount ($)	Day	Amount ($)	Day	Amount ($)
1	0.01	8	1.28	15	163.84
2	0.02	9	2.56	16	327.68
3	0.04	10	5.12	17	655.36
4	0.08	11	10.24	18	1,310.72
5	0.16	12	20.48	19	2,621.44
6	0.32	13	40.96	20	5,242.88
7	0.64	14	81.92	21	10,485.76

Starting with one penny and doubling it daily results in exponential or geometric growth. The term *exponential* comes from the fact that growth is described mathematically by an exponential expression, that is, some constant raised to a power. This means that as a population gets larger and larger, it grows faster and faster in succeeding intervals of time.

the population occurs. The curve that results looks (to the imaginative) like the letter *J*. The crash might be the result of density-dependent destruction of the environment or the abrupt change in the influence of some density-independent factor. An example of the former would be the food depletion and the accumulation of waste in a bottle of milk populated by bacteria. An example of the latter would be a first frost and its impact on flies, other insects, or annual

Biomes of the World

Biomes are the earth's major community types. Biomes are characterized by specific climatic conditions (of which temperature, rainfall, and evaporation rate are the most important), resulting in characteristic plant types. The plant types are in turn important determinants of the kinds of animals that can exist in each biome. The major **terrestrial biomes** include deserts, grasslands, tundras, coniferous forests (taigas), deciduous forests, and rain forests. A description of each of these biomes is given in the text.

Deciduous Forest in Upstate New York

Savanna in Serengeti National Park, Tanzania, Africa

Shortgrass Prairie of Eastern Colorado

Sonoran Desert in the State of Arizona

Rain Forest near the Pacific Ocean in Oregon

Desert in Full Bloom, Suguaro National Monument, Tucson, Arizona

Tundra in Denali National Park, Alaska

High mountains exhibit the same kind of biome transition pattern with altitude that is seen with latitude.

Coniferous Forest in the Rocky Mountains

Mount Rainier in the State of Washington

The **aquatic biome** (oceans, lakes, etc.) covers about 70% of the earth's surface. Wetlands or places where the aquatic biome meets various terrestrial biomes are among the most ecologically productive places on earth. Examples of wetlands are cypress swamps, estuaries, and salt marshes.

Salt Marsh near the Pacific Ocean in California

Biome Vegetation Chart

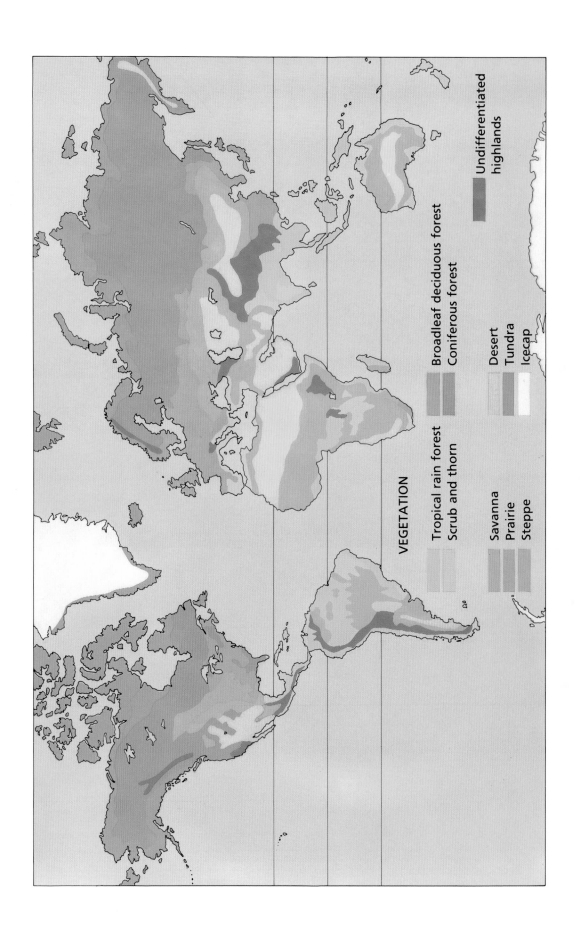

VEGETATION

Tropical rain forest
Scrub and thorn

Broadleaf deciduous forest
Coniferous forest

Savanna
Prairie
Steppe

Desert
Tundra
Icecap

Undifferentiated
highlands

plants. If the survivors are so few that the reproductive rate continues to be less than the death rate, that population is on the way to disappearance. If a sufficient number of the population survives that the reproductive rate again surpasses the death rate, that population may again go through a J-shaped rise and fall.

The Sigmoid Curve

The more common general pattern for growth of animal and plant populations is the **sigmoid (S-shaped) growth curve** (Figure 4.7). In this case limiting factors come into play to "apply the brakes" very gradually, and the population tends to stabilize around a certain level. As we have already pointed out, the *carrying capacity* of a particular environment is the maximum population size for a given species that can be supported *indefinitely*. If the population rises above the carrying capacity, it will strain the ecosystem to such an extent that the system will not be able to support it at that level, and individuals will die from lack of food or habitat, or the population will decline because of other factors that reduce the number of viable offspring. Growth sometimes takes a population well above carrying capacity initially. When this happens, the size of the population might quickly fall back toward or even below carrying capacity, and this might be followed by a period of considerable "wobble." In other cases the environment

Figure 4.7 Sigmoid Growth Rate in a Population (S-shaped Curve). Species with this pattern begin with an exponential growth rate but then respond to an environmental braking effect as their numbers approach the limits of environmental support. Sometimes such populations increase slightly above carrying capacity, but they are soon brought back within supportable limits by self-regulating density-dependent factors and fluctuate around or below this level.

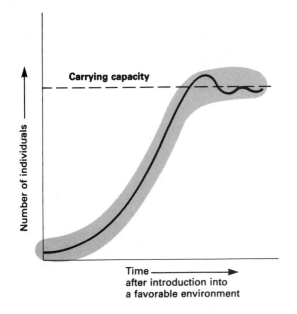

puts the brakes on ever so gradually, and population growth tapers off more smoothly as the size of the population approaches the environmental carrying capacity. In either case populations tend to fluctuate around or below carrying capacity unless some dramatic event causes a major change in the carrying capacity of that system.

There are many determinants of carrying capacity. Food supply is the ultimate regulator of many populations. Temperature seems to influence seasonal fluctuations in some populations, as does **photoperiod,** the amount of light in a 24-hour span. In many species, predation, disease, and competition may limit population growth long before food supply is endangered. The question of *ultimate* regulation is complex and controversial, and there is apparently no single answer; various factors seem to be more important for some species than others and for the same species under different circumstances and at different times.

The Concept of Niche

We are going to look at communities next, and it will help to illuminate the kinds of interactions that characterize a community if we first consider the concept of *niche* (pronounced "nitch").

Whereas the **habitat** of any given species is simply the kind of environment where one would go to find that species—its address, as it were—its niche is more complex. A species' **niche** refers to the unique, functional *role* or place of that species in an ecosystem. The ecologist Eugene Odum (1984) says that we should think of an organism's niche as its "profession," how it makes its living, how and when it gets its energy and nutrients, how and when it reproduces, how it relates to other species. According to Odum, to be complete in the description of the niche or place of a particular species, one would have to describe three things:

1. Its *physical location* within a particular habitat—its **habitat niche,** where it "goes to work"
2. Its *ecological role* within the ecosystem, e.g., the species it eats, the species with which it competes, the species that prey on it—its **trophic** or **food niche,** its "job"
3. Its *preferences* for temperature, shade, pH, humidity, slope, etc.—its **multidimensional niche,** its "working conditions"

Habitat Niche. A variety of habitat niches exist within any ecosystem. A forest, for example, can be pictured as having layers of habitat niches or

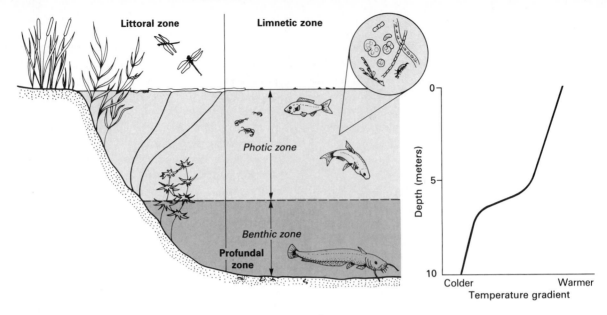

Figure 4.8 Stratification in a Lake. Differences in such factors as sunlight, temperature, and oxygen create a multitude of habitat niches or spatial niches in a lake. Terrestrial systems also exhibit stratification.

strata. Some species are found on the forest floor (wildflowers, beetles, snails). Other species are found beneath the surface (earthworms). Still others, such as lichens and algae, may occupy tree trunks. The lower branches and shorter trees provide habitats for some species (squirrels), and other species (birds) occupy the upper canopy area. **Stratification** is found in aquatic environments as well. In a lake, for example, variation in temperature, oxygen, and light cause different species to be found at different levels (Figure 4.8).

In fact, deep lakes can be divided into three zones—the littoral, limnetic, and profundal zones—on the basis of the types of vegetation and animal species found. The **littoral zone** extends from the shoreline to the farthest point where rooted vegetation is found. Cattails, water lilies, and various submerged vegetation live in this zone, along with such other organisms as frogs and insects. The **limnetic zone** is the open water area beyond the littoral zone. Its major vegetation are green and blue-green algae. Its animal species include fish and microscopic zooplankton. The limnetic zone extends downward as far as light penetrates. In a deep lake there is a third zone, the **profundal zone,** below the limnetic zone. Because there is no light penetration in this zone, its major species are decomposers and organisms that feed on dead organic matter.

Food Niches. Food niches may be demarcated by time. A hawk and an owl feed on the same food type, but one feeds predominantly in the daylight and the other is nocturnal. Thus there are equivalent

but different night and day food niches. Still other species feed at daybreak and twilight (deer, fox). Food niches may also be separated by food type. Birds that feed in the same place at the same time may occupy different niches because of what they eat—different insects, seeds, and so on. It is important to be aware of all of these subtleties because they bear on the general rule in nature that two different species cannot occupy the same niche in the same place at the same time—at least not for very long.

Multidimensional Niche. G. E. Hutchinson (1965) has suggested that the niche can be pictured as a multidimensional volume demonstrated by graphing the tolerance ranges of a species for various factors on a set of coordinates. For example, as shown in Figure 4.9, the range for two factors such as temperature and pH tolerance can be graphed. A third dimension can be added, such as shade tolerance. There are theoretically any number of such factors that form the ecological niche. Since we can graph only three dimensions, the ultimate multidimensional niche or hypervolume cannot be drawn, but the abstraction can be imagined.

Communities

Communities are groups of interacting populations. For example, a field community might consist of populations of different grasses, insects, worms, birds, and mammals interacting in various ways. Grasses provide food for certain insects and mammals; insects

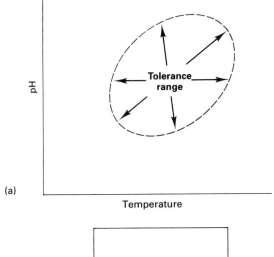

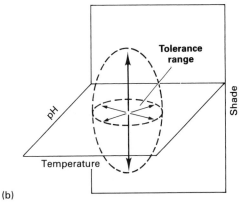

Figure 4.9   Niche as a Multidimensional Volume. The concept of niche is a complex one. Hutchinson has suggested that it can be graphed to a limited extent as a hypervolume bounded by tolerance ranges for specific factors. (a) Shown here for a given species is a graph of the tolerance range for two factors, pH and temperature. (b) A third factor, shade tolerance, adds another dimension to the graph. Many such factors help define the niche of a given species.

provide food for birds; birds prey on mammals and worms. Functionally, communities are made up of organisms with interlocking niches; each species in a community depends on certain other species in the definition of its role in the community. Though each species may not relate directly to every other species in a community, they all are interrelated. This would be analogous to the interrelated parts in an engine or the interrelatedness of the jobs in human society. A community is the living part of an ecosystem. As is true of many of the terms and concepts considered up to this point, community is an abstraction. There is a biotic community of the planet Earth. There is a biotic community in every woodlot. There is a biotic community in every aquarium.

Diversity and Stability Revisited

In Chapter 1 we broached the subject of the relationship between stability and species diversity and concluded that the relationship was fuzzy. Older environmental science textbooks nearly all had statements to the effect that the more **diversity** in an ecosystem (the more species it had), the more stable the system. Sometimes this was stated in the converse: The simpler an ecosystem, the more delicate it is. The delicacy of the Alaskan and Canadian environment (higher latitudes typically have less species diversity) was loudly touted as the main reason why the Alaskan pipeline should not have been built. Now the textbooks equivocate because over the years species diversity has not been found to be highly correlated with stability.

Odum (1984) points out that part of the problem may be semantic; there are, after all, several definitions of stability.

One kind of stability is the ability to hold fast in the face of disturbance. Another kind is the ability to return to near normal quickly after a change induced by a disturbance. Odum suggests that different kinds of stability may be mutually exclusive. He uses the phrase "inversely related." Odum ascribes **resistance stability** to California redwood communities, indicating that they are hard to change or destroy but nearly impossible to restore once they are changed. **Resilience stability,** by contrast, ascribed by Odum to California chaparral, means that although chaparral may burn easily, it recovers quickly.

Diversity may be one of several factors that improves resistance stability in nature, but it could also decrease resilience stability. Perhaps an analogy would help.

Mature redwood communities are very stable. However, once destroyed, they are for all practical purposes impossible to restore.

Consider a symphony orchestra and a three-piece jug band. Assume that one member of each group dies at random. The jug band would be devastated, but the symphony orchestra would have enough redundancy that it would sound nearly the same (because of resistance stability) as before. The remaining members of the jug band, having only the task of filling one musical vacancy, might, even allowing for training, accomplish the task in time to keep all of their engagements (resilience stability). Now assume a prolonged musicians' strike and the partial disbanding of both groups. The symphony orchestra's diversity would be a drawback this time; it might never get back to its original form. The jug band's simplicity would allow it to form again fairly easily.

If diversity improved one kind of ecosystem stability and hurt another kind, this might explain why studies have failed to show a strong positive correlation between diversity and stability.

Perhaps stability is best thought of as something ecosystems settle into. As time passes, no matter what else is true, instability is eliminated, and stability is rewarded by preservation. We should not worry nearly so much about whether or not ecosystems are simple as we should about changing them. Simple ecosystems may well be stable, after all, but arbitrarily changed ecosystems will almost certainly be less stable.

Factors Influencing Species Diversity in an Ecosystem

The Edge Effect. Although communities are sometimes depicted as distinct units, they actually merge indistinctly into one another. At the junction of two communities—a field and a forest community, for example—there is typically a zone containing species from both types of systems as well as species not found in *either* of the juxtaposed communities. This transition area is called an *edge* and is characterized by notably more species diversity and greater species density than in either of the juxtaposed communities. The mix of microhabitats and available niches is made up of the combined totals from the two systems plus some new ones created by the junction. The high level of diversity and density that occurs within these areas is called the **edge effect** (Figure 4.10).

It is also known that diversity tends to be reduced in stressed biotic communities and with a critical decrease in the size of a given type of habitat. Too many edges can lead to a decrease in diversity.

Latitude and Diversity. Diversity also tends to increase from the colder to the warmer cli-

Figure 4.10 The Edge Effect. Where two community types come together, such as a forest and a field, species in the zone between them include both forest and field species and some additional species that do not exist in either forest or field. The high level of diversity (and density) that occurs in these areas is called the *edge effect*. Estuaries where oceans and fresh water merge exhibit this edge effect.

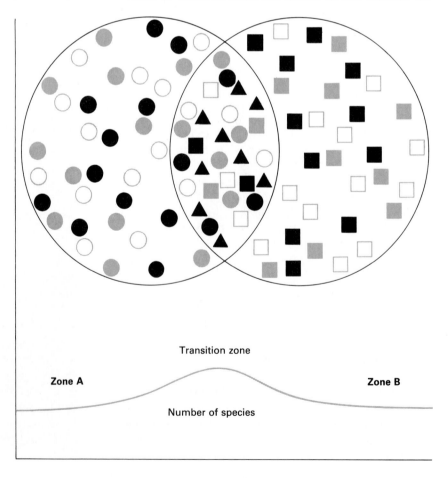

○ ◉ ● *Species in zone A*

□ ◪ ■ *Species in zone B*

▲ *Species in transition zone only*

Table 4.2   Changes in Species Abundance and Diversity with Latitude.
Numbers indicate approximate number of species.

	Florida (27° N)	Massachusetts (42° N)	Labrador (54° N)	Baffin Island (70° N)
Beetles	4,000	2,000	169	90
Land snails	250	100	25	0
Intertidal mollusks	425	175	60	*
Reptiles	107	21	5	0
Amphibians	50	21	17	0
Freshwater fishes	*	75	20	1
Coastal marine fishes	650	225	75	*
Flowering plants	2,500	1,650	390	218
Ferns and club mosses	*	70	31	11

* Data lacking.

Source: Reprinted with permission from *National Ecosystems* by W. B. Clapham, Jr. (copyright © 1973 by W. B. Clapham, Jr.) as adapted from G. L. Clark, *Elements of Ecology* (New York: Wiley, 1954).

mates (see Table 4.2). Tropical rain forests are thought to encompass two-thirds of the world's plant and animal species. One reason for this is that warmer climates encompass the ranges of tolerance of more species than cooler areas. Another reason is that many of the warm-climate ecosystems are more highly stratified—they have a greater variety of habitat niches.

Human Influences.   Given the importance of the relationships between diversity or changes in diversity and stability, the concept of diversity is especially important to humans and human activities. On the one hand, we tend to simplify ecological systems as we create cornfields, wheat fields, timber stands and front lawns. We human beings rely on only a few strains of grain for food. Though we have cultivated the strains that are most productive, we have left ourselves vulnerable to disease invasions and crop failures from climatic conditions.

On the other hand, humans also make biotic communities more complex. By creating fields separated by woodlots, a greater diversity of habitat and an enriched diversity of biotic communities may be established. Cities tend to destroy some kinds of habitats but provide others that are quite suitable for rats, pigeons, sparrows, songbirds, and other living things.

Dominance and Stability

Although a community may have many species, one or two might be the most important with regard to the nature and overall stability of the system. Such species are said to be **dominant.**

Abundance is only one of several determinants of **ecological dominance.** In a forest, insects may exist in the largest numbers, but two or three species

of trees may exert an overriding influence on the character and stability of the system. If the dominant trees were destroyed, the entire system might be drastically altered, whereas insect population changes would be less generally felt. One reason for the relative unimportance of numbers is that it takes a lot of insects to equal the biomass of just one oak. Biomass is a more meaningful expression of presence than numbers. Also, because of the second law of thermodynamics and because of the key role of plants in connecting the abiotic with the biotic, the dominance award almost always goes to plants; most terrestrial communities are named for their dominant plant species (often more than one). For example, we speak of beech-hemlock communities.

The Competitive Exclusion Principle

In the 1930s, while studying two different species of *Paramecium* in the laboratory, the Soviet biologist G. F. Gause established the principle that two species cannot occupy the same niche simultaneously (Figure 4.11). This concept is termed the **competitive exclusion principle.** If two species in the same area tend to occupy the same niche, one of three things seems to happen. If one species has a greater competitive advantage due to greater reproductive potential or some other factor, this will lead to the eventual disappearance of the other species from that area. If the habitats occupied by the two species are not completely identical, and an adjacent habitat is within the tolerance range of one of the species, that species may be forced into the adjacent habitat. A third alternative is character displacement. Two species that occupy the same area and are very much alike at first may develop differences from one another that tend to decrease competition (see Chapter 5). This divergence of characters was one of the operative principles behind what Charles Darwin saw

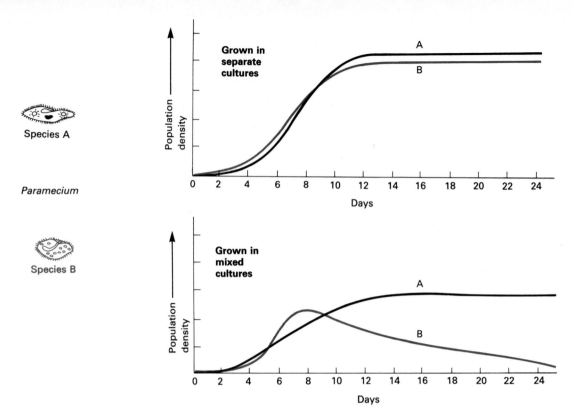

Figure 4.11 Gause's Competitive Exclusion Principle. When two species of *Paramecium* are grown in the same culture, occupying basically the same niche, one species predominates and the other gradually declines. The Soviet scientists G. F. Gause concluded that two species cannot simultaneously occupy the same niche in the same community.

Figure 4.12 Character Displacement. Species occupying the same niche may undergo physical changes over many generations that lead to the development of several niches from an initially common one. Otherwise similar species of birds in overlapping habitats drift in isolation toward characteristics that displace them from one another; differences in their bills reflect differences in types of seeds or insects eaten. This phenomenon was the basis of Darwin's observations of various types of beaks on the finches of the Galápagos Islands.

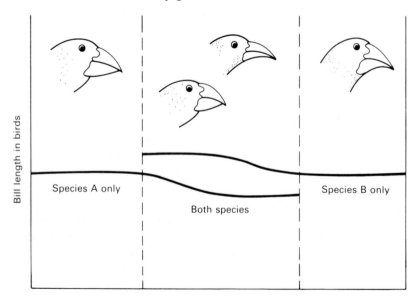

among the finches of the Galápagos Islands in the 1800s (see Figure 4.12).

Competitive exclusion as a result of the coincidence of niches was also demonstrated in a classic study in the laboratory by Dr. Thomas Park (1954) using flour beetles of the genus *Tribolium.* Park and his associates found that whenever two different species of flour beetles were put in a jar of flour, one species always thrived and the other died out. Which species survived and which one became extinct depended not on which species was the most abundant at the outset but rather on the abiotic factors of temperature and humidity. Alone, each species could survive at the tested temperature and humidity variables; together, one species did better under warm, moist conditions and the other under cool, dry conditions. Under moderate conditions, one or the other could survive, but not both.

Not all scientists agree on the significance of competition in explaining species coexistence or lack of it.

Other Interspecific Relationships

Here we consider some specific common relationships between species. As will be seen in the examples that follow, some such interactions have become **obligatory**—certain species cannot exist except in very specific interspecies relationships. In other cases we will see that some such relationships are **facultative;** that is, the species may exist either within or outside of such a relationship.

One type of interspecific interaction is termed symbiotic, which means "living together." Often **symbiosis** refers to relationships between two species that are mutually beneficial (*mutualistic*) or beneficial to one with the other unaffected (*commensal*).

In interactions under **mutualism,** both species benefit. The most common example used to demonstrate obligatory mutualism is the lichen. You often see lichens growing as a greenish-gray crusty coating on tree trunks or rocks. A lichen is composed of a fungus and an intimately associated alga. The fungus contributes moisture and minerals to the union; the alga produces food via photosynthesis. In more advanced species of lichens, the fungus does not penetrate the algal cell wall, and the relationship is indeed mutualistic. In more primitive lichens, however, the algal cell wall is penetrated, and thus a parasitic relationship results. A more consistent example of mutualism is the relationship between certain fungi and other plants. The rootlike structure of these fungi, or *mycorrhizae,* interacts with the roots of trees or other plants, enhancing mineral uptake from the soil. Mycorrhizae are important in the growth in poor soils of some trees such as pines. Another interesting example is the termite. The termite actually owes its ability to digest wood to a small protozoan that the termite carries in its gut. The protozoan secretes a substance that converts cellulose into its carbohydrate components, providing nourishment for the protozoan *and* the termite. Insect pollination of flowers is another common example of a mutualistic relationship.

Commensalism is a form of interaction in which one species benefits and the other is not affected. Epiphytes, plants that grow on the limbs of trees in tropical forests, extract no nutrients from the tree but use the branch as a point of attachment to obtain sunlight that does not penetrate to the forest floor. An orchid is an epiphyte. In another example of commensalism, a small ocean fish called a remora attaches itself to the body of a shark. The shark provides transportation, and the remora may also consume food remnants from the shark's prey.

Two classes of species interaction wherein there is benefit for one species and harm for the other are parasitism and predation. In **parasitism** the organism lives *in* (**endoparasite**) or *on* (**ectoparasite**) its host; often more than one parasite infects a single host. A "good" parasite is one that has developed a relationship with its host over long generations and will not kill its host. It must at least allow sufficient time for its host to reproduce; otherwise, in killing its host it destroys its own food supply. Consequently, a well-developed parasitic relationship may actually approach commensalism. This points up a very pertinent fact. All of our classification schemes are artificial; they are loose abstractions made on the basis of patterns observed in nature. Examples that do not specifically fit any category show only the weaknesses in our classification schemes.

Another kind of interspecific interaction is **predation.** In predation the predator is free-living and usually larger than its prey. A predator kills its prey for food. While predation is most definitely a "negative" for the *individual* caught and eaten, the picture is not so clear-cut when the focus is on the total prey

Predators play important positive roles in nature, culling the weak and imperfect of the species on which they prey.

population. Many studies show that in larger animals, predation is a mechanism for culling the imperfect young, the old, and the sickly. Deevey found in developing the life table for the Dall sheep that mostly the very young and the very old were killed by wolves.

A study of moose and wolves carried out on Isle Royale, Michigan, which was discussed in Chapter 1, shows the role of predators in population control. When there were no wolves on the island, the moose population increased dramatically, and the moose destroyed habitat in search of food. This overpopulation resulted in a population crash; many moose died of starvation. Eventually, wolves made their way to the island, and a balance was reached between the moose and the wolves, keeping both populations in check.

For many predator-prey relationships, the predator may be a relatively insignificant population control factor. To affect population size, the predator must have a significant impact either on the number of individuals able to reproduce or on those who will one day be able to reproduce. If predators take only those that are likely to die soon anyway, their effect as regulators of population size is negligible.

Generally, it would seem that in populations of *smaller animals,* predation is less discriminate, and individuals of all ages and reproductive capacity are equally subject to the predator. Consequently, predators may be one factor in regulating population size for smaller animals.

Homo sapiens is a predator of sorts, but our modern relationship to our food species, both plant and animal, comes closer to mutualism. Some plants and animals are ensured of propagation by humans, and in return humankind is supplied with a dependable food source. The development of these relationships between humans and the rest of nature is considered in Chapter 5.

Species Introductions

The practical importance of population and community dynamics in nature can best be illustrated by example. When the European hare was introduced into Australia as a game animal, it had no natural predators. The hare reproduced dramatically—perhaps close to their biotic potential—and advanced over the continent "like a swarm of locusts." No one had considered the population checks on the European hare. In 1950 a strain of virus known to be fatal to rabbits was introduced into Australia. In one year, 97% to 99% of the hare population was killed. Those that survived, however, passed on some immunity, and the population gradually increased again. Hare are still a problem in Australia.

Another episode occurred in the United States when the European starling was introduced in the 1890s to compete with the house sparrow, which had been released in the 1850s and was reproducing prolifically. As it turned out, both the starling and the sparrow populations prospered; the starling became the competitor of songbirds instead of sparrows. In addition, the starling damaged crops and eventually became a general nuisance, especially around airports.

The Distribution of Communities: Biomes

Biomes are the major types of biotic communities of the world. The distribution of various types of terrestrial biomes is primarily a function of climate and soil. From the North Pole to the equator there is a gradual transition from tundra to tropical rain forest. The same phenomenon can be observed in the transition from a high to a low altitude. Each is a "megacommunity" characterized by specific plant types. Each terrestrial biome is unique, and yet **ecological equivalents** (different species performing the same ecological function in different communities) are found in each. This is an illustration of the fact that the same fundamental patterns override the specific relationships in any subdivision of the biosphere. Some major terrestrial biomes are tundra, coniferous forest, temperate deciduous forest, chaparral, grassland, desert, and rain forest. A description of these follows. The map in the "Biomes of the World" insert shows the location of the major biomes.

Most **tundra** biome communities are located in the northern hemisphere just south of the polar ice-cap region. Tundras are cold with little rainfall and low evaporation rates. Tundras are treeless, the vegetation consisting mainly of grasses, sedges, low flowering herbs, and lichens. The roots of tundra vegetation are relatively shallow because they cannot penetrate the **permafrost** (water permanently frozen in the soil), which begins a few centimeters or more below the surface of the ground. During the warmer months, water frozen in the soil above the permafrost will thaw and, because of poor drainage and the permafrost barrier below, pool into small ponds, forming a marshy plain. Even in the warmest months, air temperatures are low—less than 10°C—but the long days of summer and the availability of water allow vegetation to flourish during this time. Tundra animal life consists of burrowers—lemmings and mice, arctic fox, and musk oxen—all of which survive year-round on the lichens and other plants. Waterfowl and caribou can also be found on the tundra during the summer months. Tundra communities are fragile as a consequence of low temperatures and short growing seasons. Any disturbance to the land by vehicles or construction can leave ruts and bare areas that take

Biomes and Land Use

Biomes have characteristics that restrict how land may be used. This raises some basic questions of land use by humans and its compatibility with natural factors. Should homes be allowed to be built in the chaparral region of southern California, a biome characterized by periodic fires? Should land in desert areas be irrigated for agricultural use?

Should wetlands be drained for agricultural use? Should homes be permitted to be built on erodible shorelines? Should the construction of homes be permitted in 50-year or even 100-year flood zones? On prime farmland? What do you think?

many years to regain their productivity because the vegetation is slow to recover.

The **taiga** is a biome dominated by evergreen, coniferous forests. The temperature range in coniferous forests is cool to cold, and there is generally more precipitation than in tundra. The soils of the taiga are acidic and deficient in minerals because minerals are leached out of soil by rainwater. Transpiration and evaporation rates are relatively low, so that the effect they have on bringing minerals to the surface is more than offset by the downward flow of nutrients. The dominant vegetation of the taiga consists of needle-leaved, cone-bearing evergreens such as spruce, fir, larch, and pine. There are also some aspen, poplar, and birch in low, moist areas. Dominant large animal species are mostly mammals like moose, caribou, elk, grizzly bear, wolverine, and beaver. The soil in the coniferous forests is characterized by a relatively thick bed of plant material that decays very slowly. Because of this, there is not much **humus** (organic layer formed from decomposing matter). There is little undergrowth because the dense foliage filters out most of the sunlight. These characteristics make the land very susceptible to erosion when trees are harvested by methods that leave the land bare.

Besides being found in northern latitudes, taiga-like and tundralike communities can be found on mountainsides in more temperate latitudes. Temperature and precipitation may vary with altitude in the same way that they vary with distance from the equator. Temperature and rainfall are major factors in determining *climax* (steady-state) communities (more on these later). Therefore, it is not surprising that in climbing a mountain one may find the same biome changes as would be present moving northward in latitude. Climbing a mountain, one may start from grassland (in the west) or deciduous forest (in the east) at the base of the mountain. At higher altitudes one will eventually reach evergreen forest resembling the taiga. At even higher altitudes, alpine communities resembling the tundra biome will be found (see Figure 4.13).

The temperate **deciduous forest** biome is characterized by a moderate climate with four distinct seasons. The dominant vegetation consists of broad-leaved hardwood trees such as beech, maple,

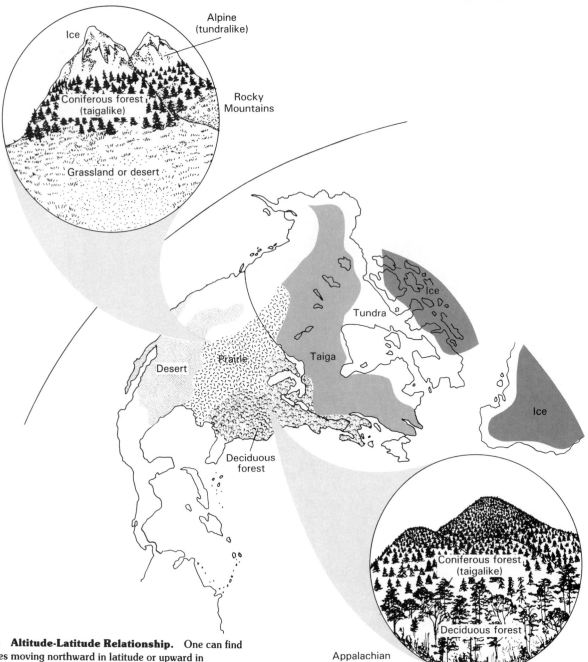

Figure 4.13 Altitude-Latitude Relationship. One can find similar biomes moving northward in latitude or upward in altitude.

oak, hickory, basswood, poplar, sycamore, and elm. In contrast to the evergreen forest, sunlight *can* penetrate to the deciduous forest floor, allowing a rich, diverse understory of shrubs, herbs, and flowering plants to flourish, particularly in the spring prior to the emergence of full summer foliage on the mature trees. Large animal species include deer and black bear. There are also a variety of smaller mammals such as fox, bobcats, raccoons, squirrels, chipmunks, and mice. Deciduous forests are also home to many kinds of reptiles, amphibians, birds, and insects. It is the temperate deciduous forest biome that has been, to a great extent, "civilized" by humans. Very few virgin timber stands remain. Forests were cut to be replaced by human settlements, cultivated fields,

roads, and other forms of development. Other once forested areas have nearly all been cut over several times to provide fuel and building materials. Good management in harvesting hardwoods can lead to sustainable yields over time, but poor techniques that result in erosion and loss of topsoil can slow or prevent a return to the hardwood climax.

The **chaparral** biome is dominated by low evergreens with waxy leaves. This biome occupies parts of southern Europe, southern Australia, southern California, Mexico, and South Africa. Chaparral tends to be fire-adapted, which means that it depends on periodic fast-burning fires to perpetuate itself and allow reproduction of low vegetation at the expense of taller trees. Chaparral is more open than deciduous forests. The precipitation rate is low in the summer months. The predominant animals are small-hooved mammals, rodents, and reptiles. The slopes of the chaparral region tend to be steep, with loose soils that are prone to erosion when disturbed. Disturbing these areas also increases runoff and can increase sedimentation in streams and flooding. In areas where prevention or control of "natural" fires has been the practice—in southern California, for example—the "unnatural" accumulation of plant material has on occasion resulted in "unnaturally" more severe fires. These have in turn resulted in slower recovery of vegetation and more extreme erosion.

The **grassland** biome is characterized by irregular rainfall, a high evaporation rate, and rich soils. Examples of grasslands are the steppes of Eurasia, the pampas of South America, and the Great Plains in the United States. In the United States, rainfall decreases as one moves from east to west. In the eastern grasslands of Illinois, Missouri, and eastern Oklahoma, grasses tend to be tall; shorter grasses predominate in the more arid western states. Low precipitation and a high evaporation rate mean that mineral nutrients tend *not* to be leached out of the soil. This plus the high organic content of prairie soils make them very rich. Native grasslands were well adapted to natural fires, which played an important role in prairie mineral cycles. Fires also serve to stabilize grassland systems by restricting the growth of trees and shrubs. Grasslands are well adapted as pastures because grass continues to grow from the base even as their upper blades are eaten. These pasturelands can be overgrazed, making them much less productive. Domesticated livestock are harder on grasslands than wild species because domestic animals tend to stay in larger groups, trampling more grass, and tend to eat fewer kinds of grass species, resulting in overgrazing of favored species and a change in the grass species mix. Use of grasslands for livestock has also resulted in campaigns against the normal predators found in the prairie, such as wolves and coyotes. This has in turn led to increases in rodent populations such as jackrabbits, ground squirrels, and prairie dogs. The original grazers, the bison, were once nearly exterminated by overhunting. **Tropical savannas** are grasslands in South America and Africa characterized by a few scattered trees amid the grasses. Loss of habitat is causing overpopulation in many of these areas as the natural species are forced into smaller and smaller areas. As cropland, grasslands have been and still are important as sources of many of our staple food plants.

The **desert** biome is characterized by very low rainfall and a high evaporation rate. There is a wide fluctuation in diurnal high and low temperatures because of a lack of humidity; humidity tends to hold and carry the heat of the day into evening and night. Characteristic desert plants include brush, cacti, and low-profile plants well adapted to the arid conditions. Desert plant root systems tend to be shallow and expansive to allow them to take adaptive advantage of the little rainfall that does occur. Consequently, vegetation is widely scattered. There are few large animals. Common desert animals include coyotes, fox, rodents, insects, snakes, and lizards. The animals, too, are adapted (behaviorally and physiologically) to lack of rainfall. Some have metabolisms that use water more efficiently; others use water sources such as dew. Many desert animals are nocturnal. Attempts have been made to cultivate desert areas by importing water and irrigating the land. The costs of this type of development are high both in terms of economics and in terms of environmental impact. Among the effects on soil, for example, is salt accumulation (see Chapter 13).

The **tropical rain forest** biome is found in Central and South America, Africa, Southeast Asia, and the East Indies. Rainfall tends to be evenly distributed throughout the year in tropical rain forests, and humidity is high. Temperatures range between 20°C and 28°C. This type of biome has the greatest diversity of both animal and plant species; plants stay green all year. Unusual plant types include the epiphytes—plants such as orchids and members of the pineapple family (bromeliads) that grow perched on the branches of taller trees where they have access to sunlight and absorb rainfall or moisture from the air. There are abundant bird species and tree-dwelling animals in the rain forest.

The soils in tropical rain forests tend not to be very fertile. Most nutrients are tied up in the biomass of the lush vegetation. Nutrient turnover is so rapid that nutrients are absorbed directly from the litter layer; root systems are not extensive. When tropical lands are cleared, nutrients tend to be lost quickly by being leached from soils in heavy rains.

The biological diversity in the rain forest is its most unique characteristic. Unfortunately, much rain forest continues to be destroyed for development purposes. The characteristic method of agriculture in these areas is slash, burn, cultivate, then move on to another area. Because the soils are not fertile, they cannot be continuously cropped. Once the native rain forest is removed, the secondary growth forest that returns does not have the same vegetative mix, probably because of the loss of nutrients as a result of leaching and lost biomass.

A **temperate rain forest** occurs in the extreme northwestern part of the United States. It is characterized by such tree species as western hemlock, western arborvitae, grand fir, redwoods, and

Sitka spruce. Overharvesting of these giant trees can destroy this biome because of the loss of nutrients tied up in the biomass of the harvested trees.

Aquatic "Biomes." Perhaps because we are terrestrial creatures, we often pay much more attention to terrestrial ecosystems than we do to aquatic ecosystems, even though aquatic ecosystems cover more than 70% of the earth's surface.

Subdividing aquatic ecosystems into biomes would be based on factors other than climate, so it is awkward to treat aquatic ecosystems the same as terrestrial biomes. In discussing all kinds of biomes it perhaps makes more sense to regard aquatic ecosystems as a *single* biome in which water is the distinctive feature. Indeed, the open waters of oceans and lakes have a similar character, the medium being water and the base of the biotic community being microscopic plants called phytoplankton. We could speak of the aquatic-plankton biome as a single entity. Where water meets land or where it lies in close juxtaposition to land (e.g., coastal areas and rocky shores, estuaries, coral reefs, rivers, streams, ponds, and small, shallow lakes), the presence of large, anchored plants and mineral and organic runoff from the land set the character of the aquatic communities. If oceans and large lakes amount to one aquatic-plankton biome, perhaps shores, reefs, rivers, streams, small lakes, and ponds could all be considered special kinds of edges where aquatic and terrestrial systems overlap or are combined. All of this having been said, however, there remains some rationale for considering rivers, lakes, streams, reefs, estuaries, and coastal regions as different kinds of ecological entities. Let us examine some of their distinctive features.

Where water is found on the surface of the earth, it is doing one of two things. It is either filling up a hole, for example, **oceans, salt lakes,** and **ponds,** or it is in the process of moving, for example, **streams** and **rivers,** either toward the ocean or to some other lower level as part of the hydrologic cycle. All freshwater ecosystems are made up of waters flowing in some net direction, but salt waters typically have no outlet to another lower level. Water enters oceans and salt lakes and typically leaves mainly by evaporation, leaving behind any dissolved minerals that may have been carried to the saline bodies of water. This accounts for why these bodies of water tend to get saltier and saltier with the passage of time.

The fresh waters found in small streams and large rivers offer the species that inhabit them a similar type of physical environment. Typically, streams and rivers contain organisms that are able to withstand the physical flowing of water and depend on the

relatively high levels of oxygen found in these bodies of water. Flowing waters are more highly oxygenated because they mix with air as they flow. Freshwater lakes tend to be wide places or holes into which streams or rivers flow and out of which streams or rivers then continue. Physically, lakes tend to be like rivers and streams with respect to their dissolved mineral content. However, they tend to be lower in oxygen because of the relative lack of mixing with air. The ecology of lakes, especially deep lakes, is unique because of the thermal stratification or layering that occurs in these bodies of water (Chapter 3).

Because of their differences in oxygen content, the human impacts on rivers and streams on the one hand and lakes on the other can be quite different. In general, the sediment put into lakes by human activities ages the lakes prematurely. Lakes (which in oxygen concentration are similar to slow-moving rivers) are more susceptible to dissolved organic matter loads (e.g., sewage) than fast-moving streams. Organic pollutants cause lakes, already low in oxygen, to be depleted to the point where they become unable to support normal aquatic life. The natural chemical composition of fresh waters can vary considerably, depending on the nature of the geochemistry and the vegetational character of the watersheds that these bodies of water drain. We will have much more to say about water pollution and the differences between lakes and flowing waters in Chapter 12.

Except for water that ends up being stored in underground aquifers, fresh waters tend to flow toward the oceans and eventually get there. Once in the oceans, the waters cannot go any lower. The only way out is by evaporation. The oceans are by no means stagnant, however; ocean currents, tides, wind, and various types of geologic action keep the waters of the oceans in motion. Oceans also exhibit different ecological zones because of the combinations of current flow, oxygenation, sunlight penetration, and the ebb and flow of tides found near their edges.

Some of the most interesting ecosystems on earth occur at the junctions of aquatic and terrestrial ecosystems. These freshwater and saltwater marshes, bogs, and estuaries (where fresh and salt water mix) can all be lumped under the heading "wetlands." Wetlands tend to be very important as breeding grounds even for aquatic organisms that otherwise inhabit the open oceans, and they tend to be among the richest ecosystems in terms of diversity and productivity because of the edge effect, discussed earlier in this chapter. These areas provide optimal combinations of water, sunlight, and nutrients for many organisms. Human beings have tended to fill in wetlands in attempts to extend land area and in this and other ways have caused many negative ecological impacts on wetlands around the globe. Many countries now have special laws protecting wetlands. The significance of wetlands, and the problems that have resulted from the loss of these systems, will be discussed in Chapter 14.

Succession

We have stated that living organisms are affected by their environment and in turn exert an influence on it. Organisms invading virgin territory may change their environment even to the extent that they make it less suitable for themselves and more suitable for other species. The natural sequence of such changes in community structure is called **ecological succession.** It occurs in amazingly regular patterns tending toward a climax or steady-state situation.

Succession is often referred to as being primary or secondary. **Primary succession** involves an unbroken sequence from rock or bare soil to climax veg-

etation. Primary succession can be found in abandoned quarries, abandoned strip mines, badly eroded slopes, and the area around recently erupted volcanoes. **Secondary succession** occurs where a climax or near-climax community is changed and succession from earlier stages begins again. Abandoned farmland is a good example of an area undergoing secondary succession. Foresters often speak of secondary growth timber where all the virgin growth has been cut down and the area has been allowed to grow to forestland again. A path or trail worn through the forest, if left untraveled for a while, would undergo secondary succession. Both terrestrial and aquatic communities undergo succession by means of a series of transitional stages called **seres** until a dynamic equilibrium is reached. Let us look at specific examples.

The transformation of rock to soil is an example of primary succession and is an inherent part of terrestrial succession (Figure 4.14). Exposed by a violent

Figure 4.14 Terrestrial Succession. Plant species alter their environment in a fairly predictable sequence, making it less suitable for their own survival and more suitable for other species, which do the same until a dynamic steady state is attained. This figure depicts the successional pattern that would occur if a patch of deciduous forest were swept clean, down to the bedrock.

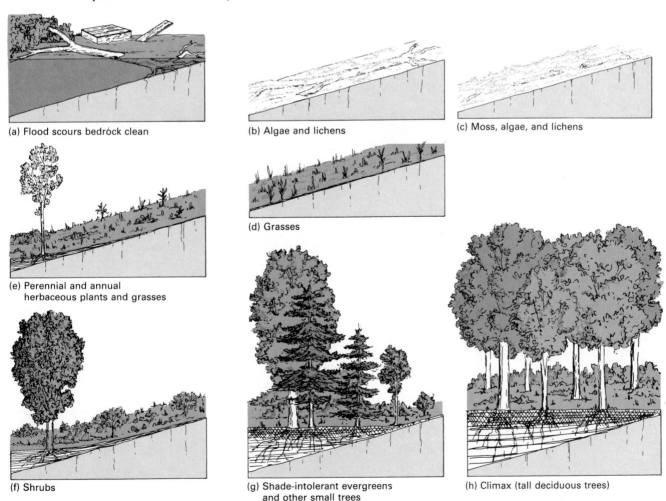

(a) Flood scours bedrock clean

(b) Algae and lichens

(c) Moss, algae, and lichens

(d) Grasses

(e) Perennial and annual herbaceous plants and grasses

(f) Shrubs

(g) Shade-intolerant evergreens and other small trees

(h) Climax (tall deciduous trees)

storm or by the draining of a lake, bare rock immediately begins to weather. Pulverized, moisture-holding material will, in time, accumulate in small low places or in cracks in the larger rock. Seeds from trees, flowers, or shrubs are likely to become trapped in these areas as well, but conditions may not be right to support most of these plants at first. Some kinds of **pioneer plants**—lichens, for example—that need little moisture *can* survive and flourish under such

conditions. So of all the potential colonizers (plants that arrive and try to become established), only certain species such as lichens may find an appropriate tolerance range. In fact, lichens are referred to as pioneer species in ecology. They prepare the way for others to follow. As they grow, lichens release organic acids that react chemically to dissolve more rock, which, together with accumulating organic matter from dead bits of lichens, creates new condi-

Figure 4.15 Aquatic Succession.
Lakes and similar aquatic communities undergo a series of changes as over time they begin to fill with sediment and organic debris. Eventually, lakes become solid land and undergo terrestrial succession. Depicted here are the successional stages following formation of a new lake by a receding glacier.

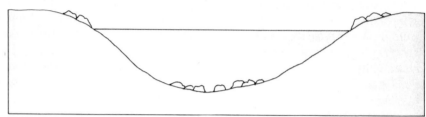

(a) Newly formed lake

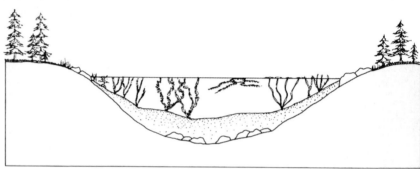

(b) Mature lake

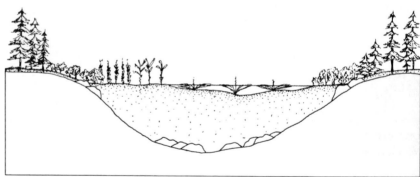

(c) Meadow/marsh

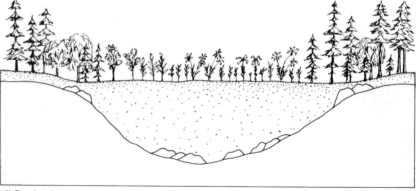

(d) Dry land

tions that may be suitable for other species. Gradually, as more water and debris accumulate in the depression and as weathering of the rock continues, moss may outdo the lichens in competition for light and other resources. Moss in turn causes still further changes in chemical and physical conditions, making the environment suitable for small plants and shrubs. As these plants provide greater shade, changes in temperature and moisture cause conditions to become more favorable for different seeds to germinate and flourish, ones that earlier would have germinated and withered. Eventually, sun-tolerant species of trees become established. As these trees provide more and more shade, shade-tolerant species of trees will dominate.

In terrestrial systems, succession is a function of the interaction of organisms and their abiotic environment. In some aquatic systems, such as ponds and lakes, the prime factor in succession is input from the outside.

Succession in lakes and ponds involves the gradual change to marsh and then solid land by the accumulation of sediment and debris on the lake bottom (Figure 4.15). Sediment and debris may be carried into the lake through runoff water that picks up materials as it travels through the watershed or drainage areas to the lake. In areas where the land surface has been disturbed, erosion is more severe, and the buildup of sediment occurs at a faster rate. In lakes that receive a lot of nitrates and phosphates, such as those receiving runoff from cattle feedlots or effluents from sewage plants, the rate of succession is increased. These nutrients result in more plant growth within the lake. When this plant growth dies and decays, this debris accumulates on the lake bottom, resulting slowly in a shallower lake.

Odum (1984) describes various patterns of secondary succession. Secondary succession in a temperate terrestrial environment is as follows. Generally, the growth of annual plants is replaced by perennials, which in turn prepare the way for grasses and shrubs. Eventually, pines and fast-growing deciduous trees develop and change the conditions to favor the growth of shade-tolerant species such as oaks or hickories. In the southeastern United States, management practices such as fire must be used on pine tree plantations to prevent succession to shade-tolerant hardwood species.

Secondary succession on grasslands is fairly well documented. The pattern involves a four-stage development from annual weeds to short-lived grass to perennial grass and finally to climax grass. A total of 20 to 40 years is required for grasslands to become reestablished from plowed ground.

Changes in vegetation, whether terrestrial or aquatic, bring with them changes in the animal species that live and feed on the vegetation. Animal species in turn influence prevailing conditions by the manner in which they help to distribute seeds, burrow into the soil, and feed on the vegetation, among other things.

A climax community is, by definition, relatively stable. The impacts that climax species have on the abiotic environment and on one another tend to be offset by other impacts and ongoing processes such that the system no longer drifts toward some other state as long as climate remains relatively constant. That is not to say that climax communities are not changing. Rather, in the climax state, a dynamic balance is achieved through self-regulating feedback mechanisms wherein minor perturbations tend to be absorbed without major alteration of the system as a whole. The impact of the disappearance of a particular plant species in the understory of a forest community due to unfavorable weather patterns in a given year may be absorbed because of other, near-equivalent plant species that are present. Changes in the availability of nutrients are accommodated by greater or lesser periods of growth. Only a major climatic disturbance such as fire, abrupt human impact, or continued chronic stress over a long period of time is likely to disrupt a climax system to move it back to some earlier successional stage.

The expected or normal climax stage for a particular region is not always reached in every location because of varying local conditions. For example, factors such as chronic erosion may prevent soil development and hence the successful growth of vegetation that requires good soil. Many of the things we humans do serve to keep ecosystems from maturing; agricultural and forestry practices (e.g., cultivation, pasturing, logging, mowing) are perhaps the most familiar examples. Human impact has also been known to cause succession to occur at faster than normal rates. A variety of human activities tend to increase siltation in lakes, causing the sequence of successional stages to occur more quickly. Normally, the time required for the transition of a lake to a climax forest may be hundreds or thousands of years.

Moving On

Many terms used in ecology are nondescriptive or even misleading. *Producers* is one that was pointed out in Chapter 1. In this chapter we have introduced some more—*climax*, for example.

If one scraped a patch of virgin Montana forest clean to the bedrock, a series of stages would result until something once again resembling a virgin forest appeared. But Montana did not always look like it does now. And it surely will not always look that way.

Population Control

Many things serve to limit the growth of populations in nature. Among the kinds of limiting factors we have reviewed in this chapter are food supplies, space or habitat, disease, predation, and inborn physiological factors that serve to reduce fertility as population density increases. Humankind is in the process of experiencing a temporary increase in limits as evidenced by the rapid increase in population in the last 100 years or so. Obviously, human population cannot continue to grow as it has for very long—some people would argue that the earth cannot really sustain the population it has now. Nature will surely impose new limits whether we do something about population growth or not. Some people say that we cannot do anything about population growth, even if we wanted to, because cultural factors are too overwhelming, and it would be impossible to get all of the peoples of the world to work on the problem cooperatively, even if they did see the problem the same way—which they don't. Should we just forget about the population problem and let nature take its course? What do you think?

There is another dimension to succession, obviously. What we have described are not climaxes in the *absolute* sense at all.

We are about to consider chemical and biological evolution. Evolution, as we shall see, is really a "supersuccession." It is the sum of the interactions between chemicals and the physical environment and then later between life forms and the environment. Evolution is succession extended to the very gradual impacts living things have on the composition of the atmosphere, on the physical state of the earth's surface, and on climate and the impact that these things in turn have on which organisms will most likely survive to reproduce.

Concepts to Remember

1. The patterns of interaction between living things are an important dimension of ecology. Interaction is the means of self-regulation of biotic communities, and self-regulation of biotic communities is the essence of ecology.
2. An interbreeding or otherwise interacting group of organisms of the same species is called a population. Certain features are unique to the population level of organization. Some examples are distribution, density, growth rates, survivorship, and age structure.
3. Many factors determine the growth rate of a population. These can ultimately be reduced to how many individuals are born, how many die, how many move in, and how many move out per unit of time. Among the determinants of these basic parameters are (1) the number of individuals (particularly females) of reproductive age, (2) the number of offspring per birth, (3) how often births occur, and (4) the survival pattern determined by predation, disease, nutrition, shelter, and living space.
4. There is no one answer to the question of what regulates population growth. Factors that influence population size change from season to season and from species to species. Among the most important factors are food supply, availability of habitat, social interactions, and climatic changes. Some factors come into play when populations reach a certain size, such as available food or spread of disease; others such

as climate have impacts not directly related to population size.

5. The maximum number of individuals of a particular species that an ecosystem can support indefinitely is called the carrying capacity. The size of a stable population usually fluctuates around or below the carrying capacity.

6. The carrying capacity of a given ecological system for a particular species may change dramatically if some severe disruption occurs such as a forest fire or the introduction of a competing or predatory species. Should a population exceed the carrying capacity of the environment, the impact on the system may permanently lower the capacity of that system for that species.

7. If we could describe everything about a species—such as how it relates to every other species in the ecosystem; what type of habitat it prefers; how, when, and what it eats; when it reproduces; and how it relates to members of its own species—we would have defined the niche of that species. Every species has a niche or job description within the ecosystem.

8. A community is an interactive group of plant and animal populations. Communities have certain characteristics that individuals and populations do not have; these include diversity, food webs, and species interactions such as competitive, predatory, parasitic, and symbiotic relationships.

9. Diversity refers to the number of different species in a community. Diversity may enhance some kinds of stability and diminish other kinds. Generally, diversity increases with the stages of succession. Diversity is greater in warmer climates than it is in colder climates. Diversity is high where two different communities overlap. Diversity is reduced when areas of a given habitat are drastically reduced in size and in stressed communities.

10. The major terrestrial communities of the world are called biomes. They include tundra, coniferous forest (taiga), temperate deciduous forest, chaparral, grassland, desert, and rain forest.

11. A natural sequence of changes (succession) in communities of living things takes place over long periods of time until a highly stable community is reached called the climax stage. Before the climax stage, the species in the transient communities interact with abiotic factors, changing the ecosystem so that it is less favorable to the current species and more favorable to new species. Aquatic ecosystems such as lakes eventually fill in to form terrestrial systems; terrestrial ecosystems usually move toward climax as one of the major biomes.

References and Further Reading

References marked with an asterisk are cited in the chapter.

* Christian, J. J., and Davis, D. E., 1964. "Endocrines, Behavior and Populations," *Science* **146**:1550–1560.

Cheater, M., 1991. "Save That Taiga," *Worldwatch*, **4**(4):10ff.

Huston, M., De Angelis, D., and Post, W., 1988. "New Computer Models Unify Ecological Theory," *BioScience* **38**:682–691.

Hutchinson, G. E., 1965. *The Ecological Theater and the Evolutionary Play*. New Haven, Conn.: Yale University Press.

Lewin, R., 1986. "In Ecology, Change Brings Stability," *Science* **234**:1071–1073.

"Managing Planet Earth," 1989. *Scientific American* **261**(3):46–175. (Special issue)

* Odum, E. P., 1984. *Fundamentals of Ecology*. Philadelphia: Saunders.

* Park, T., 1954. "Experimental Studies of Interspecific Competition: II. Temperature, Humidity and Competition in Two Species of *Tribolium*," *Physiological Zoology* **27**:117–238.

Schneider, S., 1990. "Debating Gaia," *Environment* **32**(4):5–9.

Stolzenburg, W., 1991. "The Fragment Connection," *Nature Conservancy* **41**(4):18–25.

See also the references cited at the end of Chapter 1.

Evolution and Ecology: The Emergence of the Ecosphere

5

There are several reasons for including a chapter on evolution in a book with an environmental or ecological theme. First, as was indicated at the end of Chapter 4, evolution is another kind of ecological succession.

Like ecological succession, evolution is driven by selective forces derived from the interaction among living things and between living things and the nonliving environment. Both ecological succession and evolution encompass the biotic and the abiotic. The nonliving environment also undergoes succession and evolves. As we discussed in Chapter 4, moss and other pioneer species generate the soil and other physical and chemical conditions that allow other plants to succeed them. A parallel example in evolution would be the way in which oxygen-producing organisms changed the atmosphere so as to allow oxygen-utilizing organisms to emerge and thrive. Evolutionary biology and ecology also overlap in other ways. Both, for example, have something to say about the distribution and abundance of animals and plants. Each is concerned with the influence of the other on the biology of populations.

Another reason for including evolution here is that a book on any topic should include some history. Serious students of the relationship of humans and the environment must appreciate how the natural systems on which we depend came to be and how they came to function the way they do; such students must also appreciate how the stage was set for humans by prehuman evolution and how our species was literally shaped by the environment. The history of human evolution may even reveal the roots of the environmental impact our species has today.

Still another reason for including evolution is that a grasp of the concepts of adaptation, natural selection, and evolution is essential to understanding a wide range of modern environmental problems such as resistance to pesticides and species extinction.

Finally, by considering our species in the evolutionary context, by appreciating it as the tip of a branch having origins in common with numerous other species, it should become easier to see that we are indeed an integral part of nature.

What Is Evolution?

Biological evolution is change in the genetic composition of populations resulting from the unequal reproduction of genes. The basis for evolution is *variation* and *selection*.

There is considerable variation in any group of living organisms. Not all individuals in a population of squirrels, bass, or people are exactly the same. Variation exists because of differences in environmental influences and because of individual genetic differences. Let's consider an example.

Imagine a population of mammals that in the past may have been the ancestors of the giraffe. The distance from the ground to the mouths of the members of this population ranged from 2 to 3 meters; in other words, there were some tall animals and some short ones. Suppose that during a period of two or three years there was a dry spell, much of the vegetation on which this species subsisted did not thrive, and the population had to survive by eating the leaves of a certain type of tree. Suppose that all the low-lying branches of these trees were consumed during the first year and after a time the lowest leaves that could be found were above the level of 2 1/2 meters.

Now, assuming that some of the short organisms in this population were short because they had genes that specified shortness, imagine what would happen to "short genes." Eventually, short members of this population would be unable to obtain food and would die, and the genetic packages including the genes that specified shortness would disappear. This would, of course, constitute evolution as defined at the beginning of this discussion. Perhaps long before the short organisms starved to death, they would find it more difficult to reproduce, but certainly when they reached the point of starving and dying, their chances of reproducing would be zero. So under a set of circumstances in which food was hard to reach, short organisms had a much lower reproductive rate than taller ones, and the genetic composition of that population changed. "Short genes," the genes specifying short organisms, became less common.

Try to imagine the effects of an infinite number of more subtle influences on survival in millions of species simultaneously, each resultant genetic change constituting one of the factors determining the course of evolution of every other species. The concept that

Through evolution, giraffes have become the tallest mammals. A large male may reach a height of 6 meters. Like most mammals, including humans, giraffes have just seven neck bones, though in giraffes these bones have become much elongated.

natural selection was the basis of organic evolution was first elaborated by Charles Darwin more than a century ago.

The evidence that evolution accounts for the origin of species can be found summarized in almost any textbook of general biology. The main elements of the evidence that evolution has occurred include the following:

1. Almost all vertebrates share very strong similarities in their internal and external organs and in the interrelationship among these organs.
2. All vertebrates have similar courses of embryonic development; that is, all species go through very similar stages of unfolding as embryos on their way to becoming what they will become.
3. Many vertebrates have vestigial organs. Darwin and others have pointed out that the presence of

Mini-Glossary

DNA: deoxyribonucleic acid, a chemical substance that carries genetic information.

Evolution: change in the genetic composition of a genetically isolated group of organisms as a result of certain genes or genetic combinations being reproduced more often than others.

Gene: a functional piece of DNA; a unit of heredity defined by some discrete function or potential function, usually the specification of a particular protein.

Mutation: an inheritable change in the structure of a gene.

Natural Selection: the process by which nature causes some genes and gene combinations to be reproduced more than others; the process by which the genes of the fittest organisms are selected by environmental influences for greater reproduction.

Speciation: the formation of two species from one; the process by which new species are formed; one of the consequences of divergent natural selection.

the pelvic girdle in the snake, the existence of birds that cannot fly, and the human tailbone all indicate that the current forms of these organisms are the products of evolution in which selective pressures favored the assumption of new characteristics and the loss of old ones.

4. Over time the fossil record shows "progressive" changes. An outstanding example is the horse. The geologic evidence indicates that the horse evolved over a period of 60 million years through some 30 distinct species to its present-day form (see Figure 5.1).

5. **Biogeography,** the distribution of plants and animals over the globe, shows subtle differences in species from place to place as a result of envi-

ronmental differences. This stimulated some of Charles Darwin's thinking about the origin of species. He wondered how it was that there were such differences in species and ecosystems throughout the world. The varying adaptations among the finches of the Galápagos Islands provided Darwin with his "eureka!" experience. He reasoned that if slight differences in environment from one island to another could account for the differences in the finches, the wide range of changing environments of the whole earth over millions and millions of years may have accounted for the origin of all of the earth's species.

6. It has been shown that related organisms have similar sequences of subunits in analogous DNA

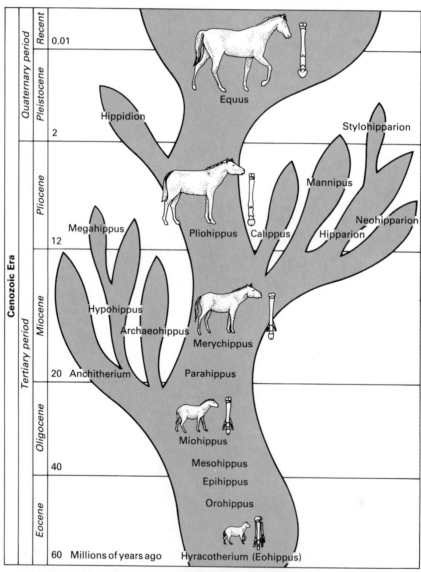

Figure 5.1 Fossil Record of the Horse over 60 Million Years. This chart offers a striking illustration of gradual evolutionary change. Note how evolution "tries" many things, only a few of which work. Many branches of horse evolution eventually died out; ten blind alleys are illustrated here. (Skeletal structure of foreleg shown next to the horses.)

and protein molecules. For example, closely related primates have very similar sequences of amino acids in their hemoglobin (a protein) molecules. The more closely related the primates, the greater the similarity.

7. Artificial selection routinely results in the evolution of special breeds and strains of both plants and animals, for example, miracle grains, short-haired terriers, and thoroughbred racehorses.

Throughout this book we discuss the principles of ecology, mainly with the present and future in mind. However, the principles of ecology have been in effect for billions of years. By the time we emerge from the other end of this chapter it should be clear that evolution is the net effect of the ecological interaction that has been going on since the beginning of time. The most profound way in which a species or a population can be influenced by the environment is through the environment's differential influence on the reproduction of organisms with different combinations of genes.

How Does Evolution Actually Work?

Darwin observed, as did some of his predecessors, that all species of plants and animals produce many more offspring than are likely to survive. He further observed that the production of offspring is followed by an intense struggle for survival, the result of which is that only a few of the offspring are able to pass on their traits to future generations. Darwin viewed the millions of eggs or seeds produced in any one year as the raw material for **natural selection,** with every facet of the environment determining which of the organisms originating as eggs or seeds would grow to maturity and reproduce. Darwin knew that there is inherited variation in all populations, in all litters, and in all clusters of eggs.

A few years after Darwin, Gregor Mendel described the principles by which characteristics are transmitted from one generation to another by the male sperm and the female egg. It was not until the middle of the twentieth century, however, that scientists confirmed that the chemical substance containing information specifying particular features and characteristics was in fact deoxyribonucleic acid (DNA). Chemical information contained in **DNA,** in units called *genes,* specifies the structure and nature of proteins, and these in turn determine the overall character and appearance of the organism. Each molecule of DNA has the ability to replicate (see Figure 5.2), or to produce, almost invariably, exact copies of itself. The copies are passed on to sperm or eggs and thereby transmit parental traits.

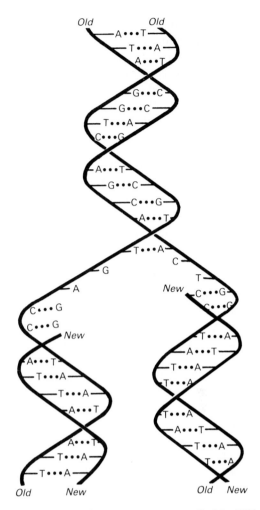

Figure 5.2 Replication of DNA. Each half of the DNA molecule serves as a template for construction of the other half. The helix unwinds and separates into halves, each half participates in forming a new complementary half, and the result is two identical double strands. This phenomenon is responsible for the inheritance of traits—why you may look like your mother or your father or have some features of each.

One important fact about DNA that explains *some* genetic variation and is therefore a factor in evolution and the origin of species is that it does *not always* reproduce exactly. Rarely, but regularly, errors, or **mutations,** occur naturally and spontaneously. The average rate of spontaneous mutation per gene in the fruit fly is about 1 in 100,000 gametes (sperm or eggs). Mutations can also be caused by X-rays and mutagenic chemicals.

Most mutations are harmful. This is reasonable. If the genes of a normally functioning organism are arranged in such a way as to produce a perfectly integrated, functioning, whole organism, random change is *most likely* to be harmful. Occasionally, such mutations are neutral; that is, they specify some change that really does not make any difference to an organism's survival. Mutations that do amount to an improvement are more likely to be reproduced. Mutations are not the only way in which variation can occur in new organisms.

Most organisms above the level of simple microorganisms are **diploid.** That is, they have *two* complete sets of genes. When the sex cells are formed, there is a random sorting of these genes such that, although the offspring will receive one complete set from each parent, it may receive any combination of genes held by the parents. Having received one set of randomly sorted genes from the father and one set from the mother, an offspring almost invariably represents a unique (except in the case of identical twins) combination of genetic material. This accounts for considerable variation in any population.

Each time one set of male and female parents produces offspring, another array of genetic combinations is presented to the world. Chance and the degree of fitness of these packages will determine which of them will survive. Darwin first pointed out that evolution works because nature is able to try countless combinations. Only a tiny percentage of the experiments turn out to be successful, but these will be preserved and will become part of the basis from which other experiments and combinations can be tried in an ever-changing environment.

Each of nature's creatures has emerged from this process and has come into being as a result of the fitness of cumulative mutations and genetic recombinations and, to some extent, chance.

Fitness for What?

The term *fitness,* as used by Darwin, refers specifically to the ability to reproduce. Although the expression "survival of the fittest" conjures up images of strong, powerful, and healthy individuals, it must be understood that strength, power, and health are important only to the extent that they are translated into the perpetuation of particular sets of genes. A healthy warrior who was constantly off fighting other warriors and eventually died by the sword would be infinitely less fit, biologically, than another individual who was not quite well enough to be accepted into the army but who stayed home and produced offspring instead. The success of any particular species can be measured only in terms of how successful that species is at keeping its genetic packages in existence, by having its offspring come into existence to survive to reproduce again. Fitness is relative; under one specific set of environmental conditions, one genetic package might be considerably more fit than another; in other environmental circumstances, the reverse could be true.

One of the best-known examples of the way in which natural selection works is in the case of the peppered moth. A shift was observed in the relative abundance of light- and dark-colored moths around Manchester, England, during the last half of the nineteenth century. It was first observed that in certain industrial areas the number of dark moths was increasing. At that time in industrial England, much economic activity was based on coal combustion. Cities like Manchester were grimy places where even the tree trunks were sooty and black from the products of incomplete combustion and coal dust. Moths are nocturnal insects; they rest during the day on places like tree trunks. With the blackening from coal dust and smoke, light-colored moths would have become increasingly more visible to predators such as birds. Under these circumstances, dark-moth genetic packages would have become less visible to predators and more apt to be perpetuated.

Such differential predation has been verified in an experiment in which light and dark moths were released into dirty and clean areas. It was found that in the clean areas, survival was significantly higher for the light-colored moths than it was in darker, dirtier areas. The selective advantage was evidently so great that it accounted for a shift of from 1% dark moths in 1850 to 99% in some dirtier areas of England at the turn of the century.

Interestingly, in recent years, with increased cleaning up of coal discharges, the percentage of light moths has begun to increase. Black moth populations are now shrinking toward England's northeast corner.

Let's consider another example, this time a hypothetical one in which a population of insects develops resistance to a particular pesticide. Suppose that a farmer, upon discovering that a barn door was covered with flies, goes to a hardware store to buy a pesticide. A plausible sequence of events for the pesticide applications is illustrated in Figure 5.3. The farmer observes that most of the insects have been killed by the first application. Among the insects originally present on the barn door, however, there may have been a few with particular genetically specified arrays of enzymes or concentrations of enzymes that enabled them to handle the pesticide just a little bit better. Suppose the pesticide in fact killed all but a very few of the most resistant types (Figure 5.3b), which also happened to be otherwise well adapted to the barnyard environment. Given the high reproductive rate characteristic of most insects, and given a few weeks, our farmer could well have been surprised to find the barn door covered once again. Although the flies looked like exactly the same kinds of insects as before, because of the selective action of the pesticide and the shifts in the genetic composition of the population that resulted from elimination of the more susceptible strains, the farmer was actually looking at a much more resistant population of insects than he saw the first time (Figure

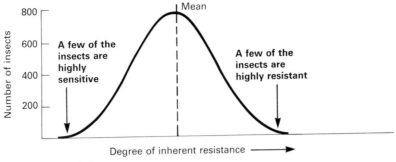

(a) Original population

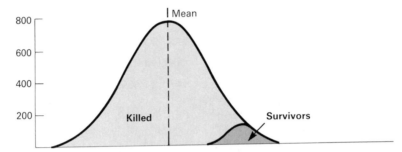

(b) Insects killed by first application of pesticide

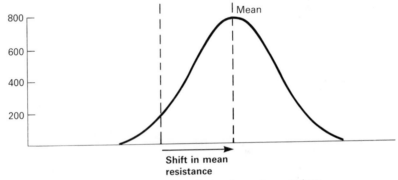

(c) Two weeks later, after the survivors have had a chance to reproduce

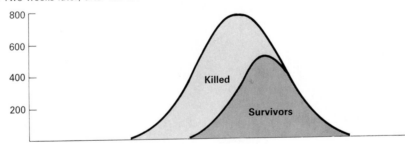

(d) Insects killed by the second application

Figure 5.3   Natural Selection of a Pesticide-resistant Strain of a Hypothetical Population of Insects.

5.3c). The farmer might find that a second spraying killed a smaller percentage than the first spraying did (Figure 5.3d) and that a third spraying three weeks later killed even fewer. Although the farmer might now go to the county extension agent and claim that the insects on the barn door had become immune to the pesticide, what in fact had happened was that the farmer had acted as an agent of selection, making it possible for pesticide-resistant genetic packages to survive. In effect, the farmer changed the genetic composition of the population toward resistance to the pesticide.

Fitness is relative and is determined by environmental circumstances. In our example it did not matter so much that in the population of flies originally encountered on the barn door there were some that could fly faster than others, some that could find food better than others, or some that could attract mates better than others. In the presence of the pesticide, the major determinant of fitness was whether or not a fly was resistant to the pesticide.

Does Natural Selection Still Operate in Human Beings?

Natural selection has operated in the past, operates now, and will operate in the future on human beings. Even though we have been able to control many of the influences in our environment, there is still considerable room for natural selection among humans. Of all potential human beings, 2% die shortly after birth; of the human beings that survive infancy, 3% die before sexual maturity; 20% never marry or have offspring; and 10% of those who do marry remain childless. On the average, about one-third of the individuals of any one generation of human beings contribute nearly 70% of the genes in the genetic packages represented by the great-grandchildren. This is significant to any consideration of humans adapting to various environmental insults, a topic that comes up occasionally in discussions about people and their relationship to the environment.

Why Discrete Gene Pools or Species Rather Than a Continuum of Genetic Packages?

Our discussion to this point raises a question: If natural selection results in incremental changes in the genetic composition of populations, why isn't there one

Is Evolution the Solution to Pollution?

Natural selection is a process whereby members of a species with traits that allow them to survive and reproduce can, over time, change the makeup of the gene pool, resulting in an organism that is better adapted to new environmental conditions. Environmental conditions are changing today. Increased pollution from the burning of fossil fuels is affecting the quality of the air. Toxics are found in soil, water, and air. Could it be that *Homo sapiens* will adapt to a more crowded, polluted world? What do you think?

continuous spectrum of organisms all the way back to the origin of life?

The answer to this question has several important dimensions. One part of the answer has to do with the concept of extinction. Most of the species that were once part of the continuous stream of life simply no longer exist. Another has to do with the concept of species. Species can be thought of as "blips," or wide places in an otherwise superthin stream of change (see Figure 5.4). The pattern of evolution has been one of a few "genetic colonizers" (variants) moving into and exploiting new niches. Typically, transitional forms were so few in number that they made little, if any, mark in the fossil record. What all of this means is that although there was indeed a continuous genetic spectrum beginning with the first life and extending to the species of today, most of the spectrum disappeared as early forms gave rise to better-adapted forms and changed the world so as to make themselves obsolete. We need to explore the notions of speciation and extinction a bit more before we move on.

For organisms that reproduce sexually, a **species** is a group of organisms that can or could interbreed and produce fertile offspring. More generally speaking, a species is a genetically related and genetically isolated group of organisms. The question of why there are species can be reduced to this one: How do genetically isolated organisms come to be? Let's consider the concept of reproductive isolation.

If a river suddenly appeared because of a shift in the earth's surface and it separated a population of any particular species into two groups (and the or-ganism could not swim), the members of the population on the two sides of the river could, over a period of time, become two separate species, unable to interbreed. With time, each of these populations would be influenced by different environmental pressures because conditions would not be exactly the same on one side of the river as they are on the other. The influences of the different environments on the two sides would cause the two populations to become increasingly dissimilar. Eventually, they might become too physically or physiologically dissimilar to interbreed even if the river dried up and they could once again associate.

There are actually a number of mechanisms by which **reproductive isolation** occurs in nature. There are physical barriers like the river in the example we just discussed, barriers that result from the disappearance of a land bridge, and habitat barriers that result from fluctuations in climate. There are a number of ways by which separation of ideal habitats by less ideal or more dangerous zones can occur.

Picture a population of primates well adapted to life in the trees. Imagine a long dry spell that results in clumps of trees becoming separated by prairielike grassland (Figure 5.5). Imagine some groups of tree-dwelling primates beginning to adapt to part-time life on the ground because of a shrinking **arboreal** (tree) habitat and the presence of food on the ground. Imagine this progressing to the point at which some primates become almost completely adapted to the ground-level niche, enabling them to move far from trees and their tree-dwelling cousins. The existence of successfully ground-adapted primates and the persis-

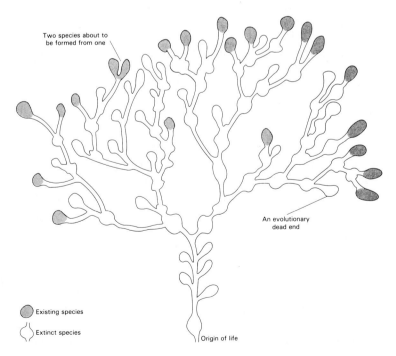

Two species about to be formed from one

An evolutionary dead end

○ Existing species

◡ Extinct species

Origin of life

Figure 5.4 Life in Search of Niches over Time. The course of evolution can be thought of as a branching stream of change over time. Along the stream are wide places—"blips"—where species flourished for a time as streamlets of genetic change resulted in good fits between organisms and environment—only temporarily in most cases.

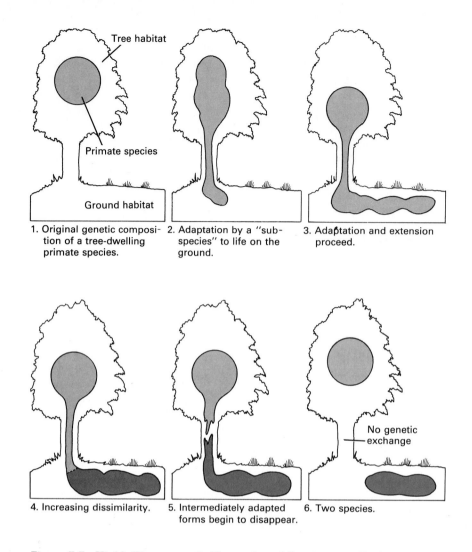

1. Original genetic composition of a tree-dwelling primate species.

2. Adaptation by a "subspecies" to life on the ground.

3. Adaptation and extension proceed.

4. Increasing dissimilarity.

5. Intermediately adapted forms begin to disappear.

6. Two species.

Tree habitat

Primate species

Ground habitat

No genetic exchange

Figure 5.5 Highly Diagrammatic Illustration of Speciation. The key to speciation and the definition of species is genetic isolation.

tence of tree-adapted primates could make it difficult for any varieties in between. Any intermediate organisms might disappear quickly because of competition from both sides.

The pattern of the appearance of new species throughout evolutionary history is best described as some combination of gradual adaptive change and **punctuated equilibrium.** The fossil record suggests that species have tended to go through long periods without changing and that these periods of equilibrium were "punctuated" by short periods of relatively rapid change. The concept of punctuated equilibrium was originally offered by Niles Eldredge, Stephen Jay Gould, and others as a substitute for the gradualist view of evolution offered, if perhaps only by implication, by Darwin. For most of the past century, biologists have held that new species arise slowly through gradual change. The fact that transitional forms tend to be absent from the fossil record has

generally been dismissed as a consequence of the incompleteness of that record. What the punctuated equilibrium view says, in effect, is that the fossil record very likely reflects the way evolution really occurred—up steps, not a gradual slope.

The logic of punctuated equilibrium is compelling. Organisms that are well adapted to a particular environment are, after all, likely to stay as they are, certainly as long as the environment does—and, according to the fossil record, even if the environment changes. Species get to be the way they are because of selection for a particular niche. Once a "fit" is achieved, stabilizing selection (see Enrichment Box 5.1) would tend to keep species as they are. Even in the face of changes in the environment, small adaptive changes in species might take a long while to appear, especially in a large population. It is more likely, many evolutionary biologists argue, that new species arise via rapid jumps that take shape in tem-

The Effects of Natural Selection

Imagine a population of a species that has a particular frequency distribution for some trait—height, weight, skin color intensity, whatever. Let's represent the population as follows:

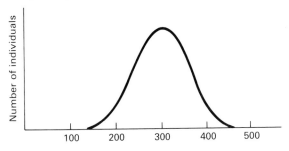

Units of some trait (weight, height, hat size, % body fat, etc.)

There could conceivably be changes in the influence of the environment that favor (1) one of the extremes, (2) the middle, or (3) both extremes (which is the same as selection against the middle).

Favoring one of the extremes results in directional selection. Directional selection results, over time, in the shift or drift of the population in the direction of the selection. This can be represented as follows:

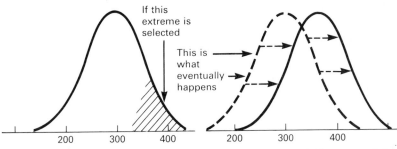

If selection favors both extremes over the middle, this is called diversifying selection. Diversifying selection eventu-

ally leads to a bimodal frequency of distribution:

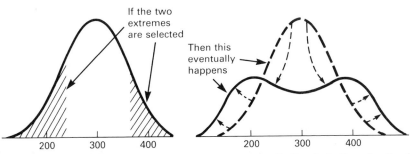

If selection favors the middle, the result is called stabilizing selection. Stabilizing selection results in a decrease in diversity in the population. Because this would tend to de-

crease the likelihood of either directional or diversifying selection, its effect is to stabilize:

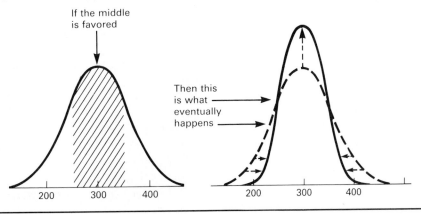

porarily genetically isolated, "splintered-off" populations. Perhaps such jumps occurring in just a few generations have the effect of matching genetic "experiments" to new or expanded niches, niches that are closely related to the original. Such splinter groups could develop quickly (for much the same reason that introduced species often exhibit explosive population growth) and perhaps eventually return to outcompete and replace the original parent population. Perhaps the tendency in some species to drive off or reproductively isolate the "deformed" or the different serves as a mechanism of speciation, driving genetic recombinants across an adaptive "landscape" where they sometimes encounter new opportunity.

One of the best pieces of evidence that evolutionary change is slow—that gene pools do in fact tend toward inertia or equilibrium—is that species have apparently tended to stay the same, even in the face of clear environmental imperatives to do otherwise. Virtually all species (more than 99%) that have ever graced the earth are now extinct. Extinction is not always a sign of failure to adapt, of course; indeed, some extinct species are extinct because they gave rise to new species that then beat them out of existence. However, most extinct species were in fact evolutionary dead-ends, their lineage extinguished because of an apparent failure to adapt to challenge quickly enough.

Extinction

Extinction means the disappearance of a species from the face of the earth forever. If extinction is bad, then things have been bad for the earth's creatures from the beginning because extinction has been the rule. Raup (1986) estimates that there have been up to 4 billion species alive at some time in the past—with only 5 to 10 million (some say more) of these alive today. This means that more than 99% of the species that have ever lived are gone. Extinction is, practically speaking, every bit as common as the origin of new species.

Superimposed on a record of relatively constant rates of extinction are episodes of mass extinction. The fossil record indicates that there have been at least six major episodes of mass extinction on land and at least four in the oceans (Figure 5.6). There is obviously some subjectivity involved in deciding how much of a rise in rate amounts to such an episode, particularly because of the "noise," or random variation, in the fossil record. Raup and Sepkoski (1986) claim that they see eight major peaks in the rate of extinction in the past 250 million years alone. Raup (1986) goes so far as to suggest that perhaps most extinctions are episodic—punctuated equilibrium in reverse, as it were. It is not hard to imagine that, like a house of cards, the disappearance of one or more key ecosystem components has tended to set off col-

96

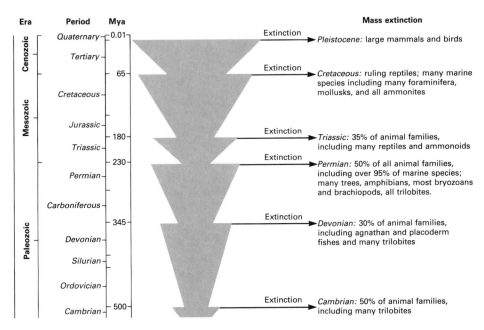

Figure 5.6   Major Periods of Mass Extinction in the History of the Biosphere.
(Mya = million years ago)

lapses of ecosystems and cascades of disappearances of variably interdependent species, even on a global scale. It is likewise easy to imagine that climatic changes may have regularly amplified such cascades of extinction by making ecosystems adapted to earlier conditions unstable. The biggest abrupt extinction ever apparently occurred at the close of the Cambrian epoch some 505 million years ago when 80% to 90% of the species and 50% to 60% of the genera disappeared. Raup and Sepkoski (1986) calculated that species disappeared at the rate of 20 families and up to 1,200 species per million years during the mass extinctions in the oceans.

There has been some recent speculation that mass extinctions have occurred regularly in the past—about once every 26 million years—and that these may be associated with regular meteor or asteroid showers. There was an especially dramatic mass extinction at the end of the Cretaceous period (at the Cretaceous-Tertiary, or C-T, boundary) 65 million years ago when a very high proportion of the oceans' plankton disappeared, along with what was left of the dinosaurs, within a relatively short time. There was apparently also a sharp loss in vegetation at the C-T boundary—flowering plants lost dominance for a time and then recovered.

Some scientists believe that the C-T boundary extinctions were caused by a collision of the earth with an asteroid. This collision is thought to have stirred up a lot of dust and to have caused a drastic decline in incident solar radiation and depletion of the ozone layer (see Chapter 10). In these and other ways

the collision caused great disruptions of the ecosphere lasting tens of thousands of years. Among the evidence for this hypothetical cataclysm is the presence of an iridium-rich deposit at the C-T boundary throughout the world, at precisely the point in geologic history when the extinctions occurred. Iridium, relatively rare in the earth's crust, is relatively abundant in meteors and asteroids. Other evidence for a major impact with some kind of extraterrestrial body at the C-T boundary includes the presence of "shocked quartz"—quartz grains that show the impact of physical shock—at the same boundary in at least several locations throughout the world.

Not everyone accepts the asteroid hypothesis; Sloan and colleagues (1986), for example (see also Briggs, 1991), believe that the end of the dinosaurs was brought on by a combination of decreased temperature, a drop in sea level, a decrease in flora, and possibly increased competition with mammalian herbivores. They and others point out that there was a progressive reduction in species diversity over a 10 million–year period leading up to the C-T boundary. Still others point out that many animals survived the boundary—the dinosaurs, in fact, were one of the few groups not to make it. Still, an asteroid collision or something like it could have been a kind of last straw, an event that brought on the collapse of an increasingly unstable biosphere.

A report in the journal *Nature* (Jansa and Pe-Piper, 1987) described a crater in the Atlantic Ocean 125 miles southeast of Nova Scotia. The crater, some 28 miles wide and 1.7 miles deep, is believed to have been caused by an object about 2 miles wide that crashed into the ocean 50 million years ago. This was 15 million years after the impact that is alleged to have brought an end to the dinosaurs; nevertheless, it does afford scientists an opportunity to assess the impact of comets or asteroids striking the ocean. Although the causes of mass extinctions will likely remain the subject of ongoing debate, it is worth noting that all of the possibilities involve some kind of environmental disruption.

It is also worth noting at this point that extinction has been one of the most important consequences, as well as one of the most important driving forces, in the evolution of life. The disappearance of a species changes the environmental setting to the same degree as the appearance of a species does. Extinction has thus been one of the regular elements of change in an ever-changing world. Over time, both the extinction of individual species or families and wholesale extinctions have created niches that have driven evolution further along.

So why should we worry about other species becoming extinct? The reason we ought to be concerned is that we humans have become one of the most important factors in species extinction. We endanger many species mainly because we eliminate their habitat, and in some cases we overexploit species for food. Some species are particularly vulnerable because they are already rare or because they tend to be more directly in our way. Chapter 14 is devoted to the impact we human beings have on our "biological resources." In that chapter we will look at how, and to what extent, we are making certain species more vulnerable to extinction. (Later on in this chapter we will look at the role humankind may have played in extinctions in the past.) In Chapter 14 we will also examine strategies—for example, captive breeding and habitat preservation—that we humans employ to keep endangered species from the brink. We will also take a selfish point of view and look at what humankind loses, or might lose, when we push a species to extinction prematurely.

How Did the Earth and Its Great Spheres Originate?

The cosmos originated 10 to 20 billion years ago. The stars, the planets, the meteors, and other assorted bits of matter are thought to have condensed from clouds of superhot gases or mixtures of atoms and subatomic particles. Some say that the clouds of gases from which the stars and planets condensed resulted from an explosion so uniquely large that it cannot really be described by using standard super-

latives, though it is sometimes referred to as the *big bang.*

Some 5 billion years after the origin of the cosmos, the earth came to be as a condensed, relatively tiny bit of matter in orbit around the sun. At first the earth was too hot to have water on its surface, but gradually, and in total conformity with physical laws, the atmosphere and hydrosphere separated from the lithosphere. Over the next few billion years the earth continued to change in form; the atmosphere changed from oxygen-free to oxygen-rich, the hydrosphere changed (it became saltier), and the lithosphere underwent changes as well. During the Precambrian period and even later, the cooling crust continued to crack and separate into pieces that drifted atop less solid substrata, colliding with one another in cataclysmic clashes that resulted in the uplifting of mountains and in various other wrinkles in the surface of the earth. Gradually, the planet became quieter and less chaotic.

Could Life Have Originated on Earth Spontaneously?

From the start the earth contained about 100 elements existing singly and in various combinations, all having certain potentials for still additional combinations. It is certain that many of the chemical entities in the earth's atmosphere, hydrosphere, and lithosphere during its first billion years combined in all sorts of combinations. It is highly likely that organic molecules were created in this early part of the earth's history. All the right ingredients were present under the right conditions—occasional electrical discharges in the form of lightning, unique and complex mixtures of chemical substances, heat, and ultraviolet radiation. As molecules absorb energy, they enter into more excited states from which they are more apt to enter into chemical combinations with other elements or molecules.

A number of theoreticians postulate that organic chemicals appeared on prebiotic earth (Azoic era) to such an extent that the oceans were converted into a weak "organic soup." Urey (1952) postulated that this organic soup may have been as much as 1% organic matter.

A few decades ago, it was demonstrated in the laboratory that in environments absolutely devoid of oxygen it is possible to produce amino acids, **polypeptides** (chains of amino acids), and nucleic acids from hydrogen, methane, nitrogen, water, and various salts by exposing these chemicals to electrical discharges or ultraviolet radiation.

Although the process has never been carried out in a laboratory, it is not considered impossible that early life originated spontaneously out of this organic soup. The first organisms may have been crystals of clay associated somehow with organic molecules (see Cairns-Smith, 1985). They would have had somehow to acquire the ability to replicate themselves out of raw organic and inorganic materials. These proto-organisms were probably like bacteria (although they would have been much simpler than the simplest of today's bacteria) and very likely were heterotrophic; that is, they probably got the energy needed for replication by breaking down the organic molecules in Urey's soup.

It is postulated that while the first anaerobic bacterialike life forms were exploiting the primordial soup, variants appeared that were somehow able to bring energy-yielding chemical reactions or light energy to bear on the assembly of *new* organic raw materials.

There are many unanswered questions in the way of a clear explanation of how life may have arisen from an organic soup. Nevertheless, the spontaneous generation of life is plausible.

Oxygen: A Waste Product and Some New Niches

The earth today has an atmosphere that is 21% oxygen. Chemically, this is an atmosphere that causes things to rust and, literally but slowly, to burn or oxidize (see Chapter 1). The atmosphere was not always this way. In fact, the earth probably began with an atmosphere containing mostly methane, ammonia, hydrogen, carbon dioxide, and carbon monoxide. Standing where you are today, 4 billion years ago, without a space suit, you would have died of asphyxiation in minutes.

According to various theoreticians, the earth's atmosphere began to increase in oxygen content as organisms perhaps similar to present-day algae appeared. By taking hydrogen from water molecules these phototrophs released oxygen as a by-product. In the Cambrian period, as a result of this kind of activity, atmospheric oxygen reached 1% of its present levels, or about two-tenths of one percent of the atmosphere (see Figure 5.7). By the end of the Silurian period, some 400 million years ago, the oxygen in the atmosphere reached about one-tenth of what it is today.

The appearance of life forms able to capture light energy and use it as a source of energy for synthesizing organic subunits was a great step forward. The fact that this activity produced oxygen as a by-product made it possible for another great step forward. The stage was set for aerobic heterotrophs,

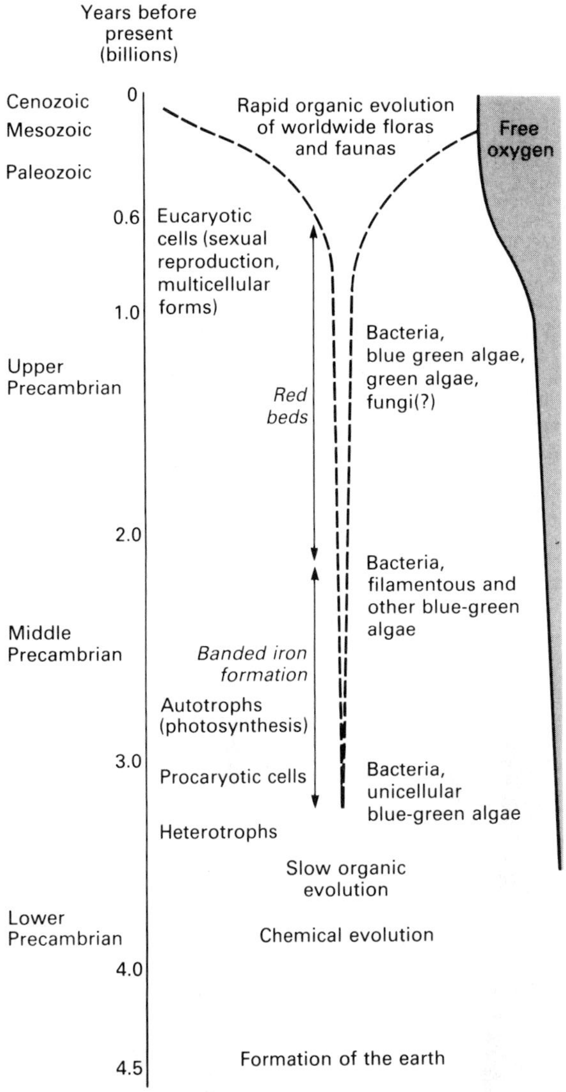

Figure 5.7 Evolution Has Been Both Chemical and Biological. Gradual chemical evolution led to the first living forms. These living forms in turn altered the makeup of the earth's atmosphere. Once sexual reproduction appeared, the resultant acceleration of recombination and evolutionary experimentation yielded much more rapid biological evolution.

organisms capable of getting their energy by oxidizing organic material using oxygen.

Could Life Originate Spontaneously Today?

What the early oxygen producers did offers a particularly striking example of how lithospheric, atmospheric, and biospheric changes all interrelate to create niches leading to new kinds of interrelationships.

We pointed out that ultraviolet light might have played a role in the formation of some of the organic molecules in the primoridal organic soup. The soup would have provided a new opportunity, one that permitted primordial heterotrophic life forms to appear and sustain themselves using the organic material as raw material. Out of these first life forms, other forms may have evolved that were capable of using light energy and produced oxygen as a by-product. Some of the oxygen was in turn chemically converted into ozone, which began to create a shield over the surface of the earth against ultraviolet radiation. This enabled photosynthetic life forms to move into areas where there was more light energy to do still more photosynthesizing. Eventually, enough of a shield was created by the production of oxygen that even the land was protected from ultraviolet radiation. This opened up still more opportunities, which would eventually be exploited, first by terrestrial plants, then by animals. Thus the evolutionary processes of the biosphere, lithosphere, hydrosphere, and atmosphere were highly interrelated.

Interrelated changes in all of the earth's spheres explain why life will probably never be able to originate again as it once did. The combinations of conditions that existed when life first originated were changed permanently by the appearance of life itself. Even if life could originate spontaneously again, newly appearing forms would have little chance in competing with the forms already in existence. They would be gobbled up, absorbed, or otherwise outdone by organisms that have had 3 to 4 billion years to adapt to a changing environment.

After the Early Life Forms

According to the fossil record, a rapid rise in the number of species began about 600 million years ago. Pre-

sumably, this was made possible by all that had gone before, the result of 4 billion years of preparatory chemical and biological evolution. Brachiopods (clamlike organisms), sponges, and gastropods (e.g., snails) appeared in great abundance rather suddenly at the beginning of the Paleozoic era nearly 600 million years ago. On land there was a flurry of plant evolutionary activity 200 million years later in the late Silurian period, which produced an enormous variety of vascular plants, providing niches that would later be filled by plant eaters. Animals diversified at a rapid rate about 570 million years ago. Dinosaurs originated from very early reptiles that appeared on earth just after the age of large land plants during the Permian period, 200 to 300 million years ago. Dinosaurs dominated the earth for well over 100 million years, but by the time the Mesozoic era closed, a majority of them had disappeared.

The Evolution of Ecosystems

One of the main points in all of what we have considered so far is that ecological relationships between organisms and the environment existed from the very first appearance of life on earth.

The sharp climb in oxygen that began about a billion years ago suggests that the earth went through an "age of producers," a time when there was more oxygen given off than consumed. A steady state was evidently reached some 100 million years or so ago because for a long time the amount of oxygen in the atmosphere has remained about the same. We saw in Chapter 1 that the net effect of respiration is the reverse of photosynthesis; as long as the two processes are coupled, there can be no net accumulation of oxygen and no net decrease in carbon dioxide. LaMont

Our knowledge of early life forms comes mostly from the fossil record. By putting together the information gleaned from great numbers of fossils, scientists can trace evolutionary trends and their rates of development.

Cole (1966) advanced the view that the 1.2 quintillion kilograms of oxygen in our atmosphere plus about another 1% of this amount dissolved in the oceans of the world reflect the amount of photosynthesis that has never been reversed by respiration. This, according to Cole and others, corresponds to only a very, very small fraction of all the oxygen that was ever produced by photosynthesis. Relatively little production has gone unmatched by decomposition, plant respiration, fire, or consumption and oxidation of plant material by herbivores. Ecosystems must have functioned much as they do now for a long, long time.

How Did Homo sapiens Evolve?

Homo sapiens now generally considers itself to be the terminal point of an evolutionary path beginning with primordial matter and extending through the appearance of life, animal forms, and the vertebrates. Vertebrates appeared in the early Ordovician period about 500 million years ago. Out of this group evolved the special class of warm-blooded vertebrates that suckle their young, called the **mammals.** *Homo sapiens* emerged from a subgroup of mammals called primates.

The stage for *Homo sapiens* had been set by all that had gone before. Dinosaurs were on their way out when a shrewlike mammal appeared in greater numbers. It must have been difficult being a small mammal during the final days of the dinosaurs. Most early mammals were believed to have been nocturnal, their days spent under cover and nights spent scurrying from cover to cover searching out the relatively enormous amounts of food that their active lifestyle and warm-blooded metabolism required. Smell would have been extremely important to such creatures.

Fossil records suggest that somewhat later in the evolution of mammals, some of the early forms took to the trees, a move that would have provided a whole new framework of selective pressures (see Enrichment Box 5.2). There would have been little advantage in having an acute sense of smell high off the ground. There would have been pressures favoring abilities to walk on larger tree branches or to grasp smaller branches. As the eons rolled by, any adaptation toward better grasping would have been a selective advantage.

Some time after the early mammals moved into the trees, a tarsierlike mammal appeared. This evolutionary line favored an adaptation that would later be important to humans. The eyes of tarsiers are up front; both look at the same target. Having two such eyes, together with the brain circuitry necessary for their coordination, made possible stereoscopic vision, the ability to perceive depth. Clearly, there is some advan-

---

Enrichment Box 5.2

Simplified Description of the Relationship between the Concepts of Natural Selection, Evolution, and Origin of Species

Natural Selection

All living things overproduce. For example, many more frog eggs are laid than will ever become egg-laying adult frogs. The environment and chance determine which offspring will live to reproduce. Disregarding pure chance (because over the long haul its effects even out and produce no bias), environmental pressures tend to go easier on the individuals with the ''best'' of certain characteristics, for example (other things being equal), the fastest rabbits, the wariest deer, the most cautious bass, the sharpest-toothed wolves, and the most discerning eagles. The result is that the genes of these most-likely-to-reproduce variants will be preserved, and those of the slow rabbits, dull-toothed wolves, nearsighted eagles, and so on will eventually be lost in a competitive world. Over the long haul this natural or environmental selection of favored gene combinations leads to evolution.

Evolution

Environmental pressures, which come in many forms (for example, temperature, light levels, availability of food), can come and go, intensify and ease up, change abruptly or gradually, change cyclically or linearly. Because of these selective pressures, the genetic composition of every species is constantly changing, sometimes quickly, sometimes slowly, usually linearly. This process, which is manifest as change in the physical and behavioral characteristics of a group of organisms, is called evolution. When it goes far enough, evolution leads to the origin of species.

Origin of Species

Imagine that somehow two groups of the same species of plants or animals become isolated from each other. The environmental pressures on the two groups would differ, since no two places or circumstances are exactly alike. The two groups would evolve away from each other. When the point was passed beyond which the two groups could interbreed and produce viable offspring, through mixed mating, the groups would technically have become separate species. Darwin imagined that all species could have come from common ancestors in just such a pattern repeated over and over and over.

tage to depth perception if one's life depends on an ability to leap from limb to limb.

Tarsierlike mammals gave rise to monkeylike and catlike creatures and, eventually, *Homo sapiens.* In our monkeylike ancestors there were many additional adaptations that eventually came to characterize humans. For example, the adoption of hand-over-hand locomotion in early tree-dwelling primates required many muscle changes and changes in the placement and orientation of internal organs. These adaptations eventually made possible other selective steps leading to the upright posture characteristic of human beings today.

Apparently, at one point in the history of primate evolution it was necessary for some of them to come down out of the trees. This move is believed to have been related to a prolonged dry spell over many parts of the habitable earth that lasted from 60 million years ago until about 4 million years ago. This period favored low forms of vegetation, such as those we now see in prairies. Forests receded in many locations, and clumps of trees became separated by large expanses of prairie or savanna. The climatic change is believed to have resulted from great continental drift—related upliftings that appeared in the surface of the earth, upliftings that produced the Sierra Nevada, the Andes, and the Himalayas (see Enrichment Box 5.3). These,

we know from our discussions in Chapter 3, would have resulted in the deflection of water-laden clouds upward, causing rain to fall on the windward side of these mountains and leaving the leeward or opposite side dry.

As the trees receded, the arboreal habitat must have become crowded, less desirable, and less able to support the organisms that had adapted to it during earlier, wetter times. Perhaps the climatic change produced a drop in the yield of fruit, and some of the tree dwellers may have found it necessary to come down to the savanna occasionally to find food there. What happened, in effect, was that as one niche began to recede, another one expanded. This is relevant to an idea we developed earlier, namely, that because of the enormous adaptive potential of life, if there is a niche to fill, it will be filled eventually because of pressures such as competition.

Adaptive features that began to take shape in tree-dwelling preprimates underwent further adaptation in the expanded niche of the savanna. The environment was giving birth to *Homo sapiens.* Adaptations that enabled a primate to grasp a limb and to move easily from one tree to another led eventually to an ability to handle tools and to perform delicate manipulations requiring hand-eye coordination. In a very real way your ability to untie a knot in your tennis shoe

Continental Drift and Evolution

Two hundred million years ago, there was just one big continent, now called Pangea, and only one enormous ocean. Sometime after the beginning of the Mesozoic era, Pangea began to break up into two continents, Gondwanaland in the southern hemisphere and Laurasia in the northern hemisphere. Gondwanaland split into several pieces, one that later became South America and Africa and other pieces that became Australia, Antarctica, and India. India later crashed into what became Asia, and the resulting wrinkles are called the Himalayas. Long after Africa separated from South America, it bumped into Arabia and provided a way for African primates to move up into Asia (about 16 or 17 million years ago).

Most of the earth's mountain ranges were formed when the continental plates rode over or beneath one another as they collided. These ranges permanently altered the climate over many parts of the earth. Continental drift also changed the ocean currents, which became another important climatic factor. These changes were an important driving force in evolution and may have been responsible for the extinction of many species. Also important was that the breakup of the original landmasses isolated many organisms. For example, the unusual animals on Australia are a result of that continent's isolation.

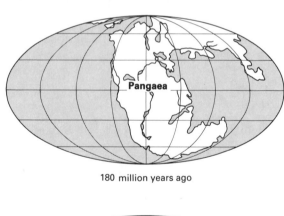

180 million years ago

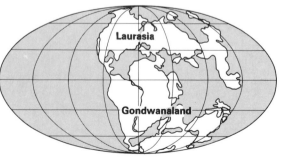

120 million years ago

is the result of pressures that favored stereoscopic vision and grasping abilities in your tree-dwelling, fruit-eating ancestors.

The fossil record of the evolution of human beings is quite incomplete, and we have much yet to learn. Tentative schemes of how humans and other primates emerged during the Cenozoic era and the relationship of modern human beings to various fossil forms are given in Figure 5.8. However the sequence went, the important thing is that from the first tool users there was strong selective pressure favoring greater and greater elaboration of the use of tools and eventually the use of energy other than muscle to power them. Some time after *Australopithecus afarensis,* about 500,000 to 1,000,000 years ago, *Homo erectus* appeared as the first species to live what the fossil record indicates was definitely a human type of existence. *Homo sapiens* emerged sometime during the next 2 million years, perhaps as recently as 100,000 to 300,000 years ago (Figure 5.8). *Homo sapiens* was the result of the competitive advantage conferred by better brains and tools.

Changes in the outward appearance of our human ancestors became quite subtle after *Homo habilis,* and it has been suggested that a Neanderthal dressed in modern clothes might pass unnoticed on

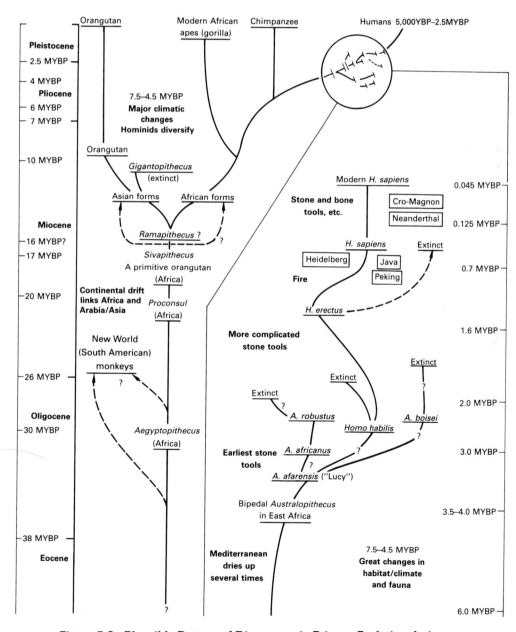

Figure 5.8 Plausible Pattern of Divergence in Primate Evolution during the Cenozoic Era and the Relative Temporal Position of Some Famous Ancestors of Modern Humans. (MYBP = million years before present)

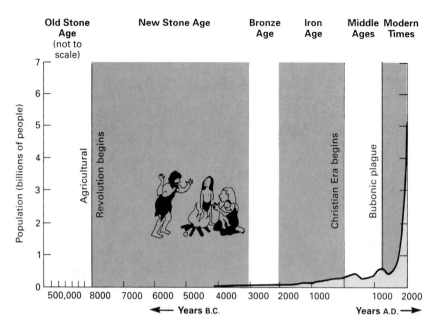

Figure 5.9 Growth of Human Population in the Past Half Million Years. The increase in numbers of *Homo sapiens* has been mostly very slow over time. If the Old Stone Age were drawn to the same scale as the rest of the figure, it would be nearly 20 feet wide. Very recently, our increase in numbers has been very fast.

one of our city streets today. The evolution of the genus *Homo* was primarily toward attributes favoring intra-species cooperation and culture, leading to a *gradual* increase in reproductive success. It was not until human beings learned to harness fossil fuels that human reproductive success began a sharper ascent (Figure 5.9). Modern life is a very recent phenomenon.

If we conceptually condense the history of humankind since the Stone Age—that is, the previous half million years—into one day, agriculture started in the last half hour—after 11:30 P.M.; Julius Caesar's reign would have begun around 11:55 P.M.; William the Conqueror would have invaded England at 11:57 P.M.; George Washington's administration would have occurred within the last half minute of the day; and humanity would have relied almost entirely on its own muscle and beasts of burden for all except the last 16 seconds.

Humans as Intelligent, Cooperative Generalists

Whereas some species are more or less completely adapted to particular niches, ours is not. Ecological adaptability seems to have been—and still to be—a characteristic of primates in general. Primates have no outstanding features they all share except the ability to survive environmental change by being relatively flexible about their needs. As a group, the primates occupy broad ecological niches. The evolutionary history of *Homo sapiens* and the sequence of adaptations that have come to characterize this species confer a little of every ability on men and women, but we are not really outstanding at any of them. Humans can swim, run,

climb, and do lots of other things. There are other organisms that can swim farther, run faster, dig deeper, and climb better; most others have sharper teeth, keener sight, and better insulation. However, no other animal can do all of the things human beings can, and no other species can make up for its shortcomings with intelligent cooperative interaction the way our species does. Some say that our species's most distinctive characteristic is intelligence. But another key feature, cooperativeness, amplified the importance of intelligence in making possible shared behavior—learning from one another and even learning from those long since dead. Because of our cooperative-interactive way of life, only one of us had to invent the wheel for all to benefit from it.

An important trend in evolution was the development of eyes at the front of the head, allowing depth perception. The ability to judge distances must have offered an advantage in the arboreal habitat. (This is a photo of a Tarsier.)

Throughout the history of evolution, niches have been created and filled as a consequence of evolution. Filling specific niches occurs slowly because they can be filled only through competition and adaptation, biological processes that take time. The success of *Homo sapiens* can be attributed to an ability to circumvent these processes through intelligence and the explosive advance of learning and invention, feeding on itself in a process that has come to be called *cultural evolution*. Cultural evolution, together with the human generalist primate heritage, has enabled our species to recognize and occupy a niche of many niches. We have become the supreme exploiters. Humans can invade any habitat and use almost anything as food. We are more than multilevel consumers; we are controllers of producers. Although we cannot convert electromagnetic energy into chemical energy (yet), we *can* make it happen. We can start, augment, or stop production; we can even alter the nature of producers.

From Hunter to Farmer and the Origins of Culture

Born out of a heritage of cooperative bands that reached high levels of organization in our primate forebears, humanity started out gathering fruits, insects, and other miscellaneous edibles. Some say that humans turned to hunting later, possibly after finding that predators killed in self-defense were also edible. In any event, the archaeological record of human campsites indicates that we started out as small bands of hunters supplementing kills with things foraged and gathered. Although there have been some complex band-level human cultures with some fairly sophisticated, cooperative divisions of labor, band-level cultures supported by hunting and gathering never got very far. Some say that this type of culture was successful enough to bring about population pressures that made agriculture necessary. At the very least, these cultures set us up to exploit agriculture once it was discovered.

People began to farm seriously about 10,000 years ago, though there were probably occasional false starts before that and perhaps a long period in which hunting and gathering were combined with agriculture in various proportions. Archaeological records indicate that humans began to farm seriously in 8000 B.C. in parts of the Near East and in Southwestern Asia and that there was full-scale agriculture in many places just after 4000 B.C.

Turning to crops for food was extremely important to the success of the species for many reasons, all of which have to do with the simple fact that plants are at the base of the food chain. As we learned in Chapters 1 and 2, as food energy moves from one level to another in the food chain, about 90% of available energy is lost with each transfer. It remains true to this

Humans are ecological generalists, highly adaptable in finding niches in ecosystems throughout the world.

day that many more people can be supported if they subsist on a basic diet of corn than if that same amount of corn were fed to hogs and the humans ate pork. In a very real sense, using plants as food eliminates the middle consumer—the herbivore, the rabbit, the hog, the cow—and the energy profit taken in the process of making meat. Plants also have fixed locations and are easier to find than animals, especially if they are planted in a particular location. Grain and beans are much more easily stored than meat, and this provides another kind of food energy cushion against hard times, something that hunting could not do as well.

For all of these reasons, agriculture permitted the establishment of stable population centers and time free from the business of subsisting. This added an enormously important dimension to the potential for human success. Because of the energy and survival advantages of agriculture, a primate that had already developed considerable intellectual powers and patterns of cooperation suddenly had much more time to think and to invent. This in turn led to greater opportunities for inventiveness and the development of technologies that produced more and more free time for more and more thinking and more and more inventions. The process accelerated sharply when humans learned to harness the energy derived from fossil fuels.

How Do Cultural Evolution and Biological Evolution Compare?

We defined biological evolution earlier, and we have now traced the process to the point at which there is apparently another kind of evolution, cultural evolution. Perhaps **cultural evolution** can be defined as changes passed on not by genetic transfer but by learning.

In comparison with the time scales we have been dealing with for most of this chapter, the time scale of cultural evolution is very short. Cultural evolution occurs very much faster than biological evolution, yet both have the same result, that of producing populations more fit in relation to a given environment. Both processes can go on simultaneously.

Flight could be considered something that confers a selective advantage and increased fitness on *Homo sapiens*. Disregarding nuclear missiles and the like, being able to fly translates into improved survival for the species through the transport of goods or of combinations of expertise to the right place at the right time. Humankind has not had to wait for biological evolution to produce human wings, however; the Wright brothers used their highly evolved intellectual capacity to *invent* wings. There are many other similar examples. So rapid are the changes that can be effected through cultural evolution and inventiveness that they outstrip the natural tendency of life—*Homo sapiens* included—to come up with responsive biological adaptive mechanisms.

The Impact of Early Humans on the Environments That Shaped Them

While *Homo sapiens* was literally being formed from the primordial dust over the long course of chemical and biological evolution, there was, as we have stated often, interaction between the forces driving evolution. Born to this heritage, humankind has had considerable impact on both its biological and its physical and chemical environment since long before there were such things as large-scale fossil fuel combustion and atomic bombs. To be sure, this impact was less in the beginning, when there were fewer of us and technological advances for the most part were not applied on a grand scale.

Because early humans subsisted by hunting and gathering, the species was controlled by relatively short control loops and simply could not hurt the environment very much. If early humans decided for some reason to cut down lots of trees or otherwise destroy vegetation, the deer and other animals they needed for food would no longer be there to sustain them. The humans would have to move on or starve long before any permanent harm could be done. Textbooks cite some notable exceptions to this, some of which have come to be known collectively as the **Pleistocene overkill.**

Early humans are believed to have improved hunting techniques during the Pleistocene epoch by using fire and advancing lines of hunters (sometimes simultaneously) to drive prey into traps or swampy areas where they could easily be killed. This was successful to the extent that it is believed to have contributed to the extinction of as much as 40% of the game species in parts of Africa. At about the same time, as humans came over the land bridge from Siberia to North America, the invaders encountered many large mammals that had never before seen humans and had not developed a fear of predators. Some anthropologists believe that as humanity advanced through North America and eventually into Central and South America, many of these species, some already endangered by environmental and habitat changes, were rendered extinct long before selection could produce more wary varieties.

As we began to develop the beginnings of today's agricultural patterns, humans began to have a more profound impact on the environment. Some of this impact was damaging, and some, though perhaps quite extensive, could be considered benign.

Scholars as early as 347 B.C. described deforestation and overgrazing around the Mediterranean Sea and how this led to the drying up of springs and erosion of soil. Both ancient and modern scholars attribute the collapse of the Babylonian Empire and other civilizations of the Middle East to such ecological disasters. There were also environmentalists in those early days. Around 40 B.C., Virgil recommended that the crops be rotated, that legumes be planted with other crops, that land be left fallow in alternate years, and that soil be regenerated with manure and ashes. Pliny wrote of human activities and how they caused alteration in local climates—for example, by changing the course of rivers and the draining of lakes. He reported that grapes and other food crops were destroyed by frosts once the temperature-moderating effects of certain bodies of water were removed.

Early agricultural humans were known to use fire to clear land for crops and to create pasturelands. This undoubtedly resulted in the destruction of forests and produced great temporary upsets in the balance of nature in and around the Mediterranean Sea and in other parts of Europe from the Middle Ages on. In Europe, great destruction of forests began around the Middle Ages as land was cleared for agriculture and as more and more lumber was used for shipbuilding.

As the human species picked up even more steam, its impact extended to some benign but nevertheless major changes in the flora and fauna of the world. Our ancestors moved cows and horses all over. Although wheat first arose in parts of the Middle East, it has been scattered by would-be farmers all over the globe. Today, some 600 million acres of land are covered by wheat most of the year, supplanting thousands and thousands of species of natural grasses, trees, and shrubs. Human horticulturists have often acted as agents of natural selection as they learned to identify good seed and to favor seeds that produced what was wanted. Humans have surely had quite an impact on the genetic makeup of a number of species.

Homo sapiens has also served as a passive factor in the evolution and development of other species. Many organisms have learned to exploit humans and the niches humankind has created. The rise of humanity was paralleled by an increase in success for a number of disease organisms that wreaked havoc on human populations throughout most of human history. Some of these diseases were transmitted by rodents that had adapted very well to human-made habitats. A number of organisms have adapted to live in harmony with humans in the habitats our species created without much of a direct negative effect on the humans. These include certain weeds, songbirds, field mice, squirrels, and other organisms like dogs that find some of the human-made niches very suitable, or at least not incompatible with their biological success.

The Question

We have just described what we consider to be a plausible sequence of events by which we humans arrived where we are. In a word, humans are the leading edge of an evolutionary strategy in which reproductive success is achieved through the intelligent use of tools to manipulate the environment to advantage.

We have recently come to realize that our powers bring serious threat as well as promise. Some of us feel confident that now that we have seen the problem, the same intelligence that brought us to this point will enable us to solve it. Others are not so sure. Now that our intelligence has gotten us into a certain amount of trouble, will we be wise enough to use it to find the solutions to get us out? That is the question.

Concepts to Remember

1. Ecological interaction has been the driving force behind both biological and geochemical evolution. In a sense, evolution is the epitome of ecological succession. Evolution is succession on million-year or billion-year time scales.
2. That humankind can be influenced by the environment is obvious if humankind is seen in terms of evolution. The changing environment literally formed us from dust in a process spanning nearly 4 billion years.
3. Genetic selection can be natural or artificial. Both result in evolution.
4. Environmental destruction is not an invention of modern *Homo sapiens;* even our Stone Age ancestors were good at it.

References and Further Reading

References marked with an asterisk are cited in the chapter.

Alvarez, W.; Kauffman, E. G.; Surlyk, F.; Alvarez, L. W.; Asaro, F.; Michel, H. V. 1984. "Impact Theory of Mass Extinctions and the Invertebrate Fossil Record," *Science* **223**:1135–1140.

Berkner, L. V., and Marshall, L. C., 1965. "On the Origin and Rise of Oxygen Concentration in the Earth's Atmosphere," *Journal of Atmospheric Science* **22**:225–261.

Bice, D. M.; Newton, C. R.; McCanley, S.; Reiners, P. W.; and McRoberts, C. A., 1992. "Shocked Quartz at the Triassic-Jurassic Boundary in Italy," *Science* **255**:443–446.

Briggs, J. C., 1991. "A Cretaceous-Tertiary Mass Extinction?" *BioScience* **41**(9):619–623.

*Cairns-Smith, A. G., 1985. "The First Organisms," *Scientific American* **252**(6):90–100.

Cloud, P. E., Jr., 1965. "Symposium on the Evolution of the Earth's Atmosphere," *Proceedings of the National Academy of Sciences* **53**:1169–1226.

*Cole, L., 1966. "Man's Ecosystem," *BioScience* **16**:243–248.

*Crick, F., 1982. *Life Itself: Its Origin and Nature.* New York: Simon & Schuster.

Darwin, C., 1859. *The Origin of Species.* Originally published November 24, 1859, in Down-Beckenham, Kent, England; this work is now available in a paperback version through Dolphin Books, Doubleday & Co., Garden City, N.Y.

Day, M., 1986. *Guide to Fossil Man,* 4th ed. Chicago: University of Chicago Press.

Dobzhansky, T., and Boesiger, E., 1983. *Human Culture: A Moment in Evolution.* New York: Columbia University Press.

Edelson, E., 1991. "Tracing Human Lineages," *Mosaic* **22**:56–63.

Eldredge, N., and Tattersall, I., 1982. *The Myths of Human Evolution.* New York: Columbia University Press.

Gould, S. J., 1982. *The Panda's Thumb.* New York: Norton.

*Jansa, L. R., and Pe-Piper, G., 1987. "Identification of an Underwater Extraterrestrial Impact Crater," *Nature* **327**:612–614.

Kerr, R. A., 1992. "Extinction by a One-Two Comet Punch?" *Science* **255**:160–161.

Lewin, R., 1991. "The Biochemical Route to Human Origins," *Mosaic* **22**:46–55.

Nance, R. D.; Worsley, T. R.; and Moody, J. B., 1988. "The Supercontinent Cycle," *Scientific American* **59**(1):72–79.

Oparin, A. I., 1953. *The Origin of Life.* New York: Dover.

*Raup, D. M., 1986. "Biological Extinction in Earth History," *Science* **231**:1528–1533.

Raup, D. M., and Sepkoski, J. J., 1982. "Mass Extinctions in the Marine Fossil Record," *Science* **215**:1501–1502.

*Raup, D. M., and Sepkoski, J. J., 1986. "Periodic Extinction of Families and Genera," *Science* **231**:833–836.

Simons, E. L., 1989. "Human Origins," *Science* **245**:1343–1350.

*Sloan, R. E.; Rigby, J. K.; Van Valen, L. M.; and Gabriel, D., 1986. "Gradual Dinosaur Extinction and Simultaneous Ungulate Radiation in the Hell Creek Formation," *Science* **232**:629–633.

Thorne, A. G., and Wolpoff, M. H., 1992. "The Multiregional Evolution of Humans," *Scientific American,* **266**(4):76–83.

*Urey, H. C., 1952. "The Early Chemical History of the Earth and the Origin of Life," *Proceedings of the National Academy of Sciences of the United States* **38**:351–363. Another classic!

Wilson, A. C., and Cann, R., 1992. "The Recent African Genesis of Humans," *Scientific American,* **266**(4):68–73.

part two

Homo sapiens in the Scheme of Natural Things

Although philosophers might never agree on some of the more mystical aspects of *Homo sapiens,* one thing is certain: Ours is an animal species, subject to the laws of nature. We have needs that must be supplied by our environment, and we also have impacts on our environment. As is the case with all other animals, our most basic needs are those for energy and minerals. But humans are different in that our needs for energy and minerals go far beyond what we eat, drink, and inhale. We have extraordinary needs for minerals—metals, for example—and for energy in the forms of electricity, heat, and the like to support all of the dimensions of civilization. As we have tried to meet the growing needs of the earth's human population, we have had extraordinary impact on the environment. Some of us worry that this jeopardizes other species; others of us, more selfish perhaps, worry that our impacts diminish the quality of human life and in fact jeopardize our own existence.

In Part Two we will consider the most limiting environmental needs of our species—our needs for food, for other forms of energy and minerals, and for water. We will examine the environmental impacts of mining, energy production, food production, and our efforts to get more water. Along the way we will consider how natural limits and impacts conspire to limit both human population growth and the quality of human life.

6

Energy in Human Affairs

Energy Up to Now: Sources and Uses

When we talked about energy in Chapter 2, we were talking mostly about the energy animals like us need, and get, from food. Now we will consider another kind of energy, the energy we humans have used and continue to use to expand and maintain our niche on earth.

Human dependence on energy goes far beyond the energy we need from food. We use energy for transportation, growing food, heating, lighting, making chemicals and other manufacturing, and energizing the tools we use in attempting to make the world more suitable for human habitation. If we were to generalize about our need for energy other than food energy, we could say that we use energy for niche expansion—to increase and sustain the area of the earth on which we live.

When the weather turns cold, many kinds of birds fly south; reptiles and other kinds of animals hibernate. Humans build fires. Rather than respond to changes in the environment, we use energy to hold

portions of the environment steady. We can live in very cold and in very hot climates because we use energy to keep the temperatures around us more or less continually suitable. When it gets dark, chickens and other birds roost. Monkeys sleep. We humans turn on lights and sleep when we are ready. Most species have to emigrate when food isn't available. We humans plant gardens and raise cattle; we ship food from one part of the world to another; we move water to areas where it is scarce; we add nutrients to infertile soils. Many species cooperate, but we humans have engineered the epitome of cooperation by making it possible to fly experts on any subject anywhere on the globe. In short, we humans use energy to make the world as we like it and then use still more energy to keep it that way.

All of this has been responsible for the success of our species. Largely as a result of our access to large amounts of cheap energy, we are thriving—in the strict biological sense—virtually everywhere on earth. The downside of this story is that it has produced too many of us too fast, and we now threaten the planet by our continued dependence on energy. Too many fires now rage as we discover that the stability of our collectively and narrowly conceived niche depends on the existence of niches suitable for other species. Our fires now threaten to change the chemistry of our planet. This is the essence of the energy crisis.

The so-called energy crisis of the mid-1970s was greatly overblown. Long lines at gas stations, home heating oil shortages, turned-down thermostats, and sharply rising energy prices made good human interest stories but reflected little more than temporary supply problems and some long overdue price adjustments. It was an economic crisis triggered by an oil embargo and prolonged by cutbacks in oil production by the OPEC cartel. In more recent years we have had other similar energy crises brought on by conflicts in the oil-rich Middle East. There will be other such crises, and these will continue to serve to illustrate how important particular kinds of energy resources have become to our species. The energy *supply* crisis will be, and should be, a continuing concern for our species, but the energy *use* crisis deserves considerably more attention than it has been getting.

To help put all of this into perspective for the material that lies ahead, consider the following points:

1. In a very brief period in the history of the world, in less than five centuries, we will have consumed virtually all of the coal, oil, and gas formed over a period of 500 million years (Figure 6.1).
2. We are running out of some of the energy sources we depend on most, and we have no assurance that we will be able to make smooth transitions to other sources of energy already available to us, let alone to those for which technology is not yet sufficiently advanced.
3. There are many uncertainties about the economic, political, social, and other ripple effects as the world goes through the crisis of passing peak production of natural gas and oil.
4. In a world in which energy is of supreme importance, there are few energy policies or plans. Because the United States uses the greatest proportionate share of the world's energy, it is especially irresponsible in not having a consistent, well-thought-out, sound energy policy. Since the early 1950s the United States has consumed more energy than it produces. Since the difference is made up by imports, this imbalance contributes greatly to the U.S. trade deficit.
5. Energy is wasted worldwide.
6. Nearly every environmental problem is directly related to the production and consumption of energy. The combustion of fuel accounts for nearly all of the *air* pollution problem; the extraction, processing, and use of fuel generates much *water* pollution. Strip-mining is an obvious example of how energy use can contribute to the deterioration of land. Energy consumption is even responsible for some less obviously connected types of pollution, such as noise and hazardous wastes.

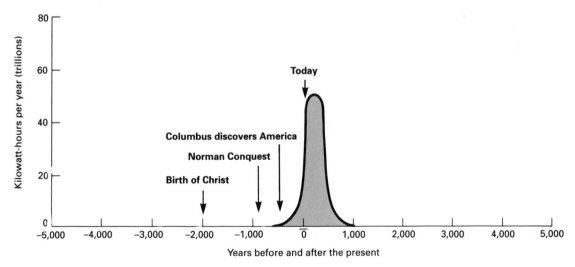

Figure 6.1 Historical and Expected Fossil Fuel Use. The use of fossil fuels will appear as just a blip even on a time scale as short as 10,000 years. After using fossil fuels for about 600 years, we are probably already on a permanent downward course in worldwide production of natural gas, at or just past the peak in production of oil, and within a few hundred years of reaching the peak in our use of coal. The "age of fossil fuels" might more aptly be called the "fossil fuel incident." The effect of the "incident" on the environment will be akin to that of a serious, noxious-fume-producing, short-lived fire aboard a spaceship.

British thermal unit (BTU): amount of heat necessary to raise the temperature of a pound of water one degree Fahrenheit.

Coal: solid combustible organic fuel formed by the anaerobic decomposition of vegetable matter. This material is composed of aromatic carbon compounds.

Consumption: the amount of energy or any other resource used in a unit of time by some unit of people. (For example, the United States consumed about 7 billion barrels of oil in 1979. This could also be expressed as more than 30 barrels per person in 1979.)

Electrical energy: electricity; energy derived from the movement of charged particles.

Fossil fuel: any organic fuel such as coal, petroleum, or natural gas formed over millions of years from what was once living matter.

Generator: a machine that converts mechanical energy into electrical energy. A generator typically consists of a coil of wire that can be made to rotate in a magnetic field. The rotation causes electrons to flow in the wire, thus producing electricity.

Kilowatt (kW): unit of electrical energy (1,000 watts) equal to about 1 1/3 horsepower.

Kilowatt-hour (kWh): unit of electrical energy (1,000 watt-hours) equivalent to the energy delivered by the flow of 1 kilowatt of electrical power for one hour. (For example, a 200-watt bulb burning for five hours will use 1 kilowatt of energy, or enough to lift 300 pounds roughly 9,000 feet into the air.)

Megawatt (MW): 1 million watts; 1,000 kilowatts.

Natural gas: naturally occurring fossil fuel gases found in coarse geologic formations, often in association with oil.

The principal chemical constituent of natural gas is methane (CH_4).

Oil: any of a large group of viscous, flammable liquids (e.g., whale oil, petroleum, vegetable oil) that are soluble in alcohol or ether but not in water. **Petroleum** is a dark-colored oil consisting of a mixture of hydrocarbons derived from once-living matter. Petroleum occurs in natural deposits scattered throughout the world and is usually recovered by drilling. The word *oil* is sometimes used as a synonym for *petroleum.*

Production: the amount of an energy or other resource mined or otherwise obtained from the earth per unit of time. (For example, the United States produced nearly 700 million metric tons of coal in 1979.)

Quad: 1 quadrillion British thermal units (BTUs). The amount of an energy resource is often expressed in terms of its heat energy content, in calories or BTUs. This makes possible direct comparisons between different kinds of energy sources. Whereas it would not be fair to compare barrels or even kilograms of oil to kilograms of coal, it would be fair to compare the heat content of a given amount of oil with that of a given amount of coal. One metric ton of coal has 0.0000000278, or 2.78×10^{-8} quads, and a barrel of oil has 0.0000000058 quads. In terms of energy content, then, a metric ton of coal has nearly five times as much energy as a barrel of oil. A quadrillion is equal to 1,000 trillion; it is a 1 followed by 15 zeros, or 10^{15} in scientific notation. One quad is thus 10^{15} BTUs.

Watt: amount of power available from an electric current of 1 ampere at the potential difference of 1 volt.

The former chair of the U.S. Atomic Energy Commission, Dixie Lee Ray, once pointed out that every American uses the electrical equivalent of 11 servants, that with electrically powered machinery one human being can do the work of 700 human beings, and that a barrel of oil has the heat energy equivalent of a person working at hard labor for two years. A direct consequence of such labor-saving access to energy is that fossil fuel is squarely responsible for delivering much of humankind from bondage to the soil. But we have traded one kind of bondage for another. Much of civilization as we now know it is dependent, directly or indirectly, on higher rates of consumption of what for now and the foreseeable future are nonrenewable, rapidly dwindling, and increasingly expensive sources of energy that cause problems in the environment. This is a crisis indeed.

The History of Energy Use: Noteworthy Recent Trends

The History of Consumption

The trend in energy consumption from the primitive era of human history to the modern day is shown in Figure 6.2. Note that during this period of time, energy consumption increased by more than 100-fold. Note also that energy consumption rose most sharply in recent times; even in the United States the days of iceboxes, outdoor privies, washtubs, chamberpots, horse-drawn carriages, oil lamps, and unheated bedrooms were simply not all that long ago.

More recent trends reflect a dramatic climb in the use of electrical energy in the American home (Table 6.1). There has been more than a doubling of total energy use in the United States every 20 to 25 years over the past century, and this pace has ex-

Table 6.1 Homes with Electricity from 1907 to the Present in the United States

Year	Homes with Electricity (%)
1907	8
1912	16
1920	35
1930	68
1940	79
1950	94
1960	99
1965–present	100

Source: Adapted from *The American Energy Consumer,* copyright © 1975, the Ford Foundation.

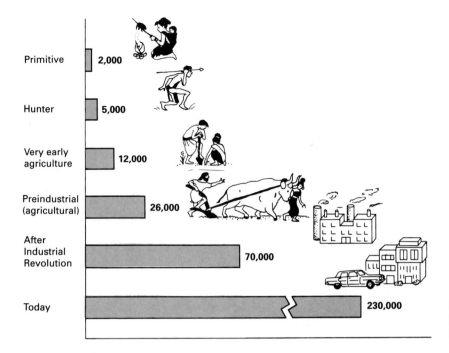

Figure 6.2 Energy Consumed at Different Stages of Human History. In kilocalories *per person* per day.

ceeded the rate of population growth for some time now (Figure 6.3). For much of the twentieth century the use of electricity, a highly desirable, somewhat inefficient form of energy, increased at nearly twice the rate of overall U.S. energy consumption, a doubling every 10 to 12 years. A similar trend is under way throughout much of the world.

Perhaps because of increased energy consciousness, but more likely because of increased energy bills, the demand for electrical energy and energy in general quite unexpectedly tapered off in very recent years, falling far below the projections made in the early 1970s. Demand for electricity did continue to increase during the ten years following the energy cri-

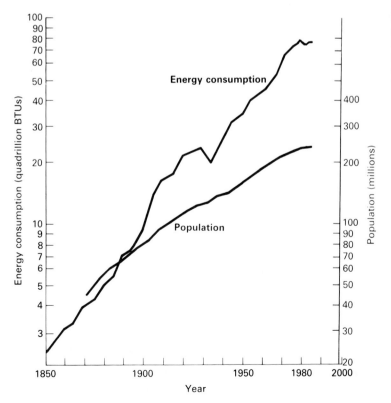

Figure 6.3 Comparison of Energy Consumption and Population Growth in the United States since the Mid-nineteenth Century. These graphs clearly indicate that per capita consumption increased considerably from the turn of this century until recently. Note that the vertical scales are logarithmic.

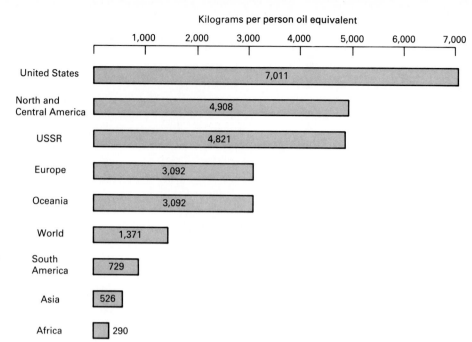

Figure 6.4 Per Capita Energy Consumption in Various Regions of the World (1988). Although energy consumption has generally increased worldwide in recent decades, the gap between the developed and developing nations has continued to widen.

sis, however, bucking the general trend of the use of energy during that period.

As for *total* energy consumption in America, after the 1973–1974 oil embargo, it fell in five of the next nine years. Energy consumption in 1981 was actually 3% below what it was in 1973. Energy consumption in 1986 was still below 1973 levels overall but new highs were being reached by the end of the 1980s.

The History of Imbalance

As illustrated in Figure 6.4, there is considerable difference in the use of energy from one country to

another, and Americans account for a particularly large share of world energy consumption. For most of the last several decades the one-twentieth of the world's population made up of Americans accounted for one-third to one-fourth of all energy consumption.

Since 1950 there have been greater proportionate increases in energy use in western and eastern Europe than in the United States, and it is expected that the gap between other developed countries and the United States will become narrower in years to come. Figure 6.5 illustrates the disparity between production and consumption within individual countries and regions of the world, resulting in a growing

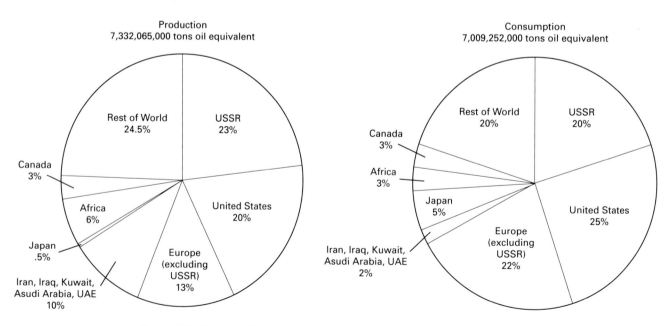

Figure 6.5 Relative Energy Production and Consumption in Various Parts of the World (1988).

Part Two *Homo sapiens* in the Scheme of Natural Things

dependence of energy-consuming countries on the energy-producing countries. We will consider the political and economic implications of this situation later in the chapter.

Energy Sources to the Present

As energy consumption has increased worldwide, changes in the major sources of energy have also occurred. In the early stages of human energy history we relied almost entirely on wood fires. A little more than 100 years ago the United States, for example, got about three-fourths of its commercial energy from wood (Figure 6.6). The increasing use of **fossil fuels** (fuels such as coal, oil, and natural gas derived from slow physical and chemical changes in deposits of accumulated dead plant and animal materials over millions of years) in more recent times has itself been marked by notable trends. These are illustrated for the United States from 1952 to 1989 in Figure 6.7. The most notable feature in these trends is the increasing dependence on natural gas and petroleum until recently.

As shown in Figure 6.7, fossil fuels currently contribute almost 90% of the energy used in the United States. Nearly half of the current energy supply for the United States is in the form of oil; another 24% is natural gas. Thus the two forms of energy that are in shortest supply are those on which the United States depends almost exclusively.

We will now summarize the history and the status of natural gas, coal, oil, and nuclear fission as energy sources. We will consider the prospects for the future for each of these sources along with solar power, nuclear fusion, hydroelectric power, and other energy options in Section B of this chapter.

Natural Gas

Natural gas has been used for a little over a century and currently provides almost one-quarter of the energy used in the United States. In the 1920s, Americans used about 800 billion cubic feet of gas annually; this had risen to 22 trillion cubic feet by the mid-1970s for an average increase of about 6% per year. Over the next ten years or so consumption fell to around 16 trillion cubic feet. Today, about two-thirds of all American homes and roughly half of American commercial enterprises use natural gas.

Because natural gas is easily stored and transported through pipes, it became extremely popular after the 1950s when large-capacity pipelines linked gas-producing regions to the big population centers of the United States. By 1974 there were already more than a quarter of a million miles of natural gas pipelines installed throughout the United States. Because natural gas yields almost no pollutants when it is burned, its use got another boost as concern for the environment intensified after the 1960s. Still another reason for the popularity of natural gas is that for most of the second half of this century, gas has been about 20% cheaper than oil on an energy-content basis. It was not until 1979 that bituminous coal became cheaper than gas on the same cost-per-BTU basis. The (pre-1979) general cost advantage of natural gas was partly due to government price controls (imposed with the Natural Gas Act of 1938) on interstate shipments. Natural gas prices have been gradually decontrolled following the Natural Gas Policy Act of 1978.

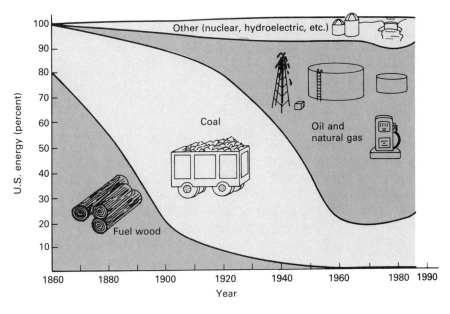

Figure 6.6 U.S. Energy Sources since 1860. Since 1860 there have been several dramatic shifts in our primary energy sources.

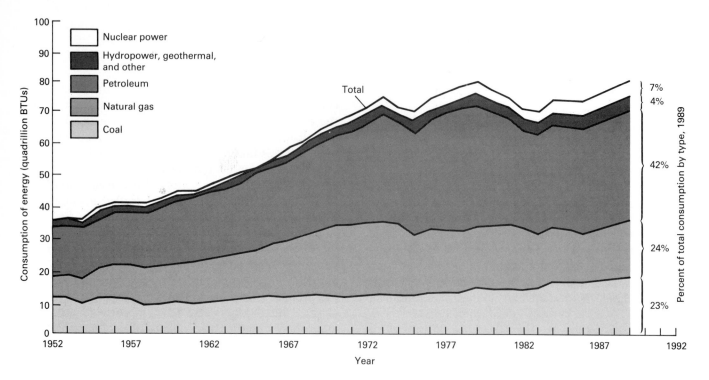

Figure 6.7 Recent Trends in U.S. Energy Consumption. Two-thirds of our energy now comes from oil and natural gas; more than 90% comes from fossil fuels, less than 5% from renewable sources.

In the 1880s the world used about 200 billion cubic feet of natural gas per year; today about 300 times as much is used as it provides almost one-fourth of the world's energy. It is a major source of energy in the United States, Canada, and the nations of the former Soviet Union. Despite the fact that the peak of natural gas production may have passed in many nations, certain countries of Europe and the Middle East are expected to increase their use of natural gas in the near future. Because of a combination of dwindling supplies, government-imposed restrictions on use, and increasing cost, the use of natural gas has declined for most of the years in the United States since its peak in 1972 (see Figure 6.7). Some observers say that although the end of natural gas may be in sight, the end may be prolonged as new discoveries are made worldwide. We will have more to say about the prospects for the future later.

Coal

Both archaeological and written records indicate that **coal** was used in Bronze Age Europe some 4,000 years ago. The Chinese were using it at least 2,000 years ago, and Hopi Indians were using coal at least 400 years before Christopher Columbus was born. Some historians say that significant coal burning began about 600 years ago, and problems with coal were recognized almost immediately—Edward I banned coal burning in London in 1306.

The rate of increase in coal use was very low at first, but the Industrial Revolution produced a sharp worldwide increase resulting in nearly a doubling of coal production every 20 years between 1860 and 1920 (Table 6.2). Coal production has risen unsteadily in the United States over the past century, but since 1850 it has increased nearly 100-fold. Figure 6.8 gives a comparative production breakdown for the leading coal-producing countries, showing that in 1988 three coal-producing nations generated over

Table 6.2 Coal Production since the Mid-nineteenth Century

Year	World Production (millions of metric tons)	Year	U.S. Production (millions of metric tons)
1860	150	1850	9
1880	320	1890	145
1900	780	1910	454
1920	1,200	1930	454
1940	1,700	1940	462
1960	2,100	1950	468
1970	2,500	1960	399
1980	3,797	1970	556
1988	4,760	1980	753
		1988	873

Sources: R. C. Dorf, *Energy Resources and Policy* (Reading, Mass.: Addison-Wesley, 1978); Energy Information Administration, U.S. Department of Energy.

Table 6.3 Uses of Coal in the United States, 1989

	Percent of Total Use
Electric utility	86
Coking coal for steel	5
Residential and commercial heating	1
Other	8
	100

Source: Energy Information Administration, U.S. Department of Energy.

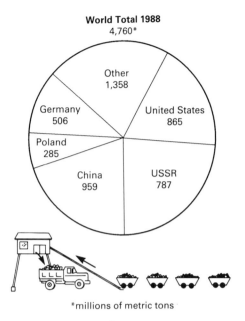

*millions of metric tons

Figure 6.8 International Coal Production. Numbers are in millions of metric tons.

half the world's total. It is noteworthy and important to later discussion that coal is used in more heavily industrial ways than other types of fuel; the ways in which it was used as of 1989 in the United States are illustrated in Table 6.3. It is also noteworthy that coal is the only significant source of fuel exported by the United States (about 12% of production was exported in 1989).

Coal today supplies a little over one-fifth of U.S. energy despite the fact that this energy source constitutes by far the largest fraction of proven U.S. energy reserves. Coal has been held to a relatively minor role in meeting American energy needs in recent decades because of environmental and other problems associated with mining it, transporting it, and burning it.

Much of the promise of the future of coal has been dimmed by the combination of (1) higher rail rates following deregulation, (2) delayed and indefinitely discontinued plans for coal conversion (synfuels) technology (related to both politics and the oil glut that began in 1982), (3) the coming of age of the acid rain issue (Chapter 10), and (4) air pollution restrictions in general. There continue to be serious safety, labor, and environmental problems with mining operations (see Chapter 17).

Petroleum

We have known about **petroleum,** or oil, for only 100 years, but it quickly became a highly pre-

ferred form of energy partly because of the ease with which it could be transported and handled.

The growth in oil production and consumption in the United States since 1920 is illustrated in Table 6.4. Notice how much faster consumption grew than production over most of this period. A comparison of trends in oil consumption for the United States and the world is made in Figure 6.9. This illustration reveals two particularly interesting facts: that the United States accounted for about 40% of world oil consumption from 1930 to 1970 and that the United States began consuming proportionally less during the 1970s. The uses of oil in the United States are shown in Figure 6.10.

The most important dimension of the oil energy situation is that we are running out. The peak in world oil production will probably occur in the 1990s, and some energy experts predict that oil supplies will begin to fail to meet demand in 20 years or so. Recent

Table 6.4 Annual Consumption and Production of Oil in the United States since 1920

Year	Consumption (millions of barrels)	Domestic Production (millions of barrels)	Imports as Percent of Consumption
1920	434	443	2%
1930	862	898	4%
1940	1,285	1,353	5%
1950	2,467	2,157	13%
1960	3,579	2,905	19%
1980	6,227	3,705	40%
1989	6,297	3,352	47%

Sources: R. C. Dorf, *Energy Resources and Policy* (Reading, Mass.: Addison-Wesley, 1978); Energy Information Administration, U.S. Department of Energy.

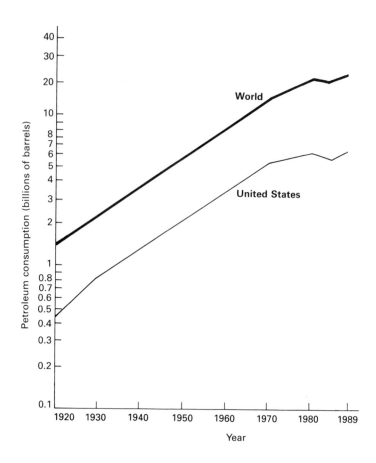

Figure 6.9 Annual U.S. and World Petroleum Consumption Rates. Note that the vertical scale is logarithmic.

trends in U.S. oil consumption are depicted in Figure 6.7 in relationship to other forms of energy.

Nuclear Power

As Enrico Fermi and his colleagues looked on anxiously, the world's first nuclear fission reaction

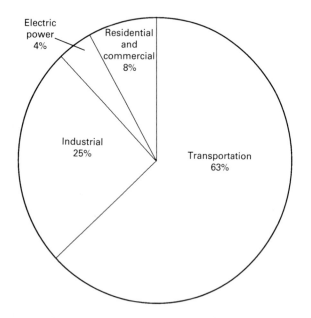

Figure 6.10 Use of Oil (Refined Petroleum Products) in the United States.

"went critical" (developed a sustained, controlled nuclear chain reaction) on December 2, 1942. More than a decade later, the first commercial-scale atomic fission reactor was turned on at Shippingport, Pennsylvania. The first *licensed* atomic power plant opened on November 15, 1957, in Vallecitos, California. Today, nuclear fission (which we will describe in some detail in Section B of this chapter) is an important energy source in many parts of the world. Although only about 6% of the world's *total* energy needs are met by nuclear fission, more than 400 nuclear power plants are generating electricity worldwide. There are more than 100 operable nuclear fission plants in the United States. In France, one of Europe's most nuclear-oriented nations, 55 nuclear power stations generate more than two-thirds of that nation's electricity. Japan has about 40 operational reactors producing nearly 30% of its electricity. In the early 1990s nuclear power provided nearly 20% of U.S. electricity (note that this is not total energy, just electricity) overall. This ranged from 7% in the Southwest to 35% in New England.

Among the factors that have sustained the proliferation of nuclear power plants are the facts that, done right, nuclear power is clean, produces no smoke, and requires very little fuel. Among the factors limiting the further expansion of nuclear power and already responsible for a precipitous decline in the orders for new plants in the United States and elsewhere are (1) economics (nuclear plants have become very expensive to build in comparison with plants that accept conventional fuels), (2) leveling off in the demand for electricity, (3) scarce and dwindling uranium deposits, (4) waste disposal problems (we have yet to come up with an acceptable long-term way of storing nuclear waste), and (5) nuclear accidents, particularly those at Three Mile Island and Chernobyl. These accidents intensified an already existing concern over the safety of atomic power and intensified negative public pressure, resulting in much altered (downward) projections for use of nuclear power in the future. We will return to this topic in Section B of this chapter.

Electricity as a Delivery Form

Electricity is a favorite form of energy largely because it can be delivered easily and directly to power appliances and machines. Trends in U.S. generation of electricity are illustrated in Table 6.5. Worldwide generation of electricity is increasing rapidly—about 5% per year during the last half of the 1980s. America generated 3.7 billion kilowatt-hours (kWh) of electrical energy (49 kWh per person) per year at the beginning of this century. By the end of the century this will likely have grown by nearly *three* orders of magnitude—1,000-fold. The generation of electricity now accounts for 36% of U.S.

Table 6.5 Electricity Generation in the United States since 1900

Year	Population (millions)	Generation (billion kWh)	Electricity Generated per Person per Year (kWh)
1900	76	4	49
1920	106	57	540
1930	123	115	935
1940	132	142	1,074
1950	152	329	2,161
1960	181	753	4,169
1970	205	1,530	7,469
1980	227	2,286	10,070
1989	260	2,779	10,629

Sources: R. C. Dorf, *Energy Resources and Policy* (Reading, Mass.: Addison-Wesley, 1978); Energy Information Administration, U.S. Department of Energy.

energy consumed, i.e. 36% of the energy used in the U.S. is used to generate electricity. We have improved the efficiency with which we can generate electricity from fossil fuels (from about 20% in 1940 to about 31% today), and this in turn reduced the relative cost of electricity over a 40-year period, beginning in the early 1930s, contributing to its popularity (Figure 6.11).

We use electrically delivered energy in many ways. Industrial uses account for about 35% of con-

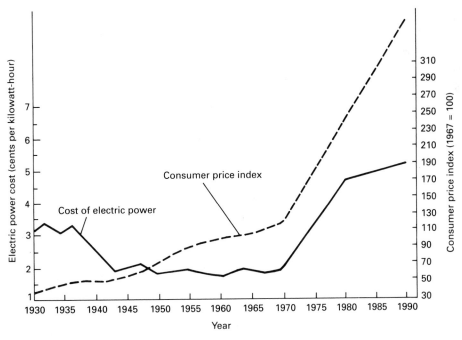

Figure 6.11 Cost of Electric Power Compared to the Cost of Living from 1930 to 1990. Electricity has become more of a bargain over the years. The relative cost of electric power fell over a period of 40 years, beginning in the early 1930s, when the consumer price index was rising. This trend reversed itself in the early 1970s. Both the cost of living and the cost of electricity have tripled since 1970.

Chapter 6 Energy in Human Affairs

119

sumption. In the American home, refrigeration and lighting top the list, together accounting for more than 22% of the total.

Electricity will likely be with us forever as a delivery form, since a great variety of sources and uses of energy can be coupled by electrical wires. Electricity is generated (to considerably varying degrees) by burning oil, coal, natural gas, wood, garbage, and trash, as well as from flowing water, wind, the earth's heat, sunlight, and nuclear fission.

Other Energy Sources

Other sources of energy make up a very small percentage of the total. We will consider these—including solar, hydroelectric, geothermal, atomic fusion, biomass, and tidal energy—in the more detailed analysis of energy sources of the future in Section B of this chapter.

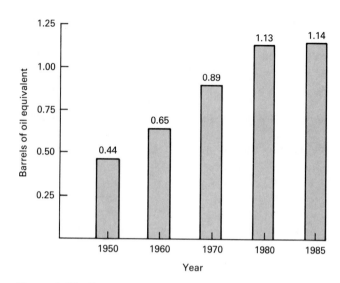

Figure 6.12   Energy Used to Produce a Ton of Grain.

Major Categorical Uses of Energy

Note the following facts about our uses of energy:

— More than 35% of U.S. energy is used to generate electricity.

— Nearly half of all the energy we consume is lost through some form of inefficiency or conversion loss as a consequence of the second law of thermodynamics (Chapter 2).

— Some 37% of the energy we use goes for residential and commercial lighting and heating and other domestic and commercial uses.

— Another 36% goes for mining, smelting, and industrial processes.

— About 6% of U.S. energy is issued as a raw material, the coal and petroleum and natural gas used in making chemicals, pesticides, plastics, and other products.

— Roughly 17% of U.S. energy is used in agriculture.

Energy in Agriculture

A century ago, the farmers of this world were, practically speaking, self-sufficient in terms of energy. The energy required for farming came from draft animals and from humans fed off the farm. Today, in sharp contrast, the world's farmers depend a great deal on external sources of energy, principally oil. According to one source, by the middle of the 1980s, the fossil energy going into farming worldwide was the equivalent of one-twelfth of the world's oil production (some 1.7 billion of 21 billion barrels). A ton of grain on the average now requires more than one barrel of oil to produce worldwide (see Figure 6.12). Most of

the increase in agricultural productivity seen worldwide in the past half century has been the direct result of energy inputs in the forms of fertilizer, mechanization, pesticides, and irrigation. Although draft animals are still used extensively in Asia and Africa, about two-thirds of the world's cropland is now estimated to be plowed with tractors. The world's tractor fleet has grown fourfold since 1950.

Perhaps the most remarkable energy trend in agriculture in the United States is the decline in human labor. During the 25-year period beginning in 1945, farm labor decreased 60% and yields tripled. Today only a fraction of one percent of the energy consumed on the American farm comes from human labor. The reason for this trend is fairly obvious: Energy has been abundant and relatively cheap during a time when labor has become prohibitively expensive. Illustrating the degree to which this is true, Pimentel and associates (1973) calculated that even by the early 1970s, while the hand planting of corn required only about one-sixtieth the energy required of machine planting, planting by hand would cost about four times more.

Sociopolitical Implications of Energy Intensiveness in Agriculture. The energy intensiveness of agriculture varies greatly around the world and is one of the reasons for the disparity in energy use from one country to another. China is a nation of more than 1 billion people, four times as many people as there are in the United States, and yet China grows nearly all of its own food on far less cropland than there is in the United States and with much less fuel energy—but with a lot more human labor. Human labor is much more plentiful and much cheaper in countries like China; machines, pesticides, and other energy-consuming practices are used comparatively less, although fossil fuel energy use has increased dramatically (100-fold) in China since 1960.

It goes almost without saying that the energy intensiveness of agriculture in the United States has had an enormous impact on the American way of life. Most Americans now live in cities working at jobs that have only a very indirect relationship to food production. The application of energy in agriculture has made it possible for this demographic shift to take place; whereas one out of ten people lived on a farm in the early 1930s, only one out of 35 people lives on a farm today. Energy intensiveness in agriculture has enabled the United States to be a largely urban society and yet be a major exporter of food for the world. Although many Americans have viewed the successes of the U.S. system of agriculture as evidence of national ingenuity and hard work, the facts suggest that much of our agricultural success can be attributed to the application of relatively cheap fossil energy. American agriculture may have grown into what will become an increasingly tight corner. Although the United States exports food, it does so at the expense of fuel that is to some extent imported from other countries.

It has been estimated that to feed the entire world with an agricultural system like that of the United States, about 90% of the world's annual energy supply would be required just to supply food. During these times of increased energy costs and dwindling energy supplies, this is a sobering thought indeed.

The Future of Energy in Agriculture. The foregoing discussion raises a number of important questions with regard to agriculture, food production, and the continuing energy crises in the world and in the United States. It has been estimated that so tightly connected is the price of oil and the price of food that if there were a fourfold increase in the price of oil, there would be a sixfold increase in the price of food. The extra increase is partly related to the second law of thermodynamics (Chapter 2). But regardless of the exact details of the economic relationships, as fossil fuels become more expensive, agricultural practices will change. Given the premise that modern-day agricultural methods are based on cheap energy, the following kinds of changes should be anticipated in the future.

We will use less chemical fertilizers and pesticides that take significant amounts of energy to produce and will begin to rely more on natural fertilizers. We will practice more scientific forms of crop rotation and soil tillage and will develop increasing respect for the soil and what we do to it—we will gradually lose the attitude that no matter what insult we inflict on

Table 6.6 Estimated Efficiency of Hauling Freight

Shipping Method	Efficiency (BTUs per ton-mile)
Airplane (all cargo)	28,610
Truck	3,420
Barge	990
Rail	1,720
Coal slurry pipeline	1,270
Oil pipeline	500

Note: Combines energy required for propulsion, maintenance, manufacturing and construction, movement of empty carrier. A ton-mile equals one ton carried one mile.

Source: Congressional Budget Office.

soil, we can restore it with some kind of energy-intensive technology.

Energy in Transportation

Transporting people and goods accounts for about 25% of the energy used in the world. The global fleet of automobiles has increased about 12-fold since the end of World War II. This has greatly increased both oil consumption and air pollution problems worldwide.

Petroleum is obviously the energy source of choice in modern transportation. Over half the oil used in the United States (about one-fourth of total energy) is used in transportation. Transportation, by itself, consumes more oil than is produced in the United States. Passenger cars consume about one-half of the total used in transport.

Americans do a lot of traveling. Almost one-tenth of all the oil consumed in the world every day is used by American motorists on their way to and from work—usually traveling alone. In America the use of cars for local travel has become exceedingly important because of our entrapment in a vicious cycle. The automobile has altered the face of America by making it possible for cities to spread out, making the automobile necessary to get almost anywhere.

A comparison of the various modes of transportation of goods and people is given in Tables 6.6 and 6.7. These data reveal that although the airplane is the most energy-intensive mode of all, the automobile

Table 6.7 Relative Energy Efficiency of Various Modes of Domestic Transportation

Transportation Mode	Efficiency (BTUs per passenger-mile)
General aviation	11,044
Airplane (carrier)	5,733
Automobile	3,498
Motorcycle	2,273
Intercity bus	1,078
Amtrak	1,765

Source: U.S. Department of Transportation.

is a close second. Far more passenger-miles are traveled by automobile than by airplane, however. Automobiles account for nearly 90% of all miles traveled by people; airplanes garner a modest 9.3%. Each year, Americans fly an average of 580 miles per person in airplanes, ride about 150 miles per person in buses and trains, travel about 120 miles on the average on inland waterways, and, as we have already indicated, travel an average of more than 6,000 miles per person in automobiles. What this suggests—and we will take up the matter in some detail in a later section on conservation—is that the automobile is an obvious target in our battle to come to grips with our energy problems.

Automobile ownership is practically universal in the United States. Beyond the necessity factor, the automobile's popularity is due in part to the tremendous amount of freedom it provides. Perhaps no other device gives people at almost every socioeconomic level (in countries like the United States) the power to overcome natural restraints on mobility. To appreciate this power, one need only reflect on the fact that we can travel at high speeds through some of the worst kinds of terrain in driving rainstorms, swirling blizzards, or intense heat, all at a comfortable 22°C with stereo music, isolated from the environment. It will be difficult at best to wrest any of this sort of freedom away. The appeal of the freedom factor is obviously a major reason why people do not use less energy-intensive forms of transportation like buses and bicycles. It seems inevitable, however, that buses, bicycles, and trains will assume an increasingly important role in the years ahead as the energy situation gets tighter and, especially, more expensive. This trend, modest though it may yet be, is being spurred by subsidies for bus travel. A great truth that may be emerging from the relatively poor results of nationwide efforts to get people away from their cars and into buses is that nothing beats the private automobile in terms of convenience. Perhaps the answer is to continue to make the auto less convenient by means of higher taxes, license fees, and parking rates.

Alternative Fuels. Because most of the world's inexpensive petroleum is in the volatile Middle East and because conventional fuels derived from petroleum are a leading cause of urban air pollution and contribute to global warming, alternative fuels and other means of propelling automobiles continue to receive serious consideration. The alternative fuels include ethanol, methanol, hydrogen, and natural gas; alternatives to the direct consumption of fuels include electric power via batteries and electric power via solar cells. We will discuss these options later in this chapter and in the chapters on air pollution.

Here are some other interesting facts about the use of energy in transportation (see Lowe, 1988):

— China has 540 times as many bicycles as automobiles.

— The United States has the second highest number of bicycles per capita of the nations of the world, but only one in 50 U.S. bicycles is used in commuting.

— A bicycle requires less than one-fiftieth of the energy per mile of an automobile.

— A 1980 study in Great Britain concluded that if just 10% of car trips of less than 10 miles were made by bicycle, that nation would save 14 million barrels of oil per year.

— Public transit has been in decline in the United States since the early twentieth century. Between 1954 and 1963 nearly 200 transit companies went out of business, and for most of the past three decades public transit has been heavily dependent on government subsidies.

— In the late 1970s there were about 300 million automobiles in the world fleet; this is expected to reach 700 million by the year 2000—almost enough autos to allow the entire population of the world to get into a car at the same time (one for every eight people).

Residential and Commercial Energy Consumption

The rise in labor-saving appliances has been largely responsible for the increasing use of energy in the American home. The comparative power consumption of various home appliances is given in Figure 6.13. American homes may have 50 or more light bulbs, and just one 100-watt light burning around the clock for a year requires the burning of about 60 gallons of oil. The use of energy for heating and cooling is also considerable; 20% of the energy used in the United States is for space heating and cooling. A majority of American households now have air-conditioning units of some kind.

The patterns of energy use in homes and commercial establishments in the United States are quite similar. The largest fraction in each case, 58% and 47% of the energy used in homes and commercial establishments, respectively, goes to space heating. Most of the remainder goes to air conditioning, water heating, lighting, and refrigeration in very roughly equal amounts.

Industrial Energy Consumption

In the United States, mining, smelting, and manufacturing activities account for 36% of total energy consumption. About 45% of the energy consumed industrially is used to generate steam, and

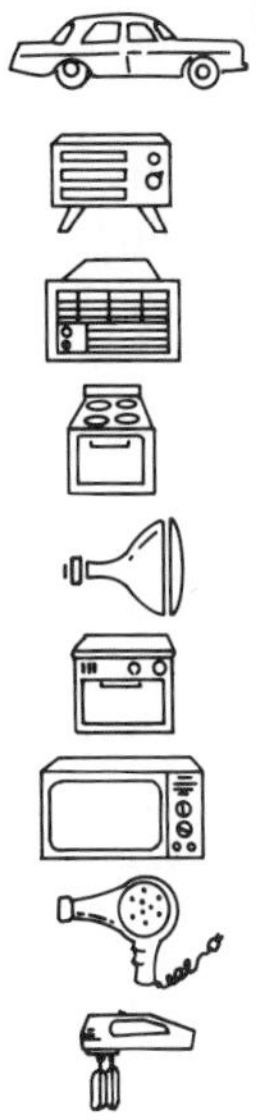

Appliance	Annual Average Consumption (BTUs per year)
Automobile	900,000,000
Home heater	180,000,000
Air conditioner	40,000,000
Stove	9,000,000
Color TV	9,000,000
Dishwasher	4,000,000
Black-and-White TV	2,000,000
Hair dryer	150,000
Egg beater	50,000

Figure 6.13 Comparative Energy Use in Automobiles and Various Home Appliances. Note that one ton of coal is equal to 25 million BTUs, one barrel of crude oil is equivalent to 5.8 million BTUs, and one cubic foot of natural gas is equal to 1,031 BTUs.

25% is used to drive motors, to light factories, and to drive chemical processes such as electroplating. Nearly all of the remainder is used to heat buildings, to heat vats of various liquids, and in other direct heating applications. In the past three decades, industrial coal consumption remained constant while industrial petroleum consumption increased almost 4% annually.

Section B

Energy Alternatives for the Future

Among the plausible energy alternatives for the future are some very well established old standbys and some untested new ideas. Coal is perhaps the most important of our old standbys.

Coal

If we find environmentally safer ways to mine, transport, and burn coal, it will become our best bet for the next several hundred years.

Although we humans have been using large amounts of coal lately (see Figure 6.14), there is still a great deal left. In the United States a little over one-fifth of current energy needs are met by coal, but coal represents about four-fifths of proven U.S. energy reserves. World reserves of coal are described in Figure 6.15, where it can be seen that the United States has more than one-fourth of the entire world's reserves.

U.S. Western Coal Reserves: Promises and Problems

Coal is not uniformly distributed. Major deposits of coal in the United States are illustrated in Figure 6.16. According to some sources, there is enough coal in the western United States *alone* to keep up with America's need for coal for 300 years at present rates of consumption.

Western coal has only recently begun to be developed, and there are several reasons for this. One is the distance from western coal to the markets where it would be consumed—the mills and population centers of the United States. However, because western coal occurs in such thick seams and because it is close to the surface, generally speaking even with shipping costs included, western coal is actually cheaper per ton than coal mined in eastern and central areas of the United States. An alternative to shipping would be to convert western coal to electricity and to send the electricity east. Although this would shift the pollution associated with burning coal away from the population centers, there would be other problems. For example, there would be problems associated with providing rights-of-way needed for transmission lines and the reduced efficiency of electrical as opposed to direct combustion heating. Another serious technological problem with western coal is the scarcity of water, needed for cooling and condensing the steam that drives the generators and for irrigation related to reclamation.

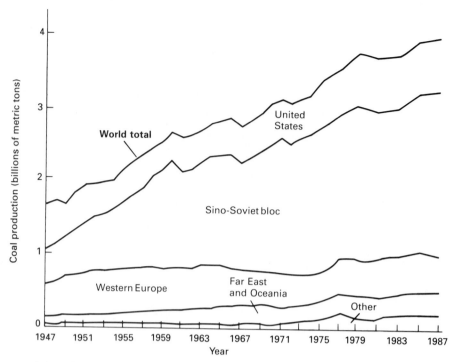

Figure 6.14 Recent Trend in World Coal Production by Various Countries and Regions.   The greatest increases in coal production during this period were in the Sino-Soviet bloc of nations. This accounted for nearly all of the increase of world production in recent times. (Note: One metric ton equals 1.102 short tons.)

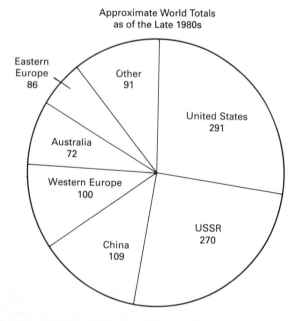

Approximate World Totals as of the Late 1980s

Figure 6.15 World Recoverable Coal Reserves.   Proved or recoverable reserves are coal deposits known to be recoverable with current technology and worth getting under current economic conditions. Annual world coal production is approximately 5 billion short tons per year (one short ton = 2,000 pounds = 0.907 metric ton), about 0.5% of the total recoverable. Thus if production continues at its present rate, existing reserves will continue to provide coal for several hundred years.

The Trouble with Coal in General

If coal is to become more of a solution to our energy problems, better ways will have to be found to reduce the environmental impacts of mining, processing, transporting, and burning it. These are just some of the problems (most are discussed further in Part Three):

— Although strip-mining practices are no longer as destructive as they once were, strip mining has a profound negative impact on land.
— Mining in some areas results in acid mine water pollution of streams and rivers.
— Coal cleaning is still done at the expense of clean water in many places.
— Sulfur dioxide from coal has increased the acidity of rain in many areas of the world.
— The burning of coal generates sulfur dioxide and particulate air pollution problems.
— The burning of coal produces carbon dioxide, which has the potential to alter the earth's climate, and oxides of nitrogen, which contribute indirectly to ozone pollution and to acid rain.
— Coal combustion also produces hydrocarbons, some of which participate in the generation of ozone and some of which cause cancer.

Part Two   *Homo sapiens* in the Scheme of Natural Things

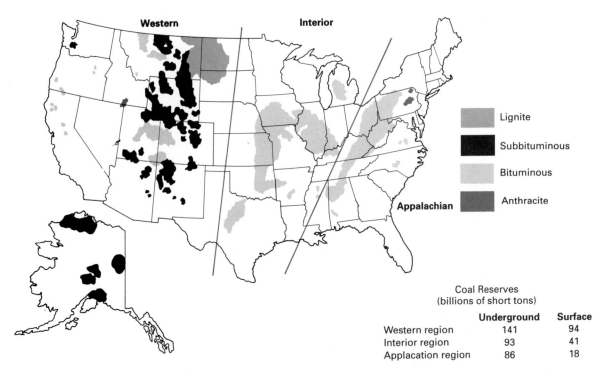

Figure 6.16 **Principal Coal Fields in the United States and Estimated U.S. Coal Reserves by Region as of 1989.** Note that the coal in Montana and Wyoming consists principally of lower-rank (higher ash, lower energy content) bituminous coal.

Coal Reserves
(billions of short tons)

	Underground	Surface
Western region	141	94
Interior region	93	41
Applacation region	86	18

___ Coal may contain radioactive materials.

___ Among the most serious problems with coal are those having to do with human health and safety. Coal mining remains one of the most hazardous of all occupations.

Despite hard work on these problems, we have a long way to go. New pollution control and coal production technologies are under development, but in all likelihood things will get worse before they get better; we will probably increase our reliance on coal faster than these problems can be solved.

Because everything must go somewhere, there will always be problems of one sort or another with coal. For example, a by-product of scrubbing (removing) SO_2 from stack effluents is a calcium sulfate sludge. A 2,000-megawatt power station burning high-sulfur coal will produce about 80,000 cubic feet of sludge per day (Eisenbud, 1979). Particulate recovery yields waste that needs disposing. Both of these forms of waste are potential land and water pollutants because they can contain heavy metals and other toxic and carcinogenic chemicals.

During the throes of the mid-1970s energy crisis, government officials called for a switch from foreign oil to domestic coal. *If all else stays the same,* an increase in dependence on coal would have the effect of making the acid rain problem worse and increas-

ing the health risks to smokers, pregnant women, the elderly, and people with respiratory and cardiac problems. According to an American Public Health Association study in 1979, if power plants were to increase coal generation of energy from 0.8 quadrillion BTUs to 1.1 quadrillion BTUs, SO_2 levels would have risen enough to kill an average of 500 more adults with heart disease and cause 16,000 more respiratory attacks among children under 5 years old each year.

There is still considerable research under way looking for ways to convert coal into other physical forms and to burn it in better ways. These alternatives are not without their own problems. Some of the problems are the same, of course. In order to liquefy coal it has to be mined in ways that have significant negative environmental impacts. New processes will inevitably mean new problems to be solved as well.

Emerging Coal Technology

A 1985 report by the Energy Research Advisory Board of the U.S. Department of Energy listed 15 categories of technologies under development for the clean use of coal. Some of the more promising and developed of these ideas are presented here.

Fluidized Bed Combustion. Among the approaches to improving the way in which coal is used as a source of energy is to convert it into forms

Alpha particle: positively charged particle consisting of two protons and two neutrons spontaneously emitted by certain radioactive materials undergoing decay.

Anthracite: hard coal; coal that burns with less smoke than softer varieties and is therefore more desirable.

Barrel (bbl): liquid measure equal to 42 U.S. gallons commonly used to measure petroleum and oil products.

Beta particle: particle equal in mass to that of an electron (negative charge) or a positron (positive charge) emitted by certain substances undergoing radioactive decay.

Bioconversion: conversion of one form of energy into another by plants or microorganisms. The conversion of solar electromagnetic energy into chemical energy in photosynthesis is a form of bioconversion, as is the conversion of plant mass into ethanol or methanol.

Biomass: literally, living or once-living material, excluding fossil fuels; in connection with energy this term applies to the use of once-living material as fuel. The most common example is wood; however, any dried plant material can be used as fuel.

Bituminous coal: soft coal, somewhat softer than anthracite.

Breeder reactor: a nuclear reactor that produces more fissionable fuel than it consumes.

Chain reaction: a sequence wherein one reaction leads to another. In atomic fission a neutron collides with an unstable nucleus, causing it to undergo fission, which in turn releases one or more additional neutrons, which cause still other fissionable nuclei to undergo fission, and so on.

Coal gasification: the conversion of coal to an energy-rich gas.

Coal liquefaction: the chemical or physical conversion of coal into a liquid fuel.

Continental shelf: the submerged shelf of land sloping away from the edge of a continent; generally defined as an area where the ocean water is less than 200 meters deep.

Core: the central part of a nuclear reactor, where the atomic fuel is located.

Critical mass: the smallest amount of fissionable material necessary to sustain a chain reaction.

Crude oil: petroleum direct from the ground, before it is refined.

Deuterium: hydrogen with an extra neutron in the nucleus; an isotope of hydrogen.

Enrichment: process by which the radioactive material in uranium ore is concentrated to about 3%, the level necessary to sustain a fission reaction.

Fission: splitting of the nuclei of the atoms of certain elements into two lighter nuclei accompanied by the release of relatively large amounts of energy.

Fusion: combining of two atoms into a single atom as a result of a collision. Because the fused mass is a little less than the mass of the two nuclei that collided, the result is the release of the amount of energy that comes from the loss of mass according to Einstein's equation, $E = mc^2$.

Gamma rays: electromagnetic radiation of the same nature as X-rays but of a shorter wavelength. Gamma rays are emitted by certain materials undergoing radioactive decay.

Heat pump: device that moves heat from colder to warmer areas; i.e., it causes heat to flow in the direction opposite to that in which it would spontaneously flow. This can be done only through the expenditure of energy.

In situ: literally, "in place"; the in situ gasification of coal, for example, amounts to the conversion of coal into gas as it sits in the original coal seam.

Lignite: very low grade of coal, intermediate in quality (energy content per pound) between peat and bituminous coal.

Liquid-metal-cooled fast breeder reactor (LMFBR): type of breeder reactor in which the coolant for the reactor core is molten sodium.

Methanol: methyl alcohol or wood alcohol (CH_3OH), the chief alcohol derived from the destructive distillation of wood. Methyl alcohol can be burned as fuel.

Neutron: subatomic particle having a mass equal to that of a proton but having no charge. When a nucleus disintegrates (fission), the resultant free neutrons can collide with other nuclei, causing them to undergo fission, provided that the neutron is moving fast enough. Neutrons are able to crash into nuclei because they are not repelled by the positive charge of the nucleus.

Nuclear fission reactor: any device in which a fission chain reaction can be started, maintained, and controlled by the regulation of dampening rods set into the core of the reactor. A fission reactor gives off heat, converting water into steam, which then drives generators.

Oil shale: sedimentary rock (not really a shale) containing the solid hydrocarbon kerogen. Processing of oil shale converts it into a gas or a liquid fuel.

Passive solar heating: heating that does not require special machines or devices other than structural features that allow sunlight in to be absorbed.

Photovoltaic conversion: direct conversion of solar radiation into electricity.

Plutonium: chemical element, all of the isotopes of which are radioactive. Plutonium 239, the most important isotope of plutonium, is an alpha particle emitter. Plutonium 239 is a fissionable material that can be made by bombarding nonfissile uranium 238 with neutrons; this is what is done in a breeder reactor.

Proven reserves: a relatively accurate estimate of the amount of a resource that can be obtained from the earth with existing technology and sold for a profit.

Rad: a delivered (absorbed) unit of radiation energy equal to 100 ergs per gram of tissue.

Refining: separation and conversion of the complex mixture of hydrocarbons in crude oil into its component fractions (for example, gasoline and heating oil).

Relative biological effectiveness (RBE): term used to compare the effectiveness of different kinds of radiation to the effectiveness of beta particles. Beta particles have an RBE of 1; alpha particles have RBEs of 10 to 20; neutron beams have RBEs of 2 to 10.

Rem: *radiation equivalent man or mammal* A rad adjusted for relative biological effectiveness (RBE): a rem is a rad times the RBE of the radiation in question. Thus 1 rem of any kind of radiation does an equivalent amount of biological damage as 1 rem of any other kind.

Resource: any naturally occurring entity, material, or space that is useful to human beings.

Secondary recovery: the extraction of oil and gas by other than spontaneous means (for example, by pumping).

Slow neutron: a "thermal" neutron with enough energy to cause unstable (fissionable) nuclei to undergo fission. Stable nuclei can be made to undergo fission only by impact with fast neutrons. The kinetic energy of a slow neutron is about equal to the kinetic energy of atoms and molecules of air at normal room temperature.

Thorium: radioactive chemical element of atomic number 90, useful as a nuclear reactor fuel. Thorium 232 is used in breeder reactors because when it captures or absorbs a slow neutron, it decays into fissionable uranium 233.

Tritium: hydrogen with two extra neutrons in the nucleus; an unstable (radioactive) isotope of hydrogen.

Uranium: radioactive element with the atomic number of 92 occurring in natural deposits throughout the world.

Roof-support technologies like this flexible roof drill improve the efficiency of coal production in underground mines. Previous technology required that pillars of coal be left to support the roof.

that can be burned more efficiently. One such method under advanced testing now is fluidized bed combustion (Figure 6.17). Fluidized bed combustion is designed both to enhance combustion efficiency and to reduce sulfur and nitrogen oxide emissions. Sulfur is absorbed by the limestone mixed with the coal in this process, and lower combustion temperatures tend to produce much less oxides of nitrogen. The major problem with this approach is that it requires the disposal of relatively large amounts of solid waste or ash.

Coal Gasification. According to various sources, oil and gas production from coal will be commercially available sometime during the 1990s. The objective of **coal gasification** is to produce a fuel from coal that can be used like natural gas in the same pipes and the same appliances. This is not a new idea. Congress appropriated $30 million in the mid-1940s to explore ways to convert coal into other forms of fuel, and coal gasification technologies have been known for some time. The main problem is that coal gasification is expensive and relatively ineffi-

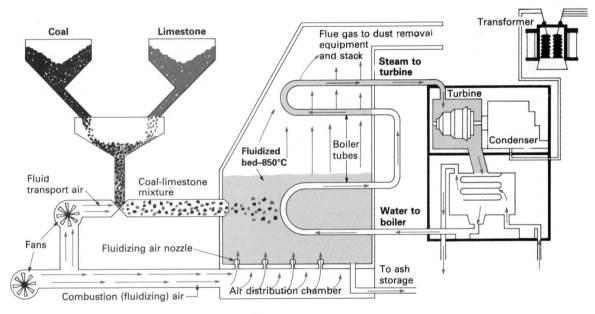

Figure 6.17 Design of a Fluidized Bed Boiler. A fluidized bed boiler is designed to minimize the pollution problems from coal relatively high in sulfur and coal that tends to slag. A number of such designs are being tested under contract with the U.S. Department of Energy.

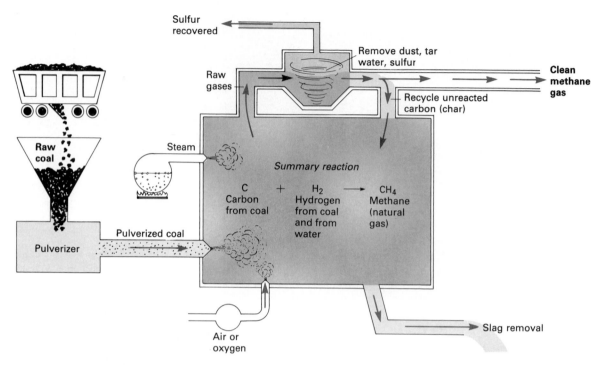

Figure 6.18   Schematic of a Process for Converting Coal into a Gas Similar to Natural Gas. This process is about 75% efficient. With recovery of the heat generated by gasification, efficiency can exceed 96%.

cient. At present there are limitations as to the kind and sizes of coal chunks that can be gasified. Such problems are all subjects of ongoing research. Nevertheless, various gasification technologies had reached the demonstration phase toward the end of the 1970s. The basic method of converting coal into gas is illustrated in Figure 6.18.

Underground or In Situ Gasification of Coal.
A unique and unusual type of energy extraction from coal is underground coal gasification. This process was actually first explored in the early 1930s in the Soviet Union. Wells are sunk into a coal seam some distance apart, the coal at the base of one well is ignited, and air is forced down one well through the coal seam into a second well. As burning takes place, a low-energy gas comes out of the well that can be captured, stored, and used.

There are problems with **in situ** (in-place) coal gasification. We do not know much about its potential for contamination of groundwater, for example. Once coal is burned, the coal seam can collapse; when this happens, water could seep in, or noxious gases could escape. Another problem is that some coal beds are not permeable enough to maintain combustion and to allow the flow of gas through the coal and up the production well. There is apparently very little information available concerning the possibility of gas leakage from such processes.

Gasification Coupled with Power Generation.
In May 1984 a combined-cycle coal gasification plant began operation near Barstow, California. This 100-megawatt demonstration plant combines gasification, desulfurization, and power generation (see Figure 6.19) in such a way that the generation of electrical power is both clean and efficient. Gasification allows sulfur to be removed, and the integration of gasification and steam generation eliminates any energy penalty (loss) otherwise associated with gasification by capturing the heat given off by the gasifier (see Balzhiser and Yeager, 1987).

Coal Liquefaction.
Coal liquefaction is intended to generate a product that can serve in lieu of petroleum. There are a number of improved processes being studied now by which coal can be liquefied. One of the key features of liquefaction is that hydrogen is used in a process called hydrodesulfurization, whereby hydrogen is combined with the sulfur and then extracted, leaving a relatively sulfur-free, low-ash synthetic fuel oil. The cost of oil derived from coal is projected to be high (although it has recently been projected to be as low as $35 per barrel; see Lumpkin, 1988) and thus may have difficulty competing with other types of liquid fuel now and in the immediate future. It is unlikely that synthetic liquid fuels will make much of a contribution to our energy supply until after the year 2000.

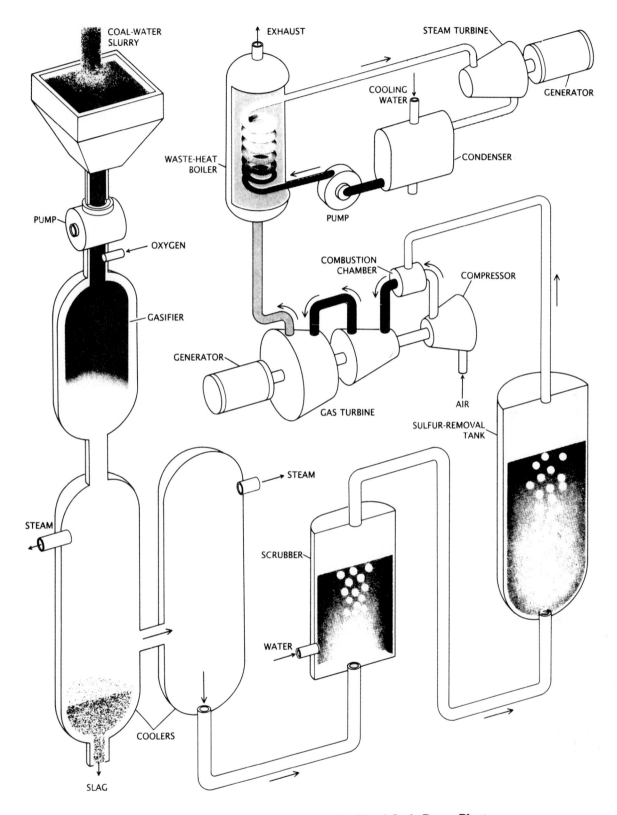

Figure 6.19 Schematic of an Integrated Combined-Cycle Power Plant.
By integrating gasification and power generation in this way, the use of energy is
more efficient than in stand-alone gasification, and there is less air pollution.

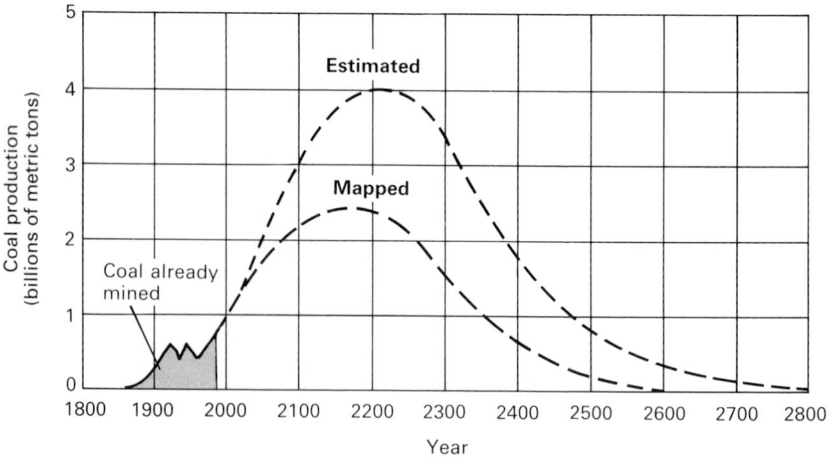

Figure 6.20 Two Estimates for the U.S. Production of Coal over the Next Several Hundred Years. This assumes production of 1 billion metric tons per year (1982 rate = 0.748 billion metric ton). The upper curve is based on an estimate of the amount of coal remaining in the United States, the lower curve on deposits already mapped. Each square of the grid represents 100 billion metric tons.

Among the environmental problems associated with coal liquefaction is the fact that toxic and carcinogenic compounds are generated as by-products of the production of coal-derived liquids (and gaseous fuels). Carcinogens have been measured in coal liquefaction process streams in pilot plants. There is some direct evidence, too. At a coal hydrogenation (liquefaction) plant that operated in West Virginia from 1953 to 1959, the incidence of skin cancer among workers was reported to be 20 times that of other white males in the United States (Morris et al., 1979). For all forms of conversion there would be a shift in solid waste generation from the site of energy release to processing and conversion plants, which would presumably be located near the mines. An environmental advantage to conversion is that sulfur can be removed somewhat more economically in conversion processes than in the cleaning or burning of coal.

The Synfuels Corporation. Recognizing that coal is America's most plentiful fuel resource but that liquid and gaseous fuels are generally more desirable, the U.S. government under the Carter administration sought to stimulate the development of "synfuels" technology for coal conversion and other energy resource development technologies. A federal Synfuels Corporation began operation in 1981 with the purpose of bankrolling innovative projects to be designed, proposed, and carried out by private industry. By 1984 this venture had supported only a few coal gasification projects and an oil shale project. A combination of erratic political winds and the oil glut that began before the synfuels program could get started apparently resulted in a mutual loss of intense interest in synfuels by both the federal government and industry.

Clean Coal Technology Program. A U.S. national program to demonstrate environmentally acceptable clean coal technologies was established dur-

ing the Reagan administration. This now supports a number of projects throughout the United States. It emphasizes near-commercial-scale demonstrations of innovative technologies for the efficient uses of coal as a fuel. The program supports an atmospheric fluidized bed combustion project in Colorado said to be operating at 35% energy efficiency and thus bringing about a 10% reduction in CO_2 emissions per unit of useful energy. It also supports projects such as pressurized fluidized bed combustion combined-cycle power systems in Ohio, West Virginia, and Wisconsin and an integrated gasification combined-cycle power system in several locations, all of which should reduce CO_2 emissions per unit of useful energy considerably more.

The Future of Coal

The future of coal is bright only because it is really the only fossil fuel resource that can contribute significantly to energy supplies in the long-range future. Worldwide there are sufficient coal reserves available to last for centuries. Despite the numerous unsolved environmental problems associated with its use, coal offers the United States its only chance for achieving energy independence in the near future before more environmentally sound alternatives come of age in the more distant future. Figure 6.20 illustrates how far we have come and how far we may progress with the coal resources in the United States, given estimated and mapped deposits.

Petroleum

Reserves

The bad news is that estimated crude oil reserves for the world are not large. Their magnitude as of the end of the 1980s is illustrated in Figure 6.21. The immediately striking thing about these data is that the Middle East has the greatest reserve by far.

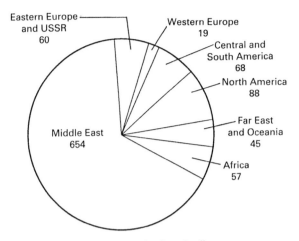

Total = 991 billion barrels of crude oil

Figure 6.21   1989 Estimated Crude Oil Reserves for the World by Region. U.S. petroleum reserves (27 billion barrels) amount to 3% of the world's total.

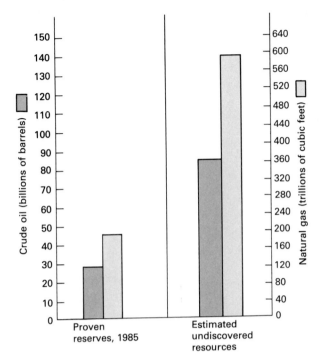

Figure 6.22   Projections of Oil and Natural Gas Resources. Recoverable conventional oil resources in the United States that are still undiscovered are estimated at 64 to 105 billion barrels; the mean estimate is 83 billion barrels. Natural gas resources are estimated at 475 to 740 trillion cubic feet.

The really important fact, however, is that we are nearing the peak of oil production worldwide. At current rates of utilization, we will come close to having consumed 80% of what is now estimated as the earth's total recoverable reserves in just one lifetime. The fact that the annual world output of oil is about 22 billion barrels (an amount equal to the total reserve of Libya) gives an idea of how fast we are reaching the "bottom of the barrel." A country-by-country account of the ratio of reserves to annual production is given in Table 6.8.

There is some disagreement about the years remaining until the United States exhausts its domestic supplies of oil. Depending on the estimator, the U.S. might run out shortly after the year 2000, or supplies could last another century. In any case, these esti-

mates, combined with data on exploratory wells being drilled and the number of these producing, point to the inescapable conclusion that U.S. supplies of oil and natural gas are rapidly diminishing (see Figure 6.22).

In the meantime, the fact that demand is expected to continue to outstrip supply by some considerable measure in the United States has serious economic implications because the difference will have to be made up by imports. Although some significant progress was made in reducing oil imports from the late 1970s to the early 1980s (8 million barrels per day in 1978 down to 5 million barrels in 1983), imports had returned to 8 million barrels per day by the end of the 1980s (see Figure 6.23 and Table 6.4). The decline in imports was a result of conservation induced by the shock of rising prices and the general increase in oil supplies that included domestic oil, which was also spurred by higher prices.

New Oil Fields, on Land and Offshore

Perhaps the most direct approach to increasing oil supplies is to find more. That this approach works is indicated by the fact that the oil field on the north slope of Alaska accounts for one-fourth of U.S. domestic production. Recent attention has turned to the

Table 6.8   Oil Reserves and Reserves-to-Production Ratio for Selected Countries in 1989

Country	Reserves (billions of barrels)	Ratio of Reserves to Production (barrels in reserve per barrel produced each year)
Saudi Arabia*	255	142
Soviet Union	59	15
Iran*	93	85
Iraq*	100	91
United States	27	9
Venezuela*	58	83
China	24	22
Nigeria*	16	23
United Kingdom	5	7

* OPEC members.

Source: Energy Information Administration, U.S. Department of Energy.

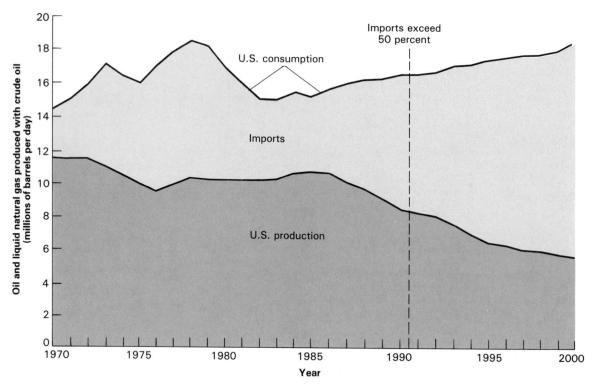

Figure 6.23 **Recent and Projected U.S. Oil Consumption and Production Trends.**

continental shelves off the Atlantic, Pacific, and Gulf coasts. In 1990, offshore oil amounted to about 15% of domestic production. This proportion is expected to rise in the future as a result of much increased offshore exploration.

The U.S. Department of the Interior estimated in 1981 that there might be as much as 27 billion barrels of oil and 163 trillion cubic feet of natural gas

To transport the oil from the Alaskan fields to major refineries, a pipeline thousands of miles long was constructed.

undiscovered in offshore deposits. But over the next few years most of the drilling off the Atlantic coast and Alaska yielded little. In 1985 the department slashed its estimates to 12.2 billion barrels of oil and 90 trillion cubic feet of gas.

In 1981, when the Department of the Interior announced a five-year-program of leasing almost the entire continental shelf for oil and gas exploration, there was much opposition to this move by environmental interests based mainly on the threat of environmentally damaging leaks and spills associated with exploration and production (see Chapter 12). The advocates of offshore oil and gas exploration carried the day, however, their main arguments being (1) that exploration impacts have been well studied and found safe and (2) that America needed the energy. The Bureau of Land Management within the Department of the Interior followed suit and opened up as many acres to oil and gas exploration in 1982 as were offered in the entire period from 1977 to 1980. A similar pattern was followed in coal and geothermal leasing.

Enhanced Recovery

Another way of increasing domestic oil supplies is to enhance the percentage of oil recovered from present oil fields. When oil wells are mentioned, gushers come to mind and gushers demonstrate that oil deposits are under pressure. If oil deposits are

Part Two   *Homo sapiens* in the Scheme of Natural Things

tapped in the right way, this pressure is enough to lift 35% to 45% of the oil out of the ground into storage tanks. Natural pressure derives from gaseous or water pressure (see Figure 6.24); oil recovery as a result of these natural forces is called **primary recovery.** It is often necessary to use pumps to get the remaining oil out of a deposit as cumulative withdrawals re-

duce the natural pressure. If natural forces have to be helped along by pumping or by injecting water or air into oil wells, this is referred to as **secondary recovery.**

Primary and secondary recovery combined often leave 40% to as much as 80% of the oil in the ground. Enhanced or **tertiary recovery** methods such as

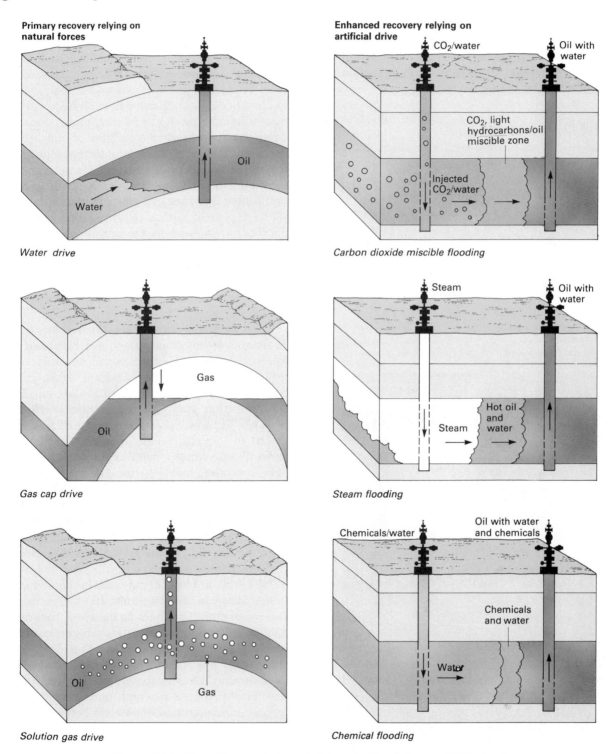

Primary recovery relying on natural forces

Water drive

Gas cap drive

Solution gas drive

Enhanced recovery relying on artificial drive

Carbon dioxide miscible flooding

Steam flooding

Chemical flooding

Figure 6.24 Three Forms of Natural Drive and Three Forms of Enhanced Recovery of Oil.

those illustrated in Figure 6.24 are being used to get at the oil that is left by conventional (primary and secondary) recovery methods. In the steam-flooding or thermal recovery method, for example, oil is heated so that it will flow out of porous rock formations more easily and is then moved out by the steam pressure. Still another method is to burn some oil in situ. In the latter method, a fire is started under one well, which then reduces the viscosity of adjacent oil, allowing it to be more easily pushed into a production well. As would be expected, oil recovered via tertiary methods is generally more costly to produce than oil obtained through primary recovery.

Oil Shale and Oil Sands

Locked in oil-bearing rock formations within a relatively small area of the high desert in Utah, Wyoming, and Colorado is an estimated 2 trillion barrels of a solid hydrocarbon equivalent of oil called *kerogen*. Deposits are found in some of the eastern states as well. The product is not oil, and the rock in which it resides is not shale, but the term *oil shale* has stuck. If kerogen can be economically melted and extracted, it will represent a resource 50 times

The Paraho Oil Shale Facility at Anvil Points, Colorado, was the site of a research project that looked at optimal retorting conditions and environmental effects of oil shale processing.

greater than the present U.S. oil reserve (Maugh, 1977).

Research is under way to find ways of getting oil from shale. One is to extract the oil from the shale in an aboveground process using chemical and physical methods. The other is an in situ approach, which may turn out to be more economically feasible. There are a number of environmental concerns. Aboveground extraction can be done by strip mining or other forms of extraction and will tend to cause water pollution and to disturb land directly. The disposal of spent shale will present other forms of the same problems. Finding a place to put spent shale will be a problem because once stripped of its hydrocarbons in processing, the shale will take up more room than it did originally because of a swelling or popcorn effect. There will also be the potential for gaseous and particulate emissions from mining, drilling, preparation, and handling of oil shale at the site of processing. No matter how it is done, the extraction of oil from shale will be a process with a relatively high capital cost, and there are a number of uncertainties about operating costs because of a need to deal with as yet unknown environmental impacts. There is also the possibility that carcinogenic and mutagenic materials will be included in the effluents from oil shale operations.

Oil sands (or **tar sands**) are deposits made up of mixtures of clay, sand, water, and **bitumen** (a tar-like, very heavy oil). Bitumen can be removed from surface-mined oil sands by steam heating, which makes the bitumen fluid enough to float out. It must then be further purified before it can be refined like crude oil. The world's largest deposits of oil sands are located in Canada, South America, and the former Soviet Union; smaller deposits are known to exist in other countries, including the United States, where most of the deposits occur in Utah. It is difficult and energy-costly to get oil out of oil sands. However, new production techniques are under development. These include new mining techniques for near-surface deposits, new drilling techniques, and new thermal production techniques.

Overall, in terms of potential for liquid fuels, the United States is fairly well off. The nation has huge reserves of coal that could be liquefied, vast deposits of oil shale, and some oil sands. The estimated total liquid fuel resource for the United States is 625 billion barrels—far in excess of the present oil reserves of Saudi Arabia. Despite this potential, very little was spent on liquid fuels research during the 1980s, the Reagan administration apparently believing that the major oil companies would do the research when the time was right. Many technical and environmental problems will have to be solved before much of the U.S. liquid energy potential can be realized.

The Price of Oil. The price of oil will have a great deal of impact on just how fast new reserves are discovered and alternative liquid fuels will come on line. The problem is that oil prices are unpredictable and tend not to be steady enough to drive new developments. In *Assumptions for the Annual Energy Outlook 1990,* the U.S. Department of Energy estimated that oil prices during 1990 would probably be $16.80 per barrel, ranging from a low of $14.40 to a high of $19.20. But the invasion of Kuwait caused prices to exceed $40.00 per barrel that year. The U.S. Energy Information Administration's best estimate was that oil would not reach $36.90 per barrel until the year 2010.

Natural Gas

Worldwide prospects for natural gas are being touted by many experts as much better than they were a decade ago. There have been improvements in exploration and production technologies and we are now going after gas as a primary target rather than as a by-product of oil production as was the case in the past. Still, natural gas, with emphasis on the word *natural,* apparently does not have much of a long-range future. Proven natural gas reserves are not very large on a worldwide scale (Figure 6.25). At present rates of consumption (approximately 70 trillion cubic feet per year), these reserves will last less than 60 years. Although the experts disagree on just how much natural gas remains to be discovered worldwide (see Masters and colleagues, 1991), the most common estimates indicate that we will eventually discover another half-century supply, again assuming current rates of consumption. Increased rates of consumption (which *are* likely) will, of course cut the lifespan of natural gas sharply.

Although natural gas production in the United States began to go up again in the late 1980s—and is expected to continue to rise until 2005—production will not likely again reach the peak production levels of the early 1970s. In other words the peak of natural gas production in the United States has passed. Each year the United States consumes about one-tenth of its proven reserves. The U.S. Geological Survey estimates that undiscovered natural gas resources will amount to about 60 percent of what has been produced already (see Figure 6.22). As supplies diminish, gas becomes harder and more expensive to find and produce. Barring any unexpected new sources, prices of natural gas are likely to move upward. America must now rely more and more on imported natural gas and on research aimed at producing pipeline-quality gas from other fuels such as coal.

It should be pointed out that the U.S. has some very large deposits of natural gas that are uneconomical to tap at present. Some natural gas and oil deposits, for instance, are held in sponge-like sand and porous rock formations. The gas can be held so securely in these "tight sand formations" that it does not flow out easily. Experiments have been performed, some even using nuclear explosives in the attempt to break up such formations, but so far these gas deposits remain untapped.

Energy from the Sun

History

Perhaps no form of energy has more inherent appeal than solar energy. Broadly defined, solar energy would include—in addition to direct radiant energy—wood, vegetation, heated surface water, and wind power. In this discussion we use a narrow definition that includes only direct solar energy. The fact that solar energy is delivered to the earth every day, whether we tap it or not, promises that we would not have to do much extra environmental manipulation to use this form of energy.

Archimedes was said to have used solar reflecting mirrors more than 2,000 years ago to set fire to ships in a Roman fleet. In the mid-1880s a French scientist developed the first known solar device able to convert water into steam to power an engine. There were some rather surprising solar technological developments in the early twentieth century. Tens of thousands of solar water heaters were sold and used in California and Florida in the early 1900s, some of them reportedly still in operation. Bell Telephone Laboratories came up with a method for converting solar energy into electrical energy nearly 30

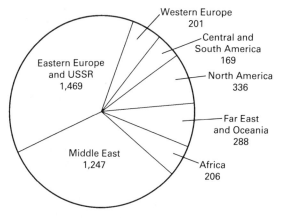

Total = 3,917 trillion cubic feet

Figure 6.25 World Natural Gas Reserves 1989. World annual marketed production is approximately 65 trillion cubic feet; U.S. production in 1989 was approximately 17 trillion cubic feet.

years ago. The world's current energy problems provide new incentives to further developments in solar technology.

Potential

Every day the sun delivers about 1400 BTUs of energy to every square foot (929 cm^2) of land in the United States. In an absolute sense we get far more solar energy than we could possibly ever use. The solar energy received by the earth in one day would take care of the world's energy requirements for 30 years at present rates of energy consumption. This dramatizes the potential of solar energy, but many factors—technological, economic, and others—stand in the way of realizing even a small fraction of this potential.

Space Heating and Water Heating

All that is needed to convert sunlight into heat in a confined space is for the sunlight to be able to pass through transparent panes and be absorbed. This simple technology is now relatively well advanced, and there are some interesting possibilities for extended application. The conversion of sunlight to heat is a straightforward, efficient, direct way of using solar energy. In terms of transmission, direct solar energy is already delivered to exactly where it is needed—everywhere—although in different amounts and, of course, not always *when* it is needed.

The solar collector can be as simple as a double layer of glass that admits light to an absorber plate painted black to maximize absorption. Air heated in such a collector can be circulated to heat a building directly, or fluid can be pumped through collectors to

Solar power provides hot water and heat for this home in Shenandoah, Georgia.

carry heat to storage depots, hot water tanks, and radiators (Figure 6.26).

There is no question that we already have the engineering capabilities to design, install, and use solar space and water heating equipment. Solar water heating is already economically competitive with electrical water heating in many parts of the United States. A research firm in New England tentatively concluded that solar water heating is really not yet commercially competitive with other forms of energy in northern states. In this study, several hundred families participated by keeping track of the energy they saved by using solar water-heating boosters. They saved 17% of the energy that normally would go into

Figure 6.26 System for Heating the Space and Water in a Home Using Sunlight.

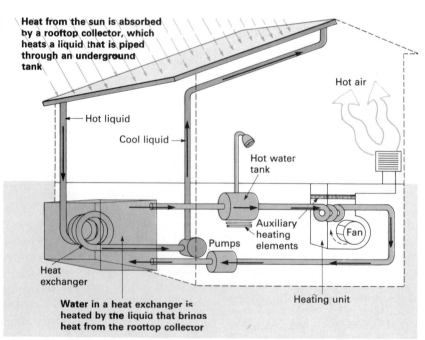

Heat from the sun is absorbed by a rooftop collector, which heats a liquid that is piped through an underground tank

Hot liquid

Cool liquid

Hot air

Hot water tank

Auxiliary heating elements

Pumps

Fan

Heat exchanger

Heating unit

Water in a heat exchanger is heated by the liquid that brings heat from the rooftop collector

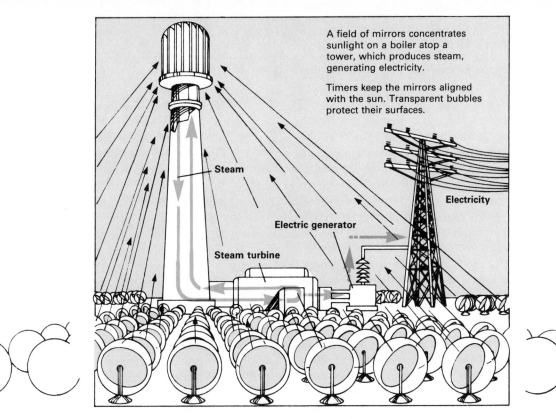

A field of mirrors concentrates sunlight on a boiler atop a tower, which produces steam, generating electricity.

Timers keep the mirrors aligned with the sun. Transparent bubbles protect their surfaces.

Electricity

Steam

Electric generator

Steam turbine

Figure 6.27 A Solar Power Plant Employing a Field of Heliostats. Each heliostat consists of a mirrorlike reflector and a control system. The controls continuously focus reflected solar rays onto a central boiler mounted on a tower. Fluid heated in the tower is then expanded through a turbine, producing electricity.

heating water, but the saving was not high enough to offset the capital construction cost in a reasonable period of time. This has already begun to change as more conventional energy sources have become more expensive and as solar capital construction becomes more standard and less expensive.

The major problem standing in the way of improvements much beyond simple **passive solar space heating** is the economics of capitalization, that is, the costs of installation and the costs of equipment and structural modifications. It will take more time to close the cost differential between solar energy capitalization and the burning of fossil fuels.

Although simple passive solar space heating can be achieved at little or no cost through appropriate designs of windows and roof overhangs, tree planting, and the opening and closing of drapes, more elaborate systems can be costly. Solar space and water heating systems such as the one illustrated in Figure 6.26 can require structures covering up to the equivalent of half the floor space of a home and cost tens of thousands of dollars. Surveys have shown, however, that many people would be willing to establish such a system if it could be shown that the result would be sharply reduced fuel bills and consequent long-term savings. As prices of fossil fuels increase, the prospects for the realization of these conditions increase along with them. The potential benefit from

solar heating is significant; the overall energy used for space heating and water heating is high; it constitutes 20% of the U.S. energy budget. Obviously, even if solar energy technology were limited to heating, it would be well worth pursuing. A high priority should be assigned to research into the design and technology of solar heating, particularly in the construction designs of office buildings and homes.

Shipments of solar collectors used primarily for heating swimming pools in the United States reached a peak of 12 million square feet in 1980 and then declined to 3.2 million square feet in 1987. Shipments of medium temperature collectors used primarily for heating water peaked at 12 million square feet in 1983, but after the expiration of the federal tax credit for such installations, shipments fell to 1 million square feet in 1987. Shipments of photovoltaic modules reached 10,000 peak kilowatts in 1988.

Conversion of Solar Energy into Electricity via Steam

An indirect application of solar energy would be to convert it into electricity. This would in effect feed solar power into existing electrical delivery and distribution systems or power grids throughout the United States and the world. One way of generating electricity from sunlight is illustrated in Figure 6.27. In this system the sun's heat is focused on a boiler, causing water to be converted into steam used to drive a turbine. Among the problems with this type of system is that it requires a several-phase conversion of one form of energy into another, and this of course means reduced efficiency. Still, because the source of energy is free, it has been estimated that **heliostats** like those in Figure 6.27 will be economically reasonable ways to capture and use solar energy. Estimated that the capital cost for construction of such facilities (with mass production of components and systems) would be comparable to the cost of more conventional means of generating electricity. The world's first solar plant like the one illustrated in Figure 6.27 opened near Barstow, California, in 1982. "Solar One," as this 10-megawatt plant is called, is now one of many solar receivers undergoing tests throughout the world.

Photovoltaic Conversion

As we have already pointed out, Bell Laboratories developed a method of converting solar energy into electrical energy more than three decades ago. This is known as **photovoltaic conversion.** Hundreds of U.S. and Soviet space vehicles and satellites have been powered by solar energy since the late 1950s. One of the very first large photovoltaic arrays, one that generated 10,000 watts of electricity, was used to power the U.S. Skylab space station in 1973.

Among the problems with generating electricity with silicon solar cells is that due to the low efficiency of present technology—30% at best—it takes a large surface area to generate the electricity needed for a single household. This, coupled with the facts that the solar cells are still expensive (although the price of photovoltaic power has dropped considerably) and that they have to be kept immaculately clean, makes large-scale implementation of this system unlikely in the very near future. Efficiency of conversion is apparently the key.

Over the next few decades, photovoltaics will become an increasingly important source of electricity. Solar-powered watches, patio walkway lights, and calculators are already well established in the United States.

Local Versus Central Conversion

Because sunlight reaches everywhere, it somehow seems perverse to collect and focus it and then ship the energy through transmission lines or some other means to places where it is to be used. We look for solar technologic developments with emphasis on systems that are not central; we look for small solar conversion units that capture and make use of energy for homes, small businesses, and office buildings directly on site. An appealing feature of this decentralized concept is that it would mean a much more stable situation; individuals would not have to depend on the

integrity of power grids that serve large expanses of territory.

The Problem of Storage and Other Research Subjects

A key research strategy for photovoltaic conversion is finding more selective and better surfaces for absorbing solar radiation and converting it into electricity. For solar space heating and water heating, the main technological problems remaining have to do with the development of better systems of storage and improvements in the architectural design of buildings to allow greater and greater utilization of this form of energy.

A common problem for all of the many applications of solar energy is what to do in cloudy periods, at times of the year when there is less sunlight, and at night. We look for significant future advances in the area of energy storage (storage of heat energy, of electrical energy, etc.). On the distant horizon, perhaps we can overcome the problem of cloud cover or night with a system of solar energy collectors in orbit that beam microwave radiation to the earth. These could be placed so as to provide a continuous supply of energy (see Figure 6.28).

Assessment

A study made by the U.S. National Science Foundation and the National Aeronautics and Space Administration in the early 1970s reached the following conclusions:

1. Enough solar radiation is received in the United States to make a major contribution to future heat and power needs.
2. There are no insurmountable technological barriers to wider application of solar energy.
3. Although solar energy is currently more expensive than some conventional types of fuel, increasing costs of other fuels will almost certainly make solar energy economically competitive in the not too distant future.
4. If developed correctly, solar energy would be available to the general public around 1980 (it was, to a limited extent).
5. Even very large scale use of solar energy would have a minimal impact on the environment.

During the 1970s the U.S. Council on Environmental Quality estimated that, given problems and the extent to which we will be able to cope with them in the near future, solar energy could have allowed the United States to meet about one-fourth of its total energy needs by the turn of the century (Carter, 1979). Current trends indicate that solar energy will not even come close to that, however; current estimates, includ-

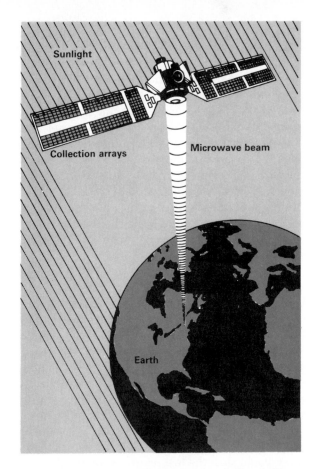

Figure 6.28   A Futuristic Way to Deal with Cloudy Days and Sunless Nights. Satellites orbiting at altitudes of 22,000 miles or more, fixed over the same point on the earth's surface, could be designed to convert sunlight into microwaves, which would then be beamed to the earth and converted into electricity.

ing those by the U.S. Department of Energy, are pessimistic. A study reported by the U.S. Office of Science and Technology Policy projected that it would take 30 to 50 years before solar sources would provide even 10% of U.S. *electricity*. Solar energy apparently will not have a major impact on our energy problem for generations. Progress is being made, however, and will likely continue at a brisk pace.

Atomic Energy

Fission

Figure 6.29 illustrates what happens when the nucleus of an atom of uranium 235 (^{235}U) absorbs a free neutron. The splitting of the uranium nucleus is called **nuclear fission,** and a number of the things produced as a result of fission are of interest to us in this book. In a later chapter we will discuss radiation emissions (alpha particles, beta particles, and gamma rays) that result from the decay of fission products and how these rays can cause radiation sickness, cancer, and birth defects. The heat energy released can be used to boil water to produce steam to drive turbines; this is the part of the fission process of most interest to us in this chapter. The fission reaction also produces free slow neutrons, which, if they happen to be absorbed by another nucleus of ^{235}U, will cause another nucleus to undergo fission, repeating the whole process. If enough uranium is present at a high enough density, a chain reaction can be generated, producing

Figure 6.29   The Products of Nuclear Fission.

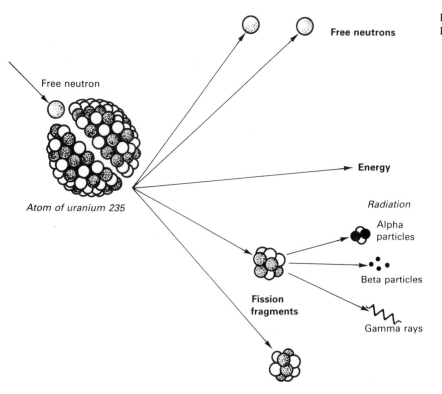

an explosive release of energy such as in an atomic bomb. At lower densities, such reactions can be controlled—made to proceed at much slower rates—as is done in a nuclear reactor.

How Does a Nuclear Power Reactor Work?

The basic design of a nuclear fission reactor is almost identical to that of any steam boiler. In a nuclear reactor the heat from fission is transmitted to water, causing it to be converted to steam, which is then used to drive turbines and generate electricity. A typical nuclear power reactor is illustrated in Figure 6.30. The rate of fission in nuclear reactors is controlled by moving dampening or control rods in or out from between rods of ^{235}U. Control rods are made of materials that absorb neutrons and thus regulate the rate at which neutrons strike and are captured by ^{235}U.

Because the world's ^{235}U reserves are quite limited even in the United States, which has a relatively abundant supply, there is considerable interest in alternatives to simple fission for the extraction of energy from atoms. One such alternative is the breeder reactor.

Breeder Reactors

The **breeder reactor** starts with fissionable fuel, but as it operates, it converts nonfissionable materials into fissionable ones. In a sense—not in the absolute thermodynamic one, of course—breeders produce more fuel than they consume, hence the term *breeder*.

Plutonium 239 (^{239}Pu) is a by-product of the interaction of neutrons with uranium 238 (^{238}U). Although (unlike its isotope, ^{235}U) ^{238}U is not readily fissionable, it can be converted into fissionable plutonium 239. This actually occurs to some extent in conventional fission reactors—reactors in which the principal reaction is the fission of ^{235}U. Breeders are actually special fission reactors in which, by allowing the fission reactor to run "hotter," fast neutrons produce more than one atom of fissionable material for every one consumed.

The vanguard of breeder reactor technology is the **liquid-metal-cooled fast breeder reactor** (LMFBR). In these reactors the breeding of ^{239}Pu is accomplished by arranging a blanket of ^{238}U around the core of a reactor in which fissionable material is undergoing fission. This is done in such a way that 1.2 or more atoms of ^{239}Pu are made from ^{238}U for every atom fissioned in the core. As the ^{239}Pu is made, it can be processed and returned as core fuel, and some can be used to fuel another reactor. LMFBRs must be cooled by liquid sodium rather than by water as in conventional reactors. The main reason is that the water would slow the fast neutrons too much, thereby preventing them from converting ^{238}U into ^{239}Pu. This

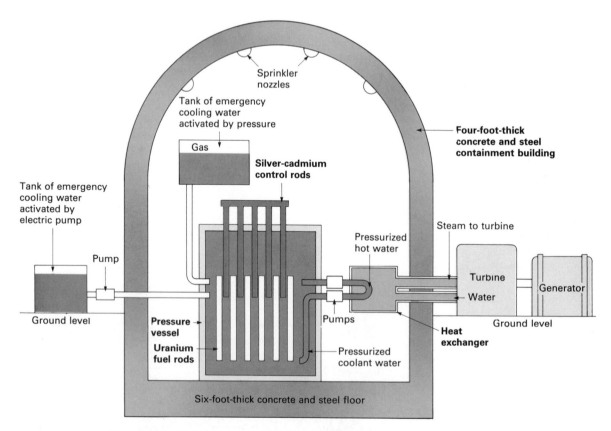

Figure 6.30   Design of a Nuclear Power Reactor.

also makes breeder reactors somewhat more efficient because liquid sodium is a better conductor of heat from the reactor core to the water that is to be converted to steam.

The principal advantage of the breeder reactor over the traditional and currently much more prevalent light water reactor is that there is considerably more ^{238}U in the world than ^{235}U. Actually, nearly all (>99%) of the uranium in uranium ore is in the form of ^{238}U. There are currently some 20,000 tons of "discarded" ^{238}U at Oak Ridge, Tennessee, and at other processing sites. This "waste" (in terms of conventional light water reactor technology) is 70% usable as fission energy in terms of LMFBR technology (see Kulcinski et al., 1979). Because a 1,000-megawatt LMFBR power plant would consume only one ton of uranium annually, that stockpile alone could supply 400 LMFBR plants for some 500 years.

Two experimental breeder reactors are operating in the United States, one in Idaho and one in Washington. Breeder reactors have been reliably producing electricity for a number of years in England and in the former Soviet Union; a French plant that began operation in 1967 ran until 1982, when it was shut down permanently because of an internal leak.

For a variety of technical, political, and economic reasons the breeder programs in both Britain and France have been slowed greatly from earlier projections. The U.S. Congress declined to continue to fund the Clinch River breeder reactor project in Tennessee in 1983, partly because the Tennessee Valley Authority said it did not need the power in the foreseeable future and at least partly because of the decline in demand for uranium as a result of the general demise of the nation's nuclear power program. Technological development continues, and while some experts believe that commercial orders for breeder reactors will be possible by the turn of the century, some economists claim that breeders will not be competitive with conventional light water reactors for another 50 years.

Consensus on the positive side of the breeder reactor is perhaps expressed best by Zaleski (1980): "With breeders . . . the world will have a major and almost inexhaustible source of energy which will allow enough time to find and develop complementary and perhaps better energy sources" (p. 137).

Opposition to the further development of the breeder in the United States runs high because of the security problems associated with plutonium. Plutonium is one essential ingredient in the manufacture of atomic bombs, and it is one of the most toxic substances produced by humankind. The potential of plutonium falling into the hands of terrorists would be greatly expanded by the increased use of breeder technology.

Atomic Fusion

At normal earthly temperatures, if two atoms were on a collision course, their nuclei would not collide because of repulsive forces that would come into play. If atoms can be made to collide in a situation in which they have high thermal or kinetic energy—that is, extremely high energy of motion at very, very high temperatures such as exist on the sun—they can be made to fuse because once the nuclei get close enough, the very strong but short-range attractive *nuclear* forces can then overcome the repulsive *electrical* forces. When this reaction, known as **fusion,** occurs between hydrogen atoms, the sum of the masses that go into the fusion product is larger than the mass of the final product (helium). The difference is converted into energy according to Einstein's equation, $E = mc^2$. (Note that it is the conversion of mass to energy that yields the energy in fission reactions also.) Our sun loses 4 million tons of mass per second in generating energy via atomic fusion. The sun is a fusion reactor.

Although we have been realizing energy benefits from atomic fusion in the sun from day one, harnessing controlled atomic fusion generated here on earth will not be easy. The trick in atomic fusion will be to create and sustain **plasmas,** matter in which the heat is so extremely intense that atoms cannot exist as atoms. Part of the problem is how to *contain* matter at such high temperatures. What on earth could be used as a container for something as hot as a star? Experiments are under way using magnetic fields to contain superhot plasmas in fusion research.

Rafelski and Jones (1987) describe an alternative form of fusion, using short-lived particles called *muons* (which can be generated artificially) to cause the fusion of hydrogen atoms rather than relying on high temperatures. The use of infrared laser beams to raise the temperature of hydrogen to 100 million degrees is described by Craxton, McCrory, and Soures (1986).

It should be pointed out here, perhaps, that there is no danger from superhot plasmas. If these fires did happen to touch the wall of a vessel, they would become extinguished. Unlike stars, the density of fusion-type plasmas is quite low; as a result, they have very little heat content. The problem is not how to keep the plasmas from burning up labs, exploding, and so on; it is how to keep plasmas hot enough.

In 1978, scientists at Princeton University reported that they were able for the first time to produce the temperatures necessary to support sustained fusion. This report was later described as a "media event" because although the temperatures reached were high, the density of the heated plasma was very, very low, and the temperature was sustained for only 15 thousandths of a second. In 1983, however, scientists at MIT achieved the minimum plasma density and confinement needed for the fusion break-even point.

This particle beam fusion accelerator in Albuquerque, New Mexico, is designed to deliver pulses of 100 trillion watts of power. It is perhaps the first machine able to ignite a controlled laboratory fusion reaction. The electrical discharges are a result of the extremely high electromagnetic fields produced during firing.

No controlled fusion reaction has yet broken even—produced as much energy as it has consumed—however the European Community's "Joint European Torus" and Princeton's Tokamak Fusion Test Reactor are moving ever closer (see Kay, 1992). The ultimate development of fusion technology may depend on a joint international effort being mounted—one that could begin operating a prototype fusion reactor during the first decade of the 21st century.

Fusion reactors would be somewhat safer and would have other advantages over fission reactions. The fusion reaction is theoretically very efficient, producing 180 times more energy than it consumes. Furthermore, fusion would have relatively little adverse environmental impact, and plenty of the fuel needed to keep the process going is available. Of the possible products of the fusion of hydrogen, for example, only **tritium,** a radioactive isotope of hydrogen, is produced as a by-product. Tritium emits no gamma rays, and its radiations are easily shielded; its **half-life** (the time required for half of any amount of radioactive material to undergo radioactive decay) is about 12 years. Tritium is dangerous, but even though it may well be very difficult to contain the large amounts of tritium that would be involved, this problem is thought by many experts to be considerably less than that of the radioisotopes produced as by-products of fission.

Although a number of fuels could be used in fusion, a likely fuel for fusion reactors of the future is heavy hydrogen or deuterium. Although it is rare, deuterium occurs as one part in 7,000 of ordinary hydrogen. One gallon of seawater contains the deuterium fusion energy equivalent of several hundred gallons of gasoline.

The advantages of fusion over fission, in summary, are as follows:

1. Fusion would provide a relatively inexhaustible source of cheap fuel.
2. Fusion would generate comparatively less serious radioactive waste disposal problems than fission.

The problems with fusion are as follows:

1. We have not solved the problem of confining the proper reaction mixture (containing 100 billion ions per cubic centimeter for one second) at the right temperature (above 44 million degrees).
2. Fusion reactors will probably be complicated, large structures without application in certain areas like propulsion; it is not likely that there will be fusion reactor atomic submarines, for example.

Kulcinski and colleagues (1979) give a more detailed comparison of the environmental impacts of fission and fusion.

The Trouble with Nuclear Energy

There are problems with the use of nuclear energy; we have touched on some of them already. The problems, all of which have to do with various real and feared impacts of radiation, can conveniently be divided into four categories, maybe five. First, there is the controversy over whether or not routine operation of nuclear power plants adds significant amounts of radioactivity to the general environment. Second, there is the very real problem of what to do with radioactive waste. Third, there is the controversy over the possibility of a core meltdown—a colossal nuclear accident. Fourth, there is the problem of security. How can uranium and plutonium be kept out of the hands of the wrong people? All of these are related to a fifth item: cost. Nuclear power plants have become very expensive to build.

Before we go into each of these problems with nuclear power, let us review some important facts about radiation.

The Effect of Radiation.
In 1927 the Nobel laureate H. J. Muller found that high-energy radiation could damage genetic material—the chromosomes of living organisms. Most of what we have found out since

then about the specific, direct effects of radiation on the human body comes from studies of survivors of the atomic bomb blasts in Japan and studies of Marshall Islanders who were exposed to radiation during atomic tests in the early 1950s. Data about chronic exposure come from X-ray technicians and radiologists who received large doses in the years before the dangers of radiation were recognized. Still other data come from uranium miners. A considerable amount of additional basic information has come from experiments with laboratory animals.

Death can occur in just a few days to a few weeks from a whole-body dose of more than 500 rems. (A **rem** is a unit of biological effect. One rem of any kind of radiation produces the same biological effect as one rad of a certain reference energy-level X-ray.) As few as 300 rems might be fatal for some people.

While we know a lot about acute effects of radiation above the levels of 150 rems, we have less information about the effects of short-term radiation in smaller doses. We do know that doses of radiation below 150 rems can have important genetic effects leading to birth defects and cancer. In a later chapter we will see that radiation may be responsible for some cancer deaths—some of these from radiation used in medicine for medically indicated reasons. A small percentage of cancers are no doubt due to background radiation in the environment. The implication is that if background radiation goes up, the number of cancers induced by radiation will also increase.

Studies of the health effects of extremely small doses of radiation are difficult, if not impossible, to conduct and interpret. As a result, the effects of very low doses of radiation are not known. There are two possibilities: (1) that below some threshold level, genetic damage can be *repaired* before it becomes "expressed," or (2) that genetic damage is directly proportional to radiation all the way down to zero. The latter is called the linear hypothesis and is favored by most experts. Most informed nonexperts believe that if the experts disagree, we should err on the safe side. But as is plainly illustrated in Table 6.9, it is impossible to avoid all radiation whether on, above, or beneath the earth.

The National Council on Radiation Protection has established the "maximum permissible doses" (MPD) for radiation workers and for the general public from artificial sources of radiation other than those related to medicine. The MPDs have been reduced over the years to their current levels of 5 rems per year for radiation workers and 170 millirems (1,000 millirems = 1 rem) per year for the average citizen. The National Academy of Sciences continues to urge that the MPDs be lowered substantially, even though actual exposure of the general public is well below the current limit.

Table 6.9 Background Radiation and Other Chronic Sources of Radiation in the United States

Source	Dose (millirems per year except as noted)
Cosmic rays:	
Sea level	41
Denver (5,000 ft)	70*
Leadville, Colorado area (10,500 ft)	160*
Mount McKinley (20,300 ft)	400*
Commercial jet (35,000 ft)	0.7 millirems per hour
U.S. average from cosmic rays	44
Gamma rays from rocks, soil (U.S. average)	26
Internal radionuclides:	
^{40}K	16
^{14}C, Ra, and decay products	2
Total	18
Grand total U.S. environmental average gonadal dose (Oakley, 1972)	88 ± 11
Human-made sources:	
Fallout (1970)	4
Nuclear power	0.003
Medical diagnostic	72
Medical radiopharmaceuticals	1
Occupational	0.8
Miscellaneous	2

* These illustrate the shielding effect of the earth's atmosphere.
Source: National Academy of Sciences.

Radiation Hazards and Nuclear Energy.
Measurements of radioactivity made in waters and areas surrounding correctly functioning nuclear power plants by state health departments, nuclear power facility operators, and the EPA have in most cases revealed no increase in environmental radiation near these plants. It has been estimated by many sources that, overall, nuclear power contributes insignificantly to background radiation. According to McBride (1978), coal contains radioactive material and radiation doses from coal-burning electrical-generating plants may be greater than those from a properly operated nuclear plant. According to one source, nuclear power installations add no more than 0.0001 millirad per year to background radiation. But there is far from universal agreement on the significance of the radiation "leak" problem.

In 1979, two major National Academy of Sciences studies were made public. The studies indicated that nuclear power would account for some 2,000 cancer deaths by the year 2000 (100 people per year). The projection was not arrived at easily or with complete agreement; 5 of the 16 commission members filed a dissenting opinion that the estimate was far too high.

Nuclear power proponents argue that the danger associated with nuclear power is greatly exaggerated. They argue that overly alarmist negative publicity has given the public a distorted view of the past record.

They claim that even educated people believe that nuclear power kills more people than automobile accidents. In actual fact, the number of deaths, including cancer deaths, related to nuclear power has so far been very small. Bernard Cohen (1983), physics professor at the University of Pittsburgh, points out that the 100 or so accidents in transporting nuclear material, many of which were front page stories, amounted to less than a 1% chance of causing a single death from escaped radiation. He points out that the historical record indicates that the risk from nuclear power to the average American is equivalent to the risk of smoking one cigarette in a lifetime, an overweight person gaining a fraction of one ounce, crossing the street one extra time every three years, or increasing the national speed limit from 55 to 55.003 miles per hour.

Despite the fine points of the controversy, there is clearly reason for concern. The fact that some radiation materials have extremely long half-lives militates in favor of a very careful and extensive consideration of all risk factors before nuclear power is allowed to proliferate beyond its present state. This becomes obvious when one considers the magnitude of the problem of what to do about nuclear waste.

The Chronic Problem of Radioactive Waste. The word chronic has a special meaning in connection with atomic wastes. **Plutonium,** one of the by-products of atomic fission, has a half-life in excess of 24,000 years. What this means is that half of the radioactivity in any given amount of plutonium will remain after 24,000 years; half of the remaining half will be around after another 24,000 years. Where do you put something you want to keep safely tucked away for 100,000 years or longer? Nuclear proponents argue that while it is true that radioactive wastes remain radioactive for thousands to hundreds of thousands of years, most of the harmful high-energy radiation will have been dissipated after a few hundred years. Obviously, even if this point is conceded, the disposal of atomic waste is a serious problem, and despite the fact that waste is currently being generated in significant amounts, the problem has not been solved. (*Significant* is a judgment word, and it refers in this context more to the nature of the waste than to its volume or weight. The U.S. Committee for Energy Awareness points out that the total volume of all of the high-level radioactive waste produced by 80 nuclear power plants up to 1984 was equal to that of a rectangular solid as large as a tennis court 45 feet deep. Most of the radioactive waste we generate is generated by the military.)

Storing spent fuel rods under water at reactor sites is generally thought to be unsatisfactory. The reactor sites are simply not permanent enough. But what then? No matter where waste is put, that place will have to remain a repository for the waste in a relatively undisturbed condition, practically speaking, for all eternity. We know more about the world than to expect that things will stay the same for very long in any one particular place even on a human time scale, let alone on the geologic scales appropriate to radioactive waste. It is not surprising that many American states have put the U.S. Department of Energy on notice that they will not welcome deep repositories or any other kinds of repositories for radioactive wastes within their borders. Between spring 1977 and spring 1980, more than 15 states passed laws that had the effect of tightly controlling or banning the disposal of radioactive waste within their borders.

This "not me" chorus undoubtedly had something to do with the nature of the Nuclear Waste Policy Act passed by both houses of the U.S. Congress in 1982 and signed by Ronald Reagan in January 1983. The bill required the Department of Energy to search the U.S. for the best geologic formations in which to hold nuclear waste. In a series of reviews, the Department of Energy was to take all relevant factors into account and narrow the list of possible locations to three in the western United States and then three more in the East. (The East/West provision was, among other things, a political compromise designed to allow the bill to pass.) After even more detailed studies of the finalists, the president was to choose a single western site first and then, a few years later, a site in the East.

In 1986, when the U.S. energy secretary announced that the three finalists in the West were Hanford, Washington; Yucca Mountain, Nevada; and Deaf Smith, Texas, he indicated that the Department of Energy was suspending the effort to find a site in the East. The announcement enraged western members of Congress, who then took steps to suspend the entire process; nevertheless, in 1987 the site at Yucca Moun-

tain, Nevada, was designated as the first repository for high-level waste.

Opponents of nuclear power have generally expressed unhappiness over the Nuclear Waste Policy Act, claiming (among other things) that safe forms of permanent storage have yet to be defined by the studies of storage techniques going on in the United States, Germany, Sweden, Belgium, Australia, Canada, and elsewhere.

Although the method is not without controversy, some scientists and scientific bodies such as the U.S. National Academy of Sciences have expressed the belief that high-level nuclear waste, blended into a glasslike material in a process called **vitrification,** can safely be disposed in deep, geologically secure depositories thousands of feet underground. The U.S. Nuclear Waste Policy Act of 1982 itself specifically calls for lowering sealed-up waste in a glasslike or ceramic form about 2,500 feet underground into geologic formations such as granite, basalt, or salt. The world's largest nuclear waste processing plant, in Aiken County, South Carolina—a plant that is now in a testing phase and that is expected in 1994 to begin converting nuclear waste into a glassy material—cost about $1 billion to build.

Among other important related developments, a chemical extraction process has been described (see Horwitz, 1987) that removes the most toxic and longest-lived radioactive materials from nuclear wastes. The process may therefore reduce the amount of waste to be dealt with by a factor of 100 to 1,000.

What about Old Reactors? Old reactors present a special kind of atomic waste problem. Reactors have a lifetime of 20 to 40 years; they wear out and become obsolete. Already, 20 plants have been closed in the Western world, 15 in the United States and 5 in western Europe. It is claimed that it will be hundreds or

In 20 to 40 years, when this plant is obsolete, what will society do with it?

even thousands of years before these plants cease being dangerously radioactive. This is a long time to resist the second law of thermodynamics. There are currently only three options—all imperfect—for dealing with closed plants: (1) mothballing (weld them shut and place them under guard), (2) entombment (bury them), and (3) dismantlement (and then do something safe with the hot parts).

The disassembling and burial of the world's first commercial nuclear reactor, the one at Shippingport, Pennsylvania, was completed recently. The reactor was cut up and shipped to Hanford, Washington, for burial at a cost of nearly $100 million. By the year 2000 there will be about 100 inactive nuclear power plants throughout the world. The licenses of about half of the currently operational nuclear power plants in the United States will expire by 2010.

Nuclear Power Plant Accidents. Quite apart from the consideration of low-level leaks of radiation from routinely operating nuclear plants is the specter of the nuclear accident. It must be made clear that when nuclear accidents are discussed, it is *not* an atomic explosion as in a nuclear bomb that is feared. Atomic power plants cannot explode like a bomb. There is some finite possibility of a meltdown, however (see Enrichment Box 6.1).

A U.S. government report published in 1975, commissioned by the Atomic Energy Commission and generally known as the Rasmussen Report (the study was directed by Norman Rasmussen) concluded that there was essentially no chance of an atomic power plant disaster. In January 1979, some two months before the Three Mile Island accident, the Nuclear Regulatory Commission (NRC) declared the Rasmussen Report misleading and unreliable and stated that the risk was probably higher.

In a report released late in 1982 (done by Sandia Labs for the NRC), the worst possible nuclear accident was projected to take more than 100,000 lives and cost $300 billion dollars. The study projected that the possibility of a meltdown and failure of all safety systems was about one in 100,000 reactor years—a 2% chance that such an event would happen before the year 2000. The study went on to point out that the consequences would be much less severe in *most* of the locations where there are functional nuclear power plants. The highest death toll would occur if such an accident occurred at Salem, New Jersey (102,000 early deaths); the greatest damage would occur if the meltdown happened at the Indian Point 3 Reactor on the Hudson River ($314 billion).

In the 1950s the U.S. Senate passed the Price Anderson Act, which exempts facilities engaged in nuclear power generation from liability for accidents and guarantees that the government will pay up to $560 million to help take care of nuclear accidents. This act did little to inspire public confidence in the safety of nuclear power. In late 1983 the NRC recommended to Congress that the limit on the utilities' liability for damages resulting from an accident at a commercial atomic power plant be lifted and that the statute of limitations be increased from 20 to 30 years.

Naturally, since nuclear engineers have long been aware of the possibility of a meltdown, there are multiple safeguards in nuclear power plants to prevent them. There are backup cooling systems; containment buildings are designed to hold fission products in

---

Enrichment Box 6.1

The China Syndrome: A Nuclear Meltdown

A **meltdown** is what happens when a nuclear reactor core cooling system and all backup systems fail. The core overheats to the extent that it melts through the floor of the reactor into the ground and heads (if the reactor is in the United States) in the general direction of China (hence the expression "China syndrome"). Actually, the molten core comes to rest a few meters down as a boiling, seething mass of molten materials. Any water coming into contact with the molten core would be instantaneously converted into contaminated steam, which could carry radioactive gases and particles over a wide area. Radioactivity would cause water molecules to split into hydrogen and oxygen. This mixture could explode, possibly further rupturing the containment building and spreading even more radioactive material around.

The following scenario has been developed and expanded as the likely or possible consequence of a meltdown in the United States.

1. An area perhaps as large as several states would become contaminated.
2. Agricultural or human activity within a several mile radius might be restricted or forbidden altogether for thousands of years.
3. Five hundred thousand people might have to be evacuated from the immediate area; outside this range, another 3.5 million people would have to restrict outdoor activities for a while to keep from receiving high doses of radiation.
4. Several thousand people might die from acute radiation exposure, and as many as 50,000 more might die later from radiation-induced cancer.
5. General panic and a hysterical state of affairs might exist for some time following an accident—perhaps leading to a shutdown of all nuclear power plants and numerous other unforeseen problems.

place; and there are neutron-absorbing materials that can be inserted into the core automatically to turn off the fission reaction.

Still, there have been some accidents. According to a report by Klaus Hopfner in *Nature* (1979), the principal environmental protection groups in Germany (Bundesverband Bürgerinitiativen Umweltschutz) claimed—on the basis of data gleaned from the files of the German Reaction Safety Society—that an incident or accident occurred in German nuclear plants every three days from 1977 to 1979. There have been many relatively minor escapes of radioactivity from plants in the United States and in other countries as well. And, of course, there was Three Mile Island and then Chernobyl.

Three Mile Island

In terms of its negative impact on the development of nuclear power worldwide and in terms of how close it came to unequivocal disaster, Three Mile Island was the worst commercial nuclear accident in U.S. history. On March 28, 1979, at Three Mile Island near Harrisburg, Pennsylvania, the failure of a faulty water pump in a nuclear reactor and then the malfunction of a valve were followed by a chain of further mechanical and human failures. The results were the release of radioactive water from the emergency cooling system into the Susquehanna River; the venting of radioactive steam into the atmosphere; the conversion of water into hydrogen gas, which raised the possibility of an explosion; and, for a time, a very real danger of a reactor meltdown. The accident released some radioactivity that would have resulted in a dose of only about 70 millirems (see Table 6.9) of radiation to an individual at the plant boundary (half the allowable annual exposure to a member of the general public). Some experts claim that there will yet be cancers, genetic defects, and other long-term outcomes. Others have emphasized that the off-site health effects will be virtually negligible. The net effect of the accident in terms of danger to health is still a subject of some controversy.

Some say that the Three Mile Island plant came close to a meltdown. Technically, it was a *partial* meltdown, in that about 1% of the metal cladding that holds the uranium fuel pellets melted.

The cleanup of the damaged reactor at Three Mile Island was still going on 12 years after the accident; the owners were trying to get permission to leave some radioactive material sealed inside for a while longer. The cost of the cleanup has been nearly $500 million to date. There was also a six-year delay in starting up the undamaged, uncontaminated adjacent reactor. (In contrast, it took only six months to restart two contaminated reactors adjacent to the one that melted down at Chernobyl.) No one was seriously over-exposed to radiation, injured, or killed at Three Mile Island, but there were some major consequences.

— President Jimmy Carter appointed an investigative commission, which found that the main problems were bureaucratic complacency regarding departures from standard safety and maintenance procedures and insufficient operator training. The report was critical of the NRC and actually recommended that it be abolished and replaced, a recommendation that was not carried out. The commission made numerous additional recommendations regarding operator training, nuclear plant siting, evacuation plans, and operating procedures, nearly all of which have been implemented.

— We have learned that even nonlethal nuclear accidents are very difficult to deal with.

— The Tennessee Valley Authority ordered a review of the design and modes of operation of all its nuclear facilities.

— There were stepped-up efforts to scrutinize construction and to block the construction of nuclear power plants in Phoenix; Sacramento; Cincinnati; Madison, Indiana, and Hannover, Germany. Some projects were discontinued.

There have been other incidents in the United States since the Three Mile Island accident. In February 1983 the Salem 1 reactor in New Jersey failed to stop a fission reaction when ordered to do so by a safety control system—something that was supposed to happen only once in a million reactor years. This was the *third* record of such an occurrence in commercial nuclear history.

The NRC now has a system for investigating nuclear power plant problems, which has been compared to the "SWAT team" (Special Weapons and Tactics team in police work) approach to solving problems. This system was implemented for the first time in June 1985, when an emergency shutdown occurred at the Davis-Besse reactor on Lake Erie in the state of Ohio.

Chernobyl

Around 1 A.M. on April 26, 1986, explosions and fires at the Chernobyl nuclear power plant in the Ukraine scattered radioactive materials over the adjacent countryside and into the atmosphere. Fires continued to burn and radioactive steam continued to escape for nearly two weeks as radioactive materials spread throughout the world. Although only 31 people died as a direct result of the initial explosions or as a consequence of intense radiation exposure during the first few months, some unknown greater number are expected to die later as a result of delayed radiation effects, including cancer and birth defects. Although most of the escaped radioactive material was short-lived, some radioactive material will remain in soil and water over a wide area of the world for some time. It is impossible to state with certainty just how many people were, and will be, exposed to how much radiation from Chernobyl for how long.

The accident started as a test of the newest of the Chernobyl plant's four reactors. During the test, safety systems were disconnected, and operating procedures were not followed, causing the reactor to become unstable. The fuel rods overheated, ruptured, and turned cooling water into steam so rapidly (and perhaps generating explosive gases) that the 1,000-ton top of the reactor was blown off. Hot radioactive fuel and burning graphite control rods were blown out into the sky.

There was much confusion following the explosion. At first local officials were uncertain of the dimensions of the tragedy they were facing. An adjacent reactor was not shut down for 16 hours following the explosion. Many firefighters, health care workers, and others received high doses of radiation at or near the accident site. The people in Pripyat (a city of nearly 50,000 people a few kilometers downwind of the power plant) and in other towns and villages within 10 kilometers of the plant were evacuated 36 hours after the explosion as radiation reached levels predetermined as requiring evacuation. The evacuation zone was later enlarged to 30 kilometers, and more than 60 small towns and villages were eventually evacuated. Hundreds of thousands of schoolchildren in Kiev (130 kilometers south of Chernobyl) were later sent on summer vacations early. A little more than a year later, 14 villages had already been resettled.

About 30 people died from radiation exposure within the first few months; they and 20 or so others were estimated to have received more than 500 rads. This group, who faced a high risk of death due to bone marrow failure, were to receive bone marrow transplants. Since many of this group also suffered thermal or chemical burns, many died of other complications such as liver failure and infection. Nearly two dozen of this group were eventually given transplants (Gale, 1986). Another 237 people who received between 100 and 500 rads were hospitalized with acute radiation sickness (Gale, 1986; Flavin, 1987). Most of the latter group (90%) are expected to survive the acute effects of radiation, although they will run a greater risk of cancer throughout their lives.

The accident also generated considerable confusion in the rest of the world. The Soviets were silent about the disaster for several days, even after suspicions were aroused in other European countries where abnormally high levels of fallout had been detected. The accident had sent radiation high into the atmosphere, where it was carried away by high-speed winds. Health-threatening levels of radiation were subsequently measured in at least 20 countries, some of these as far as 2,000 miles from Chernobyl.

Although the levels of radiation from Chernobyl that reached many parts of the world were negligible, the range of effects of such a nuclear accident is vast. This is illustrated by the fact that a layer of radioactive cesium today remains deposited in the polar ice sheet in Greenland (Davidson et al., 1987). There, a 10-centimeter-wide band containing measurable levels of radioactive cesium (^{134}Cs and ^{137}Cs) will serve as a stratigraphic marker for a long time to come. We will be able to learn more about long-range transport of materials that get into the troposphere from this marker (see Chapter 10).

At first the greatest danger was from iodine 131 (^{131}I), a radioactive form of iodine that concentrates in the thyroid gland. This threat is relatively short-term because the half-life of ^{131}I is only eight days. Dietary restrictions and the administration of iodine solutions (to block the uptake of ^{131}I) were the principal means of treatment. Most of the continuing long-range effects come from isotopes, such as ^{137}Cs, with a half-life of 30 years; strontium 90, 28 years; and plutonium, 24,000 years.

Various estimates indicate that the amounts of ^{137}Cs released by the accident were one-tenth to one-sixth of the amounts put into the environment by all nuclear testing ever done. Because the accident occurred in a populated area with other populated areas downwind, the Chernobyl ^{137}Cs dose to humans may have been 60% of the dose received by humans from all weapons tests combined (Flavin, 1987). By some estimates, as much as 50 million curies of long-lived radiation entered the environment as a result of the Chernobyl accident—many thousands of times more than in any previous nuclear accident.

To put this into additional perspective, we could point out that in the year following Chernobyl, residents of Kiev would receive an estimated 50% more background radiation than they would have had there been no accident, and people as far away as England would receive 3% to 4% increases over normal background radiation. People in the Ukraine, outside the evacuation zone, will receive an average of 0.5 rem in their lifetimes as a result of the accident—comparable to the extra radiation one would get by moving from Washington, D.C. (near sea level), to Denver (1 mile high) and living there ten years (see Table 6.9).

Precautions were taken throughout Europe after the accident. Restrictions were placed on the import and consumption of fresh vegetables and milk. In some places this was extended to fish, meat, and other foods. By one estimate, some 100 million people had to alter their diets for several months. In retrospect, many of the precautions were unwarranted. Vegetables were destroyed in Italy, for example, that had radiation levels below the safe limit. Many potassium iodide tablets taken throughout Europe probably did more harm than good.

Still, estimates by various agencies of the number of extra cancer deaths to expect over the next 70 years range from a few hundred to 100,000. The most common ballpark estimate was 40,000 extra cancer deaths worldwide (roughly half of these within a few hundred kilometers of the site) within the next 50 years. A doubling of the leukemia rate might be expected in the zone 30 kilometers around the plant for the decade that began in 1988. Such estimates are based on a United Nations benchmark set in 1977: one fatal cancer per 10,000 person-rem of radiation. The U.S. Environmental Protection Agency uses double that rate (see Marshall, 1987, and Anspaugh, Catlin, and Goldman, 1988). Ultimately, as many as 1,000 more children may be born with birth defects and up to 5,000 more with genetic defects.

There are so many variables involved that no one knows for sure just what the extended impacts of Chernobyl will be. The situation will be monitored closely for the next 50 years—and even then it will be hard to pick out the small increases over the normal numbers of cancer and other radiation-related diseases expected during this period. The area around the Chernobyl plant may have to remain uninhabited for many decades because of the long half-lives of the radioisotopes released. The cost of Chernobyl has been estimated at $3 billion, perhaps as high as $5 billion (Flavin, 1987).

Chernobyl has taught us many things and will likely continue to do so. We have learned, for example, that a nuclear accident is not simply a theoretical possibility. We have learned that the world is indeed a small place and that everything in it is connected to everything else. We have learned that the world was not prepared for nuclear emergencies. Radiation spread farther and faster than computer models said it would, and the fallout behaved more erratically than predicted. Our models will now have to be readjusted. We learned that radiation travels through food chains in inexplicable ways, ending up in greater concentrations in some food chain components than in others. The extent of medical resources needed to cope with even a limited accident like Chernobyl has made it clear that we could never mount an adequate medical response to a full-blown nuclear war.

Before Chernobyl, world consensus was in favor of nuclear power. Chernobyl changed that. In many countries, nuclear power had already been on the ropes, and Chernobyl may have delivered the knockout punch. In the United States, public support of nuclear development fell from 64% in 1975 to 19% in 1986 (Flavin, 1987).

Today the Chernobyl reactor is entombed in 300,000 tons of concrete and metal. Pripyat is a ghost town and will likely remain so for a while. Three officials who were in charge of the Chernobyl plant at the time of the accident were convicted of gross negligence and sentenced to ten years at hard labor; three subordinates received lesser sentences. The Soviet government designated the region around Chernobyl an ecological reserve to be used in the study of the impact of radiation on the environment.

Nuclear Energy as a Social Issue

Nuclear energy has become a social issue. We have all seen pictures of, heard about, or maybe even gotten involved in the protests. People have picketed, held rallies, and even placed themselves in front of cement trucks, trying to halt the construction of nuclear power plants throughout the world. There is a clear negative public attitude toward nuclear power that had already become a significant factor in the proliferation of nuclear power plants even before Chernobyl. The attitude of a large sector of the public is perhaps summed up best by Dr. Hannes Alfvén (1972), who says,

> Fission energy is safe only if a number of critical devices work as they should, if a number of people in key positions follow all their instructions, if there is no sabotage, no hijacking of the transports, if no reactor fuel processing plant or repository anywhere in the world is situated in a region of riots or guerrilla activity, and no revolution or war—even a "conventional" one—takes place in these regions. The enormous quantities of extremely dangerous material must not get into the hands of ignorant people or desperados. No acts of God can be permitted. (p. 6)

Well-informed scientists can be found on both sides of the nuclear question. Proponents cite the need for atomic energy "if our way of life is to be maintained" and the relatively good safety record of the nuclear industry. Opponents make these claims:

1. Small amounts of radioactivity may well leak into the air and water, and we do not know what

The negative public attitude toward atomic energy is often reflected in the work of editorial cartoonists.

effects these low levels of radiation will have on the incidence of diseases like cancer.

2. There is no safe, permanent system for disposal of nuclear wastes at the present time.

3. The probability of meltdowns, even if small, is intolerable.

4. We really do not have good systems for prevent-

ing the diversion of nuclear material into the wrong hands, which could lead to terrorist blackmail or even to widespread destruction.

It is generally conceded that almost anyone in possession of 5 kilograms of uranium or plutonium 239 can make a nuclear explosive. Uranium occurs natu-

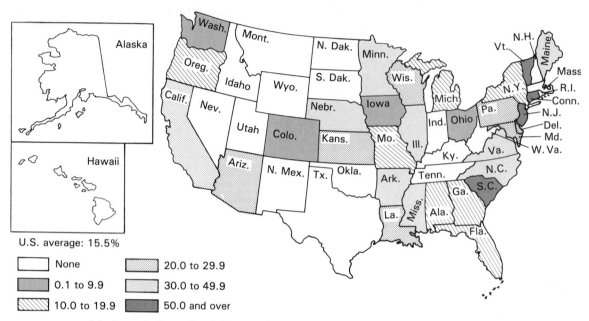

U.S. average: 15.5%

None		20.0 to 29.9
0.1 to 9.9		30.0 to 49.9
10.0 to 19.9		50.0 and over

(a) Percent of net electricity generated in each state by nuclear power in 1987.

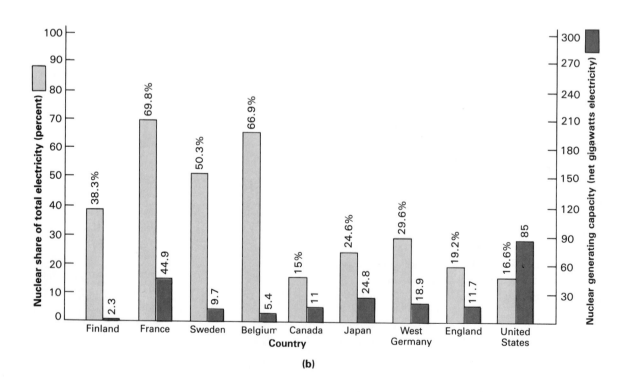

(b)

rally in many deposits throughout the world, but it must be enriched before it can be used in a bomb. Plutonium is made in nuclear reactors as a by-product of nuclear fission. For a long time the U.S. government has been reluctant to let commercial enterprises reprocess reactor fuel elements to extract plutonium from them for fear that this would increase access to plutonium by groups and individuals with mayhem on their minds. Likewise, opposition to the breeder reactor is often based on its potential for increasing the amount of ^{239}Pu available for subversive activities.

The Status of Nuclear Energy

Worldwide, some 400 nuclear plants generate electricity; 200 more are under construction. Nuclear reactors now generate about 15% of the electricity in the world. France produces two-thirds of its electricity by nuclear power, and many other European countries are heavy users of nuclear power.

Despite its setbacks, nuclear-generated electricity is the fastest-growing major energy source in the United States over the past few decades. In 1984, nuclear power replaced hydroelectric power as the second leading source of electricity. Eleven nuclear units began operation in 1987 and 1988, increasing the nuclear share of electricity production to nearly 20%. Nuclear power plants in the United States have generated the energy equivalent of more than 2 billion barrels of oil since 1973.

Yet nuclear power plant construction is at a standstill in the United States. The last order for such a plant that was not subsequently canceled was placed in 1973.

Between 1972 and 1982, fully 100 nuclear power plants under construction or on order were canceled at a total cost of $10 billion. Six more were canceled in 1983. In fact, only two of the plants ordered between 1975 and 1985 have *not* subsequently been canceled. The Tennessee Valley Authority (TVA) offered to sell eight nuclear reactors to China in April 1984, all left over from plant cancellations in 1982. The half-finished Marble Hill plant in Indiana was canceled with some $2.8 billion already spent; and after $1.7 billion had been spent, the Zimmer nuclear power plant near Cincinnati was marked for conversion to a coal-fired plant at a cost of another $1.7 billion. These actions have had a chilling effect on an industry already in trouble. There have been no new orders for nuclear plants since 1978. Although the pace of nuclear power development has slackened in some other countries as well, nuclear power abroad has flourished in comparison to that in the United States (see Figure 6.31). Why?

In its annual report to Congress in 1983, the Department of Energy cited five principal factors responsible for the demise of nuclear power in the United States:

1. Lower forecasted growth in the demand for electricity
2. High interest rates and other constraints on long-term financing
3. Reversal in nuclear power's cost advantage over coal in many places
4. Changes in and uncertainties about the regulatory climate
5. Denials of nuclear power plant certification by some states

The U.S. Office of Technology Assessment issued a report in 1984 that came to the same general conclusions. Items 2, 3, 4, and 5 on the list are related and can ultimately be traced to an overwhelming decline in public confidence in nuclear power.

Though the Office of Technology Assessment's report concluded that government regulation had little to do with the nuclear industry's problems, the NRC itself accepted some of the blame. In an April 1984 report the NRC said that it failed to screen utility companies closely enough to determine whether they were competent to deal with the complexities of nuclear power. One international study concluded that the main problem with nuclear power in the United States was poor management (see Hansen et al., 1989).

France and Switzerland have single companies or organizations developing plants. In Japan, only a few companies are involved in nuclear power plant construction and operation. Whereas the pattern in these countries has tended to generate standard designs, dozens of U.S. utilities have gotten into the act, each "inventing the wheel" in its own way.

Inexperience with nuclear power by utilities and by construction companies and changing federal regulations have no doubt contributed to the nuclear industry's problems. In Japan it takes five years to build a plant; in the United States it takes ten. Delays mean huge costs. The Zimmer plant, to give just one example, was budgeted at $240 million in 1969; by 1984 it would have cost a total of $3.1 billion to complete. Construction and operating costs for nuclear power plants are now twice that of other energy sources.

All of this raises the question, What is the future of nuclear energy in the United States?

The Future of Nuclear Energy in the United States

Public opinion and the events of recent times will preclude the headlong development of nuclear power that was once envisioned; the future of nuclear fission is nowhere near as bright as it once was. Since the plants now built have a finite life span and since no more are being built, we have a de facto policy of phasing out conventional fission technology. Perhaps the future of fission will be, as suggested by Lester (1986) and others, developed around centrally fabricated, low-power nuclear reactors of considerably lower cost and inherently safer design (see also Weinberg and Spiewak, 1984; Taylor, 1989).

The fate of the breeder reactor lies in the answer to the question, Is it worth the risk involved in making plutonium more plentiful? For both conventional fission and the breeder the question remains, What about the wastes? While nuclear fusion would have considerably fewer of these problems, its barriers are primarily technological. Perhaps fusion holds the most promise for the long-range future of nuclear power. The development of both breeder and fusion technologies will be costly, but there is now much interest in the possibility of international cooperation in the development of these technologies.

Logic may not be what dictates what will happen with nuclear energy. Unknown variables and intangibles appear to be the controlling factors, public pressure and safety considerations being two of the most important. Their influence has been variable and in the future will be determined in part by the perceived risks associated with the use of other energy sources (including the risk of dependence on foreign oil).

Ironically, while the cost of building nuclear power plants has become prohibitively high in the United States, the operating cost of nuclear power plants is about half that for plants using fossil fuels. Still, because of the uncertainties surrounding nuclear plants, utilities have been, and are likely to continue to be, reluctant to invest in nuclear power.

In summary, nuclear power has an uncertain place in our energy future. Technological breakthroughs in waste management and in fusion would improve its future, as would steps to improve public confidence. But for the moment anyway, the main questions are, What is the rush—and is it worth it? There are many different responses.

Hydroelectric Power

The electricity generated via turbines as water passes through holes in dams is really a form of solar energy. Hydroelectric power generation takes advantage of the portion of the hydrologic cycle in which the water is flowing back to its lowest point after the sun lifts it into the atmosphere by evaporation and it returns to the earth as precipitation. According to various sources, somewhat less than 20% of the world's hydroelectric capacity is currently being tapped, and it generates about 21% of the electricity produced worldwide (a little less than 7% of total energy production). Overall, about 40% of the generating capacity of hydroelectric power in the United States had been developed by 1975.

Hydroelectric power is not without negative environmental impacts, and there have been controversies over almost every dam built in the United States in the past several decades. The lakes created behind hydroelectric dams cover land, including good agricultural bottom land, and all ecosystems behind dams are altered, including the aquatic one. Deep lakes with low flows simply support different aquatic ecosystems than "wild" rivers do. Also, the development costs of hydroelectric projects are relatively high. This is partly because the life of such projects is limited by the tendency of dams to fill up with silt.

Although dams such as Hoover Dam provide multiple benefits such as flood control, water storage, improved navigation, recreation, and hydroelectric energy, they are costly and have negative environmental impacts.

Various informed sources indicate that despite the fact that a significant percentage of hydroelectric potential has yet to be developed in the United States, this will be done slowly, if at all, because (1) the major sites close to population centers have already been developed and (2) the costs of establishing reservoirs are high in relation to the benefits received. It is generally expected that the relative contribution of hydroelectric power to the total U.S. energy budget will decrease in the years ahead.

Geothermal Energy

Geothermal energy is energy derived from the heat of the earth itself. The earth has the basic structure illustrated in Figure 6.32. The central core is very hot; on the average the temperature increases about 1°C for every 100-foot drop from the surface. This is an average, however, and in some places the temperature increases much more sharply. In some locations it is very hot near the surface. The world has some 600 active volcanoes, nearly all of which are located in the "ring of fire" around the perimeter of the Pacific Ocean from Asia to the west coasts of North and South America. In these locations, molten lava and the heat from the earth's core come close enough to the surface to be useful in generating electrical energy. In the United States it is believed that Hawaii and Alaska, particularly the Aleutian Island chain, are good possibilities. Where geothermal hot spots come into contact with near-surface water, the result is hot springs and geysers, which can be tapped directly as steam. Hot water reservoirs can also be reached by drilling.

There are actually two types of subsurface geothermal fields; one is the hot-water type to which we have just referred, and the other is the dry, hot rock that could be used to heat piped-in water. Although there would obviously be the problem of keeping the hot water from cooling on the way out of the earth, it is believed that wherever there is dry, hot rock within 25,000 feet, systems are technically feasible whereby water can be injected down to the rocks, converted to steam, and then brought back up as steam to drive

Figure 6.32 Structure of the Earth. The heat in the earth's core and mantle is a source of energy that can be tapped to provide power.

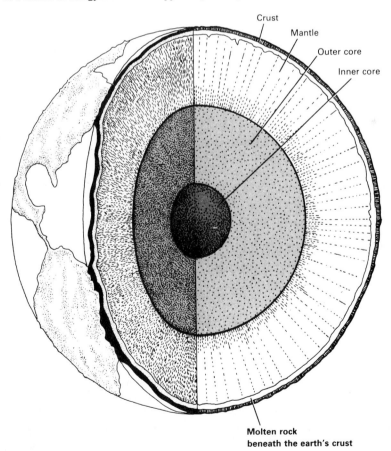

Crust

Mantle

Outer core

Inner core

Molten rock
beneath the earth's crust

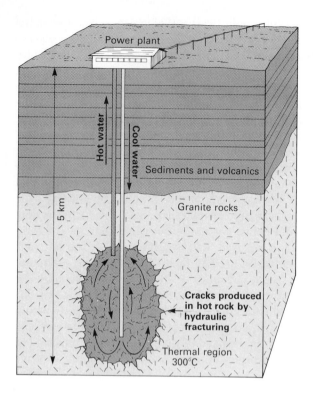

Figure 6.33 The Hydrofracturing Method of Producing Steam in Hot, Dry Rock. Water forced into the hot rocks causes cracking. Some agent (e.g., beads or grains) holds the cracks apart. Water can then be pumped through the cracks, and steam is produced to be recovered at the top and used to generate power.

Power plant

Hot water

Cool water

Sediments and volcanics

5 km

Granite rocks

Cracks produced in hot rock by hydraulic fracturing

Thermal region 300°C

turbines. One of the beauties of this suggestion is that it will require hardly any new technologies; we already have drilling rigs that can reach at least that far.

Some new technology might be needed to develop the so-called hydrofracturing method to produce porous areas in which water can be heated as it passes through (see Figure 6.33). **Hydrofracturing** is a technique in which water is injected into the heated layers of rock at 7,000 pounds of pressure per square inch, cracking the rock over a large area. The idea would then be to infuse or inject into these cracks some sort of agent that would keep the cracks open and allow water to be pumped through the cracks from the bottom and recovered as steam from the top. It is believed that such hot rock could be used to heat water for many years before cooling off and that the cost of building and operating hydrofracture geothermal plants would be somewhat less than that of conventional power plants burning coal.

There is some disagreement about the environmental problems related to geothermal energy, and controversy still rages over how extensive this source of energy really is. Axtmann (1975) reviewed some of the problems, which include the chemical (and even thermal) wastes that would be brought to the surface as geothermal power is tapped. He cites experience with New Zealand's Wairakei geothermal plant, which produces about 6 1/2 times as much heat, 5 1/2 times as much water vapor, and about half as much sulfur

pollution per unit of power produced as coal-fired electric generating plants. Axtmann admits that the plant in New Zealand is an old one and uses out-of-date technology, but his review does suggest that there may well be environmental problems in the use of geothermal energy that must be taken into account in any projection about the use of this source of energy in the future.

The significance of geothermal energy in the overall picture is somewhat uncertain at the moment. In a cooperative, U.S. government–Union Oil project at the Baca geothermal field in New Mexico there have been numerous drilling problems and problems in finding porous areas with enough hot water to supply even a single power plant. Similar problems have been encountered in La Primavera field in Mexico and in northern California (Kerr, 1991). Finding porous strata in which to reinject spent water has been a problem in many geothermal fields. Some say the hot rock is much more prevalent and much closer to the surface of the earth in many places than was previously thought. Places in the U.S. where geothermal energy is being explored include Albuquerque; El Paso; and Klamath Falls, Oregon, as well as other locations in Hawaii, Alaska, South Dakota, Louisiana, and Texas. Geothermal energy has been used for years in scattered locations such as Iceland and parts of Idaho. Although geothermal energy may well make significant contributions in these and other locations in the future, it will

154

not contribute much to the overall energy needs of the world.

Energy from Biomass

Green plants are natural solar collectors that convert solar energy into chemical energy as plant matter. Wood and other plant materials are therefore forms of solar energy. Since coal and oil are formed from plant matter, they are also forms of solar energy. Yet while wood is considered a *renewable* energy source, coal and oil are considered *nonrenewable*. Why? The difference has to do with the time required for coal and oil to form—literally millions of years. In terms of energy the important thing going on during these millions of years is that chemical energy is being *concentrated*. Terrestrial plants have a mean energy content of about 5 kilowatt-hours per kilogram dry weight; petroleum products have the equivalent of just under 12 kilowatt-hours per kilogram (Björk and Graneli, 1978); coal has the equivalent of about 9. All of this suggests a strategy of using plant mass or **biomass** as energy. The trade-off involved would be using fuel with a lower energy content (having more weight to handle) in exchange for its being renewable and available relatively immediately.

Far from being a new idea, biomass has been used as fuel for centuries, its most familiar form being firewood. In some places, biomass as firewood is still heavily exploited today. In fact, it is probably fair to say that most human beings in the world still rely on biomass for their energy needs. Biomass derived directly from photosynthesis may have a significant role to play in providing energy on a worldwide basis in the future. There have been various proposals, some in-volving the deliberate cultivation of plants to be used for fuel (**energy farming**) and some involving the use of waste or scrap biomass.

A proposal for the use of the common reed as fuel and a description of some of the spinoff benefits of large-scale reed cultivation for the purpose of generating energy are given by Björk and Graneli (1978) (see Figure 6.34). Various other common and troublesome

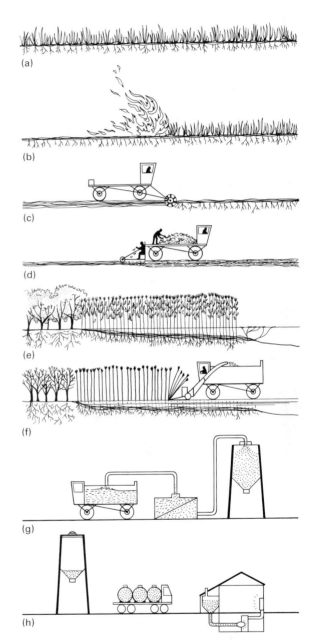

Figure 6.34 Biomass in Energy Production: A Scheme for Cultivating the Common Reed as a Source of Energy. (a) Drained lake or wetland overgrown by sedges, (b) cutting and collection or direct burning of sedge detritus, (c) destruction of sedge root and preparation of the bottom by rotocultivator, (d) planting of reed rhizome pieces, (e) establishment of reed plantation combined with wildlife management, (f) reed harvesting during the winter, (g) grinding of chopped reed stems in a mill, (h) transport of a powder for heating purposes.

If some of this municipal refuse could be used for biomass energy, two problems would be lessened—the need for energy and the problem of waste disposal.

plants have been discussed as possibilities; some of these include the water hyacinth and certain varieties of algae.

There are still other sources of biomass. Perhaps the most advantageous source is waste in municipal refuse and agricultural and forestry residues. Pimentel and colleagues (1981) have claimed that residues after forest and agricultural crop production amount to nearly one-fifth of total U.S. biomass production with a gross heat energy equivalent of 12% of the fuel consumed annually in the United States. They estimate that only about one-fifth of this is potentially harvestable and usable, but this is still equivalent to a significant fraction of our electrical energy consumption.

Alcohols from Biomass

Biomass can be burned directly, or it can be converted to other forms of fuel. Two related examples of liquid fuels derivable from biomass are *methanol* (CH_3OH) (wood alcohol) and *ethanol* (CH_3CH_2OH) (grain alcohol).

The nice thing about methanol is that it can be made from almost any kind of existing fuel. It can be made from fossil fuels as well as from a number of forms of biomass ranging from wood harvested especially for the conversion to the cellulose-based scrap in municipal, agricultural, and forestry wastes (Figure 6.35).

Ethanol is produced in other forms of bioconversion. Just as yeast makes ethanol from grape juice, a variety of anaerobic microbes, including yeast, can make ethanol from sugar (Figure 6.36) and other car-

bon compounds. Many of us have used this energy product in **gasohol,** a mixture of gasoline and ethanol.

Natural "Gas"

Methane can be generated from animal waste such as cow manure. The methane (CH_4) produced in this way can be used in the same way as natural gas. According to one estimate, if all manure produced in the United States every year by cattle and hogs were used to produce this kind of energy, it could provide 5% of our total gas consumption. Of course, there is no way that all manure could be used for this purpose. In theory, it would take about 30 head of cattle to supply the heat needs of an average household. Sewage and other wastes can also be fermented to yield usable gas. Methane produced from the anaerobic decomposition of sewage sludge is already being used to power sewage treatment plants. At the Charlotte, Michigan, wastewater treatment plant, for example, a $50,000 methane recovery system built into the plant in 1979 now yields an energy cost savings of about $11,000 per year.

Petroleum from Plants

Certain shrubs have sap containing hydrocarbons similar to those found in petroleum. Calvin (1979) claims to have demonstrated the economic feasibility of producing oil from hydrocarbon-producing species of the genus *Euphorbia*. These plants are especially appealing because they grow in semiarid areas and thus are not in competition with food crops.

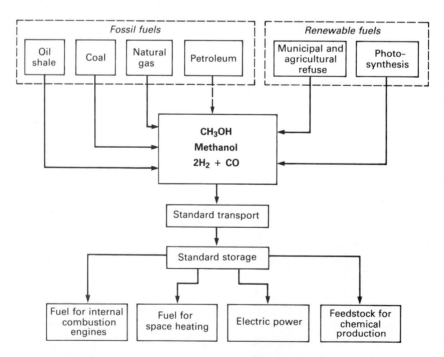

Figure 6.35 Some Sources and Uses of Methanol.

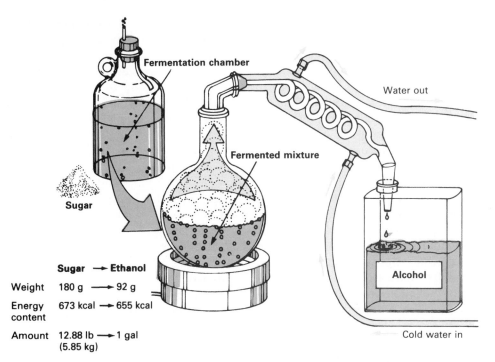

Sugar → Ethanol

	Sugar	→	Ethanol
Weight	180 g	→	92 g
Energy content	673 kcal	→	655 kcal
Amount	12.88 lb (5.85 kg)	→	1 gal

Figure 6.36. The Efficiency of Alcoholic Fermentation Producing Alcohol as a Liquid Fuel. Calculate the cost efficiency of producing ethanol from sugar by fermentation at current prices of sugar and gasoline.

Energy Balances for Biomass Conversion

It takes energy to grow biomass. In the conversion of biomass deliberately grown for energy—corn into ethanol for gasohol, for example—a key question is, Can more energy be gotten out than is put in (in the forms of fertilizer, pesticides, and gasoline)?

Chambers and colleagues (1979) conclude that if *conventional* agricultural techniques and *conventional* distillation techniques are used, the net energy balance for gasohol production is negative—more energy has to be put in than is gotten out. They acknowledge that with some generous assumptions concerning energy-efficient agricultural practices and using crop residues, there *could* be a "modestly positive energy balance." They also acknowledge that if the goal is to produce fuel that can substitute for petroleum and if more plentiful fuels such as coal provide the energy to produce gasohol, energy from biomass might not be a bad idea. Even with this, however, gasohol turns out to be a rather inefficient method of converting coal to liquid fuel.

Da Silva and colleagues (1978) reported "very favorable" energy balances for the conversion of sugarcane and sorghum to alcohol. They emphasized the importance of evaluating specific crops individually (see also Lynd and colleagues, 1991). Calvin (1979) pointed out that sugarcane is one of the world's most efficient converters of solar energy into chemical energy and that it can be converted to alcohol with

very little energy loss. He drove home his point by describing a sugar plantation as a completely energy-self-sufficient system. The waste remaining after sugar juice is extracted can be burned, yielding more than enough steam to generate all the electricity needed to run the sugar mill and the fermentation process and to generate the fertilizer required for sugarcane production.

Advantages, Problems, and Disadvantages of Biomass as Fuel

Several advantages are commonly cited for using biomass for energy. For one thing, some of the starting materials already exist, unused, in some cases posing waste disposal problems. The chief advantage is that biomass is a *renewable* energy source. Neither the use of waste biomass nor the use of biomass derived from energy farms is without problems and limitations, however.

Pimentel and associates (1978) concluded that the potential for biomass conversion from waste and energy farming is quite limited. Acknowledging that the total harvest of agricultural and forestry products amounts to 5.8×10^{15} kcal of net energy (equal to 32% of our fossil fuel consumption), Pimentel and colleagues (1984) reported that biomass-derived energy currently amounts to only 3% of the total energy consumption in the United States. However, they projected that biomass *could* provide as much as 11% of gross U.S. energy by the year 2000.

The prospect of large-scale agricultural production of biomass materials to be used as sources of gaseous and liquid forms of energy raises many important questions relative to the environment. The large amounts of land needed would raise many land use questions, not the least significant of which would be, Should land now used for food production, in a world short of food, be used to grow energy in a world short of energy? The use of marginal or otherwise currently unused lands would bring with it various environmental problems such as habitat destruction and erosion (see Chapter 13). Energy farms will tend to produce the same kinds of environmental problems found in other types of agricultural or forestry production: erosion, ecosystem simplification, ecosystem disruption, and fertilizer runoff. Collecting, transporting, processing, and ultimately burning biomass will generate air pollution and cause other pollution problems as well (see Pimentel et al., 1984).

In summary, it can be said that both biomass produced for fuel and biomass wastes can be converted into easily handled fuels such as methane, methanol, ethanol, and even powders. While this is an interesting possibility, one that might provide some fraction of our energy needs in the future, the potential of this source is not known. There have as yet been no large-scale tests of such energy systems that would indicate cost and technical difficulty. Energy farming would have much the same environmental impact as food farming and forestry, and there may well be environmental impacts yet to be discovered.

Other Energy Sources

There are two remaining kinds of energy sources. First, there are sources that have yet to be imagined or conceived. While they are sure to involve novel energy sources and novel ways of extracting energy from old sources, we obviously cannot describe them. Another category is really a miscellaneous category. It includes some old sources and some new sources, none of which are likely to provide anything more than a small fraction of our overall energy needs in the future. Three of these are wind, ocean thermal gradients, and tides.

Wind

It has long been known that fanlike structures can be used to harness the wind. Relatively recently, it has been discovered that such structures can actually drive turbines and generate electricity. The advantages of wind energy conversion are that the wind is free and does not pollute. The chief disadvantage of wind-derived energy is, as any sailor knows, that wind is intermittent and there is not much of it in many places. Environmental problems are relatively minor, but they do include visual pollution, hazards to airplanes and migrating birds, noise, and interference with television and radio. Already a number of experimental wind generators have been built, and other projects are now on the drawing boards. In California alone, there are about 14,000 wind turbines generating some 2 billion kilowatt-hours of electricity (Lenssen, 1990). In their first three years, they had produced the energy equivalent of a million barrels of oil (Moretti and Divone, 1986)—not much perhaps in light of the 6 million barrels the United States imports each day, but not insignficant either. By 1986 worldwide wind turbine sales had reached $2.5 billion. The North American share of the wind turbine market is roughly half the world total. Europe has about one-fourth, and Denmark is expected to continue to lead the European market. Other countries with substantial wind farm installations in place or projected include China, Spain, Greece, Italy, and India. India is expected to

Electrical generating equipment at the base of this 165-foot-tall vertical-axis wind turbine can generate 500 kilowatts per hour in a 28-mph wind.

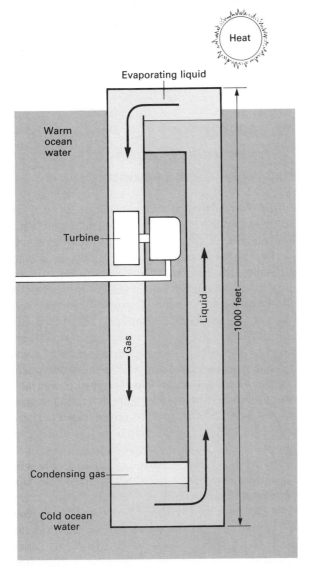

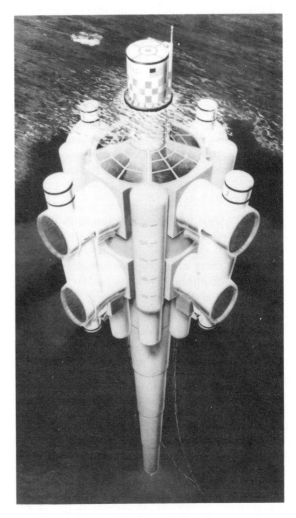

Figure 6.37 Ocean Thermal Energy. The sketch and the diagram show a device to convert ocean temperature gradients into electricity.

have about 5,000 megawatts* of capacity in place by the end of the century (Shea, 1988). Money being spent in the United States on wind turbine research is now in the tens of millions of dollars annually. Feasibility studies by the Department of Energy in cooperation with the National Aeronautics and Space Administration (NASA) have shown that prototype wind-driven generators can produce enough electricity for 30 homes in a 14-mph wind. In 1979 a 150-foot wind turbine boasting 100-foot blades able to produce 2,000 kilowatts of electricity (enough for 500 homes) at an ideal wind speed of 25 mph began operation in Boone, North Carolina. The structure, funded by the Department of Energy and NASA, cost $5.8 million.

*A watt is a unit of power like horsepower. (exerting 100 kilowatts for one hour = 100 kilowatt hours; exerting 100 kilowatts—for a year = 100 kilowatt years, etc. **Capacity** refers to a plants maximum sustainable output of power.

NASA once optimistically predicted that wind turbines would one day provide as much as 10% of the electrical power needs of the United States—perhaps as soon as the year 2000. Most estimates, however, are much lower, ranging down to a fraction of a percent.

Ocean Thermal Energy

The theoretical basis for the prospect of energy from the sea is that the oceans of the world amount to an enormous solar collector. Every day the oceans absorb the energy equivalent of hundreds of billions of barrels of oil. There are actually two ways in which the energy absorbed by the oceans can be tapped to generate electricity. One is to use the warm seawater to evaporate a fluid with a low boiling point; the other is to evaporate the warm seawater by subjecting it to a vacuum. Either way, the conversion of liquid to a gas can create pressure that can be used to drive turbines in devices like the one pictured in Figure 6.37.

The device would work as follows: In a reservoir at the top, warm water would evaporate. The increase in pressure resulting from the conversion of the liquid to a gas would push vapor past the turbine blades, rotating them as the vapor moved toward a lower, cooler area of the system, where the gas would condense into a liquid in a vapor sink. Meanwhile, more liquid would move toward the top and evaporate. This would be, in effect, a closed system. Figure 6.37 is an artist's conception of a structure proposed by Lockheed's Ocean Systems Division in Sunnyvale, California. Attached around the outside are turbine generators and pumps. The proposed structure is 250 feet in diameter and 1,600 feet long and would weigh 300,000 tons. In theory, such a structure in the right location could supply the power needed by a city of 100,000 residents.

The Department of Energy is evaluating the feasibility of this nonpolluting renewable (solar) resource. The National Academy of Sciences has looked into the matter and has determined that although the concept is technologically feasible, there are numerous developmental problems to be overcome. Because seawater is extremely corrosive, for example, corrosion will be a big technological problem.

Tidal Power

Despite the fact that there are several prototype generating stations in the world using the power of tides, this form of energy generation is not likely to be very significant in the future.

Tides are produced by the gravitational effects resulting from the interaction of the earth, the moon, and the sun. The magnitude of tides varies in cycles depending on the relative positions of the earth, moon, and sun. Tidal power is harnessed by taking advantage of the moving of water in and out of bays much like the generation of power in a hydroelectric dam.

According to various sources, the total potential rate of energy extraction from tides in shallow seas (where the tides are greatest and where the energy is most accessible) is just a little over a billion kilowatts. Only a small fraction of this could be harnessed with existing technology. The chief advantages of tidal power are that it is clean and free; the chief disadvantages are that (1) it is expensive to build and maintain the structures to harness the free energy and (2) building dams across harbors or inlets would create navigational problems.

After indicating that the potential of waves, tides, currents, and salt and temperature gradients *could* provide energy on a "significant scale," Isaacs and Schmitt (1980) concluded that the most important relationship of the oceans to power needs will continue to be the use of seawater for cooling and possibly the use of the sea floor for disposal of nuclear wastes. This fairly accurately sums up the prospects of ocean-derived energy—at least as it stands now.

Electricity and the Prospects of Improved Electrical Energy Storage and Transmission

Although electricity, as such, is not a source of energy, it has an important place in the consideration of energy alternatives for the future. The demand for electricity in the United States is likely to grow at 2% to 3% annually. The Electric Power Research Institute estimates that even if the rate of growth is low, the United States will need the equivalent of 250 new large coal or nuclear power plants by the year 2010 (see Cruver, 1989). Many of the world's developing countries are short of electrical power at present, and deficits in electrical power capability inhibit economic growth in these countries. Demand for electrical power in Egypt is expected to triple from the present 6,000 megawatts by the year 2000; capacity is not expected to keep pace. Demand exceeds supply in the Philippines by about 8%, in Bangladesh by 20%, in India by 15%, and in Pakistan by 25% (Cruver, 1989). Demand will certainly increase throughout the world for the foreseeable future.

Because of the energy lost in the production, transmission, and use of electricity (through design inefficiency and because of the laws of thermodynamics), much attention will be given to improving the efficiency of electrical energy generation, transmission, and use in the future. A particularly significant technological hurdle is the storage of electricity. Storage of electricity is important for two reasons. One is that electricity generated via some of the newer forms of energy—solar energy and wind, for example—is available only intermittently. Quite simply, if solar- and wind-derived electricity are to go far, provision will have to be made for energy needs at night and when the wind is not blowing. A second need for storage is related to the facts that demand for electricity is intermittent and some types of electrical generation produce electricity best at relatively constant rates; nuclear energy is a notable example.

Electricity as such cannot be stored. To be stored, electrical energy must first be converted into some other energy form such as chemical energy—in a battery, for example. Another alternative would be to pump water uphill and to allow it to come back through turbines during periods of peak demand. Because of the second law of thermodynamics, much energy will be lost as these transfers take place, however.

Hydrogen as Stored Electricity

Electricity can be used to split water molecules (H_2O) into hydrogen (H_2) and oxygen (O_2) gas. Hydrogen generated in this way can be saved and used later as a fuel, in effect amounting to a way of storing electrical energy.

Hydrogen is a flammable material that can exist as a gas or a liquid under various conditions of temperature and pressure. Hydrogen gas has about one-third fewer BTUs per cubic foot than natural gas; however, liquid hydrogen has a much higher energy content than gasoline. Thus if a way were found to produce it cheaply, hydrogen would represent a useful form of energy, one that would be adaptable to many of the ways in which we use energy today.

It should be no surprise that it takes more energy to produce energy in the form of hydrogen than we would get if hydrogen were burned. If we were to substitute electrically derived hydrogen only for natural gas, for example, the demand for hydrogen would require four times as much electrical power as is now generate in the United States.

A major disadvantage to the use of hydrogen is that it is explosive. The major advantage of hydrogen is that it pollutes very little when it is burned; it simply reverts back to water when combined with oxygen, sometimes producing small amounts of nitrogen oxide. Because it can be transported as a gas or a liquid, hydrogen is relatively inexpensive to transport, and it can be much more efficiently transported than electricity.

Electrical Transmission in the Future

Another major problem with electrical energy is the inefficiency of transmission lines. Although it is now technologically and economically feasible to transmit electricity over great distances, transmission lines disturb the environment, they are not aesthetically pleasing, and they may have negative effects that are yet unfathomed (see Slesin, 1987). As we discussed earlier, over half of the energy used to generate electricity is lost in generation and transmission inefficiency. In the future we will look for more underground transmission of electrical energy and for the development of materials and technologies whereby electricity can be transmitted with reduced loss of energy. We also look for ways of generating electricity more efficiently.

Conservation in Our Energy Future

We learned two fundamental principles about efficiency of energy utilization by living systems in Chapter 2. One of these principles is that the shorter the energy chain, the more energy there is available to the elements in the chain. More energy, you will recall, is available to primary producers than to consumers. Another principle we illuminated in Chapter 2 was that energy efficiencies are variable. Some living things are more efficient than others.

Extending these fundamental principles to the human condition, we find that the use of energy in the form of fossil fuels by civilization is subject to the second law of thermodynamics just like energy use by the primary and secondary consumers in an ecosystem (Figure 6.38). Also, some of the uses we make of energy are more efficient than others. Efficiency is thus central to the issue of energy utilization by human beings. The purpose of this section is to consider the possibility, plausibility, and potential significance of energy conservation—using less energy and using it more efficiently (see Enrichment Box 6.2).

Energy Efficiency Around the World

A comparison of the efficiencies of energy use throughout the world suggests that the potential for energy conservation in some countries is very great. The United States has long been the world's outstand-

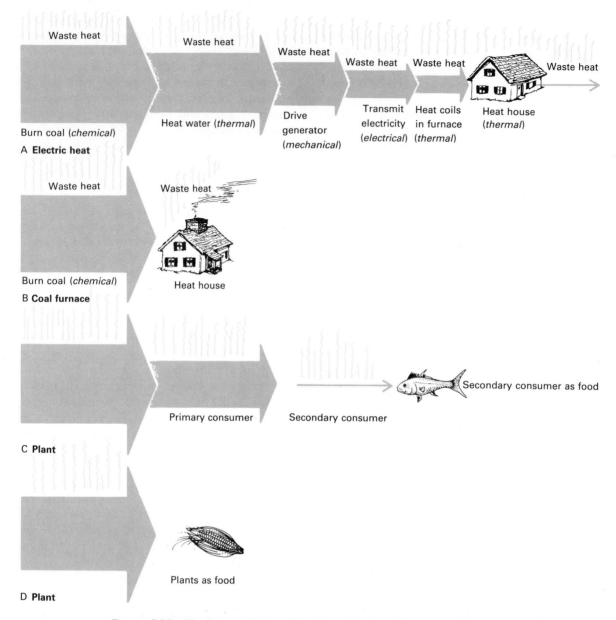

Figure 6.38 The Energy Chain. The shorter the chain of energy transfer, the fewer energy conversions and the less energy waste. This is true for ecosystems and for the uses that civilizations make of energy. Which is better in terms of heating a house, A or B? Which is better for providing food energy, C or D?

ing example of energy inefficiency. Many nations of the world achieve standards of living comparable to that in the United States on just a little more than *half* the amount of energy used in the United States per capita. The Swedes, for example, use less than two-thirds of the energy per capita and half the energy per dollar of gross domestic production used by Americans (Johansson et al., 1983). Reviewers of the energy situations in the United States and Sweden suggest that the Swedes have been able to accomplish greater efficiency because of higher energy prices, which encourage more efficient energy use. The Swedes have more

effective energy conservation construction codes and other incentives for energy conservation in housing. Johansson and colleagues (1983) project that even with a 50% increase in the consumption of goods and services in the future, Sweden will be able to reduce energy consumption by 35% through the use of even more efficient transportation and building technologies. Gasoline and other fuel taxes also have a strong energy-conservative effect on Swedish transportation. There are still more factors responsible for energy efficiency in Sweden. These bear looking into to see how they might apply to other societies.

Some progress has been made in the United States in recent years. According to Shepard (1991), Americans are now spending about $100 billion less each year than would have been true if the nation had remained at 1973 levels of consumption. This thanks to investments in more efficient vehicles, appliances, and buildings. The cumulative savings from efficiency gains in the U.S. since the first Arab oil embargo now total about $1 trillion. Far more astounding than this is the potential for even greater gains. Shepard calls for a system of "feebates," fees levied on energy-wasteful products and rebates for efficient ones.

Conservation Potential

There are many areas in which the potential for conserving energy is very great. Among these are lighting, home heating, home air conditioning, appliances, packaging, transportation, industry, and agriculture. Significant amounts of energy could be saved in every category in which energy is used.

Lighting. During the coal miners' strike in early 1978, many institutions reduced lighting in all buildings, disconnecting every other fluorescent light, as an energy conservation measure. Some energy was saved, but the amazing thing was that no one seemed to mind. This raised questions about how much light is actually needed. It is difficult to get at the truth.

Some critics claim that the lighting industry has pushed for standards of lighting in construction that are two to three times higher than that required for good vision. The critics claim that largely because of lobbying, lighting standards for schools have been increased from 30 foot-candles (a foot-candle is a measure of brightness) in the early 1950s to more than

double that by the mid-1970s. Where we have been able to find estimates of what is needed for good vision, it seems that 20 to 25 foot-candles is considered adequate for schools. If this is true, it is doubly profound, because more than 10% of all electricity generated in the United States goes for lighting. Eight million tons of coal must be burned each year to provide lighting in the United States.

According to Reisner (1987), it costs about $1 per square foot to illuminate office space in the United States; the energy used is equivalent to that produced by 100 nuclear power plants. Reisner further claims that much of this is unnecessary, given new lighting source and lighting control technologies and actual lighting needs.

If nothing else, the lighting question makes one wonder how lighting needs are decided. How do we go about deciding how much light we actually need for different purposes in different places? Do highly motivated special-interest groups have the most to say about such things? Obviously, it takes money to buy, install, and power lighting systems; one would think—naively, perhaps—that economic pressure would quickly identify the borderline between good vision and low costs. But then again, perhaps not. Perhaps when government becomes involved by establishing standards for light levels in all public buildings, the economic competitive factors that would normally favor efficiency no longer operate very well.

Politics aside, better lighting efficiency is possible even if light levels are kept constant. A 40-watt fluorescent bulb gives off more light than a 100-watt incandescent bulb, and it uses about half the energy. In other words, fluorescent light gives off three to four times more light per unit of energy input than incan-

---

descent light. Light bulbs are now available that fit into regular incandescent light sockets but that are actually fluorescent lamps.

Heating and Cooling. Heating and cooling are the biggest consumers of energy in the home. Approaches to conservation here include these tactics:

1. Tightening up existing homes
2. Adapting to lower temperatures in winter and higher temperatures in summer
3. Developing improved systems for heating and cooling homes
4. Improving the design of newly constructed houses to conserve energy more efficiently
5. Using landscaping and external factors to improve the conservation of energy

For existing homes the obvious strategies are insulation, storm windows, and weather stripping. A 1/4-inch gap at the bottom of a 36-inch-wide door provides the same potential for escaping heat as a 9-square-inch (3-inch by 3-inch) hole in the wall. It is estimated that fuel bills can be cut almost 4% by just adding weather stripping to the average existing home.

Computer simulation research and actual measurement have shown considerable energy savings by reducing temperature in homes during the winter—maybe as much as 3% for every Fahrenheit degree the thermostat is turned down in a cold climate.

The Heat Pump. There are a number of relatively new ways of heating and cooling homes; one of these is the **heat pump.** The heat pump is a device that collects heat from air and pumps it into or out of a house or a building. In winter, such pumps actually collect heat from outside and pump it in; in the summer they do the reverse. The ability of the heat pump to pump heat in decreases as the outside air temperature drops, and at 0°C a balance is reached at which supplemental heating must be used to heat the house. However, 80% of U.S. heating occurs in the temperature range of 0°C to 18°C. In the solar-assisted heat pumps that are now being developed, solar energy is used to lower the break-even point.

Construction Codes. The community of Davis, California, is often cited as an example of energy efficiency. This community has a number of ordinances that promote the conservation of energy through building codes. The codes limit the proportion of window to wall area, favor houses that come with trees, set insulation standards, and the like. According to one source, these ordinances reduced electric consumption in Davis by 18% during a time when electric

consumption increased 8% nationally. Housing design changes obviously have a great deal of conservation potential. When we speak of design, we mean things like the amount of insulation in walls, the ratio of window to wall space, and the design of overhanging roofs (Figure 6.39) as well as the use of more efficient heating systems.

Trees and Things. Deciduous trees located near a house shade the house from the sun's rays during the summer, decreasing the air-conditioning burden. Leaves fall off in the autumn, allowing the sun's rays through during the winter to help heat the house. This effect can be accentuated if a house has a properly designed overhang.

The use of evergreen trees in landscaping can also make a difference. It takes twice as much fuel to heat a house when the outside temperature is 0°C and the wind is blowing at 12 mph as it would for the same temperature with a wind of only 2 mph. The windbreak effect of evergreen trees can really make a difference.

Appliances. A significant amount of energy is used in dishwashers, gas and electric stoves, televisions, stereos, and hot water tanks. There is therefore a great deal of conservation potential in appliance design. It is estimated, for example, that as much as half of the gas consumed in a gas stove is used in the pilot light. Refrigerators have long been made with hot motors at the bottom, even though it is well known that heat rises.

Commerce. The advances in home design just discussed are applicable to commercial buildings as well. Changes have already begun to take place in big building design. Manhattan's 59-story Citicorp Tower is an example. Although it has an aluminum and glass exterior, the aluminum panels contain twice the insulation found in similar older buildings; the glass is double and is coated to reflect the rays of the sun. In a federally funded demonstration project it was shown that if lighting were cut from the usual level, which requires 5 watts per square foot, to a lower level requiring only 1.65 watts by focusing light to where it is needed and not providing it everywhere at once, a significant amount of energy could be saved. The results of this experiment have been confirmed in a number of other studies.

Industry. Industry uses about a third of America's energy. Perhaps big industry can be expected to be the bellwether of energy conservation, being the most sensitive and quickest to react to economic pressures. In 1977, according to one source, American industry reduced its energy consumption by

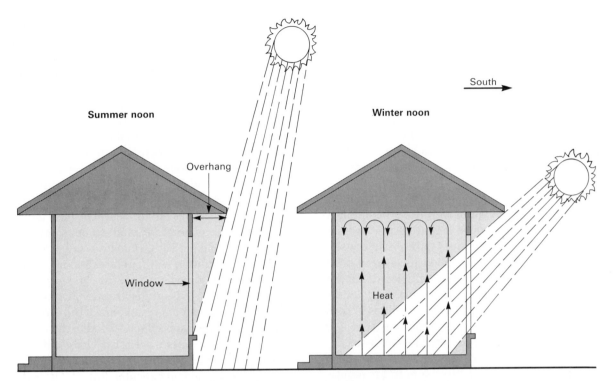

Figure 6.39 Using Housing Design to Save Energy. The sun is higher in the sky during the late spring and summer than it is in the winter. In the northern hemisphere, a south-facing house with a properly designed overhang can keep the sun's rays from coming through the south-facing windows during the summer but let them in during the winter, when they are more welcome. In the southern hemisphere, the overhang is needed on north-facing windows.

4% over the previous year while turning out nearly 6% more products. Many conservation steps were taken in industry during recent periods of energy cost escalation simply to save money. In the past when natural gas and other forms of energy were not expensive, it apparently did not pay to try to save. Now it pays. The 31% jump in the price of natural gas from 1975 to 1976, for example, was a big incentive indeed.

It is interesting that recycling has proved to be an industrial energy saver. According to one source, producing copper from recycled copper requires only a tenth of the energy as starting from ore. For aluminum the saving is even greater. The net energy savings from the use of scrap in making steel is less impressive. Hannon and Brodrick (1982) claim that although increased use of scrap would reduce energy use, it is not economical mainly because of the high volatility of scrap prices. They further claim that even if all steel were made from scrap, the energy savings would be only 6%. Others say that the savings would be much greater, however.

Packaging. Consider the following:

__ One source estimates that the Oregon law providing that all bottles be redeemable for a deposit saves enough annually to heat 50,000 homes for an entire winter.

__ A 100-watt bulb could give light for five hours on the same amount of energy needed to make a single disposable can or bottle.

__ It takes the output of several nuclear power plants to produce the energy needed to make the throwaway containers used in the United States each year.

Although there are some energy-based arguments that favor throwaway containers (one being that owing to the weight difference, a truck can carry twice as much beverage in aluminum cans as in returnable glass bottles), there is little doubt that in the overall analysis, throwaway packaging contributes to our energy problem.

It is generally conceded that in the United States, nearly all goods are overpackaged. There must be a significant energy conservation potential in the manufacturing and transport of packaging materials (see Chapter 7).

Electricity. Only about 30% of the heat generated in an electric power plant makes its way to the consumer. The rest is lost through the inefficiency of energy transformation and transfers. Some of this loss is unavoidable, but some is not. Earlier in the chapter we reviewed some of the ways being explored to improve the efficiency of the generation and transmission of electricity.

As we will see in Chapters 9–11, the efficient use of electrical energy is relevant to a host of environmental issues. Electrical utilities in the United States, for example, now produce nearly 70% of U.S. sulfur dioxide emissions, 20% of the greenhouse gases, and 50% of all nuclear waste.

A particularly troublesome aspect of the conservation of energy in association with the production of electricity in the United States is that electric utilities have not had sufficient incentives to promote conservation. The rules of their operation have been such that they make money when they sell electricity, not when they help to conserve it. Fuel adjustment clauses in the regulation of electric rates allow utilities to pass along fuel costs directly—so that fuel costs do not affect profits. Utilities thus have little incentive to conserve fuel or even to purchase fuel at the lowest price. These problems have pointed to the need for regulatory reform, and various proposals are under consideration and review in several states and at the national level. To reward utilities that can deliver electrical power more efficiently, a number of states, including Maine, Nevada, and Ohio, and the federal government are considering ways to encourage "least-cost planning" in the regulation of electrical utilities. The orientation has to change from selling kilowatt-hours to selling heat and light. It will be interesting to see how this goes.

Transporation. About a fourth of the energy consumed in the United States is used in transportation. Obviously, the potential for conservation here is significant. The system of personal transportation in the United States is outlandishly inefficient in every way. Considering the objective of getting people from one point to another, nearly all of the energy consumed in the automobile is wasted. Should we abolish the automobile immediately? This is obviously not practicable or possible, so we must address ourselves to a long-range strategy for finding and using substitutes for the automobile.

To begin with, perhaps as much as half of the gasoline consumed in the United States is consumed in trips of less than 3 miles—trips within walking distance or easy range of a bicycle. Second, gas mileage is far poorer than need be because U.S. cars are still too big and unnecessarily powerful. There are already cars on the road, after all, that can go 60 miles on a gallon of gasoline. Apparently, economic and legal incentives have already begun to make a difference, however. Gas mileage in 1977 U.S. automobiles re-

portedly improved as much as 30% over that in 1974 models. Some 1979 and 1980 cars got as much as 70% better gas mileage than their counterparts made in 1974, and mileage has been getting better ever since (see Figure 6.40).

While many observers expect to see continued improvement in the efficiency of the automobile, many of us advocate a long-range solution in which we gradually do away with automobile transportation as we know it today. Automobile driving should be both allowed to and made to become very expensive. Normal economic pressures should be accentuated through systems of taxation such as we will discuss later in the chapter. Clearly, much of what we get from the automobile we could get with some less convenient form of mass transportation. Although we acknowledge that the concept is somewhat simplistic given present-day attitudes toward mass transportation, we cannot help but think that even if a small part of the large fraction of the gross national product related to the automobile were diverted to mass transit, the United States could construct and operate one of the most efficient, pleasant, desirable, and spectacular transit systems in the world within a short time.

Agriculture. We have already alluded to a number of energy-conservative practices that might be adopted in agriculture. There may be complex reasons for it, but Amish farmers use far less fuel energy (but much more human energy) per pound of produce, obtaining almost the same yield per acre as their non-Amish counterparts. According to data gathered by investigators from San Diego State University, Eastern Illinois University, and Columbia University, Amish farmers in Pennsylvania used much less energy to produce milk than their neighbors on adjacent farms. The same study showed that the Amish produced 115 bushels of corn per acre using organic fertilizer, while their neighbors, who used much more energetically expensive chemical fertilizers, got 165 bushels per acre. Though other studies have reached similar conclusions, we do not think the definitive studies have been done showing realistically how much energy could be saved by changing agricultural practices in the United States. Because the use of energy in agriculture is significant, energy conservation in agriculture must be explored further.

Conservation and the Energy Crisis

As energy became generally more expensive in the U.S. during the 1970s and 1980s, we learned that energy conservation was more than a theoretical possibility. A lot of energy has been saved since 1974; much less oil, coal, gas, and electricity has been used than was expected.

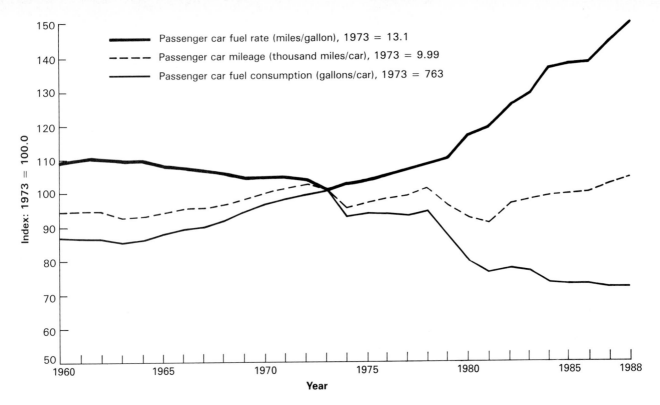

Figure 6.40 Average Annual Passenger Car Mileage, Fuel Use, and Fuel Efficiency, 1960–1988. The vertical axis can be read as the percentage of 1973 values shown in the figure.

There were several reasons for this, actually, improved efficiency being only one. Other important contributors included slower growth in the economy than was expected and changes in industrial output away from energy-intensive industries such as steel. Industrial demand for electricity rose only 20% during the 1970s. It remained virtually flat from 1979 to 1982, when it actually declined, and it did not rise substantially until the late 1980s. Overall, conservation of electricity during the 1970s and 1980s virtually eliminated the need for a "couple of dozen" new power plants in the United States (Reisner, 1987). As a bot-

---

For Further Thought . . . 6-2

What Should You Pay for a Gallon of Gas?

Gasoline prices are cheaper in the United States than in most other places. For example, in European countries the price can be several times higher. Because of the nature of a market economy, the price of oil simply does not reflect the fact that it is a finite, nonrenewable resource. During the 1973–1974 oil embargo, the price of gasoline in the United States rose significantly. One of the consequences was that gasoline consumption decreased. People were more conservative in their use of gasoline, opting to walk or ride bikes on short trips. They used public transportation more, they shared rides, and they consolidated trips to get the maximum done on the least amount of gas. They also tended to purchase more efficient cars. All of this indicated that raising the price of gasoline may be an effective method of "forcing" conservation. But increasing gasoline prices would also work hardships on the people who may be least able to afford higher gasoline costs. The cost of public transit would increase, small businesses dependent on transportation would suffer, and the cost of home delivery and home services would rise. The impact of all these things would be greatest on the poor. At the same time, those who

use the most gasoline would pay the most in absolute terms. Those who could afford the luxury of a big gas-consuming car would have to pay for it. Some people suggest that money from a sizable gasoline tax could be used somehow to offset additional costs to the working poor and to fund other worthwhile programs. Is increasing the price of gasoline by taxation to encourage conservation a good idea? What do you think?

tom line, covering all forms of energy conservation by all types of users, energy consumption per unit of economic output was an impressive 27% lower in 1986 than it was in 1970 (Energy Information Administration, 1990). According to Reisner (1987), if the United States had not experienced the conservation that it did, it would have used one whole "OPEC-output-worth" more energy in 1987 than it actually did. During a time when energy alternatives such as solar energy and wind were saving a few hundred million dollars in fuel costs, conservation was saving billions.

Considerable potential for energy conservation remains. In 1987 the U.S. Congress passed the **National Appliance Energy Conservation Act** requiring manufacturers to achieve energy efficiency targets. The act was modeled after a similar piece of legislation in California, where refrigerators use a third less energy than do those on the national market (Reisner, 1987). Some estimate that this law can potentially save $28 billion in electricity by the year 2000 (with about $4.7 billion in extra manufacturing costs) and will save as much energy as that produced by 22 large nuclear power plants (Reisner, 1987). The Energy Research Advisory Board recently projected a possible further U.S. energy efficiency gain of 20% to 30% by the year 2000.

Section C

The Economics and Politics of Energy

Our use of energy in the future will not be determined solely on the basis of environmental considerations. At best, environmental factors will be given more and more weight in what will no doubt continue to be a struggle involving economic and political considerations in all environmental matters. Those who care most about the environment cannot afford to ignore economics and politics. These must be understood and dealt with as if they were as real as ecosystems.

Dollars and Watts: The Economics of Energy

Economics is, among other things, the science of how value is determined. Economists concern themselves with the factors that go into determining the value of commodities such as energy. It will become apparent throughout the remainder of this book that economics is a major dimension of each and every environmental problem. Here we wish to explore briefly some economic dimensions of the energy problem. (Some of the fundamental principles of economics are covered in detail in Chapter 19. It might be useful to read the section on supply and demand in that chapter before going on with this section. See also Enrichment Box 6.3.)

At the philosophical end of the spectrum of questions we will consider are the following: How basic is energy as a commodity in our value system worldwide? Is money really a symbol for energy? Is money really only a means of trading in energy, such that a dollar spent is equivalent to a unit of energy consumed? A related question is, What is the relationship between economic growth and the consumption of energy? Other specific and somewhat more practical—and perhaps more answerable—questions we will explore

---

The Role of Demand in Energy Production

Energy production is governed by many factors, not simply the amount of fuel left in the ground. To be sure, production limits are ultimately set by the amount of a resource to be had. But demand plays an important role in determining production along the way until the last "drop" is gone (assuming we could ever get to that point, which we cannot).

Simply stated, unless somebody wants some, no natural gas will be produced no matter how much there is. More accurately, no natural gas will be produced unless someone wants to *buy* some. The price of gas then becomes a major determinant of production because the more people are willing to pay for it, the harder the explorers will look for it, and the more production companies are willing to spend to get it out. This becomes complicated because the factors involved all influence one another. High prices reflect high demand, but

they also tend to decrease demand. To make a long story short (at least for the time being), whenever there is a downward turn in production, it might be equally accurate to say that "gas is harder to get" or "demand is down because prices are too high."

Market forces are related to production in such a way—that is, with a little lag built in—that boom-and-bust cycles are sometimes generated. A U.S. oil production boom following the price increases in the 1970s was followed by a period of overproduction made worse by declining demand (conservation and fuel switching). This was in turn followed by a period of declining prices, a drop in domestic production, and a drastic drop in exploration. The decline in price was followed by an increase in demand, then an increase in price, and so on.

---

are the following:

1. To what extent does the price of energy determine the use of energy?
2. What effect have higher energy prices had on world economics?
3. What impact does the regulation of energy prices have on the use of energy?

Is Money a Symbol for Energy?

It has been pointed out that when people double their income, they just about double their use of energy. Hannon (1975) has gone so far as to suggest that energy be made the coin of the realm. He thinks of energy as a much more realistic basis for wealth than the gold in Fort Knox. Energy, according to Hannon, is the basis for all value in our economic and legal system, and therefore we should embrace this reality and structure our financial systems accordingly. While there is some truth to this, the fact that countries having similar standards of living consume quite different amounts of energy per capita suggests that the relationship between economics and energy is perhaps not so tight. There are apparently ways for dollars to be spent in which a "bigger energy bang can be gotten for a buck." The relationship between gross national product (GNP) (the value of all goods and services produced by a nation per year) and energy consumption in the United States has been rather close over most of the period since 1920, but there has been a discernible slow but steady increase in the efficiency of the GNP. From 1973 to 1989 there was a 50% **real growth** in the GNP (growth after adjustment for inflation) and less than a 10% increase in energy consumption (Energy Information Administration, 1990). This indicates clearly that although GNP and energy consumption are related, they are not rigidly coupled (see Figure 6.41).

To What Extent Does the Price of Energy Affect Its Use?

This is somewhat the reverse of the question raised in the preceding section. As is true of much that we have discussed so far, the picture is not crystal

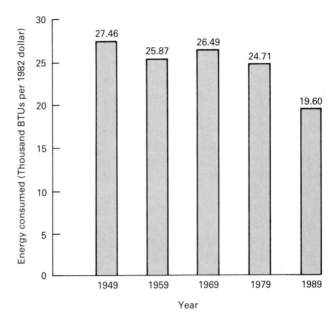

Figure 6.41   Relationship between Gross National Product and Energy Use, 1949–1989.

clear. In the extreme it is obvious that there *is* a relationship between how much energy costs and how much of it is used. Studies have shown that the better off an individual is, the more energy he or she tends to use. Americans whose incomes are low tend to own fewer energy-consuming devices. The appliances that they do have tend to be models that consume less energy; low-income people have fewer frost-free refrigerators and automatic as opposed to nonautomatic washers. Although certain data can be interpreted to mean that energy consumption does not rise quite as rapidly as income, some who have looked at these data suggest that not all the energy consumed in producing certain goods and services is accounted for; that is, we do not have good data on how much indirect energy is consumed in producing the things well-off people tend to have. The truth seems to be that although cost or price is a major factor controlling energy consumption, the connection is less than absolute.

Mini-Glossary

Consumer price index: the cost of any array of typical goods (market basket) expressed as a multiple of the cost of the same or equivalent items in some base year. For example, if a list of items costs $251 in 1983 and the equivalent items cost $100 in 1961, the consumer price index would be said to have risen by 151% from 1961 to 1983.

Efficiency: yield divided by input, usually expressed as a percentage. (For example, if 10 quads of coal were converted to electricity and then to 1 quad of electric heat, the entire process could be said to be 1/10 or 10% efficient.)

Gross national product (GNP): the total value of goods and services produced by a nation in a one-year period.

Per capita: per person.

Oil prices held constant from 1952 to 1972 and then gradually began to outdistance inflation. During the period when oil prices were increasing substantially, U.S. dependence on oil actually increased even though the price of oil was passed along to consumers. Apparently, even with the increase in price, people felt they could derive more economic good—more value, if you will—from the product they purchased than it cost them to buy it. This suggests that over the short term and up to some unknown point, at least, energy cost has little impact on how much is used.

What Are Some of the Effects of Energy Price Regulation?

The U.S. government has long regulated the price of energy transported across state lines and energy imported from foreign countries. The effect of regulation has been to hold prices low for certain domestic energy commodities, in effect shielding consumers from the realities and full costs of oil and natural gas. At one point, domestic gas and oil prices were regulated such that foreign oil cost four times as much. Some observers have argued that this has been harmful and has tended to aggravate and accelerate the onset of energy crises. Clearly, if economics is to come into play so as to favor conservation and to slow down our consumption of energy, the marketplace must be allowed to indicate to consumers the true price of energy.

It can be argued easily that price regulation brings on shortages. To keep the price low for a particular commodity is in effect promoting the use of that substance because people derive more economic value than they must pay in terms of dollars or in trade. Not only does this favor the using up of the resources in question, but it also retards the development of alternative resources because a primary stimulus for the development of alternatives is economic pressure. This is not to say that an abrupt departure from price regulation would be without problems. Because of the inertia of the way we have done things in the past, abrupt deregulation has a disproportionate impact on the poor. The poor suffer most from a rapid rise of the cost of fuel. Even so, all things considered, the price of energy should not be regulated. Other solutions must be found for the problem of the impact of energy costs on the poor.

What Impact Have Rising Energy Prices Had on the World Trade Balance?

The United States has had a serious balance of payments deficit in recent decades largely because of its dependence on foreign oil. While this is a direct problem for the United States and other user countries, it has also created ripples that have affected the economics of the world as a whole. We cannot help but wonder what sorts of economic horrors lie in wait for us if the imbalance becomes even greater. We have already seen the effect this can have on world political stability.

The Hidden Costs of Energy

Our consideration of the economics of energy would not be complete without some consideration of the fact that not all of the costs of using energy are reflected in the price of energy. We opened this chapter with the declaration that the world's most serious energy problems are the negative environmental impacts (costs) of using energy. We will consider these in detail in later chapters.

Another major dimension to the hidden costs of energy is defense. According to various estimates, in 1989 alone, the U.S. Department of Defense spent between $15 billion and $54 billion to protect oil supplies in the Persian Gulf. During 1991 "Operation Desert Storm" added several tens of billions of dollars in extra protection costs. According to Hubbard (1991), the smallest of these estimates adds about $23.50 to the actual cost of each barrel of oil imported into the U.S., thus subsidizing the consumption of Middle East oil in the U.S. and throughout the world. Other hidden costs which we will examine in Chapter 11 include pollution-related health care costs.

The Politics of Energy

Energy and World Politics

The fact that recoverable world energy reserves are far from evenly distributed ensures that energy reserves and resources will continue to be an important factor in world politics. Even though the United States has a considerable fraction of the world's total recoverable energy reserves, it has had a negative balance of trade in fossil fuels since the late 1950s. Many industrialized countries of the world, notably countries of western Europe and Japan, have had relatively stable levels of energy production since 1960 while undergoing considerably increased consumption during the same period. Japan now imports nearly every drop of petroleum it consumes and nearly three-quarters of its coal.

Some time ago, the oil-producing nations came to a collective realization that it was not in their own economic or political interests to deplete their oil reserves quickly in response to the demands of the industrialized nations and that it would be better for them to prolong the period of depletion by raising prices and imposing taxes on their energy resources instead. This presents another important dimension to the politico-economic side of the world's energy picture.

The U.S. Domestic Politics of Energy

The Reagan and Bush administrations, apparently believing that the answer to U.S. energy problems is deregulation and opening up federal lands and offshore areas to the recovery of traditional energy sources, have not followed the Carter administration's lead in subsidizing the development of solar energy or even coal conversion technologies by industry. The Reagan administration took the United States government out of the conservation picture almost completely, offering a 1983 budget that cut funding for conservation programs by 97%. The administration also proposed the liberalization of coal, gas, and oil leasing policies on federal lands, *including wilderness areas.* Understandably, this provoked much opposition and debate.

Special-interest groups in the United States traditionally speak of the need for a sensible energy policy, adding, implicitly or explicitly, *only if it does not hurt them or threaten them in any way.* Obviously, the United States cannot break away from old patterns and chart a new course without some rather widespread sacrifices. Blame for the fact that the United States has had no policy at all for so long rests squarely on the shoulders of the American people. It has been said by many that the American public made the key decisions that led to U.S. withdrawal from Vietnam and that this is how most major decisions are made. Because of their short terms in office, nearly all elected officials tend to respond to perceived public sentiment on almost any issue. Why doesn't the public feel that an energy policy is needed? Apparently, many Americans have for too long believed that there is no energy problem. Numerous surveys done in the thick of the energy problems of the late 1970s determined that almost half of the people interviewed were not aware that the United States had to import oil to meet current energy needs. The U.S. system of government fails—by design— to get very far ahead of public opinion. This may have served the country very well in the past, but the future is likely to bring problems for which such a response time may be critically too long.

The Problem of Lag. It takes time to develop alternative sources of energy. The first nuclear reactor in the United States went critical in 1943, and by the mid-1970s, nuclear energy still accounted for only about 1% of the nation's energy consumption. It takes five to eight years to open a new coal mine. A nuclear electric plant can take more than ten years to put into operation. It can take as long as 12 years to make a new oil field productive. This lag is part of the reason why crisis management, management by public opinion, and the lack of a positive natural energy policy are less than ideal in responding to energy crises.

An energy bill was passed by the 95th Congress in October 1978. Some important provisions of that energy bill were the following:

1. Prices of newly discovered natural gas—not old or stored natural gas—were permitted to rise 10% per year until 1985, when the price controls were to have been lifted altogether. (The deadline was later extended.)
2. New industrial and new utility plants were required to use coal or fuels other than oil or natural gas. Plants already existing before the bill took effect were required to switch to alternative fuel supplies by 1990.
3. Utility commissions were required to consider revamping utility rate structures in favor of conservation, for example, establishing lower prices during off-peak hours.
4. Utilities were required to provide their customers with information about energy savings, and provisions had to be made for the public to be able to borrow money—to be paid off through utility bills—to pay for conservation improvements. Grants and government-backed loans were to be available to families. Mandatory efficiency standards were authorized for many types of home appliances.

Several factors seemed to stand in the way of implementing stronger measures than those in the original (1978) energy bill. One of the factors is institutional uncertainty, the fuzzy relationship between the government and the energy production establishment. There is a great contrast, for example, between the relationship that exists between government and the nuclear power industry and that between government and the coal industry. A second factor is performance uncertainty, the unknowns associated with future energy sources and what they will and will not be able to do. Third, there is uncertainty about future demand.

The U.S. House of Representatives passed a new Energy Bill (HR 776) on May 27, 1992. Major provisions of the bill would ease licensing of natural gas pipelines and nuclear power plants, overhaul regulations governing electrical utilities, promote alternative (non-gasoline) fuels for motor vehicles, and mandate greater energy efficiency. The new bill would also restrict offshore oil drilling. The House bill is similar to a

Senate bill (S 2166) passed earlier and the two bills were being reconciled in Conference Committee as we went to press.

The Need for a United States Energy Policy

The U.S. government must play more of a major role in energy affairs. The responsibility for doing so clearly falls to the federal government under directives having to do with general welfare, interstate commerce, and a host of other specific constitutional provisions. The federal government must force decisions that consider the big picture and the long-range picture —something that market forces are rarely able to do very well.

The U.S. government must establish improved national energy policies that encompass (1) the use of federal lands, (2) regulation of electric utilities (the government owns an electric company, TVA), (3) the strategic petroleum reserve and other aspects of energy bearing on national security, and concerns about air pollution and climate change. Indeed, some of these items will require an international approach to regulation. At issue relative to item 1 is the fact that the U.S. government owns areas—the outer continental shelf plus about one-third of the land area of the United States—estimated to contain over half the nation's undiscovered energy resources. Should areas such as Alaska's National Wildlife Refuge be opened to energy exploration? Environmentalists say leave it alone. Oil companies and the U.S. Department of the Interior argue that exploration and production can coexist with the refuge. All of this is to say that the "market" simply will not solve our energy problem if the problem is properly broadly defined. The federal government is already involved in mine safety, mine regulation, and regulation of interstate commerce, including energy prices, energy import regulation, and pollution control. It should get back into the conservation promotion business; it should develop a bipartisan national energy plan, and this plan should have major provisions for the continued support of energy research. Government has an important role in the support of energy research and development. The current research program of the U.S. Department of Energy is extensive—more than $2 billion.

There is a substantial role for government in the exploration of new sources of energy and improved ways of using energy. The role is essentially one of speeding up research that might have been conducted later in response to immediate economic forces. The Atomic Energy Commission subsidized the building of atomic power plants well before such plants were economically competitive. Problems with that venture notwithstanding, the government should continue to fill this role in the future.

Our Energy Future: Projections Regarding Energy Use and Energy Sources

Future Trends in Energy Use

The fact that the future cannot be known applies to energy consumption as it does to everything else. Throughout the mid-1970s, many estimates were made of future trends in energy use. Nearly all of them forecast continued increasing universal consumption. Although energy use will most likely continue to grow throughout the world for the foreseeable future, the rate of increase in countries like the United States has already fallen off sharply. After decades of growth in energy consumption in excess of 7% annually, energy consumption in the United States actually fell in four of the eight years following the Arab oil embargo of 1973–1974. In 1981, total energy consumption was 3% below 1973 levels. Many factors, including the depressed economy, higher energy prices, temporary shortages, and conservation laws and programs, combined to cause this downturn, but it was not foreseen. There is no way to tell just what will happen in the future. A number of factors are working against further decline in energy use. The U.S. population will continue to grow, and our consumption of energy may well become increasingly inefficient, owing mostly to an expected switch to electrical heating of homes. Some observers predict that Americans may well be using more than 100% more fuel by the turn of the century. But who knows?

Future Energy Trends

U.S. Department of Energy projections of world energy sources are given in Figure 6.42. For the United States specifically, there will continue to be a predominant dependence on petroleum and natural gas through the end of the century and beyond. Coal use is expected to increase significantly. Solar, geothermal, and wind energy will contribute less than 5% of the energy used in the nation well beyond the end of the century.

These are some of the other things we should expect to see in our energy future:

___ Increased use of automobile fuels other than gasoline (see Enrichment Box 6.4.)
___ Development of inherently safer reactors for nuclear power
___ Much more emphasis on conservation
___ Increased dependence by many countries on foreign oil
___ Much more emphasis on efficiency

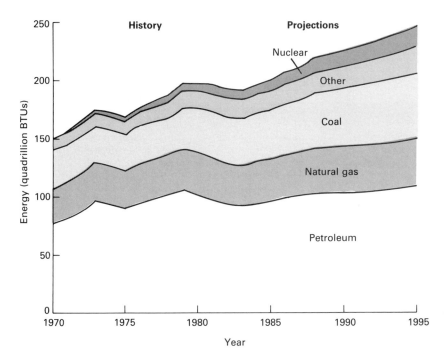

Figure 6.42 Energy Sources Projected through 1995 (Not Including Those Countries under Communist Control as of 1990—Data Not Available).

___ Increasing concern about the greenhouse effect and ozone depletion

___ Significant improvements in photovoltaic conversion and other new energy technologies

___ Greater use of energy despite improved efficiency—some estimates put world population at 9 billion people within 50 years (see Chapter 8).

The use of electricity will increase in the world overall in the next few decades. Many countries will be catching up to the United States in terms of electrically powered technology; in the United States, consumption of electricity is expected to move from the present 25% of total energy consumption toward 50%. Transportation costs and environmental considerations will combine with increased costs of oil and natural gas to cause the increased dependence on coal to take the form of electricity. Despite the bleak nuclear power situation in many countries, it is very likely that nuclear power will provide much of the energy we use in

---

Enrichment Box 6-4

Sunraycer, a solar-powered car, averages nearly 42 miles per hour across nearly 1,900 miles of Australia, winning the very first transcontinental race of solar-powered vehicles. Designed and built by GM Hughes Electronics, Sunraycer was powered by 7,200 solar cells, each the size of a large postage stamp and having an efficiency of 16.5% in converting sunlight into electricity.

Table 6.10   Comparative Estimated Short-Term and Long-Term Value and Potential Environmental Impacts of Future Energy Resources

Energy Source	Prospect for Providing a Significant Fraction of the World's Energy Needs		Relative Environmental Impact	Spheres Significantly Affected	Type of Impact and Chapter in Which Covered
	Short Term (1990s)	Long Term (beyond 2000)			
Coal					
(in general)	very good	very good	high	air, water, and land	
Deep mining	very good	very good	high	water and air	Deep mining produces acid mine water problems (Chapter 12)
Strip mining	very good	very good	high	air, water, and land	In addition to above, strip mining can have significant negative impact on land (Chapter 13)
Liquefaction and gasification	poor	good	high	air, water, and land	Burning of coal produces hydrocarbon products, some of which are carcinogenic (Chapter 18)
					Liquefaction and gasification process may constitute additional source of carcinogens
					Mining, especially deep mining, is associated with occupational dust disease such as pneumoconiosis (Chapter 10)
Oil					
Wells	good	fair	moderately high	water	Marine pollution from offshore wells and ocean transport (Chapter 12)
Shale/sands	poor	good	high	air, water, and land	Solid waste problem from spent shale (Chapter 6)
					Processing oil shale may yield carcinogens (Chapter 18)
					Combustion of hydrocarbon products produces mutagenic and carcinogenic chemicals (Chapter 18) and contributes to ozone pollution (Chapters 9–11)
Natural Gas	good	fair	low		
Nuclear					
Fission	fair	uncertain	potentially high	principally land but also air and water	Problems are related to radioactive waste disposal and, to a much lesser extent, escape of radioactive materials from power plants (Chapter 18) Possibilities also exist of colossal accidents (reactor core meltdown) and theft of plutonium that could be used by radical groups or individuals to make nuclear weapons
Fusion	poor	unknown	low to moderate	water	Impact problems limited to tritium control
Geothermal	poor	poor	moderate	water	
Solar	poor	fair to good (space heating)	low	land (use)	
Wind	poor	fair	low	land (use)	
Biomass	poor	fair	moderate to high	land, water	Energy farms would have problems in common with other types of agriculture (Chapters 12, 16)
					Burning biomass will generate hydrocarbons and potentially carcinogenic compounds (Chapters 9, 10, 18)
Hydroelectric	fair to poor	fair to poor	low to moderate	water	Dams change flow; floods lowland (Chapter 12)
Conservation	good	good	very low	indoor air	None

Note: Environmental impacts of current energy sources are described in chapters on pollution in Part Three. The impacts of many new energy sources can only be speculated on at present.

the future. Nuclear revival in the United States will be slow at best. It will take decades to build and test pilot plants of the smaller, inherently safer kind and to construct commercial plants following that. It will also take some time for the nuclear power industry to earn public confidence.

Our best hope for the future is to improve energy efficiency worldwide. Only improved efficiency can reconcile the need for more energy and environmental concerns. Energy efficiency can simultaneously stretch existing energy resources; slow climate change; reduce the environmental impact of exploration, production, and consumption; and buy more time to develop additional energy resources.

A summary of our short-range and long-range energy alternatives is presented in Table 6.10.

Future Energy Sources and Global Warming

In Chapter 10 we will consider the greenhouse effect—the problem of global warming. There we will see that fossil fuels—specifically, the carbon dioxide that they add to the atmosphere as they are used—are at the root of this potentially serious global environmental problem.

The threat of global warming makes it all the more imperative that we find more efficient fossil fuel technologies and other, non-carbon-dioxide-generating energy sources.

As we have pointed out, many of the energy alternatives that do not produce carbon dioxide have either limited potential or are as yet undeveloped. Still, on a worldwide basis, hydroelectric power amounts to 21% of all electricity produced—displacing more than 500 million tons of carbon that would otherwise be emitted. If wind power and solar power development were ac-

celerated such that they reached 150,000 megawatts by the year 2010, these could in combination displace more than 70 million tons of carbon emissions per year. As we move toward other such energy sources, we should be aware that conservation has much greater potential than alternatives to fossil fuels in reducing carbon dioxide buildup. In the interest of global climate stability, we simply must continue to invest in energy-efficient buildings and in more fuel-efficient automobiles. It has been estimated that in the United States, appliance efficiency standards already enacted will displace over 300 million tons of carbon by the end of the century.

What Must Be Done?

The technologically advanced countries of the world must do two things very soon:

1. Reduce their degree of dependence on other countries in the interest of world stability, both political and economic
2. Prepare for transition to new energy sources working on *both* the supply and demand sides of the problem

Reducing demand is a straightforward task. The problem is that conservation has no constituency, and many governments have policies that hinder conservation. Working on the supply side is less straightforward. So what is the best course?

It would clearly not be wise for any nation or the world as a whole to put all of its eggs into coal, nuclear power, solar power, or any other basket. It is important that we explore and exploit all options, including the emotionally endangered nuclear option. This exploration will bring risk as well as potential new energy sources.

Lovins (1978) and others have advocated what has become labeled the "soft energy path" for the

Cost-conscious businesses like this bank are already using soft energy technology—solar panels.

future. The **soft energy path** is cast as an alternative to the so-called hard energy path of trying to do more and more of what we have been doing in the face of increasing economic, social, political, and environmental resistance. The soft path would be based on a reordered match-up between the sources of energy and their end use. Strategically, the soft path would move us through a period of much more efficient energy use toward a system based on the decentralized generation of *renewable* energy sources. Specifically, soft energy technologies, including such things as solar heating, biomass conversion, and small-scale hydroelectric plants, would come into prominence over a 50-year period during which the strategy would be to improve the efficiency (including conservation) and cleanliness of traditional fossil fuel sources of energy.

Is There Room for Optimism?

Although there seems to be little doubt that adequate energy at a noneconomically, nonenvironmentally devastating cost will be among the most difficult and serious problems for *Homo sapiens* in the next half century, perhaps there is room for some optimism. We find at least three sources of hope.

First, the prospects of improving our lot through conservation are good, for these reasons:

— Much greater efficiency is already in reach either with currently available technology or with technology that is now on the drawing board. Up to 1% improvement in efficiency per year is possible.
— High energy costs will stimulate technological innovation.
— Growth in energy consumption is not rigidly coupled to prosperity.

Second, much of humankind is *not* dependent on fossil fuels. Most of the people in the developing nations (which comprise 75% of humanity) are only marginally involved with modern fuels and electricity. It is ironic that many developing nations have tried to match economic models based on nonrenewable resources, although more and more of them have begun to see this as a two-edged sword. For these people and these nations there is still time and increasing incentive to tie future development to improved uses of renewable energy sources.

The third source of optimism is expressed in a report issued in early 1979 by the U.S. National Academy of Sciences Committee on Nuclear and Alternative Energy Systems:

It is important to keep in mind that the energy problem does not arise from an overall physical scarcity of resources. There are several plausible options for an indefinitely sustainable energy supply, potentially accessible to all the people of the world. The problem is in effect a socially acceptable and smooth transition from gradually depleting resources of oil and natural gas to new technologies whose potentials are not now fully developed or assessed and whose costs are generally unpredictable. This transition involves time for planning and development on the scale of half a century. The question is whether we are diligent, clever, and lucky enough to make this inevitable transition an orderly and smooth one. (p. 89)

The report went on to say that the prospect of our being diligent, clever, and lucky enough will be based on social and political factors. Is there room for optimism? Do you think that the resolution of the conflicting values and special interests we have touched on in this chapter will be orderly and smooth?

Concepts to Remember

1. Energy as food is fundamentally important to human beings in the same way that it is for all organisms. But human civilization requires several orders of magnitude more energy than its members require in food. Civilization allows us to sustain a much larger population than could be sustained otherwise, but civilization brings an enormous extra dependence on energy.

2. The energy crisis of the 1970s was caused by temporary supply problems and overdue price adjustments, not by any absolute shortage of energy. It did, however, point up the importance of energy and helped to bring the broader, more fundamental, and continuing crisis into clearer focus.

3. Supplies of oil and natural gas are dwindling, and we should expect continuing problems in making transitions to other sources.

4. Energy use accounts directly for many of our environmental problems. This will be made clear in Part Three of this book.

5. The use of all forms of energy has increased dramatically in the United States during this century, electricity leading the way as a delivery form.

6. America and many other developed countries use more energy than they produce. This situation, which is not expected to improve, creates balance-of-payment problems and other political and economic tensions. It also creates national security problems.

7. Fossil fuels—oil, coal, and gas—at present account for more than 90% of the energy used in the United States.

8. Residential use (mainly heating), transportation, industry, and commerce are the four main energy use categories. Personal use amounts to about 37% of the total; business, industry, agriculture, and government use the rest.

9. Coal is the only energy resource currently available to the United States in large amounts. However, increased dependence on coal would mean an intensification of the environmental problems associated with the use of coal.

10. At current rates of utilization, we will have come close to using 80% of the world's total petroleum reserves in just one lifetime.

11. The peak of natural gas production in the United States has apparently passed.

12. Solar energy will not have much of an impact on U.S. or world energy needs for quite a few generations.

13. The immediate future prospects of nuclear energy are poor; the long-range future is uncertain. Nuclear power currently provides about 6% of world energy needs. Although the promise of breeder and fusion technology is great, the real and especially the *perceived* threat of nuclear power appears to be even greater.

14. Geothermal energy may have some local significance but will contribute little to our overall energy needs in the intermediate future.

15. Conservation and increased efficiency offer the greatest potential for impact on world energy problems in the foreseeable future.

16. Americans use far more energy per unit of gross national product than other countries with comparable standards of living. This indicates that the U.S. standard of living need not be compromised in any fundamental way by energy conservation.

17. The cost of energy has some impact on its use but not as much as might be expected. This indicates that energy is undervalued and suggests that there may be some room for inducing conservation by increasing taxes or otherwise passing along to consumers more of the actual cost of using energy.

18. Regulation of energy prices to keep costs low has the effect of promoting the use and even the waste of energy.

19. All governments need sound energy policies, and because of the nature of energy use and impact, the responsibility falls to the national governments.

20. In the next quarter century, coal, oil, and natural gas will continue to supply the bulk of our energy. Our task will be to conserve; to perfect new, environmentally sound, energy use technologies; and to make smooth transitions to the substitutes for gas and oil, whatever they may be.

21. We need continuous sources of energy, but this need must not be allowed to supersede the need for a functional, clean, and healthy environment.

22. There is room for optimism. We are *not* going to run out of energy; the problem will continue to be to learn to use it, in whatever form, without harming the environment.

Marshal Taylor Case

Marshal T. Case is the senior vice-president for education for the National (U.S.) Audubon Society. The Audubon Society is a well-established organization whose purpose is to advance conservation and environmental protection through research, education, and political action. The work of the society is accomplished by a professional staff supported by a large grass-roots membership network consisting of more than 500 local chapters nationwide. Mr. Case supervises the work of 75 employees in 12 geographic locations, raising some $5 million annually for the Audubon Society's educational programs. His division develops regional and national educational initiatives such as Audubon Adventures, a curriculum-based school club program begun in 1984, now reaching some 552,000 children, including 125,000 minority students. Among Mr. Case's division's international initiatives is a weekly radio show, *Youth Audubon News,* broadcast by the Russian State Radio and TV, and other specials on Russian television. Other international efforts include a student exchange program and the translation of Audubon curriculum materials into other languages. The Audubon Society is currently developing a Spanish-language edition of Audubon Adventures for countries in Central and South America. Mr. Case's group oversees ecology camps and workshops—some internationally—for teachers and other professionals. Mr. Case serves as dean of the Audubon Society's own college, the Institute for Expedition Education. Through the Audubon Society's Education Division, Mr. Case also oversees six regional environmental education centers and coordinates educational programs in all other Audubon departments and divisions. Mr. Case says that the best aspect of his work is the opportunity it gives him to have a

long-lasting, far-reaching, positive impact on attitudes toward the environment through his work with children.

Mr. Case has a Bachelor of Science degree in science education and wildlife management. His previous work includes service as director of the Cape Cod Museum of Natural History and as executive director of the Connecticut division of the Audubon Society. The basic qualifications for the kind of work Marshal Case does today include strong administrative and fund-raising skills and a solid background in the natural sciences and education.

References and Further Reading

References marked with an asterisk are cited in the chapter.

* Alfvén, H., 1972. "Energy and the Environment," *Bulletin of the Atomic Scientists* **28**(5):5–15.
* Anspaugh, L. R.; Catlin, R. J.; and Goldman, M., 1988. "The Global Impact of the Chernobyl Reactor Accident," *Science* **242**:1513–1518.
* Axtmann, R. C., 1975. "Environmental Impact of a Geothermal Power Plant," *Science* **187**:795–803.
* Balzhiser, R. E., and Yeager, K. E., 1987. "Coal-fired Power Plants for the Future," *Scientific American* **257**(3) (September): 100–107.
* Björk, S., and Graneli, W., 1978. "Energy, Reeds, and the Environment," *Ambio* **7**(4):150–156.
* Calvin, M., 1979. "Petroleum Plantations for Fuel and Materials," *BioScience* **29**:533–538.
* Carter, L. J., 1979. "Policy Review Boosts Solar as a Near-Term Energy Option," *Science* **203**:252–253.
Carter, L. J., 1987. "Nuclear Imperatives and Public Trust: Dealing with Radioactive Waste," *Issues in Science and Technology* **3**(2):46–61.
* Chambers, R. S.; Herendeen, R. A.; Joyce, J. J.; and Penner, P. S., 1979. "Gasohol: Does It or Doesn't It Produce Positive Net Energy?" *Science* **206**:789–795.
* Cohen, B. L., 1983. *Before It's Too Late: A Scientist's Case for Nuclear Energy.* New York: Plenum Press.

* Craxton, R. S.; McCrory, R. L.; and Soures, J. M., 1986. "Progress in Laser Fusion," *Scientific American* **255**(2)(August): 68–79.
* Cruver, P., 1989. "Lighting the 21st Century," *The Futurist* **23**(1):29–34.
* Da Silva, G. J.; Serra, G. E.; Moreira, J. R.; Concalves, J. C.; and Goldenberg, J., 1978. "Energy Balance for Ethyl Alcohol Production from Crops," *Science* **201**:903–906.
* Davidson, C. I.; Harrington, J. R.; Stephenson, M. J.; Monaghan, M. C.; Pudykiewicz, J.; Schell, W. R., 1987. "Radioactive Cesium from the Chernobyl Accident in the Greenland Ice Sheet," *Science* **237**:633–634.
Dukert, J. M., 1983. "Analyzing Future Energy Use," *Exxon USA,* 4th quarter, pp. 24–29.
* Eisenbud, M., 1979. "Reassessing Our Environmental Imperatives," *Exxon USA,* 2d quarter, pp. 8–11.
Energy Information Administration, U.S. Department of Energy, 1989. *Energy Facts, 1988.* Washington, D.C.: Government Printing Office.
* Energy Information Administration, U.S. Department of Energy, 1990. *Annual Energy Review, 1989.* Washington, D.C.: Government Printing Office.
* Energy Research and Advisory Board, U.S. Department of Energy, 1985. *Clean Coal Use Technologies.* Washington, D.C.: Government Printing Office.
* Flavin, C., 1987. *Reassessing Nuclear Power: The Fallout from*

Chernobyl. Worldwatch Paper 75. Washington, D.C.: Worldwatch Institute.

Flavin, C., 1988. "How Many Chernobyls?" *Worldwatch,* January-February, pp. 14–18.

* Flannon, B., 1975. "Energy Conservation and the Consumer," *Science* **189**:95–102.

* Gale, R. P., 1986. "Chernobyl: Biomedical Consequences," *Issues in Science and Technology* **3**(1):15–20.

Goldman, M., 1987. "Chernobyl: A Radiobiological Perspective," *Science* **238**:622–623.

Hamakawa, Y., 1987. "Photovoltaic Power," *Scientific American* **256**(4)(April): 87–92.

* Hannon, B., and Brodrick, J. R., 1982. "Steel Recycling and Energy Conservation," *Science* **216**:485–491.

* Hansen, K.; Winje, D.; Beckjord, E.; Gyftopoulos, E. P.; Golay, M.; and Lester, R., 1989. "Making Nuclear Power Work: Lessons from Around the World," *Technology Review,* February-March, pp. 30–38.

* Hopfner, K., 1979. "West Germany: A Nuclear Incident Every Three Days," *Nature* **281**:418.

* Horwitz, E. P., 1987. "New Radioactive Waste Treatment," *The World and I,* March, pp. 58–72.

* Hubbard, H. M., 1991. "The Real Cost of Energy," *Scientific American* **264**(4) (April):36–42.

Hubbert, M. K., 1974. *U.S. Energy Resources: A Review as of 1972.* Washington, D.C.: Senate Committee on Interior and Insular Affairs.

* Isaacs, J. D., and Schmitt, W. R. 1980. "Ocean Energy: Forms and Prospects," *Science* **207**:265–273.

* Johansson, T. B.; Steen, P.; Bogren, E.; Frederiks, R., 1983. "Sweden beyond Oil: The Efficient Use of Energy," *Science* **219**:355–361.

Johnson, W. A.; Stoltzfus, V.; and Craumer, P., 1977. "Energy Conservation in Amish Agriculture," *Science* **198**:373–378.

* Kay, W. D., 1992. "The Politics of Fusion Research," *Issues in Science and Technology* Winter (1991–92):40–46.

* Kerr, R. A., 1991. "Geothermal Tragedy of the Commons," *Science* **253**:134–135.

Kreith, F., and Meyer, R. T., 1983. "Large-Scale Use of Solar Energy with Central Receivers," *American Scientist* **71**:598–605.

* Kulcinski, G. L.; Kessler, G.; Holdren, J.; and Hafele, W., 1979. "Energy for the Long Run: Fission or Fusion?" *American Scientist* **67**:78–89.

Lennons, J., and Malone, C., 1991. "High-level Nuclear Waste Disposal and Long-term Ecological Studies at Yucca Mountain," *BioScience* **41**(10):713–717.

* Lenssen, N., 1990. "California's Wind Industry Takes Off," *Worldwatch,* 3(July-August), pp. 38–39.

* Lester, R. K., 1986. "Rethinking Nuclear Power," *Scientific American* **254**(3)(March): 31–39.

* Lovins, A. B., 1978. "Soft Energy Technologies," *Annual Review of Energy* **3**:477–517.

* Lowe, M. D., 1988. "Pedaling into the Future," *Worldwatch,* July-August, pp. 10–16.

* Lumpkin, R. E., 1988. "Recent Progress in the Direct Liquefaction of Coal," *Science* **239**:873–877.

* Lynd, L. R.; Cushman, J. H.; Nichols, R. J.; and Wyman, C. E., 1991. "Fuel Ethanol from Cellulosic Biomass," *Science* **251**:1318–1322.

* Marshall, E., 1987. "Recalculating the Cost of Chernobyl," *Science* **236**:658–659.

* Masters, C. D.; Root, D. H.; and Attanasi, ■. ■., 1991. "Resource Constraints in Petroleum Production Potential," *Science* **253**:146–152.

* Maugh, T. H., 1977. "Oil Shale Prospect on the Upswing," *Science* **198**:1023–1027.

* McBride, J., 1978. "Radiological Impact of Airborne Effluents of Coal and Nuclear Plants," *Science* **202**:1045–1050.

* Moretti, P. M., and Divone, L. V., 1986. "Modern Windmills," *Scientific American* **254**(6)(June): 110–118.

* Morris, S. C.; Moskowitz, P. D.; Sevian, W. A.; Silberstein, S.; and Hamilton, L. D., 1979. "Coal Conversion Technologies: Some Health and Environmental Effects," *Science* **206**: 654–662.

Penney, T. R., and Bharathan, D., 1987. "Power from the Sea," *Scientific American* **256**(1)(January): 86–92.

* Pimentel, D. L.; Hurd, E.; Bellotti, A. C.; Forstar, M. J.; Oka, I. M.; Sholes, O. D.; and Whitman, R. J., 1973. "Food Production and the Energy Crisis," *Science* **182**:443–449.

* Pimentel, D.; Nafus, D.; Vergara, W.; Papaj, D.; Jaconetta, L.; Wulfe, M.; Olsvig, L.; Frech, K.; Loye, M.; and Mendoza, E., 1978. "Biological Solar Energy Conversion and U.S. Energy Policy," *BioScience* **28**:376–382.

* Pimentel, D.; Moron, M. A.; Fast, S.; Weber, G., et al., 1981. "Biomass Energy from Crop and Forest Residues," *Science* **212**:1110–1115.

* Pimentel, D.; Fried, C.; Olson, S.; Schmidt, K.; Wagner-Johnson, K.; Westman, A.; Whelan, A.; Foglia, K.; Poole, P.; Klein, T.; Sobin, R.; and Bochner, A., 1984. "Environmental and Social Costs of Biomass Energy," *BioScience* **34**:89–94.

* Rafelski, J., and Jones, S. E., 1987. "Cold Nuclear Fusion," *Scientific American* **257**(1)(July): 84–89.

* Reisner, M., 1987. "The Rise and Fall and Rise of Energy Conservation," *Amicus Journal* **9**(2):22–31.

Roberts, L., 1982. "Ocean Dumping of Radioactive Waste," *BioScience* **32**:773–776.

* Shea, C. P., 1988. "Harvesting the Wind," *Worldwatch,* March-April, 12–17.

* Shepard, M., 1991. "How to Improve Energy Efficiency," *Issues in Science and Technology.* Summer, 85–90.

* Slesin, L., 1987. "Power Lines and Cancer. The Evidence Grows (Elf Fields)," *Technology Review* **90**:52–59.

Smil, V., 1984. "On Energy and Land," *American Scientist* **72**:15–21. Describes how switching from fossil fuel to renewable energy will change our patterns of land use.

Sperling, D., 1989. *Alternative Transportation Fuels: An Environmental and Energy Solution.* Westport, Conn.: Quorum Books.

Spinrad, B. I., 1988. "U.S. Nuclear Power in the Next Twenty Years," *Science* **239**:707–708.

* Taylor, J. J., 1989. "Improved and Safer Nuclear Power," *Science* **244**:318–325.

United Nations Statistical Office, 1990. *Production, Trade and Consumption of Commercial Energy.* New York: United Nations.

U.S. Department of Energy, 1989. *The Secretary's Annual Report to Congress, 1988–1989.* Washington, D.C.: Government Printing Office.

U.S. Department of Energy, 1990. *Clean Coal Technology: Project Sites.* Washington, D.C.: Government Printing Office.

U.S. National Academy of Sciences Committee on Nuclear and Alternative Energy Systems, 1978. "U.S. Energy Demand: Some Low Energy Futures," *Science* **200**:142–152.

* U.S. National Academy of Sciences Committee on Nuclear and Alternative Energy Systems, 1979. *Energy in Transition: 1985–2010.* Washington, D.C.

Vergara, W.; Hay, N. E.; and Hall, C. W., 1990. *Natural Gas: Its Role and Potential in Economic Development.* Boulder, Colo.: Westview Press.

* Weinberg, A. M., and Spiewak, I., 1984. "Inherently Safe Reactors in a Second Nuclear Era," *Science* **224**:1398–1402.

* Zaleski, C. P., 1980. "Breeder Reactors in France," *Science* **108**:137–144.

7 Mineral and Water Resources

Minerals: Essential Nutrients of Civilization

Early humans needed chemical nutrients just as all living things need them. They had to have calcium for bones and teeth; iron for hemoglobin; and hydrogen, oxygen, phosphorus, potassium, iodine, nitrogen, sulfur, copper, zinc, molybdenum, and magnesium for their basic structure and function. Later, *Homo sapiens* began to rely on inorganic chemical substances for purposes other than food. They beat elements into spears and plowshares, so to speak. In doing so, *Homo sapiens* began to use water and certain minerals in ways designed to help get the chemical elements needed as food (Figure 7.1). Humans have mined minerals and transformed them into fertilizers, farm machinery, and other equipment. We extended food production into dry areas as metals derived from minerals became the components of water pipelines, pumps, and irrigation systems. Shipping food into places where it could not be produced was made possible by minerals that had been transformed into transportation systems. And so it went.

The expanding use of minerals and the use of energy to power mineral-derived machinery long ago brought us to the point at which civilization is absolutely dependent on nonfood mineral (and energy) resources. Minerals other than those humans eat have become essential to human civilization.

Just as the human need for energy extends well beyond the energy (calories) that individual humans must eat every day, the need for minerals extends beyond minimum daily requirements printed on the sides of cereal boxes. It is the "minimum daily requirements" of civilization that will be our focus in this chapter.

Resources and Reserves: Some Definitions

Despite the fact that, practically speaking, elements are neither created nor destroyed, minerals are generally regarded as nonrenewable resources. **Nonrenewable resources** are those not naturally regenerated at rates comparable to their rates of depletion. We learned in Chapter 3 that the great spheres differ in chemical composition, in terms of both the types and the amounts of chemicals that

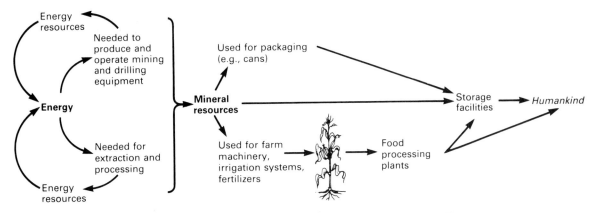

Figure 7.1 The Expanding Role of Minerals in Getting Food to Humans.
Humans have long used minerals to improve access to food. Human societies have become nearly as dependent on food-getting systems made from minerals as the individual person is on food. We also use minerals in other ways—to provide shelter, for example, which also improves our prospects for survival and gives us a higher quality of life.

comprise them. The most abundant elements in the earth's crust are listed in Table 7.1. Whereas the chemicals that make up the atmosphere and hydrosphere tend to be relatively well dispersed, minerals in soil and rock (the lithosphere) vary from one area to another; they tend to occur in deposits of varying richness. These deposits are the result of physical-chemical phenomena like crystallization that produce a given mineral. The solid state of the earth's crust keeps such deposits from scattering once they have been formed. The deposition of most minerals occurs slowly, in geologic time frames. For some minerals the conditions that caused or permitted a given combination of elements to become concen-

Table 7.1 Abundant Elements in the Earth's Crust

Element	Symbol	Percent by Weight	Percent by Volume
Oxygen	O	46.6	93.8
Silicon	Si	27.7	0.9
Aluminum	Al	8.1	0.5
Iron	Fe	5.0	0.4
Calcium	Ca	3.6	1.0
Sodium	Na	2.8	1.3
Potassium	K	2.6	1.8
Magnesium	Mg	2.1	0.3

Note: Figures have been rounded to nearest one-tenth.
Source: B. Mason, *Principles of Geochemistry,* 3d ed. (New York: Wiley, 1966), table 3.4, p. 48.

For Further Thought . . . 7-1

Future Generations: Should We Care?

As we have learned, some natural resources are for practical purposes nonrenewable. We use these nonrenewables daily. Gasoline from oil is one of the main ones. For some of us, natural gas for heating and cooking is another. There are those who believe that all of us have an ethical obligation to consider future generations in our use of nonrenewables, that we should promote conservation and use of substitutes to save some nonrenewables for the future. Others feel no such obligation. They believe that each generation will find new technologies allowing them to maintain a high standard of living. This point of view is based on the idea that necessity is the mother of invention. Do you think you have a responsibility to leave the next generation with some reservoir of nonrenewable resources? Do you think each generation should fend for itself? How far ahead should we look in setting public policy on natural resource use? Should policies on finite resources be different from policies on renewable resources? What do you think?

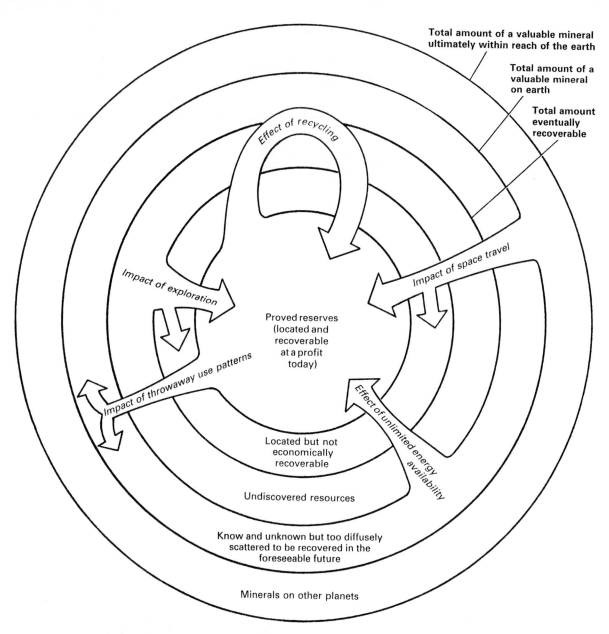

Figure 7.2 The Difference between Reserves and Various Categories of Resources. Technically speaking, even the minerals on other planets can be considered resources.

trated in mineral deposits may no longer exist. This is why mineral ores are considered nonrenewable. Once they are mined and used, they are gone as ores.

Mineral resources are the minerals that are important to human beings. **Mineral reserves** are mineral deposits that are recoverable at a profit under existing or imminent conditions of technology and economics. A schematic relationship between mineral resources and reserves is shown in Figure 7.2. It should be noted that linear use (extract, use, and discard) tends to *decrease* reserves; discoveries and new recovery technologies tend to *increase* them; recycling tends to hold them constant.

Supply and Demand, Resources and Reserves

The relationship between resources and reserves changes with supply and demand. As supply goes down and/or as demand goes up:

1. Reserves become more valuable—cost goes up.
2. The search for new deposits intensifies because of the economic incentive.
3. Resources or deposits that had previously been uneconomical may become economical and thus shift into the category of reserves.

Element: a substance composed of just one kind of atom. There are only 106 different elements; in various combinations they make up all the materials in the universe.

Mineral: a chemical element, a specific inorganic compound, or a specific complex aggregation of elements and compounds. (*Examples:* iron—an element, asbestos—a mixture of compounds)

Ore: a mineral-containing substance or mixture from which that mineral can be extracted. For example, bauxite can be mined and processed to extract the aluminum it con-

tains. If the amount of aluminum per given volume is high, it is a *high-grade ore;* if the amount is low, it is a *low-grade ore.*

Reserve: deposits of a mineral that have been located and are recoverable under existing or imminent economic and technological conditions.

Resource: the total amount of a mineral on the earth (or potentially in the universe). Includes material whose location may or may not be known and material that may not now be economically or technically recoverable.

---

4. The search for more efficient technologies of mineral extraction intensifies.
5. The search for substitutes intensifies.
6. Recycling and reuse become more desirable practices.
7. Conservation increases, and wasteful practices decrease.

As the limits are reached and the gap between supply and demand makes cost too high for a particular mineral, cost serves as a feedback influence causing demand to level off. Demand may actually fall off as people decide to do without or to settle for less desirable substitutes (where substitution is possible). All of the foregoing is grossly simplified, but it should illustrate some of the complexities of the human relationship to natural resources, which we will explore in this chapter.

Minerals and Limiting Factors

If we accept the premise that the total amount of nearly every chemical element on earth is constant and that all minerals can be synthesized from the elements, the ultimate limiting factor in obtaining minerals is energy. One cubic kilometer of surface rock contains 200 million metric tons of aluminum, over 100 million metric tons of iron, 800,000 metric tons of zinc, and 200,000 metric tons of copper. If we had unlimited energy, trace elements could be accumulated from such rock or even from air, soil, and water. We could mine anything—even landfills—and we would never run short. But we do not have unlimited energy.

The U.S. Bureau of Mines reported a decrease in the consumption of certain minerals per unit of output from 1972 to 1982. Consumption of copper declined at a 1% average annual rate, and consumption of manganese, tin, and zinc declined at an average annual rate of 5%. This decline was attributed to

changes in production and use as a result of the 1973 oil shortage and the consequent increasing cost associated with the energy-intensive operations of mining and processing these metals.

Problems are arising today because many of the highly concentrated deposits have been mined out, leaving more low-grade (less concentrated) deposits to be mined. Because needed minerals are now only to be found in more dispersed **low-grade deposits,** more energy is needed to mine and concentrate them. The National Academy of Sciences has indicated that by the year 2000, the grade of copper ore available at that time will likely require three times as much energy to yield one ton of copper as employed today. In the 1980s one large copper mine operated with only 5.8 pounds of copper recovered per ton of ore. Industry accounts for nearly 40% of all fuels and electricity used in the United States, and of this, as much as 50% is used in the extraction and processing of minerals. This percentage is likely to increase in the future.

World Trends and Imbalances in the Use of Mineral Resources

We have been using up our mineral resources at faster and faster rates, particularly since World War II. Some of the increased rate of consumption can be attributed to increases in population, but increases in per capita consumption have had a greater effect. From 1948 to 1978, population in the United States increased 49%, while the use of nonfuel minerals increased significantly more than that: ferrous metals 83%; phosphate rock, lime, and salt, 235%; and aluminum 650%. However, some studies show a trend toward decreasing use of certain minerals per dollar of gross national product and even a leveling out of per capita consumption. These materials include steel, cement, paper, aluminum, ammonia, chlorine, and ethylene. This trend is attributed to the development and use of substitute materials (use of plastic in automobiles, for example), design changes that result in more efficient use of materials (thinner aluminum cans), saturated markets (most Americans have major appliances), and new products with relatively low mineral content (such as VCRs and computers). Generally, projections through the year 2000 indicate a sustained positive rate of increase in total consumption of 3% to 5% for most nonfuel minerals. However, there is some indication that for most metals, world consumption since 1980 has not changed significantly.

It is important to note that there have been changes over time in the things we have used in support of civilization. There has been a relative increase in the use of nonrenewable materials (including minerals) since 1900. There has also been an increase in the use of petroleum-based materials. The amount of petroleum-based plastics by weight produced in the United States is several times greater than the amount of aluminum and all other nonferrous metals combined.

It seems worth noting that some nations of the world use up mineral resources more quickly than others. Without question, industrialized nations are the major users of nonfuel minerals. According to the World Commission on Environment and Development,* developed countries, with about 26% of the world's population, account for 79% of the world's consumption of steel and 86% of consumption of other metals. Currently, the United States alone, with 5% of the world's people, uses more than 33% of the world's mineral resources. Every American uses over 40,000 pounds of new minerals each year. In a lifetime a U.S. citizen will use about one-half ton of lead, one-half ton of zinc, 2 tons of aluminum, and 45 tons of iron and steel.

Where Do We Stand Today?

Inventory

Even if the rate of consumption remains static, the world is within 50 years of exhausting its reserves of lead, mercury, silver, sulfur, tin, and zinc. A mineral is considered to be depleted or used up when 80% or so of its known reserves have been mined. Total depletion of a mineral resource deposit is an unrealistic end point because of economics. (Prices rise so sharply as supply diminishes that certain uses of a commodity can no longer be justified.) In any case, if a particular mineral were indeed critical, the

* This commission was set up at the behest of the General Assembly of the United Nations. Its report is often referred to as the Brundtland Report, after the commission chair, Gro Harlem Brundtland of Norway.

most severe impact that would result from running short would be reached as supplies became very tight, well before the end.

Since there is no way to know for certain whether future rates of consumption will rise, drop, or stay the same, all that predictors can do is state and explain the assumptions on which their predictions are based. The assumption that future rates will remain constant produces a **static reserve index** (0% increases in the rate of consumption). If one assumes some degree of exponential (faster and faster) increase in the rate of utilization (a given percentage each year), the result is an **exponential reserve index.**

The Trouble with Inventories

Estimating resources and reserves is difficult because so many factors are involved. Finding new deposits would extend the life of a given mineral, as would increased recycling of products that contain the mineral. The development of new technologies to rework old deposits or to work known deposits previously considered too poor to be economical could increase the supply of a given mineral. The rate of use of a mineral could change because of the availability of a substitute material (which would result in a decrease in use) or the expansion of uses for a mineral (which would result in an increase in use). Consequently, it is often more realistic to talk about several alternative estimates of how long a mineral may be available based on specific assumptions about changes in rates of use and available reserves.

Increasing Our Resource Reserves: Promise and Problems

Finding New Deposits

Many human beings are scouring the globe looking for new mineral deposits. All sorts of technologies are employed, ranging from prospectors with picks to satellites. Using satellites to identify areas of likely deposits is termed **remote sensing.** Since many deposits have been found already, what is left to find will be harder and harder to locate. The law of diminishing returns has been operative for some time. Deposits that are being found now are of generally lower quality than those we have tapped in the past. Recently, increasing attention has been given to the hitherto least explored portions of the earth—the ocean floor and Antarctica.

Mining the Deep Sea

It is already known that concentrations of nickel, cobalt, manganese, and perhaps copper are located on the ocean floor in sufficient quantities to be mined economically. Nearshore mining of some minerals has been going on for several years. The minerals include sand and gravel, calcium carbonate, phosphorites, barites, sulfur, and tin. The development of a deep-sea mining industry has been slow. Some problems relate to the need to develop better technologies for mining through several thousand meters of salt water; some relate to the lack of discovery of "rich" deposit sites.

Some problems with the development of a deep-sea mining industry are related to the political uncertainties involved—who owns the deep sea and its contents? There have been attempts to reach international agreement on deep-sea mining (see Enrichment Box 7.1). The industrialized countries are concerned with ensuring supplies for themselves; developing countries seek also to be included in the benefits of mining international waters.

Today the mining of minerals is much cheaper on land than it is under the oceans. Someday, however, deep-sea minerals may be cheaper, leaving countries with land reserves in a major slump. Some international agreement will be needed to protect the economies of countries with major land reserves and to make the transition easier than it might otherwise be.

The mining of deep seabed materials has long-range potential. Because of the need for technology development and because of still uncertain legalities, deep-seabed mining is not likely to have a great impact on resource availability in the near term. Should development come, only a few types of minerals are likely to be commercially recoverable.

Antarctica

Since 1959, Antarctica has been managed according to the Antarctic Treaty. Under that treaty, Antarctica is to be maintained for peaceful uses only and for scientific investigation. International cooperation in projects is encouraged. But the debate over management of the continent is resurfacing in light of interest in the development of mineral resources there. To date, the only mineral resources that have been discovered that are of suitable concentrations for recovery are coal and iron. Because these are available in less extreme climates and closer to markets, economics dictates against their current exploitation. According to a report by the Office of Technology Assessment, it will likely be 30 years before any mineral exploitation occurs in Antarctica. The report cites geologic, economic, environmental, and political constraints.

Since the 1959 agreement had no provision relative to extraction of mineral resources and was in fact a voluntary moratorium on exploration, further international agreement by the treaty nations has been sought. Some nations such as France and Australia supported making Antarctica a world ecological preserve or world park where all commercial mining would be banned. In 1988 the treaty nations worked to draft an agreement called the Convention on the Regulation of Antarctic Mineral Resource Activities. The purpose of the agreement was to establish a mechanism to regulate mining. There was concern that this agreement would in effect end the voluntary moratorium on mining and encourage commercial exploration. The agreement was ultimately rejected.

Enrichment Box 7.1

UN Conference on the Law of the Sea

In the summer of 1980, after eight years of meeting, the UN Conference on the Law of the Sea appeared to have reached an agreement on who controls the ocean and its resources. The treaty does the following:

1. It establishes a 12-mile territorial limit for coastal nations and a 200-mile exclusive economic zone (EEZ) in which the coastal nation controls all fishing, marine life, and mineral rights.
2. It prescribes fines for nations that pollute the sea.
3. It establishes an international seabed authority to oversee the mining of deep-seabed minerals. The authority can set policy consistent with the treaty and license companies to mine outside the EEZs. The authority is authorized to collect taxes on the minerals produced. The authority can also do its own mining. Revenues from its mining and tax collection go to the United Nations to be used for developing nations (de-

veloped nations will lend the authority money to support its initial mining activities).

On December 10, 1982, 119 nations signed the treaty in Montego Bay, Jamaica. The United States did not sign the treaty, nor did several other industrialized nations. Nevertheless, President Reagan, by executive order on March 10, 1983, established the U.S. Exclusive Economic Zone (EEZ), extending U.S. jurisdiction over minerals found within 200 nautical miles of the U.S. coast and that of its island territories.

In 1984 the United States, along with Belgium, France, West Germany, Italy, Japan, the Netherlands, and the United Kingdom, did sign the Provisional Understanding Regarding Deep Seabed Matters to serve as a mechanism for resolving conflicting claims in deep seabed mining. (Belgium, France, Italy, Japan, and the Netherlands are also parties to the Law of the Sea Treaty.)

Instead an agreement to ban mineral and oil exploration in Antarctica for at least the next 50 years was adopted. The agreement was signed by 24 countries and was made part of the 1959 Antarctic Treaty. The ban could be lifted after 50 years by a two-thirds majority vote of the treaty parties.

Finding Better Ways to Extract Minerals

Research financed by private industry and government is under way that would develop improved technologies for mineral extraction. Improved efficiencies would have the effect of improving the cost–to–selling price ratio for marginal deposits. Such deposits might then move into the category of reserves (see Figure 7.2). What do we mean by improved extraction methods? We refer mainly to reduction in the number of dollars needed to extract a unit of resource—dollars per kilogram of iron, let's say. We mean reduction in the amount of energy required to extract the mineral, reduction in the environmental impact of mining and processing, and reduction in hazards in the workplace, all of which are involved in calculating the cost of extracting minerals.

However, according to the report of the World Commission on Environment and Development, the rise in consumption of materials in developing countries will continue in order to meet essential needs of growing populations. Greater efficiency in material extraction and production will help to lessen the impact, but environmental problems relating to mineral use are likely to intensify.

Problems with the Strategy of More

As resources become harder to find and harder to extract, more and more energy must be used in the finding, mining, and extraction of minerals. This shifts part of our mineral resource problem into the category of the energy resource depletion problem addressed in Chapter 6. In turn, the use of more energy generates more pollution (see Chapters 6, 9, 10, 11, 12, and 13). There are also a number of direct environmental impacts to consider.

The mining of low-grade mineral deposits not only requires much more energy per ton but also requires the disposal of much more **overburden** (the rock layers overlying the mineral that must be removed) and **tailings** (what is left after the mineral is extracted). The area of land disturbed is greater per ton of mineral extracted. Extra processing adds still more pollutants to the air and water. Impacts on health and worker safety are additional factors to be considered in calculating the trade-offs for mineral supply. The environmental impacts of mineral resource extraction and use are summarized in Figure 7.3.

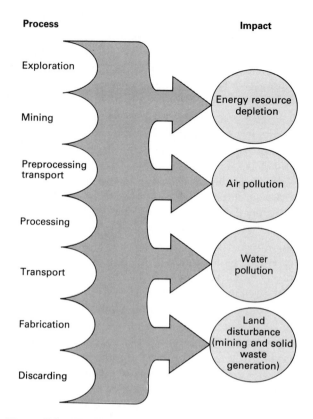

Figure 7.3 The Relationship between Mineral Utilization and Environmental Problems. Many stages in the extraction and use of minerals have considerable real or potential negative impact on the environment. Although there are likely to be more nonpolluting technologies in the future, it is generally true that the lower the grade of a mineral ore, the greater the potential negative impact from extraction and processing on the environment.

Elsewhere in this text (Chapter 16), we will review in detail the many environmental problems associated with errant (wasted) mineral resources. Mercury, nickel, sulfur, chromium, lead, and cobalt escape into the environment to poison fish, birds, and people. There is also the problem of solid waste, which can include aluminum, iron, copper, and nickel.

Let's take a look at alternatives to solving mineral problems, other than finding more minerals.

Strategies to Reduce Rates of Mineral Consumption

Repeated Use of Minerals: Recycling

An environmentally ideal solution to the problem of mineral supply would increase the availability of a given mineral while using less energy, disturbing less land, and polluting less air and water. Table 7.2 shows some of the beneficial environmental impacts of using recycled materials instead of virgin materi-

Table 7.2 Environmental Benefits Derived from Substituting Secondary Materials for Virgin Resources (in percent)

Environmental Benefit	Aluminum	Steel	Paper	Glass
Reduction of:				
Energy use	90–97	47–74	23–74	4–32
Air pollution	95	85	74	20
Water pollution	97	76	35	N/A
Mining wastes	N/A	97	N/A	80
Water use	N/A	40	58	50

N/A = Data not available.

Source: Based on Robert Cowles Letcher and Mary T. Sheil, "Source Separation and Citizen Recycling," in *The Solid Waste Handbook,* ed. William D. Robinson. Copyright © 1986 John Wiley and Sons. Used with permission.

als. There are savings in energy consumption, diminished air and water pollution, less mining wastes, and a decrease in water use.

If recycling increased by 50% in the steel industry and tripled in the paper industry, energy equivalent to 500,000 barrels of oil could be saved daily. This amounts to more than the daily energy output of 14 nuclear power plants. If the amount of paper recycled were tripled, $750 million could be saved annually in disposal costs.

Though the recycling of paper, metals, and glass is already an established part of the economic system, it is not widely practiced.

Many manufacturers do see recycling as a way of obtaining cheaper raw materials. For example, glass can be melted and re-formed at lower temperatures than the virgin constituents of glass. This saves fuel costs. Also, as steel mills have become more efficient, there is less scrap steel available. Consequently, steel cans become an alternate source of scrap steel. Producing aluminum from used cans requires only 5% of the energy needed for production from raw materials. This is a significant fuel and cost savings.

The question then arises, Why don't we recycle more? The answer has many facets. Basically, relatively rigid institutional forces encourage use of raw materials and discourage use of scrap or waste materials. In the United States, institutional policies such as tax breaks for depletion of a mineral resource, higher transportation rates for scrap material than for virgin materials, and federal mining and forestry leasing programs were established at a time when development of resources was a national priority and resources were plentiful. The institutional policies have carried over into our patterns of using convenience products and throwaway materials. We have not yet changed to adjust to the current circumstances in which wise use of resources and resource conservation should be national priorities.

A shift to an economic system emphasizing secondary (as opposed to virgin) materials industries would be no simple matter (see Figure 7.4). Most industrial machinery is designed for processing virgin materials. The collection and sorting of used materials for recycling requires costly labor. Most products have not been made with recycling in mind; consequently, the complex mixtures of metals in automobiles require expensive separation techniques and so are poor reserves. Products that are made basically of one mineral—such as aluminum or steel—are actually rich reserves, but such products tend to be hard to find. As in any system, changing one cog—such as the way materials are used—disrupts the operation of the interlocking wheels of civilization.

Lack of markets for products with recycled content is another major detriment to recycling. Governments are major purchasers of goods. Government procurement policies that mandate recycled content in the goods purchased can help to establish markets. As demand for goods with recycled content increases, more recycling facilities come on line, and consumers are further encouraged to recycle. We will discuss recycling in greater detail in Chapter 17.

Small-scale community recycling is just the beginning. It is vital that industry also make resource recovery a priority.

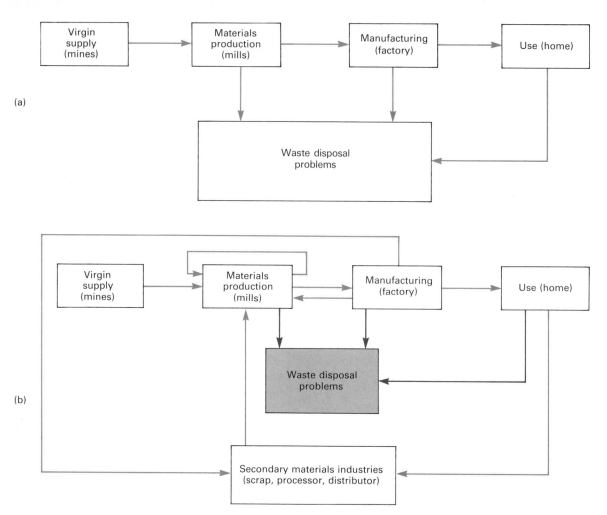

Figure 7.4 Secondary Materials Industry in the Grand Scheme of Material Flow. (a) The most common pattern for materials use has been the one-way flow from virgin materials to product to disposal. (b) The development of a secondary materials industry changes the pattern to a system of cycles.

Extended Use of Products: Antiobsolescence

In a society in which obsolete and old-fashioned things are collected once a week and buried, a shift toward products designed to last longer would decrease the demand for the materials from which those products are made. Such a shift will not come easily. Obsolescence is partly related to the fact that cheaply made and rapidly worn-out products are not expensive in the short run, as are more soundly made products. There is a real demand for such inexpensive things.

Even if this could be overcome, major economic changes and adjustments would be required by any movement toward antiobsolescence. Labor-intensive activities would replace energy- and mineral-intensive activities. For instance, the production of fewer cars each year would decrease the use of minerals and the energy required to transport and process them, but the work force would have to shift from assembly work into repair and maintenance. Cars would cost more, since manufacturers would be selling fewer and better products. Perhaps not many people could afford cars. Repercussions would spread to the steel industry, sales, and other related enterprises. Imagine this sort of domino effect over the entire spectrum of manufacturing! Still, this approach deserves consideration.

Substitutes: Let Them Eat Cake?

Another approach to solving mineral shortage problems is to work toward decreased demand beyond that resulting from rising prices.

First of all, technology can find substitutes for scarce minerals. In specific instances we have already seen a shift from copper to aluminum and from iron and steel to aluminum. Chromium on automobiles has largely been replaced by aluminum and plastic. There might well be more alternative materials that can be modeled to fit the function of the mineral in short supply. The creation of new types of materials has been an increasingly popular alternative in the twentieth century. However, new synthetic materials such as plastics have created other problems, such as that of disposal. Also, these materials have been based on petroleum.

What about substitutes that are not only in greater supply but are also less environmentally harmful? The advantages must be weighed along with numerous other considerations on a case-by-case basis. A shift from metal or plastic to a renewable resource—wood—is possible in some cases. However, forests are already being used faster than they are produced. At the moment, renewable wood happens to be a lot more expensive than plastic made from nonrenewable petroleum. Policies and practices governing renewable resources must be examined closely if demand is to shift in their direction.

We humans live mostly for the moment. We could say that economic natural selection should already have brought us to the point at which we are using the best mineral (in terms of suitability for the job and low cost). However, new substitutions can be found only through painstaking research. Once found, it takes a while for substitutes to be placed into the manufacturing system—machines and whole factories have to be retooled. Now, though, as equipment and processes in industries are upgraded and replaced, we must look for environmentally sound substitutes for scarce minerals and for minerals that hurt us as we use them.

Minerals and World Politics

The uneven distribution of mineral deposits throughout the world adds a political dimension to the whole question of mineral supply. Figure 7.5 reveals U.S. mineral requirements and the amounts imported from other countries. Given the essential nature of most of these minerals and their significance in the U.S. economy, they are important factors in U.S. foreign policy considerations (see Figure 7.6). For some strategic minerals, that is, minerals indispensable to defense and industry, the United States is 80% to 100% dependent on foreign sources. The U.S. Office of Technology Assessment cites chromium, cobalt, manganese, and platinum group metals as the "first tier" of metals of strategic importance to the United States. Manganese is used in steel and cast-iron production; cobalt is used to harden steel and in

making jet engine parts and cutting tools; platinum is used in the manufacturing of air pollution control and telecommunications equipment and as a catalyst in the chemical industry; chromium is needed in making stainless steel. In 1985, South Africa and the Soviet Union produced 55% of the world's manganese, 62% of its chromite (chromium ore), and 93% of its platinum group metals. Zaire and Zambia provided about two-thirds of the world's cobalt.

We are all aware of the dependence of the United States on imported oil from the OPEC cartel. Although it is generally thought that a cartel for non-fuel minerals is unlikely because of the erratic distribution patterns of the deposits and because of the potential for substitutes, mineral resources are factors in our foreign policy. For example, in 1966 the United States was party to a UN action placing economic sanctions against Rhodesia (now Zimbabwe). Imports from that country were banned. In 1971, however, the United States reversed its position on the sanctions. Low stockpiles and concern that we would become dependent on the Soviet Union for an essential mineral contributed to this policy reversal. In 1977 the United States joined the boycott again after stockpiles were replenished and improved technology and increasing exports by other countries made supplies more available.

A related problem arises when supplies are curtailed for reasons other than an embargo. For example, in 1978, internal problems in Zaire prevented ample production of cobalt. Prices soared, and there were major disturbances in the marketplace. A strike at South Africa's Impala Platinum Holdings mine in December 1985 sent platinum prices from an average of $291 per ounce in 1985 to $425 per ounce by early March of the following year.

In 1986 the U.S. Congress passed the Comprehensive Anti-Apartheid Act, which banned the importation of gold coins and other materials from government-owned sources in order to pressure the South African government to abandon its apartheid policies. However, the act contains provisions that allow exceptions for materials designated strategic for the economy or the defense of the United States that are unavailable from other reliable sources.

Developed countries are depending more on developing countries for minerals. In 1959, developing nations provided 19% of the imports consumed by developed nations. By 1981 that had increased to 30%.

Several approaches have been taken to address the political realities of mineral supplies. Since 1939, the United States has maintained a national defense stockpile in order to have a given number of years' supply of strategic minerals on hand at all times. The policies associated with use of the stockpile have

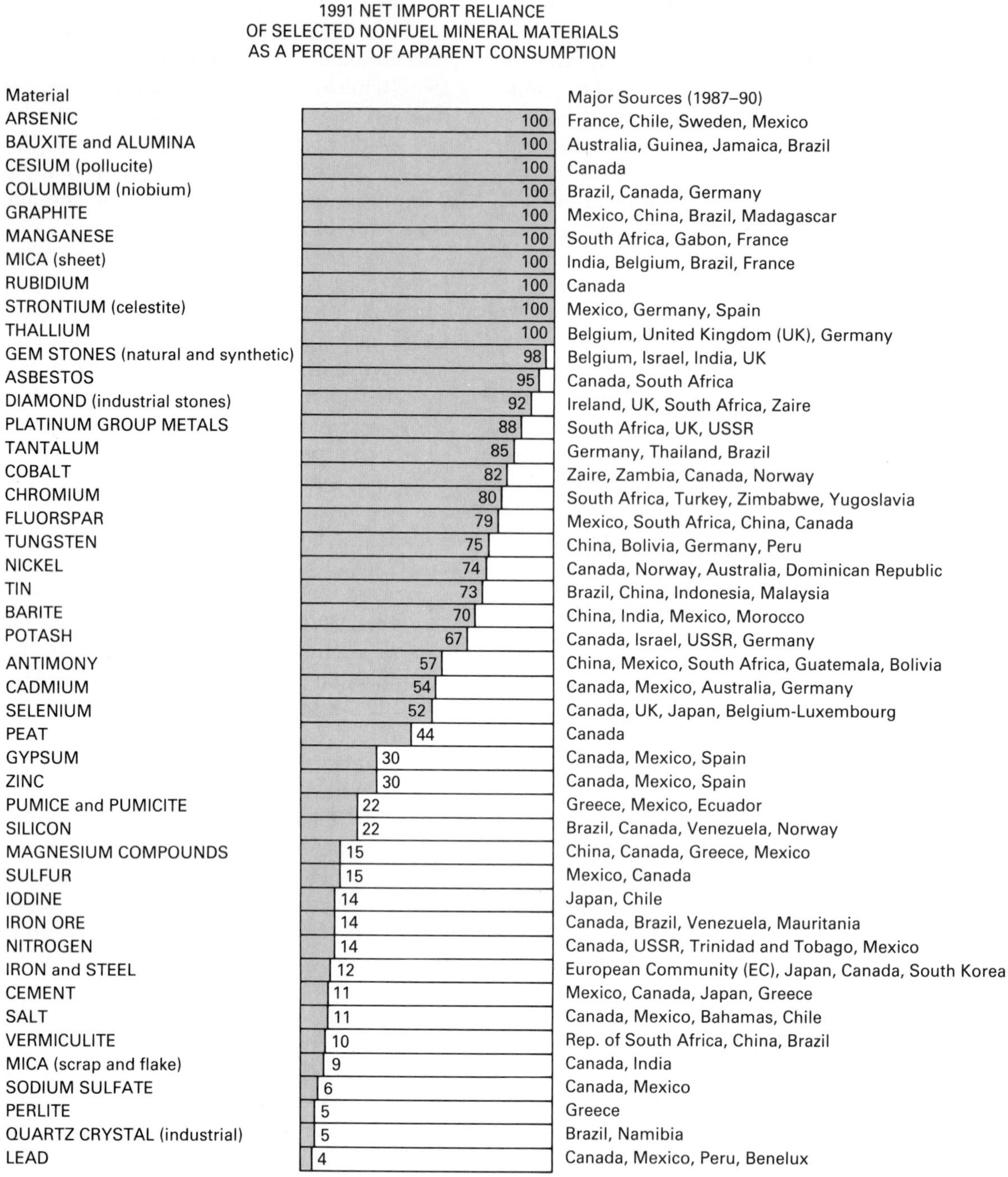
1991 NET IMPORT RELIANCE
OF SELECTED NONFUEL MINERAL MATERIALS
AS A PERCENT OF APPARENT CONSUMPTION
Material
Major Sources (1987–90)
ARSENIC 100 France, Chile, Sweden, Mexico
BAUXITE and ALUMINA 100 Australia, Guinea, Jamaica, Brazil
CESIUM (pollucite) 100 Canada
COLUMBIUM (niobium) 100 Brazil, Canada, Germany
GRAPHITE 100 Mexico, China, Brazil, Madagascar
MANGANESE 100 South Africa, Gabon, France
MICA (sheet) 100 India, Belgium, Brazil, France
RUBIDIUM 100 Canada
STRONTIUM (celestite) 100 Mexico, Germany, Spain
THALLIUM 100 Belgium, United Kingdom (UK), Germany
GEM STONES (natural and synthetic) 98 Belgium, Israel, India, UK
ASBESTOS 95 Canada, South Africa
DIAMOND (industrial stones) 92 Ireland, UK, South Africa, Zaire
PLATINUM GROUP METALS 88 South Africa, UK, USSR
TANTALUM 85 Germany, Thailand, Brazil
COBALT 82 Zaire, Zambia, Canada, Norway
CHROMIUM 80 South Africa, Turkey, Zimbabwe, Yugoslavia
FLUORSPAR 79 Mexico, South Africa, China, Canada
TUNGSTEN 75 China, Bolivia, Germany, Peru
NICKEL 74 Canada, Norway, Australia, Dominican Republic
TIN 73 Brazil, China, Indonesia, Malaysia
BARITE 70 China, India, Mexico, Morocco
POTASH 67 Canada, Israel, USSR, Germany
ANTIMONY 57 China, Mexico, South Africa, Guatemala, Bolivia
CADMIUM 54 Canada, Mexico, Australia, Germany
SELENIUM 52 Canada, UK, Japan, Belgium-Luxembourg
PEAT 44 Canada
GYPSUM 30 Canada, Mexico, Spain
ZINC 30 Canada, Mexico, Spain
PUMICE and PUMICITE 22 Greece, Mexico, Ecuador
SILICON 22 Brazil, Canada, Venezuela, Norway
MAGNESIUM COMPOUNDS 15 China, Canada, Greece, Mexico
SULFUR 15 Mexico, Canada
IODINE 14 Japan, Chile
IRON ORE 14 Canada, Brazil, Venezuela, Mauritania
NITROGEN 14 Canada, USSR, Trinidad and Tobago, Mexico
IRON and STEEL 12 European Community (EC), Japan, Canada, South Korea
CEMENT 11 Mexico, Canada, Japan, Greece
SALT 11 Canada, Mexico, Bahamas, Chile
VERMICULITE 10 Rep. of South Africa, China, Brazil
MICA (scrap and flake) 9 Canada, India
SODIUM SULFATE 6 Canada, Mexico
PERLITE 5 Greece
QUARTZ CRYSTAL (industrial) 5 Brazil, Namibia
LEAD 4 Canada, Mexico, Peru, Benelux

Source: Bureau of Mines. *Mineral Commodities Summaries 1992,* p. 3.

Figure 7.5 Imports as a Percentage of Total U.S. Consumption, 1991. Because minerals and mineral deposits are not equally distributed worldwide, there is a political dimension to mineral consumption. the *net import reliance* is imports and exports plus adjustments for government and industry stock exchanges. *Apparent consumption* is U.S. primary and secondary production plus net importandt reliance.

been controversial. Throughout its existence, various administrations have sought to sell off parts of the stockpile for budgetary reasons or to influence the world market.

Currently, some European countries and Japan have entered into international commodity agreements with mineral-producing countries. Producer countries bargain for premium prices and technical assistance. User countries are willing to give generous terms to prevent supply disruptions. All of this, of course, affects the mineral market system worldwide and strains political allegiances. Energy will continue

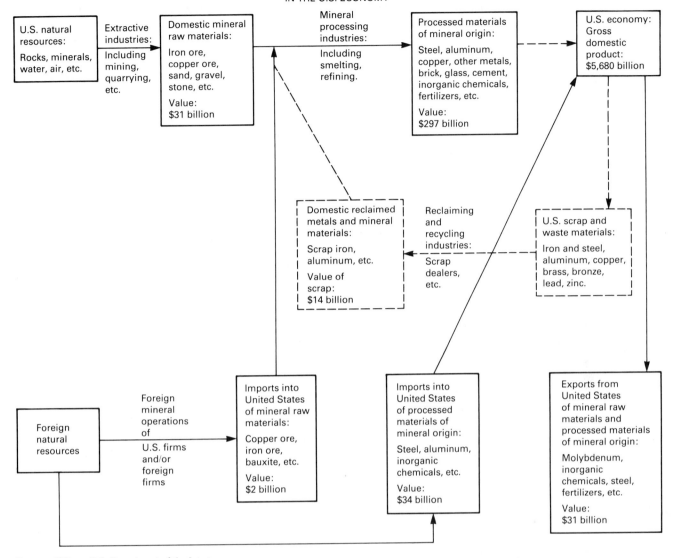

Bureau of Mines, U.S. Department of the Interior

Source: Bureau of Mines
Mineral Commodities Survey 1992 p. 2.

Figure 7.6 The Role of Nonfuel Minerals in the U.S. Economy. Because
nonfuel minerals are a significant factor in the U.S. economy, the need for mineral
imports and the need to retain and expand mineral exports has an impact on foreign
policy decisions. This is true for all governments.

to be one of the biggest issues in world politics in the future, but mineral supplies will be close behind. Water will be another continuing problem.

Section B

Water Resources

Just as minerals are the essential "nutrients" for modern industrial civilization, water has been civilization's essential solvent. Water has been the vehicle

for extending the range available to humans for settlement and development. Water has turned deserts into farmland by irrigation; water has provided major transportation routes; water has been used to cool people, float boats, and carry off the garbage and sewage wastes of concentrated populations of people; water has even been used to produce energy.

Bringing water to arid or semiarid regions opens the way for agriculture, additional settlements, and eventually business and industry—all of which become dependent on the continued flow of artificial streams. Obviously, extending water resources to areas that are otherwise not suitable for large human settlements carries with it the risk of overextension and other problems.

Like mineral shortages, water shortages can be solved only by increasing supply and/or decreasing demand. In this section we examine world water distribution and consider the approaches that have been used and those that have been proposed for increasing water supply and decreasing demand.

Worldwide Water Resources

There are 4×10^{20} gallons (1.5×10^{21} liters) of water above, on, and in the earth, but it is not evenly distributed. Water "deposits" occur as rivers, lakes, streams, and oceans. Water is also found in the soil as groundwater and is bound up as ice in glaciers. Figure 7.7 shows the earth's reservoirs of water in proportion to one another. If our entire supply of water were equal to 55 gallons, the oceans would comprise over 53 gallons (97%). Of the remaining 2 gallons, more than 1 gallon would be tied up in glaciers and ice caps. Much of the rest of the water would be found in underground aquifers or in the soil (only about 50% of this is within 1 mile of the surface, however, and only about 25% can be extracted with current technologies). Less than 1 fluid ounce would be left for surface waters such as lakes, rivers, and inland seas or estuaries. So although there are 400 quintillion gallons of water in the ecosphere, only a small part is directly available to human beings. At any one time, only about five of every 100,000 gallons

of water in the world supply is in motion as precipitation, running streams, or atmospheric vapor. The rest is stored underground, in lakes, in glaciers, or in the oceans. Water may be held or stored in the ground for thousands of years, in a lake for 100 years, or locked in a glacier for 40 or so years. Eventually, however, stored water will find its way back into the hydrologic cycle.

Precipitation plays a major role in water supply, since it is precipitation that replenishes or **recharges** reservoirs that have been diminished through evaporation, runoff into the oceans, or human withdrawals from the ground or the surface. If the water loss or withdrawal exceeds the rate of water recharge in a particular area for very long, the water supply is obviously in jeopardy.

Figure 7.8 shows the final disposition of the water that falls on the United States annually as precipitation. The average annual rainfall for the United States is 30 inches, or about 4,200 billion gallons per day (bgd). About 70% of this (2,750 bgd) is lost in evaporation before it can be "used"; approximately 1,450 billion gallons per day becomes runoff into surface bodies of water or seeps into the soil to become groundwater.

A similar diagram could be drawn for any country in the world. According to one calculation, there is enough water in the world in theory to sustain 20 billion people. If that is the case, what is the reason for water shortage problems? The answer lies, obviously, in the fact that water, like minerals, is not evenly distributed. There are water-rich and water-poor countries. There are water-rich and water-poor

Figure 7.7 A Drop in the Barrel.
Though the earth contains vast water resources, only a small part of the water in the hydrosphere is available at any one time for use by people. This figure indicates the world's distribution of water, treating the whole supply as if it were 55 gallons. The primary sources of water for human use are from groundwater, freshwater lakes, and rivers.

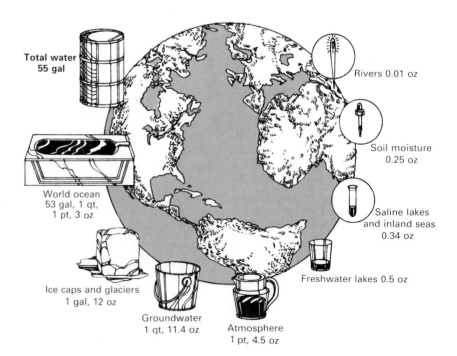

Total water 55 gal

Rivers 0.01 oz

Soil moisture 0.25 oz

World ocean 53 gal, 1 qt, 1 pt, 3 oz

Saline lakes and inland seas 0.34 oz

Freshwater lakes 0.5 oz

Ice caps and glaciers 1 gal, 12 oz

Groundwater 1 qt, 11.4 oz

Atmosphere 1 pt, 4.5 oz

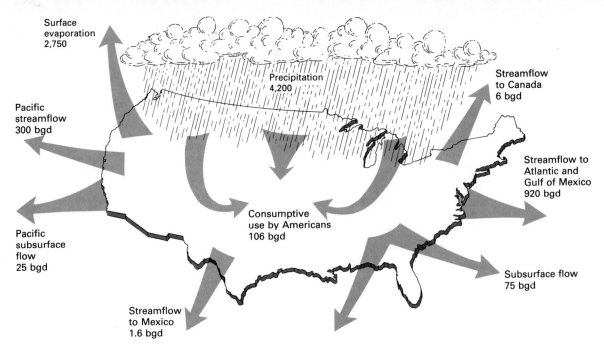

Figure 7.8 U.S. Water Budget. What happens to the rain that falls daily on various parts of the United States? Most (70%) of the 4,200 billion gallons per day evaporates back into the atmosphere; the remainder drains to surface and underground reservoirs.

regions within the same country. Before examining some of the major water supply problems, let us first discuss some fundamental terms relating to water supply.

Water Supply and Water Budgets

What is involved in determining a water budget for a region? Figure 7.9 is a diagram of the water budget for a given region. **Renewable water sup-**

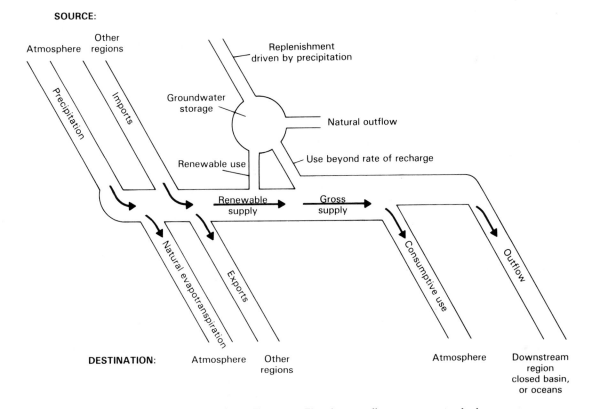

Figure 7.9 Regional Water Budget. This diagram illustrates very simply the factors that influence the available water supply in a given region.

ply is water that results from precipitation and imports (human-made transfers and diversions from other regions). Tapping groundwater is another way of making water available to the region. This water adds to the **gross supply** available, but this may not be renewable if it is withdrawn faster than it is replaced by precipitation. Water is lost to a region by natural evaporation, by exportation to other regions, by consumptive use, and by outflow into a downstream region. The rate of evaporation is a function of climatic factors such as temperature and humidity. Water can be exported out of a region by a variety of diversion methods such as canals and channels. Consumptive use of water and instream water use demand more in-depth explanations.

Withdrawals versus Consumption. In most uses of water, its quality is altered; either its temperature is changed, or materials are added to the water. If such water is returned to its original quality, it can be described as having been *used* without being *consumed*. Used (but not consumed) water can be used again. After water is used to carry away waste, the wastewater can be treated and returned to a surface or groundwater reservoir and made available again for other uses. Water used in agriculture, however, is not immediately recyclable; 60% or more of the water used in irrigation is consumed in evaporation and transpiration and thus is not immediately available as surface or groundwater. Industry is the largest withdrawer of water, but agriculture accounts for the largest consumptive use by far. We shall see how the inordinate consumption of water by agriculture is leading to significant problems in water-short areas.

Withdrawals versus Instream Use. There are also *nonwithdrawal* uses of water. These are called flow uses, on-site uses, or instream flow uses. Such uses include navigation, commercial and sport fishing, habitat protection, dilution of wastewater, and recreational or aesthetic uses. Instream uses have for many years been taken for granted. However, as surface supplies become more and more overextended, the need to allow for a minimum instream flow to protect instream uses has taken on new importance. For example, in 1987, as a result of drought in the northwestern United States, federal officials had to curtail providing irrigation water to farmers in the Yakima Valley so as to ensure sufficient instream water for migrating salmon. As water supplies become more and more committed, such competition will increase.

The renewable supply of water in a region must meet the instream needs for water as well as the needs for consumptive use. Consequently, the renewable water supply is not available *in toto* for consumptive use. However, the ratio between consumptive use and renewable water supply can be a useful index, in that it can be indicative of the long-term viability of current water use patterns. If consumptive use is approaching the renewable supply, the region is in danger of overextending its use of its water resource. If consumptive use exceeds renewable supply, groundwater is being used at a rate faster than it is being recharged to meet existing water needs. Such a situation cannot exist for long.

Water Use

Water is used for many purposes throughout the world. These purposes can be broadly defined as domestic (cooking, cleaning, bathing), agricultural, and industrial (manufacturing and cooling). About 70% of worldwide water demand is for agriculture, 25% for industry, and less than 10% for residential and municipal uses. Global water demand will continue to rise as population increases and as agriculture and industry expand.

It has been calculated that by the end of the twentieth century, if population continues to grow and climate stays the same, the amount of water available per person will decline by 24%. Countries with growing populations will experience a more significant decline: Kenya an estimated 50% drop; Nigeria 42%, Bangladesh and Egypt 33%. It is estimated that currently 86% of the world's rural population (about 2 billion people) lacks an adequate supply of clean water.

Worldwide, the major consumptive use of water is for agriculture. Over 70% of the water withdrawn globally goes to agriculture. However, only about 37% of all irrigation water is actually taken up by crops. The majority of it is lost to evaporation and runoff because of inefficient application methods.

A prime example of what can happen when water is withdrawn at a rate faster than it can be replenished is demonstrated by the fate of the Aral Sea. The Aral Sea, located between Kazakhstan and Uzbekistan in the former Soviet Union, was at one time the fourth largest lake by area in the world. Between 1960 and 1987 its area decreased by 40%; there was a 13-meter drop in its water level. Salinity levels rose, and over 40 million tons of salt from the dry seabed are now dispersed over surrounding farmlands annually by the wind. This dramatic change is the result of withdrawal of water from the rivers that feed the lake, withdrawals primarily for crop irrigation. Although restoration of the Aral Sea has become an important goal, the adequacy of efforts to restore it remain to be seen.

Table 7.3 describes water scarcity problems in several countries. According to a report by the World-watch Institute, chronic water shortages in the 1990s are predicted in northern China, northern Africa, Mexico, parts of India, the Middle East, and the western United States.

It should be noted that water shortages can be of two types. Shortages may result from low supply or from poor water quality. Chemical spills have caused alerts requiring boiling of water and curtailment of water use until the toxic materials passed through or became sufficiently diluted. Thus even if the supply is sufficient, water of the quality needed for human consumption and municipal and industrial uses may not be available. Quality and quantity are aspects of the same problem of water supply. We will discuss water quality problems in detail in Chapter 12.

Table 7.3 Water Scarcity in Selected Countries and Regions

Country or Region	Observation
North and East Africa	Ten countries are likely to experience severe water stress by 2000; Egypt, already near its limits, could lose vital supplies from the Nile as upper-basin countries develop the river's headwaters.
China	Fifty cities face acute shortages; water tables beneath Beijing are dropping 1 to 2 meters per year; farmers in the Beijing region could lose 30% to 40% of their supplies to domestic and industrial uses.
India	Tens of thousands of villages throughout India now face shortages; plans to divert water from the Brahmaputra River have heightened Bangladesh's fear of shortages; large portions of New Delhi have water only a few hours a day.
Mexico	Groundwater pumping in parts of the valley containing Mexico City exceeds recharge by 40%, causing land to subside; few options exist to import more fresh water.
Middle East	With Israel, Jordan, and the West Bank expected to be using all renewable sources by 1995, shortages are imminent; Syria could lose vital supplies when Turkey's massive Atatürk Dam comes on-line in 1992.
Former Soviet Union	Depletion of river flows has caused the volume of the Aral Sea to drop by two-thirds since 1960; irrigation plans have been scaled back; high unemployment and deteriorating conditions have caused tens of thousands of people to leave the area.
United States	One-fifth of total irrigated area is watered by excessive pumping of groundwater; roughly half of western rivers are overappropriated; to augment supplies, cities are buying farmers' water rights.

Source: S. Postel, 1989. *Water for Agriculture: Facing the Limits.* Washington, D.C.: Worldwatch Institute, p. 25.

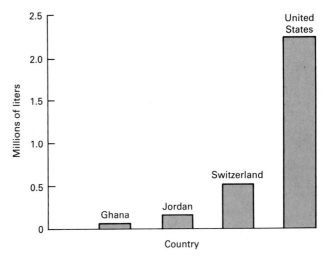

Figure 7.10 Comparison of Annual Per Capita Withdrawals of Water. Per capita consumption of water varies dramatically from country to country. The United States withdraws more water per capita than any other country.

Water Demand

Although the amount of water needed per person per day for biological survival purposes (in food and water) is less than half a gallon, actual water use per person varies dramatically. Domestic water use may range from 76 to 270 liters per person per day for European countries to as low as 2.5 liters per person per day in some developing countries. Figure 7.10 shows a comparison of annual per capita consumption for all uses (domestic, agricultural, and industrial) for several countries. The United States is the biggest user of water of all the industrialized nations. A citizen of France or Germany uses about one-third the water of a U.S. citizen. Figure 7.11 indicates the domestic water consumption of the average U.S. family in a day.

Note that lawn watering and toilet flushing are the leading domestic activities in volume of water used. Water is also used in numerous industrial and agricultural ways. When these other uses are averaged over the entire population, per capita use increases to more than 2,000 gallons per day.

To illustrate this indirect water use, let's consider a 12-ounce can of root beer. Figure 7.12 shows that the water required to produce the can, from mining to manufacture of a single finished can, amounts to about 63 liters.

Figure 7.13 shows trends in water use in the United States from 1950 to 1980. Offstream water withdrawals include water withdrawn for public supply (supplied to domestic, public, commercial, and industrial users from public supplies), rural supplies (self-supplied domestic and livestock), irrigation, and self-supplied industrial. From 1950 to 1980, total wa-

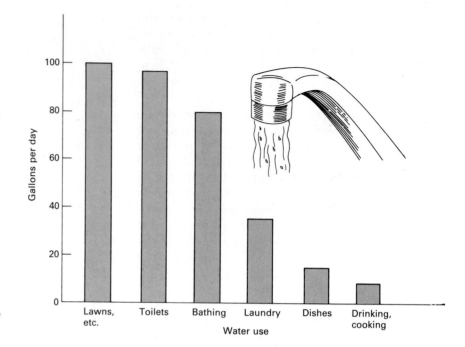

Figure 7.11 Estimated Daily Use of Water by a U.S. Family of Four. Total usage per day will run in excess of 300 gallons. Per capita usage can be figured by dividing each number by 4.

Figure 7.12 Water Needed to Make a Can. To calculate the water input per unit of product made, we need to consider not just water used in the manufacturing process itself but also water used in the mining or harvesting of the raw materials used in the manufacturing process. How many other products do you use daily that directly or indirectly have used or consumed water?

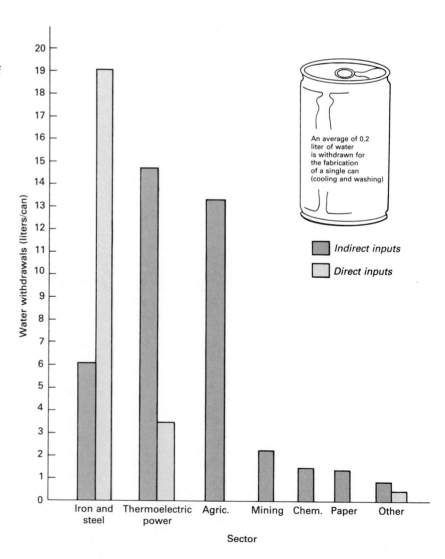

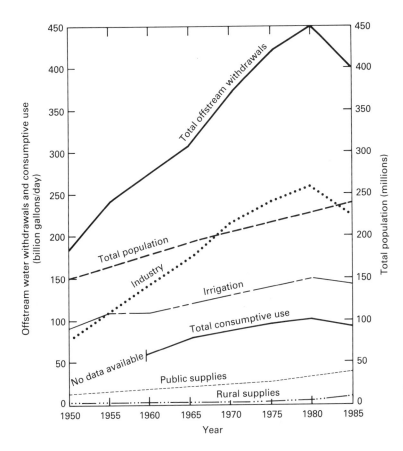

Figure 7.13 Trends in Withdrawals, Consumptive Use, and Population. What conclusions can you draw about changes in per capita use over these years?

ter withdrawals increased by 2 1/2 times, while the population increased only 1 1/2 times. Per capita use for all uses grew from 1,200 gallons per day to 2,000 gallons per day. Although total water withdrawals continue to increase, the rate of increase appears to be slowing. Increased demand may be triggering more efficient use.

In the United States, about 50% of the water withdrawn for domestic use is groundwater; in the ru-

ral parts of the nation, about 95% of the domestic supply is groundwater.

Groundwater

There is 13 to 20 times as much water in the ground as is available on the surface. This water may be contained in the soil or in **aquifers**—areas underlain by impermeable rock (unconfined) or areas between two impermeable rock layers (confined or **ar-**

For Further Thought . . . 7-2

Choices about Water

Water is in very short supply in some areas of the United States and is becoming increasingly scarce in many other areas. In many areas with access to groundwater, annual water use exceeds recharge. There are many competing uses for water. Crops can't grow without it, and we have major agricultural operations in areas with limited rainfall. Commercial and industrial uses of water are critical, and jobs depend on them. We have grown dependent on water to flush our wastes and water our gardens and lawns. We have found that if we don't maintain a certain volume of water in streams, some aquatic species die, and pollutants may not be sufficiently diluted. On what criteria should allocation of scarce water resources be based? What kinds of uses of

water should be restricted when supply begins to fail to meet demand? Some people believe that we should restrict development on the basis of availability of water resources. What do you think?

tesian) (see Figure 7.14). About one fourth of this water is extractable by using today's technology.

Groundwater reservoirs are recharged by rainfall; however, with increasing demands on groundwater supplies, the risk of pumping more water than is recharged in any given time increases.

When withdrawals exceed recharge, several things may happen. The water table levels drop. The

water table is the upper surface of an unconfined aquifer. A drop in the water table may result in subsidence (lowering) of the ground surface into the voids left by reduced amounts of groundwater. Subsidence permanently diminishes the size of the aquifer.

Another problem with the overuse of groundwater occurs in coastal areas. Overuse of ground-

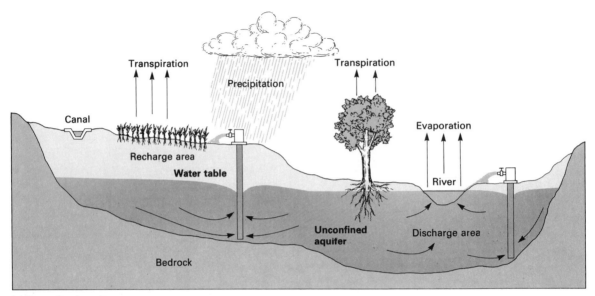

(a) Unconfined aquifer

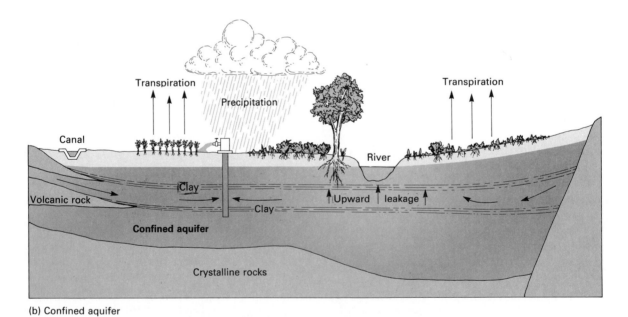

(b) Confined aquifer

Figure 7.14 Groundwater. Rainwater infiltrating the soil collects in aquifers as groundwater. The upper surface of an unconfined aquifer is the water table. The water table shifts upward when water accumulates after rain, recharging the aquifer; it shifts downward as water is withdrawn by plants or human activities, as it flows to lower levels, or as it is discharged to the surface.

water supply and the consequent lower water tables result in the intrusion of salt water into the water table, making the groundwater brackish. **Saltwater intrusion** can be a groundwater problem in any coastal area.

Other problems arise when cities begin to rely on groundwater supplies. Urbanization, which covers soil with asphalt, causes increased runoff and less percolation of rainfall into the land. This decreases recharge and lowers the water table.

Table 7.3 describes problems of overuse of groundwater in several countries, including China, Mexico, and the United States. The United States is one of the few countries to have tried to assess its overpumping of groundwater. It has been determined that 72% of the total acreage irrigated in Texas is done so by overpumping of groundwater. This is true of 57% of the acreage in Kansas, 46% in Oklahoma, and 41% in Arizona. According to the Worldwatch Institute, about one-fifth of the irrigated acreage in the United States is watered by overpumping of groundwater. Depletion problems in Texas, California, Kansas, and Nebraska date back to the early 1980s.

The impact of overpumping is readily apparent in the Ogallala Aquifer.

The Ogallala Aquifer. The Ogallala Aquifer underlies the High Plains states from Kansas and Nebraska to Wyoming, including Texas, New Mexico, Oklahoma, and Colorado. The agricultural economy of the Texas High Plains has been built from the overdrafting of the Ogallala Aquifer.

Since the 1950s, water tables have dropped by as much as 100 feet. The water level continues to drop 2 to 5 feet a year. In Gaines County, Texas, the cost of pumping 1 acre-foot of water increased from $1.50 to $60.00 in a decade. (An acre-foot is the amount of water it would take to cover 1 acre 1 foot deep; it is approximately 325,000 gallons.) The cost increase is due to increased energy prices and increases in the amount of energy required to pump the water up. Some farmers in the area have stopped irrigating their crops because of increased pumping costs as a result of the lowering water table.

Over the years, federal policies have encouraged use of water from the Ogallala Aquifer by providing price supports for water-intensive crops like cotton. Even now, High Plains farmers are eligible for a depletion allowance on pumped groundwater from the Ogallala Aquifer. In Kansas alone this amounts to benefits estimated at $1 billion. A congressional study of the Ogallala Aquifer indicates that the aquifer can supply the Plains states with water for the next 40 to 50 years. The problem is, What happens after that? As other demands compete with agriculture for water, how will limited supplies be allocated?

Water Rights and Water Economics

Water economics has had an interesting history. Throughout early history, water was free. There were no charges for using water, diverting water, or depositing waste in bodies of water.

It is interesting to note that although water was free, the rights to water developed differently in the eastern and western parts of the United States. In the East there evolved the **doctrine of riparian rights** (rights relating to the bank of a watercourse), which said that a landowner adjacent to a stream had the right to the water in the stream undiminished in quantity and quality. This placed responsibility on upstream users and was directed toward protecting private rights in streams and lakes. In the West, where water was scarce, the old adage "first come, first served" applied. The **doctrine of appropriation** states that the right to water is acquired when water is diverted from the watercourse for a "beneficial" use, and in all cases the water right that was acquired first takes priority over those acquired at a later time. "Beneficial" was interpreted as economic; thus diversions for economic purposes were the only uses that carried with them recognized rights. Instream uses of water or diversions for wildlife or other "uneconomic" purposes were not recognized historically as beneficial uses. It is likely that in both the East and the West, sophisticated accounting systems will have to be developed in order to ensure that public water needs are met, even at the expense of private needs.

Today the cost of water varies with the system of acquisition, purification, and transportation. Pricing is generally not strictly proportionate to the amount used. In fact, in many places, the more an industry uses, the cheaper the rate. Not surprisingly, this subsidization of water has created situations wherein it is cheaper to draw clean, fresh water than to reuse the old.

A prime example of water subsidization is the U.S. Bureau of Reclamation program for agriculture. The water supplied to farmers subsidized by the Bureau of Reclamation costs around $10 an acre-foot in Phoenix, Arizona. The cost of unsubsidized water is much higher, more than $100 an acre-foot. What is even more ironic is that 33% to 45% of the acreage irrigated by water subsidized by the Bureau of Reclamation is used to produce surplus crops. The subsidies for a 960-acre farm can range anywhere from one to several million dollars.

For domestic use, charges may be based on volume, but they are often based on such things as number of taps, property frontage, or even property value. Overall, our systems of pricing developed when water shortage was not a concern. With increasing water shortages and continued population and industrial growth, the need to increase supply or decrease demand arises. A change in the way we charge for water would be one mechanism for influencing water use. There are others.

Throughout the world, governments use various economic incentives to encourage industry to locate in certain areas. These may indirectly subsidize water use (and energy and other raw material use) and be counteractive to sound conservation practices. Investment tax breaks, low-interest loans, depreciation allowances, and noncompliance fees are all examples of economic instruments employed by governments that can affect resource use.

Water Conflicts

According to the World Commission on Environment and Development, global water use doubled between 1940 and 1980 and is likely to double again by the year 2000. It is estimated that two-thirds of the projected water use will go to agriculture.

With growing demand in urban areas for water, conflicts between agriculture and domestic and industrial uses are inevitable. The World Commission lists several examples of disputes over water in North America (Rio Grande), South America (Rio de la Plata and Paraná), South and Southeast Asia (Mekong and Ganges), Africa (Nile), and the Middle East (Jordan, Litani, Orontes, and Euphrates).

As with minerals, water policies have significant political implications. In the spring of 1990, Turkey announced that it was going to cut off water temporarily from the Euphrates to Syria and Iraq. Turkey has been building a series of dams to allow it to control the waters of the Euphrates that flow through its borders. Although Syria draws most of its water from the Jordan River Basin, it already rations its water and cuts off water for several hours a day in some areas. Egypt is totally dependent on the Nile for its water but has no control over the river's headwaters, most of which are in Ethiopia. The ability to use water as a political weapon will no doubt affect foreign policy in many areas.

Because agriculture is such a major consumer of water, water disputes within a given nation are likely to pit agriculture against domestic and industrial users. In such a conflict, it is generally thought that agriculture will lose out to other uses. The reasons are predominantly economic. According to Reisner (1986), agriculture accounts for 85% of all water used in California, yet agriculture comprises only 2.5% of the total California economy. In more rural New Mexico, 92% of the water goes to agriculture, but agriculture accounts for only 18% of the economy. According to a Worldwatch Institute report, planners in China figure that water used for industrial purposes generates 60 times the economic value it would if used for agriculture.

This does not necessarily mean that agricultural productivity will suffer. It does mean that more efficient irrigation techniques will need to be employed. This will be discussed further in Chapter 8.

But unlike minerals, water conflicts are going to be not with distant nations but with neighbors. The Great Lakes Charter is an example of neighborly cooperation in water management.

Regional Water Management

In July 1982 the U.S. Supreme Court, in *Sporhase* v. *Nebraska*, declared that water is an article of interstate commerce and therefore a state cannot ban the export or diversion of water unless water conservation is tied to a clear public health or welfare purpose. Limitations to water use must be imposed equally on all users, instate and out of state. This ruling was one factor that led to the establishment of the Great Lakes Charter, a consortium of the states and Canadian provinces bordering the Great Lakes. The members of the charter have agreed to work together to establish programs in the individual states and provinces to manage and regulate the diversion and consumption of water in the Great Lakes Basin. A set of management principles has been agreed on, and the transformation of these into programs is the next step. For example, Wisconsin and Illinois have passed legislation requiring permits for any diversions or consumptive uses greater than 2 million gallons per day. This regional approach to management of water resources is a model for other regions to consider.

Water Marketing

Another approach to managing water use has developed in the western United States. Since water is a commodity of commerce, the concept of marketing water or water rights is becoming more common. Farmers with rights to a given amount of water could sell the water they do not use to another user. For example, by improving irrigation methods and conserving water, a farmer could have an "excess" of water to sell to a municipality. The economics provide conservation incentives to both buyer and seller. Table 7.4 shows some examples of water rights sales in the western United States.

A related phenomenon, termed "water ranching," has developed. Agricultural land is purchased primarily for its water rights. According to a report of the Worldwatch Institute, over 230,000 hectares of

Table 7.4 A Sampling of Water Rights Sales in the Western United States, 1986–1989

Buyer	Seller	Volume (acre-feet per year)	Cost ($ per acre-foot)	Purpose
Westminster, Colorado	Local irrigators	272	6,176	Municipal use
Westpac Utilities, Reno, Nevada	Urban homeowners on formerly irrigated land	2,000	2,000	Municipal use
Phoenix, Arizona	McMullen Valley Farm	30,000	1,017	Projected municipal needs after the year 2000
Albuquerque, New Mexico	Local irrigators	1,360	1,000	Projected municipal needs after the year 2020
Nevada Waterfowl Association	Truckee-Carson Irrigation District	35	214	Wetlands and waterfowl protection
Central Utah Water Conservancy District	Local irrigators	85,000	164	Municipal use in lieu of a previous plan to drain wetlands for their water

Source: S. Postel, 1989. *Water for Agriculture: Facing the Limits.* Washington, D.C.: Worldwatch Institute, p. 29.

farmland had been purchased for this purpose. The land may be farmed until residential or industrial water demand warrants the shift of water use from agriculture to these other uses. At that point, the agricultural function of the land will cease.

Increasing Supply: Old Ways and New

The most common method used for increasing water supply has been diverting water to where it is needed. Also, streams have been dammed to control water fluctuation, and groundwater supplies have been tapped by means of wells. In recent times, new methods have been tried or proposed for increasing the supply of fresh water. These include desalination of seawater, melting of glaciers, and seeding of clouds. Each of the ways of increasing supply carries with it some ecological and other implications.

Diversions

Homo sapiens has often seen fit to base its technology on models found in nature. It is not surprising then that one of the major ways of redistributing water is through human-made diversions. Artificial streams such as channels or pipelines have been constructed to take water to where it is wanted.

The western United States provides several examples of major diversion projects developed to carry water to the growing population centers in the western states. The California State Water Plan outlines a

This open, lined section of the Los Angeles Owen River Aqueduct runs approximately 40 miles, providing water to the Los Angeles area.

system of diversions from the Colorado River and other streams to the north. The Central Arizona Project is designed to pipe water 250 miles from the Colorado River to cities and land areas of central Arizona.

Diversion projects are major undertakings often subsidized by federal taxpayers. In earlier times, when water supplies were plentiful, diversions were rarely opposed. Today, diversion projects are mired in political, social, and environmental problems. Shortages are widespread, and people with water are becoming more and more possessive.

Northern Californians oppose plans to divert water from northern California to southern California, for example. There are, in fact, real problems with giving up water apart from the obvious one of not keeping enough. Lower supplies in the donor area eventually mean higher water prices there. Diversions may affect instream flows and species in the area from which volumes of water are taken. Likewise, future industrial development may be threatened. Then what about droughts? When these occur, which users—direct or diversion—get how much water?

Also, diversions do not necessarily have only positive effects at the receiving end. Dry areas into which water is brought usually have high evaporation rates. In normally arid regions, evaporation pulls water from deep in the soil to the surface. As the water evaporates, the salts it carries are deposited on the surface of the land. This may lead to an accumulation of salts that leaves the land unfit for further agricultural uses. The San Joaquin Valley in California provides a good example of this very phenomenon; once a fertile valley, it is now endangered by salt accumulated from irrigation waters. A similar problem has occurred in California's Imperial Valley. Worldwide, about 1 million hectares annually are damaged by salinization. Fully 36% of India's irrigated land is damaged by salinization, 27% in the United States, and 20% in Pakistan. We will discuss the implications of this further in Chapter 8.

Then, too, water availability spurs population growth and further development, requiring even more water. A large portion of the people in the world are totally dependent on "engineered streams" or diversions for their water needs. The future management of these areas is of vital importance as prolonged droughts bring increasing potential for catastrophe.

Dams

Another method for increasing water supplies in a given area is to impound water in dams. Dams have been around as long as the beaver and probably from the first time a tree fell across a stream, causing a pileup of debris and a backup of water. Human-made dams are large-scale versions with controls for varying the water level.

Dams serve human needs in several ways. First, dams collect water, hold back the excess water during flooding, and release it gradually during low-flow periods. This means that the floodplain below the dam is now available for farming and development with less risk of flooding. Second, the large pool of water in the reservoir provides a constant supply and continuous flow for human consumption, farm irrigation, and industrial uses. Third, the pool has recreational potential. A dam may also generate electricity.

But there are ecological problems with dams. Damming produces a large pool area in the river or tributary. As the moving water enters the pool area, it slows down. This causes sediment to settle out. Eventually, silt buildup makes reservoirs practically nonfunctional because of the decrease in volume of the reservoir.

Another problem is that some plants and animals that flourish in streams are not adapted to and will not survive in reservoir impoundments. Lake flora and fauna will take over the niches. Furthermore, the land area flooded by the reservoir is no longer available for land uses; in some cases this has meant relocation of the human and other inhabitants. The potential for new human inhabitants with different skills in the area served by the reservoir may well be increased, so a sort of "niche" exchange takes place as a dam is built. Such exchanges can have far-reaching negative as well as positive effects.

A classic example is the case of the Aswan Dam in Egypt. In 1960, construction was begun on the dam that was to turn the Sahara into a rich farming area. However, a number of major adverse side effects have become apparent since the dam's completion in 1970. Lake Nasser, the reservoir produced by the dam, is filling with silt at a faster rate than was expected. The nutrients carried by the Nile are enriching the lake and causing problems with eutrophication. Before the dam was built, the nutrients and silt were deposited annually downstream in the Nile Delta. Because this source of fertilizer for the delta has been cut off, the productivity of the delta area has decreased. The newly irrigated Sahara is also plagued with problems of increasing salinity. Another hazard that was overlooked relates to human health. Diseases involving vectors that breed in standing water are on the rise.

In the United States, various governmental entities are taking action to reverse some of the ill effects of major dams. The Tennessee Valley Authority is adding devices such as baffles and weirs to its dams to increase the oxygen concentration in the river in order to support an adequate fish population. Experiments on Norris Dam near Knoxville have proved successful. Several other dams have already been retrofitted, with the possibility of more being retrofitted in the future. At some dams, attempts are being made to regulate the water temperature downstream by changing the water intake upstream. In areas with fish and plant species adapted to warm water, drawing water through the intakes from warm surface water protects native fish and plant species. Where species are adapted to cooler temperatures, intake waters can be drawn from cooler deeper waters. A change of a few degrees can make a significant difference. The Bonneville Dam is modifying its turbines to protect salmon and steelhead trout. Control of water flow over the dams to lessen erosion below the dam is a management technique that is also being employed in some places.

Glaciers and Ice Mining

What is the potential for using the fresh water tied up in glaciers for human water needs? Proposals have been made to pull icebergs to major coastal cities and mine them for water. The logistics of such a project are probably feasible; however, there are a number of unanswered questions. What effect would the ice have on the surrounding waters near the coast? How would it affect climate? When the ice was replaced by water in the arctic area, less sunlight would be reflected, and more would be absorbed. Would this warm the arctic waters enough to cause melting of the ice remaining there? Every action may have enormous environmental repercussions because, after all, everything is connected to everything else.

Desalination

Almost 97% of the earth's water is salty. Hence it is logical to consider the feasibility of using salt water for human needs. Salt water can be used directly for power plant cooling along coastal areas. However, it is too salty to be used for most industrial processes, agriculture, or drinking. Before seawater can be used for these purposes, the salt must be removed. This is known as **desalination.**

There are three methods by which salt can be removed from seawater: distillation, electrolytic and osmotic separation, and freezing.

1. In distillation, the water is boiled or heated, and pure water is collected as it evaporates. The salt is left behind. Proposals have been made that would tie nuclear power plants to distillation plants. Waste heat from the power plant would provide the energy for heating the water.
2. In electrolysis, positive and negative electrodes attract the charged salt and leave the water

away from the electrodes less salty. In a process called reverse osmosis, seawater is pushed under pressure through a membrane that allows the water to pass through but not the salt.
3. In freezing technology, the salt forms crystalline pockets as the ice freezes.

As predicted by the second law of thermodynamics, all of these methods require energy.

Although the technology for desalination exists, it is now used only in small-scale plants. Desalination is economical in arid regions where there are no other water sources to compete with it. There are several thousand desalination plants operating worldwide making over 3 billion gallons of seawater potable each day. About 60% of this desalination occurs on the Arabian peninsula.

Even if cost could be brought into a more competitive range, there is another problem. What do we do with all of the salt? Every million gallons of fresh water that is produced from seawater leaves about 150 tons of salt behind.

Cloud Seeding

Another source of fresh water is the clouds. **Cloud seeding** is used to cause rain. Chemicals are sprinkled into the clouds to serve as surface agents on which water can condense.

How successful cloud seeding actually is has been difficult to verify scientifically. There is no sure formula for producing rainfall at will from cloud seeding. Investigations have been going on since the late 1940s and continue today. Cloud seeding poses interesting legal questions. Who owns the water in the clouds? If clouds are seeded, they may drop their water in Washington State rather than in the Rocky Mountains. There is only a given amount of water in the condensation phase; if it is used in one place, it will simply not be available to users downwind.

Decreasing Demand

In the discussion of minerals we noted that the demand for any one mineral could be shifted by the development of substitutes. There could be other mineral deposits in more abundant supply or new synthetic materials. Other methods for decreasing demand involve longer-lasting mineral products and lifestyle adjustments.

In the case of water, most of these alternatives are not available. There are no substitutes to which demand can shift. There is nothing like water. However, demand can be decreased by improving the efficiency of water use. This would be directly comparable to recycling and somewhat comparable to the development of longer-lasting products in the case of minerals. Let us look at some examples for improved water resources management.

Water Resource Management

We can conveniently classify water uses as agricultural, industrial, and domestic, and there is room for improved water resources management in all of these areas.

Irrigation, being a major consumptive use, can be executed much more efficiently. Irrigation can be carried out via underground pipelines, which decrease the amount of evaporation and consequent water loss. Drip and sprinkler irrigation systems conserve water. Lined trenches for water transportation prevent seepage loss. Plant hybrids requiring less water have been developed. Improved timing of water deliveries, avoidance of overdeliveries, and control of weeds or competing vegetation focus water on the desired crop. All of these can increase the agricultural output per unit of water used.

As water resources become scarce and prices increase, industry will find more efficient ways to control its water use. Industry has already begun to reuse its water, particularly water used primarily for cooling. Leakage and water pressure controls are other ways of decreasing industrial uses.

Every flush of a toilet ordinarily carries 4 to 7 gallons of water with it. Devices are on the market now that cut down on the volume of water used in each flush. These devices are a little more sophisticated than the brick-in-the-tank technique, though the brick still works. (In the early 1970s, environmentalists encouraged people to put a brick in the toilet tank to take up some water space; when the tank refilled after each flush, not as much water was needed to fill the tank to the desired water level.) Devices for showers and faucets are available that reduce water flow. San Francisco now requires that toilets use no more than 2.5 gallons per flush and that showerheads permit a flow of no more than 2.5 gallons per minute.

Changes are also gradually occurring in the system of charging for water use. Individual metering of homes is increasing. Incentives for water use at off-peak times is an interesting possibility. Several states have approved the use of waterless sewage waste systems in which human wastes and kitchen wastes are used to form compost. The U.S. Army Corps of Engineers has calculated that universal metering in New York City would curb water use by 50 million gallons per day, domestic conservation devices would save 85 million gallons per day, and leakage controls would save another 25 million gallons per day.

Water quality is another issue to be considered. For the most part, the same quality of water is used for all activities, whether domestic, industrial, or agricultural. This again is a reflection of earlier times when water was abundant and free. In the home, the same quality of water is used for sewage disposal as for drinking. Because of shortages, sequential use of water may become the pattern at home and work. Human wash water can be used to clean the car and then to flush sewage wastes. There is, in fact, a growing market for waste water (see Vaughan, 1992). This leads to another mechanism for decreasing demand, alteration of lifestyle.

Bluegrass Versus Rock

Lavish and inefficient use of water resources is a feature of the American standard of living. It seems to have developed from the axiom that "cleanliness is next to godliness." Use of waterless systems for sewage wastes and reuse of domestic water will require fundamental adjustments. Learning to appreciate the beauty of a rock garden instead of a carpet of green front lawn—especially in places like the Southwest—will require changes in values and attitudes. Setting priorities among essential uses and distinguishing them from luxuries will have to be done as water becomes an increasingly scarce commodity.

Institutional Adjustments

For change to occur, institutional policies and practices in the area of water resources management must be altered. There must be economic incentives to conserve. This can be brought about by a change in pricing structure. Some method of compensating for water *not* used is a possibility—just as we have compensation for land not used—and for providing this unused water to a water user in need. A clearinghouse for information on tested conservation practices and devices would be valuable. A major reason that water-conserving techniques are so little used is lack of knowledge of their effectiveness. A public education program is needed. Efforts have to be made to curtail government water subsidies to water users. Irrigation systems are often subsidized by public funds in Third World countries as well as in the United States.

The relationship between water quality and water quantity needs to be fully understood. Shortages stem from overloading water supplies with sewage or industrial wastes. Why use water of drinking quality to flush toilets or even for cooling processes? Methods for sequential water uses need to be explored—with, of course, all caution to prevent negative health effects. Related to water quality is good land management to prevent sedimentation, increased siltation, and the filling of water reservoirs.

The Future

Humans and human civilizations depend on water and minerals. Interestingly, water and mineral resources are interlinked. Water is used in the extraction, processing, and shipping of minerals. In the future, as poorer grades of minerals are mined, the percentage of our water resources used in mining and industry is expected to increase significantly. Water and mineral resources have been exploited, and the time has now come—or may even have passed—for them to be better managed. There are ecologically sound ways to approach the management of these essential resources, but they have yet to be implemented. Water and mineral resource problems will be among our most important environmental problems in the future.

Concepts to Remember

1. Access to mineral resources has become as crucial to modern civilization as food is to the individual person.

2. Mineral *resources* are minerals that are important to human beings. Minerals that are accessible under existing economic and technical conditions are called *reserves*. The relationship between resources and reserves changes with the relationship between supply and demand.

3. Nonrenewable resources are those being used up faster than they are regenerated. A small number of countries, the so-called industrialized nations, use up a disproportionate share of the world's mineral resources.

4. Although the life of our mineral reserves can be extended by finding new deposits, mining the sea, finding substitutes, and developing more efficient extraction processes, none of these strategies makes as much ecological sense as consuming less.

5. As highly concentrated deposits of minerals are depleted, lower-grade deposits are mined, requiring more energy for mining and processing per unit of usable mineral produced, resulting in more environmental disturbance.

6. Rates of mineral consumption can be reduced by recycling, by extended use of products (antiobsolescence), and by substituting other materials. Various institutional factors can work against the large-scale adoption of these practices, including tax policies, subsidies, and transportation pricing.

7. The uneven worldwide distribution of mineral deposits adds a political dimension to mineral supplies. The foreign policies of industrialized nations reflect, at least in part, their mineral supply needs.

8. Access to water has extended the range of land available to humans for settlement and development. Many people worldwide now depend on the availability of water supplied by human engineering projects.

9. Only a very small percentage of the total water in, on, and above the earth is accessible to humans at any one time as fresh water.

10. Usable water reservoirs are recharged by precipitation. If water is withdrawn faster than it is replenished, the water supply is in jeopardy. In many places, water withdrawal exceeds recharge. Lowering of the water table can lead to other problems such as saltwater intrusion.

11. Water withdrawn for *use* may or may not be *consumed*. Water used for cooling will be returned to a water reservoir and be immediately available for another use; much of the water used for irrigation will be consumed by evaporation.

12. Instream uses of water are also important and can be jeopardized by excessive withdrawal. Instream uses include navigation, recreation, support of aquatic life, and waste dilution.

13. Water rights have developed differently in the eastern and western United States. In the eastern United States, where water is relatively plentiful, laws tend to protect the rights of all users along the stream course. In the West, with scarcer water resources, the doctrine of appropriation or "first come, first served" has prevailed. The marketing of water rights is becoming more common.

14. Methods for increasing water supply have included the building of dams and diversion projects, desalination of ocean water, cloud seeding, and ice mining. All of these bring major environmental impacts directly or as a result of increased energy use.

15. Decreasing demand for water can be achieved by greater efficiency in its use.

16. With increasing water scarcity, water conflicts are likely. Agriculture will compete with domestic and industrial uses. Nations, and regions within nations, will compete for available resources.

Marat Khabibullov

Marat Rafikovich Khabibullov, Ph.D., is an assistant professor and vice-dean for international programs in the School of Ecology at the Kazan State University in the Tatar Autonomous Republic of Russia. Dr. Khabibullov teaches biology and ecology courses related to his interests in herpetology, evolutionary ecology, wildlife protection, and community ecology; he also serves as the international program coordinator for the Business Unit of the school of ecology. The Business Unit of the School of Ecology is an organization through which the faculty engage in fee-for-service project work for local governments and private companies springing up throughout the Tatar Republic. The Business Unit functions much like consulting firms found throughout the world engaging in such activities as ecosystems analysis and environmental impact assessments. Part of Dr. Khabibullov's job is to arrange partnerships between Russian environmental companies with companies in North America. Dr. Khabibullov recently spent ten months with Entrust Canada, Inc., an environmental management consulting firm in Calgary, Alberta, to learn about how such firms operate. Dr. Khabibullov is very much interested in the development of his country's environmental policies and administrative regulations designed to protect nature.

Dr. Khabibullov says that he never has a "typical" day. His routine usually includes some teaching and some form of dealing with government bureaucracy arranging exchanges of faculty and students. Much of his time is spent organizing and participating in meetings and mediating arrangements between the Business Unit and the environmental work to be done. He spends at least several hours each day writing and reading.

The big challenge that Dr. Khabibullov and the republics of the former USSR now face is to make sure that a regard for nature and a clean environment are built in to the emerging new order in that part of the world. Today, environmental disasters and serious problems of land degradation, air quality, and water quality cover one-fifth of the former Soviet Union. The new independent states must find a way to move from

the environmentally ineffective command and control mode of their former strong central government to a more effective environmental policy mode to fit within the decentralized, market economy now under development.

Dr. Khabibullov has a doctorate in zoology from the Institute of Zoology of the Academy of Sciences of the USSR in Leningrad. His Ph.D. dissertation was based on field studies of the reptiles of Kugitangtau in eastern Turkmenistan. He has served as a visiting lecturer on environmental topics at the University of Wisconsin at Madison, the University of Oregon, Western Washington University, Cornell University, and Dartmouth in the United States; McGill University in Montreal, Canada; and the University of Siena in Italy. In addition to Russian, Dr. Khabibullov speaks and writes in English; he also speaks French, Italian, Polish, and Tatar.

References and Further Reading

References marked with an asterisk are cited in the chapter.

Abelson P., 1991. "Desalination of Brackish and Marine Waters," *Science* **251**(4999):1289.

Anderson, C., 1990. "Recycling," *Governing* **3**(11):5A–30A.

Bureau of Mines, U.S. Department of the Interior, 1992. *Mineral Commodity Summaries.* Washington, D.C.: U.S. Government Printing Office.

Dixon, J. A., 1989. *Dams and the Environment.* Washington, D.C.: World Bank.

Frederick, K. D., 1986. "Watering the Big Apple," *Resources* (82):2–4.

Gottlieb, R., 1988. *A Life of Its Own: The Politics and Power of Water.* Orlando, Fla.: Harcourt Brace Jovanovich.

Leopold, L. B., 1990. "Ethos, Equity, and the Water Resource," *Environment* **32**:16–20ff.

Maurits La Rivière, J. W., 1989. "Threats to the World's Water," *Scientific American* **261**(3):80–94.

Micklin, P. P., 1988. "Desiccation of the Aral Sea: A Water Management Disaster in the Soviet Union," *Science* **241**:1170–1175.

Morgan, J. D., 1991. "Strategic Materials for the Future," *Futures Research Quarterly* **7**(4):5–19.

O'Mara, G., 1988. *Efficiency in Irrigation.* New York: World Bank.

Pollock, C., 1987. *Mining Urban Wastes: The Potential for Recycling.* Washington, D.C.: Worldwatch Institute.

Postel, S., 1985. "Thirsty in a Water-rich World," *International Wildlife* **15**:32–37.

Postel, S., 1989. *Water for Agriculture: Facing the Limits.* Washington, D.C.: Worldwatch Institute.

Reisner, M., 1988–1989. "The Next Water War: Cities versus Agriculture," *Issues in Science and Technology* **V**(2):98–102.

*Reisner, M., 1986. *Cadillac Desert.* New York: Viking Press.

Rogers, P., 1986. "Water: Not as Cheap as You Think," *Technology Review* **89**:30–43.

Smith, R. T., 1988. *Trading Water: An Economic and Legal Framework for Water Marketing.* Washington, D.C.: Council of State Policy and Planning Agencies.

Vaughan, R. J., 1992. "Reclaiming Western Water," *Governing* **5**(6):76–77.

World Commission on Environment and Development, 1987. *Our Common Future.* New York: Oxford University Press.

Young, J. E., 1992. "Mining the Earth," *State of the World 1992* Lester Brown New York: W.W. Norton & Company.

8 Population, Food, and Hunger

Section A

Population in Perspective

Once upon a time, a great many people were invited to a banquet. Even though none of the guests had any idea who was giving the banquet or why they had received invitations, they all made plans to attend.

It came to pass that on the night of the banquet, more guests appeared than were anticipated. Perhaps some of the guests had failed to comply with the RSVP request, or perhaps there was some failure of coordination among the planners, inviter, and cooks. At any rate, the result was too little food for the number of guests that appeared. To make matters worse, the problem did not become apparent to the kitchen and serving staff until they had already distributed standard portions to the people at the first few tables. After that point the waiters adjusted by serving smaller and smaller portions until those at the last few tables were actually underfed.

Naturally, there was a mixed reaction to the whole state of affairs. In the kitchen the collective impression was: *There are too many people out there!* In the dining room the predominant impres-

sion was: *There isn't enough food here!* Of course, those who had been underfed felt much more strongly that the problem was lack of food.

We introduce this chapter on population and hunger with this story to emphasize at the outset that the concept of "overpopulation" is relative (Figure 8.1). If the problem is that there are too many people, the problem is not completely stated until the following questions are answered: Too many people in relation to what? To available food? To nonrenewable resources? To the assimilative capacity of the biosphere? To personal preference?

Overpopulation?

Three notable characteristics of overpopulation among animals in general are mass suffering, a high rate of premature death, and deterioration of the environment. Some intelligent people argue that such conditions characterize *Homo sapiens* at this very time. Other intelligent people argue that these problems are not simple consequences of numbers of people; they are a result of the way people do things. They say that the earth is far from overpopulated.

208

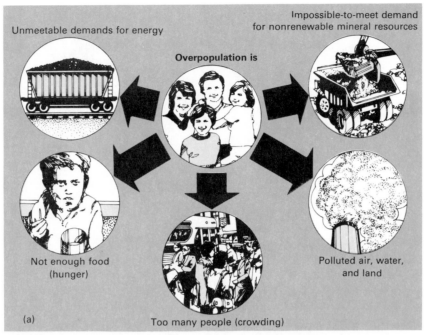

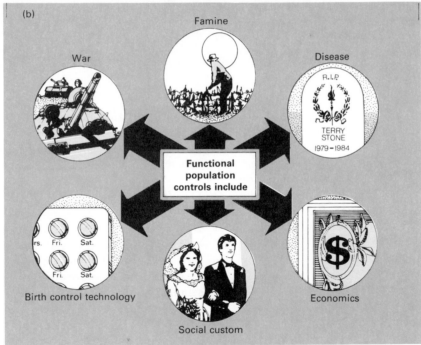

Figure 8.1 Overpopulation Is More than a Big Number. (a) The carrying capacity of the earth is related not only to numbers of people but also to impact of their lifestyles. A few people with a highly consuming lifestyle can have a much greater impact on the ecosphere than a larger number of people with a simpler lifestyle. (b) Likewise, population control is much more than birth control technology. Economic, cultural, and religious factors also come into play.

Throughout the world, 15 to 20 million people die each year because of malnutrition or starvation. Is this a problem of overpopulation or a problem of the inequitable distribution of food and wealth?

It has been calculated that one U.S. citizen has an impact on the biosphere equal to that of 25 citizens of India. Thus we would need to multiply the population of the United States by 25 to compare it to the population of India in terms of effect on the biosphere. For other comparisons, other relationships might apply. For example, with 5% of the world's population, the United States consumes over 33% of

The population problem is not just a question of numbers. The various environmental impacts of the world's cultures, customs, and lifestyles must be considered in dealing with it.

the mineral resources of the world. By the single standard of mineral resource consumption, the population of the United States is six or more times larger than actual numbers indicate.

The point? One's concept of overpopulation depends on personal values, societal standards, the lifestyle to which one is accustomed, or the lifestyle one thinks *everyone* should have. The population problem is more than a problem of numbers.

Limits to Growth

The enormous potential for human population growth has been known and discussed for hundreds of years. In the sixteenth century, Thomas More specified in *Utopia* "that the city neither be depopulated or grow beyond measure." In the eighteenth century, Robert Wallace wrote, "Under a Perfect Government . . . mankind would increase so prodigiously that the earth would at last be overstretched and become unable to support its numerous inhabitants." Thomas Malthus (1766–1834) started a controversy that is still raging with his observation that population will always expand to some limiting situation in which misery is the general condition. Malthus sets forth an explanation of why, in his estimation, "the power of population is indefinitely greater than the power in

the earth to produce subsistence for food." The basis of his argument is his hypothesis that population increases geometrically whereas subsistence increases only arithmetically (see Chapter 4).

War, Famine, and Disease

Historically, the three most important factors serving as checks on human population have been death by war, famine, and disease. Throughout history, although human population was increasing gradually, there were many "booms" with subsequent "crashes" attributable to one or more of these factors.

A war affects population in two ways. Directly, it kills people. Indirectly, it sets up conditions in which famine and disease can occur.

Famine and disease may also result from other factors. One of the most notorious famines occurred in Ireland in 1845. The success of the potato as a crop in Ireland (beginning around 1588) contributed to a population increase from 2 million in 1687 to over 8 million in 1841. It is estimated that in 1845–1846, when the potato blight occurred, over 1 million people died of starvation or related diseases; many more people moved to other countries. Today, Ireland's population fluctuates between 4 and 5 million.

The outstanding example of disease as a population control factor was the fourteenth century's

Black Death, the plague during which one-quarter of the population of central Europe died. England's population was cut in half by the plague between 1348 and 1379.

War, famine, and disease are still at work today, smoldering and flaring with an ever-increasing potential for impact on human numbers. A quick examination of almost any newspaper or weekly newsmagazine will corroborate this point.

Self-regulatory Mechanisms

Do other factors limit growth in *Homo sapiens*? Scientists have found factors other than lack of food, predation, and disease that serve as limits to population growth in various animal species. For example, the males of many species stake out territories. Territoriality limits the numbers that can breed in a given area (see Chapter 4). Physiological effects of crowding lead to hormonal changes in some animals, which result in diminished reproductive potential and even death (see Chapter 15).

How much of this applies to humans is not known, but many scientists hold that such self-regulation is not physiologically innate in humans. They believe that in our species, similar regulatory control is achieved by culture and social custom. Lack of food may, among other things, trigger the social regulation of population growth. Various other socioeconomic factors are also important. The scientific and industrial revolutions seem to have led to population decrease rather than increase. In an agrarian society, children are unquestionably assets—as workers. In an industrial setting this is not necessarily so. Children may in fact reduce mobility and make accumulation of goods more difficult. This is the basis of an anti-Malthusian theory stating that poverty increases breeding, whereas plenty decreases it.

Physical Laws and Population Growth

J. H. Fremlin (1964) once calculated the absolute limit of the earth's population. He assumed that if all political, sociological, medical, and other human problems resulting from an expanding population could be solved, the final limit would be a physi-

---

For Further Thought . . . 8-1

Population and Hunger

There are too many people in the world to feed. There is not enough food available to feed the people in the world. These two statements reflect two different views on the root cause of hunger in the world. One side believes that the world is overpopulated and that the resources are not available to feed all people adequately, that we are reaching the limits of our productivity, and that the answer is to control population growth. The other side feels that we have not begun to reach the limits of productivity. There is food, but our distribution and political systems have failed us. This view frames the problem as one of technological and social shortcomings, not resource shortcomings. What do you think?

cal heat limit. He projected that world population would be absolutely (physically) limited after it reached the point at which the earth was covered over its entire surface by a single 2,000-story building. The upper 1,000 floors would be devoted to the machinery to keep the lower 1,000 going. Ducts and wires, elevator shafts, and conveyor belts would leave 3 to 4 square yards for each individual in the lower 1,000 stories. A plate on the roof would be kept at the melting point of iron to radiate away all of the earth's heat. This world would support 100 quadrillion (1×10^{17}) people. On the bright side, Fremlin points out, such a population might have as many as 10 million William Shakespeares. The dark side is not hard to imagine. Fremlin's exercise points up the absurdity of the ultimate extension of always trying to solve the problems brought on by increasing population and never coming to grips with the fact that population growth itself may be a problem.

Models of Growth and Limits to Growth

The first practical attempt to develop a model for how the systems of the earth interact was made by Jay Forrester (1971). On the basis of this model, another group headed by Donella and Dennis Meadows attempted in 1972 to project the factors that would eventually check human population growth. Although the assumptions and methodologies used in this projection continue to be controversial, some of the factors involved in checking population limits and industrial growth were defined, and a first step was taken toward expressing their interrelationships in a model.

Using a computer, the Meadows team attempted to show the interactions of five factors affecting the ability of the world system to continue intact. These factors were population growth, agricultural production, nonrenewable resource utilization, industrial output, and pollution. Conservative assumptions

about future technological advancements were made and included in the formula. Depending on the assumptions that were made about changing rates of growth, food production, resource development, and environmental deterioration, one or another of these factors proved limiting to the continuation of the world system over specific periods of time. For example, in the "standard run," which assumed "no great changes in human values" or "in the functioning of the global population/capital system as it has operated for the last one-hundred years," nonrenewable resource depletion was projected to cause the collapse of the world economic and political system well before the year 2100. When the assumption was made that twice as many nonrenewable resource reserves are available as were used in the standard run, system collapse was again projected before 2100, but in this run the limiting factor was pollution. No doubt other models attempting to predict the future of global society will emerge and be scrutinized and disputed. Although a model is only as good as the accuracy of the assumptions on which it is based, modeling does help to clarify the factors to be considered in assessing the vulnerability of the world system and defining the variables—some of which are under the control of world decision makers (also see Meadows, 1992).

Demography

Demography is the study of the vital and other social statistics of human populations. A study of certain key indices over time can be used to project future trends in human population growth. Let's look at some of these significant statistical factors.

Population Growth Indices: A Review

As we discussed in Chapter 4, growth rate is a function of birth rate, death rate, and net migration.

Mini-Glossary

Age-specific fertility: births per year per 1,000 women of a given age.

Crude birth rate: births per 1,000 people per year.

Crude death rate: deaths per 1,000 people per year.

Demographic transition: the shift from high fertility and mortality to the low fertility and mortality pattern that characterizes industrialized nations.

Demography: the study of vital and other social statistics.

Developed countries: industrialized nations generally characterized by high gross national product and low birth rates.

Developing countries: nonindustrialized nations, also called *less developed countries* and *Third World nations*, characterized by low gross national product and high birth rates.

General fertility rate: births per year per 1,000 women aged 15–44.

Net growth rate: rate of natural increase plus percentage change from net migration.

Net migration: immigration minus emigration (may be a negative value).

Rate of natural increase: birth rate minus death rate divided by 10 (expressed as a percentage).

Birth rate and death rate can be expressed in terms of births and deaths per 1,000 persons per year. In 1990, the Census Bureau counted about 248,710,000 people in the United States, there were 4,179,000 births and 2,162,000 deaths. The birth rate was thus 4,179,000 divided by 248,710,000/1,000, or 16.8 births per 1,000 people, and the death rate was 2,162,000 divided by the same figure, or 8.7 deaths per 1,000. Ignoring migration, the growth rate for that year was 16.8 − 8.6 = 8.2 per 1,000 people. This is the *rate of natural increase*. It is usually expressed as the *annual percentage increase*. A rate of 8.2 per 1,000 is 0.82 per 100, so the natural rate of increase in 1990 was 0.82%.

To determine the actual *net growth rate* for a given year, one would also have to take migration into account. Assuming a net U.S. in-migration of 700,000 in 1990, or a percentage increase of 0.28%, the actual rate of population increase was 0.82 + 0.28 = 1.10% in 1990.

These birth and death rates are crude because they do not provide certain relevant information. A less crude, more information-rich term is *fertility rate*. The **general fertility rate** is the number of births per year per 1,000 women ages 15–44 (the reproductive group); it is used to measure changes in overall fertility in a population. For predictions of longer-range trends, other fertility factors may be used. The **age-specific fertility** rate is the number of births per year per 1,000 women at a given age, say, 28. By comparing such specific rates over a period of time, general trends in population growth can be determined.

The rate of natural increase is a function of fertility and of the number of women of reproductive age. If the fertility rate remains constant but there are more women entering the reproductive age, there will be more babies. This illustrates the importance of age structure within a population (see Chapter 4). Generally speaking, developed countries are characterized by a bell-shaped profile, which indicates a stable population (see Figure 8.2). Developing countries exhibit an age structure like a broad-based pyramid. In a developing country, 40% to 50% of the population may

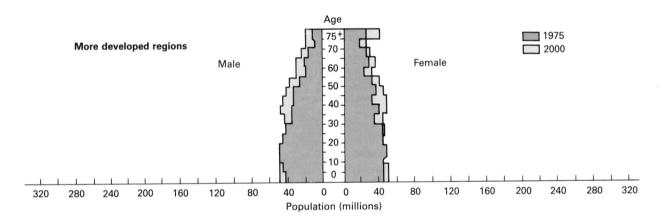

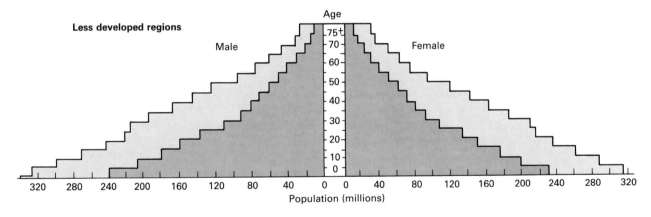

Figure 8.2 Age-Sex Composition of the World's Population, 1975 and 2000. These two pyramids of age structure of populations typify the difference between developed and developing countries. The broad base of developing countries indicates a proportionately large number of young people, which means potentially explosive population growth. The profile of the developed countries indicates a decreasing birth rate and a stable population.

be 15 years of age or younger. Such countries have an explosive potential for further growth should death rates decline quickly (see Chapter 4).

Age structure is related to the survivorship curves we discussed in Chapter 4. The human survivorship curve has changed dramatically with improvements in and accessibility of medicine. The decrease in infant mortality in both developing and developed nations has contributed significantly to the population increases of the past several decades.

Zero Population Growth

If crude birth rate is equal to crude death rate for a long enough time, the result is **zero population growth** (ZPG). For human beings, an average of 2.1 children per female is the theoretical replacement rate, the rate at which ZPG is eventually reached (if immigration is disregarded). The extra tenth of a percent over 2—the rate one might expect to be the replacement rate—takes into account deaths prior to reproduction, the fact that fewer females are born than males, and other factors.

In any growing population, it may take a number of years for growth to stop, following the achievement of 2.1 births per female. This will be true, for example, in a population in which the proportion of women of reproductive age is increasing. In other words, replacement rates must be maintained for some time before ZPG can be attained.

Figure 8.3 indicates that if the *world* attained replacement rate fertility between 2000 and 2005, the *total* population would continue to increase by 2.6 billion people and would eventually stabilize around 8.5 billion. If replacement rate were not reached until around 2020, world population would stabilize eventually around 10.7 billion. If the world did not attain fertility replacement rate until the middle of the twenty-first century, population would grow to 13.5 billion because of the lag resulting from the age structure of the world population.

Lag is extremely important in determining the meaning and effects of changes in population growth rates. Population control is a long-range proposition; it cannot be achieved instantaneously at the time of an impending crisis.

Historical Perspectives and Trends

Now that we are familiar with some of the terms and indices used to gauge population trends, let us look at the historical growth of human population. Two major factors permitted substantial human population increases. One was the increased ability to produce food through agriculture. The second was the substantial lowering of the death rate achieved through improved sanitation and medicine.

World Population

World population reached 1 billion in 1850, 2 billion in 1930 (80 years later), 3 billion in 1960 (30 years later), 4 billion in 1975 (15 years later), and 5 billion in 1987 (12 years later). Most projections put world population around 6 billion in the year 2000 or a little before. The momentum of population growth is immense.

Figure 8.4 shows world population growth over time. It is impossible to gauge from this figure whether we are following a J-curve or the first part of a sigmoid curve. A J-shaped curve is characteristic of species that increase exponentially and then, because of the lack of density-dependent feedback control, overshoot carrying capacity and go through a population "crash" (see Chapter 4). In the S-shaped pattern characteristic of other species, population growth levels off gradually and stabilizes somewhere below carrying capacity. Some experts feel that our progression along an S-shaped or a J-shaped curve is still a matter of collective choice. Others fear that carrying capacity has already been seriously exceeded and that this will lead to a considerable wobble or even a crash in the future.

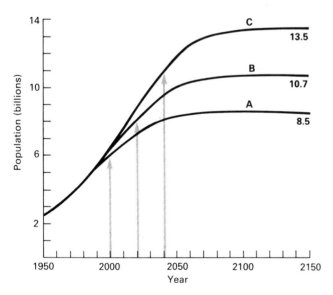

Figure 8.3 World Population Growth Projections. Even if a replacement rate of fertility could be reached immediately worldwide, total population would continue to grow for many years because of the broad base of young women already born and about to enter their reproductive years. This graph indicates three projections for ultimate world population, based on when replacement rate is achieved (see arrows). For example, even if the world attained replacement fertility between 2000 and 2005 and the population would then stand at 5.9 billion, the world population could be expected to continue to rise another 2.6 billion before stabilizing at 8.5 billion.

214

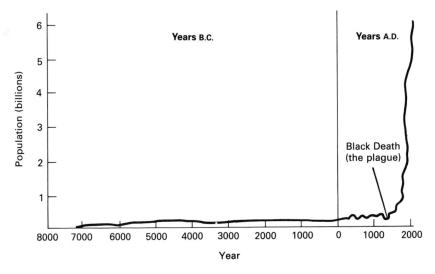

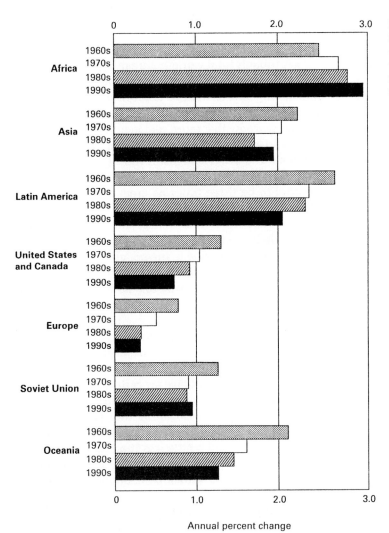

Figure 8.4 What Goes Up Might Come Down. The growth of human population has been dramatic in the past 150 years. Providing an ever-increasing number of persons with clean air, water, nourishing food, and shelter is becoming more and more difficult. Will we follow a J-shaped curve or an S-shaped curve as time goes on?

At an annual percentage growth rate of 1%, the human population will double in 70 years; at 2%, the doubling time is 35 years; and so on. Average annual growth rate for the world is about 1.8%. This growth rate has a population doubling time of 39 years. However, this rate varies greatly from country to country. Average annual growth rates run around 2.0% in developing nations and 0.6% in developed regions. Figure 8.5 shows annual growth rates by continent from 1960 to 1990. The highest growth rate

Figure 8.5 World Population: Annual Growth Rate by Continent, 1960 to 1990. This bar graph shows that the annual population growth rate has decreased worldwide since 1960 except for Africa. These figures represent averages for the continents; subregions may reflect growth rates above or below this average.

Annual percent change

is in Africa. At the current rate, the population of Africa will double in 28 years. Some subregions of Africa are growing even faster. Kenya and Zambia, with almost a 4% annual growth rate, will double population in only 17 years.

Developed Versus Developing Countries

Since 1950, the fertility rates in developed countries have decreased significantly. Many of the developed countries have reached replacement fertility levels, and in Germany, death rates now actually exceed birth rates. For all developed countries the fertility rate is just about at the replacement rate. Should these rates continue, these countries will achieve stable populations or even experience population declines. Although no one knows for certain, it is generally thought likely that once low fertility rates are reached, they will be maintained around replacement levels, with periodic fluctuations. The population problems facing many of the developed countries in the future will stem from increased immigration, urbanization, and wide differences in income.

What about the developing countries? Figure 8.5 shows that annual growth rate has also declined in many of the developing countries since 1960. Although fertility rates in some developing countries are on the decline, there is in the age structure of the populations of these countries great potential for continued growth.

Almost 90% of current world population growth can be attributed to the developing countries. It has been predicted that of the 6 billion people expected to occupy this planet by the year 2000, fully 5 billion will live in developing countries. According to a U.S. government study, developing countries had 66% of the world's population in 1950, had 72% in 1975, and are likely to have 79% in the year 2000. Table 8.1 reflects the annual population increment for some slow- and rapid-growing countries and regions at about 1990 growth rates.

Population problems in developing countries include those associated with urbanization and distribution of goods, as in developed countries, but in developing countries these problems will continue to be overshadowed by poverty, hunger, and the need to increase economic output faster than population growth. Historically, population growth in the developing countries has far outstripped resource development and improvement in the quality of life. It has been predicted that simply to maintain present, sometimes very low standards of living in developing countries, economic output must double by the year 2000.

Table 8.1 World Population Growth by Selected Countries and Regions

Region	Population Mid-1990 Estimate	Population Growth Rate	Annual Increment
	(million)	(percent)	(million)
Slow-Growth Regions			
Western Europe	159	0.2	0.3
North America	278	0.7	1.9
Former Soviet Union	291	0.9	2.6
Australia	17	0.8	0.1
China	1,120	1.4	15.7
Japan	123	0.4	0.5
Rapid-Growth Regions			
Southeast Asia*	455	2.1	9.5
Latin America	447	2.1	9.4
India	853	2.1	17.9
Western Asia†	132	2.8	3.7
Africa	661	2.9	19.2

* Principally Burma, Indonesia, the Philippines, Thailand, and Vietnam.

† Principally Iraq, Syria, Turkey, and Saudi Arabia.

Sources: Worldwatch Institute; Population Reference Bureau.

Updated figures: Population Reference Bureau, 1990 Population Data Sheet.

World Trends

Although total world population will continue to increase for many decades, most population growth *indices* are on the decline and are expected to continue to decline. Crude birth and death rates, gross and net reproductive rates, and general fertility are all expected to drop. The world birth rate appears to be declining at a faster and faster rate. The drop in birth rate in developing countries from 1970 to 1977 was reported to be three times as great as the drop from 1950 to 1970. This dramatic drop in birth rate in developing countries has, since the mid-1960s, exceeded their continuing declining death rate, resulting in an overall decline in the population growth rate (see Figure 8.6). In 1987 the world birth rate did increase slightly due primarily to a jump in birth rate in China. We will discuss this phenomenon later. It is important to keep in mind that even with these promising trends, population growth overall will continue well into the twenty-first century because of the lag effect. Table 8.2 reflects estimates of where population size is likely to stabilize for certain countries.

It appears that trends toward urbanization will continue. It is estimated that by the year 2000, 75% of the population in South America, over 40% of Africa's population, and a little less than 40% of the people in Asia will live in cities. Today 20% of the world's people live in a city with a population of 100,000 or greater.

According to the World Commission on Environment and Development, Third World cities alone

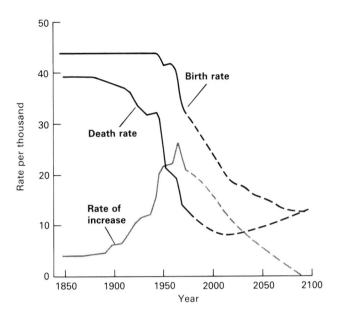

Figure 8.6 Birth Rate, Death Rate, and Rate of Population Increase in Developing Countries, 1850–2100. Since the mid-1960s, the rate of increase in population in developing countries has decreased significantly owing to a dramatic decrease in birth rate. The dashed curves are projections. The increase in death rate shown for the mid to late twenty-first century is due to the large number of older people in the population at that time.

Table 8.2 Projected Population Size at Stabilization, Selected Countries

Country	Population in 1988	Estimated Size of Population at Stabilization	Change from 1988
	(million)	(million)	(percent)
Slow-Growth Countries			
China	1,087	1,571	+ 45
Former Soviet Union	286	377	+ 32
United States	246	289	+ 17
Japan	123	128	+ 4
United Kingdom	57	59	+ 4
Former West Germany*	61	52	− 15
Rapid-Growth Countries			
Kenya	23	111	+382
Nigeria	112	532	+375
Ethiopia	48	204	+325
Iran	52	166	+219
Pakistan	108	330	+205
Bangladesh	110	310	+181
Egypt	53	126	+138
Mexico	84	199	+137
Indonesia	177	368	+108
India	817	1,700	+108
Turkey	53	109	+106
Brazil	144	298	+107

* Former West Germany has a negative birth rate.

Source: Worldwatch Institute, 1987. *State of the World*; Population Reference Bureau, *1988 World Population Data Sheet.*

could grow by 750 million people between 1985 and 2000. This means increased needs for urban infrastructure to provide clean water and sanitation, food and health services, shelter, transportation, and education. Even now problems of inadequate facilities, overcrowding, and disease trouble many urban areas. To achieve the additional infrastructure levels that will be needed would require an enormous development effort and would be extremely costly. There is little likelihood that the means for such an effort are available.

Population in the United States

Figure 8.7 shows the historical changes in U.S. population growth. We would like to draw attention to several important points. There was a decrease in the *rate* of population growth in the 1930s, a time of economic depression. A "baby boom" followed World War II and is reflected in the sharp rise in the population growth rate prior to 1950. Population growth since the 1970s appears to be slowing down, but the full effect of the baby boom—as the boom babies have babies—is yet to be seen (Figure 8.7b).

Total births in the United States have risen since 1975 as a result of this large baby boom generation entering their childbearing years, but the *birth rate* has remained low. It is thought that this boom generation is either delaying childbirth or will have a much lower fertility rate, resulting in a continuation of the population growth rate decline. Some think there is little or no reason to expect an increase in fertility rates.

Since the 1950s the birth rate has declined significantly more than the death rate. This decline accounts for a slowdown in the rate of population growth. The fertility rate in the United States is also on the decline. The net population growth rate in the United States has fluctuated around 0.9% in the recent past.

Legal immigrants account for 25% of present U.S. population growth (see Figure 8.8). Illegal aliens add another significant but unknown percentage.

The U.S. Census Bureau has projected U.S. population growth on the basis of high (2.7), medium (2.1), and low (1.7) fertility rates. The Census Bureau estimates that the earliest the population size will stabilize or decline under any of the estimates is around the year 2030.

U.S. Population Distribution Trends

Where are the people? In 1900, approximately 40% of the people in the United States lived in urban areas; in the 1980s, approximately 75% lived in **standard metropolitan statistical areas** (SMSAs) (urban areas of 50,000 or more together with the county containing the urban area and adjacent counties with

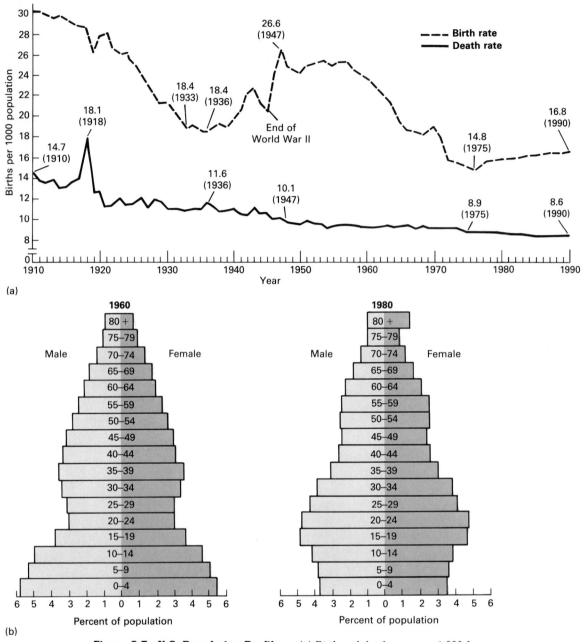

Figure 8.7   U.S. Population Profiles.   (a) Birth and death rates per 1,000 from 1910 to 1990. Birth rates slowed in the 1930s, then picked up again after World War II. A leveling-out period began in the late 1970s. Its duration will depend on the reproduction patterns of the baby boom children now in their reproductive years. (b) U.S. population by age and sex, 1960 and 1980. The baby boom bulge has had a significant impact on the economics of American life, from education to job seeking. Even more changes are likely as this predominant portion of the population enters the middle and late years.

employment ties). Nearly 50% lived in 36 metropolitan areas with populations of 1 million or more.

The center of population in the United States has been moving steadily westward. The 1990 census places it about 10 miles northwest of Steelville, Missouri, reflecting a continuing population shift to the south and west.

U.S. Immigration Policies and Trends

Immigrants are "aliens"—that is, non-U.S. citizens—admitted to the United States for residence of one year or longer. As far back as 1882, Congress limited immigration into the United States by enacting the Chinese Exclusion Law (repealed in 1943). In 1917, Congress passed an act barring certain peoples from

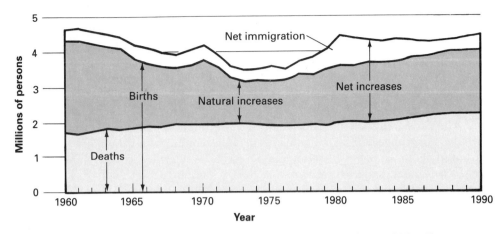

Figure 8.8 Components of U.S. Population Change, 1960 to 1990. This graph depicts the various factors that affect population growth (births, deaths, and migration) and their relationships from 1960 to 1990.

Asia and the Pacific Islands. In the 1920s, Congress enacted a quota law consisting of complicated formulas for determining the number of immigrants to be allowed from various countries. In 1965 the quota law was abolished, and 170,000 was set as the annual number of immigrants allowed from the eastern hemisphere with no more than 20,000 coming from any one country. A limit of 120,000 was imposed for the western hemisphere. Exemptions were allowed in both cases for relatives of U.S. citizens.

The Immigration Act of 1965 emphasized family reunification and needed skills and professions rather than national origin as immigration criteria. Congress passed legislation in 1990 to increase the number of legal immigrants to about 700,000. The emphasis is on allowing unrestricted immigration of those who have family who are permanent residents of the United States (as is currently the case for U.S. citizens) and

allowing a greater influx of immigrants with needed skills.

Above and beyond its quotas, the United States also provides asylum for refugees. This figure varies depending on world events. The shiploads of "Marielitos" from Cuba in 1980 added over 100,000 persons to the U.S. population. In a "normal" year, however, refugees may average only 100,000 to 150,000.

Between 1945 and 1965, some 43% of all U.S. immigrants came from Europe. From 1966 to 1970 this figure was reduced to just over 33%, with another third coming from Canada and Latin America and the remaining third from western India, Asia, and Africa. This pattern of increased migration from Asia and South America has continued in recent years.

People do not just immigrate to the United States; some also emigrate. About 125,000 to 150,000 persons leave the United States each year. Although

Most people in the United States live in urban areas having relatively small amounts of green space.

numbers vary, substantially more migrants come into the United States than leave annually.

The number of illegal aliens is growing. In many years, more illegal aliens were expelled than legal immigrant aliens were admitted. Actual numbers of illegal aliens entering the United States are unknown. Estimates range from 150,000 to 250,000 a year, but may be much more. The Census Bureau has estimated that the United States may have 3.5 to 6.0 million undocumented immigrants annually. The National Academy of Sciences has estimated 2 to 4 million. In 1986, Congress enacted legislation to address the issue of illegal aliens primarily by making it illegal for employers to hire undocumented workers.

Population Growth and Economics

What is the relationship between population growth and economic growth? What impact can birth control programs have on the overall economic development of a country? Is a growing population a requirement for continued economic growth?

Economic Growth and Population

As early as 1972, the President's Commission on Population Growth and the American Future reached the following conclusions:

> We have looked for, and have not found, any convincing economic argument for continued national population growth. The health of our economy does not depend on it. The vitality of business does not depend on it. The welfare of the average person certainly does not depend on it.

The study cited the fact that because, in the past, periods of rapid population growth have also been periods of rapid economic expansion, we tend to assume that the two go hand in hand: the more people, the more consumption of goods and services, the higher the GNP. The report went on to suggest that "the diminished burden of providing for dependents and for the multiplication of facilities should make more of our national output available for many desirable purposes" such as human resource investments and expenditures for *qualitative* improvements and environmental objectives.

In the Third World, population growth characteristically has the effect of *decreasing* the economic welfare of the individual or, at best, greatly retarding improvement in individual welfare. Many studies show that developing countries would make more progress faster by investing in slowing population growth rather than accelerating industrial growth. The *World Development Report 1984* states that rapid population growth is impeding development. In other words, over time, the cost of implementing birth control programs is recovered many times over as extra income per capita for the population.

In 1984 at the UN Population Conference in Mexico City, the U.S. administration challenged the position that population growth plays a role in retarding economic development in the Third World. Instead, it proposed that more people mean higher consumption and are thus a spur to economic development and that a free-market system is what is needed.

At the request of the Agency for International Development (AID), the National Research Council undertook to study the issue further. The council's report, released in 1986, looked at the effect that slower population growth has on exhaustible resources, renewable resources, health and education, distribution of income, and cities. In addition, it examined the question of whether a couple's fertility behavior imposes costs on society at large. Termed "revisionist" by some, the report emphasized that factors other than population growth influence economic development and that the adaptability of individuals, governments, and markets mitigates the negative effects of population growth. For example, in the area of exhaustible natural resources, there are economic and market conditions that reduce their use long before they are depleted (higher costs, substitution, etc.; see Chapter 7). Growing populations significantly affect renewable resources held in common. Overuse of forests, habitat destruction, and overfishing are examples. The report concluded, however, that "on balance . . . slower population growth would be beneficial to economic development for most developing countries," although quantifying these benefits is difficult. Moreover, family planning programs can improve the individual lives of people in developing countries and thus decrease any economic burdens on society that may result from growing populations.

In sum, the report seemed to reinforce the fact that the relationship between population growth and economic development is a complex one.

Although some gains are being made in efforts to slow population growth in developing countries, absolute increases in total numbers due to the lag effect will keep per capita GNP levels low for many decades. Predictions are that the gap between GNP in developed and developing countries will continue to widen.

Economic Birth Control

In industrialized societies where the rate of economic development has exceeded the rate of population growth, it has been observed that childbearing tends to decrease as income rises. In modern American society, in which both parents generate income, children may be viewed as more of an economic liability than an economic asset. While it might be argued that the answer to the world population problem is to give everyone a chance at a good-paying job and a comfortable life, this strategy would generate enormous environmental pressures. As we have pointed out often, affluent countries such as the United States have a far greater impact on the global ecosystem via increased pollution and increased resource depletion than many less developed countries.

Population Growth and Culture

One of the most significant factors in population growth is culture. Unless technology and population control programs take social customs into account, their chances of success are slim.

Determining Family Size

There are cultural reasons for high fertility in the Third World countries. They include high infant mortality, the importance of family and kin in subsistence, education, and care of the elderly (all functions that fall to other institutions in industrialized countries), and the importance of a male heir. All of these tend to encourage more childbearing. An important contributor to the present population problem is the fact that cultural norms have not yet adjusted to the decrease in infant mortality. Previously, higher birth rates were required to compensate for early death. The problem is that birth rates have not yet fallen in many countries in proportion to the decline in infant mortality and death rates in general. Old ways fade slowly.

A number of things operate to keep birth rates low in the developed countries, including these:

1. A lower infant mortality rate
2. Compulsory education, which reduces or eliminates economic productivity by children
3. Mechanization, which diminishes the need for human labor
4. Social programs for the elderly
5. The advancement of birth control technology
6. Shifts in social status symbols

Status has become associated with affluence and occupation, not family size; women are finding satisfying roles other than child rearing.

Birth Control:
Methods and Programs

Egyptian papyruses dating back to 1900 B.C. show that contraception was known even then. However, information on contraception has not always been easily available. The subject was taboo for centuries. One of the earliest American advocates of disseminating birth control information was Margaret Sanger (1883–1966). Sanger's experiences as a nurse convinced her of the need to provide contraceptive information to people who wanted it. Objections from religious organizations and the failure of the medical profession to acknowledge the need made her task an uphill fight. (The American Medical Association gave its first endorsement of birth control in 1937.)

Table 8.3 shows the major forms of birth control that are available today. Prescriptive methods are those that require prescription by a doctor; nonprescriptive methods are available over the counter. The effectiveness of any birth control device varies among individuals; forgetting to take a pill or improper placement of a diaphragm can result in method failure. Use of more than one method, such as condoms combined with contraceptive foam, can increase the overall effectiveness of the methods. Each method has advantages and disadvantages. Negative health side effects have been documented from the use of some methods.

Another method, in addition to those listed, is the ancient practice of coitus interruptus (withdrawal of the penis from the vagina prior to ejaculation). This is not a very effective practice; some sperm may be released prior to ejaculation, and a good deal of self-control is required.

Breast feeding delays the onset of ovulation in a new mother for a period of time. In Third World countries, breast feeding, though unreliable as a form of birth control, can play a significant role in the spacing of pregnancies. However, breast feeding has been reduced in Third World countries by the promotion of canned formula, which has resulted in other problems as well. Formula prepared from contaminated water is one example.

Manufacturers of birth control devices are becoming much more aware of their responsibility to alert the public to the potential side effects of their products. Court cases against a U.S. manufacturer of a common intrauterine device by users who became sterile have caused most IUDs to be withdrawn from the U.S. marketplace. Consumer advocates express the need for people to be able to choose among contraceptives with full knowledge of the potential risks. Concerns have been raised that the constant threat of lawsuits will make contraceptive devices less available in the United States and will hamper research.

New Technologies

Technological approaches to contraception are varied. Researchers continue to look for methods that are safe, reliable, convenient, and reversible. Recent efforts have centered on the development of a vaccine to immunize a female against sperm, against certain

Table 8.3 Methods of Birth Control

Type	Description	Relative Effectiveness
Prescriptive Methods		
The pill	Series of pills containing hormones similar to those that naturally regulate the menstrual cycle	Two pregnancies per 100 women using the pill during the first year
Intrauterine device (IUD)	Small piece of shaped plastic inserted in the uterus	Five pregnancies per 100 women using the device during the first year
Diaphragm	Shallow rubber cap with a ring rim that fits securely in the vagina to cover the cervix (entrance to the uterus)	Nineteen pregnancies per 100 women using the device during the first year
Hormonal implant (Norplant)	Capsules containing a hormone that inhibits ovulation are implanted in the upper arm. Hormones are released gradually over time. Effective up to 5 yrs.	Less than 1 pregnancy per 100 women in the first year and 1 pregnancy per 100 women in the fifth year
Mifepristone (RU 486)	A drug in pill form that blocks the action of progesterone	Four pregnancies per 100 women
Nonprescriptive Methods		
Vaginal contraceptive sponge	Doughnut-shaped sponge that fits over the cervix and is impregnated with a spermicide	Nine to 11 pregnancies per 100 women using the device during the first year
Condom	Sheath of thin rubber or animal tissue that covers the erect penis during intercourse	Ten pregnancies per 100 couples using the device during the first year
Contraceptive foam	Foam containing a chemical that stops sperm, placed in the vagina to cover the cervix	Eighteen pregnancies per 100 women using the device the first year
Fertility awareness method (natural family planning or rhythm method)	System designed to help a woman determine ovulation and thus the fertile period in the cycle by monitoring body changes	Twenty-four pregnancies per 100 women using the method during the first year
Permanent Methods		
Sterilization (vasectomy, tubal ligation)	Surgery to sever the reproductive tubes to prevent release of sperm or egg	Less than one pregnancy per 100 couples in the first year following the operation on one of the partners

Source: Based on A. L. Schirm et al., "Contraceptive Failure in the United States: The Impact of Social, Economic, and Demographic Factors," *Family Planning Perspectives 14*(2):68.

hormones needed to maintain pregnancy, or even against her own eggs. Pills for males are also under investigation, as is the use of sex hormones to halt sperm production.

A promising contraceptive technique especially for developing countries is the vaginal ring. According to the UN Fund for Population Activities, these devices may be in widespread use by the year 2000. The World Health Organization is developing the vaginal ring. This device, worn continuously for a three-month period, releases hormones in the same amounts as taking a daily pill would.

Hormonal implants are another relatively new contraceptive technique. The hormonal implant is a device inserted under the skin that releases a hormone commonly found in birth control pills. It is usually implanted in capsules about the size of a matchstick in the upper arm during a simple office procedure involving a local anesthetic. In newer forms the hormone is encased in a shell that also dissolves, making removal unnecessary. The implant is usually effective for up to 5 years. Because no daily or preintercourse preparation is required, this can be a very effective birth control mechanism. More than 50,000 women in over 40 countries have used such implants. Use in the United

States was approved in December 1990, under the trade name Norplant. Some drawbacks with the technique are irregular bleeding, cost, and potential long-term health risks.

Other promising technologies are those that would easily and reliably pinpoint ovulation, thus making natural family planning methods more reliable. A home test using urine or saliva that would measure hormonal changes is feasible, but a concerted research effort would be necessary to perfect the technology.

A very controversial birth control technology is the drug mifepristone, commonly referred to as RU 486. RU 486 is a drug that blocks the action of the hormone progesterone. The production of progesterone is essential to the maintenance of a pregnancy; without it, the uterine lining is shed. Because progesterone is important both for implantation and for maintenance of the uterine wall after implantation, drugs that block progesterone may be contraceptive (preimplantation) or abortifacient (postimplantation).

Currently RU 486 is approved for use in France, Great Britain, and China. In France it is generally used to terminate a confirmed pregnancy during the first seven weeks. A small dose of another drug, prostaglandin, is given in conjunction with RU 486. This drug causes uterine contractions to help expel the fertilized egg and uterine lining. Contrary to popular belief, RU 486 is not a pill that can be taken in the privacy of one's home. Because of some side effects, medical supervision is required. Of women in France choosing to interrupt a pregnancy in its early stages, 25% to 33% choose RU 486 because it does not involve a surgical or invasive procedure. Anesthetic is not required.

Distribution of RU 486 in the Netherlands and Sweden is under consideration. Protests of use of RU 486 in the United States by opponents of abortion have delayed legal availability of the drug in this country. Investigation continues into the potential use of RU 486 as a birth control rather than an abortive drug, as theoretically it could be used to halt an egg's release or prevent implantation.

When contraceptive methods fail, the problem becomes one of how to handle an unwanted pregnancy. Counseling programs address the alternatives and their psychological and long-term implications. Among the current alternatives are abortion, adoption, and single-parent families.

As we have already seen, the motivation to limit births apparently comes fairly automatically to individuals in highly industrialized societies. The question we would like to explore now is how motivation can be influenced in other ways. A related question is, *Should motivation be influenced?*

Government Involvement in Population Control

There is power in numbers. Italy under Mussolini and Germany under Hitler attempted to increase the birth rate by taxing bachelors, giving married people preferential job treatment, reducing taxes for large families, and granting marriage loans. France developed a family allowance scheme in the 1930s; the amount of the allowance increased with the size of the family. Sweden enacted a program in the 1930s based on assisting families who really wanted children. The program provided services rather than monetary allowances.

In the United States, federal influence on birth rates has been indirect. During certain periods, military draft policies have encouraged early marriage and provided incentives for childbearing. Tax breaks are given for each added dependent. One of the few federal laws dealing with contraception was the Comstock Act of 1873, named after Anthony Comstock, founder of the Society for the Suppression of Vice. This act prohibited the importation, transportation in interstate commerce, and mailing of any article whatever for the prevention of conception. This law was not repealed until January 8, 1971, although it was widely ignored long before that.

The effectiveness of government population programs may be questionable, but the precedent they set for government involvement in procreation is important. In recent years the key questions have had to do with limitation rather than promotion of reproduction.

Proponents of government intervention in population control cite examples of government intervention in other personal, individual matters. The government permits citizens only one spouse at a time, enforces compulsory education, and may require that critically ill children be treated regardless of personal or religious objections.

Incentives and Coercion

Table 8.4 summarizes some of the possible ways in which governments might influence population control. These proposals include involuntary fertility control, educational campaigns, and legal and social reforms. Not many of them have ever been implemented or even seriously considered. The proposals vary widely in their scientific readiness, political viability, administrative feasibility, economic feasibility, ethical acceptability, and presumed effectiveness.

Two solutions consistently emerge as effective tools in population control. One is family planning programs. The other is improvement in the status of women.

A world fertility survey conducted between 1977 and 1984 showed that if all of the women who said they did not want any more children were actually able to prevent further pregnancies, the number of births would decline by 27% in Africa, 33% in Asia, and 35% in Latin America. Studies indicate that family planning programs account for anywhere from 10% to 40% of the observed declines in birth rates.

Where women's status has improved, birth rates are declining. The report of the World Commission on Environment and Development states unequivocally that social and cultural factors are the predominant determinants of fertility. It singles out the status of women as the most important of these factors. The commission stresses that policies that offer economic incentives and disincentives for bearing children must also have a component aimed at improving the status of women. Employment opportunities, education, and higher age at marriage are all important.

China: A National Policy on Birth Control

More than one out of every five persons in the world is Chinese. Because of this, world population trends are significantly affected by what is happening in China. The decline in world birth rate in the 1970s coincided with the initiation of family planning efforts

Table 8.4 Alternatives for Controlling Birth Rates

Proposal	Scientific Readiness	Political Viability	Administrative Feasibility	Economic Capability	Ethical Acceptability	Presumed Effectiveness
Extension of voluntary fertility control	high	high on maternal care, moderate to low on abortion	uncertain in near future	maternal care too costly for local budget, abortion feasible	high for maternal care, low for abortion	moderately high
Establishment of involuntary fertility control	low	low	low	high	low	high
Intensified educational campaigns	high	moderate to high	high	probably high	generally high	moderate
Incentive programs	high	moderately low	low	low to moderate	low to high	uncertain
Tax and welfare benefits and penalties	high	moderately low	low	low to moderate	low to moderate	uncertain
Shifts in social and economic institutions	high	generally high, but low on some specifics	low	generally low	generally high, but uneven	high over long run
Political channels and organizations	high	low	low	moderate	moderately low	uncertain
Augmented research efforts	moderate	high	moderate to high	high	high	uncertain
Family planning programs	generally high, but could use improved technology	moderate to high	moderate to high	high	generally high, but uneven, on religious grounds	moderately high

Source: B. Berelson, ''Beyond Family Planning,'' *Science 163* (1969):540.

in China. The increase in the world's birth rate in 1987 coincided with a jump in birth rate in China. China has taken definitive steps in its birth control efforts.

In the 1970s, China initiated strong birth control policies with an emphasis on "late, sparse, and few"— later childbearing, longer spacing between children, and fewer children. The program was successful; the fertility rate dropped from 5.93 births per woman in 1970 to 2.66 births per woman in 1979. However, a more stringent one-child policy was later implemented. This was done because it was thought that even with the decline in fertility rate, the number of women entering the reproductive years would have continued to cause huge increases in population and make economic development gains much more difficult.

The new policy, "one couple, one child," has been controversial. It has consisted of an educational campaign promoting one-child families, a system of incentives and disincentives to comply, and—perhaps its most controversial aspect—abortion to prevent the birth of a second child conceived outside of the plan. Although policy and major direction are set at the national level, policy implementation is at the local level, carried out by birth-planning committees. Consequently, the specific ways in which the program has

been implemented have varied from area to area. In some localities, forced sterilization is a component of the program. Educational campaigns have emphasized child care, nutrition, and conduct, as well as the virtues of a one-child family. Incentives have included access to better housing, food supplies, and cash; disincentives have included fines, demotion, and termination of work bonuses. The system has had other reported effects, including the killing of baby girls and deformed babies at birth in order to be eligible for another child. By 1984 the fertility rate had dropped to 1.94, slightly below the replacement rate.

In 1984 and 1985, action was taken at the national level to relax the family planning policy by expanding conditions under which couples could have a second child. Also, reforms were made to increase voluntary participation. The result of this relaxation of the program has been a much larger increase in births than expected. The number of babies born in 1986 was 1.6 million more than the number that had been set by the national plan. The fertility rate in 1991 was 2.25. The population in China is projected to reach 1.2 billion in 1995, and 1.3 billion in the year 2000. Part of the increase is due to the increased number of women entering childbearing age. Also, more relatively wealthy rural Chinese are willing to pay the fines in

A society has many ways—some subtle, some obvious—to influence people's motivation to have or not have children. In China, billboards are part of a campaign to convince couples to have just one child.

order to have more children. Ironically, some of the economic reforms initiated in China have inadvertently encouraged larger families. Individual farmers can now sell their own crops at a profit, so larger families with more helping hands are economically advantageous.

According to the Worldwatch Institute, one alternative for China now is to establish a rural social security system. Such a system would ensure that needs are met in old age. In such a system, government subsidies and programs would replace the need to have children to provide security in old age. Also, improvements can be made in the opportunities provided for women.

It is likely that China will renew its efforts to reduce family size. China's future depends on the degree of success it has in its family planning policy.

India: A National Program of Birth Control

We often read of famine and disasters in India, but their impact on population growth has been negligible. A great famine occurred in Bengal in the 1940s when India's population was 310 million; in 1951 the population was 360 million, and in 1971 it was 548 million.

India was the first country to initiate a national family planning program in the 1950s (although it was not pursued vigorously until 1965). In the mid-1970s, India was the first nation to advocate compulsory sterilization as a national policy.

India's family planning program is credited with about a 50% reduction in all births, due primarily to its stress on sterilization. Although the Indian government's policy of compulsory sterilization resulted in dramatic increases in sterilizations in 1976, this policy was one of the prime factors leading to the overthrow of Indira Gandhi's government in the late 1970s.

Prospects for the Future

Where population control is concerned, no action is indeed action. Doing nothing is an endorsement of the status quo. The question is not whether population growth in the Third World will slow down but how? At stake in the population question are basic human freedoms, which will be diminished either by the increasing regulation that will be necessary to care for a growing population or by the use of coercive measures for population control. It is an ecological reality that the population growth of *Homo sapiens* will be checked somehow. Should we let it happen naturally, or will human intervention to check population growth be less traumatic? Should we work on means or on motivation? Shall we educate or legislate? Shall we provide incentives or penalties?

Section B

Food, Hunger, and Nutrition

Despite all the advances in human knowledge that have taken place, somehow or other we have failed to solve the very basic problem of hunger. Even in the United States, up to 20 million Americans may go hungry at least some part of each month.

What is the essence of the food, hunger, and nutrition problem? Is it the need for increased production? Is it the need for better distribution of food? Is it related to food habits? Is it a problem of a population expanding too fast? Is it a problem of poor technology or the inappropriate transfer of technology? Is it a problem of priorities that place food production and agricultural research behind other developments? Is it a problem of lack of overall economic development? Is it a problem of culture? Is it a problem of politics and trade? The answer to all of these questions is yes.

Even in the United States, a land of plenty, some people are going hungry.

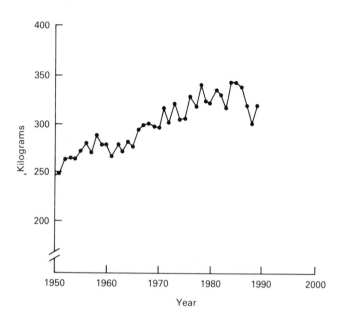

Figure 8.9   World Grain Production Per Capita.
Although grain production has been able to keep up with population growth, there is some indication that this will not continue to be the case. Weather and government policies had a role to play in the 1987–1988 decline shown here. It is uncertain how much of the decline can be attributed to short-lived phenomena and how much is due to diminishing returns on chemical and energy inputs and land degradation.

Food Production: Current Situation

Where do we stand after several decades of intense effort to increase worldwide food production? After World War II, technological advances such as chemical fertilizers and hybrid crops resulted in major increases in food production worldwide. From 1950 to 1973 the worldwide production of grain more than doubled. While population was growing at a rate of about 1.9% annually during this period, grain production was increasing at a rate of over 3% annually. The year 1973 is a landmark because it was the year that energy prices began to rise dramatically. Because many of the agricultural practices responsible for the major increases in food production up to that time were energy-intensive (increased use of fertilizer, pesticides, gasoline-powered farm equipment), the high cost of energy was a major factor in slowing the increase in per capita food production. Since 1973, world grain production has increased less than 2% annually, while the world's population has been growing at about 1.7%.

Figure 8.9 shows the trend in worldwide grain production per capita. The peak of 345 kilograms (759 pounds) occurred in 1984. Several factors contributed to the decline in 1987 and 1988. They include droughts in the United States and China, monsoon failure in India, and the implementation of government programs designed to idle cropland. This is in addition to the fact that limits are being reached in terms of return on increased agricultural inputs of fertilizer and energy, agricultural land is being lost to other uses, productivity is declining from poor maintenance of agricultural lands, and new agricultural technologies have proved not to be easily implemented on a widespread basis.

It is extremely difficult to determine what percentage of the decline is the result of environmental degradation. The Worldwatch Institute has made an attempt to quantify the additional annual loss of world grain output as a result of environmental degradation. It calculates a loss of 14 million tons of grain annually: 9 million from soil erosion; 1 million from waterlogging of soils and salinization of irrigated lands; 2 million from a combination of loss of soil organic matter from burning cow dung and crop residue, shortening of the shifting cultivation cycle, and compaction of soil from heavy equipment; 1 million from air pollution; and 1 million from flooding, acid rain, and increased ultraviolet radiation from loss of the ozone layer.

Another significant indicator of how well we are doing in meeting the demand for food is how nations overall are becoming more or less self-sufficient in food production (see Enrichment Box 8.1) (Table 8.5 reflects trends in this regard). Although there are other exporter countries, including Argentina and Brazil in recent years, the United States and Canada remain the major grain exporters today. Should any prolonged climatic change or other phenomenon cause crop failure or even decreased production in North America, this too could change. It almost changed in 1983 and again in 1988. In 1983 the United States had the smallest corn harvest since 1974. A combination of factors brought this about.

Regional Perspective on Agricultural Development

Sometimes in reviewing worldwide statistics on agriculture, we fail to recognize regional variations. Below is a summary of some basic conditions relating to agricultural development by region.

Africa

__ A drop in per capita food output of about 1% a year since the beginning of the 1970s
__ A focus on cash crops and a growing dependence on imported food, fostered by pricing policies and foreign exchange compulsions
__ Major gaps in infrastructure for research, extension, input supply, and marketing
__ Degradation of the agricultural resource base due to desertification, droughts, and other processes
__ Large untapped potential of arable land, irrigation, and fertilizer use

West Asia and North Africa

__ Improvements in productivity due to better irrigation, the cultivation of high-yielding varieties, and higher fertilizer use
__ Limited arable land and considerable amounts of desert, making food self-sufficiency a challenge
__ A need for controlled irrigation to cope with dry conditions

South and East Asia

__ Increased production and productivity, with some countries registering grain surpluses
__ Rapid growth in fertilizer use in some countries and extensive development of irrigation
__ Government commitments to be self-reliant in food, leading to national research centers, development of high-yielding seeds, and the fostering of location-specific technologies
__ Little unused land and extensive, unabated deforestation
__ Growing numbers of rural landless

Latin America

__ Declining food imports since 1980, as food production kept pace with population growth over the last decade

__ Government support in the form of research centers to develop high-yielding seeds and other technologies
__ Inequitable distribution of land
__ Deforestation and degradation of the agricultural resource base, fueled partly by foreign trade and debt crisis
__ A huge land resource and high productivity potential, though most of the potentially arable land is in the remote, lightly populated Amazon Basin, where perhaps only 20% of the land is suitable for sustainable agriculture

North America and Western Europe

__ North America the world's leading source of surplus food grain, though the rate of increase in output per hectare and in total productivity slowed in the 1970s
__ Subsidies for production that are ecologically and economically expensive
__ Depressing effect of surpluses on world markets and consequent impact on developing countries
__ A resource base increasingly degraded through erosion, acidification, and water contamination
__ In North America, some scope for future agricultural expansion in frontier areas that can be intensively farmed only at high cost

Eastern Europe and the former Soviet Union

__ Food deficits met through imports, with the former Soviet Union being the world's largest grain importer
__ Increased government investment in agriculture accompanied by eased farm distribution and organization to meet desires for food self-reliance, leading to production increases in meat and root crops
__ Pressures on agricultural resources through soil erosion, acidification, salinization, alkalization, and water contamination

Source: World Commission on Environment and Development, *Our Common Future* (New York: Oxford University Press, 1987) pp. 121–122.

The Reagan administration initiated a farm program in 1983 to reduce grain surpluses in storage in the United States. The program was designed to award surplus stored grain to farmers who voluntarily kept their farmland out of production that year. The so-called payment-in-kind (PIK) program resulted in a loss in production of about 2.2 billion bushels of corn.

That same year a severe drought in the Corn Belt in the United States resulted in the loss of another billion bushels of corn. The combined impact of these events reduced world carryover stocks of grain to the lowest level experienced in several years. In 1988, the drought in North America and Asia again reduced world grain stocks to low levels.

Table 8.5 World Grain Trade by Region, 1950–1988 (in millions of metric tons)

Region	1950	1960	1970	1980	1988*
North America	+23	+39	+56	+131	+119
Latin America	+1	0	+4	−10	−11
Western Europe	−22	−25	−30	−16	+22
Eastern Europe and USSR	0	0	0	−46	−27
Africa	0	−2	−5	−15	−28
Asia	−6	−17	−37	−63	−89
Australia and New Zealand	+3	+6	+12	+19	+14

Note: A plus sign indicates net exports; a minus sign, net imports.

* 1988 figures are estimated.

Source: L. R. Brown and J. E. Young, "Growing Food in a Warmer World," *Worldwatch,* November-December 1988, p. 32.

The world fish catch was nearly twice as great in 1970 as in 1950 but turned downward for several consecutive years in the early 1970s. Per capita, the fish catch has changed little since 1970. There is concern that overfishing of the oceans is occurring and that pollution of the oceans is also taking its toll.

Factors that in the past allowed for dramatic increases in food production are not likely to continue to result in additional major increases. These increases were the result of the development of high-yield crops, energy-intensive technologies including pesticides and fertilizers, and irrigation. In the future, increases will occur, but in many cases with diminishing returns (more energy and effort put in per unit of food produced). The less developed countries have a poor outlook for improved diets. According to the World Bank, the number of malnourished people in the world could rise to 1.3 billion by the year 2000.

The Potential and Limits of the Land

There are only three ways by which more food can be made available to the world market. One is by bringing new lands into production; the second is by increasing the yield on already productive land; the third is by improving the use of existing food supplies

New Lands

There seems to be some general agreement that increased food supplies will not come from bringing into production land that has not been used for crops before. As long ago as 1973, the United Nations' Food and Agriculture Organization (FAO) announced that most of the 11% of the world's land that is suitable for cultivation was already in use. Even more telling is that the world area in grain production has actually diminished by 7% since 1981. The former

Soviet Union, China, and the United States have continually decreased the amount of land cultivated for crops; for example, since 1978, land area devoted to grain in the former USSR has declined by 13%; it has diminished by the same amount in China since 1976. Although the overall volume of grain produced has not declined in China, it has in the former Soviet Union. The lands removed from cultivation were marginal lands that went into production in the mid-1970s when grain prices were high.

The Food Security Act passed by the U.S. Congress in 1985 has a conservation component designed to take erodible land out of crop production and discourage cropping of grassland and draining of wetlands for crops. More than 10 million hectares has so far been taken out of production under this program.

In the mid-1970s, overexpansion of cropland did not provide a long-term, sustainable base for increasing crop production. It will not be a major factor in the future. Africa and Latin America have the greatest potential for cultivating new lands. But cultivation of tropical forests has not proved to be a sustainable alternative (see Chapters 13 and 14).

Eventually, on a worldwide scale, the law of diminishing returns will render a strategy of new lands ineffective—if it is not so already. Land under cultivation is projected by some to increase by only 4% by the year 2000. But if this does happen, it will be more than offset by the daily loss of top-quality farmland by conversion to other uses, by loss of topsoil, and by conversion of farmland into desert by improper management (see Chapter 13).

Desertification

One way to increase food production is by irrigation. In that regard, again some limits have been reached. The amount of irrigated land in the United States and China is declining. It is anticipated that India is the only country likely to increase its irrigated area significantly over the rest of the century.

Irrigation brings some of its own problems. In the United States, about one-fourth of the irrigated cropland is watered by overuse of groundwater. Water tables are declining by 6 to 48 inches per year. Another problem is irrigation of dry areas where evaporation is high, resulting in salt accumulation.

Almost 230 million people live on lands that are relatively barren because of human activities; in West Africa the Sahara expands southward an estimated 10 kilometers a year. In addition, almost 125,000 hectares of cropped land is lost annually because of poor irrigation practices in arid lands, resulting in soil waterlogging and salinization (salt accumulation). This process of improper management of arid or semiarid land to the extent that it is no longer suitable for range or cropland is called **desertification.** It is a serious and growing problem that results from climate and human activities. Extensive irrigation, poor soil drainage, deforestation, and urban development are all contributors to soil erosion, salinization of topsoils and irrigation water, and destruction of native vegetation. According to the World Commission on Environment and Development, about 30% of the earth's land suffers from some degree of desertification. Almost 20% of the arid lands in South America, Asia, and Africa have been severely affected. In the United States the primary problem areas are the San Joaquin Valley in California, the Wellton-Mohawk Irrigation District in Arizona, the Santa Cruz and Pedro river basins in Arizona, the counties of Kiowa and Crowly in Colorado, and the High Plains in Texas.

Actual global losses to desertification are estimated to run about 6 million hectares annually—an area about the size of Maine. This includes 3.2 million hectares of rangeland, 2.5 million hectares of rain-fed cropland, and 125,000 hectares of irrigated farmland. Ironically, the trend toward desertification may accelerate as population growth puts additional pressures on land for food, energy, and shelter.

Increasing Yields

What about the prospects for increasing yields from existing farmland? Whereas 1 hectare of land supported 2.6 people in 1970, a hectare will have to support four people in the year 2000.

One way to increase yield is to use fertilizer. Figure 8.10 shows how fertilizer input per capita substituted for grain area per capita between 1950 and 1986. Average yield per hectare increased from 1.1 tons in 1950 to 2.3 tons in 1986. This was the result of increased irrigation and improved crop varieties, as well as increased fertilizer use. To maintain 1987 consumption levels in the year 2000, given a projected population of over 6 billion by then, will require at least a 25% increase in average grain yields.

Eventually, the law of diminishing returns will prevail here also, however. The annual growth rate of fertilizer use has dropped worldwide from 6% to 3%. The United States is already reaching the limit. A ton of fertilizer added to the U.S. Corn Belt 20 years ago increased the world grain harvest by 15 to 20 tons; today the same amount of fertilizer would result in only a 6- to 10-ton increase (Brown, 1988).

Most of the potential for increased yields through fertilization is in the developing countries. However, energy requirements for fertilizer production will keep the costs of fertilizer high and out of reach of those who need it most. Even in some developing countries there are indications that yield per unit of fertilizer is declining, just as in the United States. In addition, many developing countries are the very ones where, even with sufficient fertilizer,

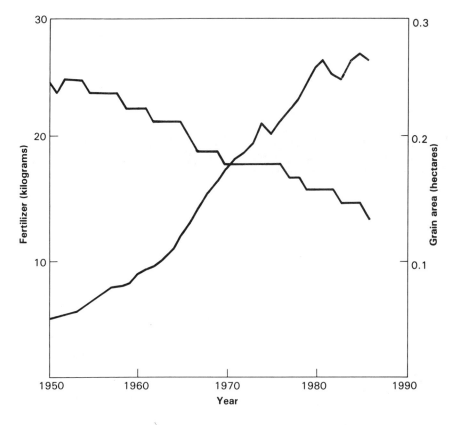

Figure 8.10 World Fertilizer Use and Grain Area Per Capita, 1950 to 1986. This graph shows dramatically how fertilizer use increased and substituted for less grain area per capita in the world from 1950 to 1986. Thus more grain could be harvested per hectare to feed a growing world population.

Enrichment Box 8-2

How Many People Can We Feed?

Some researchers have assessed the "theoretical" potential for global food production. One study assumes that the area under food production can be around 1.5 billion hectares (close to the current level) and that average yields could go up to 5 tons of grain equivalent per hectare (as against the present average of 2 tons of grain equivalent). Allowing for production from rangelands and marine sources, the total "potential" is placed at 8 billion tons of grain equivalent.

How many people can this sustain? The present global average consumption of plant energy for food, seed, and animal feed amounts to about 6,000 calories daily, with a range among countries of 3,000–15,000 calories, depending on the level of meat consumption. On this basis, the potential production could sustain a little more than 11 billion people. But if the average consumption rises substantially—say, to 9,000 calories—the population carrying capacity of the earth comes down to 7.5 billion. These figures could be substantially higher if the area under food production and the productivity of 3 billion hectares of permanent pastures can be increased on a sustainable basis. Nevertheless, the data do suggest that meeting the food needs of an ultimate world population of around 10 billion would require some changes in food habits, as well as greatly improving the efficiency of traditional agriculture.

Source: World Commission on Environment and Development, *Our Common Future* (New York: Oxford University Press, 1987), p. 99.

water may be the limiting factor. But water is scarce, and irrigation also requires energy. Irrigated lands have been growing by less than 1% annually worldwide since 1980, compared to 4% growth in the 1950s and 1960s. Brown and Young (1990) state that the irrigated area per person has declined by 8% since 1980. They suggest that gains in newly irrigated areas will be offset by losses from desertification, lower water tables, and silt buildup in reservoirs.

The ultimate limit to increased yields is the photosynthetic efficiency of plants. Once that limit is reached, increased fertilizer or water has little or no impact (see Enrichment Box 8.2).

Yields can also be increased by growing more crops on the same land in any given year. Varieties of crops with shortened growing seasons are now available. However, more crops will require more resource and energy inputs.

We cannot escape the fact that increases in food production in the past were made possible in part by energy (see Figure 8.11 and Chapter 6). The world's simultaneous food and energy crises present a profound dilemma indeed.

The Green Revolution

In the mid-1960s a technological breakthrough in the development of improved pure-line varieties of rice and wheat resulted in such an optimistic outlook for food production that it was called the **Green Revolution.** New wheat varieties were shorter and stiffer; this allowed them to respond to fertilizer without growing tall and falling over from their own weight. They had proportionately more grain than straw. The varieties appeared to be more adaptable to diverse environments and showed some resistance to diseases and insects. They required shorter growing seasons and matured over wide ranges of day length. The prognosis looked good. Between 1967 and 1969, miracle grain-based food production increased by 27% in India and Pakistan.

In the early 1970s, however, some of the promise began to fade. For one thing, the new strains performed best with large amounts of fertilizer, even though without fertilizer their yields exceeded those of traditional varieties. The wheat required much water. The new varieties were genetically very similar to one another, and this increased pest susceptibility. In the early stages of the "revolution," the miracle varieties were highly subsidized, so the high costs of production were somewhat masked. Even so, the new varieties often were not economically available to the poorest rural people. Socially, because the Green Revolution technology tends to use machines instead of human labor, it presents problems in many countries for the farm labor market and the millions of landless laborers in rural areas. Good technology out of the context of social and economic norms cannot solve the problem of food production in developing countries.

The technology of the Green Revolution *has* increased crop yields. Nevertheless, adoption of the high-yield varieties by Third World countries has not come as quickly or as broadly as was originally predicted. Currently, high yield wheat and rice is grown on about 35% of the grain area in Asia and the Middle East, a little over 20% of Latin America, and 1% of Africa. The low usage in Africa is due to its generally poor soils and low rainfall; the inability to pay for fertilizer, pesticides, and irrigation; and the traditional growing of crops other than wheat and rice,

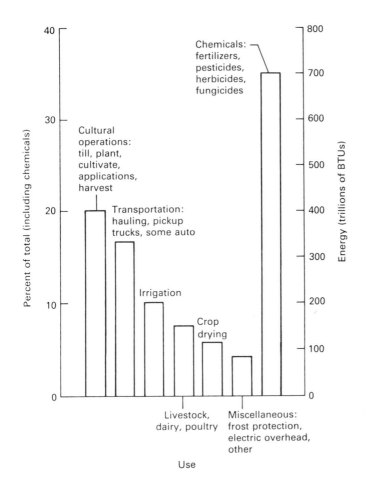

Figure 8.11   Energy in Agriculture.   Modern agriculture is energy-intensive. Energy is used directly in running farm machinery, but the largest single energy consumers in modern agricultural practices are fertilizer and herbicide production (see also Chapter 6).

such as maize, sorghum, millet, cassava, yams, and sweet potatoes.

A New Role for Traditional Agriculture

Along with adjustments in the direction of the Green Revolution, there is a growing awareness of the value of examining traditional agricultural practices as another way of increasing yields on existing land. In many parts of the world, farming has taken place on the same land for many, many years without depleting the land resource. In the United States and throughout the world, the principles of **sustainable** (also referred to as **regenerative, low-input,** and **organic**) **agriculture** are being studied.

Worldwide, traditional agricultural practices are characterized by few external inputs (chemicals, irrigation); the effective accumulation and cycling of nutrients; diversity in cropping; and protection of the soil. They are generally based on sound ecological principles and operate in concert with natural ecological cycles and systems. With few external inputs, the cost of implementing traditional practices is low. Another advantage is that these practices are familiar to the farming population.

In the Third World, the advantages of traditional methods are many. Socially and culturally, farmers are familiar with them. They do not require expensive inputs the way many of the high-yield varieties of crops do. Traditional methods require what is local and at hand. They are sustainable over the long run. But traditional practices alone will not produce the yields necessary to feed the people of the Third World. More research to enhance these practices is needed.

A good example of traditional agriculture is the practice of agroforestry. Used in the Sahel region of Africa, it could be applied in other regions as well. Acacia trees are grown in the crop fields, interspersed between the sorghum and millet crops. Over the years the farmers there have come to realize that the yields around the acacia trees were the highest. Ecologically, what is happening is that the acacia trees fix nitrogen, draw nutrients from the deep soil to the surface, and improve the soil texture.

Legume-based crop rotations are common in traditional agriculture to build up nitrogen reserves in the soil, contrary to the modern monoculture technique, which requires artificial fertilizer inputs. However, it has been found that small amounts of artificial fertilizer can be used along with organic fertilizer to increase yields above those obtained using either organic or synthetic fertilizer alone.

The traditional farming method in many Third World countries has been *shifting cultivation*, a practice by which an area is cleared and farmed for a while, and then, as yields decrease, a new area is cleared and the old area is allowed to regenerate. In the practice called "alley cropping," planting crops between rows of nitrogen-fixing trees rather than totally clearing a field, the need for a fallow or regenerative period is eliminated; the trees remain to fix nitrogen while the crop grows.

Traditional agriculture in the Third World has its limitations. The genetic potential for high yields is limited, soils are often poor, and the practices that enrich and protect the soil are often abandoned under population pressure that pushes for higher yields per hectare. Without question, however, the enhancement of traditional methods holds a key to widespread adoption of sustainable and higher-yielding techniques by poor farmers in the Third World. More needs to be done in this area.

Improving the Use of Existing Supplies

What kinds of changes would improve the percentage of food produced that actually reaches and is consumed by people? Less meat production and more direct consumption of grain, better preservatives and storage techniques, and better distribution systems are all possibilities.

Meat versus Grain for Human Consumption. The second law of thermodynamics tells us that far more people could be supported on grain than on the meat of grain-eating animals. Though this suggests a simple partial solution to the hunger problem, it is not really so simple. The economics of current demand worldwide will probably hinder a voluntary decline in meat production. The pressure for food is coming not just from growing populations in the Third World but also from rising demands for meat in industrialized nations.

The increased buying power of the OPEC nations will result in greater food imports. The consumption of grain in affluent countries is proportionately four to five times that of poorer countries owing to the use of grain for poultry and livestock to produce meat, milk, and eggs. Almost 40% of the grain consumed in the world is eaten by livestock. In the United States, 70% of its grain consumption goes to livestock, about 20% in China, and 2% in sub-Saharan Africa and India.

Livestock protein that is produced by the conversion of forage crops and rangeland proteins that are *not* usable by humans can actually increase the amount of protein available to human beings. The problem is that the remaining livestock protein is produced from feed that *is* usable by humans. Effi-ciencies are such that 5 kilograms of vegetable protein suitable for human consumption but fed to livestock instead results in only 1 kilogram of livestock protein for human consumption. It should be noted that chickens convert more plant protein to animal protein per unit consumed than cattle do. The net effect, however, whether grain is fed to chickens or to cattle, is a decrease in calories for human consumption. Almost 30% of the world's supply of protein suitable for human consumption is fed to livestock. A forage-fed-only system for livestock would release around 135 million tons of grain for human consumption, an amount capable of feeding 400 million people—a large number but less than 10% of the world population. Worldwide, one-third of the fish harvest is used to feed animals other than humans.

Increased demand for U.S. grain, the need for grain exports to help the balance of payments, and the resulting greater costs of grain-fed beef are all likely to result in a decrease in the production of grain-fed livestock. Increased forage feeding of livestock could use lands not suitable for crop production. Forage-fed cattle also have a lower fat content, less waste, and a higher protein content—producing healthier meat for consumers. There is also a detrimental side to the forage feeding of cattle, however. The demand for low-grade beef for the fast-food industry is vast. Tropical forests are being cut to create pasture areas to supply this market. We will discuss the implications of this further in Chapter 14.

Storage and Preservation

An unbelievable amount of grain intended for human consumption is lost to rodents, insects, molds, and fungi. Rodentproof storage bins would greatly increase India's food supply. Proper ventilation and improved preservation technology would help to prevent the loss of grain to molds.

Food from the Sea

Since ancient times, humans have harvested food from the seas. Some have calculated that 200 billion metric tons of living matter are produced by green plants in the ocean each year. This is at least as great as plant production on the land. The FAO estimates the maximum sustainable yield from the oceans to be about 100 million tons.

The commercial harvest of fish from the sea reached almost 100 million tons in 1989. The continued increase in fish catches can be attributed to improved fishing techniques, the fishing of more remote areas, and the harvesting of less desirable fish species (Lenssen, 1989) (see Figure 8.12). There are, however, some problems in the way of achieving

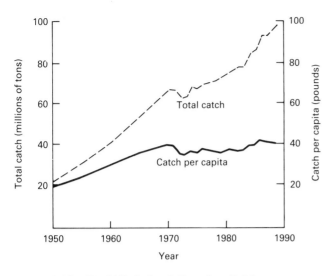

Figure 8.12 World Fish Catch Trends. Fish have historically provided a major source of protein in developing countries. However, there are some indications that significant increases in fish catches are unlikely in the future.

greater yields. Numbers of various species of fish have peaked and declined due to overfishing. Lower catches per boat and a dominance of smaller fishes are indicative of the fact that certain fish species in certain areas are being overharvested, that is, being harvested faster than the supply can be replenished. Over 40 of the 280 fish stocks monitored by the FAO are classified as overexploited or depleted (Lenssen, 1989).

Artificial production of fish through **aquaculture** in both fresh and salt water is growing worldwide at the rate of 6% to 8% a year. Finfish, shellfish, and algae are all raised by aquaculture. According to Bardach (1987), about 85% of the mollusk crop is raised via aquaculture, and about two-thirds of the algae used for food condiments and other products. Aquaculture is a versatile technology because it can be done in small ponds as well as on a commercial scale.

Freshwater aquaculture in the United States has centered on the cultivation of finfish. The channel catfish industry and the rainbow trout industry are two examples of successful aquaculture in the United States. In Hong Kong, 18% of the agricultural land is covered with fish ponds. Sewage-fed aquaculture is practiced in China, India, Thailand, and Vietnam, where fish are cultivated on algae that grow in wastewater lagoons.

A variation on fish farming is ocean ranching. In this approach, scientists manipulate conditions in an impounded area of the coast to stimulate production—seeding an area with eggs or larvae, improving habitat by providing food shelter, and discouraging predators. Good ecological information on the species to be harvested will be the key to the future success of ocean ranching.

Although total marine productivity may in fact equal total terrestrial productivity, marine food resources are scattered over a much larger surface and

The oceans are rich sources of food. We must take care not to harvest them faster than they can renew themselves.

have a third dimension—depth. This makes harvesting difficult. About 1 billion liters of seawater would have to be processed to yield 1 ton of plankton. Given the energy involved, larger fish can concentrate protein more efficiently than humans, a fact that might save us from plankton dinners.

Half of the fish in the sea are found in regions where there is a natural upwelling of nutrients from the ocean depths. What about fertilizing the ocean, increasing plankton, and then harvesting fish higher up the food chain? The problem again is diffusion and economics. It would take a large amount of energy to pull nutrients from the ocean bottom or to produce enough fertilizer otherwise to have a significant effect.

Research is now under way to improve food production in low-nutrient seas by imitating the operation of the coral reef (see Adey, 1987). Coral reefs are systems of high productivity; understanding exactly how their nutrients are obtained and used could provide leads to enhance production in low-nutrient areas of the sea without nutrient enrichment.

Long aquatic food chains yield calories available for human consumption less efficiently than land production does. It has been suggested that it takes as much protein in the form of plankton to produce one annual world catch of fish as is contained in 40 world harvests of wheat or 75 world harvests of rice.

As we will see in Chapter 12, the protection of high fish concentrations near shorelines and estuaries is extremely important in sustaining yields, yet it is these areas that are most susceptible to pollution. The need to travel farther and farther out to sea for a fish catch results in increased energy consumption. Another problem is that while fish provide more protein per gram than many other foods, access to seafood inland is limited in poorer countries where transportation and storage services are not available.

According to the findings of the World Commission on Environment and Development, a world catch of 100 million tons in the year 2000 is well short of predicted demand. On a per capita basis, the oceans may contribute less to human nutrition in the year 2000 than they do now. This is especially disturbing when one notes that 40% of the animal protein supply in developing countries comes from fish, compared to 25% for the world as a whole (Lenssen, 1989).

Food and Health

Three-quarters of a billion people in the world do not have enough to eat. In some cases they are **undernourished,** which means that they do not have sufficient intake of calories. Many are **malnourished;** that is, they lack sufficient protein. Proteins are complex substances composed of basic units called amino acids. Humans require different types of amino acids. Vegetable or animal protein entering the digestive tract is broken down into individual amino acids, which are then absorbed and later reassembled to form human body protein. Most plant protein is typically deficient in one or more of the amino acids necessary for humans. Since different plants are deficient in different amino acids, a proper mix of plants can yield a sufficient variety of amino acids. Malnutrition is most common in cultures in which the diet consists predominantly of plant material of only a few different types.

Caloric intake quite obviously varies from country to country and from one income level to another within the same country. The groups most vulnerable to nutrition problems are the very young, the very old, and pregnant and nursing women.

The Food and Agriculture Organization (FAO) and the World Health Organization (WHO) have calculated the amount of food a person needs to carry on daily life activities. A study by the World Bank looked at populations in the Third World in relation to this standard. In 1980, fully 340 million people had less than 80% of the standard and 730 million people had less than 90% of the standard. Although the *percentage* of people below the 90% level dropped from 40% to 34% between 1970 and 1980, the *number* of hungry people increased from 680 million to 730 million.

Marasmus is a disease caused by insufficient calories and protein; it affects children less than a year old most acutely and is characterized by progressive emaciation, extreme loss of weight, and wasting away. We are all familiar with pictures of children suffering from **kwashiorkor,** a disease caused by a lack of protein. Kwashiorkor affects predominantly 12- to 36-month-old children. It is characterized by an enlarged liver, accumulation of abdominal fluid contributing to a "pot belly," skin discoloration, retarded growth, loss of hair, and the accumulation of fluid in other tissue. Both marasmus and kwashiorkor result in apathy and apparent mental slowness.

Nutrition in pregnant women is important to both mother and child. Lack of proper nutrients during the first six months of pregnancy can result in mental defects in the child because of impairment of brain cell development. Malnutrition in the sixth to seventh month may affect the child but may be reversed, as only brain cell size is stunted. Generally, poorly nourished mothers give birth to higher-risk premature and underweight babies.

Many diseases are caused by a lack of vitamins. Some important ones are **beriberi** (lack of thiamine), **scurvy** (a vitamin C deficiency), **pellagra** (a niacin deficiency), and **rickets** (a lack of vitamin D and/or calcium). Lack of vitamin A causes a drying of the eye membranes and may lead to blindness. Anemia results

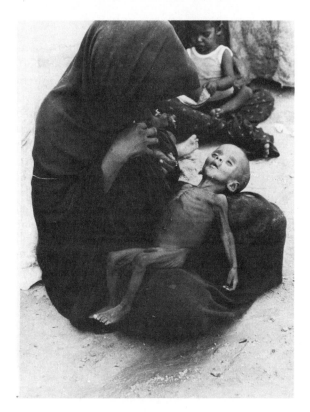

In some parts of the world, even after a drought has eased, poor nomad families who have lost their cattle have no economic means to feed themselves. It is the children who suffer the most.

from a lack of folic acid, vitamin B_{12}, or iron. A lack of iodine in a pregnant woman may lead to a mentally deficient child.

Overall poor nutrition lowers the resistance and makes one more susceptible to infection and disease. The 1985 report of the Physician Task Force on Hunger in America notes that some clinics in poor areas of the United States are reporting cases of marasmus, kwashiorkor, and vitamin deficiency diseases.

Food Distribution and Economic Development

The hungriest people in the world are usually the poorest. The demand for food by the hungry is simply not recognized in a worldwide economic system in which food is a commodity—it is bought and sold and only rarely given away. What this means is that the most logical strategies for eliminating hunger are aimed at economic development, giving the hungry the means to buy or produce food and not just giving them food.

The Politics of Food

The importance of economics and politics to food distribution became quite clear in the late 1970s when President Carter imposed an embargo on grain sales to the Soviet Union (intended as a rebuke for Soviet intervention in Afghanistan). This forced a decline in demand, which, along with the decline in grain exports resulting from the strengthening dollar, led to huge surpluses and plummeting grain prices. U.S. grain surpluses were such in 1982–1983 that the Reagan administration paid farmers for keeping land out of production not in dollars but in surplus grain, which in turn could be sold in the marketplace. Such is the capriciousness of politics and economic demand in a world in which hunger is rampant.

U.S. aid to developing countries amounts to less than 0.2% of the gross national product. The goal set by the United Nations for aid to developing countries by developed countries is 0.7% of GNP. Industrial countries average only 0.34% of GNP. But there are variations. Norway, for example, contributes 1% of its GNP.

As far back as 1978, the U.S. Commission on Hunger recommended basing foreign policy toward developing countries on the primary goal of alleviating hunger and developing self-sufficient agricultural programs in those countries. The commission was explicit in its insistence that the United States should not try to feed the world but rather to assist it in learning to feed itself. Besides foreign trade in agricultural products, the United States has programs to assist developing countries via foreign economic assistance and technical aid.

That the "help them feed themselves" approach works is exemplified by India. Once written off as hopeless, India has increased food production in recent decades to the extent that it began to reach food self-sufficiency by 1980. Increased yields are attributed to nonfood aid and technical assistance, which resulted in increased use of fertilizer and more land under irrigation. The Indian government provided price supports for crops and fertilizer to encourage production in India by locals.

The World Commission on Environment and Development supports efforts to develop self-sufficiency in agriculture in developing countries. In addition, the commission concludes that world food strategies must also contain provisions that conserve resources and that provide for sound livelihoods for the working poor. The report acknowledges the role of government intervention in agricultural activities in both industrial and developing countries. It finds government policies generally lacking. Rather than being ecologically oriented, most policies are focused on the short term. In addition, they tend to apply nationally rather than to reflect regional needs and differences. For example, subsidizing a particular crop nationwide means that certain crops may be grown in an area more suitable for a different crop. In addition, some policies actually encourage degradation of the agricultural resource base with incentive structures that result in overuse of the land.

Appropriate Technology

Any strategy to transfer technology to less developed countries must take into account the prevailing social, economic, and cultural patterns. In other words, the technology to be transferred must be appropriate and realistic. High-protein food that people will not eat will not solve a protein deficiency problem. High crop yields from intricate mechanized farming techniques may not be transferable to small family farmers who cannot get machinery. Food that does not reach the hungry because of bureaucratic red tape has no impact on the hunger problem. As we saw in Chapter 6, agricultural practices that are highly energy intensive will be less and less useful to poorer nations, which are simply unable to afford fossil fuels.

Since a good amount of land that is unsuitable for cropping *is* suitable as pasture, this land has considerable potential for increased animal production in developing countries. Developing countries already have large numbers of cattle, buffalo, sheep, and goats. What has not been done is maximizing their full potential for work, food, and fertilizer through appropriate methods that are compatible with the prevailing customs and practices. Animal production requires resources that are abundant in most developing countries—forage crops and human labor.

Another example of the application of appropriate technology is the promotion of aquaculture in China, where demonstration projects raising carp have been immensely successful. One variety of carp eats waste vegetation from farms or grass cut along the side of the pond; other varieties eat phytoplankton; others feed on the organic matter on the pond bottom. Human and animal excrement is fermented to remove pathogens and then used to fertilize the pond. In a complete cycle, humus from the bottom of the pond is returned to the fields as fertilizer. These aquaculture projects are conducted under the auspices of the Pearl River Fisheries Research Institute in Guangzhou (Canton). Representatives from other Asian countries have been trained in these techniques, which are simple, basic, low in energy and capital input, high in output, and appropriate for the needs of many developing countries.

In the past, developing countries have been somewhat suspicious of the concept of appropriate technology, labeling it a method to keep developing countries from entering the high-technology markets of the West and sharing a similar standard of living. Such suspicions are being somewhat allayed by successes in using appropriate technology, especially when the government of the receiving nation is involved in identifying and promoting appropriate agricultural technology. For example, the Indian Institute of Science supports the ASTRA (Application of Science and Technology to Rural Areas) project. This project has done research not just on agriculture but on rural energy systems, housing technology, water projects, and the processing of agricultural wastes and residues. In addition, a system of international agricultural research centers is distributed worldwide in countries such as Mexico, the Philippines, Colombia, and Nigeria; these centers are dedicated to increasing yields and protein levels in staples grown in the Third World.

Research

Where is science headed in improving world food supplies? To date, the greatest advances in increasing crop yields have come from using fertile soils in favorable climates. Research since the Green Revolution has focused on the need to address some of the problems associated with the high-yield varieties of crops such as decreased resistance to pests. However, future research must look at improving yields in areas with less than ideal soil and climate. This means developing varieties of crops that can better tolerate saline soils, acidic soils, and soils with high levels of toxic minerals, as well as varieties that are drought-tolerant or flood-tolerant.

Because of all the analysis of the Green Revolution's successes and failures, the direction of research has broadened to include considerations such as the applicability in tropical areas of small-scale, low-capital, low-machinery solutions as well as more technological approaches.

The international agricultural research centers that have been developed are coordinated by the Consulting Group on International Agricultural Research (CGIAR). They are studying 21 food crops (out of 3,000 plants that have been used for food throughout history) and focus on issues such as the conservation of genetic resources and maintenance of pest resistance in high-yield crops. They have recently made a move to include the sustainability of agriculture and the protection of the environment as goals in their research. The International Rice Research Institute (IRRI) in the Philippines, one of the first research groups involved in the Green Revolution, is shifting its focus to varieties of grain that could be profitable to poor and marginal farmers. A promising technology includes the growing of the fern *Azolla microphylla* in rice paddies. Blue-green algae that fix nitrogen proliferate along with the fern and add nitrogen to the system. The International Institute of Tropical Agriculture in Nigeria has found a leguminous African shrub that may be useful in adding nitrogen to African rice fields. Fungi that help enhance phosphorus uptake by roots of crops have been discovered in Colombia. Although the research centers conduct only a small fraction of all ongoing research, their policies are influential in directing governmental research.

Scientists are looking at food substitutes for livestock, freeing grains for human consumption. Finding ways of decreasing nitrification (the conversion of nitrate to nitrite), which releases volatile nitrogen gases from the soil, to maintain fertility longer is another goal of agricultural research. Losses by denitrification and volatilization may be as much as 10% to 30% of the nitrogen applied. Another problem arises here, however: The known nitrification inhibitors are also long-lived, generally toxic substances.

Use of perennials in crop production is under study. Perennials would eliminate the need for annual tilling and would thus be a major factor in erosion control. Plants that could use nitrogen from the atmosphere are the subject of biogenetic research.

Research on nutrition and diet is also going on. We know less than we need to know in this area. For instance, the possible efficacy of breast feeding as a means of improving nutrition and providing some immunity against disease has been overlooked. Instead, prepared formula, which is costly and can become contaminated by impure containers or impure water, has been promoted in the developing nations.

Certain techniques such as remote sensing of the environment via satellite may be used to provide information that will make crop predictions more reliable and might help to stabilize the economics of food. These same techniques have proved helpful in determining appropriate irrigation schedules to prevent too much or too little watering. Weather control, although controversial, will no doubt continue to be studied.

Biotechnology

After the Green Revolution, the new light on the horizon has been **biotechnology.** Biotechnology refers to a variety of biological techniques, the most common one being bioengineering. Bioengineering involves the ability to transfer genes, and hence desired traits, from one organism to another. For example, drought resistance, salt tolerance, and pest resistance could be engineered into staple grain crops. Grains could be adapted to grow in poor soils or in other unfavorable conditions. The potential is enormous. There are, however, some constraints.

In the area of plant bioengineering, in particular, developments have been slow (see Young, 1990). For one thing, the genes in crops must be mapped; that is, the specific gene must be matched with the trait it carries. The transfer of "foreign genes" to crops has proved difficult. Once a gene is successfully transferred, conventional methods of growing the plants must be used to evaluate the trait and, if all is found to be satisfactory, to produce the new crop or altered plant for dissemination. This could take anywhere from 5 to 15 years.

Research groups such as the CGIAR have not made much of a commitment to plant biotechnology. Large U.S. biotechnology companies do not necessarily gear their research to the needs of the developing countries. Luxury crops rather than staples are likely to bring greater profits and thus receive greater attention. There has been a focus on the development of herbicide-resistant plants. This provides a good market for herbicide producers, but in developing countries, where labor is plentiful, weed control is more cheaply done by hand.

Biotechnology has been successfully applied in certain developing countries. In Dalat, Vietnam, 30 families are growing potatoes using tissue cultures from the International Potato Center in Peru. With an average initial cost of $250, the families earn $100 to $120 a month from their tissue-cultured potatoes. These families got their start from information provided by a Vietnamese agricultural extension official.

There are drawbacks to bioengineering, however, and selecting applications requires great foresight. Janzen (1987) points out that genetic engineering could eliminate the one safeguard against further destruction of the tropics. Plants engineered to the environmental conditions there would make agriculture economically viable where now it is not—but at the same time would further endanger tropical forests (see Chapter 14).

Many concerns must be worked out. Large biotechnology companies want to patent their new bioengineered organisms and fear that developing countries are less likely to honor these patents. Developing countries prefer home-based labs and are pursuing these. In addition, they are skeptical that outside firms will take their indigenous crops and seeds, create improved strains, and then try to sell them back to the developing countries at high prices or take over their export markets.

There are many questions to be answered. Should there be greater emphasis on food commodities over nonfood commodities (tobacco, cotton, biomass for energy conversion) in allocating research dollars? Can the protein content of such staples as rice,

maize, yams, and cassava be improved? Should we teach more tropical agriculture to foreign exchange students? Can family farming survive, or are agribusiness and market farming the only way to go? A key question is, How sustainable is current agricultural output? Numerous ecological, economic, and social questions also need answers.

A total systems approach including ecological, technical, and social considerations seems to be surfacing. As Wendell Berry (1974) states:

> A healthy farm culture can only be based on familiarity; it can only grow among a people soundly established upon the land; it would nourish, and protect a human intelligence of the land that no amount of technology can satisfactorily replace.

In Closing

The problems of population, food, and hunger will be with us for some time. Although fertility rates in the developing world have been declining since the late 1960s, total populations will continue to increase significantly because of the large numbers of women of prereproductive age. Currently, almost 90 million people are added to the world each year. Although new environmentally appropriate technologies and economic development strategies will help to improve food distribution and to alleviate hunger, the ultimate solution to the food problem is to limit the demand for food. Some fear that by continually finding the means to produce more food, we are short-circuiting one of the mechanisms of population control. If we fail to see the ultimate need for control of population, we will find ourselves up against the ecological carrying capacity of our planet.

Concepts to Remember

1. Historically, war, famine, and disease have served as checks on human population growth. Today, by and large, the mechanisms that serve to regulate population growth in humans are economic, social, and institutional.
2. Demography is the study of the vital and other social statistics of human populations. Indices such as birth rate, death rate, rate of natural increase, and fertility rate can be used to project trends in human population growth.
3. If constant age-specific death rates (e.g., constant infant mortality rates) are assumed, the age structure of a population and its fertility rate are the two key factors in predicting its long-range growth trend. When

an expanding population's fertility rate falls to the replacement rate, it might still take a number of years for the population to stop growing, that is, to reach zero population growth (ZPG). Although the U.S. fertility rate has been at or below the replacement level for some years, ZPG will probably not be reached until at least the year 2030.
4. The lag effect in stabilizing population growth is highly significant. At an annual growth rate of 2%, the world's population would double in 35 years. Currently, world population growth is about 1.8% annually. Even though fertility rates are generally on the decline worldwide, total population will continue to increase for

many decades. This means continuing strain on food resources from the land and sea and on other life-supporting resources.

5. While nearly all nations of the world have experienced a declining death rate, the industrialized nations have had a greater decline in birth rate than the developing nations. A shift from high fertility and high mortality to low fertility and low mortality is called the demographic transition. The industrialized countries have made it all the way through such a transition; many of the developing countries have not.

6. Worldwide, especially in developing nations, governments are becoming involved in population control programs. As long as population is increasing rapidly, it is difficult to make improvements in the standard of living per capita; the gap widens between the haves and the have-nots.

7. The world or a nation can be overpopulated in relation to available food, nonrenewable resources, the assimilative capacity of the biosphere, or the quality of life desired.

8. Since the mid-1970s, food production worldwide has increased less than 2% annually—barely ahead of population growth. Per capita consumption of food varies considerably from country to country. People may be undernourished (lacking sufficient calories) or malnourished (lacking sufficient protein).

9. Mechanisms for bringing more food to the world market are limited and will require expensive energy inputs.

10. Improper management of cropland is reducing the amount of land available for agriculture on a sustained basis. Desertification of cropland is a worldwide problem.

11. The Green Revolution, technological advances made in the development of high-yield varieties of rice and wheat, has improved yields, but not to the extent originally predicted. Part of the problem lies in the fact that the new grains cannot be easily obtained or grown by developing countries on a large scale because of their high cost and need for mechanization.

12. Grain-fed livestock reduces the calories available to feed people. Forage-fed livestock increases the calories available to feed people. Complex cultural and economic factors prevent these two facts from being exploited on a large scale to increase human food supply.

13. The hungriest people in the world are usually the poorest. From an economic perspective, food is a commodity to be bought and sold. U.S. aid to developing countries amounts to less than 0.2% of the U.S. GNP. The United Nations has set a goal of 0.7% of the GNP of wealthy nations to be given to developing nations.

14. The long-term policy in food distribution should be to provide developing nations with the skills and resources to grow their own food. Any strategy to transfer technology to developing countries must take into account prevailing social, economic, and cultural patterns if it is to succeed.

15. Research to improve world nutrition and food supply includes developing crop varieties that will grow in saline or acidic soils and drought- and flood-resistant strains. Food substitutes for grain-fed livestock and methods of preventing nitrogen loss from soils are other areas of study.

16. Bioengineering offers numerous possibilities for improving crop varieties. It has been slow in developing in the area of improvements in plants. There are both technical and political problems that need to be addressed.

17. Adaptation of traditional farming methods is a promising area for future research. Traditional farming has evolved in a manner that relies on natural ecological principles and cycles. It tends to be more sustainable than chemical-intensive forms of agriculture. However, traditional methods alone do not produce the high yields needed to feed a growing population. Enhancements can and should be made to increase yields and productivity in ways that are accessible to small farmers in poor countries.

18. Some current crop production is using up groundwater supplies faster than they are recharged, or the crops are produced on land with serious erosion problems. Such production is not sustainable over the long run. The amount of current food supply being grown in a manner to be sustainable over the long run is not known.

19. Although new environmentally appropriate technologies and economic development strategies will help to improve food distribution and to alleviate hunger, the ultimate solution to the food problem must include limiting the demand for food.

20. It has yet to be determined whether the human population growth curve is J-shaped or S-shaped.

Michael R. Sanders

Michael Sanders is a senior resource specialist with the Boulder County (Colorado) Division of Parks and Open Space. Mr. Sanders supervises the Resource Management Program of Boulder County's park system, which encompasses some 20,000 acres. He conducts research projects on such topics as human-wildlife interactions, rare plant communities, forest ecology (fire and insect ecology), fisheries development, and environmental education. He assembles and manages collections of plant and animal species for various studies, evaluates the uses of trails and other recreational facilities, and gets involved directly in delivering environmental education programs. He gives about 15 to 20 presentations per year. His administrative duties include the preparation and monitoring of the division's budget and supervision of the Resource Management Program staff. Currently, Mr. Sanders is conducting a study of the interaction between humans and mountain lions along Colorado's front range. He is gathering information that will be used in making decisions about potential land acquisitions, wildlife management practices, and the development of recreational facilities. Mr. Sanders says that the thing he likes most about his work is that he gets to spend about 80% of his time in the mountains of Colorado—even his worst days are great. He says that even though natural resource work is highly competitive, getting into it is well worth the effort.

Mr. Sanders has a bachelor's degree in vertebrate zoology and a master's degree in resource management. Most of the earlier part of his career was spent with the National Park Service in Yellowstone National Park, the Grand Canyon, the Everglades, the Great Smoky Mountains, and Glacier National Park. Mr. Sanders indicates that a bachelor's degree is the minimum educational background needed for the work he does; a master's degree is preferable. Anyone aspiring to the kind of work he does would need a working knowledge of natural systems and some expertise and experience in resource management.

References and Further Reading

References marked with an asterisk are cited in the chapter.

*Adey, W. H., 1987. "Food Production in Low-nutrient Seas," *BioScience* **37:**340–348.

*Bardach, J., 1987. "Aquaculture," *BioScience* **37:**318–319.

Brough, H., 1989. "Skin-deep Contraception," *Worldwatch*, September-October, pp. 10–11.

*Brown, L. R., 1988. "The Growing Grain Gap," *Worldwatch*, September-October, pp. 10–18.

Brown, L. R., 1989. "Feeding Six Billion," *Worldwatch*, September-October, pp. 32–40.

Brown, L. and Young, J., 1990. "Feeding the World in the Nineties," *State of the World 1990*, New York: W. W. Norton.

Brown, L. R.; Durning, A.; Flavin, C.; French, H.; Jacobson, J.; Lowe, M.; Postel, S.; Renner, M.; Starke, L.; Young, J., 1990. *State of the World.* Washington, D.C.: Worldwatch Institute.

Brown, L. R., 1992. "World Grain Takes a Spill," *Worldwatch*, May-June, pp. 35–36.

Budd, W. W., 1992. "What Capacity the Land," *Journal of Soil and Water Conservation*, **47**(1):28–31.

Durning, A., 1990. "How Much Is 'Enough,'" *Worldwatch*, November-December, pp. 12–19.

Durning, A. and Brough, H. B., 1991. *Taking Stock: Animal Farming and the Environment.* Washington, D.C.: Worldwatch Institute.

*Forrester, J. W., 1971. *World Dynamics.* Cambridge, Mass.: Wright-Allen Press.

*Fremlin, J. H., 1964. "How Many People Can the World Support?" *New Scientist* **415:**285–287.

Gibbons, A., 1990. "Biotechnology Takes Root in the Third World," *Science* **248:**962–963.

Hamilton, D. P., 1990. "RU 486: More than an Abortion Pill," *Technology Review* **93**(4):18–19.

Hoffman, C. A., 1990. "Ecological Risks of Genetic Engineering of Crop Plants," *BioScience* **40:**434–437.

Jacobson, J. L., 1989. "Baby Budget," *Worldwatch*, September-October, pp. 21–31.

Jacobson, J. L., 1991. "India's Misconceived Family Plan," *Worldwatch* **4**(6):18–25.

*Janzen, D. H., 1987. "Conservation and Agricultural Economics" (Letter), *Science* **236:**1159.

Kasperson, J. X., and Kates, R. W., 1990. "Overcoming Hunger in the 1990s" (Special Issue), *Food Policy* **15:**274–358.

Keyfitz, N., 1989. "The Growing Human Population," *Scientific American* **261**(3):119–126.

*Lenssen, N., 1989. "The Ocean Blues," *Worldwatch*, July-August, pp. 26–35.

*Meadows, D. H.; Meadows, D. L.; Randers, J.; and Behrens III, W. W., 1972. *The Limits to Growth: A Report for the Club of Rome.* New York: Universe Books.

Meadows, D., 1992. "Beyond the Limits," *The Amicus Journal*, **13**(4):27–29.

*National Research Council, 1986. *Population Growth and Economic Development: Policy Questions.* Washington, D.C.: National Research Council, National Academy of Sciences.

Pimentel, D.; Dritschilo, W.; Krummel, J.; and Kutzman, J., 1975. "Energy and Land Constraints in Food Protein Production," *Science* **190**:754–761.

Pimentel, D.; Oltenacu, P. A.; Nesheim, M. C.; Krummel, J.; Allen, M. S.; Chick, S., 1980. "The Potential for Grass-fed Livestock: Resource Constraints," *Science* **207**:843–848.

Postel, S., 1989. *Water for Agriculture: Facing the Limits.* Washington, D.C.: Worldwatch Institute.

*President's Commission on Population Growth and the American Future, 1972. *Population and the American Future.* New York: New American Library.

Richards, C., and Gold, R. B., 1990. "RU 486: Medical Breakthrough Held Hostage," *Issues in Science and Technology* **VI**(4):74–78.

Speth, J. G., 1990. "Toward a North-South Compact for the Environment," *Environment* **32**:16–20ff.

Starets, Richard L., 1992. "The Ethics of Immigration: National Interests and Global Concerns," *Insights on Global Ethics,* **2**(1):6ff.

Torrey, B. B., and Kingkade, W. W., 1990. "Population Dynamics of the United States and the Soviet Union," *Science* **247**:1548–1552.

Ulmann, A.; Teutsch, G.; and Philibert, D., 1990. "RU 486," *Scientific American* **262**(6):42–48.

Ward, G. M.; Sutherland, T. M.; and Sutherland, J. M., 1980. "Animals as an Energy Source in Third World Agriculture," *Science* **208**:570–573.

Westoff, C. F., 1986. "Fertility in the U.S.," *Science* **234**:554–559.

Wolf, E. C., 1987. "Raising Agricultural Productivity," in *State of the World, 1987.* New York: Norton.

*World Commission on Environment and Development, 1987. *Our Common Future.* New York: Oxford University Press.

*Young, J. E., 1990. "Bred for the Hungry?" *Worldwatch,* January-February, pp. 14–22.

part three

The Impact
of Human Activities
on Health
and the Environment

Human beings have many kinds of negative impacts
on the environment. We pollute, and we change the
environment in other ways that ultimately hurt us. In
this part of the book we will examine our
environmental impacts from every angle. We will
look at the nature of pollutants, sources of
pollutants, effects of pollutants, pollution control,
pollution trends, costs of controlling pollutants, and
costs of not controlling pollutants. Pollution affects
all of the great spheres. In separate chapters we will
look at air pollution (Chapters 9–11), at water
pollution (Chapter 12), and at the ways in which we
harm land and soil (Chapter 13). Throughout these
chapters we will consider the ways in which our
pollution hurts us directly by affecting our health
and the ways by which we hurt ourselves no less
significantly by diminishing the ability of the
environment to support us—by harming other
species and by changing the earth's climate, for
instance.

We also have direct impacts on the biosphere.
We introduce species and we eliminate species,
sometimes upsetting natural balances. We destroy
rain forests and other habitats, thereby jeopardizing
species whose roles in the stability of the planet are
as yet unknown. We will examine our direct effects
on other species in Chapter 14, and we will look at
how we directly affect one another—human to
human—through noise in Chapter 15.

The contamination of the environment by
hazardous and especially toxic materials is a very
serious problem, which we will consider in Chapter
16. As we look at the ways in which hazardous and
toxic materials contaminate all of the great spheres,
we will be reminded once again that everything is
indeed connected to everything else. In Chapter 17
we will examine the problem of waste disposal, and
in Chapter 18 we will look at the relationship
between cancer and the environment, illuminating in
general the relationship between environmental
contamination and disease.

9 Air Pollutants and Their Sources

The atmosphere is a mixture of gases that forms a layer about 400 kilometers (250 miles) thick around planet earth. The bottom 16–19 km (10–12 miles) is the most important part of the atmosphere in terms of weather and other aspects of biogeochemical cycles (Chapter 3). The lowest 600 meters (2,000 feet) of the atmosphere constitutes nearly all of the atmospheric portion of the ecosphere (Figure 9.1; see also Figure 1.2).

Normal air contains about 78% nitrogen and 21% oxygen, the remaining 1% being made up of carbon dioxide and several other trace gases. Both carbon dioxide and oxygen are absolutely vital for nearly all living systems; for most living things, oxygen is the most immediately important part of the nonliving environment. Human beings can live for weeks without food, for several days without water, but for only minutes without oxygen.

The average adult human exchanges about 16 kilograms (about 35 pounds) of gases per day, about six times the weight of food and water consumed. In a lifetime a human being exchanges many millions of cubic feet of air in hundreds of millions of breathing cycles. This is one of the principal reasons why the quality of air is so important.

Curiously, air does not seem to be considered to be as much a resource as water. What is worse, we tend to think of air as the *absence* of something. You must have often heard such naive statements as "This box is empty!" "The room didn't have a thing in it," and "As far as we could see, nothing." Many of the problems we address in these next three chapters stem from this kind of attitude toward air.

Air *is* a resource, and air quality is something that must be preserved. Just as there are watersheds, there are airsheds. Although airsheds are far less distinct entities than **watersheds** (regions drained by a river or other body of water), an **airshed** can be defined as the land area that contributes to the air—and the things found in the air—that flows over or through a particular geographic area. Fairly regular wind flow patterns and land contours are what give airsheds their definition. As is true of watersheds, airsheds do not necessarily coincide with political or other artificial boundaries. This presents various sociopolitical problems, which we will consider in Chapter 11.

The Air Pollution Problem Defined

In Chapter 6 we compared the impact of what we called the "fossil fuel incident" to a flash fire aboard a spaceship. This analogy is particularly relevant to

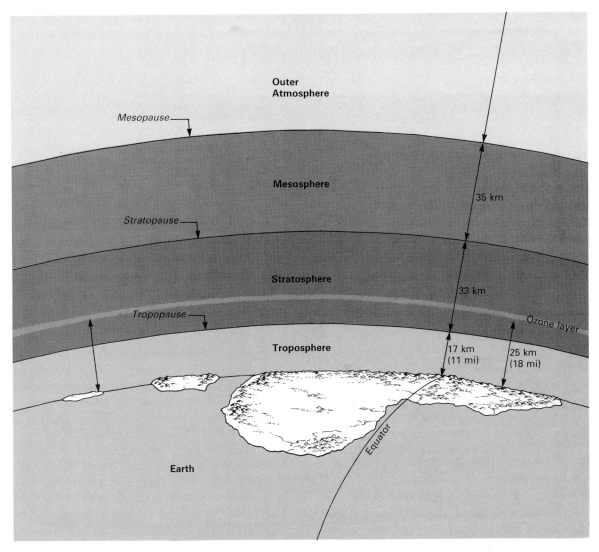

Figure 9.1 **Major Subdivisions of the Atmosphere.**

the next several chapters. Fires kill most often not by burning living things but by generating smoke and fumes that choke them. And so it is with fires aboard *Spaceship Earth*. Fumes from our fires foul our atmosphere and kill people, trees, and crops directly. The fumes of our fires and our technologies have also had, and continue to have, effects that may in the future convert food-producing areas into dust bowls, melt polar ice caps, flood coastal areas, and increase ultraviolet, cancer-causing radiation at the earth's surface. Contamination of the atmosphere (air pollution) is a very serious ecological and human health problem.

The concept of air pollution that we will use has mainly to do with the things that humans add to air. Although we will discuss natural sources of air contaminants, our principal focus will be on the extra measure of deterioration in the quality of the air that comes from the things we humans put into the air.

Air pollution is thus the transfer of harmful amounts of natural and synthetic materials into the atmosphere as a direct or indirect consequence of hu-

Concentrated human activity is what causes the air we breathe to become polluted.

This photograph taken in 1910 shows the effects of over 400 years of weathering on this grotesque figure decorating Lincoln Cathedral in England.

man activity. Air pollution is the dust, gas, and droplets we stir up in doing what we do as human beings.

Air pollution is a complex problem because a pollutant can be any of a number of chemical substances existing in gaseous, liquid (aerosol), or solid form (roughly 90% of the weight of all pollutants in the air is gas, however). Furthermore, pollutants can be added to the air directly (**primary pollutants**), or they can be created in the air (**secondary pollutants**) from other pollutants under the influence of electromagnetic radiation from the sun.

The effects of pollutants vary considerably because of differences in their concentrations and their chemistry. Some are far more toxic than others, and some have far greater impacts than others on materials and ecosystems.

Air pollutants have many different kinds of effects on humans and other parts of the natural world.

Air pollution can have a psychological impact; it can make the environment unpleasant. Air pollutants can erode statues and painted surfaces, cause things to become soiled, and damage property in other ways. Air pollution can impair human health and the health of other organisms. Air pollution can also alter climates and the chemistry of lakes, streams, and soil. All of these things make air pollution profoundly important in the most basic ecological sense.

Because there are established ways of preventing many kinds of pollutants from getting into the atmosphere, air pollution is more of a sociopolitical and economic problem than a scientific one. Scientifically, the challenge seems to be to define more clearly the relationships between and among specific pollutants, specific sources, and specific impacts. The problem with this is that the effects of some pollutants are subtle; even profound effects are sometimes separated from exposure by long time lags and may

In 1984, only 74 years later, atmospheric pollution had worn the figure to a barely recognizable remnant.

be difficult to pin down. Sociopolitically, the problem is related to our relative inability to deal with subtle impacts, our failure to appreciate cause-and-effect connections that are not obvious, and our reluctance to focus political attention on difficult chronic problems. Our sociopolitical system (maybe humankind in general) seems to be designed to respond best (if not only) to abrupt *unmistakable* disasters.

We will now consider some classic air pollution disasters. We do this reluctantly because we hesitate to dignify these episodes by calling them disasters and seeming to give them preeminence over the much more important *chronic* air pollution disaster.

The real impact of air pollution is not so much the isolated alerts or incidents that kill people or make them ill for a day or two as it is the chronic, day-to-day pollution that leads to lung disease and acid rain.

The disasters we are about to consider were really meteorological or toxicological accidents. They were episodes that attracted attention because people died as a direct effect of bad air. But as we shall see, most of those who died were older people or people with preexisting respiratory or heart conditions. Most of those who died were pushed beyond their physiological limits of compensation and adaptation.

We wonder what fraction of the people who died during these *acute* air pollution episodes were first weakened by chronic background air pollution. How many were brought to the point of high risk by exposure to air pollution over an entire lifetime?

A Short History of Some Notable Air Pollution Disasters

In the history of air pollution there have been some notable historic events. These episodes are instructive in that they demonstrate three important things:

1. Many air pollutants are poisons.
2. Weather, topography, and air pollution are very closely related.
3. Pinpointing the exact relationship between air pollutants and health can be quite difficult.

Meuse Valley, Belgium, 1930

In early December 1930, while the world was in the throes of the Great Depression, stagnant weather conditions and a cold mist covered most of the northern European country of Belgium. The mist was particularly concentrated along the Meuse River Valley, an industrial valley not unlike the Ohio River Valley and parts of the Mississippi River Valley in the United States. Because the air was not moving, the valley held the combustion products of its factories and mills; these pollutants, together with water in the air, turned the valley air into a stronger and stronger solution of poisons.

Several thousand people became quite ill simultaneously, from one end of the valley to the other, and there were approximately 60 deaths over and above the number normally expected in a three-day

period. Practically all of the deaths resulted from acute heart failure. Coughing, shortness of breath, and vomiting were among the reported symptoms. It was noted that most of the deaths were among elderly people and those with preexisting heart conditions and respiratory diseases. But many young, healthy people became ill. Cows, rats, birds, and other animals were also noticeably affected.

Though notable, the Meuse Valley episode received little attention in 1930; air pollution was not yet a common concept. Today very little is actually known about this disaster or about the relationship between the health effects and pollutants that were present in the air. As often happens in the history of humankind, when such lessons go unheeded, they are repeated.

At the same time of year, 18 years later, another disaster occurred in a Monongahela River Valley town in western Pennsylvania, under almost identical weather conditions.

Donora, Pennsylvania, 1948

Western Pennsylvania is cool in late October, and autumn rains can bring a chilling dampness. On such a day in 1948, in the small industrial valley town of Donora, the winds died down and the air became still. Like a large soup tureen, the valley held the emission products from factories and trains, and the "soup" was incubated together with people for several tragic days. Nearly half of the people of Donora were stricken with severe eye, nose, and throat irritation, chest pains, and labored breathing. Twenty people died over and above the number expected during that period. Here, too, animals as well as people got sick. Here, too, people with preexisting illnesses— most asthmatics, nearly everyone with heart diseases, and most others with diseases like chronic bronchitis and emphysema—suffered serious illness. But a world caught up in the postwar recovery had little room for air pollution in its order of priorities. Air pollution was apparently still not yet a concern.

London, 1952

Four years after the episode in Donora, a disaster of a slightly different sort struck London, England. A high-pressure mass dominating the weather picture for four or five days caused the air to stagnate and allowed pollutants to build up. Within a week there were almost 3,000 more deaths than expected, and an additional 1,200 to 1,500 excess deaths occurred in the weeks following the episode. During the four or five days in which stagnant conditions persisted, hospital admissions were 40% greater than normal. Sickness claims filed with health insurance systems more than doubled. Supposedly, home fire-place (coal) smoke mixed with the moisture of the London fog generated a deadly sulfuric acid mist.

A similar disaster occurred again in London in 1956. This time there were about 1,000 deaths above normal, even though the conditions that caused the problem lasted only about 18 hours.

There were some significant differences between the London disasters and those of Donora and Belgium. In London in 1952 the problem was believed to be precipitated by a highly synergistic combination of water vapor, carbon monoxide, sulfur dioxide, and tar—all of which are generated by the combustion of sulfur-containing coal in home fireplaces. Much of the blame for the disaster in Belgium was initially and is still placed on sulfur dioxide in factory smoke; however, some scientists think that gaseous fluoride compounds were the cause of the problem. Whereas some investigators stated that sulfur dioxide was one of the main ingredients in the Donora disaster, others think that zinc ammonium sulfate may have been a major cause. The most generally accepted conclusion about the Donora case in particular is that *several* toxic agents were responsible, none of which would have caused the problem alone.

We draw attention to this disagreement to suggest that if it was difficult for scientists and physicians to identify the cause of disasters involving a large number of *sudden* deaths, it must be even more difficult to identify the cause or causes of illnesses that are *chronic*—diseases like cancer, lung disease, and heart disease. These chronic diseases develop gradually and insidiously with vague, inconspicuous symptoms until, at some point, respiratory and cardiovascular systems are compromised to the extent that any extra stress can push them beyond their limits of compensation. When death finally occurs, it may have no *apparent* connection to air pollution. Or, as we have just seen, the death of a person with a chronic disease could be blamed on a single acute air pollution episode.

We turn now to the nature of particular air pollutants and the difficulty of linking each of these, singly or in combination, to specific sublethal effects on health and quiet, less dramatic disasters.

The Properties of Air Pollutants

Some General Considerations

The major air pollutants are those produced in significant amounts and those having documented health or other environmental effects. The chemical composition and characteristics of some of the most important air pollutants are given in Table 9.1. While these pollutants all have effects, sources, and control strategies in common, each is chemically unique.

Mini-Glossary

Acid rain: rain that is more acidic than normal because it contains sulfuric acid and nitric acid derived from oxides of sulfur (SO_x) and nitrogen (NO_x) in the atmosphere. Acidity can also come out of the atmosphere as dry deposition (as particulate matter). Acid rain is sometimes included under the more general label *acid deposition,* which includes the deposition of solid acid-forming materials, gaseous acid-forming materials, and all forms of acid precipitation.

Acute: immediate, brief, and severe—in reference to the *duration* of exposure or to the *effects* of pollutants, those effects that follow exposure more or less immediately as a direct reaction to exposure. (For example, acute exposure to even low levels of ozone may result in acute damage to the lungs.)

Adsorption: the adherence of a gaseous substance or a substance in solution onto the surface of a solid.

Aerosol: a gas that contains suspended solid particles or droplets of liquid able to stay suspended in air because of their very small size (usually less than 1 micrometer in diameter).

Air pollution: the presence of contaminants in the air to such a degree that the air causes problems or the use of air as a resource is impaired.

Air pollution episode: a striking surge of high levels of air contamination resulting in notable problems such as discomfort, illness, or even death.

Ambient air: outside air, the air around us.

Chronic: long-lasting or long-term in reference to either *duration* of exposure or *effect* of exposure to a pollutant. (For example, chronic exposure to even low levels of ozone can result in permanent scarring of the lungs, i.e., chronic lung disease.)

Convection: the movement of a gas or a liquid upward as a result of heating, which causes a decrease in the density in the gas or liquid and makes it rise. Convection currents in air tend to disperse air pollutants as the pollutants are carried up and away from the surface of the earth.

Dust: solid particles suspended temporarily in the air.

Emission: discharge of a pollutant from some source into the environment.

Fly ash: gas-borne solid particles resulting from the combustion of fuel and other materials.

Particulate: (adjective) in small pieces, as in particulate matter; (noun) a small particle of solid matter or a droplet of liquid of a size that allows it to remain suspended in air.

Photochemical process: chemical or physical changes brought about by the action of sunlight.

Photochemical smog: collection of harmful materials in the air resulting from the action of sunlight on nitrogen oxides, hydrocarbons, and other chemicals in the air. Automobile exhaust is the major source of photochemical smog in urban areas.

Pollutants: in a strict general sense, a pollutant is anything that changes air, water, or any other resource in some way such that use of the resource is impaired.

Pollutant Standard Index (PSI): a calculation by formula of the degree to which air quality relates to the standards set by the EPA for each of the major pollutants (see Chapter 11).

ppm (parts per million): the number of parts of a given substance in a million parts of a mixture by volume (see Enrichment Box 9.1).

Smog: term that combines the words *smoke* and *fog,* coined originally in Los Angeles to characterize a visible combination of smoke and fog. One type of smog is due to a high concentration of smoke particles and fly ash. Photochemical smog is the result of the interaction between nitrogen oxides and hydrocarbons under the influence of sunlight.

Smoke: a combination of solid and liquid particles under 1 micrometer in diameter emitted from burning materials.

Synergism: a phenomenon in which the effect of a combination of materials is greater than the sum of the separate effects of the individual substances.

Thermal inversion: an atmospheric meteorologic condition in which a layer of warm air acts like a lid to trap a layer of cold air beneath it. This frustrates the normal convection of air upward as the surface of the earth is heated; the air and any pollutants being vented into it are trapped.

Table 9.1 Molecular Composition and Characteristics of Major Air Pollutants

Pollutant	Composition	Characteristics
Sulfur dioxide	SO_2	Colorless, heavy, water-soluble gas with a pungent, irritating odor
Particulates	variable	Solid particles or liquid droplets including fumes, smoke, dust, and aerosols
Nitrogen dioxide	NO_2	Reddish brown gas, somewhat water soluble
Hydrocarbons (and other volatile organic compounds)	variable	Many and varied compounds of hydrogen and carbon
Carbon monoxide	CO	Colorless, odorless toxic gas, slightly water soluble
Ozone	O_3	Pale blue gas, fairly water soluble, unstable, sweetish odor
Hydrogen sulfide	H_2S	Colorless gas with a very offensive "rotten egg" odor, slightly water soluble
Fluorides (e.g., hydrogen fluoride, HF)	variable	Pungent, colorless, water-soluble gases (hydrogen fluoride)
Nitric oxide	NO	Colorless gas, slightly water soluble
Lead	Pb	Metallic, can exist in a variety of chemical compounds with different characteristics
Mercury	Hg	Metallic, can exist in a variety of chemical compounds with different characteristics

Trace Concentration Units in Perspective

The terms parts per million, parts per billion, and parts per trillion are used often in discussions of air and water pollution. This table will help to put these terms into perspective.

Unit	1 part per million (ppm)	1 part per billion (ppb)	1 part per trillion (ppt)
Length	1 inch/16 miles	1 inch/16,000 miles	1 inch/16,000,000 miles (a 6-inch leap on a journey to the sun)
Time	1 minute/2 years	1 second/32 years	1 second/320 centuries
Money	1¢/$10,000	1¢/$10 million	1¢/$10 billion
Weight	1 ounce salt/31 tons potato chips	1 pinch salt/10 tons potato chips	1 pinch salt/10,000 tons potato chips
Volume	1 drop vermouth/80 fifths gin or 1 drop vermouth/50 liters gin	1 drop vermouth/500 barrels gin or 1 drop vermouth/50,000 liters gin	1 drop vermouth in a pool of gin covering the area of a football field 43 feet deep (about 50 million liters) or 1 drop vermouth in 520 tank cars each holding 30,000 gallons of gin
Area	1 square foot/23 acres or 1 m^2/ 1 km^2	1 square inch/160-acre farm	1 square foot in the state of Indiana
Action	1 bogey stroke/3,500 golf tournaments 1 lob/1,200 tennis matches	1 bogey stroke/3.5 million golf tournaments 1 lob/1.2 million tennis matches	1 bogey stroke/3.5 billion golf tournaments 1 lob/1.2 billion tennis matches
Quality	1 bad apple/2,000 barrels	1 bad apple/2 million barrels	1 bad apple/2 billion barrels
Rate	1 dented fender/10 car lifetimes	1 dented fender/10,000 car lifetimes	1 dented fender/10 million car lifetimes

Source: Adapted from a compilation by Dr. Warren Crummett of Dow Chemical Co.

Oxides of Sulfur

A number of oxides of sulfur (SO_x) have deleterious environmental effects.* The most notable, and the one on which we will focus in this section, is sulfur dioxide (SO_2). Coal-burning electrical power plants are blamed for producing most of the sulfur dioxide problem in the United States. On the average, 70% of the sulfur dioxide in the air over U.S. cities comes from these utilities.

Fuels vary greatly in their sulfur content. High-sulfur coal from certain locales might have as much as 5% sulfur (**low-sulfur fuels** are those with less than 1% sulfur content). Natural gas contains only trace amounts (see Enrichment Box 9.1) of sulfur; this is why, when controls were first placed on sulfur dioxide emissions, many plants, factories, and power-generating stations switched from coal to natural gas. The energy crisis of the mid-1970s and the increased

cost of low-sulfur fuels caused a reversal of the trend toward use of gas and oil and stimulated a search for new solutions to the sulfur dioxide problem.

Sulfur dioxide is itself a poison, but it can also react with ozone, hydrogen peroxide, water vapor, and other substances in the atmosphere to form sulfuric acid (H_2SO_4). Sulfuric acid is one of the strongest acids known; it is able to corrode limestone, metal, and clothing, and it has a devastating effect on delicate respiratory tissue. Sulfuric acid derived from sulfur-containing air pollutants is a major contributor to the acidity in acid rain.

In terms of amounts emitted into the air and toxicity, sulfur dioxide may be the most toxic and dangerous air pollutant for the United States as a whole. For every ton of high-sulfur (say, 4%) coal burned without benefit of modern smokestack scrubbers, as much as 160 pounds (63 kg) of sulfur dioxide is released into the air. U.S. power plants, factories, and other sources have been emitting around 20 million metric tons of sulfur oxides per year since well before 1940. (Emission *trends* for oxides of sulfur and other air pollutants will be presented in Chapter 11.) In the early 1980s, more than one-fourth of the

---

* The subscript x is sometimes used to designate all of the oxides of a pollutant. SO_x is a symbol for all of the oxides of sulfur, for example, sulfur dioxide (SO_2) and/or sulfur trioxide (SO_3). NO_x is a symbol for all of the oxides of nitrogen, for example, nitric oxide (NO) and/or nitrogen dioxide (NO_2).

American people lived in areas where emission densities of oxides of sulfur exceeded 100 tons per square mile per year. Largely as a result of the Clean Air Act, only about 2% of the U.S. population now lives in counties where SO_x levels are *occasionally* an acute threat to health. The main problems with the oxides of sulfur now appear to be acid deposition (including acid rain) and impairment of visibility by sulfates formed from SO_2.

We will discuss the health effects of sulfur dioxide in more detail in Chapter 10. For the moment, let us put the health effects of SO_2 into perspective vis-à-vis its concentrations in polluted air. The following figures were compiled from various publications of the U.S. Environmental Protection Agency. (See U.S. Environmental Protection Agency, 1982c, for a detailed analysis of studies of both human and animal exposure.) (See also Enrichment Box 9.1.)

— Atmospheric background	0.2–0.4 ppb
— Nonindustrial city air	0.01 ppm
— EPA's 24-hour primary standard (see Chapter 11)	0.14 ppm
— Asthmatics begin to experience distress	0.5 ppm (for 1 minute)
— Odor threshold	0.5–1.0 ppm
— Level at which even normal people experience bronchial spasms	1.0 ppm (for 1 hour)
— Impaired lung function, occupational limit	5.0 ppm (per 8-hour day)
— Water-logged lungs (**pulmonary edema**) and permanent damage	20.0 ppm

How well we do in controlling the SO_2 problem will depend on the energy strategy we adopt. If the use of coal increases faster than the application of SO_2 controls, there will be an increasing sulfur dioxide pollution problem. Energy conservation will also make a great deal of difference.

Oxides of Nitrogen

When air is fed into a combustion mixture, particularly when the combustion is occurring at a temperature above 2,000°F, the oxygen and nitrogen present in the air as O_2 and N_2 combine to form nitric oxide ($N_2 + O_2 \rightarrow 2NO$). Nitric oxide is not thought to be very harmful and does not do much damage because it cannot readily dissolve in water or in tissue. However, through the action of sunlight, nitric oxide can combine with oxygen to form nitrogen dioxide ($2NO + O_2 \rightarrow 2NO_2$). Nitrogen dioxide is a reddish brown toxic gas that has considerable environmental impact. Nitrogen dioxide is similar to sulfur dioxide; through various reactions with substances in the atmosphere, nitrogen dioxide is converted into inorganic nitrates, peroxyacetyl nitrate (PAN), and an equivalent of sulfuric acid called nitric acid (HNO_3). Nitric acid may do even more harm to materials and to health than the oxidant NO_2, and it is implicated in the formation of "acid fogs" observed in southern California.

Natural agents like soil bacteria produce far greater *total* amounts of the oxides of nitrogen than humanity does by its fires. The problem with the oxides of nitrogen that humans generate is that they are generated in and around cities, where they reach harmful concentrations, sometimes 10 to 100 times greater than those found in rural areas.

Nitrogen dioxide is produced in the combustion of coal, oil, natural gas, and motor vehicle fuel and wherever temperatures are high enough to cause atmospheric nitrogen and oxygen to combine. Some 20 million metric tons of nitrogen oxides are emitted annually in the United States, though only the people of southern California are exposed to occasionally unhealthful accumulations (Conservation Foundation, 1984). As was true of SO_x, NO_x is an important part of the acid precipitation problem affecting many locations throughout the world.

As might be expected, the health effects of the oxides of nitrogen are similar although not identical to those of oxides of sulfur. For oxides of nitrogen (NO_x), again from the World Health Organization (1977), the U.S. Environmental Protection Agency (1982a), and other sources:

— Atmospheric background	4 ppb
— City air background	<80 ppb
— EPA's air quality standard (annual mean)	0.05 ppm
— Increased respiratory rate in rats	0.8 ppm (over a few hours)
— Pungent odor noticed	1–3 ppm
— Increased airway resistance in humans	2.5 ppm (1 hour)
— Occupational limit	5 ppm (8-hour day)
— *Reversible* increase in airway resistance	5 ppm (10 minutes)
— Pulmonary edema, fatal	100–150 ppm (1 hour)

Note that the effects of NO_x (and all other pollutants) vary both with concentration and with duration of exposure. (See U.S. Environmental Protection Agency, 1982a, for more detailed information.)

Because of the role of ultraviolet radiation in converting nitric oxide into nitrogen dioxide, there is a daily pattern in nitric oxide and nitrogen dioxide concentrations in cities (see Figure 9.2).

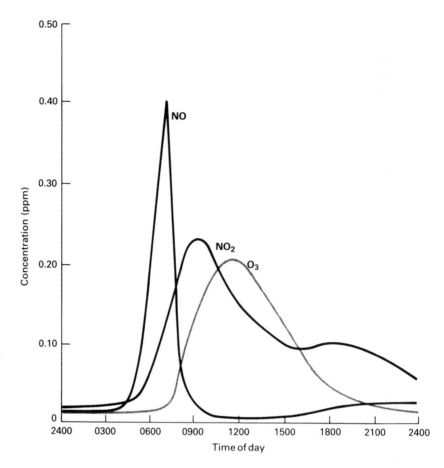

Figure 9.2 Typical Daily Pattern of Nitric Oxide, Nitrogen Dioxide, and Ozone Levels. Shown here is the pattern in Los Angeles on July 19, 1965. NO levels rise with morning traffic. A few hours later NO_2 levels rise as NO_2 is generated from NO. Still later, ozone levels rise as ozone is generated from NO_2, hydrocarbons, and sunlight. NO all but disappears, first as it is converted to NO_2 and then as it reacts with ozone. Ozone levels eventually drop as ozone reacts with the NO generated by afternoon rush-hour traffic.

Gaseous Hydrocarbons and Other Volatile Organic Compounds

Hydrocarbons form a miscellaneous category; the term **hydrocarbon** has come to mean any compound composed of carbon and hydrogen. The properties of hydrocarbons vary over a wide spectrum of chemical reactivity. Some hydrocarbons—for instance, certain **polycyclic hydrocarbons** (hydrocarbons that occur in multiple-ring structures; see Figure 18.11)—may have a considerable direct effect on humans by virtue of chemical **carcinogenicity** (ability to cause cancer).

Aside from this sort of direct, chemically specific problem, atmospheric hydrocarbons or the more general category of volatile organic compounds (VOCs), which are emitted at the rate of 18 million metric tons

Among the pollutants entering our air from automobile exhaust are lead, nitrogen dioxide, volatile hydrocarbons, carbon monoxide, and particulate matter. The hydrocarbons and nitrogen oxides may react with sunlight to form smog.

per year in the United States, are a problem largely because they participate in the formation of ozone.

Human-produced sources of ozone-generating hydrocarbons include unburned gasoline and evaporated solvents, especially from refineries. Figure 9.3 illustrates a sequence of petroleum production, refining, distribution, and use and indicates numerous points in this sequence at which hydrocarbons escape into the atmosphere.

Humans produce only about 15% of total global atmospheric hydrocarbons, but again the problem is accentuated by the pattern of concentration of human emissions in cities and industrial centers and, to some extent, by the nature of the specific hydrocarbons emitted.

Carbon Monoxide

Since carbon monoxide poisoning has been a common method of suicide over the years, nearly everyone is aware that an idling automobile produces considerable amounts of this pollutant. Human beings contribute only about 10% of the carbon monoxide load dumped into the atmosphere, and nearly all of this comes from incomplete combustion of fuels—largely in the automobile. Naturally produced carbon monoxide comes mixed with methane and other substances in marsh gases and other gases emitted from decaying material. Carbon monoxide also escapes from forest and grass fires and volcanoes. Some is formed through chemical reactions in the upper atmosphere.

In keeping with the pattern already described in our consideration of specific pollutants, human-generated carbon monoxide is a problem because most of what we generate is dumped into the areas in which we also live and breathe. In cities, 95% to 98% of the carbon monoxide in the air is from human sources, and the levels of CO are many times higher than the average levels in the natural world. Overall, carbon monoxide is the single most abundant pollutant known to affect human health that we vent into the atmosphere.

Whereas background levels of carbon monoxide average about 0.1 mg/m^3 of air (see Enrichment Boxes 9.2 and 9.3), concentrations may reach levels of 80–150 mg/m^3 in heavy traffic. Prolonged exposure to concentrations as low as 58 mg/m^3 (50 ppm) can impair judgment and reflexes. Such levels can affect vision, produce headaches, and exert strain on the heart. This last effect is a result of the fact that the oxygen-carrying capacity of the blood is reduced by carbon monoxide when it binds to hemoglobin; this is carbon monoxide's most important health effect. Carbon monoxide is otherwise relatively innocuous, with

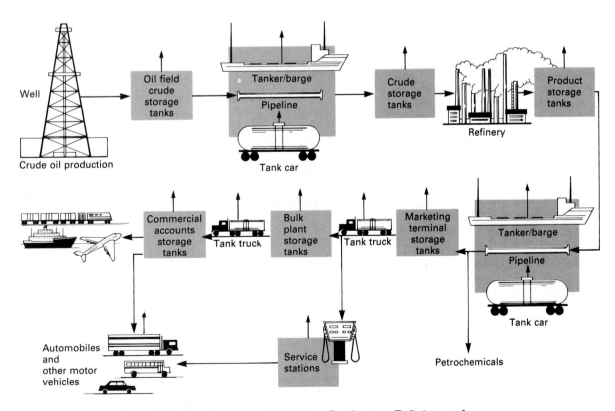

Figure 9.3 The Patterns of Petroleum Production, Refining, and Distribution. This diagram shows sources of evaporation of volatile organic compounds.

Microgram per Cubic Meter

This is an important unit of expression for the concentration of a pollutant in air. A microgram (μg) is one-millionth of a gram (there are 454 grams in 1 pound) or one-thousandth of a milligram. A sheet of typing paper weighs about 4 grams, and one-fourth of a sheet weighs about 1 million micrograms. The paper covered by a period having a diameter of 0.5 millimeter and an area of 0.196 mm^2 ($\pi r^2 = 3.14 \times 0.25^2 = 0.196$) would represent 1/76,428 of the area of a fourth of a sheet of paper (140×107 mm = 14,980 mm^2). The paper under a period would therefore weigh about 13 micrograms, or 1/76,428 of a gram.

A cubic meter (m^3) is about the size of a typical desk. Thus if you scattered the molecules that make up the paper under a period throughout a volume of air equal in size to a desk, you would have a concentration of 13 μg/m^3.

---

none of the effects on *materials* that the oxides of nitrogen and sulfur have.

The technology for controlling carbon monoxide emission involves adjustments in combustion processes to allow more complete combustion of fuels. For example, carbon monoxide can be decreased by increasing the air-to-fuel ratio in the internal combustion engine. This solution is not quite as straightforward as it sounds however.

Carbon Dioxide

Carbon dioxide is a colorless gas (at common temperatures and pressures) with a very faint odor. It is a relatively minor normal component of the atmosphere (0.03%), but it plays a major part in the carbon cycle discussed in Chapter 3. Carbon dioxide is produced in respiration and fermentation as sugars and other foods are oxidized. Plants use carbon dioxide as a starting material in photosynthesis. Carbon dioxide is

Parts per Million (ppm) and Micrograms per Cubic Meter: The Relationship

Sometimes the concentration of a pollutant is expressed in terms of micrograms (μg) per cubic meter (m^3). Sometimes the concentration is expressed in terms of parts per million (ppm). The expression μg/m^3 is a weight-per-volume expression; it indicates how much all of a particular pollutant in a unit volume of air weighs. The ppm expression defines a volume-per-volume relationship. It is an expression of the fraction of a unit volume of air occupied by a pollutant.

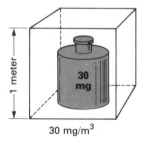

30 mg/m^3

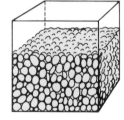

If half of any given volume of air were taken up by pollutant molecules, the pollutant would be at a concentration of 500,000 parts per million.

It so happens that a **mole** (the molecular weight of a compound expressed in grams) of any gas takes up 22.414 liters at standard conditions of temperature (0°C) and one atmosphere (760 mmHg) of pressure. One-fifth of a mole would occupy one-fifth of 22.414 liters, and so on. What this means is that the conversion of ppm to μg/m^3 is a function of the molecular weight of a particular compound. There is no constant that can be generally used to convert ppm to μg/m^3. There is a formula, however:

If You Have		To Get
To convert ppm to μg/m^3:	multiply by	$\dfrac{\text{molecular weight}}{0.0224}$
To convert μg/m^3 to ppm:	multiply by	$\dfrac{0.0224}{\text{molecular weight}}$

one of the basic end products of the burning of wood, coal, tobacco, leaves, paper, and other carbon-containing materials. When carbon dioxide is generated in this way, it is generally thought of as a much *more desirable* end product than the more toxic combustion product, carbon *monoxide*.

Relative to the other air pollutants, carbon dioxide is unreactive and is not really considered a pollutant in the most common sense of the word because it has no direct health effects. Although it is slightly soluble in water, producing weakly acidic carbonic acid, carbon dioxide makes this list of pollutants because of neither its chemical reactivity nor its solubility. It makes the list because as a constituent of the atmosphere, carbon dioxide absorbs infrared radiation, keeping some of the earth's heat from being radiated quickly into space. Carbon dioxide has a large share of the responsibility for the "greenhouse effect" (see Chapters 2, 6, and 10).

The relationship between the facts that (1) CO_2 has been increasing for some time, at least partly as a result of fossil fuel combustion, (2) the earth does appear to be getting warmer, and (3) even slight warming might upset delicate energy balances (Chapter 2), melt polar ice, and cause ocean levels to rise, inundating coastal cities, means that carbon dioxide is an air pollutant in every sense.

Photochemical Oxidants (Ozone)

The story of how Los Angeles came to be nearly synonymous with photochemical smog is an interesting one. At first the problem of watering eyes was thought to be due to particulate matter; later it was attributed to sulfur dioxide. Reductions in the emission of these substances had little effect on smog, however. Later, the notion that hydrocarbons were responsible for the problem came into vogue, and the source was believed to be mainly refineries. Control programs for refineries also proved ineffective. Eventually, the suggestion was made that the villain was the automobile, that the combination of hydrocarbons and oxides of nitrogen in automobile exhaust and sunshine leads to the formation of ozone and other photochemical oxidants. Although we will speak mainly of ozone here, a host of products are generated in the atmosphere through the interaction of nitric oxide, sunlight, and hydrocarbons (see Enrichment Box 9.4). These include ozone, peroxyacetyl nitrate (PAN), and acrolein (there are other products as well). Of these, ozone is the one chosen for measurement as an indicator of the presence of the family of photochemical oxidants that make up photochemical "smog." This is because ozone may account for as much as 90% of the oxidant chemicals in smog.

Although oxygen molecules can absorb ultraviolet radiation directly, causing them to split into two oxygen atoms ($O_2 = 2O$) that eventually go on to form ozone ($O + O_2 = O_3$), this occurs to a significant extent only high in the atmosphere. Short-wavelength ultraviolet radiation capable of doing this does not reach the earth's surface. Another mechanism is necessary to create the highly reactive atomic form (O) of oxygen and then ozone in the air that people actually breathe. It turns out that nitrogen dioxide is a very efficient absorber of the ultraviolet light that does reach the earth's surface. As NO_2 absorbs such radiation, it is broken down (**photolyzed**—split by light) into NO and O. In a subsequent reaction, O combines

---

Enrichment Box 9.4

Oxidation, Oxidants, and Photochemical Oxidants

For our purposes we can define **oxidation** generally as the *loss of electrons*. Thus when an atom or a molecule loses some electrons, it can be said to have undergone oxidation or to have been *oxidized*. That substance that takes the electrons is referred to as an *oxidizing agent, oxidizer,* or *oxidant*.

The oxidation of hydrogen can be depicted as follows:

$$H_2 - 2e^- = 2H^+$$

Everything must go somewhere, of course, and when something like hydrogen loses electrons, something like oxygen has taken them:

$$O + e^- = O^-$$

The gain of electrons is called **reduction;** the substance that does the gaining is said to be *reduced*. Thus oxidizing agents become reduced as they oxidize; conversely, *reducing agents* are oxidized. Oxidation and reduction are always coupled, so the examples just given can also be expressed as follows:

$$H_2 + O_2 \rightarrow H_2O_2$$

or

$$2H_2 + O_2 \rightarrow 2H_2O$$

In both cases, hydrogen is oxidized, and oxygen is reduced. Both are called **oxidation-reduction reactions.**

High energy wavelengths of light (e.g., ultraviolet) can cause the formation of rather powerful oxidants (chemicals that can readily take electrons from other chemicals).

Ozone is a powerful oxidant; because of the way it is produced, the term *photochemical oxidant* applies. Any oxidizing agent created in a photochemical reaction is technically a photochemical oxidant.

---

with O_2 to form O_3 or ozone (see Figure 9.4a). Fortunately, the NO produced by the initial photolytic reaction can react with O_3 and cause a reversion to NO_2 and O_2. The interrelationships of these reactions are shown in Figure 9.4a.

As suggested in Figure 9.4a, if no other factors were involved, ozone would break down as quickly as it is formed by reacting with NO. However, there are apparently two pathways other than the one shown in Figure 9.4a by which NO can revert to NO_2:

$$RO_2 + NO \longrightarrow NO_2 + RO \qquad (A)$$

where R is a hydrocarbon or other organic radical (**radicals** are atoms or groups of atoms that are chemically important constituents of molecules; they

Figure 9.4 (a) The Basic Chemical Reactions by Which Ozone Is Formed in the Lower Atmosphere. (b) The Role of Hydrocarbon Radicals (R) in Ozone Accumulation.

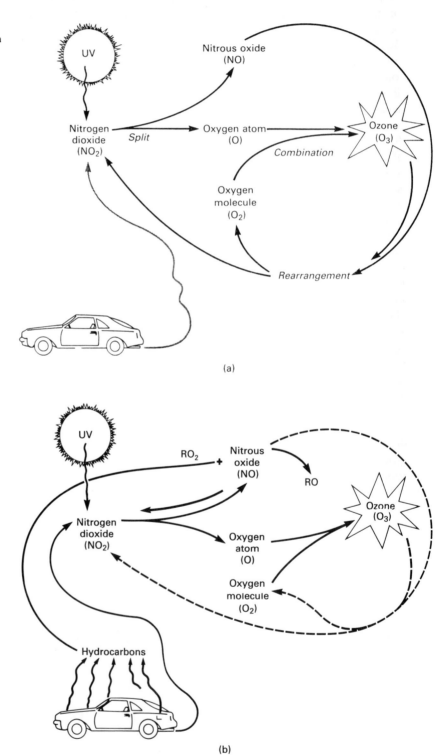

give the molecule certain reactive characteristics), and

$$2NO + O_2 \longrightarrow 2NO_2 \qquad (B)$$

Note that both of these reactions spare ozone from reacting with NO and thereby allow ozone to build up if it is being actively generated (see Figure 9.4b). Reaction B is actually too slow to spare much ozone. Reaction A, however, is fast enough to spare ozone. It is this reaction that becomes increasingly important in the presence of hydrocarbons.

Be aware that the chemistry involved is quite a bit more complicated than is suggested by these generalizations, and the details of the mechanisms involved are still somewhat controversial. The generalized reactions do, however, describe the basic relationship between hydrocarbons and ozone.

Although we are most concerned with the millions of tons of volatile organic compounds emitted via human activity every year, it may be that volatile organics emitted from trees and other vegetation are also factors in the generation of surface-level ozone. It has been shown, for example (Sandberg, Basso, and Okin, 1978), that in the San Francisco Bay, summer violations of federal ozone standards correlate well with precipitation during the preceding two winters. The hypothesis is that hydrocarbons released from growing trees and vegetation participate in the formation of ozone. The original haze of the Great Smoky Mountains is believed to have been the result of an interaction between sunlight and forest-derived organic compounds in the air. Today most of the grayish haze associated with these mountains is apparently caused by sulfates derived from human activity.

The reaction process that produces photochemical smog is such that the ozone concentration peaks 40–60 km downwind (the actual distance depending on topography, wind speed, and other meteorologic conditions) of major sources of the reaction ingredients. Ozone concentrations in the air entering New York are about equal to the concentration in the air leaving New York (Cleveland and Graedel, 1979); however, metropolitan New York appears to be the source of the starting materials that are responsible for the higher levels of ozone in Connecticut. This obviously has some significant implications for air pollution control, which we will consider later.

Ozone as a Problem. Some people feel the effects of ozone when it is present at only 0.001 ppm. At 0.05–0.1 ppm, some people experience impaired eye muscle coordination and a drop in visual acuity. Pulmonary edema can be produced in human test subjects at less than 1 ppm.

According to the U.S. Environmental Protection Agency (EPA), the probable median health effects on which the ozone standard of 0.12 ppm was based are that 0.15 ppm produces chest discomfort, irritation of the respiratory tract, and reduction of pulmonary function; at 0.17 ppm, asthma, emphysema, and chronic bronchitis are aggravated; and at 0.18 ppm there is reduced resistance to bacterial infection. These figures are medians, meaning that at these points half of all people exposed to these levels will experience the effect and half will not. The 0.12-ppm standard (see Chapter 11) is considered by the EPA to provide an adequate margin of safety.

The allowable concentration of ozone in industry (for eight-hour exposures) is 0.05 ppm. However, in some industries, concentrations of ozone can occasionally reach nearly 1 ppm; this level is also sometimes reached in acute air pollution episodes in places like Los Angeles. At times there are relatively high concentrations of ozone in cabins of airplanes flying above 30,000 feet. Above 80,000 feet, ozone concentrations in (outside) air reach 12 ppm, which is lethal even when exposure time is very short.

As a powerful oxidizing agent, ozone can damage crops and other vegetation and can cause the premature deterioration of rubber, fabrics, and other materials (see Chapter 10). By promoting the formation of light-scattering particles, ozone and other oxidants contribute to the kind of lessened visibility that makes it difficult to see across the Los Angeles basin.

Atmospheric Ozone as a Protector in Jeopardy. Ironically, ozone is also an important, useful constituent of the atmosphere because of its role in protecting life from ultraviolet radiaton (see Chapters 5 and 18). The ozone in the stratosphere screens out all but a fraction of a percent of the harmful radiation having wavelengths shorter than 340 nanometers (a **nanometer** is one-billionth of a meter). The problem is that certain human-produced pollutants tend to remove stratospheric ozone via a catalytic cycle involving chlorine atoms. We will consider the ozone shield and the ozone hole in Chapter 10.

Particulate Matter

Particulate pollutants, which account for about 5% of the weight of all air pollutants, make up a miscellaneous category. The term *particulate* itself implies a single kind of gritty entity. However, particulate pollution has multiple components, including sulfate salts, sulfuric acid droplets, salts of metals (like lead or oxides of iron), dust from finely divided particles of carbon or silica, liquid sprays and mists, and a host of uncataloged substances.

The size of the particles that make up particulate matter is an important characteristic. Individual particles are measured in units called **micrometers** (μm), 1 million of which add up to 1 meter. Particulates range in size from 0.005 μm to about 100 μm. Although natural dusts constitute half the total mass of particulate matter in the atmosphere at any one time, this dust has a relatively small impact because it tends to be coarse. Being heavy, these particles settle out of the atmosphere quickly and otherwise do not get to delicate lung tissue. **Fine particulate matter** (less than 2.5 μm in diameter as defined by the EPA) is generally considerably more hazardous to human health than **coarse particulate matter.**

Although rain generally tends to clean particulate matter out of the air, it is not very effective in removing pieces smaller than 2 μm in diameter. Particulates in this category tend to remain suspended and, depending on turbulence and wind conditions, can be transported over long distances. Very fine particles behave almost identically to gases.

Although particles as large as 15 μm in diameter can reach the nonciliated portion of the lung and even the alveoli (see Chapter 10), particulate matter with diameters less than 10 μm reach the alveoli with greatest efficiency. (The peak in efficiency of alveolar deposition for mouth and nose breathing occurs in the range 2–4 μm.)

Because of their small size (80% are less than 2 μm), transportation-derived particulates have an important impact on health even though they make up only about 1% of the particulate load in the atmosphere.

Particulates occur as sprays, mists, and dusts from spraying and grinding activities, land clearing, and highway building. Soot and fly ash are emitted from electrical power plants and factories. Significant amounts of particulates also come from forest fires and agricultural fires. Secondary particulate particles can be created in the atmosphere by the reaction of gases producing a solid or droplets or when one substance acts as a nucleus onto which other materials condense to produce new chemical entities. Hydrocarbons, for example, can react with oxidants in the atmosphere to produce peroxide radicals, which, through chemical chain reactions, eventually form large organic molecules, which in turn join together to form small droplets or solid particles.

Once a particle gets inside the lungs, its effect depends on its chemical nature; and, as we stated, there are many chemically different particulates. Particulates do, however, have several general effects on humans and the environment. First, some of them are actually forms of dirt, which simply make things dirty. As they become absorbed onto the surface of materials, particulates can promote rust and corrosion by attracting water. Sulfates and sulfuric acid are particularly **hygroscopic** (tending to take up water). Particulates such as carbon can also cause complicated health effects by carrying into the lungs various gaseous substances absorbed on their surfaces. The cancer-causing chemical **benzopyrene** has been found to be associated with particles of soot in just such a way.

Since the advent of tougher clean air laws in the United States in the late 1960s, particulate pollution has been *relatively* effectively controlled. There was a 58% decline in particulate emissions from 1970 to 1982. In the mid-1980s, only 25% of the U.S. population was exposed to levels of particulates that begin to have health effects. Most of the progress made in keeping particulates out of the air since the early 1970s was made by removing the large particles from large stationary sources, for example, power plants and mills. Some additional progress in the early 1980s was attributable to decreased industrial activity in an economic slump.

Initially, the size of suspended particulate matter was disregarded in setting air quality standards; in 1987, however, standards were based on particles less than 10 μm in diameter (see Chapter 11).

Metals: Lead and Mercury

Lead. Contamination of the environment with lead has increased dramatically since the beginning of the industrial age (Figure 9.5). Lead is fairly well known as an environmental problem related to its use in paint. This has been a serious problem in rundown old housing where very young children eat leaded paint chips, apparently because they taste sweet, and suffer lead poisoning. Lead has also been a problem in occupational settings, for example, in and around lead smelters and where lead-based solder is used. The general problem of atmospheric lead contamination comes largely from the automobile and the high-compression engine.

Tetraethyl lead was introduced as a motor fuel additive in the 1920s as a means to slow gasoline combustion to reduce engine knock and wear. Nearly all of that lead got back into the environment through the air. That leaded gasoline has been the main source of atmospheric lead was made clear by the impact of unleaded gasoline in the United States. According to the Conservation Foundation (1984), there was a 64% decline in the average concentration of atmospheric lead between 1977 and 1982—paralleling a 68% decline in the use of leaded gasoline in the United States over roughly the same period.

Because the automobile has been and still is a major source of lead pollution (leaded gas continues to be used in certain parts of the world), urban air on the average worldwide has concentrations of lead nearly 50 times the level in rural air and nearly 300 times that found in remote areas (Nriagu, 1990). The lead content in the air and in the blood of people in remote reaches of Nepal is substantially lower (3.4 μg/dl in adult

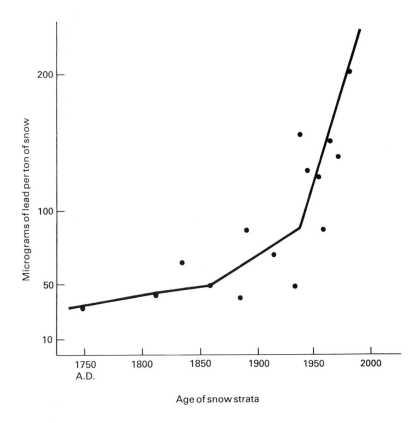

Figure 9.5 **Changing Lead Content in the Snow of the Ice Sheet in Greenland over the Past Two Centuries.** The first sharp rise in the curve probably reflects the lead stirred up by the Industrial Revolution. The second, sharper rise, in 1950 or thereabouts, undoubtedly reflects the lead used as an antiknock ingredient in gasoline.

males) than it is in the industrialized areas of the world (the U.S. normal range is 15–25 μg/dl) (Piomelli et al., 1980). Typical body burdens of lead are some 500 times higher today than they were in people living before the industrial age (Marshall, 1984). (Ancient Romans may have been exceptions. Their bones typically have high lead levels, and lead poisoning linked to lead food and beverage vessels used by the upper class is believed to have contributed to the eventual fall of Rome.)

Also paralleling the decline in use of leaded gasoline in the United States and the general decline in per capita lead consumption (in the resource use sense), a survey by the National Center for Health Statistics and the Centers for Disease Control involving 27,801 people aged 6 months to 74 years showed a 38% decrease in blood lead levels from 1978 to 1981 (Conservation Foundation, 1984). According to a report by the National Academy of Sciences, surveys of children living in large cities in recent decades have commonly revealed blood lead levels of 40–60 μg per deciliter of whole blood. This is one-fifth to one-third of what it would take for a doctor to write "lead poisoning" on a chart.

Human beings acquire lead mainly by ingestion and inhalation. The way in which humans are exposed to lead is important because not all routes are equally open to lead. Roughly half of inhaled lead is absorbed; a smaller (variable) fraction of the lead ingested in food

is absorbed. After lead gets into the blood, it is gradually excreted, and some of it may be stored in bone. Depending on the rates of storage and excretion relative to the amount absorbed, blood levels may reach the limits of toxicity. Hormone changes and stress may cause the lead in bone to be released, causing surges in blood lead and health problems in individuals who previously carried blood lead burdens near threshold. The specific biochemical, physiological, and health effects of lead will be considered in Chapter 10.

Mercury. Most of us were unknowingly introduced to the concept of mercury poisoning by Alice on her way through Wonderland. The Mad Hatter was apparently a parody of hatters in Lewis Carroll's day, who did tend to go mad as a consequence of the mercury (mercuric nitrate) used in the curing of furs and felt used in making hats.

Today as in Alice's time, a major source of atmospheric mercury is the natural degasing of the earth's crust. This produces between 25,000 and 125,000 tons of mercury per year. Geothermal steam used for power production contains significant amounts of mercury, which escapes into the atmosphere as cooling tower exhaust. Mercury emissions from geothermal plants are comparable to the emission of mercury (a contaminant in coal) from coal-fired power plants. In industry, mercury is used primarily in paper, chemical, and paint manufacturing. From these industries and from agricultural sources (pesticides and fungicides), mercury escapes into the air and soil and reaches human beings by a number of routes (see Figure 9.6). The concentration of mercury in urban air averages about 4 nanograms per cubic meter—10 times that of remote areas (Nriagu, 1990). Mercury escapes into the environment as a vapor, as a solute in water, and as a solid particulate in various chemical forms having different degrees of toxicity. Pure metallic mercury is relatively harmless in comparison to, say, mercuric chloride ($HgCl_2$), which is a very deadly poison. The ingestion of as little as 1 gram (1/28 ounce) of mercuric chloride can cause death. The specific biochemical and health effects of mercury will be described in Chapter 10.

Radioisotopes

We have chosen to cover the details of the sources and effects of radioactive pollutants, both those transported in water and those transported in air, in Chapters 6 and 18. Here we emphasize only the following points.

Most of the radiation to which humans are exposed comes from natural (background) sources (see Chapter 6). We receive about 100 millirems of radiation per year from natural sources. We receive another 100 or so millirems on the average from artificial sources; most of this comes from medical and dental X-rays. We receive about 5 millirems per year from artificially made atmospheric pollutants derived from activities such as nuclear weapons testing. Fallout containing the radioisotopes plutonium 239, strontium 90, and others has added 10% to the dose of radiation we humans receive in a given year. The U.S. EPA indicates that nuclear power plants contribute less than 0.01 millirem to the radiation each American receives from air, water, and soil (Chapter 6).

Other Air Pollutants

There are pollutants other than those described here; some of them are included in Table 9.1. We will discuss still other pollutants in greater detail in other chapters. Asbestos, for example, will be discussed in Chapter 18. Asbestos continues to be important because hundreds of thousands of buildings throughout the United States and other countries were coated with it for fireproofing and to improve acoustics. It was not until 1973 that the EPA began to restrict the use of materials containing fractions of asbestos exceeding 1%.

Fluoride compounds, such as those released into the atmosphere from the smelting of iron and aluminum and the processing of phosphate fertilizers, can

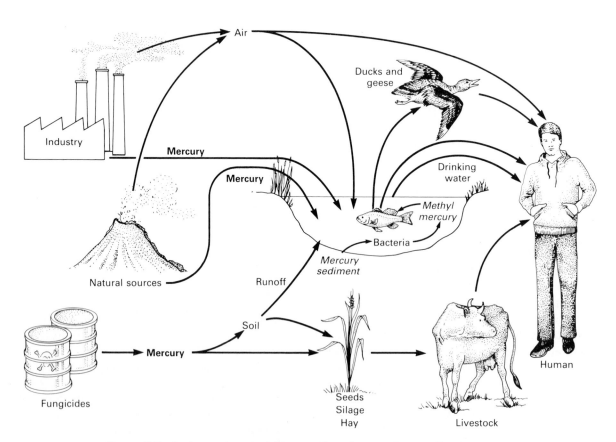

Figure 9.6 Pathways by which Mercury Can Get into Human Tissues through Water, Air, Soil, and Food Chains.

severely damage plants and have been reported to adversely affect human as well as animal health. Fluorine is a very reactive oxidant that is capable of impairing enzymes and other biologically important chemical substances. The biological effect of fluorine gas and gaseous fluoride compounds that has received most attention is its ability to affect the calcification of bones and teeth. A little fluoride helps to strengthen teeth (hence fluoride toothpaste and fluoridated water), but too much can cause disruption of phosphocalcium crystal formation and produce mottled teeth. In excess, fluoride actually decalcifies bones and teeth, making them brittle.

Air Toxics

Beyond the general kinds of pollutants found in air almost anywhere are toxic substances called **air toxics** that get into the air only from particular kinds of industry or from certain types of accidents at the local level. Perhaps the most serious industrial accident was the escape of toxic gas from a Union Carbide plant in Bhopal, India, in 1984 that killed several thousand people and injured many thousand more. More commonly, we hear about less serious railroad tanker accidents and tanker truck accidents that require people to be evacuated. Although such dramatic accidents obviously get considerable attention, the more routine emissions of highly toxic chemicals from chemical facilities are also a serious problem.

The Sources and Causes of Air Pollution

Only part of the air pollution problem comes from the *amount* of pollutants that get dumped into the air. There are also important considerations having to do with space, time, and circumstances. We would like to bring these aspects into sharper focus.

Space is part of the air pollution problem because most air pollutants are released into the very air we breathe. Residential areas or cities occupy about 1% of the land area in the United States, and it is the air over this 1% that receives emissions in highest concentration (see Enrichment Box 9.5).

Time is part of the problem because the rate of polluting is important. In every major city there is a monumental exodus of automobiles from the suburbs in the morning and the reverse late every afternoon. The result is a sharp peak of effluent release twice each day. There are other types of such surges of effluent—output at rates that exceed the ability of the environment to disperse or to neutralize the pollutants.

Still other temporal pollution patterns are the result of patterns in *circumstance*. For reasons related to weather, for example, the "ozone season" occurs in the summer. As we mentioned earlier and as we will discuss in the last portion of this chapter, circumstances of weather and topography are important factors in air pollution. Keep space, time, and circumstance in mind.

Natural Sinks and Overloading

Before we get deeply into sources, we would like to crystallize a concept touched on indirectly in several places earlier. To begin with, it is obvious that at any one time the contaminant level in the atmosphere falls far short of the cumulative amounts that are released into the atmosphere. This indicates that there are *sinks* for various pollutants; something happens to sop them up.

According to Schlesinger (1979), sinks for atmospheric pollutants (and other types) fall into three broad categories:

1. Spontaneous chemical transformation—e.g., oxidation
2. Microbial degradation in soil and in water
3. Physical processes such as solubilization and sedimentation

The degree to which these processes operate varies from place to place, from pollutant to pollutant, and from time to time. In other words, the amount of time any chemical remains in the environment is highly variable and depends on the various chemical, physical, and biological features of the environment and on the chemical in question.

The main removal mechanisms for atmospheric particulates are physical; particles are cleared by either dry or wet deposition. Wet deposition refers to the ability of rainfall, for example, to sweep particles out of the atmosphere. Dry deposition is of primary importance in the troposphere (see Figure 9.1), where sedimentation or settling out increases as particles get bigger.

Atmospheric gases are removed by both physical and chemical processes. Gases either are adsorbed onto solids or undergo chemical reactions in the air. A summary of the sinks for various selected atmospheric gaseous pollutants is given in Table 9.2.

Natural Sources of Air Pollution

The rest of nature puts more things and greater quantities of bad things into the air than humans do. Sulfur dioxides, oxides of nitrogen, carbon monoxide, methane, hydrocarbons, and particulates emanate from volcanoes, swamps, forests, natural fires, and the action of wind on soil not covered by vegetation. Volcanoes contribute great amounts of dust and particulate matter as well as noxious gases. Organic decay in

Air Pollution in Cities throughout the World

Urban air pollution is a problem throughout the world. The sources of the problem vary with the state of development and other factors in each country, but the problem is serious in developed and developing countries alike.

Air pollution has been a serious problem in the highly developed countries of North America and Europe since the beginning of industrial age; automobiles have become an increasingly larger part of the problem in these countries in the last half of the twentieth century. As these countries have struggled to clean up the air in their own cities, the problem of transboundary pollution has become increasingly serious. A case in point is that of the United States and Canada.

Air pollution has been one of the greatest sources of strain in U.S.-Canadian relations for more than a half century. In the early 1920s, complaints by American farmers that fumes from a smelter in British Columbia were injuring crops led to a settlement and eventually to the establishment of a principle of international law that one country must not pollute the air of another. In the 1980s, the issue changed direction with Canada objecting to the sources of acid rain originating in the United States (see Chapter 10). There has been some progress on this issue recently, but the two nations are still working to resolve it. The world should expect these two nations to lead the way in the development of a model for environmental cooperation badly needed throughout the world.

Some of the most serious examples of urban pollution can now be found in countries that have undergone rapid recent development. Taiwan is an interesting case in point because its notably rapid development, coupled with the fact that it is a small, densely populated island, underscores the nature of the air pollution problem. The total number of vehicles in Taiwan is approaching 10 million—mostly motorcycles of the two-cycle, more highly polluting kind. To a visitor it can seem that all of these vehicles are on the streets at the same time! Taiwan has only recently become serious about air quality. It has upgraded what had been a bureau into a full-fledged environmental protection administration reporting directly to the executive government council. The equivalent of hundreds of millions of U.S. dollars are now being spent annually by the government and by industry for air pollution control.

The centrally planned economies of the former Soviet Union and the so-called East Bloc nations of Europe offer another interesting set of examples. Although in theory central planning is supposed to be sensitive to environmental considerations, in practice environmental considerations were lost in the dust of rapid, post–World War II industrialization. The result is that the industrial cities of Poland, eastern Germany, Czechoslovakia, Hungary, Romania, and republics of the former Soviet Union are among the most highly polluted in the world. Before its dissolution, the Soviet government identified more than 100 cities—with some 50 million people in all—in which air pollution levels exceeded maximum permissible levels by as much as tenfold. The Polish Academy of Science recently indicated that a third of the people of Poland—mostly in the coal and steel belt of Krakow and Silesia—live in areas of "ecological disaster." One report (Cherfas, 1990) indicated that 92.6% of the people of Leipzig suffer some sort of health problem caused by oxides of sulfur; 85.7% suffer from problems caused by dust. According to Ember (1990), the uncontrolled burning of (highly polluting) brown coal provides 70% of the energy in eastern Germany; one-third of the population of Hungary lives in areas where air pollution does not meet international standards; and as a result of these and related problems, infant mortality rates are rising, life expectancy is decreasing, and cancer rates are rising throughout eastern Europe. Not surprisingly, environmental activism and interest in the environment are on the rise throughout eastern Europe. There are now many "green" groups active in every one of the former Soviet republics. A conference on environmental education held in the Soviet Union (before its dissolution) was attended by several hundred enthusiastic educators eager to seize on *glasnost* as an opportunity to begin to solve environmental problems.

---

tropical forests and other places puts a variety of gaseous compounds like methane into the atmosphere, and forests contribute various organic compounds including ketones, aldehydes, and other complex hydrocarbons, which, as we have already discussed, can participate in the generation of ozone. Somehow the fact that there is a lot of naturally occurring air pollution seems to offer little comfort in light of the seriousness of the highly concentrated, largely *human-derived* air pollution problem in our cities today.

Humans and Air Pollution: Technology as a General Source

In *The Closing Circle,* Barry Commoner (1971) made the point that modern technology is more costly, is more energy-consuming, and tends to pollute more than the technologies it has replaced over the years. Beyond the impression that we do lead somewhat better and happier lives today, close inspection seems to reveal that we have to pay more, in terms of energy and deterioration of the environment, for equivalent units of economic good produced than we did in decades past. Commoner offers aluminum as an example; he points out that aluminum takes more energy to produce than steel. Although there are unarguable advantages to using aluminum in packaging, construction, and other areas, the fact is that to make an aluminum beer can requires about six times more energy and generates several times more pollution than to make a steel one.

Electrical power generation is responsible for a significant fraction of the air pollution problem. As we saw in Chapter 6, even though our increasing use of electrical power has unquestionably added to the qual-

Table 9.2 Sinks for Selected Atmospheric Gaseous Pollutants

Pollutants	Sink	Estimated Residence Time
CO	Uptake by soil and conversion to CO_2 by microbes	0.09–2.7 years
CO_2	Dissolves in oceans; taken up by plants in photosynthesis	2–10 years
H_2S	Oxidation to SO_2	0.08–2 days
SO_2	Precipitation scavenging then oxidation to sulfate particles	20 min–7 days
NO	Atmospheric oxidation to NO_2	4–5 days
NO_2	Precipitation scavenging and oxidation to nitrate	3–5 days
Hydrocarbons	Oxidized to CO_2; absorption on soil then microbial degradation; photochemical degradation	1.5–2 years (methane)
Ozone	Photochemical reaction (reversion to O_2) in atmosphere	2 hr–3 days

Source: R. B. Schlesinger, "Natural Removal Mechanisms for Chemical Pollutants in the Environment," *BioScience 29* (1979):95–101.

ity of life, the improvement-to-power ratio has remained rather small. The automobile, which by now you may have recognized as a major villain in this chapter, offers another particularly profound example of the fact that sometimes it is not what we do that matters, or even how much of it we do, as it is the way we do it.

For many years, pollution by automobiles increased much faster than the number of automobile miles traveled in America. Various design changes in the automobile were responsible for this. Escalation in the power of internal combustion engines in the 1950s and 1960s was achieved through engineering based on increases in engine displacement and in the compression of the gasoline-air mixture before ignition. This produces more power per stroke, but it also results in higher combustion temperatures and, consequently, greater quantities of NO_x and other pollutants. There are other examples, to be sure, but the foregoing will serve to illustrate that part of the air pollution problem has to do simply with the way we do things.

Let us now take a look at some general categories of specific human generators of air pollution in preparation for considering source-based control strategies in Chapter 11. A summary of the sources of major pollutants is given in Table 9.3.

The major contributors of air pollution, in decreasing order, are transportation, electric power gen-

eration, industry (including fuel combustion and materials processing), forest and agricultural fires, and incineration. Each source of pollution involves the rapid burning of some kind of fuel.

Transportation and Air Pollution: The Mobile Sources

Human-generated air pollution sources can be divided into mobile and stationary sources. **Mobile sources** are automobiles, buses, trains, airplanes, and other fossil fuel–powered modes of transportation. **Stationary sources** include factories, incinerators, and other kinds of nonmobile sources. There is a practical reason for making such a division; it has to do with the differences in the pollution control problems presented by these two categories. Mobile sources tend to be much smaller, much more plentiful, and much more widely dispersed than stationary sources and are therefore more difficult to monitor.

Automobiles and trucks (highway vehicles) are the main mobile source problem with respect to air pollution because (1) to carry the same load, trucks and cars emit about six times as much pollution as railroads and (2) automobiles produce many more times the amount of pollution per gallon of fuel con-

Table 9.3 Sources of Major Air Pollutants

Sulfur dioxide	Combustion of coal, oil, and other sulfur-containing fuels; petroleum refining, metal smelting, paper making
Particulate matter	Fuel combustion, industrial processes, construction, forest fires, refuse incineration, automobile traffic
Nitrogen dioxide	Produced by combinations of atmospheric nitrogen and oxygen at high combustion temperatures such as those in automobile engines; also a by-product in the manufacturing of fertilizers
Volatile organic compounds (e.g., volatile hydrocarbons)	Motor vehicles—evaporation from gasoline tanks and carburetors; industrial processes involving solvents
Carbon monoxide	Combustion of fuel, e.g., gasoline
Photochemical oxidants	Produced via complex photochemical reactions in the atmosphere involving hydrocarbons, nitrogen dioxide, and sunlight
Hydrogen sulfide	Various processes in many kinds of chemical industries; oil wells, refineries
Fluorides	Fertilizer manufacture, ceramics manufacturing, aluminum smelting
Lead	Combustion of leaded gasoline; solder, lead-containing paint; lead-smelting operations
Mercury	Paper, chemical, paint manufacturing; pesticides; fungicides

Fuel Economy Standards: Do We Need Them?

Although it is a well-established and convenient form of transportation in many countries, the automobile is a major contributor to air pollution and consumes finite resources. Technologically, it is possible to build extremely energy-efficient cars. Proponents of high, mandatory automobile fuel economy standards for the United States point to the benefits of reduced dependence on foreign oil, improved balance of payments, and reduced greenhouse gas emissions. Opponents claim that new standards will require greater downsizing, resulting in less safe vehicles, higher vehicle prices, and a loss of competitiveness on the part of U.S. automakers. The U.S. Office of Technology Assessment says that without new standards, fuel economy is unlikely to improve significantly. Should new fuel economy standards be imposed? What do you think?

sumed than diesel-powered trucks or trains or jet aircraft. Overall, highway vehicles generate roughly 10 times more air pollutants than other mobile sources. We will ignore for the moment the facts that the building of highways takes many times the energy needed to lay railroad track on a per-mile basis and that railroads take up less right-of-way than highways do.

Disregarding the relative toxicity of pollutants, automobiles alone produce about two-thirds of the *weight* of air pollutants in the United States that are human-derived, and this fraction can reach 90% in certain cities. In the United States, transportation accounts for over two-thirds of the carbon monoxide emissions, over one-third of the volatile organic compounds, about half of the oxides of nitrogen, less than 5% of the oxides of sulfur, and about 17% of the particulates (mostly diesel engines). Motor vehicles continue to contribute most of the hundreds of thousands of tons of lead that are introduced into the atmosphere each year.

Stationary Sources (Electric Utilities, Industry, Agriculture, and Construction)

The burning of coal and other fuels is the principal cause of air pollution coming from stationary sources in many countries. Coal combustion in electrical utilities and in smelting generates oxides of sulfur, oxides of nitrogen, hydrogen fluoride, carbon monoxide, and particulates such as carbon, silica, aluminum, and iron oxides in addition to assorted hydrocarbons and metals such as lead, mercury, cadmium, selenium, vanadium, and zinc. Electric utilities together with other industrial, commercial, and residential stationary fuel combustion sources in the United States contribute one-third of particulate emissions, about

80% of SO_x emissions, half the NO_x, 10% of volatile organics, and less than 10% of CO emissions.

Solid Waste Disposal and Incineration

Packaging of products ranging from food to toys contributes significantly to the air pollution problem. First, pollution is generated in making all packaging. Then, after a package is opened and the product is removed, the package must be disposed of, often by incineration. Americans have accepted the extra cost that packaging has brought to various goods almost without realizing that they pay for packaging at least twice. The second time is the cost to health and the environment and the real cost of solid waste disposal as packaging is picked up, transported, and burned, releasing its combustion products into the atmosphere. Incinerators emit carbon monoxide, aldehydes, hydrocarbons, particulates, and a variety of

Stationary sources of air pollution, such as power plants, are easier to monitor than mobile sources, which tend to be small and dispersed.

miscellaneous gases. The shift in composition of packaging from paper to plastics has brought about changes in the character of the emissions from incinerators. Among the things that now appear in significant amounts in effluents from burning refuse are chlorine and chlorides. The chlorides are derived from plastics made from polyvinyl chloride and other chloride-containing components of plastic.

Leaves and refuse burned in residential areas obviously also contribute to the air pollution problem. Many American cities have banned the burning of leaves because it adds particulates and assorted noxious gases to the atmosphere, usually at a time of year when there are stagnant conditions. Leaf burning offers a particularly poignant example of people causing deterioration of a resource through the acceleration of a natural process to the point at which products accumulate more rapidly than they can be dispersed. An ecological paradox here is that leaves could be composted (many cities now have leaf collection and composting programs) and used to enrich garden soil in the same way that peat moss adds to the quality of soil. Thus in addition to polluting the air, burning the leaves deprives the leaf burner of the moisture-holding, soil-improving qualities of compost.

Incineration and open burning account for roughly 5% of particulate emissions, 3% of volatile organics and CO, and negligible fractions of total SO_x and NO_x emissions in the United States. Emission fractions unaccounted for come from forest fires and other miscellaneous sources.

The Influences of Weather and Other Synergistic Factors on Air Pollution

There is no city on earth where the level of air pollution is harmful on a continuous basis, although Mexico City, Los Angeles, Leipzig, and several other cities come close. Variation in pollution is a result of irregularity in the emission of pollutants and a number of factors that determine where pollutants go and whether or not they will accumulate. Weather and meteorologic conditions relevant to the pollution problem include light, humidity, temperature, wind, thermal inversions, and rain. Other factors include topography and interactions between chemical pollutants (Table 9.4). Biological magnification of certain pollutants is still another factor, one that depends on the nature of the food chains in a particular location (see Chapters 2 and 16). As we will see later, the effect of pollutants on health varies not only with the levels of pollutants themselves and their interaction but also with a number of host factors such as age, the

Table 9.4 Summary of Factors Affecting the Distribution of Air Pollutants

Increase in Factor	Pollution*	Typical Effect
Precipitation	−	Cleanses the air (but brings acid rain)
Humidity	+	Dissolves many pollutants; renders pollution more visible
Fog	+	As above
Sunshine	+	Initiates oxidation
Wind velocity	±	Reduces pollution near the source but disperses it faster and more widely
Wind direction from contaminating source	+	Causes greater contamination
Increasing temperature with increasing height	+	Provides less dispersion
Barometric pressure	+	Reduces wind and hence dispersion
Height of emitting source	−	Enhances dispersion and dilution of contaminant
Mountains, hills	+	Break the force of winds but promote light winds
Valleys	+	Trap pollutants
Plains	−	Provide greater dispersion
Distance from contaminating source	−	Greater dilution

* A minus sign means less pollution; a plus sign means more pollution.

Source: Adapted from G. L. Waldbott, *Health Effects of Environmental Pollution,* 2d ed. (St. Louis: Mosby, 1978). Used with permission.

presence of respiratory or cardiovascular illness, genetic predisposition, degree of physical activity, and level of stress.

Synergism Defined

When the effects of two things are **synergistic,** they have an impact together that is greater than the sum of their separate impacts. A particle of asbestos becomes harmful when it reaches the alveoli of the lung. Fortunately, most particles are caught in the upper respiratory tract and are carried back out on a layer of mucus fanned by tiny cilia (Chapter 10). Smoking interferes with the action of the cilia, paralyzing them. Smoking thus increases the chances that asbestos or other particulate material will reach and remain in the delicate lower respiratory tract. As will be discussed in Chapter 18, medical researchers found that asbestos workers who smoked had far more lung cancers than nonsmokers or non–asbestos workers and had far more cancers than smokers who were *not* asbestos workers. They also found that nonsmokers who worked with asbestos had actually no greater risks of developing lung cancer than did other nonsmokers. Asbestos and cigarette smoking are thus synergistic. The combination of the two things is much more effective in producing lung cancers than the sum of the separate effects of the two apart. Nobody knows what the asbestos does exactly, but asbestos is known to be able to stimulate the production of certain enzymes, one or more of which might be able to convert one or more of the chemicals in cigarette smoke into carcinogens. Such an effect would not be apparent in the absence of tobacco smoke.

There are many other examples of **synergism.** Gaseous pollutants that are highly reactive or extremely soluble and that would normally react and be absorbed in the upper respiratory airway can reach deep into the lungs by being adsorbed onto the surface of particulates and carried into the lungs. Sulfur dioxide exposure in the presence of certain particulates is a more serious problem than sulfur dioxide exposure alone. Similarly, sulfur dioxide with water vapor is much more toxic than the same levels of sulfur dioxide in a dry environment. This is because sulfur dioxide can dissolve in water and then be carried into the lower respiratory tract as a mist.

There is also *negative synergism,* or **antagonism.** During the air pollution disaster in London in the 1950s, animals housed in dirty pens did not seem to suffer nearly as much as animals housed in clean pens. The reason for this is believed to have been the neutralization of sulfur dioxide by the ammonia being released from waste materials in the dirty pens. The combination of ammonia and sulfur dioxide is actually less harmful than the effects of either ammonia or sulfur dioxide alone.

In a survey conducted in the early 1960s, the United States government found that nearly 40 different substances not present in natural air *were* present in the air of many American cities. The potential for synergistic interactions between atmospheric contaminants (and even normal constituents) makes impossible any simple determination of the cause-and-effect relationship between a particular health or environmental effect and a single air pollutant. Attempts to sort out individual effects often end up in a tangle because of the many complex chemical interactions that are possible.

Another Reason to Complain About the Weather

Normally, as the sun peeks over the horizon in the morning, the temperature of the lower atmosphere is warmest near the ground and drops slightly with distance from ground level. (Normally, every 1,000-foot increase in altitude brings a temperature drop of about 5.4°F. In metric terms, there is a 1–3°C drop for every 100-meter rise.) As the sun begins to beat down,

the ground becomes warm and heats the air immediately adjacent to it. As it is heated, surface air becomes less dense and begins to rise. Normally, the air mass cools as it rises, but the rate of cooling usually does not quite allow the temperature (or, more accurately, the density) of the rising air mass to match that of the ambient air at each new height that it reaches. The air mass might continue rising for 11–16 km (7–10 miles), all the way to the end of the troposphere (Figure 9.1), before its density matches that of the surrounding air.

This phenomenon by which dirty air can be moved away from the earth's surface vertically is blocked whenever there is a temperature inversion. A temperature inversion or **thermal inversion** is a condition wherein temperature actually increases with altitude within a layer in some part of the troposphere. In other words, there is a layer of warm air over some cold air. The effect of this is that surface-heated air rises through the lower levels only until it reaches the inversion layer. There the rising air is suddenly *not* warmer (less dense) than the air above or around it, and so it stops rising. Convection is thus halted, and air is effectively trapped, together with any pollutants it carries, as long as the inversion persists. Figure 9.7 depicts such a thermal inversion, showing the pattern of temperature change with height.

Meteorologists say that thermal inversions can come about in several ways. In a **radiation thermal inversion** or **ground inversion,** the ground is cooled at night by thermal radiation and cools the air above it but not the air farther up; this causes an inversion. This is a normal nighttime phenomenon usually erased by the morning sun. If the ground air is cooled too much or if, because of clouds, the sun cannot reverse an inversion, it hangs on.

Another kind of inversion is called a **subsidence inversion.** In this case, upper air masses (in high-pressure areas) compress lower air masses such that the upper portion of the lower air mass is compressed more and thus heated more. This sets up high-pressure doldrums, which might last several days or longer. Stationary highs are rather common over the northern continents during the winter months. They are also fairly common in late summer over the eastern United States, creating air stagnation conditions. Stationary highs are common over the West Coast of the United States practically all year long.

If a city generates pollutants (they all do) and if it is located in an area with prolonged thermal inversions, which block vertical air movement, it will almost invariably have a serious air pollution problem. These factors converged on Donora, Pennsylvania, in 1948; they converge on Los Angeles almost constantly. Los Angeles is unique in that it has very frequent inver-

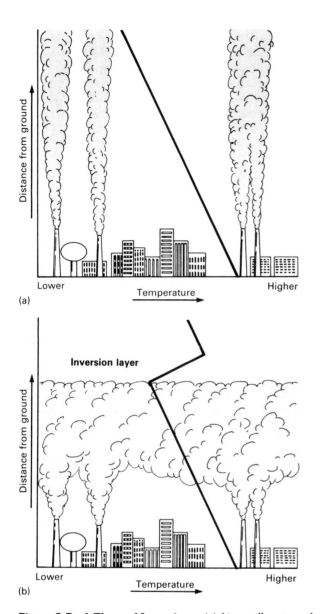

(a)

(b)

Figure 9.7   A Thermal Inversion.   (a) Air usually gets cooler with altitude; temperature drops 1–3°C for every 100 meters in elevation. (b) Sometimes layers of air occur in which this relationship is inverted. Inversion layers act like lids; the air that rises from below (as the sun warms the ground and the air just above it) stops when it reaches the inversion layer. Warm air rises only as long as the air around it is cooler. Note: Diagonal line represents relative temperature.

sions. It has about 250 days each year when its air is unhealthful. In 1981, Los Angeles had around 110 days of air quality alerts and 140 days of air quality warnings (see Chapters 10 and 11).

The Creation of Pollutants in the Atmosphere

We have already discussed how ultraviolet radiation from the sun would have a much more intensely adverse effect on life on earth if it were not for an ozone shield high in the atmosphere. The effect of ultraviolet radiation on living things is that this type of radiation can be absorbed by DNA (Chapter 5) and other chemicals that make up living things. Chemicals that absorb energy become activated and are more likely to react with some other chemical and become changed. Ultraviolet radiation and even light energy can cause pollutants to react with one another and thus change them from one form to another. This adds considerable complexity to the air pollution problem. Several of the most important examples of radiation-induced chemical reactions in the atmosphere were discussed earlier.

Temperature and Wind

Temperature affects the formation, action, and interactions of pollutants in several ways. Part of the influence of temperature on pollution is related to the fact that chemical reaction rates tend to increase with temperature. Also, the warmer the air, the more water the air can hold. Many air pollutants dissolve in water to form acids. This effect is at least partially offset (as an air pollution problem, anyway) when water vapor in the atmosphere condenses and returns to earth as rain, sleet, or snow, cleaning the atmosphere.

Temperature is also important because of its effect on wind (Chapter 2). Wind cleans by diluting and dispersing. This can be a great boon to the residents of the city that generates pollutants, but it can be a problem for people who live in small towns and cities downwind. Wind can also bring in pollutants. According to some reports, smog generated in Los Angeles has been linked to ill effects in places 200 miles away. Air pollution from New York can be carried by west winds through Connecticut and as far as Massachusetts. Some parts of Connecticut have the highest ozone concentrations in the Northeast, largely as a consequence of the pollutants carried there from New York.

Topography

The disasters we considered in the first part of this chapter illustrate the importance of topography in air pollution. Cities at the bottoms of topographic depressions or nestled in mountains are especially prone to dangerous levels of pollution when inversions put lids on these depressions. But mountains can also help to *create* inversions. This happens because heat is radiated from the tops of mountains faster than it is from valleys. As air next to cool mountain slopes is cooled, it pours down the sides of the mountains and fills the valley up with cool air, aggravating and often creating inversion conditions that are favorable to the buildup of air pollution.

The Effects of Pollutants on the Environment and on Human Health

In Chapter 10 we will see that if chemists have a great deal of difficulty keeping track of pollutants in the atmosphere under all kinds of variable conditions, environmental and biomedical scientists often have the even more difficult task of trying to link specific pollutants to specific effects. Part of the reason this is so difficult is that in addition to variations in dose and length of exposure, the actual impact of a given pollutant on a human being depends on the human being, age and genetic heritage, whether or not other illnesses are present, degree of physical activity, and stress. Likewise, the effects of air pollutants on an ecosystem depend on the ecosystem—its species composition, the chemistry of the soil, and other variables. In Chapter 10 we will look at effects of air pollutants on people, places, and things.

Concepts to Remember

1. Although there have been some particularly notable, even spectacular, acute air pollution disasters, these are insignificant in comparison to the impact that air pollution has on all of us over many years of *chronic* exposure.

2. The properties of air pollutants are highly variable. Their chemical properties are summarized in Table 9.1. (Biochemical and health effects are covered in Chapter 10.)

3. The oxides of sulfur and the oxides of nitrogen have similar chemical properties. Beyond their harmful effects on lung tissue, both of these classes of air pollutants also contribute to the acid rain problem and damage materials.

4. Many hydrocarbons are important individually as toxic substances (see Chapter 18); hydrocarbons are important as a class of compounds in the generation of ozone (summarized in Figure 9.4).

5. Carbon dioxide is *not* a human health problem. Its standing as a pollutant stems from its contribution to the "greenhouse effect."

6. Carbon monoxide is toxic by virtue of its ability to compete with oxygen for hemoglobin. It does not damage materials.

7. The health effects of ozone and its effects on materials are derived from its strong oxidizing properties.

8. Particulate matter is a chemically miscellaneous category of pollutants. The size of suspended particles is a major determinant of their health effects. Large particles settle out quickly; generally, only particles less than 10 μm in diameter can stay suspended in the air. Particles less than 1 μm in diameter are difficult to filter and are most likely to get by air defenses and to reach the lungs.

9. There are natural sinks for air pollutants, but these can be overloaded.

10. The *rate* at which pollutants are emitted, the size of the *space* into which they are emitted, and the *circumstances* under which they are emitted are all critical determinants of pollutants' impact.

11. Our technology is a source of many pollutants.

12. *Stationary* and *mobile* are the two categories of sources of air pollution from the point of view of control strategy.

13. Most air pollution (in terms of weight) comes from automobiles and other forms of transportation in the U.S.

14. Weather has a particularly profound effect on air pollution. The effects of wind, rain, and other factors on air pollutants are summarized in Table 9.4.

15. Inversions limit the vertical movement of polluted air; mountains limit horizontal movement.

16. Although air contaminants are generated from natural sources in significant amounts on a global scale, these are of little importance in relation to human-generated air pollutants in cities, where emissions are highly concentrated.

Lynn M. Hodges

Lynn M. Hodges is Environmental Education Section manager for the Tennessee Valley Authority, headquartered in Norris, Tennessee. The TVA is a federal agency established in the 1930s to provide electricity and to foster economic development and sound natural resource management practices in a seven-state region in the southeastern United States. Mr. Hodges's job is to promote environmental education throughout the Tennessee River Valley and beyond. The purpose of the unit he leads is to improve environmental awareness, knowledge, and citizenship skills among teachers, students, and citizens. Mr. Hodges's unit is staffed by ten professionals operating out of Norris and works through a TVA-supported network of some 16 environmental education centers based in colleges and universities throughout the Tennessee Valley. Through programs at the university-based centers, elementary and secondary school teachers are helped to build environmental concepts into their teaching. The Environmental Education Section of the TVA supports the development of environmental and energy education materials (multidisciplinary teaching materials and videos), which are then made available to teachers through workshops held throughout the network. Mr. Hodges describes his work as team building, quality control, and marketing. He puts together the teams that develop educational materials and establishes production standards and mechanisms ensuring the usefulness and quality of the materials. He spends a great deal of his time traveling throughout the valley and beyond, touching bases at each of the environmental education centers and meshing the work of his agency with that of other federal agencies and environmental groups throughout the United States. He has represented the TVA on the U.S. Department of Education's Federal Interagency Committee on Education and in the Alliance for Environmental Education, made up of some 120 private, nonprofit organizatons.

Mr. Hodges has a Bachelor of Arts degree, and when he graduated from college, he was certified to teach English and mathematics. He began his career as a teacher and first became involved in environmental education through his participation in curriculum development projects at the TVA's Land Between the Lakes. Before he assumed his present position, Mr. Hodges served as a state consultant in environmental education for the Kentucky Department of Education and completed a Master of Arts degree in education. He also served on the Environmental Education staff of the TVA's Land Between the Lakes. He became manager of the TVA's Environmental and Energy Education Program in 1981.

References and Further Reading

References marked with an asterisk are cited in the chapter.

Benedicto, C. M., 1988. "The EPA (Taiwan) Attacks Air Pollution," *Free China Review,* March, pp. 32–37.

* Cherfas, J., 1990. "East Germany Struggles to Clean Its Air and Water," *Science* **248:**295–296.

* Cleveland, W. S., and Graedel, T. C., 1979. "Photochemical Air Pollution in the Northwest United States," *Science* **204:** 494–496

* Cleveland, W. S.; Kleiner, B.; McRae, J. E.; and Warner, J. L., 1976. "Photochemical Air Pollution: Transport from the New York City Area into Connecticut and Massachusetts," *Science* **191:**179–181.

* Commoner, B., 1971. *The Closing Circle.* New York: Bantam Books.

* Conservation Foundation, 1984. *State of the Environment: An Assessment at Mid-decade.* Washington, D.C.: Conservation Foundation.

* Ember, L. R., 1990. "Pollution Chokes East-Bloc Nations," *Chemical and Engineering News,* April 16, pp. 7–16.

* Marshall, E., 1984. "Senate Considers Leaded Gasoline Ban," *Science* **225:**34–35.

* Nriagu, J. O., 1990. "Global Metal Pollution," *Environment* **32**(7):7–11, 28–32.

* Piomelli, S.; Corash, L.; Corash, M. B.; Seaman, C.; Mushak, P.; Glover, B.; and Padgett, R., 1980. "Blood Lead Concentrations in a Remote Himalayan Population," *Science* **210:**1135–1136.

* Sandberg, J. S.; Basso, M. J.; and Okin, B. A., 1978. "Winter Rain and Summer Ozone: A Predictive Relationship," *Science* **200:**1051–1054.

* Schlesinger, R. B., 1979. "Natural Removal Mechanisms for Chemical Pollutants in the Environment," *BioScience* **29:**95–101.

* U.S. Environmental Protection Agency, 1982a. *Review of the National Ambient Air Quality Standards for Nitrogen Oxides: Assessment of Scientific and Technical Information* (EPA 450/5-82-002). Washington, D.C.: U.S. Government Printing Office.

U.S. Environmental Protection Agency, 1982b. *Review of the National Ambient Air Quality Standards for Particulate Matter: Assessment of Scientific and Technical Information.* Washington, D.C.: U.S. Government Printing Office.

* U.S. Environmental Protection Agency, 1982c. *Review of the National Ambient Air Quality Standards for Sulfur Oxides: Assessment of Scientific and Technical Information* (EPA 450/5-82-007). Washington, D.C.: U.S. Government Printing Office.

U.S. Environmental Protection Agency, 1984a. *Revised Evaluation of Health Effects Associated with Carbon Monoxide Exposure: An Addendum to the 1979 Air Quality Criteria Document for Carbon Monoxide.* Washington, D.C.: Office of Health and Environmental Assessment.

U.S. Environmental Protection Agency, 1984b. *National Air Pollutant Emission Estimates, 1940–1982.* Research Triangle Park, N.C.: Office of Air Quality.

* World Health Organization, 1977. "W.H.O. Environmental Health Criteria for Oxides of Nitrogen," *Ambio* **6:**290–292.

Other general air pollution references are given in Chapters 10 and 11.

10

The Effects
of Air Pollution

This chapter is about the effects of air pollutants on people, places, and things. More specifically, it covers the effects of air pollutants on human health, trees and crops, fish and other animals, climate, lakes, ecosystems, statues, paintings, buildings, fabrics, rubber gaskets, and other material things. We will begin with human health—for several reasons: (1) The effects of air pollution on human health are significant and relatively direct; (2) we know more about its effects on us than we do about its effects on climate and on ecosystems—effects that may ultimately prove to be even more significant for our species in the long run; and (3) most air pollution laws and control strategies, which we will consider in Chapter 11, are based primarily on human health considerations.

The Effects of Air Pollution on Human Health

An impressive amount of evidence links air pollutants to respiratory and other diseases in humans. In the first part of this chapter, we review the effects that air pollutants have on cells and tissues, and we look at data linking air pollution and specific pollutants with

diseases like emphysema, chronic bronchitis, lung cancer, dust diseases, and other respiratory diseases. In the middle of the chapter is a special section on **epidemiology,** an investigative approach to medicine that looks at patterns of disease in groups or populations in relationship to various environmental factors. Let us begin our consideration of the health effects of air pollutants by reviewing the basic anatomy and physiology of the respiratory tract and the built-in defense mechanisms against environmental insult.

The Respiratory Tract and Its Defenses

The basic anatomy of the respiratory tract, the lungs, and the interface of the junctures of pulmonary blood vessels (capillaries) and air sacs (**alveoli**) is presented in Figure 10.1. The *alveolar sacs,* shown at the ends of tiny air ducts called *bronchioles,* serve collectively as the organ of gas exchange with the environment. In human beings the total surface area of the alveoli is approximately that of a tennis court. An area this large is required for the oxygen–carbon dioxide exchange serving the trillions of cells that make up the human body. (You may wish to consult a textbook on human physiology to review the physiology of gas exchange.)

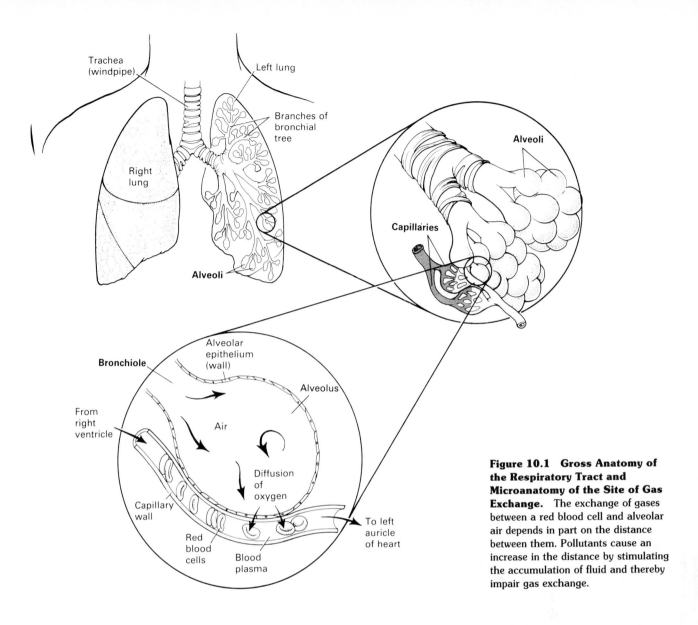

Figure 10.1 Gross Anatomy of the Respiratory Tract and Microanatomy of the Site of Gas Exchange. The exchange of gases between a red blood cell and alveolar air depends in part on the distance between them. Pollutants cause an increase in the distance by stimulating the accumulation of fluid and thereby impair gas exchange.

Mini-Glossary

Alveolus: a very small, thin-walled sac in the vertebrate lung. The alveoli are the sites of gas exchange between blood and air.

Chlorofluorocarbons: a family of chemically stable compounds made up of hydrocarbons to which chlorine and fluorine atoms are attached.

Cohort: in epidemiology, any defined group of people under observation (for example, those being checked periodically for the development of disease).

Enzyme: a protein or a principally proteinaceous molecule that catalyzes a chemical reaction, facilitating conversion of a chemical substance from one form into another.

Epidemiology: an investigative approach to disease that looks for the factors that account for the frequency and patterns of disease within defined populations.

Hepatotoxin: a liver poison; any chemical substance that causes injury to the liver.

Macrophage: any of the large cells found throughout the body that are capable of engulfing and digesting invad-

ing particles or bacteria. (Literally, *macro* = "big," *phage* = "eater")

Mutagen: any substance capable of altering cellular DNA such that a mutation results.

National Academy of Sciences: a nongovernmental organization of scientists and engineers established by the U.S. Congress in 1863. The NAS serves as an official adviser to the U.S. government in matters of science and technology.

National Research Council: a body established by the National Academy of Sciences in 1916 to coordinate the activities of scientists and engineers in government, industry, and universities.

Phagocyte: any of the cells of the human body that serve as a means of defense by engulfing invading bacteria.

Substrate: the substance that enzymes act on. Enzymes usually react with only specific substrates.

Vinyl chloride: a chemical substance that is polymerized to yield polyvinyl plastic.

Imagine the difficulty of keeping a warm, moist area as large as a tennis court free of bacteria, mold, and dust! The respiratory tract contains many effective built-in defense mechanisms against both particulate and gaseous air pollutants. These defenses are spread out along the passage leading from the nostrils to the most delicate portion of the respiratory tract, the alveoli.

The first line of defense is the nostril, especially the nasal hairs just inside. The hairs function like the air filters found in automobiles, air conditioners, and furnaces. Most *large* particles are caught in this first part of the respiratory tract and end up in the alimentary canal after being swallowed or are blown into handkerchiefs. Even some *gaseous* pollutants do not travel farther than the nasal passages because they are quickly absorbed. At high concentrations, some of these very same pollutants can cause damage in the upper respiratory tract, but here irritation and damage are less problemmatic than they would be in the alveoli or the delicate bronchioles.

The sense of smell is another important defense mechanism. When we get a whiff of a pungent chemical, we reflexively turn away, and the brain insists that it be taken away. This is an often overlooked defense mechanism—perhaps because not all dangerous airborne substances smell bad.

Particulate matter can be caught in the mucous membranes lining the respiratory tract from the nose into the upper portions of the bronchial tree. Particulate matter thus entrapped is eventually blown out or swallowed and passed into the gastrointestinal tract, from which it is either excreted directly or absorbed and then excreted by the kidneys. The washout defense mechanism, mucus, can be stepped up in the face of especially severe environmental insult. Tears and running noses (both of which serve to wash irritants away) are often reactions to irritating gases, as are the cough and the sneeze. Both of the latter irritation-initiated reflexes are also parts of the air pollution defense system. They are quite a bit more dramatic than the normally smooth, quiet, and unnoticeable action of the respiratory cilia.

Lining the upper and lower portions of the respiratory tract are cells with tiny hairs that beat in a coordinated rhythmic pattern so as to move toward the outside anything that becomes caught on them (Figure 10.2). These cilia carry mucus and trapped particles up and out of the respiratory tract to where they can be expelled or swallowed. Cigarette smoke retards the action of the cilia lining the respiratory tract and thereby acts as an important cofactor in the damage caused by many air pollutants. Smokers deprived of the action of the cilia are left only with a cough, which they sometimes suppress with lozenges.

Particles that make their way to the ends of the bronchioles into the alveoli must face the **alveolar macrophages.** These *phagocytic* (engulfing) cells are the last line of defense. They engulf particulates and work to digest them or to move them back up the respiratory tract to be carried out by cilia (ciliated cells do not extend down into the alveoli of the lung). This

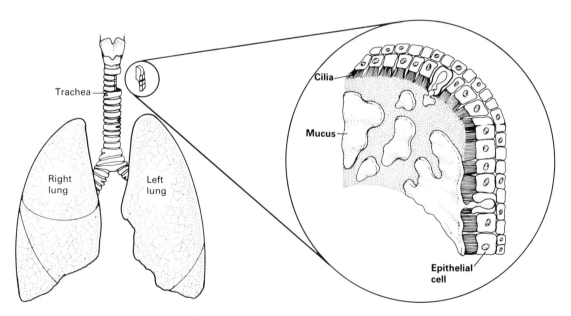

Figure 10.2 The Ciliated, Mucus-covered Epithelium of the Human Airway. The respiratory tract is lined, except for the lower reaches, with cells bearing tiny hairlike cilia that undulate in a coordinated fashion, carrying debris caught in the mucus up and out of the respiratory tract.

system of phagocytes is particularly good at disposing of invading bacteria (which macrophages can also kill), but the macrophages do not do very well with certain kinds of particulate material. As a result, certain kinds of particulate matter can become lodged in the lung more or less permanently. Smoking is also significant at the alveolar macrophage level; it has been shown that cigarette smoke can inhibit the action of these scavenger cells. A summary of the body's total "air defense" system is given in Figure 10.3.

Once they dissolve in tissue fluids and reach cells, the effects that pollutants have depend on their chemical properties (see Table 10.1) and consequent biochemical action (Table 10.2). Some pollutants are so reactive that they do their damage in the first cells they encounter. Pollutants that are less reactive can be absorbed into the bloodstream and carried in an active form to organs of detoxification and excretion like the liver and the kidneys. Some may not be excreted or detoxified very rapidly and may accumulate in certain organs or tissues (for example, strontium 90 in bone, radioiodine in the thyroid, DDT in fat, lead in bone), where they may have cumulative, latent effects (see Chapters 16 and 18).

The Effects of Pollutants on Enzymes and Other Vital Parts of Cells.

To have a health effect, pollutants must influence some crucial structural or functional substance in the body. Many of the effects of pollutants are the result of their ability to damage vital cell structures or to interfere with enzyme function.

Cells can do what they do only because of constituent enzymes that promote most biological reactions. In many enzymes, metals are integral parts of their structure. Zinc as a pollutant can interfere with enzymes that normally contain magnesium if zinc becomes substituted for magnesium in those enzymes. Metals and other kinds of pollutants can destroy enzymes by binding to them indiscriminately, causing various structural changes. Enzymes function only so long as they have a particular molecular shape. Anything that reacts with an enzyme can distort the enzyme's shape such that it can no longer combine with its substrate. Then vital reactions fail to occur. With functionally impaired enzymes, cells die.

Pollutants that eventually react with structural components of the cell membrane or with other parts of the structural and chemical makeup of a cell can

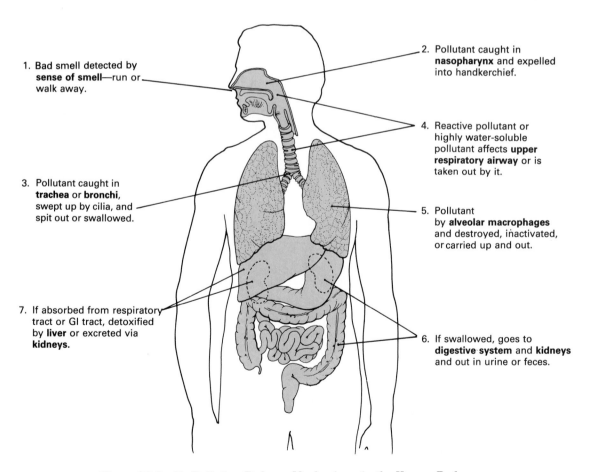

1. Bad smell detected by **sense of smell**—run or walk away.

2. Pollutant caught in **nasopharynx** and expelled into handkerchief.

3. Pollutant caught in **trachea** or **bronchi**, swept up by cilia, and spit out or swallowed.

4. Reactive pollutant or highly water-soluble pollutant affects **upper respiratory airway** or is taken out by it.

5. Pollutant by **alveolar macrophages** and destroyed, inactivated, or carried up and out.

6. If swallowed, goes to **digestive system** and **kidneys** and out in urine or feces.

7. If absorbed from respiratory tract or GI tract, detoxified by **liver** or excreted via **kidneys.**

Figure 10.3 Air Pollution Defense Mechanisms in the Human Body.

Table 10.1   Chemical Reactivities of Air Pollutants

Pollutant	Effects
Sulfur dioxide	Through various reactions in water, this water-soluble, rapidly diffusing, acid-forming oxidizing agent can form sulfurous acid and sulfuric acid, both of which react with organic matter as well as with metals and other materials.
Particulate matter	Airborne particles vary in properties, depending on their chemical nature. Droplets of sulfuric acid behave as sulfuric acid, for example. Solid particles can absorb various chemicals onto their surfaces.
Nitrogen dioxide	This oxidizing agent can react in water to form nitric acid. Being less soluble in water, NO_2 tends to travel farther into the respiratory system. Nitric acid is a very powerful oxidizing agent capable of reacting with nearly all metals and most organic compounds, particularly lipids. NO_2 is also involved in the production of ozone in the atmosphere.
Hydrocarbons (volatile hydrocarbons sometimes included in a larger group of volatile organic compounds)	These combine with many kinds of compounds; products have various reactive properties. For example, carcinogenic hydrocarbon compounds can combine with DNA, causing mutations and cancer. Volatile hydrocarbons and other organic compounds also participate in atmospheric reactions generating ozone.
Carbon monoxide	Though only slightly soluble in water, CO is extremely dangerous because it has a greater affinity for hemoglobin than oxygen does.
Ozone (and other photochemical oxidants)	Ozone is an unstable, highly reactive oxidizing agent able to combine with many organic compounds in cells and tissues as well as with rubber and other materials.
Hydrogen sulfide	H_2S behaves like a weak acid and can also act as a reducing agent, making it mildly reactive with various organic compounds and metals.
Fluorides	These highly reactive reducing agents can combine with many otherwise nonreactive inorganic and organic compounds.
Nitric oxide	NO has little direct effect, being insoluble in water (tissue), but it can be converted to nitrogen dioxide (NO_2).

also result in cell death. A chemically damaged cell membrane may allow essential cellular constituents to escape or allow things normally kept out to get in. In either case the cell is likely to die. Pollutants that react with DNA, though they may most often kill the cell, can also cause mutation and cancer.

Tissue Reactions to the Death of Cells

The number of cells killed or altered will determine how generally felt a reaction to a pollutant will be. Cells die all the time in small numbers. The body's cells are replaced completely every six or seven years.

However, when lots of cells die in an organ system in a short time, there may be drastic local and bodywide (**systemic**) effects.

As cells become damaged and die, they release substances (such as histamine) that cause capillaries to **dilate** (expand) and become more permeable to water. As this happens, fluid is able to escape from the capillaries, causing localized tissue fluid accumulation. This is part of the reason why a wound or a bruise swells. Bruises become warmer because the blood supply to that area is increased. All of this serves to concentrate repair mechanisms in the injured area.

Table 10.2   Biochemical Effects of the Major Air Pollutants

Pollutant	Effects
Sulfur dioxide	SO_2 tends to be absorbed quickly and acts in the upper respiratory tract; it reacts with cellular constituent chemicals, e.g., enzymes. Sulfuric acid (H_2SO_4) lowers pH, impairing enzyme function, and destroys various functional molecules.
Particulate matter	Effect varies depending on the nature of the particles. Carbon particles and others cause scarring of the lungs via complex walling-off and fibrogenic reactions. Particles carrying absorbed mutagens lead to damaged DNA in the lung and elsewhere.
Nitrogen dioxide	Direct effects include the oxidation of cellular lipids; some nitric acid–mediated effects are similar to those described for H_2SO_4.
Hydrocarbons (and volatile organic compounds in general)	Some can react with constituents of cells; e.g., the carcinogenic hydrocarbons like benzopyrene (see Chapter 18) react with DNA, causing mutations.
Carbon monoxide	CO competitively inhibits the combination of oxygen and hemoglobin.
Photochemical oxidants (e.g., ozone)	These agents oxidize cellular constituents.
Heavy metals	Heavy metals can take the place of metals that are normal parts of enzymes or otherwise bind to enzymes, rendering them inactive. Other effects depend on the chemical nature of particular metals.

But these mechanisms can go too far. Secretion of fluid in an irritated bronchiole, up to a point, can help wash out a pollutant; after that point, the accumulated fluid impairs gas exchange and causes respiratory distress.

Another outcome of the injury and death of cells is the scar. The body responds to the death of cells by replacing them with scar tissue, which serves to reinforce the injured area. Scars are good for many kinds of wounds, but if scars develop beyond a certain point in organs that must be perfused (continually washed through by blood or air), such as the liver and the lung, these organs become less elastic and less functional. A less resilient lung or liver cannot be as well perfused with air or blood, respectively. As a consequence, lung and liver functions decrease. Impaired lung perfusion causes stress on the right heart and generalized stress on body tissues as the cardiovascular and respiratory systems' ability to deliver oxygen to the tissues decreases.

The buildup of scar tissue takes time (resulting from repeated injury), whereas the localized inflammation and swelling just described is an immediate effect of pollutants. We refer to scar formation as a **chronic effect** and to the irritation reaction as an **acute effect**. What air pollutants ultimately do to a human being physiologically is summarized in Table 10.3.

Major Classes of Air Pollution—Related Diseases

The following is one of many possible classifications of diseases related to the general problem of community exposure to air contaminants:

1. Pulmonary irritation and acute impairment of lung function (breathing)
2. Cancer
3. Structural changes
4. Systemic toxicity (e.g., lead poisoning)
5. Suppression of host defense mechanisms, leading to increased susceptibility to infection
6. Other types of reduced tissue oxygenation (e.g., carbon monoxide asphyxiation)

This classification scheme is far from neat, since the categories overlap. However, it will serve as a useful framework within which we can consider some of the specific health effects of air contaminants.

Acute and Chronic Irritation Diseases and Impaired Lung Function

With even brief exposures to the oxides of sulfur (e.g. 5 ppm for a few minutes), to nitrogen (dioxide), or to ozone, breathing and gas exchange can be impaired because of **edema** (tissue fluid accumulation), mucus

Table 10.3   Acute and Chronic Physiological Effects of Major Air Pollutants on Human Beings

Pollutant	Effects	
	Acute	Chronic
Sulfur dioxide	Gives rise to irritation reactions, which cause capillaries to dilate and exude fluid; this leads to tissue fluid accumulation and swelling (edema), bronchial spasms, and shortness of breath. General physiological reaction to SO_2 is similar to allergic asthma, i.e., with impaired pulmonary function via increased airway resistance, impaired lung clearance, and increased susceptibility to infection.	Contributes to and aggravates lung diseases like chronic bronchitis, pulmonary fibrosis via irritation leading to decreased pulmonary function and increasing stress on the heart.
Particulate matter	Depending on nature and size, can cause irritation, altered immune defense, or systemic toxicity.	Depending on the nature and size of the particles, can cause decreased pulmonary function and stress on the heart.
Nitrogen dioxide	Incompletely understood, although cell membrane disruption appears to be the principal reason for respiratory tract edema.	Cell membrane damage and acid-induced irritation leading to or contributing to diminished pulmonary function and right-heart stress.
Carbon monoxide	Asphyxiation, heart and brain damage, impaired perception.	Increased red blood cells (polycythemia) in blood, leading to increased resistance to blood to flow; weakness, fatigue, headaches.
Photochemical oxidants (e.g., ozone)	Decreased pulmonary function and right-heart stress.	Emphysema, fibrosis, right-heart failure, aging of lung and respiratory tissue.
Hydrocarbons	The primary harm of hydrocarbons is in their participation in ozone production. Cancer is one direct primary effect of some organic compounds.	

Smoking in Public Places

Controversy continues over the right of individuals to smoke in public places. For some people, exposure to smoke is a health hazard triggering asthmatic and bronchial attacks. Recent studies indicate that exposure to smoke can present a serious risk to the health of nonsmokers. Many people find cigarette smoke harmful and offensive. Smokers believe that they have a right to smoke in public places. They believe that restrictions are not put on other air pollutants such as perfumes or sprays that can also cause allergic reactions in others. They believe that smoking prohibitions are an infringement on their rights. Should smoking be abolished in all public spaces? What do you think?

production, and bronchospasms (see Figure 10.4) secondary to irritation and inflammation. Bronchospasms, edema, and mucus production impair air flow, decrease lung capacity, and decrease rates of gas exchange between blood and alveolar air because of the extra fluid the gases have to pass through. All of these tend to reduce the amount of oxygen delivered to the tissues throughout the body and tend to make the heart work harder. Chronic lung irritation, whether induced by air pollutants, cigarette smoking, living with cigarette smokers, allergic reactions, or all of the above, may lead to chronic conditions and ultimately to permanent structural alteration of the lungs.

Human beings who have asthma, chronic bronchitis, or emphysema get these diseases principally through various combinations of hereditary predisposition, smoking, and occupational exposures, helped along by exposure to air pollutants in the general environment. However they may come by their lung diseases, these individuals, along with children and the elderly, are among the groups of people highly susceptible to air pollutants who are taken into account in establishing air quality health standards (see Chapter 11). Increased airway resistance has been reported in asthmatics with exposures to 0.1 ppm NO_2 for an hour. The same effect in healthy adults might not occur until NO_2 levels reach 2.5 ppm for several hours. With SO_2, asthmatics can show increased airway resistance at 0.1–0.5 ppm for a few hours or even minutes, whereas it might take up to 5 ppm for the same amount of time to produce equivalent effects in normal individuals. For a quarter century, bronchitis and emphysema rates have doubled every five years in the U.S. In the mid-1980s, Americans were dying of emphysema, bronchitis, and bronchial asthma at the rate of 50,000 per year. Today emphysema remains the fastest-growing cause of death in the United States. Emphysema, chronic bronchitis, and related chronic lung conditions are among the most significant causes of disability now compensated by social security.

Figure 10.4   Diagrammatic Representation of the Bronchial Muscle Spasms Caused by Irritation.

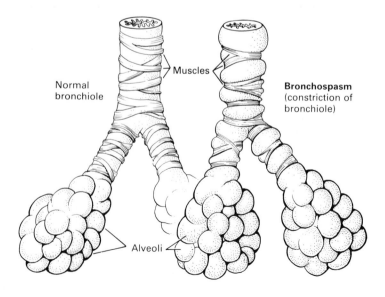

Aggravated Asthma. Asthma is an allergic reaction in which the membranes of the bronchioles are irritated by inappropriately severe reaction to foreign materials such as pollen. The disease is marked by narrowing of airways caused by bronchospasms (see Figure 10.4), edema of bronchial linings, and oversecretion of mucus. There are various forms of asthma: **Extrinsic asthma** is precipitated by external factors; **intrinsic asthma** can be precipitated, apparently in the absence of external factors, by such things as emotion and exercise. In an asthmatic attack, air literally becomes trapped in the lungs because it cannot be expelled. Breathing may, in severe cases, become so impaired as to result in oxygen starvation and death. Air pollutants in the class of irritants are known to aggravate asthma; pollutants (such as certain particulate pollutants) may also be involved in the precipitation of asthma attacks.

Chronic Bronchitis. The problem in chronic bronchitis is inflammation and edema of the linings of the bronchi, resulting in increased mucus production and chronic coughing. The cough and continued mucus secretion cause more irritation and more mucus secretion and coughing, ultimately leading to the actual destruction of small bronchioles. Cigarette smoking and air pollution in industrial environments seem to be the major causes of chronic bronchitis. Air pollution in the more general environment can make matters worse by impairing clearance of mucus, impairing defense mechanisms leading to infection, and adding to the irritation. Chronic bronchitis affects about one in five American males between the ages of 40 and 60.

Emphysema. As if it were not bad enough by itself, chronic bronchitis can lead to emphysema. Bronchospasms and accumulated mucus can in time conspire to trap air in the alveoli. The obstruction acts like a one-way valve allowing air to come in during deep inspiration but keeping it from being expelled. Eventually, the walls of overinflated alveoli rupture, and several alveoli coalesce into fewer, bigger sacs. This irreversible phenomenon leads to emphysema. Because small sacs have larger surface-to-volume ratios than larger sacs, emphysema amounts to a structural condition in which the surface area of the lung is reduced, resulting in reduced breathing efficiency (see Figure 10.5).

Cancer

Among the things that can be present in polluted air are certain chemical substances called **carcinogens,** which are able to react with DNA in such a way as to cause cancer. Inhaled carcinogens may cause cancer in respiratory tissues directly, or they may be

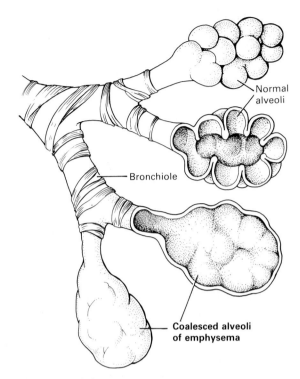

Figure 10.5   Diagrammatic Representation of Emphysema.   The walls of the alveoli break down, coalescing individual alveoli into larger sacs having an overall smaller total surface area. The effect is to decrease the ability of the lung to exchange gases.

picked up by the bloodstream and cause cancers in other body tissues.

Some substances found in polluted air are well-established carcinogens such as benzopyrene; others are just beginning to be identified as carcinogenic agents. Ozone, for example, has been shown to be associated with increased lung cancer in experimental animals exposed to levels of ozone equivalent to those found in Los Angeles during a smog alert (Hassett et al., 1985). It is not known if ozone itself is carcinogenic or if it acts as a cofactor in some way. Cancer and the environmental causes of cancer will be considered in detail in a later chapter; we will not discuss it further here.

Dust Diseases

We have already discussed how irritants in contaminated air may eventually help to bring about structural impairment and loss of lung function. The **pneumoconioses,** or dust diseases, are caused by a special class of air pollutants that bring about structural changes in the lungs. The term *pneumoconiosis* applies to *all* dust diseases of the lungs. Two examples of such diseases are **silicosis,** in which the active ingredient is silicon dioxide, and anthracosis, or **black lung,** in which the active ingredient is coal dust. Through a mechanism that is only poorly understood, dust in the lungs causes a response in which fibrous tissue is deposited around dust particles, creating "macules." The dusts that cause the pneumoconioses (see Table 10.4) are relatively inert and tend to stay in place over many, many years. The fibrotic reaction eventually decreases the function of the lungs; this in turn puts stress on the heart. Whereas pneumoconioses tend to cause diffuse lung changes, certain agents such as beryllium and hair spray produce granular knots or lumps called granulomas (see Waldbott, 1978). Granulomas also have an adverse effect on pulmonary function.

Asphyxiation

Carbon Monoxide. The chief environmental problem with carbon monoxide is that it combines with hemoglobin, competing with oxygen. In an atmosphere with more than 10 ppm carbon monoxide, even in the presence of normal amounts of oxygen, the blood can be rendered unable to carry sufficient oxygen to tissues. If as little as 5% of the hemoglobin in the blood is carrying carbon monoxide, the oxygen-carrying capacity of the blood is reduced to the extent that performance tests begin to reveal physiological impairment.

Because the affinity of carbon monoxide for hemoglobin is more than 200 times that of oxygen, there is some doubt that even extremely low levels of carbon monoxide are safe. As yet, there is no experimental evidence, or even any theoretical basis, for establishing a threshold level for carbon monoxide problems. The impairment of oxygen transfer by carbon monoxide is probably a continuum that ranges from severe impairment and death at very high concentrations down to no impairment at trace levels.

It appears that most people suffer no problems when inhaled air contains concentrations below 10 ppm of carbon monoxide. At 100 ppm, however, nearly all humans experience headache, dizziness, and impaired perception. At 300–400 ppm even for a few minutes, vision is impaired, and there may even be nausea and abdominal pain. Exposure to 1,000 ppm carbon monoxide for less than an hour is fatal. It is important to note that the key parameter is the percentage of blood hemoglobin in the form of *carboxyhemoglobin* (COHb). This percentage is related to the level of CO in ambient air, the duration of exposure, and the ventilation rate (which increases with exercise). In terms of carboxyhemoglobin, 3% to 4% COHb has been associated with reduced work efficiency in some people. Studies have demonstrated decreased vigilance, altered perception, and decreased manual dexterity at COHb levels of 5% or greater.

Groups at special risk from CO exposure are fetuses and young infants, the elderly (especially those with heart or lung problems), and individuals with heart disease, lung disease, or anemia (including sickle-cell anemia).

Hydrogen Sulfide. Another major asphyxiant, hydrogen sulfide, is sometimes encountered in high concentrations in sewers and in various occupational settings. This toxic gas impairs tissue oxygenation indirectly by paralyzing the breathing control center in the brain. Although hydrogen sulfide's characteristic "rotten egg" odor gives it away at low concentrations, at higher concentrations it rapidly paralyzes the sense of smell and prevents any warning. Hydrogen sulfide affects neurons, apparently by decreasing their ability to conduct impulses.

Systemic Toxins

Chemicals picked up from the environment by the lungs may cause disease in organs other than those of the respiratory tract. Air pollutants that get into the bloodstream via the lungs to cause problems in various other parts of the body (Figure 10.6) are called *systemic toxins*. Lead and mercury are prime examples of systemic toxins that can be picked up from the air.

Lead. As a heavy metal, lead causes first biochemical, then physiological, then health effects as a consequence of its reactions with enzymes, other proteins, and other chemical constituents of cells. Apparently, two of the most lead-sensitive types of tissue are brain cells and the bone marrow cells that normally give rise to red blood cells. Interaction between lead and the brain—particularly the developing brain—takes the forms of mental retardation, lowered IQ, and behavioral abnormalities. In red blood cell

Table 10.4 Agents That Cause Scar Formation in Lung Tissue

Material	Disease Designation
Inorganic fibers and dusts	
Crystalline silica	Silicosis
Asbestos	Asbestosis
Talc	Talcosis
Coal (pure)	Anthracosis (black lung)
Kaolin	Kaolinosis
Graphite	Graphite lung
Organic fibers and dusts	
Cellulose	Bagassosis
Cotton	Byssinosis (brown lung)
Flax	Byssinosis (brown lung)
Hemp	Byssinosis (brown lung)
Metallic fumes	
Tin oxide	Stannosis
Iron oxide	Siderosis
Beryllium oxide	Berylliosis

Source: From *Chemical Contamination in the Human Environment* by Morton Lippmann and Richard B. Schlesinger. Copyright © 1979 by Oxford University Press, Inc. Adapted by permission.

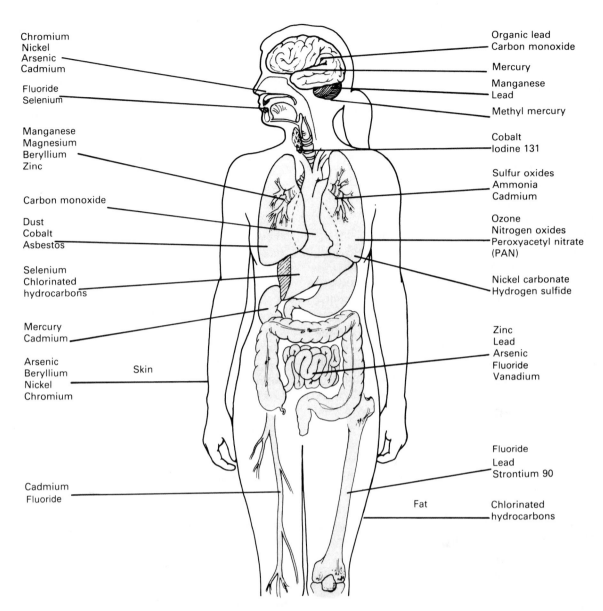

Chromium
Nickel
Arsenic
Cadmium

Fluoride
Selenium

Manganese
Magnesium
Beryllium
Zinc

Carbon monoxide

Dust
Cobalt
Asbestos

Selenium
Chlorinated
hydrocarbons

Mercury
Cadmium

Arsenic
Beryllium
Nickel
Chromium

Skin

Cadmium
Fluoride

Organic lead
Carbon monoxide

Mercury

Manganese
Lead

Methyl mercury

Cobalt
Iodine 131

Sulfur oxides
Ammonia
Cadmium

Ozone
Nitrogen oxides
Peroxyacetyl nitrate
(PAN)

Nickel carbonate
Hydrogen sulfide

Zinc
Lead
Arsenic
Fluoride
Vanadium

Fluoride
Lead
Strontium 90

Fat

Chlorinated
hydrocarbons

Figure 10.6 Targets of Air Contaminants.

precursors, lead inhibits one of the key steps leading to the synthesis of hemoglobin. This manifests itself as microscopically stippled red blood cells and the appearance in the urine of two chemicals (delta-aminolevulinic acid and coporphyrin III) that would normally have been incorporated into hemoglobin. The ultimate manifestation of the blood effect is an impaired ability of the blood to carry oxygen to the tissues—a kind of anemia.

Lead has other effects. Both benign and malignant tumors have been produced in rats and mice exposed to lead acetate (a lead salt). Lead can also affect the kidneys by damaging renal tubules.

Until very recently it was thought that a blood lead level of 50 μg/dl (40 μg/dl in children), where a reduction in blood hemoglobin began, was the health effect threshold, but newer evidence indicates that the threshold may be much lower. The banning of leaded gasoline in the United States was based on observations that brain wave changes occur in children with blood lead levels as low as 15 μg/dl and that lower IQs, mental retardation, and behavioral changes have been associated with 40–60-μg/dl levels in preschool children.

It was estimated that the proposed reduction would cut in half the amount of impaired mental development in children being caused by lead in 1984; some 50,000 children would be saved from brain damage in only two years. The EPA estimated that the IQs of 4 million children could be raised by 2.2 points if lead were removed from gasoline altogether. Earlier, the Director of the U.S. Center for Environmental Health estimated that there would have been 80% fewer cases of lead toxicity among children if lead had been eliminated from gasoline entirely between 1977 and 1981.

In 1985, the EPA moved the date after which lead could no longer be put into gasoline up to 1988. The reason was findings in at least two studies that low levels of lead in the environment are also linked to high blood pressure. Originally, lead was believed to affect primarily children; now it looks like it has a significantly negative effect on adults as well. The EPA acknowledges that eliminating the use of lead as a gasoline additive will cost refiners over half a billion dollars, but the agency also points out that the actual cost of the lead ($608 million) would be saved. Moreover, the action would save American consumers $914 million in car engine maintenance, $187 million in fuel economy, and $600 million in health costs, for a total of $1.71 billion (Sun, 1985).

A number of research questions remain concerning the effects of lead. We need to know more about the relative effects of organic and inorganic lead compounds, and we need to know more about how much lead we get from air, diet, and drinking water. We also need to know more about the environmental variables that can affect human responses to lead.

Mercury. Mercury kills cells by causing the denaturation of proteins. In episodes of acute mercury poisoning, mercury tends to kill cells of the organs with which it comes into contact and thus impairs those organs. When mercurial compounds are eaten, for example, they affect the digestive tract and the kidneys (because some of the mercury is absorbed from the gastrointestinal tract and is filtered from the blood by the kidneys). Chronic exposure causes lesions of the mouth and skin and neurological problems. Air pollution is more important than ingestion to the *chronic* kinds of mercury toxicity.

The classic symptoms of poisoning by mercury vapor are irritability, excitability, loss of memory, insomnia, tremor, and gingivitis (gum disease). Such symptoms do not occur below an average mercury concentration in air of 0.1 mg (100 μg) per cubic meter, although loss of appetite and psychological problems have been reported to occur at levels slightly below 0.1 mg/m^3 (WHO, 1978). Adverse health effects in workers have not been known to occur below the level of 0.05 mg (50 μg) per cubic meter of air (WHO, 1978). The effects of methyl mercury in adult human beings become noticeable in sensitive individuals when levels of mercury in the blood reach 20–50 μg per 100 ml.

Much more research on mercury and its effects is needed, particularly with respect to various *forms* of mercury and its conversion products in air, food, and water and with respect to dietary intake and the dose-response relationship for certain health effects of mercury.

Mercury as a water pollutant will be discussed in Chapter 12.

Metal Fume Fever. Fumes of certain metals —particularly manganese and zinc but also oxides of antimony, arsenic, cadmium, cobalt, copper, iron, lead, magnesium, mercury, nickel, and tin—can cause temperature elevations in people who breathe them. **Metal fume fever,** as this is called, is also accompanied by a dry throat, chest constriction, fatigue, headache, back pain, nausea, and muscle pain. Although primarily an occupational problem, metal fume–related health problems have also been reported in people living adjacent to metal processing plants.

The Effect of Air Pollution on Resistance to Disease

A number of studies have shown that one of the most important effects of air pollution on health may be that it makes individuals more susceptible to infection. This possibility has been raised by many experiments with animals. In such studies, one group of animals is exposed to a pollutant while another is not; each group is then equally challenged with an infectious microorganism of some sort—the disease organism that causes pneumonia, for example. In one study it was shown that one-third of the animals *not* exposed to nitrogen dioxide came down with pneumonia and died when challenged by a causative disease organism, but *all* of the animals previously exposed to nitrogen dioxide for just a short time died of pneumonia.

Colfin and Blommer (1967) showed that if mice were exposed to irradiated exhaust (irradiated to mimic the effects of solar radiation), they had much greater death rates due to subsequent challenges by streptococcal pneumonia than controls (mice not subjected first to the exhaust). The study used concentrations of auto exhaust fumes that were somewhat *lower* than those found in heavily polluted city air. Experimental evidence like this indicates that many pollutants—for example, metals, organic pesticides, and gaseous pollutants—can in fact impair our immune defenses and make us more susceptible to infectious disease (Caren, 1981).

Air Pollution and Feeling Good

According to some reports, people do not function very well on smoggy days—they suffer from impaired efficiency. For some people, at least, pollution brings a decrease in the feeling of well-being. Studies have shown that absenteeism, accident, and suicide rates are all higher during unfavorable weather and on overcast and smoggy days. Some people believe that smog is at least partly responsible for some of these effects. Certain studies have suggested that smog leads to personality changes and accentuates traits such as forgetfulness and irritability. Such effects seem to be most pronounced on the first day of a pollution episode.

Since, as the World Health Organization defines it, **health** is a state of complete physical, mental, and social well-being and not merely the absence of disease or infirmity, effects on these kinds of conditions must be included in any discussion of the health effects of pollutants. Although feeling good may be more subjective than a disease like emphysema, not feeling good may have indirect impacts that equal or surpass the negative impact of a specific disease. For instance, the California Department of Public Health reports that as the level of photochemical oxidants rises, so does the accident rate. As another example, carbon monoxide at low levels may not have much of a direct impact on mortality rates, but in its negative effects on perception, particularly among people driving automobiles at high rates of speed, it may ultimately have considerable impact.

Epidemiology: A Population-based Approach to Environmental Disease

The Place of Epidemiology in Understanding Disease

As pointed out by Mausner and Bahn (1974), there are three fundamental ways of studying human disease:

1. The basic science approach (studies using experimental animals, biochemicals, etc.)
2. The clinical approach (studies of sick people and volunteer experimental subjects)
3. The epidemiological approach

The *basic science* approach is concerned with every detail of the causes (**etiology**) and the biochemical, biological, and physiological steps involved in the progression (**pathogenesis**) of the disease in question. What is the causative factor? How does it get in? Where does it go? How is it metabolized? How is it excreted? What tissues are affected? What happens to the factor in water? In air? What exactly does it do at first? What happens then?

The *clinical* scientist is more directly concerned with the impact on the individual human. What measurable changes occur in the exposed human that might be used to detect the presence and determine the extent of disease? What are the physiological

symptoms? What will arrest the disease? What will relieve the symptoms?

The third approach, the community, population, or *epidemiological* approach, examines the relative frequency of disease within a defined population and within subgroups of that population. What kinds of people are getting sick? Is the disease more prevalent among the young? The old? Factory workers? Does the disease exhibit a pattern that indicates that it is communicable?

Another way of saying all of this is that epidemiology is one of several important ways of studying the relationship between environmental contaminants and disease. First, animals, plants, living cells, and even biochemicals can be exposed to the environmental contaminants under investigation to see what happens. Second, human volunteers can (under some circumstances) be exposed to see what happens to people directly. Third, the patterns of relationship between disease and exposures can be systematically observed in large groups of people. In the real world, all three approaches are almost always employed in concert. For example, epidemiology might discover that people who work with a particular chemical have relatively high rates of cancer. Laboratory scientists might then expose experimental animals to a suspected causative agent to confirm or refute the connection experimentally. If and when a connection is confirmed, clinicians would begin to look more carefully for the symptoms of disease among people who have been exposed to the agent. The order of the involvement of the three approaches is variable; any one might initially lead to the suspicion that there is a connection between specific agents and a disease.

In 1974, Dr. John Creech, a physician in Louisville, Kentucky, observed that an unusual number of vinyl chloride workers had died of a rare liver cancer (hepatic angiosarcoma). Subsequent epidemiologic checks of death certificates confirmed the significance of this relationship. Almost simultaneously, scientists in Italy confirmed that vinyl chloride caused cancer in animals. Laboratory scientists and clinicians throughout the world then went to work investigating just how vinyl chloride "caused" liver cancer and how the cancer might be detected, diagnosed, and treated.

For reasons that may not become entirely clear until after Chapter 18, we have singled out epidemiology for emphasis here. Epidemiology is a bit more abstract and less straightforward than experiments with animals or people. The most important reason for mentioning it here is its special importance in connection with environmental diseases. Epidemiology can help to make up for the facts that animals do not always react the same as humans to environmental contaminants and humans cannot ethically be deliberately exposed to any amount of suspected cancer-causing agents, for example. In some cases, epidemiology—which is really a way of collecting data from exposures that are going on anyway—may be the only way to document the effect of a particular contaminant on human beings.

What Do Epidemiologists Do?

Epidemiologists study the patterns of disease (incidence over time, spatial distribution, etc.) within defined groups of people (a county, a state, a metropolitan area, etc.) and the relationships of these patterns to *other* patterns within that group (e.g., occupational patterns, residential patterns, who drinks water coming from what sources). Epidemiologists try to match up patterns of disease with other patterns in an effort to identify possible cause-and-effect relationships. More often, epidemiologists try to *confirm* or *refute* match-ups already under suspicion. For example, after the initial observation that an unusually large number of vinyl chloride workers died of liver cancer, epidemiologists went to work to assess just how much greater the risk of liver cancer is to vinyl chloride workers with varying degrees of historical exposure. Another way of defining **epidemiology** is that it is the study of patterns of disease within *defined* populations and the factors that are responsible for or influence those patterns.

The key word in this definition is *defined*. Perhaps the most easily overlooked aspect of the epidemiological approach is that it is dependent on *rates* of disease. For example, 300 cancer deaths in a year would be high in a group of 6,000 people; the same number might be normal in a group of 100,000 people and low in a group of 1 million. The point is that the number of people with a certain disease can have meaning only in terms of the number of people who might have developed the disease. Further, this is only meaningful in terms of the rate of disease in some standard or control population. Time is another important dimension here. Epidemiologists can work with *morbidity* expressed in terms of time, that is, the percentage of people who have the disease at one particular point in time or the number who are diagnosed within a given period of time (e.g., per year). If **mortality** (death) is used, this must also be qualified in terms of some defined period of time.

In terms of our earlier example, if someone came up to you and reported finding three cases of liver cancer, you might reply, "So what? A lot of people die of liver cancer each year." But if someone came up to you and reported the discovery of three cases in two years of a rare form of liver cancer in a single group of 1,200 American vinyl chloride workers, you might be moved to ask the key question, "How prevalent or how common is liver cancer among people in general?" Suppose that someone came forward and reported finding three cases of a certain type of liver cancer among a few hundred vinyl chloride workers in a single plant and added that only 25 or 30 cases of this type of liver cancer had been reported in the whole world that year. In this instance, you could be fairly sure that an occupational disease had been discovered.

Although the vinyl chloride–hepatic angiosarcoma relationship was discovered in just such a way, differences in relative rates are usually not nearly so impressive. Epidemiologists usually work with small relative differences in rates of disease between the group of people exposed to some agent and those not exposed.

In epidemiological investigations, precise mathematical analyses are required, and extreme care must be given to the methods by which disease is observed, measured, and cataloged in well-defined groups. Great caution must be exercised to ensure that apples are always being compared to apples. Chemical workers may have significantly greater or less risk of developing certain diseases than people in general. But these rates might be identical to rates in groups of non–chemical workers with similar distributions of age, sex, race, and socioeconomic status. Let us consider the concept of controls in a bit more detail.

Control Groups

As pointed out by Lave and Seskin (1979), if an association is found between exposure to air pollution in cities and a lung disease such as lung cancer, there would be four possible explanations, assuming no bias in the way the sampling of city and noncity dwellers and smokers and nonsmokers was carried out:

1. The observation is a false, chance occurrence resulting from the way the sampling was done.
2. Air pollution causes or contributes to lung disease.
3. Lung disease causes or contributes to air pollution.
4. Some third factor is responsible for *both* lung disease and air pollution.

To conclude that air pollution is a cause of or a contributing factor in lung disease, the other three explanations have to be ruled out. Generally, the first possibility is ruled out by repeating the study a number of times, and the third possibility is usually ignored. It is the fourth possibility that gives epidemiologists the most trouble, and this is where controls come in. The selection of a control group is intended to minimize the possibility that other factors might line up on one side or the other side of the comparison being made. If cities have high proportions of smokers, old people, and males and if old people, smokers, and males have high rates of lung cancer, it could be falsely concluded that cities cause lung cancer.

Age offers a particularly good case showing that great care must be taken to compare apples with apples in epidemiology research. Since cancer is largely a disease of the relatively old, investigators must be sure that a difference in cancer rates between two populations is not a result of the fact that one population has proportionately more old people in it. This does not mean that comparisons cannot be made between populations having different age profiles; it does mean that if such comparisons are made, the data has to be *age-adjusted*. Typically, this is done by comparing the age-specific rates of the two populations in some way. If a particular population is being compared to a large standard population, say, that of the whole United States, then the age-specific rates for the standard populations can be used to calculate an "expected" number of cancers for the number of people in each age bracket of the population being evaluated. This expected number of cancer cases can then be compared to the number actually observed, any difference or ratio then being subjected to a statistical test of significance. This indirect method or other methods of making mathematical adjustments in populations being compared can also be applied to sex, race, and other factors that might influence rates of cancer in populations.

Epidemiologists work very hard to find control groups that are matched for as many other factors as possible. The effects of benzopyrene on lung cancer would be studied best (epidemiologically) by comparing the rates of lung cancer in a city that has high benzopyrene levels with the rates in a city that has low levels but is otherwise "identical."

Obviously, no two cities or groups of any sort are identical, and this is the curse with which all epidemiologists must live. The task then amounts to controlling confounding variables to the extent possible. Among the most common variables controlled in epidemiological studies are age, sex, race, and socioeconomic status.

Retrospective and Prospective Epidemiology

Epidemiology has the advantage of being able to look forward and backward. Suppose that we were concerned that a certain type of skin cancer was possibly caused by exposure to a certain pesticide. One of the first steps we might take would be to identify as many people as possible in whom the cancer was diagnosed within a certain (large enough) period of time. These would then be matched with an equal number of similar people who have never had the cancer and do not have it now. We could then look into the histories of the two groups and determine the proportion of those *with* the cancer who were exposed to the pesticide versus the proportion of those *without* the cancer who were exposed to the pesticide. If it turned out that a very high proportion of those with the cancer in question were exposed to the pesticide and relatively few of those in the control group had ever been exposed to the chemical, we might hypothesize that there was a connection. This type of backward-looking study (**retrospective study**) is called a *case-control study*.

Our second step might be to watch groups of exposed and nonexposed people over time to see how many of each group develop skin cancer, for example. This would be a forward-looking or **prospective study**. Prospective studies start with exposure and look into the future for disease; retrospective studies generally start with disease today and look back for exposure (see Table 10.5). Retrospective studies may also look back for both exposure and disease.

The Trouble with Controls and the Trouble with Epidemiology

We mentioned earlier that there is no such thing as a perfect control group. A consequence of this and some other limitations inherent in the epidemiological approach is that conclusions derived from epidemiological investigation are far from absolute. This much may be obvious to the sophisticated reader who knows that absolute answers really do not come even from much more precise and more completely controlled laboratory investigations. Epidemiologists can be fairly exacting in establishing the degree of correlation between a disease and a suspected etiologic agent. But even when correlations are nearly perfect, nothing can really be inferred *absolutely* in terms of cause and effect.

In science, as in courtrooms, association is considered circumstantial evidence, although "guilt" can eventually become established through consistent association. Proof of the sort that comes from elegantly constructed laboratory experiments does not happen in epidemiological investigations. Proof usually emerges only gradually. Often it does not emerge fast enough. This is one of the major reasons why despite large volumes of circumstantial evidence, pollution control and the relationship between smoking and health continue to generate controversy.

Epidemiological Evidence that Air Pollution Is Indeed a Real Health Problem

Much of the general information we have about the relationship between air pollution and disease has come from epidemiological studies. Some of these studies have focused on air pollution in general rather

Table 10.5 Comparison of Retrospective and Prospective Studies of the Relationship between Disease and Causative Agents

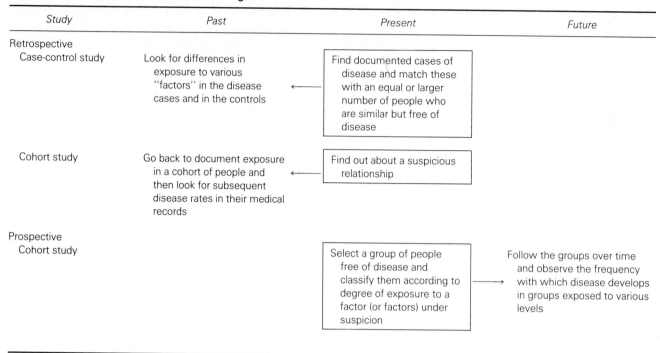

Study	Past	Present	Future
Retrospective Case-control study	Look for differences in exposure to various "factors" in the disease cases and in the controls ←	Find documented cases of disease and match these with an equal or larger number of people who are similar but free of disease	
Cohort study	Go back to document exposure in a cohort of people and then look for subsequent disease rates in their medical records ←	Find out about a suspicious relationship	
Prospective Cohort study		Select a group of people free of disease and classify them according to degree of exposure to a factor (or factors) under suspicion →	Follow the groups over time and observe the frequency with which disease develops in groups exposed to various levels

Source: J. S. Mausner and A. K. Bahn, *Epidemiology: An Introductory Text* (Philadelphia: Saunders, 1974).

than on specific pollutants because of the difficulty of sorting out specific pollutants in the air where people live. In an early study reported in the *British Journal of Preventive Social Medicine* (Fairbairn and Reid, 1958), for example, postmen who lived in northeastern London, the most heavily polluted part of the city, had a higher rate of respiratory disease, greater frequency of earlier retirement, and greater numbers of deaths from chronic lung disease than postmen who lived in other areas of the city. In another study it was determined that in Erie County, New York, children living in areas of the county with low levels of pollution were hospitalized for asthma at a lower rate per 100,000 people than children in more heavily polluted areas of the same county.

An American Lung Association pamphlet describes another such study showing that of the 38,200 deaths in Nashville, Tennessee, over a 12-year period, more people died of breathing ailments in parts of the city that were heavily polluted than in less polluted areas. The same pamphlet also describes a British study based on observations of a cohort of 3,866 children followed from birth to the age of 15. This study determined that there was a definite relationship between the rates of lower respiratory tract infection and life in high-pollution environments.

We will discuss cancer at length in a later chapter, but it is appropriate to mention here that one of the best-documented connections between pollution and health has to do with the relationship between air pollution and lung cancer. While acknowledging that cigarette smoking is the main culprit, the National Research Council has blamed the rising tide of lung cancer at least partly on the quality of the air over American cities. Lung cancer is the single greatest cause of cancer deaths in cities. In some cities the incidence of lung cancer is twice that in rural regions. The link between lung cancer and smoking may account partially for this; that is, there may be differences in the smoking habits of city folk and rural dwellers. It is more likely, however, that the increased risk of lung cancer comes from the combination of cigarette smoking and the unhealthful elements of city air or exposure to other things characteristic of cities. Lave and Seskin (1970), who evaluated a number of studies done in England comparing rural and urban air, claim that there is significant evidence of an association between air pollution and lung cancer. Their report went further, however, suggesting that for both males and females the incidence of more than 16 different kinds of cancer was higher in cities than it was in rural areas.

Episodic studies have been done attempting to find correlations between days of high air pollution levels and admissions to hospitals for respiratory problems. Close correlations have generally been found between bad air days and respiratory diseases, including bacterial infection, influenza, bronchitis, allergic disorders, and cardiovascular diseases.

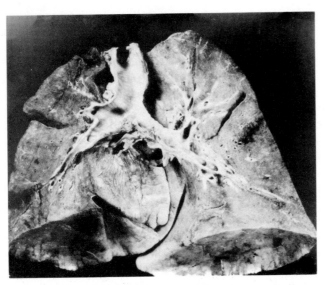
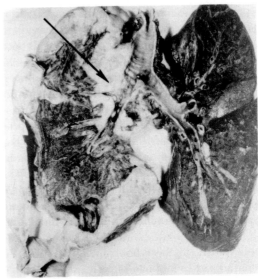

The link between cigarette smoking and lung cancer is clear. At left are the healthy lungs of a nonsmoker; at right are the lungs of a heavy smoker with lung cancer (the arrow indicates the area of cancer).

Some studies have attempted to relate respiratory disorders to more specific air pollution measurements. In New York City, young children were found to have greater respiratory disease problems in areas or on days characterized by high levels of particulate matter and carbon monoxide in the environment. In a classic Chicago study of air pollution and its relationship to disease, it was determined that serious respiratory disease or aggravation of disease in individuals with chronic lung disease—especially in older age groups—was much higher when sulfur dioxide levels were high. This relationship was found to exist on the day on which the exposure occurred, and it was even stronger for disease problems observed on the day following exposure (Carnow et al., 1969).

In their comprehensive look at the literature concerning the relationship between air pollution and disease, Lave and Seskin (1970) concluded that abatement of air pollution could bring about a 10% to 15% decrease in death due to heart disease. They said that there is a strong relationship between death due to bronchitis and air pollution and that mortality due to bronchitis could be reduced up to one-half in certain places by eliminating air pollution episodes and reducing levels of pollution overall. They further suggested that a 10% overall decrease in particulate matter would decrease the infant death rate overall by 0.7%, the neonatal death rate by 0.6%, and the fetal death rate by 0.9%. The National Research Council has estimated that if air pollution were reduced in and around U.S. cities by half, the lung cancer rates might be cut by as much as 20%. EPA criteria reports and staff papers document numerous epidemiological studies linking air pollution and human disease. Refer to recent EPA publications for an up-to-date account.

The Effects of Air Pollution on Other Animals, Plants, Microorganisms, and Ecosystems

We presented the effects of air pollution on human health first, not because this is necessarily the most important effect of air pollution on the planet but simply because they relate most directly to the organisms reading these pages and because they are among the most direct kinds of effects. Another important problem may well be the effect that air pollution has on life forms in general. Unfortunately, difficult as it is to establish tight connections between human health and pollution, the effects of pollution on the balance in nature and on ecosystems are considerably more difficult to assess.

The Effects of Air Pollution on Animals Other than *Homo sapiens*

Studies *have* shown harmful effects of air pollution on animals; not surprisingly, most of these involved species in which humans have an economic interest. Fluorides, for example, have been shown to cause more worldwide damage to farm animals than any other kind of pollutant. We described the health effects of fluorides on humans in Chapter 9. In other animals, similar effects have been documented by

Lillie (1970). Most fluoride problems in animals have occurred among livestock pastured in and around aluminum smelters. Waldbott (1978) described several cases of lead poisoning in animals allowed to graze near lead smelters. Other air pollutants that have been shown to have an adverse effect on farm animals are selenium, molybdenum, and mercury. Though it has not been well documented, we must assume that wild and domestic animals are affected by air pollutants in the same general ways as humans.

The Effects of Air Pollution on Plants

Damage to plants by air pollutants has been known since the beginning of the twentieth century, but what were once only isolated episodes have become more general and more widespread. A list of air pollutants and the plants they affect is given in Table 10.6.

According to reports published by the EPA, the air pollutants most responsible for plant damage are oxidants such as ozone, sulfur dioxide, acids derived from the oxides of both sulfur and nitrogen, and various particulates. Plants that are susceptible to these kinds of pollutants include vegetables, fruits, other kinds of agricultural crops, grasses, shrubs, trees, and commercial flowers. The susceptibility of plants to pollutants is influenced by such variables as temperature,

wind, light intensity, soil fertility, and relative humidity. High soil moisture and atmospheric humidity actually intensify the damage.

A particularly poignant example illustrating the extent to which pollutants can affect crops was the severe damage to the potato crop on the eastern shore of Virginia in the summer of 1971, which was linked to oxidant pollution (Marx, 1975). This was a profound episode in that up to half the potato crop was lost that summer. In the United States, ozone is believed to be responsible for 90% of all air pollution–related crop damage, amounting to billions of dollars each year (Skarby and Sellden, 1984). Discussion of the economic impact of air pollutants on crops will be extended in Chapter 11.

Air pollution affects plants in many of the same chemically fundamental ways that it affects animals—by the oxidation of cellular constituents and enzymes, for example. Some of the primary symptoms that appear in plants are **chlorosis** (yellowing, which reflects an impaired ability by plants to manufacture chlorophyll) and the destruction of leaf tissue. As with animals, the influence of pollutants on plants extends to weakening them to the point at which they are more susceptible to attack by infectious agents or other natural enemies. Oxidants like ozone, for instance, are believed to weaken ponderosa pines to the extent that the trees are more susceptible to the western bark

Table 10.6   Effects of Air Pollutants on Plants

Chemical	Symptom	Sensitive Plants*	Examples of Concentration for Sensitivity
Chlorine	Bleaching, leaf tip and margin browning, dropping of leaves, yellow spots	Radish, alfalfa, peach, buckwheat, corn, tobacco, oak, white pine	Radish, 1.3 ppm
Fluorides	Leaf tip and margin yellowing (chlorosis), dwarfing, leaf abscission, decreased yield	Gladiolus, tulip, apricot, blueberry, corn, grape, blue spruce, white pine	Gladiolus, apricot, 0.1 ppb
Nitrogen oxides	Brown spots on leaf, suppression of growth	Azalea, sunflower, mustard, tobacco, pinto bean	Pinto bean, 3 ppm
Sulfur dioxide	Bleached spots on leaf, chlorosis, suppression of growth, early abscission, reduced yield	Barley, pumpkin, alfalfa, cotton, wheat, lettuce, apple, oat, aster, zinnia, birch, elm, white pine, ponderosa pine	Alfalfa, barley, cotton, 0.3 ppm
Ozone	Reddish brown flecks on upper surface of leaf, bleaching, suppression of growth, early abscission, premature aging	Alfalfa, barley, bean, oat, onion, corn, apple, grape, tobacco, tomato, spinach, aspen, maple, privet, white pine, ponderosa pine	Tomato, tobacco, 0.05 ppm
Other oxidant gases, e.g., peroxyacetyl nitrate (PAN)	Glazing, silvering, or bronzing of lower surface of leaf	Pinto bean, mustard, oat, tomato, lettuce, petunia, bluegrass	Petunia, lettuce, 0.2 ppm
Unsaturated hydrocarbons, e.g., ethylene	Leaf abscission, dropping of flowers, loss of flower buds, chlorosis, suppression of growth	Orchid blossom, carnation blossom, azalea, tomato, cotton, cucumber, peach	Orchid, 0.005 ppm Tomato, 0.1 ppm

* Certain varieties of these plants are sensitive.

Source: From *Chemical Contamination in the Human Environment* by Morton Lippmann and Richard B. Schlesinger. Copyright © 1979 by Oxford University Press, Inc. Adapted by permission.

beetle. The pine dies of complications of bark beetle infection rather than the oxidant directly—but dead is dead. There are other examples of disease in plants that are aggravated by oxidant injury. Air pollutants may also be responsible for some other subtle effects on plants such as impaired reproduction and germination through increased mutation rates (Marx, 1975).

Some observers regard damage to crops by air pollutants as an enormous problem with worsening implications for the future. To others this is only a remote concern. Most farmers see the damage due to SO_2 as only one of a number of problems with their crops. The effects of things like SO_2 are in fact difficult to distinguish from the effects of insects, nutrient imbalances, and weather. For an extended general discussion of the effects of acid deposition on vegetation, including agricultural crops and forests, see the various EPA criteria documents and staff papers specific for each pollutant. See also the upcoming section on acid rain.

Air Pollution as Fertilizer?

Most fertilizers contain phosphorus, nitrogen, and potassium but no sulfur, even though sulfur is the fourth most important plant nutrient. In the early 1950s it was shown that plants absorb sulfur directly from the atmosphere and incorporate it into the materials that make up the plant. This has led to the suggestion that agricultural crops in the United States may have become dependent on the sulfur in air pollution. It is conceivable that where soils are locally deficient in sulfur, atmospheric inputs may in fact help to fertilize plants. Given the many negative impacts of sulfur oxide air pollution, this is not much of an argument in favor of air pollution. It does perhaps suggest a future use of the sulfur scrubbed from smokestacks.

Plants as Early Warning Systems

Some plants are very sensitive to pollutants (Table 10.6), and it has been suggested that plants can serve as warning systems for chemicals in the air, much as canaries were once used in coal mines. One can easily imagine a system in which the plants that are most sensitive to a particular kind of pollutant are planted in particular locations and checked periodically as indicators of the recent history of air pollution in that area. Although this would provide only after-the-fact information, an advantage of such a biological assay is that it would integrate and cumulate the effect of pollution on living things.

The Effects of Air Pollution on Ecosystems

Because they are large, indistinct, and complex, real ecosystems are difficult to study. Because ecosystems are subject to often unfathomed natural varia-

tions, it is even more difficult to determine the specific effects that particular pollutants have on them. Pollutants never occur alone in the real world—and never under controlled conditions.

Clearly, to affect an ecosystem, a pollutant must have some effect on the structure and function of the ecosystem. Among some of the things that might be watched for are changes in rates of decomposition, in biogeochemical cycles, in species diversity, in community structure, in primary productivity, and in ecosystem productivity—to name just a few of the more obvious examples. The state of our ability to study ecosystem-level effects of pollutants is summarized in a volume edited by Sheehan and colleagues (1984). In this book, specific cases in which the above-named parameters were evaluated are also summarized. The objective of such studies is to be able to identify early indications of ecosystem injury due to particular kinds of perturbants such as acid precipitation. Sadly, it appears that so far we recognize the damage much too late and only after it has become quite severe.

Acid Rain: What Goes Up Must Come Down—Somewhere, Sometime, in Some Form

The earliest published reference to acid rain was made by an English chemist, Robert A. Smith, in 1872, in a book titled *Air and Rain: The Beginnings of a Chemical Climatology* (Cowling, 1982). Modern scientific awareness of acid rain began with agricultural scientist Hans Egner in the late 1940s and limnologist Eville Gorham and meteorologists Carl Rossby, Christian Junge, and Erik Eriksson in the 1950s. A comprehensive picture was developed by Svante Oden, a Swedish soil scientist, beginning in the late 1960s, providing the outline of the acid rain picture we have today.

In the early 1970s it became widely noticed that lakes without any known source of acid (such as coal mine seepage) in Canada, the northeastern United States, and Scandinavian countries were becoming increasingly acidic and that fish were disappearing from them. Acidification of lakes was reported at about the same time in England, Brazil, and Scotland (Reisner, 1977). Acid from the sky seemed to be the only explanation, and monitoring demonstrated that rainfall and even snow did indeed carry notable acidity. The acid rain issue was born.

We appear to be stuck with the term *acid rain* even though the problem is really a more general *acid deposition* problem. Acid or acid-forming materials may be deposited from the air in the form of snow, sleet, fog, or even gases or dry particulate matter—as well as in the form of rain.

Acid rain should have been no surprise. Everything, and there are no exceptions, must go somewhere. For more than a century we have been burning enormous quantities of coal and oil and smelting enormous quantities of ore. Coal (and to a lesser extent oil) and the ores of many metals (e.g., copper, nickel, lead, and zinc) contain sulfur. In the presence of oxygen at high combustion temperatures, sulfur compounds are oxidized to become sulfur oxides (SO_x). Combustion at high temperatures in power plants, smelters, and automobiles also creates oxides of nitrogen, mostly as atmospheric nitrogen combines with oxygen. In Chapter 9 we described how the oxides of both sulfur and nitrogen undergo chemical reactions in air to form some of the most powerful acids known.

Gravity keeps the things in the atmosphere from moving off into space, and this leaves only two possibilities: Either sulfur and nitrogen compounds come down in some form, most probably as acids, or they are all still up there somewhere. Almost all has probably returned to earth as dry deposition or acid precipitation somewhere downwind of where it was formed.

Why Acid Rain Now?

We have already cited a published reference to acid rain dated 1872, but there was acid rain long before there were dinosaurs. As long as there has been carbon dioxide in the atmosphere, some of it has dissolved in rainwater to form carbonic acid (see Chapter 3). Even "pure" rainwater can have a pH of 5.6 or even lower. Over the eons, nature, without any help from humankind, has even made acid rain from the oxides of sulfur vented by volcanoes and from other gases vented into the atmosphere from natural sources. Acid soils have been around since before there were terrestrial plants. There is nothing new about any of these things.

What is new is that human beings have been adding significantly to acid deposition at faster and faster rates in many parts of the world. Forests in North America and Europe may now be receiving as much as 30 times more acidity than they would if precipitation fell from or through clean air (Postel, 1984). In terms of sulfur alone, eastern North American rainfall has been shown to have 2 to 16 times the sulfate content of rainfall downwind of unpopulated areas (Galloway, Likens, and Hawley, 1984). Wet deposition of sulfate in China has been shown to be 7 to 130 times that of remote areas of the southern hemisphere (Galloway et al., 1987).

Apparently, acid deposition has rather recently begun to cause obvious damage to the animals and plants in susceptible waters and soils where rain falls most heavily. Intensification of the acid rain problem from the 1950s through the 1970s in the United States

(Figure 10.7) is probably related to some combination of several factors.

First of all, most of the coal burned by electric utilities has been burned relatively recently. Much oil and gasoline has been burned at high, nitrogen dioxide–generating temperatures in recent decades (see Figure 10.7; see also Chapters 6 and 11). Although total SO_x emissions have not changed much since the turn of the twentieth century, enormous amounts of SO_x were already being emitted in 1900. For more than 100 years, human generation of SO_x has roughly equaled natural emissions. Some perspective is offered by Postel's (1984) observation that the International Nickel Company (Inco) smelter in Sudbury, Ontario, generates more than twice the sulfur that Mount St. Helens did in its biggest sulfur-emitting year. In West Germany, NO_x emissions rose 50% between 1966 and 1978. Since the 1950s, NO_x emissions have roughly doubled in the United States and tripled in Canada (Postel, 1984). A major point here is that the acid rain problem does not correlate strictly with the amount of coal used; it correlates better with the type of use, particularly the methods of combustion, and with the accelerated, high-temperature combustion of oil (Patrick, Benetti, and Halterman, 1981).

A second contributing factor can reasonably be presumed to be the tall-stack strategy adopted in the early 1970s and late 1960s (Figure 10.7). We used to be concerned only about the local human health effects of air pollutants like SO_2. At first we thought that we could deal with SO_2 by using tall smokestacks to lift the pollutants up and over the populations surrounding power plants. We thought that the pollution would be

Tall smokestacks were built to solve air pollution problems by lifting pollutants high into the atmosphere to be diluted. This practice is now known to contribute to air pollution.

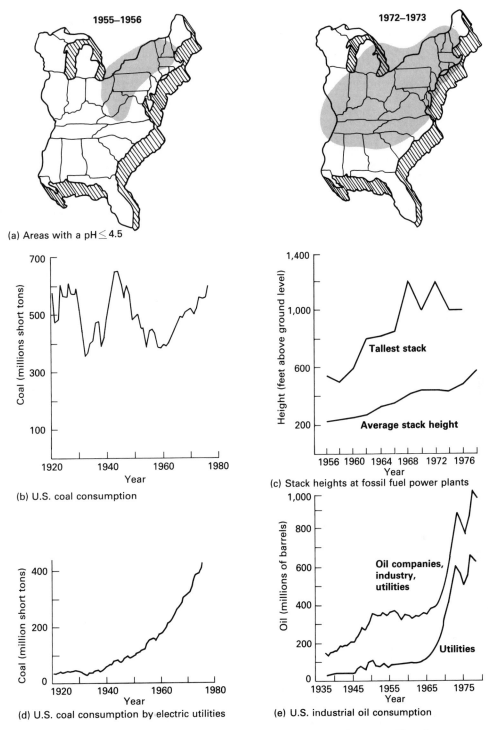

(a) Areas with a pH ≤ 4.5

(b) U.S. coal consumption

(c) Stack heights at fossil fuel power plants

(d) U.S. coal consumption by electric utilities

(e) U.S. industrial oil consumption

Figure 10.7 Trends Related to Acid Rain. (a) Areas in the eastern United States that had rainfall with a pH of 4.5 or lower (1950s and 1970s); (b) overall coal consumption; (c) stack heights at fossil fuel–burning power plants; (d) coal consumption by electric utilities; (e) oil consumption. Conclusion: It is not just coal consumption that causes acid rain.

sufficiently diluted by the time it came back down that no health standard would be violated (see Chapter 11). Only recently have we begun to pay more attention to other impacts such as acid rain.

Finally, the recent overt appearance of damage from acid rain may, in part, be a consequence of an intensifying "conspiracy" of environmental insults. Acidity from air pollution, oxidation by air pollutants, insects, drought, and other factors may simply have ganged up on forests. Nature has a finite capacity to absorb insult, and perhaps limits are being exceeded in the most sensitive of lakes and forests suffering the most intense combinations of insults.

The acid rain problem itself now boils down to a few basic questions: Just how far can the elements of acid rain travel? What exactly are the effects of acid deposition on lakes, soil, crops, forests, animals, and ecosystems? What level of emission control makes sense in light of damage and the cost of emission control? What factors other than the burning of coal contribute to acid deposition? Just what is the best technology to keep the oxides of sulfur and the oxides of nitrogen out of the air?

The Effects of Acid Rain

In the eastern provinces of Canada, acid rain has apparently rendered hundreds if not thousands of lakes fish-free and has had a considerable impact on other forms of aquatic life. In this same area the $24 billion forest industry is jeopardized by acid rain. Crops and other plants have apparently also been affected. In Norway, acid rain is believed to have rendered many southern lakes and rivers nearly fish-free. Sweden is reported to have some 3,000 to 4,000 so-called dead lakes—lakes without fish, frogs, or lily pads, where algae are the only visible form of life. Some 10,000

additional Swedish lakes have been damaged. In recent years the growth of spruce has declined 50% on Camels Hump in Vermont; there has been a 40% decline in the growth of dozens of other species. Perera (1978) links decreased hardwood growth to acid rain in New Hampshire, and acid deposition is believed to be a major factor if not the whole story in widespread forest death in Germany.

And so it goes worldwide. The principal effects of acid deposition appear for the moment to be rather devastating on aquatic and forest communities (see Figure 10.8). A bit more is known about the former than the latter—perhaps because, with forests, acid rain is only part of a more complex air quality problem.

Effects of Acid Deposition on Lakes and Aquatic Communities. There are several specific mechanisms by which acid rain is thought to influence aquatic systems. An obvious mechanism is that involving pH itself. In Chapter 3 and elsewhere we discussed the concept of tolerance limits, and we mentioned pH specifically. Living things have optimal pH levels and pH limits. Departure from near-optimal pH means suboptimal reproduction, growth, and survival. Acid rain can change the pH of lakes directly and through the soil acids it mobilizes as it runs over and through soil on its way to streams and lakes. In aquatic systems, acidification can also cause toxic metals

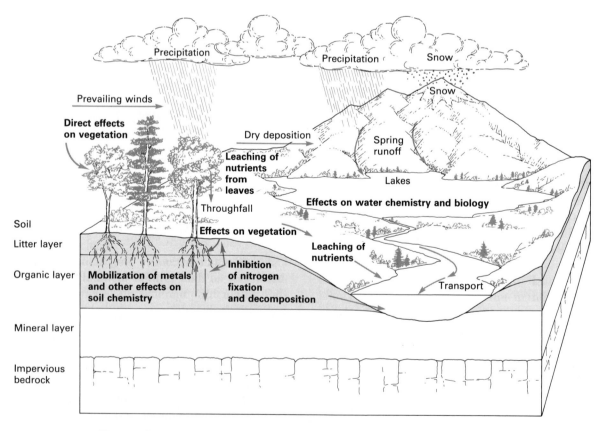

Figure 10.8   Sources and Effects of Acid Deposition.

(e.g., aluminum and mercury) to be leached from sediment—in a lakebed, for example—and from soil as acid water percolates through it. Acidified water may even be able to leach metal from pipes on the way to faucets in houses.

It is difficult at best—impossible at worst—to connect specific causes and effects in a large, complex ecosystem such as a lake. For example, when lakes are found that appear to have suffered from acidification, not enough is usually known about how the lake was before it became acidified. To overcome this problem, Schindler and colleagues (1985) studied a whole, poorly buffered, small lake in Ontario. Acid was deliberately added to the lake, and then the lake was carefully studied as the pH fell from 6.8 to 5.0 over an eight-year period. The observers noted dramatic changes in the lake's food web: elimination of key organisms at pH values as high as 5.8, changes in the phytoplankton species, cessation of fish reproduction (it was projected that the lake would become fishless in about a decade following the point at which the pH reached 5.4), disappearance of bottom-dwelling crustaceans, and the appearance of filamentous algae (such as had been reported in dying lakes in high-acid-deposition areas). They concluded that the changes were caused by changes in acidity and not by secondary effects such as aluminum toxicity.

What kills fish in acidified lakes? Mayer, Multer, and Schreiber (1984) report that at pH 3 or lower, fish tend to die of suffocation as a consequence of excess mucus on their gills. These researchers postulated that fish may also die of complications of heavy-metal poisoning resulting from the mobilization of the metals by the acidity. In either case, fish are particularly sensitive after overwintering or the rigors of spawning. Well before lethal levels of either acidity or metals are reached, species may disappear as a consequence of declining fertility or decreased resistance to stress.

Not all bodies of water are equally sensitive to acid deposition. Acidity is a measure of the effective concentration of hydrogen ions (see Chapter 12). These ions come directly from the dissociation of compounds that have hydrogen ions to donate (e.g., $H_2SO_4 \rightarrow 2H^+ + SO_4^{2-}$). But hydrogen ions can be tied up by various compounds and chemical reactions called *buffering reactions* (Figure 10.9).

If soils or bodies of water have buffers in or around them, they can maintain their pH in the face of continual hydrogen ion loading. Limestone, for example, acts like an antacid tablet. The principal ingredient in limestone is calcium carbonate ($CaCO_3$); in an antacid tablet the ingredient is usually sodium bicarbonate ($NaHCO_3$). In both of these cases the carbonate forms carbonic acid in the presence of hydrogen ions, and this is in turn converted to carbon dioxide and water (see Chapter 3).

There may even be other ways in which the effects of acid rain can be neutralized. It has been suggested, for example, that lakes and streams with a lot of organic material may suffer less from metals because organic material is able to bind the metals, reducing their toxicity (see Patrick, Benetti, and Halterman, 1981).

The susceptibility of a body of water to acidification due to acid deposition, then, depends on five factors:

1. The amount of acid-contributing material that falls into the watershed
2. The path that the material takes on its way to the body of water (over the surface or filtered through soil or bedrock)
3. The ability of the soil to contribute additional acid when flushed by acid-containing precipitation
4. The buffering capacity of the soil
5. The rate at which buffers are released to the soil from underlying bedrock

All of these factors vary tremendously from place to place, even from slope to slope around the same lake. Soils that can buffer acid precipitation can ultimately lose this ability if demand for buffering exceeds the weathering rates by which buffers are made available. Figure 10.10 illustrates the most sensitive areas of North America.

Surges may be a problem even in lakes of relatively low sensitivity. Even if a lake is eventually able to buffer the acidity and neutralize the metals it receives over the course of a year, aquatic life could be damaged severely by the surges in acidity and toxic metals that occur during heavy rains or spring thaws. Melting snows, for example, could bring a cumulative winter's

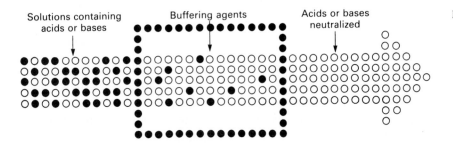

Solutions containing acids or bases Buffering agents Acids or bases neutralized

Figure 10.9 Buffering.

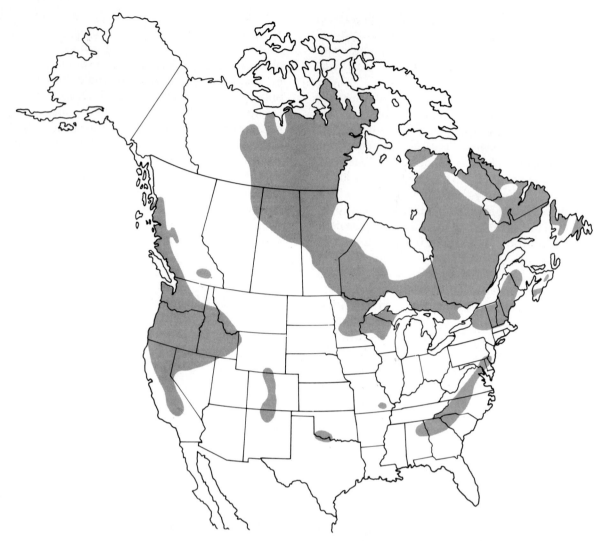

Figure 10.10 Areas of North America with Lakes Particularly Sensitive to Acid Deposition.

deposition to a lake right at the crucial time of egg laying, fertilization, or hatching.

The aquatics subgroup of a joint U.S.-Canadian working group on acid rain concluded that when acid deposition reaches 20–30 kg per hectare per year, moderately sensitive lakes begin to exhibit chemical and biological changes. This group found consistent evidence that effects were negligible below 17 kg per hectare per year and that changes were long-term when deposition of sulfur oxides exceeded 30 kg per hectare per year. Allegedly because of political pressure, this group came up short of recommending that emission controls be targeted to achieve a 20 kg per hectare per year limit (Roberts, 1983). Reducing deposition to the 20-kg limit would require roughly a 50% decrease in emissions, and it now appears that deposition must be limited to something in the range of 9–14 kg per hectare to protect the most sensitive aquatic ecosystems (Schindler, 1988).

In some parts of the world, lime is being put into acidified lakes. Although this strategy works—it has been shown to return aquatic life to some semblance of normal—it has been described as about as effective as distributing a few antacid tablets during an epidemic of chronic hyperacidic indigestion. Liming lakes might work up to a point, but it makes better sense in the long run to minimize the insult.

There is reason to believe that once acid deposition is stopped, even lakes in sensitive areas will heal themselves. Schindler and colleagues (1985) suggest that biological processes contribute more to neutralization of waters than the geochemical processes.

Effects of Acid Deposition on Terrestrial Systems. Despite the fact that acid rain has been linked conclusively to damage to aquatic ecosystems, similar connections for terrestrial systems have yet to be clearly documented. Forests are in trouble around the world, but there is apparently more to it than acid rain.

The extent of what appears to be air quality–related forest destruction worldwide is staggering. In Germany, Austria, Czechoslovakia, Poland, Yugoslavia, Hungary, Romania, the Netherlands, Belgium, France, Switzerland, Sweden, and the United States, some 7 million hectares, or 16.5 million acres, are experiencing forest death. Symptoms have been reported in Russia, Canada, Great Britain, Spain, and Italy as well. Overt damage in central Europe seems to be well ahead of damage in the United States (see Table 10.7). Only on a few mountain peaks in New England does damage rival the widespread destruction in the forests of Germany, Poland, and Czechoslovakia. Forests with damage linked to air pollutants cover an area of central Europe equivalent to half the size of Austria (Postel, 1984).

Forest death, or **Waldsterben** as it is known in Germany, is a frightening phenomenon. Parts of the Black Forest look like scenes of evil conjured up in fairy tales. Conifers and deciduous trees alike stand stark and leafless, dead, some covered with shrouds of lichens and the fruiting bodies of decomposers. The forest floor is littered with bonelike dry, brittle, peeling branch fragments and trees uprooted by the wind. Other trees are alive but deformed; abnormal tufts of needles sprout at the tips of their crooked branches. Whatever is killing the trees is not very discriminating; Norway spruce, Scotch pine, larch, silver fir, and at least some species of beech, birch, ash, alder, maple, and oak are among the dead.

Table 10.7 Estimated Forest Damage in Selected European Countries, 1988 (in millions of hectares)

Country	Total Forest Area	Estimated Damage Area	Estimated Damage Percentage
Czechoslovakia	4.58	3.25	71
Greece	2.03	1.30	64
United Kingdom	2.20	1.41	64
Germany	10.32	5.13	51
Norway*	5.93	2.96	50
Poland	8.65	4.24	49
Bulgaria	3.63	1.56	43
Switzerland	1.19	0.51	43
Finland	20.06	7.82	39
Sweden	23.70	9.24	39
Spain	11.79	3.66	31
Austria	3.75	1.09	29
France	14.44	3.32	23
Total for Europe†	140.96	49.65	35

* Conifer forest only.

† Does not include Turkey or any of the former Soviet Union except Estonia and Lithuania.

Source: Worldwatch Institute data.

That there was a problem with forests in Germany became generally apparent in 1980. A 1982 survey revealed that about 4% of the forests' 600,000 acres were affected. In 1983 in Bavaria, 46% of the spruce showed damage; 60% of the pine area and 78% of the white fir area were also damaged (Plochmann, 1984). By 1985 the cost of the forest death in West Germany was estimated at $1 billion (Hinrichsen, 1987).

Recently, signs of dramatic declines in U.S. forests have been noted. In many parts of the United States, a long-term downward trend in the radial growth of many kinds of trees began in the 1950s. In recent years, however, certain areas of the southern Appalachian Mountains went from exhibiting no damage to 39% of red spruce being 50% defoliated (Holden, 1986).

Although there are differences between what is happening to forests in Europe and in the United States, there are also some striking similarities. In both locations the most severely affected areas tend to be on the windward slopes of mountains often enshrouded in clouds or fog. Also, in both locations the forests appear to have been under stress for some time. Tree ring thickness has been reduced since about 1970 in the eastern United States; in Europe, diameter growth began to fall in some species as early as 1960. It is believed that the damage in both locations is caused by a variety of factors, including these:

1. Ozone and other photochemical oxidants. There has been an upward trend in these pollutants on both continents over the past two decades. Levels of ozone known to damage trees and other vegetation are common in Germany and in many places in the United States. Ozone causes injury directly (as an oxidant), resulting in defoliation and consequently reduced photosynthesis and reduced growth. Ozone may also leach nutrients such as calcium and magnesium from leaves.

2. Sulfur dioxide. According to Hinrichsen (1987), the effects of sulfur dioxide are evident in the complete mortality of vegetation in a zone of some 1,865 square kilometers surrounding the single largest source of sulfur dioxide in the world, the Sudbury, Ontario, Inco smelter.

3. Deposition of nitrogen compounds. These are believed to alter the physiology of trees. One possible effect is that this causes them to continue growing at the expense of physiological preparation for winter.

4. Acid precipitation. This is believed to affect trees by leaching potassium, calcium, and magnesium from leaves and needles and/or by soil acidification (see Ulrich, 1990) and acid mobilization of aluminum and other toxic substances.

5. Heavy metal deposition. Elevated levels of lead, cadmium, copper, and zinc have been found in forest litter and soils throughout the Appalachian Mountain chain. All of these are toxic and may have many kinds of deleterious effects on trees.

6. Organic chemicals. There are many organic chemicals in polluted air that could conceivably interfere with tree hormone systems and in other ways affect growth.

All of these are believed to be operative in the damage to forests seen in Europe and the United States. Ozone is fairly generally believed to be the most important of these on both continents. In their study of the effects of acid rain and ozone on photosynthesis, Reich and Amundson (1985) showed that "any increase in ozone reduced photosynthesis" in all species of tree they tested—and that this was in turn related to reduction in growth or yield. They observed no such effect with acid rain and observed no interaction between simulated acid rain and ozone—there was no effect of acid rain on the effect of ozone, for example. This is not to say that acid deposition has no effect on forests. Among the plausible possible effects of acid rain on forests suggested by studies thus far are (1) acidification of soil, (2) leaching of nutrients from the soil or from leaves, (3) mobilization of toxic metals in the soil, (4) inhibition of nitrogen fixation in soil, and (5) stimulation of growth directly by nitrates, forcing growth late in the year and making trees more susceptible to freezing. An example of the kinds of complications that preclude simple explanations is that nitric acids and nitrates may act as fertilizers in the summer and as poisons in the winter. Complicating the picture still further is that some forest species are much more susceptible than others to pollutants, particularly acid deposition. It will take many years of research to sort out the relative importance of all of the relevant factors involved in *Waldsterben*.

Perhaps the fact that it is now possible to monitor the health of forests from space will help somewhat. There are scanners on Landsat satellites that can acquire from orbit data on light reflected from squares 185 km on a side. Scientists who specialize in such remote sensing hope that such data can be used to identify unique spectral signatures (particular patterns of reflected light frequencies) associated with the response of vegetation to various stress agents—not only permitting the mapping and monitoring of the decline of a forest but also perhaps eventually identifying fingerprints of the effects of particular injurious agents.

Effects of Acid Rain on Wildlife Other than Fish. The effects of acid rain on wildlife other than fish are largely unknown. Some species—frogs and toads, for example—rely on the aquatic environment for part of their life cycle. It is conceivable that acid rain can affect other types of animals with even less direct connections to aquatic environments. Water birds, for example, could possibly ingest harmful levels of metals mobilized from sediments by acidity, but there is no documentation yet. As an indication that there are metals to be mobilized in and around bodies of water, it is noteworthy that bald eagles can become lead-poisoned by eating waterfowl that have swallowed or been wounded by lead shot. A few years ago, a U.S. district court ordered the Fish and Wildlife Service to ban hunting in parts of several states unless the states agreed to require the substitution of steel for lead in shotgun shells.

The forest declines in both North America and Europe have an obvious connection to the wildlife that use the forests for food or shelter.

Acid Rain as an International Issue

The acid in acid rain apparently can come from far away. Airborne sulfate originating in England and elsewhere in northwestern Europe has been linked to increased acidity in the rain coming down in Scandinavia (Ottar, 1977). Some of the acid rain in New England and New York is believed to originate in coal burned along the Ohio River. American oxides of sulfur and nitrogen end up in Canadian acid rain; to a lesser extent, Canadian oxides end up in American acid rain.

Until now the connections between acid rain and its sources have been derived primarily via computer models and by correlations between emission and deposition. Only recently have atmospheric scientists begun to follow tracer substances put into the atmosphere to see just how they travel. Such studies tend to support computer models (see Blair, 1984). But even as we wait for more direct confirmation, transboundary pollution has already raised some delicate political issues throughout the world.

In August 1980 the United States and Canada signed an agreement to work on transboundary pollution, but the United States balked at following through. The Reagan administration was at first reluctant to concede that air pollution was a cause of acid rain and then continued to be reluctant to impose the cost of emission control on industry. Although a joint scientific working group was formed, the group released its report a year late with considerable disagreement as to the degree of emission control needed. Canadian officials accused the American government of stalling and even blatant interference with the work group (Roberts, 1983). Critics of the Canadian position pointed out that Canada has an economic stake in driving up the price of American electricity; that the Sudbury, Ontario, plant is still the largest generator of

SO_2 in the world; and that Canada's SO_2 emissions per capita are greater than those of the United States.

Problem Status and the Need for Action

Acid deposition is acknowledged as a real problem worldwide, though there remains some controversy as to its seriousness. The problem has been worsening everywhere, and it will undoubtedly continue to worsen until something is done. Greater reliance on coal—which is running ahead of the development of clean-coal technologies (see Enrichment Box 10.1)—will likely accelerate aggravation of the problem. The United States has been quibbling for more than a decade over the measures that should be undertaken to combat acid deposition. The issue went untackled for so long partly because it was seen at first as a regional issue. Now, however, even though the problem is most serious in the East (see Figure 10.11), areas sensitive to acid deposition are known to exist in Minnesota, Wisconsin, Michigan, and large parts of the West.

In 1990 a report on a $500,000 study (the National Acid Precipitation Assessment Program) mandated by the U.S. Congress a decade earlier was released (see Roberts, 1991). The report indicated that the problem of power plant and automobile emissions is not the crisis that many have claimed it to be. But the report went on to conclude that controls on acid precipitation would indeed benefit water, fish, trees, and buildings. The director of the study indicated that there appears to be a near consensus, scientifically, that acid rain is a moderate environmental problem. People on both sides of the issue thus found comfort in the report.

One of the most important scientific conclusions of the study was that because the conversion of the oxides of sulfur and nitrogen into acid compounds is limited by the availability of oxidants such as hydrogen peroxide, hydroxy radicals, and ozone, the relationship between emissions of NO_x and SO_x and acid rain is probably nonlinear; that is, a reduction in emissions of 50% would not necessarily translate into a 50% reduction in acid rain. Here are some other findings:

__ Data from 150 monitoring sites in the eastern United States indicate that the severity and geographic extent of acid deposition actually decreased over the decade of the 1980s.

__ The average acidity of precipitation in the eastern United States is pH 4.4, compared to 5.0 in natural rainwater.

__ Cloud water in the eastern United States is 5 to 20 times more acidic than precipitation, and trees above 4,500 feet elevation are bathed in these clouds about a third of the time during the growing season.

__ At least 1,200 (4.2%) of U.S. lakes are acidic; the sulfate ion, used as an indicator of human-made and most likely atmospheric sources, was responsible for most of the acidity in the acidic lakes of New England and 73% of those in the upper Midwest. The report went on to say that current deposition levels are such that most lakes highly susceptible to acidification have already been acidified and that few additional lakes would become acidic as a result of acid deposition even over the next 50 to 100 years.

__ Most U.S. forests appear to be unaffected by acid rain; there may, however, be a problem with high-elevation red spruce in certain parts of the East. It may be that cloud water acidity and ozone may conspire to cause red spruce to be less able to deal with winter injury.

Enrichment Box 10.1

Clean Coal Technology

The Clean Coal Technology Program is a cost-sharing program sponsored by the U.S. Department of Energy. As of the early 1990s, some three dozen projects designed to find improved methods of combating pollution from coal were being supported. One such project is a pressurized fluid bed combustion combined-cycle demonstration project at the Ohio Power Company's Tidd plant near Brilliant on the Ohio River. The technology being developed here is expected to reduce sulfur oxide emissions by 90% and nitrogen oxide emissions by half or more, at the same time operating at much improved thermal efficiencies. Such technology will enable utilities to conform to the Clean Air Act even when using high-sulfur coal and at the same time reduce the generation of carbon dioxide per unit of useful energy—and thus also help to solve the global warming problem. Another clean-coal project, this one under development by Encoal Corporation, a Shell Mining Company subsidiary near Gillette, Wyoming, will convert subbituminous coal into a higher-grade solid fuel and at the same time produce a low-sulfur liquid fuel. Major reserves of the starting material for this process are available in Wyoming, Alaska, North Dakota, Montana, and Texas.

By some estimates, clean-coal technologies could cut acid emissions by half or more in the United States. Energy conservation has an even greater potential for reducing acid rain.

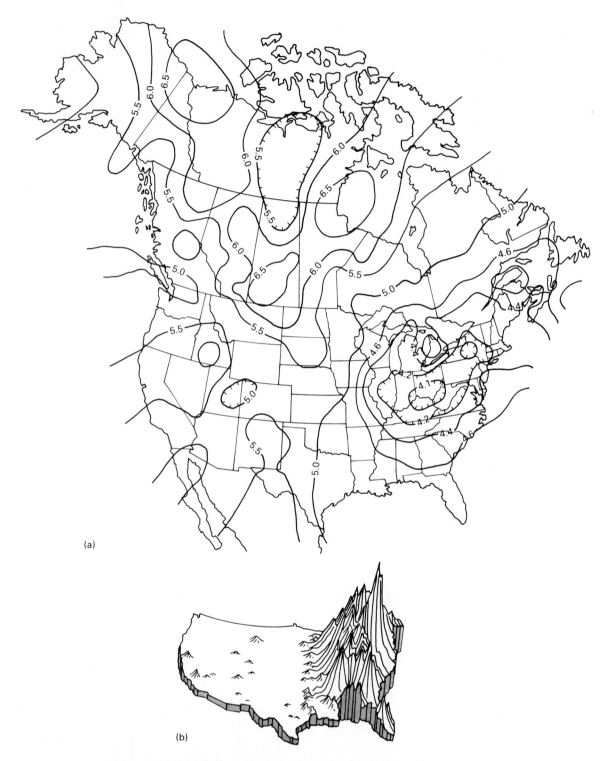

(a)

(b)

Figure 10.11 Where Acid Rain Falls and Will Fall in the United States and Canada. (a) Average annual pH of rainfall in 1980; (b) particulate sulfate pollution predicted for the United States in 1990.

___ The oxides of neither sulfur nor nitrogen are apparently responsible for forest decline or for crop damage; *ozone* was found to be a significant problem with respect to crop damage.

___ Acid deposition is responsible for the deterioration of building materials, statues, and paint; acid deposition is also responsible for a deterioration of visibility (adding difficulty to see through it) in many places.

The U.S. Clean Air Act of 1990 mandates that coal-fired power plants reduce sulfur oxide emissions by one-half and also calls for sharp reductions in the emission of nitrogen oxides. Many other countries have passed laws requiring reduced emission of these starting materials for acid deposition. More controls are needed in countries that have only begun to abate sulfur and nitrogen oxide emissions, and many other things need to be done (see Enrichment Box 10.1). We should obviously stop wasting energy; and we should adopt recycling as a way of life. According to Postel (1984), one-third of Canada's copper is made from scrap rather than ore, and this alone keeps about 1 million tons of sulfur dioxide out of the air each year. Each ton of paper made from waste paper reduces energy use by one-third to one-half and reduces air pollution by as much as 95%. Producing aluminum from scrap rather than ore cuts nitrogen oxide pollution by 95% and sulfur dioxide by 99%. However we go about it, we must continue on the strategic course of reducing sulfur oxide and nitrogen oxide emissions even as we work to determine the limits of harmfulness.

The acid rain issue is a classic case of ecology at odds with the private-market, financial-analysis view of economics (see Chapters 11 and 19; see also Hitzhusen and Hemphill, 1984). As with other similar issues, the control costs are far more easily figured than the cost of no control. Achieving the degree of control that will bring us close to protecting all but the most sensitive aquatic systems will cost as much as $20 billion per year in the United States alone. No one can say for sure just how much the damage related to acid deposition is worth. No one knows for sure how much momentum the damaging effects have. No one knows how much of the damage is relatively irreversible.

Should we pay billions of dollars every year in control costs to save the fish in thousands of lakes and streams—and maybe crops and trees? Should we protect all of the lakes and streams, or should we set less stringent, less costly standards and save all but the most sensitive? What about related consequences of sulfur and nitrogen oxides other than acid rain? Where do they fit in?

Politicians have a very difficult time dealing with issues and questions such as these because action on

pollution issues might well carry them too far from the immediate concerns of their constituents. Also, the most appropriate actions can be translated into large costs to special-interest groups; the translation into general welfare is much less clear-cut.

The Effects of Air Pollution on Materials

The Bavarian Monuments Protection Office in Germany has documented an alarming deterioration rate of buildings and monuments made of stone and plaster and even of painted glass works dating from the Middle Ages. Particularly striking examples are the statues of the Augsburg Cathedral. These badly damaged statues are literally covered by a black coating that contains large amounts of gypsum (a product of the reaction between limestone and sulfuric acid). Heating systems and power-generating stations have been identified as the source of the sulfur dioxide that is believed to cause the problem.

Pollutants affect health because they are reactive—they can and do react with almost every living and nonliving thing. The oxides of sulfur can react with aluminum to form aluminum sulfates and thus corrode aluminum surfaces. Dilute acid solutions formed from a variety of air pollutants can etch limestone and even marble. In its reaction with limestone, sulfuric acid (as well as other sulfur compounds) causes gypsum ($CaSO_4 \cdot 2H_2O$) to be formed. When this occurs on surfaces, coatings such as those just described are formed; when gypsum forms in cracks, the expansion

Pollutants react with almost every material. On this statue of Thomas Jefferson at Columbia University in New York, the dark coating was formed by atmospheric pollution reacting with the stone of the statue.

Table 10.8   Effects of Major Air Pollutants on Materials

Chemical	Primary Materials Attacked	Typical Damage
Carbon dioxide	Building stones, e.g., limestone	Deterioration
Sulfur oxides	Metals	
	Ferrous metals	Corrosion
	Copper	Corrosion to copper sulfate (green)
	Aluminum	Corrosion to aluminum sulfate (white)
	Building materials (limestone, marble, slate, mortar)	Leaching, weakening
	Leather	Embrittlement, disintegration
	Paper	Embrittlement
	Textiles (natural and synthetic fabrics)	Reduced tensile strength, deterioration
Hydrogen sulfide	Metals (silver, copper)	Tarnish
	Paint	Leaded paint blackened due to formation of lead sulfide
Ozone	Rubber and elastomers	Cracking, weakening
	Textiles (natural and synthetic fabrics)	Weakening
	Dyes	Fading
Nitrogen oxides	Dyes	Fading
Hydrogen fluoride	Glass	Etch marks, opacity
Solid particulates (soot, tars)	Building materials	Soiling
	Painted surfaces	Soiling
	Textiles	Soiling

Source: From *Chemical Contamination in the Human Environment* by Morton Lippmann and Richard B. Schlesinger. Copyright © 1979 by Oxford University Press, Inc. Adapted by permission.

of gypsum crystals causes the limestone to crumble away.

Hydrogen sulfide can combine with silver to form silver sulfide and tarnish silverware. Ozone can cause the oxidation of fabrics directly, causing them to age faster. A summary of the material effects of air pollutants is given in Table 10.8. Economic impacts will be considered in Chapter 11.

The Effects of Air Pollution on Climate: The Greenhouse Effect

We mentioned the role of carbon dioxide in the so-called greenhouse effect in Chapters 6 and 9. We will reiterate a few points and expand our consideration here.

Carbon dioxide and other constituents of the earth's atmosphere are transparent to visible light but are relatively opaque to long-wave, infrared radiation (see Chapter 2). This is the basis of the greenhouse effect. Greenhouses, even unheated ones, are much warmer inside than the air outside. The reason for this is that glass will transmit light but is relatively opaque to infrared radiation (Figure 10.12). Everyone is familiar with the heating of automobiles by the sun. The interior of an automobile standing for a time in the sun will be much warmer, summer or winter, than the air outside, even if the engine and the heater have been off for hours.

So it is with the earth and its atmosphere. The temperature of the earth's surface is determined, in part at least, by the presence in the atmosphere of infrared-absorbing molecules like water, carbon dioxide, and ozone. There are also a number of trace gases in the atmosphere such as methane, ammonia, sulfur dioxide, and certain chlorinated hydrocarbons that also have an effect on the overall heat-retaining properties of the earth's atmosphere. The combination of water vapor, ozone, carbon dioxide, and trace gases behaves like glass in a greenhouse. The heat these materials absorb and hold accounts for the earth's surface temperature.

In the past century, human activity has added about 360 billion tons of carbon dioxide to the atmosphere, increasing the overall concentration of this gas in the atmosphere about 25%. The difference between what we have generated and what has actually been added to the atmosphere is that amount of CO_2 that has dissolved in the oceans or has been absorbed by plants and animals in biomass. Theoretically, this atmospheric increase in carbon dioxide should have caused an average rise in the earth's temperature of about 1°F; this is not far from what may have actually occurred. We are still venting billions of tons of carbon dioxide into the atmosphere each year (Figure 10.13). Measurements show that the average carbon dioxide content of the atmosphere has risen more than 9% since 1958. The current rate of increase is 1 ppm per year, equivalent to 2.3×10^{15} grams of carbon. If this rate continues, there will be some 660 ppm of CO_2 in the earth's atmosphere by the middle of the twenty-first century. This is twice as much as there was in 1900 (Woodwell et al., 1983).

Not all of the increase in atmospheric CO_2 in recent years has come from the burning of coal and oil. Stuiver (1978) claims that from 1850 to 1950, land use practices reduced terrestrial biomass some 7% and that this has contributed to increased atmospheric CO_2. The earth's plants do, after all, exchange 100 billion metric tons of carbon as CO_2 each year; any net increase in biomass would mean a net consumption of CO_2, and any net decrease would mean a net release.

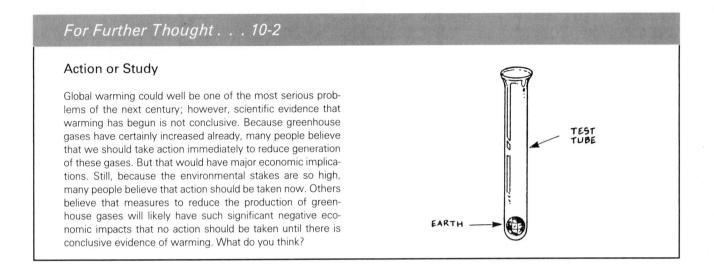

Action or Study

Global warming could well be one of the most serious problems of the next century; however, scientific evidence that warming has begun is not conclusive. Because greenhouse gases have certainly increased already, many people believe that we should take action immediately to reduce generation of these gases. But that would have major economic implications. Still, because the environmental stakes are so high, many people believe that action should be taken now. Others believe that measures to reduce the production of greenhouse gases will likely have such significant negative economic impacts that no action should be taken until there is conclusive evidence of warming. What do you think?

Superimposed on the gradual increase in carbon dioxide levels in the earth's atmosphere (Figure 10.13) is a breathinglike pattern generated by the seasonal uptake of carbon dioxide through growing, followed by the demise of vegetation. Levels of carbon dioxide are highest at the end of winter and lowest at the peak accumulation of biomass at the end of the growing season. This suggests that the amount and types of vegetation that cover the earth have some effect on atmospheric carbon dioxide levels and that such

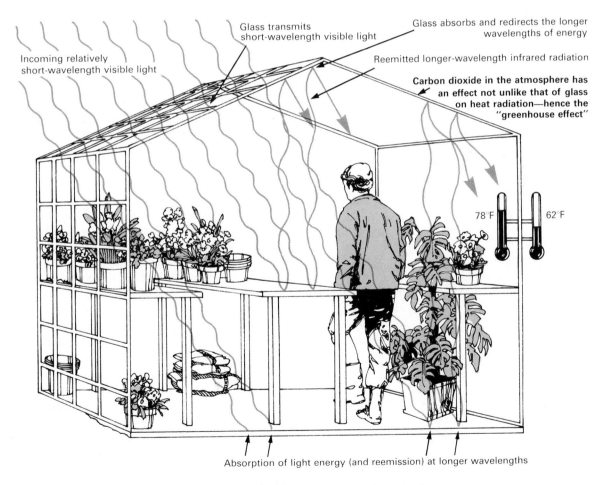

Figure 10.12 The Greenhouse Effect—in a Greenhouse. Constituents in the earth's atmosphere—mainly carbon dioxide—do for the earth's atmosphere and surface what the glass does for a greenhouse.

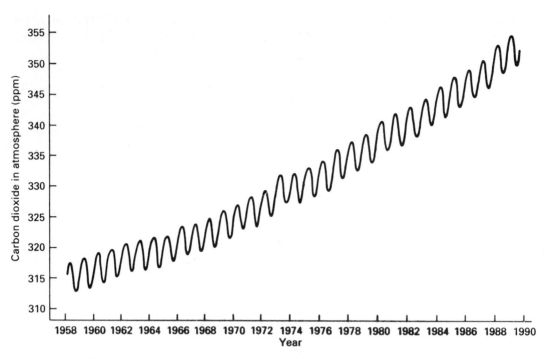

Figure 10.13 Trends in Atmospheric Carbon Dioxide. These trends have been measured since 1958 at the Mauna Loa Observatory on the island of Hawaii by Charles D. Keeling of the Scripps Institute of Oceanography. The curve shows the monthly average concentration of carbon dioxide and seasonal oscillations caused by removal of carbon dioxide by photosynthesis during the growing season in the northern hemisphere and the subsequent release during fall and winter.

things as deforestation and development would tend to cause a shift upward in atmospheric carbon dioxide levels (see Houghton, 1987).

Whatever the relative importance of the sources of increased atmospheric carbon dioxide, atmospheric CO_2 levels are increasing, and the levels of other "greenhouse gases" are increasing as well. A World Meteorological Organization panel of experts indicated that trace gases such as methane, chlorofluorocarbons, nitrous oxide, and ozone may have an impact on global warming equal to that of carbon dioxide (although the trace gases are less abundant, they are more efficient in trapping heat). Atmospheric methane has been increasing about 1% per year since 1950, and based on an analysis of bubbles trapped in the Greenland ice sheet, methane levels have apparently doubled in the past few hundred years after many thousands of years of stability. The coincidence of the rise occurring along with the rise in human population suggests human involvement, probably by means of rice paddies, the clearing of forests, and the abolition of methane sinks caused by pollutants. Ozone and nitrogen oxide trends were described earlier; the chlorofluorocarbons will be discussed in another context shortly.

Though the rise of greenhouse gas levels seems unarguable, it is not yet clear what the climatic signif-

icance of this has been thus far and what the significance is likely to be in the future. Has the greenhouse gas increase already caused a significant increase in the earth's temperature? What are the likely consequences of any such increases? Will the earth's temperature continue to increase if the levels of these gases in the atmosphere continue to increase?

It will not be easy to answer these questions. For one thing, we seem to be putting antagonists into the air simultaneously. Whereas CO_2 tends to hold heat in and to increase the temperature of the earth, *particulate matter* tends to reflect solar energy and bring about a cooling of the earth. For another thing, warming tends to increase cloud cover, and cloud cover tends to cause cooling by reflecting incoming energy. Another part of the problem is that fairly dramatic fluctuations in the earth's atmospheric chemistry and average temperature occur naturally, and it is not really clear how much of a relative effect humans have had or will have on climate. In 1883, to cite one classic example, a volcano 700 miles west of Bali spewed enormous quantities of dust and debris all the way into the stratosphere (Figure 9.1), where it remained for several years. For several annual cycles thereafter, the earth was cooler because much of the normal solar flux was blocked out. Recorded history also indicates that in the seventeenth century, volcanic eruptions caused a series of little ice ages in which glaciers advanced in some places almost as far as they did in the big ice ages. Even if it were clear that the earth's average temperature has increased over the past 100 years or so, some scientists argue that this could not be linked with any certainty to the increase in greenhouse gases.

But it is not at all clear that the earth is warming. Some experts say it is (see Houghton and Woodwell, 1989; Jones and Wigley, 1990), and others say it may be (see Etkins and Epstein, 1982), but there is simply too much fluctuation in temperatures from place to place worldwide to be sure (see Lindzen, 1990). Most actual temperature measurements over time on land and in oceans, measurements of ice cover, and other things that could indicate a temperature trend fail to reveal a clear signal that change is taking place. There are, however, some indications that warming may be underway.

A recent report (Schindler et al., 1990) covering two decades of climatic, hydrologic, and ecological records for an area of lakes in northwestern Ontario showed that the ice-free season had increased by some three weeks and that the average air and lake temperatures have increased by 2°C.

Anomalies in the upper 100 meters of cold Alaskan permafrost measured by Lachenbruch and Marshall (1986) suggested to them that the arctic surface has warmed from 2°C to 5°C in the preceding 100 years—amounting to still more circumstantial evidence of the greenhouse effect. This observation is particularly intriguing because most models indicate that greenhouse-induced warming will be greatest in the Arctic.

Atmospheric scientists use modeling techniques (see Enrichment Box 10.2) to try to predict the relationships between changes in atmospheric composition and planetary temperature. As pointed out by Schneider (1987), the processes that make up a planetary climate are too large and too complicated to be reproduced physically in any laboratory experiments, but they can be simulated with the help of computers. In order to construct a mathematical model of climate, certain assumptions must be made, and these typically draw heavily on historical trends. Unfortunately, although we do have fairly good information going back over millions of years about *gross* conditions and trends, information about more minor trends and transitions is sketchy at best. Despite the fact that mathematical models generally fall far short of being able to predict reality, they are useful in predicting some of the plausible, and even the most likely, consequences of reasonable assumptions. Hence they are clearly better than extrapolations made from a few isolated observations. We use short-term models all the time to predict the weather, and we all know from experience that

these may be useful, but they are not always accurate. Imagine, then, the difficulty of predicting on the scale of millions of years when such variables as motions of the earth's crust, variations in the earth's orbit, transfer of heat from the surface of the oceans to deep waters, and the extent of the polar ice sheets cannot simply be assumed to be constant (as they are when meteorologists generate five-day forecasts).

Models are only as good as the assumptions on which they are based. The problem is that there is usually considerable uncertainty about every key element in a model. In the case of global warming, for example, there is even some disagreement as to how fast the carbon dioxide content of the atmosphere will increase as we continue to burn fossil fuel. Had the rates of increase in fossil fuel combustion from the mid-1950s until 1973 held at 4.3% per year, the atmospheric concentration of CO_2, according to one model, would have doubled in less than 60 years and was estimated to cause an eventual rise in global mean temperature of 3.0°C (more than 5°F). If recent, less dramatic trends hold, however, the doubling and the warming will take longer. The current carbon dioxide concentration in the atmosphere is 350 ppm; methane is 1.7 ppm, and according to ice core analysis, concentrations have not been this high any time during the past 160,000 years (La Brecque, 1989). Two facts that are disturbing in this connection are that (1) there does appear to be a correlation between the atmospheric concentrations of greenhouse gases (carbon dioxide and methane) and the recurrence of ice ages as indicated by gas analysis of ice cores going back nearly 200,000 years, and (2) even though it is not yet clear that temperature is rising, geologic history indicates that temperature changes can occur very abruptly (La Brecque, 1989).

In mid-October 1983, both the National Academy of Sciences (NAS) and the EPA released reports on CO_2-induced global warming. The EPA report predicted temperature increases of 4°F by the year 2040 and 9°F by 2100 (27°F in polar regions), with a 1- to 2-foot rise in sea level as early as 2025. This was projected to cause inundation of low places in Charleston, South Carolina; Galveston, Texas; and other coastal cities and to alter rainfall patterns in the Midwest. The NAS report reached basically the same conclusion, predicting a doubling of atmospheric CO_2 sometime late in the twenty-first century with a 7°F rise in global temperatures. It is noteworthy that the waxing and waning of the ice ages of the past involved temperature changes of only 11–12°F (Kerr, 1983).

It does not take much of a change in average temperature to cause major changes in the physical world. Mathews (1987) claims that even a 3°F rise will put global climate beyond the range of all human experience.

Modeling

Before there were computers and supercomputers, the only way to study complex phenomena like weather and climate was to use some form of a direct approach. Theories had to be tested directly in the real world or in some semblance of the real world isolated in a laboratory. Often, theories had to be modified following each experiment and then retested over and over. This approach often ran up against cost limitations and was typically very slow to get answers. Computers have made it possible to build a kind of loop into the scientific process that takes much of the drudgery out of science. Basically, high-speed computers have made it possible to use mathematical models based on a theory and then to run computations based on the models. Scientists can run such models over and over, allowing them to plug in different variables and values for variables to assess the effect of the changes on the outcome. Refined in this way, theories can then be given more nearly final checks in the real world.

Supercomputers are called supercomputers because they can handle trillions of numbers and solve intricate mathematical equations rapidly. A model of Los Angeles smog, for example, might have to deal with the interactions among 50 different chemicals at more than 10,000 points in the Los Angeles basin, handling a half million or more simultaneous equations. The technology of supercomputing is now striving for the ability to perform a trillion mathematical operations in a single second. We expect this barrier to be surpassed before the end of the century. Supercomputer technology has come none too soon. We have long been able to accumulate data much faster than we have been able to analyze it, and this gap has been widening. An earth observation system to be launched in 1998 will be sending back 1 trillion bytes (binary "words") of information per week. For comparison, the Library of Congress (some 15 million books) contains about 10 trillion bytes. Supercomputers will have to be made even much faster than they are now in order to handle and analyze this magnitude of data flow.

Another significant advance in the computer world is the development of high-speed networks that enable computers to be connected to one another more effectively. This will enable researchers in different locations and in different disciplines to work more effectively together.

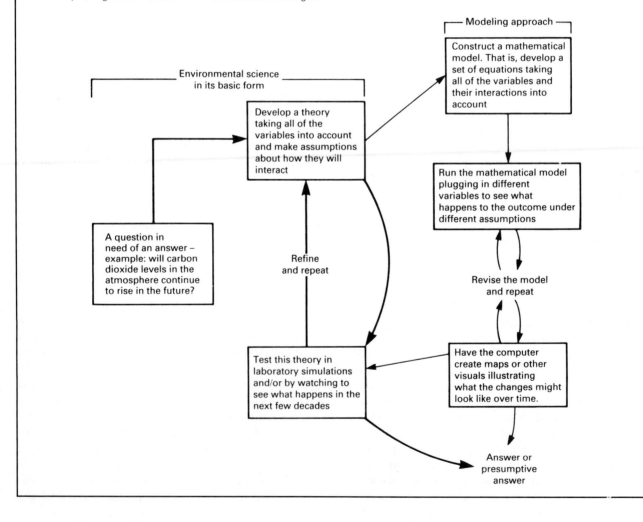

Scientists disagree on the seriousness of the buildup of CO_2. Some see the climate changes that may be induced by the buildup of carbon dioxide and other greenhouse gases (and any temperature changes this may induce) as rather minor perturbations superimposed on much more profound natural waxing and waning of ice ages and warm periods. It is not even possible to attribute the average 1°F rise in the earth's temperature over the past 100 years to carbon dioxide, they say. Variations in the earth's orbit and orientation (called the Milankovitch variations because M. Milankovitch suggested them in the 1920s and 1930s) are believed to have controlled the timing of 100,000-year ice ages during the past million years.

As for the danger of further buildup, some scientists have postulated that even if warming does start, it will quickly beget more clouds than normal, and this will tend to make the earth cooler—the combined system acting as a kind of planetary thermostat. Idso (1984) suggests that even if CO_2 levels build, and even if this has a negative effect on farming in some areas, such effects will be offset by increased photosynthesis (CO_2 uptake), since CO_2 is a limiting factor known to increase photosynthesis. However, others point out that even though increasing CO_2 levels does increase production in experimental systems, there is no direct support for the notion that this applies on a global scale, where most plant communities are already limited by the scarcity of water and nutrients. Kramer (1981) points out that we particularly need to know how forests will react to increased CO_2 because they carry out about two-thirds of global photosynthesis.

What should we do about this problem? There are some very definite uncertainties about the rates of global warming as a function of the amounts of greenhouse gases in the atmosphere. There are also uncertainties about the tightness of other linkages between particular causes and particular climatic effects. Still, it seems prudent to act now. Scientific consensus is that major climatic changes are a *highly probable* consequence of the buildup of carbon dioxide and other gases in the atmosphere. Therefore, it is highly probable that there would be major rearrangements and disruptions in the ecosphere as some places become hotter and dryer and other places become warmer and wetter. Carbon dioxide comes from fossil fuel combustion and, to some extent, from tropical deforestation. The reduction of fossil fuel combustion would be a good idea even if there were no such thing as the greenhouse effect (see Chapter 6). Reduced fuel consumption for whatever reason would reduce emissions of carbon dioxide and other greenhouse gases such as ozone and oxides of nitrogen. Likewise, there are many other good reasons why we should slow tropical deforestation (see Chapter 14). For reasons that will become obvious, there has already been an

international agreement to reduce worldwide emissions of chlorofluorocarbons. It seems logical that we should extend these strategies on an international basis—perhaps, as suggested by the head of Canada's Atmospheric Environment Service, via a "global law of the atmosphere"—even as research continues. The United Nations in 1988 set up the Intergovernmental Panel on Climate Change (IPCC). This panel was charged with assessing the effects of climatic change and evaluating strategies to deal with global warming. In 1990 the United States proposed a system of greenhouse gas emission credits that would allow each country to release only a specified amount of gas—and the credits could be bought and sold. This and other action steps must be further considered and adopted—even before it becomes crystal clear that the earth is heating up. At the Earth Summit held in Brazil in June of 1992 (see Chapter 19), a world treaty on climate change was signed in which the nations of the world pledged to work toward reduction of "greenhouse" gas emissions. Terms of the treaty which had been negotiated prior to the summit contained no binding goals or timetables for reducing carbon dioxide emissions; each nation is to generate its own action plan. A number of nations have already pledged to reduce carbon dioxide emissions to 1990 levels by the year 2000.

According to an EPA report (Lashof and Tirpak, 1989), the increase in greenhouse gases can indeed be slowed down. The report suggested that sharp cutbacks in the use of fossil fuels coupled with a significant increase in solar power, nuclear power, and biomass energy over the next century, plus planting new forests worldwide, cutting back sharply on the production of chlorofluorocarbons, installing gas capture systems in landfills worldwide, developing new ways of producing rice and meat, could lead greenhouse gas buildup to level off in the twenty-second century and in the process reduce global warming to less than half of what it would be otherwise (see Marshall, 1989).

Effects of Air Pollution on the Ozone Shield

As suggested by much of our discussion about ozone (Chapters 5 and 9 and earlier in this chapter), we have every reason to be ambivalent about this photochemical oxidant. When it is present at the surface of the earth, it amounts to a problem, yet its presence in the lower stratosphere (see Figure 9.1) is essential for health and life. This amounts to a kind of irony in that one of the most important air pollution problems is *increased* ozone, while another important air quality problem—the one we will consider here—is *decreased* ozone.

Ozone in the stratosphere serves as a shield against harmful solar radiation. It absorbs, and thus effectively filters out, almost all of the high-energy, ultraviolet (UV) radiation from the sun. Ultraviolet electromagnetic radiation (see Figure 2.3) has enough energy to change the chemicals that absorb it. Among the chemicals that will absorb UV radiation are ozone, proteins, and nucleic acids such as DNA (see Chapter 5). If ozone does not absorb UV radiation high in the atmosphere, this radiation reaches the surface of the earth, where it changes the chemistry of living things. Particularly problematic is the fact that UV radiation is mutagenic; some of the changes that UV radiation can cause in DNA result in cancer.

There is concern now about the ozone in the stratosphere for two very good reasons. First, there is not much ozone up there; if all of it were compressed into a pure ozone layer, it would be only a few sheets of paper thick. Second, there is evidence that stratospheric ozone is decreasing and that certain human activities are responsible. By some estimates, a 1% drop in stratospheric ozone could increase by 3% the amount of ultraviolet radiation reaching the surface of the earth. This in turn could cause a 2% to 5% increase in the incidence of skin cancer (see Cohn, 1987). There is also reason for concern that increased surface ultraviolet radiation could have adverse effects on crops and marine plankton among other things (see Smith and colleagues, 1992).

When we speak of stratospheric ozone levels, we speak in terms of *average* levels. Ozone levels are not static in the stratosphere. There are seasonal and other periods to the photochemical production and destruction of stratospheric ozone. Production, for example, parallels variations in solar radiation. Ozone and ozone-scavenging chemicals are moved into and out of the stratosphere with normal variations in atmospheric mixing. We have good reason to believe that in recent years there have been abnormal declines in average stratospheric ozone levels. We also have good reason to believe that this is linked mostly to chlorofluorocarbons.

Any chemical or process that can cause the disappearance of ozone and can reach the stratosphere is a serious threat. Nuclear war, because of the things it could put into the stratosphere, is one such threat to the ozone shield. Likewise, supersonic air transport has at times caused concern, although the oxides of nitrogen (see Chapter 9) that enter the atmosphere through air traffic are now generally thought to be less serious than once feared. Today the biggest part of the problem is chlorofluorocarbons.

Chlorofluorocarbons (CFCs), or freons, are a class of chemical compounds made up of carbon, chlorine, and fluorine. They were discovered in the 1930s. Because of the low boiling points of the CFCs, and because they are nonreactive, nontoxic, noncaustic, noncorrosive, and nonflammable, they are in many ways ideal for use as aerosol can propellants, as solvents (cleaning agents) in electronics manufacturing, as plastic foam–blowing agents, and particularly as refrigerants. For most of the time since the 1930s, most of the CFCs manufactured were used as refrigerants in the United States. Unfortunately, CFCs are now implicated in the destruction of the earth's ozone shield.

The nonreactivity of CFCs allows them eventually to reach the stratosphere, where they interact with ultraviolet radiation as follows:

$$CF_2Cl_2 + UV \longrightarrow Cl + CF_2Cl$$

The Cl then reacts with ozone to form chlorine monoxide and oxygen:

$$Cl + O_3 \longrightarrow ClO + O_2$$

Chlorine monoxide reacts with atomic oxygen in the stratosphere and generates another molecule of oxygen:

$$ClO + O \longrightarrow Cl + O_2$$

The Cl then starts the whole sequence over again, and the net effect is the conversion of ozone into molecular oxygen. The process continues with the chlorine sometimes bound up in other reservoir molecules temporarily and continues until one of the water-soluble reservoir molecules, such as hydrogen chloride, drifts low enough to be washed out in rain. Meanwhile, each chlorine atom has the potential to eliminate tens of thousands of ozone molecules. The problem may be especially urgent because of the potential for a very long lag effect. Most of the CFCs ever manufactured and subsequently released into the atmosphere have been released since 1975. Even if CFC production stopped today, the removal of ozone would continue for "at least another century" (Rowland, 1987) because some chlorine compounds persist for decades.

Two satellites (*Nimbus 4* and *Nimbus 7*) collected data from 1970 through 1979 showing that at an altitude of 25 miles (40 km), ozone was decreasing at the rate of 0.5% per year. NASA satellite data showed that ozone in the stratosphere declined 3% overall from 1978 to 1984 (Cohn, 1987).

In a report released in March 1982, the National Research Council estimated that there would be a 5% to 9% reduction in stratospheric ozone late in the twenty-first century if CFC production continued at its current pace.

Solomon and colleagues (1984) measured levels of chlorine monoxide in the stratosphere (about 30 km) and determined its daily variation. They concluded that the chlorine already in the stratosphere will result in an ultimate decrease of 3% to 5% in stratospheric ozone.

A National Academy of Sciences report released in 1984 indicated that stratospheric ozone was unlikely to fall more than 4%. This prediction was considerably less than the 18.6% decline predicted in a 1979 report. The difference is attributed to better understanding of what goes on in the atmosphere and not to a decline in the use of chlorofluorocarbons. A 4% reduction in stratospheric ozone will, other things staying the same, mean a significant increase in skin cancer rates. In a 1987 EPA study it was estimated that if CFCs increase by 5% per year, 40 million Americans will get skin cancer over the next fifty-five years, and 800,000 will die of it.

Ozone Holes and Beyond. In 1985 a group of British atmospheric scientists reported that the springtime amounts of ozone in the air over Antarctica had dropped 40% between 1977 and 1984 (Stolarski, 1988). Other studies ultimately confirmed that a region of the antarctic atmosphere about the size of the United States was apparently rapidly losing its ozone. There was in effect an **ozone hole** over Antarctica.

Why a hole over the Antarctic? So far, it looks like the problem may be due to a combination of meteorology and CFCs. Chlorine monoxide levels have been found to be elevated in the ozone hole in the spring, compared to levels at middle latitudes. A good possible explanation for this is that antarctic weather somehow causes the release of chlorine from protective reservoir molecules in the spring (Molina et al., 1987; Stolarski, 1988; Toon and Turco, 1991). In any case, since the first report of an ozone hole over Antarctica the magnitude of the ozone-depletion threat has run ahead of world efforts to deal with the problem—as new, more alarming data has been collected.

In 1988, an international Ozone Trends panel analyzed 17 years of ground-based ozone measurements and concluded that the northern mid-latitudes were losing ozone at a rate of 1–3% per decade. In early 1991, NASA's satellite-borne Total Ozone Mapping Spectrometer indicated that ozone losses were on the order of 4–5% per decade. Not only have ozone-depleting, man-made chemicals continued to build up, but the 1991 eruption of Mount Pinatubo in the Philippines has injected natural ozone-depleting chemicals into the upper atmosphere. Apparently sulfuric acid aerosols from Mount Pinatubo (and other sources) react with nitrogen oxide and keep it from reacting with chlorine and thus keep it from blocking the effect of chlorine on ozone.

In February of 1992, the U.S. National Aeronautics and Space Administration reported the discovery of "alarming" levels of ozone-destroying chemicals over Canada, the United States, and Europe. Using NASA's Upper Atmosphere Research Satellite, a team of researchers reported finding antarctic-like levels of chlorine monoxide in January over Scandinavia and northern Eurasia (including the cities of London, Moscow, and Amsterdam). A second study, using NASA aircraft in flights over Canada and northern New England, found that atmospheric levels of chlorine monoxide and bromine monoxide were the highest ever measured (even in flights into the antarctic ozone hole) in the upper atmosphere.

Scientists are fairly solidly united in the position that the chemicals responsible for destruction of the ozone layer should be phased out as quickly as possible. This will not be easy, nor would it be immediately effective (see below). The market value of CFCs in the United States alone is more than $750 million per year, and the estimated value of goods and services tied to CFCs is about $27 billion (Cohn, 1987). The best approach to the issue is obviously international, but this raises the additional issue of the differential stages of development of the nations around the world. Countries just beginning to acquire refrigeration equipment are obviously loath to stop making the cheapest, most effective refrigerants available, CFCs.

In the late 1970s the United Nations Environment Program called on governments worldwide to reduce the use and production of CFCs. In 1981 the UN began negotiations that dragged on until 1985, when the Convention for the Protection of the Ozone Layer was signed by 20 countries. This called for a freeze on CFC use at 1986 levels and for a 50% reduction eight to ten years later. In 1987, most of the world's industrial nations agreed in the so-called **Montreal Protocol** to cut the use of CFCs in half by the year 2000. In 1990, as concern grew based on new findings, this was amended to a cut of 20% by 1993, and 50% by 1995 and eliminating CFC production altogether by 2000. For a copy of the Montreal Protocol as amended in the 1990 meeting see, "Environmental Effects of Ozone Depletion: 1991 Update" available from the United Nations Environment Programme, P.O. Box 30552, Nairobi, Kenya.

In February of 1992 the United Nations Environment Program issued a report outlining the consequences that could result from a 5 to 10 percent ozone depletion by the year 2000. The consequences included an additional 300,000 cases of skin cancer each year worldwide and 1.6 to 1.75 million more cases of cataracts. Citing new forecasts of a growing ozone hole over the Northern Hemisphere, the Bush administration announced in early 1992 a speedup in phasing out chemicals that damage the ozone layer. The United States will now phase out production of chlorofluorocarbons by the end of 1995 rather than by 2000 as agreed to in the Montreal Protocol.

Even if all the nations of the world eventually agree to a halt in CFC production, *today,* it would be a while before the situation would improve. Because of the amount of CFCs already in the atmospheric circulation and because of their stability, CFC-induced ozone depletion is expected to continue for several decades. It is estimated that even if CFC emission is reduced by 95% in the next 10 years, it would still take a *century* for atmospheric levels of CFCs to return to the levels present in 1980.

The ozone crisis stands as a clear case showing that human activity can indeed have a significant negative impact on global ecology. Because it now appears that what we eventually do about this problem will probably be too little too late, it should serve as an important lesson concerning the "sister" problem of global warming. The problems of ozone depletion and global warming have the same problematic features of lag and relative irreversibility.

The greenhouse effect and the destruction of the ozone shield may or may not cause any great peril in our lifetimes, but the reality of the threats of greenhouse effect–induced weather disruptions, rising sea levels, and stratospheric ozone–induced increased incidence of skin cancer cannot be brushed aside. To do so would be like sailing at high speed through uncharted rocks with no one at the controls. Because they are associated with both increased global warming and ozone depletion, we must reduce the production and use of CFCs. It already seems plausible that less harmful substitutes for today's CFCs will be found. Some are already being used that break down in the lower atmosphere, and there are fluorocarbons that contain no chlorine (Rowland, 1987). Beyond this, we must reduce our dependence on fossil fuels, we must halt deforestation, and we must promote reforestation

worldwide. We obviously need more research on all of these topics to determine what else might be done and how fast we should proceed.

These environmental issues, more than any others, make the point that all humans, national boundaries notwithstanding, live in the same world.

Air Pollutants and Visibility

According to the EPA, impairment of visibility is "perhaps the most noticeable and best documented effect of particulate matter in current U.S. atmospheres" (U.S. EPA, 1982). Visible smoke plumes are obvious examples of this problem. Urban and even multistate regional haze is another dimension for those of us who value visual range, the color of the sky, and the ability to see stars at night. In socioeconomic terms, the fact that air pollution has reduced visibility in recent decades in the United States (particularly in the East) has considerable impact on air travel, property values, and psychological well-being. Many different air pollutants are involved in reactions that yield haze-generating fine particles in the atmosphere. These include ozone, sulfur dioxide, and nitrogen dioxide. Both the reactions and the fate of the hazes they form are influenced a great deal by meteorologic conditions including wind, rain, sunlight, temperature, and humidity.

The very same pollutants that reduce visibility may also affect climate via reduction of net solar radiation, enhanced cloud formation, and fog formation. Carbon dioxide is not the only pollutant that affects climate.

Concepts to Remember

1. The respiratory tract defenses against air pollutants include nasal hairs; ciliated cells lining the upper airway; moist, mucus-coated, contorted airways; coughs and sneezes; the sense of smell; and alveolar macrophages. These defenses can be overwhelmed.

2. To have a health effect, a pollutant must get at and react with some structural or functional element in cells. Contact with cells may be direct or indirect. Some pollutants are absorbed into the bloodstream and are carried to sensitive organs, where they do their damage.

3. Two of the most general consequences of air pollutants for tissue are scar formation and

tissue fluid accumulation. The latter results from blood vessel dilation that is secondary to cellular damage, as in a bruise.

4. The major classes of air pollution–related diseases are pulmonary irritation, dust diseases and other structural changes, systemic toxicity, asphyxiation, and cancer.

5. Air pollution can make an individual more susceptible to infections and other diseases that are seemingly unrelated to air pollution.

6. Epidemiological or population-based investigation is one of the principal ways by which we learn about the connections between air pollutants and human disease. Epidemiology is the study of patterns of disease

in populations in relation to environmental exposures and various other patterns within those populations. Epidemiological investigations tend to produce circumstantial rather than hard evidence.

7. Air pollutants affect living things other than humans in the same fundamental ways that they affect humans—through chemical interference.

8. Acid rain is principally a result of the venting of oxides of nitrogen and oxides of sulfur into the atmosphere. Acid rain is part of the more general problem of acid deposition, which includes dry deposition as well as wet.

9. Acid rain affects living systems through pH changes directly, by causing the mobilization of toxic metals such as mercury and aluminum,

and possibly by other mechanisms currently under investigation.

10. Carbon dioxide is classified as an air pollutant because human activity is increasing the amount of this gas in the atmosphere; this trend is expected to cause global warming via the greenhouse effect.

11. Ozone in the stratosphere serves as a shield protecting the earth's surface from harmful ultraviolet radiation. Chlorofluorocarbons such as freon are apparently causing the catalytic destruction of stratospheric ozone; the ozone hole over the Antarctic is thought to be a special case of such destruction. Increased surface UV radiation may result in increased incidence of skin cancer and other ecological problems.

Environmental Career Profile

Gary J. San Julian

Dr. Gary J. San Julian is vice-president for research and education of the National Wildlife Federation. The National Wildlife Federation's purpose is to educate, inspire, and assist individuals and organizations to take action to conserve wildlife and other natural resources. Dr. San Julian's work involves the development and staging of educational programs for adults and children throughout the United States. He works with various governmental natural resource agencies on educational projects. He likes his work very much because of the opportunity he has to change policies and attitudes. The National Wildlife Federation's wildlife camps and "conservation summits" provide educational adventures that in turn affect many others when the participants go back to their schools

and communities. Dr. San Julian has a Bachelor of Science degree in wildlife biology, a Master of Education in agricultural education, and a Ph.D. in wildlife biology. He has worked previously as a schoolteacher, an agricultural extension wildlife specialist, and a college professor. Dr. San Julian counts his experience in education and resource management as the most valuable in the work he does now.

The challenge for Dr. San Julian and the National Wildlife Federation will continue to be to help all Americans to understand the place of wildlife and the importance of protecting its habitats as critical dimensions of environmental protection.

References and Further Reading

References marked with an asterisk are cited in the chapter.

Ausubel, J. H., 1991. "A Second Look at the Impacts of Climate Change," *American Scientist* **79**:210–221.

Beardsley, T., 1992. "Add Ozone to the Global Warming Equation," *Scientific American,* March, 29.

*Blair, G., 1984. "Chasing Acid Rain," *Science* **84**:64.

*Caren, L. D., 1981. "Environmental Pollutants: Effects on the Immune System and Resistance to Infectious Diseases," *Bio-Science* **31**:592–596.

*Carnow, B. W.; Lepper, M. H.; Shelelle, R. B.; and Stamler, J., 1969. "Chicago Air Pollution Study: SO₂ Levels in Acute Illness in Patients with Chronic Clinical Pulmonary Disease," *Archives of Environmental Health* **18**:768–776.

*Cohn, J. P., 1987. "Chlorofluorocarbons and the Ozone Layer," *BioScience* **37**:647–650.

*Colfin, D. L., and Blommer, E. J., 1967. "Acute Toxicity of Irradiated Auto Exhaust," *Archives of Environmental Health* **15**:36–38.

*Cowling, E. B., 1982. "International Aspects of Acid Deposition," in "Acid Rain: A Water Resources Issue for the 80's," ed. R. Herrmann and A. I. Johnson. *Proceedings of the American Water Resources Assn., International Symposium of Hydrometerology.* Bethesda, Md. pp. 3–12.

*Etkins, R., and Epstein, E. S., 1982. "The Rise of Global Mean Sea Level as an Indication of Climate Change," *Science* **215**:287–289.

*Fairbairn, A. S., and Reid, D. D., 1958. "Air Pollution and Other Factors in Respiratory Disease," *British Journal of Preventive and Social Medicine* **12**:94–103.

*French, H. F., 1990a. *Clearing the Air: A Global Agenda.* Washington, D.C.: Worldwatch Institute.

French, H. F., 1990b. "You Are What You Breathe," *Worldwatch,* May–June, pp. 27–35.

*Galloway, J. N.; Dianwu, Z.; Xiong, J.; and Likens, G. E., 1987. "Acid Rain: China, United States, and a Remote Area," *Science* **236**:1559–1562.

*Galloway, J. N.; Likens, G. E.; and Hawley, M. E., 1984. "Acid

Precipitation: Natural versus Anthropogenic Components," *Science* **226**:820–830.

Gillis, A. M., 1991. "Why Can't We Balance the Globe's Carbon Budget?" *BioScience* **41**(7):442–447.

Gornitz, V.; Lebedeff, S.; and Hansen, J., 1982. "Global Sea Level Trend in the Past Century," *Science* **215**:1611–1614.

*Hassett, C.; Mustafa, M. G.; Caulson, W. F.; and Elashoff, R. M., 1985. "Murine Lung Carcinogenesis Following Exposure to Ambient Ozone Concentrations," *JNCI* **75**:771–775.

*Hinrichsen, D., 1987. "The Forest Decline Enigma," *BioScience* **37**:542–546.

*Hitzhusen, F., and Hemphill, R., 1984. "The Economics of Acid Rain," *Socioeconomic Information,* Cooperative Extension Service Number 667, May. Columbus, Ohio: The Ohio State University Press.

*Holden, C., 1986. "Forest Death Showing Up in the United States," *Science* **233**:837.

*Houghton, R. A., 1987. "Terrestrial Metabolism and Atmospheric CO_2 Concentrations," *BioScience* **37**:672–678.

*Houghton, R. A., and Woodwell, G. M., 1989. "Global Climate Change," *Scientific American* **260**(4):36–44.

*Idso, S. B., 1984. "Carbon Dioxide and Climate: Is There a Greenhouse in Our Future?" *Quarterly Review of Biology* **59**:291–294.

*Jones, P. D., and Wigley, T. M. L., 1990. "Global Warming Trends," *Scientific American* **261**(8):84–91.

*Kerr, R. A., 1983. "Carbon Dioxide and a Changing Climate," *Science* **222**:491–492.

Kerr, R. A., 1992. "Pollutant Haze Cools the Greenhouse," *Science* **255**:682–683.

*Kramer, P., 1981. "Carbon Dioxide Concentration, Photosynthesis, and Dry Matter Production," *BioScience* **31**:29–33.

*La Breque, M., 1989. "Detecting Climate Change," *Mosaic* **20**(4):3–17.

*Lachenbruch, A. H., and Marshall, B. V., 1986. "Changing Climate: Geothermal Evidence from Permafrost in the Alaskan Arctic," *Science* **234**:689–696.

*Lashof, D. A., and Tirpak, D. A., 1989. *Policy Options for Stabilizing Global Climate* (Report to Congress, Office of Policy, Planning and Evaluation). Washington, D.C.: U.S. Environmental Protection Agency.

*Lave, L. B., and Seskin, E. P., 1970. "Air Pollution and Human Health," *Science* **169**:723–733.

*Lave, L. B., and Seskin, E. P., 1979. "Epidemiology, Causality, and Public Policy," *American Scientist* **67**:178–186.

*Lillie, R. J., 1970. *Air Pollution Affecting the Performance of Domestic Animals: A Literature Review* (Agricultural Handbook No. 380). Washington, D.C.: U.S. Department of Agriculture.

*Lindzen, R. S., 1990. "Some Remarks on Global Warming," *Environmental Science and Technology* **24**:424–426.

*Marshall, E., 1989. "EPA's Plan for Cooling the Global Greenhouse," *Science* **243**:1544–1545.

*Marx, J. L., 1975. "Air Pollution Effects on Plants," *Science* **187**:731–733.

*Mathews, J. T., 1987. "Global Climate Change: Toward a Greenhouse Policy," *Issues in Science and Technology,* Spring. **3**(3):57–68.

*Mausner, J. S., and Bahn, A. K., 1974. *Epidemiology: An Introductory Text.* Philadelphia: Saunders.

*Mayer, K. S.; Multer, E.; and Schreiber, R. K., 1984. *Acid Rain: Effects on Fish and Wildlife.* Washington, D.C.: U.S. Department of the Interior.

*Molina, M. J.; Tso, T.; Molina, L. T.; and Wang, F. C., 1987. "Antarctic Stratospheric Chemistry of Chlorine Nitrate, Hydrogen Chloride, and Ice: Release of Active Chlorine," *Science* **238**:1253–1257.

Oden, S., 1968. *The Acidification of Air and Precipitation and Its Consequences in the Natural Environment* (Ecology Committee Bulletin No. 1). Stockholm: Swedish National Science Research Council. English-language edition published by Translation Consultants, Ltd., Arlington, Va.

*Ottar, B., 1977. "International Agreement Needed to Reduce Long-Range Transport of Air Pollutants in Europe," *Ambio* **6**:252–269.

*Patrick, R.; Benetti, V. P.; and Halterman, S. G., 1981. "Acid Lakes from Natural and Anthropogenic Causes," *Science* **211**:446–448.

*Perera, F. P., 1978. "Fine Particles in the Atmosphere," *Natural Resources Defense Council Newsletter* **7**(2):1–16.

*Plochmann, R., 1984. "Air Pollution and the Dying Forests of Europe," *American Forests* **90**(6):17–21.

*Postel, S., 1984. *Air Pollution, Acid Rain, and the Future of Forests* (Worldwatch Institute Paper No. 58). Washington, D.C.: Worldwatch Institute.

Rahn, K. A., and Lowenthal, D., 1984. "Elemental Tracers of Distant Regional Pollution Aerosols," *Science* **223**:132–139.

*Reich, P. B., and Amundson, R. G., 1985. "Ambient Levels of Ozone Reduce Net Photosynthesis in Tree Crop Species," *Science* **230**:566–570.

*Reisner, M., 1977. "It's 1977, Why Don't We Have Cleaner Air?" *Natural Resources Defense Council Newsletter* **6**(1):2–3.

*Roberts, L., 1983. "Acid Rain Clouds U.S. and Canadian Relations," *BioScience* **33**:418–422.

*Roberts, L., 1991. "Learning from an Acid Rain Program," *Science* **251**:1302–1305.

Rosswall, T., 1991. "Greenhouse Gases and Global Change: International Collaboration," *Environmental Science and Technology* **25**(4):567–573.

*Rowland, F. S., 1987. "Can We Close the Ozone Hole?" *Technology Review* **90**:50–59.

Rowland, F. S., 1991. "Stratospheric Ozone in the 21st Century: The Chlorofluorocarbon Problem," *Environmental Science and Technology* **25**(4):622–628.

*Schindler, D. W., 1988. "Effects of Acid Rain on Freshwater Ecosystems," *Science* **239**:149–157.

*Schindler, D. W.; Beaty, K. G.; Fee, E. J.; Cruikshank, D. R.; DeBruyn, E. R.; Findlay, D. L.; Linsey, G. A.; Shearer, J. A.; Stainton, M. P.; Turner, M. A., 1990. "Effects of Climatic Warming on Lakes of the Central Boreal Forest," *Science* **250**:967–970.

*Schindler, D. W.; Mills, K. H.; Malley, D. F.; Findlay, D. L.; Shearer, J. A.; Davies, I. J.; Turner, M. A.; Linsey, G. A.; and Cruikshank, D. R., 1985. "Long-Term Ecosystem Stress: The Effects of Years of Experimental Acidification on a Small Lake," *Science* **228**:1395–1396.

*Schneider, S. H., 1987. "Climate Modeling," *Scientific American* **256**(May (5)):72–81.

Schwartz, S. E., 1989. "Acid Deposition: Unraveling a Regional Phenomenon," *Science* **243**:753–763.

*Sheehan, P. J.; Miller, D. R.; Butler, G. C.; and Bourdeau, P., eds., 1984. *Effects of Pollutants at the Ecosystem Level.* New York: Wiley.

*Skarby, L., and Sellden, G., 1984. "The Effects of Ozone on Crops and Forests," *Ambio* **13**:68–72.

*Smith, R. C.; Prezelin, B. B.; Baker, K. S.; Bidigare, R. R.; Boucher, N. P.; Coley, T.; Karentz, D.; MacIntyre, S.; Matlick, H. A.; Menzies, D.; Ondrusek, M.; Wan, Z.; and Waters, K. J., 1992. "Ozone Depletion: Ultraviolet Radiation and Phytoplankton Biology in Antarctic Waters," *Science* **255**:952–958.

*Solomon, P. M.; de Zafra, R.; Parrish, A.; and Barrett, J. W., 1984. "Diurnal Variation of Stratospheric Chlorine Monoxide: A Critical Test of Chlorine Chemistry in the Ozone Layer," *Science* **224**:1210–1214.

*Stolarski, R. S., 1988. "The Antarctic Ozone Hole," *Scientific American* **258**(Jan. (1)):30–36.

*Stuiver, M., 1978. "Atmospheric Carbon Dioxide and Carbon Reservoir Changes," *Science* **199**:253–258.

*Sun, M., 1985. "EPA Accelerates Ban on Leaded Gas," *Science* **227**:1448.

*Toon, O. B., and Turco, R. P., 1991. "Polar Stratospheric Clouds and Ozone Depletion," *Scientific American*, June, 68–74.

*Ulrich, B., 1990. "*Waldsterben:* Forest Decline in West Germany," *Environmental Science and Technology* **24**:436–441.

*U.S. Environmental Protection Agency, 1982. *Review of the National Ambient Air Quality Standards for Nitrogen Oxides: Assessment of Scientific and Technical Information; Review of the National Ambient Air Quality Standards for Particulate Matter: Assessment of Scientific and Technical Information; and Review of the National Ambient Air Quality Standards for Sulfur Oxides: Assessment of Scientific and Technical Information*. Washington, D.C.: U.S. Government Printing Office.

*Waldbott, G. L., 1978. *Health Effects of Environmental Pollutants*, 2d ed. St. Louis: Mosby.

*Woodwell, G. M.; Hobbie, J. E.; Houghton, R. A.; Melillo, J. M.; Moore, B.; Peterson, B. J.; Shaver, G. R., 1983. "Global Deforestation: Contribution to Atmospheric Carbon Dioxide," *Science* **222**:1081–1086.

*World Health Organization, 1978. *Air Quality in Selected Urban Areas in 1975–1976* (WHO Publication No. 41). Geneva: World Health Organization.

11

Control of Air Pollution

In this chapter we turn our attention to the economic, legal, and technical aspects of air pollution control. The scope of this book does not permit us to undertake a review of the details of air pollution control in even a fraction of the countries of the world. Air pollution control is fairly highly developed in some countries, under early development in some, and nonexistent in still others. Air pollution and the approaches being followed to deal with it in each country have aspects that are unique. The economy of each nation is certainly unique. Volumes could be (and have been) written about air pollution control in any of the highly industrialized and automobilized countries of the world. Of necessity, then, we set out here only to illuminate the most important *general* aspects of the technical, legal, philosophical, and economic dimensions of air pollution control. How much does it cost to control air pollution relative to the cost if air pollution is not controlled? What is the basis in law that requires governments to protect people from the effects of air pollution? What are some of the legal strategies used in clean air laws? What do these laws regulate? What are some of the ways in which air pollution can be controlled through technology? Is technology the only way to solve the problem, or is

it but a minor element in a large array of possible strategies?

As we explore the answers to these and other questions, we will rely heavily on the United States for purposes of illustration. The United States has a long history of air quality problems and of air pollution control legislation, going back far enough to permit an assessment of the effectiveness of its laws. But our focus on one country should not impair the usefulness of this chapter as a starting point for a thorough consideration of air pollution control in any country or state.

As we move on, we will consider both the costs associated with air pollution control and the costs of not controlling air pollution. The costs of not controlling pollution can be thought of as the dollar value of the negative effects of air pollution discussed in Chapters 9 and 10. We looked at the technical aspects of the *effects* of air pollutants in Chapter 10; here we look at the technical aspects of air pollution *control*. We also look at air pollution control laws and how well they have worked so far. All of these things are closely related, of course; economic and technical factors have much to do with the kinds of laws we enact and how well they work.

The Economics of Air Pollution

Our resolve to control air pollution is determined in part by cost. Perhaps it would be more accurate to say that our resolve is determined by our perception of the difference between the cost of controlling pollution and the cost of not controlling pollution. Assessing the cost of not controlling air pollution is difficult because it is subjective; assessing the cost of air pollution control is relatively straightforward. Let us start with the cost of control.

The Cost of Air Pollution Control

Expenditures for pollution control in the United States reached and passed $100 billion per year during the last few years of the 1980s. A total of $115 billion was spent on capital investment and operating pollution control equipment in 1990 by federal, state, and local governments and private entities and this amounted to 2.1% of real gross (U.S.) national product. Pollution control costs as a share of U.S. GNP have more than doubled since 1972 and spending in 2000 is likely to be triple the 1972 levels (Council on Environmental Quality, 1991). Figure 11.1 illustrates the proportion of total pollution control costs borne by the U.S. EPA, state governments, local governments and private entities for 1972, 1980, 1987, and projections for 2000. Part b of Figure 11.1 shows that *air* pollution control specifically amounts to about 30% of the total cost of pollution control. Figure 11.2 illustrates emission control costs for automobiles alone, in the United States. Obviously, a lot of money has been spent, is being spent, and will likely continue to be spent to control air pollution.

As costs have increased, the debate has intensified over the need for air pollution controls and the laws requiring them. The specter of lost jobs adds further complexity to the debate. Pollution control costs can and do result in higher prices and thus lower demand for products. Diminished demand for consumer goods in turn leads to lost jobs. The closing of plants unable to meet emission standards also means lost jobs. Some 155 plants closed in the United States between 1971 and 1984 all due at least in part to the cost of meeting emission standards. These shutdowns alone amounted to more than 30,000 lost jobs (U.S. EPA, 1984). Of course, since someone must make, install, and maintain pollution control equipment, pollution control also means a certain increase in jobs—different jobs. The U.S. EPA estimates that the net impact of pollution control legislation in the United States has been an increase, not a decrease in employment. Nevertheless, to answer the tough questions being asked about the need for air pollution control, we need more and better information about the costs of not controlling pollution. But whereas the costs of pollution control are rather easily translated into dollars, doing the same thing with pollution control benefits is much more difficult, and the results are more controversial.

The Cost of Air Pollution Damage

To date, there have been two relatively thorough attempts to establish the costs of air pollution damage in the United States. The first was a report by Barrett and Waddell published by the EPA in 1973 titled, *The Cost of Air Pollution Damage: A Status Report.* The second was published by A. M. Freeman in 1982, based on a report for the Council on Environmental Quality. Although these studies are now somewhat dated,

---

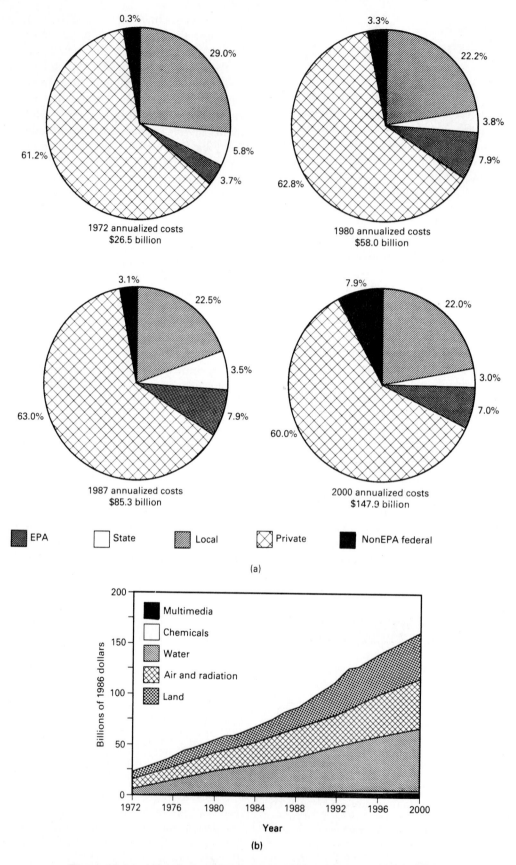

Figure 11.1 a) Total annualized costs of U.S. pollution control by funding source for 1972, 1980, 1987, and projections for 2000, in 1986 dollars. b) Trends in total annualized cost of U.S. pollution control by medium.

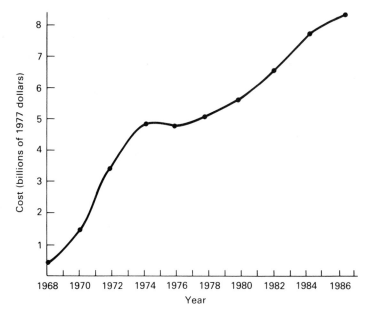

Figure 11.2 Emission Control Costs for Light-Duty Passenger Cars. The curve of increased air pollution control costs on American automobiles has been steep since 1968. The figures given here include annualized capital costs, maintenance costs, and fuel consumption penalties (reduced fuel consumption economy caused by air pollution control devices).

there have been no more recent, comparably comprehensive studies published. We will review these two studies in some detail because they serve as very good illustrations of the complexities of such studies and they offer interestingly different approaches to the evaluation of air pollution control benefits. Both reports looked at the impact of air pollution in terms of the costs associated with health and with damage to vegetation, materials, and residential property. The interesting difference between the two reports is that whereas Barrett and Waddell estimated the total cost of the impact of air pollution, Freeman estimated the dollar benefit of an assumed 20% improvement in air quality that occurred between 1970 and 1978.

We should point out that there have been and continue to be many studies focusing on specific individual categories of impact such as health effects (see Horst et al., 1984; U.S. EPA, 1988b; Hall et al., 1992; Krupnick, and Portney, 1991). Many such studies are cited throughout the text and others are included among the references at the end of this chapter. It would be a good class exercise to find, compile, and review those studies released since this book was published.

All studies of the kind described here are based on a set of assumptions concerning such things as the cost of being sick, the cost of missing work, the value of a human life, and the relationships between pollution and sickness or material damage. Literally hundreds of specific assumptions had to be made in coming up with the estimates that came from these studies. Refer to the original reports for a description of these sets of assumptions and the justifications used for making them.

Barrett and Waddell concluded that the annual economic impact on vegetation, materials, residential property, and health due to air pollution in 1968 was approximately $16.1 billion. Freeman estimated that the benefits of air pollution control enjoyed ten years later amounted to $21.4 billion (Figure 11.3). Freeman assumed 1978 economic conditions, 1978 population, and an improvement in air quality of 20% between 1970 and 1978. He concluded that the annual economic benefit of the Clean Air Act was actually somewhere between $4.6 and $51.2 billion dollars in 1978. The size of this range suggests that the data base is poor and that he really did not have the basis for an exacting determination of the benefits of clean air (or conversely, the cost we pay if we do not keep the air clean). We should emphasize that Freeman did not estimate the total cost impact of dirty air; he estimated what it was worth to Americans in 1978 to have air 20% cleaner than it was in 1970.

Freeman's numbers obviously cannot be directly compared to those of Barrett and Waddell simply by adjusting for inflation. Nor can the estimates be made comparable by multiplying by 5. The reasoning that if 20% of the impact of pollution costs x dollars, all of it should cost 5x dollars is not legitimate because the benefit of removing the first 20% is not necessarily the same as that derived from removing the next 20% and so on.

Although all of the studies that have been done do not agree and the data we have are far from precise, it is safe to conclude that there is a cost associated with dirty air and that the cost is highly significant. We will now examine briefly some of the specific categories of impact. Again our purpose is to illustrate the kinds of

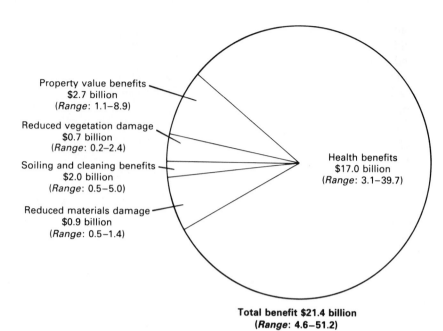

Property value benefits
$2.7 billion
(*Range*: 1.1–8.9)

Reduced vegetation damage
$0.7 billion
(*Range*: 0.2–2.4)

Soiling and cleaning benefits
$2.0 billion
(*Range*: 0.5–5.0)

Reduced materials damage
$0.9 billion
(*Range*: 0.5–1.4)

Health benefits
$17.0 billion
(*Range*: 3.1–39.7)

Total benefit $21.4 billion
(*Range*: 4.6–51.2)

Figure 11.3 Benefits of Pollution Control. One estimate of the economics of air pollution control benefits enjoyed in 1978. Numbers here express the 1978 dollar value of an assumed 20% lower level of air pollution brought about by clean air laws.

things that go into assessing the economic impact of air pollution on health, vegetation, and materials.

Health Costs of Air Pollution

Barrett and Waddell's estimates of the overall impact of air pollution on health with respect to selected diseases are given in Table 11.1. They estimated that the health costs of air pollution exceeded $6.06 billion per year in the late 1960s. Their data was derived from one of the earliest attempts to estimate the cost of damage to health from air pollution, published by Ridker in 1967. Table 11.1 illustrates the need to include many *indirect* factors in determining the health cost of pollution damage, that is, the cost of absenteeism due to sickness as well as the cost of premature death (wages lost) and the cost of treatment.

As indicated in Figure 11.3, Freeman concluded that the 1978 health benefit of the 20% reduction in air

pollution that occurred between 1970 and 1978 was somewhere between $3.1 billion and $39.3 billion (with a best estimate of $17.0 billion).

A 1979 EPA report estimated the total annual impact of air pollution on health was $5 billion to $16 billion in increased mortality (earlier death) and $36 billion in sickness (health care and lost work).

The Economic Impact of Air Pollution on Vegetation

Summarizing a host of studies of the economic impact of air pollution on vegetation, Barrett and Waddell concluded that the value of the air pollution damage that occurred in 1968 was $120 million. They noted that their estimate included only the major pollutants and their effects on only agricultural, horticultural, and forest crops. A 1969 report by the Stanford Research Institute (SRI) estimated a $113 million annual vegetation loss due to air pollution. Food, crops,

Table 11.1   An Early Estimate of the Cost Associated with Selected Air Pollution–related Diseases in the United States

	Costs Associated with Selected Diseases (in millions of dollars)							
Type of Cost	*Cancer of the Respiratory System*	*Chronic Bronchitis*	*Acute Bronchitis*	*Common Cold*	*Pneumonia*	*Emphysema*	*Asthma*	*Total*
Premature death	518	18	6	N.A.	329	62	59	992
Treatment	35	89	N.A.	200	73	N.A.	138	535
Absenteeism	112	52	N.A.	131	75	N.A.	60	430
Total	665	159	6	331	477	64	257	1,957

Source: L. B. Barrett and T. E. Waddell, *The Cost of Air Pollution Damage: A Status Report* (Research Triangle Park, N.C.: National Environmental Research Center, U.S. Environmental Protection Agency, 1973), p. 9.

ornamentals, noncommercial forests, and parks were considered. Heintz, Hershaft, and Horak (1976) released a comprehensive study of the early 1970s. Adjusted to 1978, it gave a range of loss of $1.2 billion to $12.2 billion per year for plant damage due to photochemical oxidant pollution. To this Freeman added $500 million as the cost of damage to vegetation due to SO_x pollution (including estimated plant damage from acid rain).

Freeman assumed a 20% improvement in oxidant pollution between 1970 and 1978; he assumed *no* improvement in SO_x pollution for the same period and concluded that we were enjoying a benefit of $200 million to $2.4 billion in vegetation saved by air pollution control in 1978 (see Figure 11.3).

At a 1987 research conference it was estimated that (1) ozone costs U.S. farmers between $1 billion and $5 billion a year in lost crop production, (2) ozone levels of 0.05 ppm can reduce peanut, soybean, cotton, and wheat yields by as much as 12%, and (3) as much as 10% of the nation's crops may be lost to all forms of air pollution each year (Seabrook, 1987).

The Economic Impact of Air Pollution on Animals

No thorough study has ever been conducted of the economic impact of air pollution on animals. Notable episodes of damage to animals by air pollution have been highly localized, and any economic consequences were relatively unimportant on a national scale.

The Economic Impact of Air Pollution on Materials

Barrett and Waddell cited a number of reports, most notably one by Uhling (1950) in which it was estimated that in the 1950s, corrosion due to air pollution would cost the United States up to $5.4 billion annually. In still another report published by the Rustoleum Corporation, annual corrosion losses due to air pollution were estimated to be in the neighborhood of $7.5 billion in the 1960s.

Freeman concluded that air pollution damage to materials amounted to $2.4 billion to $7.2 billion in 1978. Again assuming a 20% improvement in air pollution between 1970 and 1978, he projected a 1978 annual benefit (in the form of undamaged materials) of $500 million to $1.4 billion.

Soiling is a specific kind of effect of air pollution on materials. It has to do mainly with the settling out of particulates or the absorption of particulates on fabrics and structures like windowsills. The cost of cleaning up made necessary by soiling was not included in the estimates referred to so far, but the results presented in one (early) soiling study are given in Table 11.2. This

Table 11.2 Cost of Air Pollution–related Soiling. Summary of a report by H. Beaver in London. The results of more recent studies are summarized in U.S. EPA criteria documents and staff papers listed in Chapters 9 and 10.

Category	Cost (millions of dollars per year)
Direct	
Laundering	70
Painting and decorating	84
Soiling and depreciation of buildings (other than houses)	56
Corrosion of metals	70
Damage to textile and other goods	147
Indirect (loss of efficiency)	280
Total	707

Note: The cost per person was $14 per year in nonpolluted areas and $28 per year in polluted areas.

Source: L. B. Barrett and T. E. Waddell, *The Cost of Air Pollution Damage: A Status Report* (Research Triangle Park, N.C.: National Environmental Research Center, U.S. Environmental Protection Agency, 1973), p. 35.

table shows that a number of things must be included in any consideration of the economic impact of soiling. The study cited in Table 11.2 revealed that the costs of keeping things clean in polluted areas of England were twice those in relatively unpolluted areas.

The Impact of Air Pollution on Property Values

Barrett and Waddell also estimated the cost to society that comes from decreases in residential

The cost of cleaning soiled buildings is one of the many costs to be considered in comparing the relative economics of pollution and pollution control.

property value due to pollution. Again, they cite a number of studies in which estimates were made of various kinds of damage, pointing out that each study involved a number of specific assumptions. They concluded that air pollution's impact on residential property values amounted to $3.4 billion to $8.4 billion annually by 1968. The method used by Barrett and Waddell was to estimate how much people were willing to pay to live in an unpolluted area. This was then converted into the estimated economic impact of incremental amounts of pollution.

In his estimate, Freeman added a special adjustment for the decrease in willingness to pay as air quality got better; this compensated for the fact that it is more important to people to remove the first half of the pollution in the air than it is to remove the remaining half. Freeman estimated that the 1978 annual property value benefits derived from a 20% improvement in air quality from 1970 to 1978 amounted to $1.1 billion to $8.9 billion. This implies that the residual impact in 1978 amounted to $5.5 billion to $44.5 billion from mobile and stationary source pollution combined. Freeman notes that the impact on property values is also counted in other categories (for example, deterioration of paint is a material impact as well as a property value impact). Accordingly, he counted only 30% of the property value impact in his grand total (see Figure 11.3).

In our discussion and in the reports by Barrett and Waddell, Freeman, and others, many qualifications are given in the cost estimates. Many things are left out of estimates simply because there is very little if any basis on which to make a dollar cost estimate. For certain materials this problem is profound indeed. What about the deterioration of materials of historical or aesthetic significance? What about rare paintings and books, for example? How would one assess the value of a beautiful vista that can no longer be appreciated because of air pollution? What about reduced visibility as a highway safety factor? What about the depressing psychological effect of air pollution? How does one assign a value to the deterioration in life that comes from the presence of noxious odors—without regard to the effect they might have on health?

The Economic Impact of Air Pollution on Ecosystems
We are a long way from being able to assess the economic impact of air pollution on the smooth functioning or probability of the sustained functioning of an ecosystem. We only poorly understand the effects, never mind assigning a value to their impact. Although it will be difficult to fit *ecosystems* into our *economic* systems, we must continue to search for ways of doing exactly that.

Determining the Cost-Benefit Ratio for Air Pollution Control

We now have an idea of the costs and benefits of controlling pollution. That is really all we have. It is obvious that we will never have all of the exact information we need, but we must start nonetheless to obtain more and better economic information. We need it in order to assess the cost-benefit ratio for controlling air pollution.

Figure 11.4 contains four generalized curves illustrating some plausible relationships between the cost of uncontrolled pollution, the cost of pollution control, the benefit of pollution control, and the value of retained resources. The only thing certain about the curves is the direction of their slopes. Whether the curves should be straight lines or not remains to be seen, although there may be more of a basis for defining some curves than for others. If we had *accurate* curves of the sort presented in Figure 11.4, we would have a better handle on the air pollution problem. We would have completed the next necessary step toward solutions, that step being the determination of the nature and the degree of the problem in economic terms.

It takes more than statements that something is bad to bring about correction. As difficult or nearly impossible as the task may be, we must be able to express badness in terms of dollars or some other commonly understood unit and to do this with good data. Only then will it be possible to get corrective action that will subsequently withstand ever-present economic pressures.

We call attention again to Figure 11.4, which shows that the cost of removing or stopping pollutants increases in an exponential way with the percentage of the pollutant removed. In other words, it may be relatively easy to remove the first few tons of a pollutant from the air, but it would be impossible to remove the last few molecules.

Another curve in Figure 11.4 illustrates how societies might perceive the benefit of pollution control or how much pressure there might be to spend money to do something about pollution. The curve shows that while the public might want very much to do something about the worst of the pollution, it would be increasingly less willing to pay for removing the last bits of pollution. One economic concept has to do with the fact that curves such as the two we have just described intersect at some point, the point at which the cost of controlling pollution matches the price people are willing to pay for controlling it. The point of intersection would obviously depend on such things as the technology available for pollution control and the

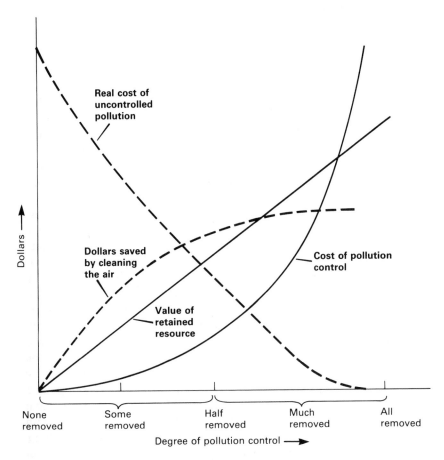

Figure 11.4 Cost versus Benefit. The graph shows some of the generalized relationships between the cost of controlling pollution and the cost of not controlling pollution. Note that the curve for the cost of not controlling pollution and the curve for pollution control costs intersect. This occurs at the theoretical point beyond which in a "macro" sense, it is not worth cleaning the air any further.

magnitude of the real and perceived effects of air pollution.

Air Pollution and the Law

The modern history of clean air legislation in the United States began with the Motor Vehicle Act of 1960, which was eventually followed by the Clean Air Act of 1970. The 1970 Clean Air Act was really a compilation of amendments to earlier legislation, and the Clean Air Act has itself been amended three times, once in 1974, 1977, and 1990. The Clean Air Act of 1970 was tighter than earlier legislation in that it required the establishment of standards for ambient air and set up somewhat more restrictive criteria for granting delays and variances in achieving these standards. The act also for the first time specified that the federal government would determine the best available technologies to be used in achieving performance standards for industrial plants, automobiles, and other sources of air pollution. The act also established federal control of particularly hazardous substances like asbestos.

The **Environmental Protection Agency (EPA)** was established in 1970 as an umbrella agency

in charge of the environmental programs previously scattered through many branches of the U.S. government. The Clean Air Act gave the EPA responsibility

The EPA has set air quality standards, but there is no guarantee that they will be met in all cities. The smog problem in Los Angeles, caused by frequent atmospheric inversions and an automobile-oriented culture, will not be easy to clear up.

for implementing the provisions of the act and certain regulatory and enforcement powers. Although many of the responsibilities and authorities are intentionally delegated to individual states, many other authorities are ultimately retained by the EPA. The act empowers the EPA to take any necessary action to stop or correct any air pollution emergency. The EPA can also levy fines, as can state or local environmental control agencies.

The original Clean Air Act implied that areas of the United States that were cleaner than the standards would not necessarily be allowed to be made dirtier. However, it was later recognized that there had to be a compromise between the desire to preserve cleanliness and the development of relatively undeveloped areas of the country. Using its broad authority, the EPA established three classes with regard to just how dirty air cleaner than the standard would be allowed to become. In Class I areas, the quality of air could be degraded slightly, moderate deterioration would be allowed in Class II areas, and dirtiness would be permitted to reach the national standard in Class III areas. Class III regions are so designated only upon appeal and only when it has been shown that the designation will optimize the economic growth and environmental objectives of the people affected.

Air Pollution Standards

The Clean Air Act of 1970 required that the EPA set ambient air quality standards with a margin of safety such that the most sensitive people would suffer no adverse health effects. These standards were to be the goals in cleaning up the air, and they would be the limits beyond which clean air would not be allowed to deteriorate. Using the data available at the time, in April 1971 the EPA identified six pollutants as requiring a national ambient air quality standard. These six

were particulate matter, sulfur dioxide, carbon monoxide, photochemical oxidants, nitrogen dioxide, and hydrocarbons. They were singled out because they were known to have effects on human mortality and morbidity and to have effects on vegetation, material, visibility, and other things that would fall within the realm of public welfare. Studies that served as the basis for setting these standards originally are described in the respective criteria documents issued by the EPA.

The Making of a Standard

Because it was recognized that setting a standard at almost any level would generate controversy over whether that standard was too high or too low, the procedure in the 1970 legislation for setting a standard was rigorous and involved multiple reviews. The standard-setting process, in summary, went as follows:

1. Review all available data on the health and environmental effects of a pollutant or conduct studies to generate data.
2. Prepare a *criteria document* summarizing all relevant data followed by a *staff paper* (by the EPA's Office of Air Quality Planning and Standards) in which the key studies of the criteria document are evaluated and the critical elements to be addressed in setting the standard are identified.
3. Conduct an open review of the staff paper and follow this up by staff recommendations for a standard to the EPA administrator (these recommendations are published in the *Federal Register*).
4. Publicize the administrator's decision regarding a standard in the *Federal Register* for the purpose of seeking comments from the general public, and invite individuals and organizations to submit comments within a certain time period.

Mini-Glossary

Ambient air quality standard: federal limit for a pollutant in ambient air that serves as a target in local air quality improvement or protection programs. The *primary standard* protects public health; the *secondary standard* protects public welfare. The federal government establishes air quality standards, but stricter standards may be established by state governments.

Emission standard: limit on the amount of a pollutant that can be legally discharged into the environment from a particular source. Under the Clean Air Act of 1970, emissions from *existing* sources are controlled by the states under state implementation plans approved by the EPA. The federal government retained control over *new* sources (construction begun after 1972), establishing maximum emission standards for new plants built in any state.

Environmental Protection Agency (EPA): the federal agency charged with the enforcement of all federal regulations having to do with environmental pollutants. The address is 401 M Street N.W., Washington, DC 20460.

Indirect source: a facility, road, building, structure, or other installation that causes air pollution to be generated through associated mobile source activity such as automobile, bus, and truck traffic. (Examples of indirect sources are a sports complex and a shopping center.)

Ringlemann charts: a series of charts that simulate various pollution densities by presenting different degrees of darkness used for estimating the opacity (density) of smoke rising from stacks and other sources.

Variance: permission by a legal body for a company or an individual to operate outside the limits prescribed in a law or standard, usually to allow the person or company to develop the means to bring its activity into compliance.

How Safe Is Safe?

In setting limits on environmental pollutants, should the purpose be to protect the most sensitive individual or the majority of the people with average sensitivities? Should the worst-case scenario always be used as a yardstick? The more stringent the regulations, the greater the cost to implement them. Money spent on increased regulation is not available for other uses in the economy. Yet people with sensitivities have no means of protecting themselves from pollutants in the common atmosphere. What criteria do you think should be used in setting air quality standards?

5. Promulgate a final standard in the *Federal Register* to become effective and enforced after a specified length of time.

National Ambient Air Quality Standards

In the early 1970s the EPA issued air quality standards for the pollutants described above. Lead has since been added to the list, and hydrocarbons have since been deleted. Standards for each of the pollutants were established in two categories, primary and secondary. Primary standards are set with public health in mind. The limit established as a **primary standard** is intended to identify a safety limit with respect to human health. The **secondary standard** is intended to take in all nonhealth effects of a pollutant and is a kind of public welfare standard. The secondary standard, where different, is more stringent.

Current primary and secondary national ambient air quality standards are given in Table 11.3. Also given in that table are (1) the levels of each of the major pollutants set as limits by the American Conference of Governmental Industrial Hygienists (for exposures in a 40-hour workweek in occupational settings) and (2) the levels of each pollutant considered to constitute a hazardous or emergency situation.

It should be noted that there are pollutants in Table 11.3 other than those for which primary and secondary standards have been set. It should also be noted that the *ozone* standard originally set in 1971 was 0.08 ppm for a one-hour average. In June 1978 the EPA proposed to raise this to 0.10, but in 1979 the standard was actually set at 0.12 ppm for a high hourly average.

Of 105 urban areas in the United States in 1978, only Honolulu and Spokane, Washington, met the health standards for ozone set in 1971. Raising the standard to 0.12 ppm meant that 20 cities rather than two had clean air as far as photochemical oxidants go. The EPA estimated that the annual cost of meeting the new standard for ozone would be $4.5 billion compared to $6 billion for the 0.08-ppm standard.

Setting standards is only one of the steps in the program for reaching and maintaining an appropriate quality of air nationwide specified in the Clean Air Act of 1970. Implied in the setting of the standard is that there have to be ways established for measuring the amounts of pollutants in ambient air, plans devised for bringing existing air quality levels into compliance with the standard, and ways of enforcing compliance.

Measuring and Monitoring Pollutants in the Air

Measuring air pollutants is no simple matter. First, methods for making measurements must be identified, established, and refined, and there must be systematic ways of sampling air so as to obtain the most accurate representation of environmental concentrations in a given large area. Under the provisions of the Clean Air Act, the EPA specified the methods to be used to measure air pollutants. Local agencies charged with monitoring air pollution must place measuring instruments in appropriate locations and determine—if measurement is not to be continuous—how often to take readings.

Once it has been determined that the air in a particular region falls short of meeting clean air standards, control strategies (state implementation plans) must be devised by the states (and approved by the EPA) to bring that air into compliance, for example, by setting and enforcing limits on how much of each pollutant can be discharged from pollution sources. **National ambient air quality standards (NAAQS)** are levels of specific pollutants that the EPA says we must stay below or strive to get below. **Emission standards** are limitations on the amount of a pollutant that may be discharged from specific sources.

According to present U.S. laws, authority to enforce NAAQS and emission standards is to be dele-

Table 11.3 National Ambient Air Quality Standards and Occupational Limits of Exposure for the Major Air Pollutants

Pollutants	Toxicity Rank	Primary Standard (to protect human health)	Secondary Standard (to protect public welfare)	Limits of Exposure Set for 40-Hour Week by the American Conference of Governmental Industrial Hygienists	Pollutant Levels Considered Hazardous
Sulfur dioxide	3	80 $\mu g/m^3$ annual arithmetic mean (0.03 ppm); 365 $\mu g/m^3$ (0.14 ppm) maximum 24-hr avg.*	1,300 $\mu g/m^3$ maximum 3-hr avg. (0.05 ppm)	5 ppm	2,100 $\mu g/m^3$ 24-hr avg.
Particulate matter of less than 10 μm in diameter†	4	50 $\mu g/m^3$ annual geometric mean; 150 $\mu g/m^3$ maximum 24-hr avg.*	equal to primary	—	875 $\mu g/m^3$ 24-hr avg.
Nitrogen dioxide	5	100 $\mu g/m^3$ annual arithmetic mean (0.05 ppm)	equal to primary	5 ppm	3,000 $\mu g/m^3$ 1-hr avg.
Hydrocarbons‡ (excluding methane)§	8	160 $\mu g/m^3$ (6 to 9 A.M.) (0.24 ppm)	equal to primary	varying limits set for certain chlorinated and other derivatives	—
Carbon monoxide	9	10 mg/m^3 maximum 8-hr avg. (9 ppm)*; 40 mg/m^3 (35 ppm) maximum 1-hr avg.*	no secondary standard	50 ppm	46,000 $\mu g/m^3$ 8-hr avg.
Ozone	1	235 $\mu g/m^3$ maximum 1-hr avg. (0.12 ppm)	equal to primary	0.1 ppm	1,000 $\mu g/m^3$ 1-hr avg.
Hydrogen sulfide	7	none		10 ppm	
Hydrogen fluoride	2	none		3 ppm	
Metals	6	—	—	50 $\mu g/m^3$ to 150 $\mu g/m^3$	—
Lead	—	1.5 $\mu g/m^3$ maximum quarterly avg.	equal to primary	0.15 mg/m^3	

* Not to be exceeded more than once per year.

† In the original standards, *total* suspended particulate matter was the basis for the standard; new standards promulgated in 1987 are based on particles having diameters of less than 10 μm as the indicator pollutant.

‡ Since deleted; non-health-related standard used as a guide for ozone control.

§ A natural pollutant produced in swamps via anaerobic decay.

gated to the states by the EPA, but the EPA retains the authority to step in if its administrator learns that a state is not being strict enough in fulfilling its responsibility. The EPA also requires states to establish performance standards for new sources.

New source performance standards are emission standards for different classes of new factories and plants. Such standards are to be based on the best available pollution control technology for controlling emissions from plants and other pollution sources of varying technological types; they are expressed in terms of maximum amounts of pollutant emitted per unit of activity, for example, tons of product produced. Performance standards may be different for different kinds of emission sources to the extent that they may be different for plants producing the same product by different processes. Standards of performance established by the EPA for asphalt plants, concrete plants, refineries, sewage treatment plants, iron and steel plants, lead smelters, and certain other sources are described in the March 8, 1984, *Federal Register.* Other standards have been published since.

The Bubble Policy

In 1979 the EPA adopted a *bubble policy* with respect to emission control. Before this policy was adopted, emission control regulations were applied to every individual source of air pollution within an industrial plant. For the purposes of emission control, the bubble policy treats an entire plant as if all of its emissions came from a single port or stack emerging from an imaginary bubble surrounding the entire plant. Under this policy a facility that has several sources of pollution may be brought into compliance (with a state implementation plan) by a larger-than-required reduction in emission from one stack within the imaginary bubble and a less-than-required reduction in some other stack within the same bubble. Such flexibility is intended to allow a more cost-effective mix of pollution control measures without compromising clean air.

Enforcement: Giving Meaning to Air Quality Standards

Most of the enforcement of the Clean Air Act has been delegated to the states. As noted, the law required the states to submit implementation plans describing how emissions would be reduced and what steps would be taken to meet, maintain, and enforce the NAAQS. Such plans were to be approved by the EPA, and following approval, the states had to come up with a timetable for reaching the air quality standard.

At the federal level, violation of emission or performance standards can bring fines. The law authorizes the EPA to shut down plants during especially severe episodes. These emergency powers were used by the EPA for the first time in late 1971 when a serious air pollution episode occurred in Birmingham, Alabama. Penalties whereby a noncomplying firm is charged an amount equal to the savings gained from delaying compliance have put more teeth into the enforcement effort.

The Air Toxics Program of the EPA

One section of the U.S. Clean Air Act deals with toxic contaminants of air other than the so-called **criteria pollutants**—pollutants for which NAAQS have been established. Section 112 was designed to deal with pollutants that can get into the air and have toxic effects where, when, or as they are emitted—for example, in a train derailment or an industrial accident, or chronically as in an industrial leak or from a wood-burning stove—but may never reach measurably significant levels in air, generally speaking. Section 112 requires the EPA to identify such chemicals and mixtures and to develop strategies to control their emission.

The **air toxics** problem is clearly a difficult one. Toxic air pollutants are released through a wide variety of routine processes of life, and accidental releases of toxic chemicals occur every day. U.S. factories reported total emissions of 1.3 million tons of hazardous materials as recently as 1987, and this included more than 100,000 tons of carcinogens. Perhaps the most serious accidental release of toxics that has ever happened was the escape of toxic gas from a Union Carbide plant in Bhopal, India, in 1984, killing several thousand people and injuring many thousand more.

EPA progress in fulfilling its obligations under Section 112 was slow at first, but the agency was prodded by public watchdog groups and by its own studies that showed that the air toxics problem was indeed significant. In one of its studies, for example, the EPA found that the routine release of toxic substances into the air was causing some 1,300 to 1,700 deaths from cancer per year. In June 1985 the EPA announced a comprehensive strategy to deal with the air toxics problem. Elements of this air toxics strategy include (1) faster publication of national emission standards for hazardous air pollutants, (2) referral of regulation of certain air toxics to state and local agencies, (3) technical support to states monitoring air toxics, (4) support for a uniform right-to-know law (see Chapter 16), and (5) development of guidelines for community accidental release programs.

By March 1988 the EPA had issued decisions to regulate or not to regulate 36 pollutants under the air toxics program; 35 additional pollutants were then under review. The 1990 amendments to the Clean Air Act identified 189 substances and compounds that *must* be considered hazardous air pollutants.

We will consider air toxics further as part of the general problem of hazardous materials in Chapter 16.

Compliance Trends

The U.S. trend in bringing air pollution or air quality into compliance with the standards set in 1971 has been generally downward; that is, there has been a reduction in emissions and there has been an improvement in air quality. Despite increases in population and increased economic activity, anthropogenic emissions of criteria air pollutants (sulfur dioxide, nitrogen oxides, ozone, particulate matter, carbon monoxide, and lead) have declined in the U.S. since 1976. Still, as of 1992, 100 urban areas in the United States had air pollution levels exceeding federal standards. Figure 11.5 illustrates ambient air quality and emission trends in relationship to U.S. National Ambient Air Quality Standards for the period 1981 through 1990. Longer-range emission trends are illustrated in Figure 11.6.

Progress with emission control and ambient air quality levels of some pollutants has been remarkable. Most of the steep reduction in lead emissions is attributable to the phasing out of leaded gasoline. Tailpipe emissions on new cars have been reduced by nearly 97 percent since the early 1970s and half of the remainder is expected to be eliminated by 1995 (Kinnear, 1991). Ozone has remained problematic. As illustrated in Figure 11.5, ambient concentrations of criteria air pollutants are, on the average, below the standards with the notable exception of ozone. In the early 1990s, about 67 million people in the United States are routinely exposed to ozone concentrations that exceed the standards set by the Clean Air Act. At the core of the problem of the slow progress in dealing with ozone levels in American cities, according to the U.S. National Research Council, is that emissions of volatile organic compounds have been chronically underestimated. The NRC indicates that new approaches to controlling VOCs are needed (Stone, 1992).

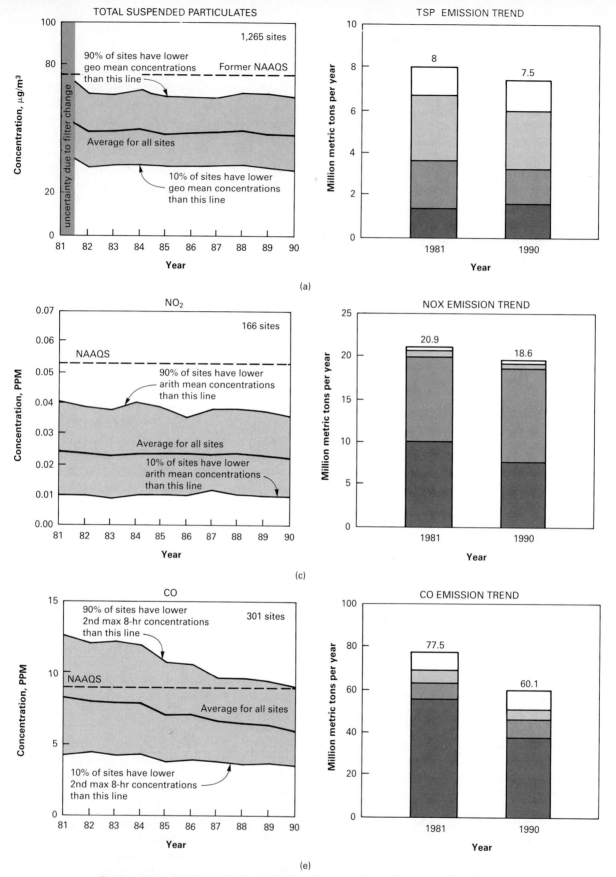

Figure 11.5 Ten-Year Ambient Air Quality Trends and Emission Trends in the United States. Plotted in each panel are average measurements for the number of sample sites shown. The right-hand panel in each set shows emissions by source for 1981 and 1990. (*Source: EPA, 1991, National Air Quality and Emission Trends Report 1990.*)

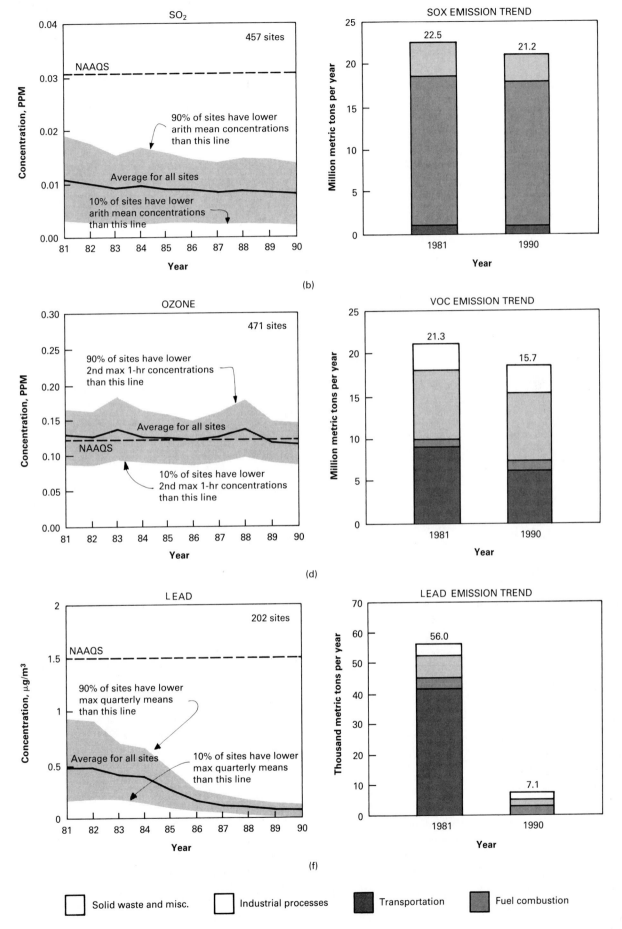

SO₂

OZONE

LEAD

SOX EMISSION TREND

VOC EMISSION TREND

LEAD EMISSION TREND

(b)

(d)

(f)

☐ Solid waste and misc.   ☐ Industrial processes   ■ Transportation   ■ Fuel combustion

Chapter 11   Control of Air Pollution

323

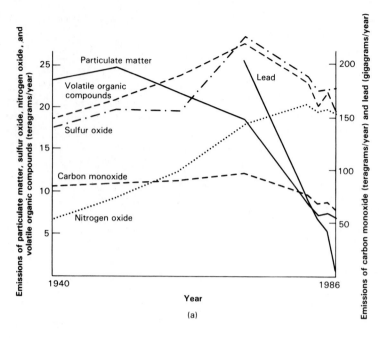

(a)

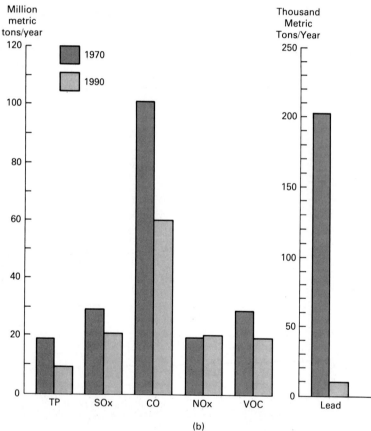

(b)

Figure 11.6 (a) Air Pollutant Emission Trends from 1940 to 1986. One teragram equals 10^{12} grams (1 million metric tons); one gigagram equals 10^9 grams (1,000 metric tons). National emission trends are not necessarily consistent with local emission trends. Also, because of other influences—meteorologic conditions, for example—national emission trends may not be consistent with national air quality trends. **(b) Comparison of 1970 and 1990 Emissions of Criteria Air Pollutants in the United States.** Lead clearly has had the most impressive decline; only the nitrogen oxides have not shown some improvement.

It should be noted here that most monitoring of air quality in the United States is conducted in urban areas and therefore does not reflect levels of air pollution that nonurban areas are experiencing. It should also be noted, since atmospheric emissions have impacts over broad geographic areas extending to the global, that global atmospheric concentrations of carbon dioxide and other greenhouse gases are increasing.

Monitoring of NAAQS and emission trends is made possible through the existence of a uniform network, the State and Local Air Monitoring System

(SLAMS), and a subset of this established in 1979 called the National Air Monitoring System (NAMS). Both systems monitor different numbers of high-pollution sites for different pollutants.

The Air Pollution Index and Air Pollution Alerts

Before 1976, air pollution indices varied from city to city in the United States. In 1976 the Council on Environmental Quality established a task force to come up with standardized air pollution indices to be used across the country. The task force recommended that an index be established on the basis of five principal pollutants: sulfur dioxide, carbon monoxide, total suspended particulate matter, oxidants, and nitrogen dioxide. The **pollutant standard index (PSI),** illustrated in Table 11.4, assigns the value of 100 to a concentration of a pollutant when it is at the NAAQS level set by the EPA. If a pollutant reaches half the level of the NAAQS, its PSI is reported at 50. At higher levels of the index, the numbers have the same kind of significance in terms of human health. For example, a PSI of 500 for any one of the five pollutants means that that level of that pollutant brings significant risk or significant harm. Another key feature of this index now in general use is that each pollutant elevated above the NAAQS is reported separately; any pollutant exceeding the value of 100 is reported. National trends with respect to the Pollutant Standard Index are illustrated in Figure 11.7.

Table 11.4 Nationwide Standardized Air Pollution Index

PSI Value	Air Quality Level	TSP (24-hr), $\mu g/m^3$	SO$_2$ (24-hr), $\mu g/m^3$	CO$_2$ (8-hr), mg/m^3	O$_3$ (1-hr), $\mu g/m^3$	NO$_2$ (1-hr), $\mu g/m^3$	Health Effect Descriptor	General Health Effects	Cautionary Statements
500	significant harm	1,000	2,620	57.5	1,200	3,750			
400	emergency	875	2,100	46.0	1,000	3,000	hazardous	Premature death of ill and elderly. Healthy people will experience adverse symptoms that affect their normal activity.	All persons should remain indoors, keeping windows and doors closed. All persons should minimize physical exertion and avoid traffic.
300	warning	625	1,600	34.0	800	2,260	hazardous	Premature onset of certain diseases in addition to significant aggravation of symptoms and decreased exercise tolerance in healthy persons.	Elderly and persons with existing diseases should stay indoors and avoid physical exertion. General population should avoid outdoor activity.
200	alert	375	800	17.0	400*	1,130	very unhealthful	Significant aggravation of symptoms and decreased exercise tolerance in persons with heart or lung disease, with widespread symptoms in the healthy population.	Elderly and persons with existing heart or lung disease should stay indoors and reduce physical activity.
100	NAAQS	260	365	10.0	160	—	unhealthful	Mild aggravation of symptoms in susceptible persons, with irritation symptoms in the healthy population.	Persons with existing heart or respiratory ailments should reduce physical exertion and outdoor activity.
50	50% of NAAQS	75†	80†	5.0	80	—	moderate		
0		0	0	0	0	—	good		

* 400 $\mu g/m^3$ was used instead of the O$_3$ alert level of 200 $\mu g/m^3$.
† Annual primary NAAQS.
Source: Council on Environmental Quality.

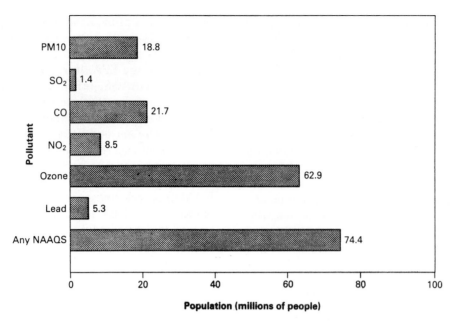

Figure 11.7 Numbers of Americans in 1990 Breathing Air with Pollutant Levels above U.S. Ambient Air Quality Standards. Ozone is clearly the biggest problem.

Stay Indoors?

Sometimes during air pollution alerts, people are advised to stay indoors and to minimize physical activity. Human studies have shown that a subject's level of exercise during ozone exposure is directly related to the magnitude of diminished lung function. Restriction of physical activity would obviously reduce gas exchange and lessen exposure of lung tissue to ozone and other damaging pollutants.

The reason for advising people to stay indoors is perhaps less obvious. Some recent evidence indicates that there may be an ironic twist to this advice. The advice is based on observations that for a variety of complicated reasons, pollutant levels are usually much lower indoors during outdoor air pollution episodes. This is due partly to the fact that there are things in indoor air that react with pollutants and neutralize them. The irony is that indoor pollution may be more of a problem than outdoor pollution *most of the time*. People spend anywhere from 70% to 90% of their time indoors and another 5% to 10% of their time in an automobile or other vehicle. This is significant in light of the fact that formaldehyde, nitrogen dioxide, asbestos, pesticides and other such highly toxic substances, carbon monoxide, radon, particulate matter, and various other components of tobacco smoke have been implicated as airborne health hazards in many homes and other buildings.

Formaldehyde apparently escapes from plywood, particle board, and certain types of foam insulation. The type of insulation known to emit formaldehyde was installed in more than 100,000 homes per year during the late 1970s. Gold (1980) reports that levels of formaldehyde high enough to cause dizziness, rashes, nosebleeds, and vomiting have been measured in some homes.

Nitrogen dioxide is emitted from gas stoves. Particulates come from tobacco smoke, cooking, and aerosol sprays. Carbon monoxide comes from cars in garages, stoves, and other sources of incomplete combustion. Tobacco smoke contains particulate matter, carbon monoxide, and assorted carcinogens. It has long been known that cigarette smoke has health effects on nonsmokers. If it can be inferred from the 90% of the time people spend indoors that 90% of all cigarettes are smoked indoors, American buildings receive the combustion products of 500 billion cigarettes each year.

A double irony to all of this is that the problem is getting worse because of efforts to save energy. Weatherstripping and other means of tightening up buildings reduce the rate of air exchange and increase the buildup of pollutants indoors. The EPA once estimated that there would be 20,000 more cases of cancer (mainly of the lung) each year from radon exposure if ventilation rates were cut in half (Gold, 1980).

Radon. Radon is an invisible, odorless, tasteless, radioactive gas found in nature. It comes from the natural radioactive decay of uranium and is different from other decay products in that it exists as a gas—a fact that allows radon to become part of the air. The uranium that gives rise to radon is commonly present in soils and rocks. If the radon gas is generated in the open, it quickly becomes diluted to insignificant levels as it mixes with air. Problems arise when radon is released in confined places such as basements and crawl spaces. In areas in which there are significant amounts of radon in the soil, the accumulation of the gas in a house or other structure will depend on its construction (mainly the degree of ventilation). Radon can move through the soil and enter a house directly through cracks in basement floors, or it can enter well water and escape into the home as the water is used. There is indirect evidence that radon increases the risk of developing lung cancer—the more radon, the higher the concentration, and the longer the exposure, the greater the risk. The evidence comes from the cancer experience of workers exposed to various amounts of radon in their work. Interestingly, studies have generally failed to show a strong correlation between radon levels and lung cancer rates in areas where radon levels are notably higher than average. In one such study (Cohen, 1989), the lung cancer–radon relationship was examined in U.S. counties where the lung cancer incidence rate was either high or low. Although in theory the radon levels in the high-rate counties should have been three times that in the low-rate counties, the actual findings were that the high counties had half the radon levels of the low counties. Such studies may be confounded in part by the fact that there can be great differences in radon levels from house to house and in part because of the overwhelmingly greater significance of smoking on lung cancer rates. Nevertheless, it has been estimated that 5,000 to 20,000 lung cancers are caused by radon each year in the United States. The cancer is believed to be caused as radon undergoes decay and releases radiation that has the capability of altering genetic material. Two sources of information about radon have been published by the U.S. Environmental Protection Agency. One of these (U.S. EPA, 1987) is a summary of methods found to reduce high concentrations of radon in houses.

Other potential air pollution problems in the home include various household products containing asbestos, molds, and toxic materials. Asbestos, well established now as a lung tissue carcinogen, is still found in many homes as a fireproof insulator and noise-reducing material. In many older homes, asbestos is present in the form of tape on heating duct joints. Asbestos fibers get into the air whenever asbestos-containing materials are disturbed. Molds can be found in almost every home, growing in damp places like shower stalls, air conditioners, basements, and the soil of potted plants. Such molds are allergens for many people, and they cause cold, flu, or allergy-type symptoms. Mold control strategies include using dehumidifiers to dry the air, using disinfectants, and changing filters regularly. Pesticides, which are toxic, be-

come part of the air in homes when they are used to control household pests such as termites (see Chapter 16). Other toxic substances get into home air by means of cleaning materials and as solvents found in products such as paints. The risk in using household air toxics such as these can be minimized by carefully following the directions on the container.

Bringing in the outdoors seems to help. Plants help to remove certain indoor air pollutants, such as carbon monoxide, nitrogen oxides, and toluene. Experiments at a NASA laboratory found that philodendrons and spider plants are particularly good at getting rid of indoor air pollutants.

It is ironic that the wood stove, in a way a symbol of an ecologically sensitive lifestyle, has become an ecological problem in many U.S. cities. There are well over 10 million wood stoves in use in the United States, and nearly a million more are purchased each year. Denver alone has more than 250,000 wood stoves and fireplaces. Even wood stoves with nearly perfect ventilation add particulate matter and other pollutants to indoor air. Also, wood stoves are becoming increasingly responsible for much outdoor air pollution; they are already a significant part of the problem in Denver. Although wood is a renewable energy source and is in other ways environmentally sound, wood burning generates both carbon monoxide and particulate matter (smoke) as well as other pollutants.

The matter of indoor air pollution raises an important point. Indoor pollutants should be taken into account in epidemiological studies that attempt to measure the relationship between health and air pollution. It is not simply what can be measured in outdoor air that people are being exposed to; indoor exposure may have an overriding effect. To be technically correct in assessing the health effect of exposure to whatever pollutant, the levels of that pollutant should be measured in all of the places where people spend time and then their exposure weighted accordingly in relation to the amount of time they spend in each place. This is actually done by having representative people carry samplers around with them all day.

1977 Amendments to the Clean Air Act

As the Clean Air Act was implemented, it became obvious that some of its provisions had to be tightened and others had to be relaxed. Some of the reasons for the 1977 amendments included poor fuel consumption efficiencies in cars equipped with air pollution control devices, failure of the auto industry to reach the target dates set by earlier versions of the act, failure of many regions of the country to achieve the primary standards by 1975 and secondary standards by 1977 as required in the original legislation, and pressure to postpone implementation of transportation control plans required of urban communities.

The amendments to the Clean Air Act became effective on August 9, 1977. Among the things the amendments did were extend deadlines for achieving certain emission standards and require states to submit revised plans for attaining air quality standards.

Effectiveness of the Clean Air Act and Its Amendments Through the 1980s

In some respects the Clean Air Act has been successful; in other respects, accomplishment has fallen short of expectations.

On the plus side, much of the trend of increasingly dirty air before 1970 has been halted and reversed; standards have been set, and progress toward achieving these standards has been significant. As we pointed out earlier, emissions of most pollutants have been reduced by various control technologies and regulations (see Figure 11.6). Reductions in particulate emissions are attributed to the installation of control equipment in industry. Sulfur dioxide emissions have been reduced through the use of low-sulfur fuels, scrubbers, and new combustion technologies despite increased electric power generation. Hydrocarbon levels were reduced significantly in many places as a direct result of emission controls on highway vehicles (Figure 11.8). In short, most areas of the United States have better air than they had in 1970.

On the negative side, progress has been slow; many of the amendments to the Clean Air Act in 1977 simply extended the deadlines for reaching clean air goals, and by the mid-1980s it was obvious they would have to be extended once again.

As for the negative side of the bigger picture, it is estimated by the EPA that some 150 million Americans are breathing unhealthy air today. The American Lung Association estimates that air pollution in the United States costs some $40 billion annually in health care costs and in lost productivity. Figure 11.7 shows that the number of people living in U.S. counties with air quality levels above the primary NAAQS was still quite significant in 1990.

The Clean Air Act amendments of 1977 specified that cities had to meet the NAAQS by 1982; in 1983, because many cities didn't achieve this goal, the deadline was extended five years to December 31, 1987. Despite the threat of loss of federal aid, many cities missed this deadline as well.

Ozone and carbon monoxide continue to be the two most stubborn problems in the effort to control air pollution, particularly in the northeastern United States and in southern California. Twenty-five northeastern cities did not make the December 1987 deadline. Although emission controls on automobiles have helped, there remains the problem of multiple

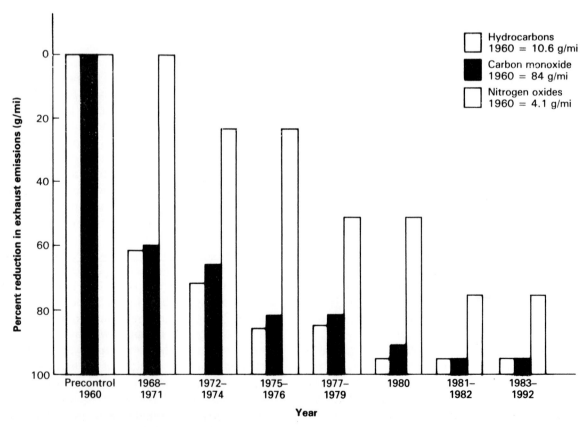

Figure 11.8 Trends in the Reduction of Automobile Emissions in the United States.

miscellaneous sources of ozone-generating organic compounds—quick-drying paint, bakery exhaust, and the like. Automobiles are still a big part of the problem. Vehicle inspection is being required in more and more cities, and these and other additional air pollution controls will likely be implemented in some form in the future.

To cite one specific example, the Los Angeles basin exceeded the ozone standards about half the time in 1987—and still does. Some 60 other cities are still in violation of the standard about 30 times a year. Although many of the latter were projected to meet the standard by 1992 (see Cone, 1987), southern California cities will have to resort to even more stringent air pollution emission control laws under the amendments to the Clean Air Act passed in 1990.

It was necessary to amend the Clean Air Act yet again because Americans were still breathing dirty air a decade after the 1977 amendments. At least part of the reason for this was that as a narrowly focused piece of legislation, the act did not regulate such clearly relevant things as planned obsolescence, transportation inefficiency, and the wasting of energy in other ways. In these respects, the 1990 amendments may also fall short. Perhaps other crises will make it more generally obvious that we need laws that are ecologically comprehensive—laws able to overcome social inertia in significant ways—to sacrifice current "sacred cows." We must continue to address both what we do and how we do it.

The New U.S. Clean Air Bill

After a political stalemate that lasted more than a decade, amendments to the Clean Air Act were signed into law in 1990. If fully implemented and enforced, the new law will require that automobiles emit 40% to 60% less pollution, industry must cut toxic chemical emissions by 90%, cleaner-burning gasoline must be provided to cities with the most serious smog problems, and coal-burning utilities must cut sulfur oxide emissions by half. Many of the bill's provisions will not take effect until the end of the 1990s, and many cities are given up to 15 years to meet air quality standards. Much depends on the administrative regulations to be issued by the EPA. The original Clean Air Act, after all, was to have delivered clean air to all American cities by 1987; almost 100 cities have yet to meet these standards. The new law has four main titles or sections, covering (1) attainment and maintenance of air quality standards, (2) motor vehicles, (3) toxic air pollutants, and (4) acid deposition. Other titles address permits, stratospheric ozone depletion, enforcement, research, and job loss benefits. It is estimated that the new law will cost some $25 billion annually. A summary of the provisions in the law follows (for further details, see Pytte, 1990):

1. *Planning.* The law requires each state to develop an implementation plan to include setting enforceable emissions limits, providing for the means to acquire air quality data, controls for interstate air pollution, and requirements that stationary sources monitor emissions, among other things.

2. *Urban air quality standards.* Smog is to be reduced by 15% in the first six years, then by an additional 3% annually until air standards are met. The most serious problem cities will have longer; Los Angeles has 20 years. New emission controls are required on both mobile and stationary sources.

3. *Motor vehicles.* Hydrocarbon emission limits are to be reduced by 40% (to 0.25 gram per mile). Nitrogen oxide emissions must be reduced 60% to 0.41 gram per mile. The law calls for cleaner-burning gasoline in cities with the worst smog problems and a gradual phasing-in of alternative fuels for automobiles (see Enrichment Box 11.1). It also requires stronger warranties on emission control devices.

4. *Air toxics.* Industry is required to install the "maximum achievable control technology" to reduce the emissions of some 189 toxic chemicals (including many carcinogens) by 90% by the end of the 1990s. The law also requires the EPA to establish pollution control programs sufficient to reduce the cancer risk by 75% by 1995 and creates a special chemical safety board to deal with chemical release accidents. The law also requires the EPA to work with the National Academy of Sciences to review "risk assessment methodology used by the Environmental Protection Agency to determine the carcinogenic risk associated with exposure to hazardous air pollutants."

5. *Acid rain.* Coal-burning electrical utilities are required to reduce sulfur oxide emissions by half—some 10 million tons; nitrogen oxide emissions must be reduced by 2 million tons from 1980 levels. This section of the law has a provision whereby under a national cap of 8.9 million metric tons of sulfur dioxide annually by the year 2000, each utility company will be given a limit that it can exceed only by buying "rights" to additional sulfur dioxide emissions from another company.

6. *Ozone depletion.* The law calls for the halt of CFC production by the end of the twentieth century and calls for the phase-out or restriction of production and use of these and other ozone-depleting substances by various deadlines.

Penalties in the new law prescribe up to $25,000 per

Alternative Fuels

Because most of the world's inexpensive petroleum is in the politically volatile Middle East and because conventional fuels derived from petroleum are a leading cause of urban air pollution and contribute to global warming, alternative fuels and other means of propelling automobiles continue to receive serious consideration. The alternative fuels include ethanol, methanol, hydrogen, and natural gas; alternatives to the direct consumption of fuels include electric power via batteries and via solar cells.

— *Methanol* produces less carbon dioxide than gasoline, but the actual impact on CO_2 generation depends on how the methanol is made. If made from coal, the CO_2-generating impact of methanol is at least as high as that of gasoline, possibly twice as high. Methanol also gives off lower emissions of hydrocarbons, the precursors of ozone.

— *Ethanol* produces less CO_2 than gasoline, but its overall impact on carbon dioxide depends on the energy source used in distillation and in the energy intensiveness of the corn crop or other source material.

— *Compressed natural gas (CNG)* burns more efficiently than gasoline and thus will reduce CO_2 emissions, but if it leaks into the atmosphere, methane (natural gas) is also a greenhouse gas. CNG (compressed natural gas) is already used fairly widely in parts of Europe and Canada. CNG must be stored under pressure and is generally much more difficult to handle than gasoline.

— *Propane* is an efficient, "high-octane" fuel, but it requires modifications in automobile engine design.

— *Electricity* can be attractive as a propellant for motor vehicles, depending on the means by which the electricity is generated. One of the downsides of the use of electricity is that recharging can take hours.

Whatever the potential impact of alternative fuels or changes in the means of propelling vehicles, changes in the design of cities and in transportation habits could have a far larger, potentially positive environmental impact. In any case, governments must play a major role in the development of alternate fuels because the market alone is not likely to cause them to be developed. Start-up barriers for both producers and users are large, and the market does not on its own consider such things as global warming, air pollution–induced health problems, and other indirect economic costs or even energy security. The amended U.S. Clean Air Act does, in fact, require that certain numbers of vehicles able to use cleaner fuels be sold in California, it gives the administrator of the EPA the authority to require urban buses to use alternative fuels, and it directs the administrator of the EPA to issue regulations governing the formulation of gasoline.

day in civil penalties and criminal prosecution of executives of companies that knowingly endanger health.

Air Pollution Control Strategies and Technologies

As established earlier, many American cities still exceed primary standards more than 100 days each year. A little of this problem could be blamed on the need for better control technologies, but the primary reason for the failure to achieve clean air in the United States has been the failure of strategies other than technological ones. There are obviously a number of very fundamental technological *and* nontechnological approaches to the control of air pollution, all of which are being used to some degree. These might be enumerated as follows.

1. *Alternatives.* A number of kinds of alternatives could be used: (a) We could change our lifestyles to require less of the energy that brings pollution

Mini-Glossary

Afterburner: an air pollution control device that keeps undesirable organic compounds from escaping by burning them.

Bag house: an air pollution control device that traps particles by forcing air containing dust and other particles through large filter bags usually made of fiberglass.

Catalytic converter: an air pollution control device that removes organic compounds from a stream of exhaust by completing their oxidation into carbon dioxide and water in the presence of a catalyst.

Cyclone: an air pollution control device that removes heavy particles by causing the particle-containing gas to swirl and removing the particles through centrifugal force.

Electrostatic precipitator: an air pollution control device that removes particles by putting an electric charge on them and then attracting them to an electrode of opposite charge.

Scrubber: an air pollution control device that uses a liquid spray to remove aerosol and gaseous substances from an airstream. The principle is that the gases are removed by absorption, by dissolving in the liquid, or through chemical reaction with a chemical in the liquid spray.

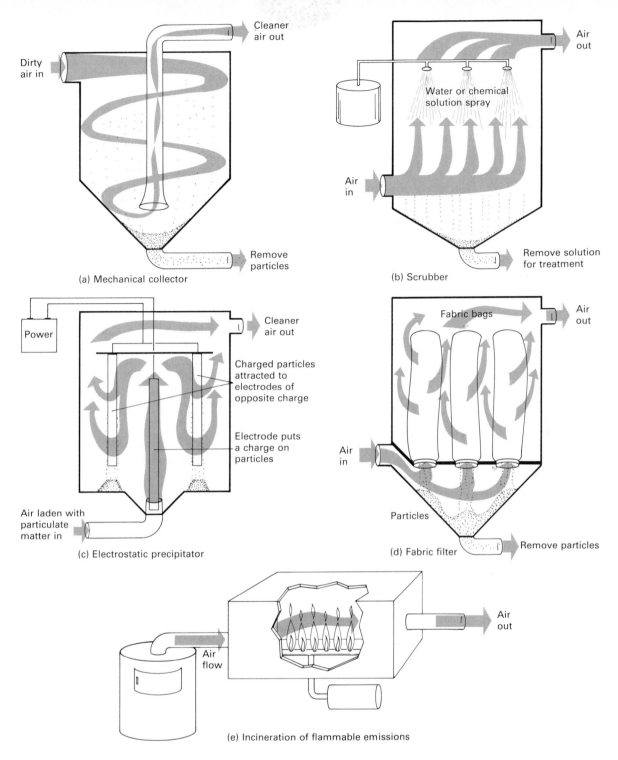

(a) Mechanical collector

(b) Scrubber

(c) Electrostatic precipitator

(d) Fabric filter

(e) Incineration of flammable emissions

Figure 11.9   Common Air Pollution Control Devices.

with it, (b) we could find alternative ways to achieve our desired lifestyle by finding different technologies that are less polluting, or (c) we could use alternative sources of energy that do not have as much pollution as a by-product.

2. *Reduction.* We could continue to do the same sorts of things but less of them. For instance, we could switch to still smaller cars, modify our technology to make cars still more efficient, or make our overall expenditure of energy more ef-

ficient by insulating houses (and finding ways to neutralize indoor pollutants).

3. *Remove pollutant-generating materials from fuels.* An example of this kind of technological fix would be to remove sulfur from coal before it is burned.

4. *Keep pollutants from escaping into the air as fuels are burned.* This can be done by putting catalytic converters in automobiles and installing scrubbers (see Figure 11.9).

5. *Remove the pollutant from ambient air.* We include this absurd strategic element for the sake of completeness (we say absurd though it has been proposed that small concentrations of pollution-scavenging chemicals could be added to the air over cities on days of high pollution) (Maugh, 1976). The second law of thermodynamics tells us that it would be more efficient to approach the problem of minimizing air pollution by getting at the offending material before it is scattered throughout the environment.

6. *Change the affected receptors to protect them.* For example, we could coat statues with protective materials, breed resistant plant species, and add lime to lakes.

The range of strategies and plans adopted around the country for controlling pollutants have included most of this spectrum. Implemented and planned approaches involving the automobile, for example, have included the following:

1. Reduction in size, horsepower, and compression ratios of the internal combustion engine
2. Improvements in the technology of controlling the emission of pollutants from automobiles
3. Reduction of the number of automobiles by improving mass transit, encouraging carpooling, restricting downtown parking, and establishing bicycle paths
4. Periodic inspections of automobiles
5. Replacement of the gasoline engine with electric or steam engines

Strategies for reducing sulfur dioxide pollution by power plants have included the following:

1. Substituting low-sulfur fuels for high-sulfur fuels
2. Using smokestacks high enough to poke through inversion layers (as we discussed earlier, this is a misguided strategy; it results in acid rain downwind)
3. Chemically scrubbing sulfur dioxide and other compounds of sulfur from the effluent going up stacks
4. Removing sulfur from the coal and oil and gas before it is burned—in addition to reducing the demand for electric power

Let us look at each of the approaches to air pollution individually and see which of them seem to hold the most promise. We will editorialize a bit; see if you agree.

Changing Lifestyles

If the environmental movement up to this point has done anything at all, it seems to have revealed at least one great truth: that appeals to human reason to change lifestyles and reduce consumption are exercises in energy-wasting rhetoric. Perhaps human beings respond only to physical or financial pain. Even then it might be argued that humans can perceive relationships between pain and pollution only when the pain is immediate. Unfortunately, it appears that we must rely on some combination of law, significant economic pressure, and education to bring about necessary, pollution-reducing changes in lifestyle. A law that requires that cars get at least 30 miles per gallon and an economic system that requires that the cost of pollution control be passed on to consumers are obviously much more effective than simply saying "please." In this regard we think it somewhat ironic that the *energy crunch,* which resulted in legislation that weakened the Clean Air Act, may—because of the expense of energy—ultimately prove to be the most important effector of clean air by bringing about reductions in fuel consumption.

Alternative Technologies

Alternative solutions often bring alternative problems. The Clean Air Act legislation in the early 1970s brought about a gradual but profound switch from coal to natural gas, natural gas being a much cleaner fuel. Replacing gasoline-powered engines with electric engines, for another example, would simply change the source of pollution from the automobile to the plants where electricity is generated. Such a switch might at least make the pollution more controllable by consolidating source locations. Furthermore, legislation would probably be required to effect a switch to electric automobiles. Our technology can already produce electric automobiles, we would have to get used to stopping every so often to be recharged, and we might not be able to go as fast. Most people are not likely to switch to this kind of alternative willingly and on the basis of appeals to reason.

Perhaps the key element in the strategy having to do with alternative technology is *efficiency.* Most air pollution comes from the combustion of fossil fuels. Obviously, any gain we make in efficiency would have a direct payoff in terms of reducing the pollution problem. Reducing energy consumption would reduce the air pollution problem.

Alternative Energy Sources

Most of what we had to say about different sources of energy for our society was presented in Chapter 6. Suffice it to say here that not all of the ways of producing energy pollute the air to the same degree. Each form of energy generation has its unique problems and lends itself in slightly different ways to the control of air pollution. This must be taken into account in the consideration of various solutions to our air pollution problem and our energy crises.

Cleaning Fuel

It is possible to remove inorganic sulfur from coal. The basic technology for doing this already exists, but only recently have the pressures of economics, law, and scarcity of fuel combined to focus technological attention on more economical ways of doing such things. The point we wish to make here is that it is possible to reduce air pollution by taking chemicals like sulfur out of the fuel. Naturally, this costs money.

Trapping Effluents

The technologies for cleaning effluent gases extend from "tightening up" (no open tanks, no leaky valves or joints) to keeping things like hydrocarbons from leaking into the environment to chemically or physically scrubbing gaseous or particulate pollutants from the effluents going up stacks and out exhaust pipes.

It is worth noting here that many toxic and hazardous materials, including air pollutants, are valuable resources that are simply out of place. So it would make sense to approach pollution as an expensive loss. In other words, there would be valid reasons for pollution control even if pollution did not harm people and materials. Many firms have found that by being forced to tighten up, they actually save money by preventing the loss of valuable materials.

Several kinds of particulate control devices are illustrated in Figure 11.9. It should be noted that each of them produces a product that must be disposed of in some way so as not to cause a substitute environmental problem. The scrubbers shown in Figure 11.9 can effectively remove gaseous pollutants from the air. Highly water-soluble gases, for instance, can be removed in water sprays or in chemical sprays that neutralize the pollutants. Here too there must be ways of disposing of the resultant solutions. Another type of scrubber is a combustion chamber or afterburner. Hydrocarbons can be removed from gases in such devices through combustion.

The control of automobile emissions is now effected largely by the catalytic conversion of pollutants to relatively harmless reaction products.

The state-of-the-art catalytic converter uses metals such as platinum, palladium, and rhodium. Platinum and palladium act as catalysts in the conversion of hydrocarbons and carbon monoxide in automobile exhaust into carbon dioxide and water. Rhodium, a reducing catalyst, converts oxides of nitrogen into nitrogen gas and oxygen. As early as 1976, fully 85% of the automobiles sold in the United States were

At top, fumes pour from the basic oxygen furnace stack at Bethlehem Steel Corporation's plant in Bethlehem, Pennsylvania. At bottom, the same stack emits almost no fumes after installation of an electrostatic precipitator control system.

equipped with catalytic converters. According to the 1976 report of the Council on Environmental Quality, catalytic converter technology, together with the trend toward smaller cars and better gas mileage, made 1976 cars 13% better than the 1975 models and 30% better than the 1974 fleet in terms of reduced air pollution. As indicated earlier, tailpipe emissions on new cars *today* are only 3%, or less, of those on early 1970s models.

Although improved and more economical control technologies are needed, the most important limiting factors in cleaning up the air appear to be sociopolitical and socioeconomic.

In Summary

Scattered throughout this chapter were defined and implied tasks for science and technology. Some things we need to learn, and other things we must learn to do more economically. Perhaps the most glaring things we do not know are the effects that low levels of pollutants alone and in combination have on human beings. Perplexing problems in this regard remain in many areas, including these:

1. Varying sensitivity to pollutants among test species and human beings

2. The dynamic chemistry of pollution under the influence of sunlight
3. The varying stability of pollutants
4. The chronic, slow, insidious nature of air pollution-related diseases
5. The variable biological magnification of different pollutants (see Chapter 16).

Although these will translate into many hours of scientific work, far more of the task of dealing with air pollution falls to institutions other than science and technology.

Studies can be designed, and laborious, elegant, time-consuming, expensive studies can and surely will add to what we know as facts. Research can pinpoint health effects, establish more precise thresholds (if there are thresholds), and differentiate very toxic from not-so-toxic pollutants. Both political and economic value judgments will obviously be required not only because our understanding will always be short of complete but also because there will always be at least a little room for choice and interpretation.

In closing, we would like to recall one simple fact and point to the connection between this chapter and the earlier chapters on energy and biogeochemical cycles. The fact is that *there will always be pollution.* We have defined pollution as the sum total of the adverse effects we have on our environment, and so, as with all environmental problems, the task is to learn more and to work to diminish the effects. A connection between this chapter and Chapter 3 is that pollutants are constituents of biogeochemical cycles. Air pollution is a result of our kicking up the lithosphere too fast and furiously for the ecosphere to handle it. When this happens in nature, pollution *becomes* the environment, and we have to live in our own waste.

There is a profound connection between these chapters on air pollution and the chapter on energy. Air pollution is largely a phenomenon related to the combustion of fossil fuels. The statement that we may have to choose between energy and the environment simply is not true. What we have to choose between is the waste and socially costly uses of energy and the environment. As more of us recognize this simple fact, significant things should begin to happen more quickly.

Concepts to Remember

1. The cost of controlling air pollution is significant. Some $300 billion was spent to control air pollution in the United States between 1978 and 1987, nearly all of it because of the Clean Air Act. Annual costs of all forms of air pollution control amount to just over 1% of the gross national product.
2. By some measures, more jobs have been gained than lost because of the Clean Air Act. Things in general cost a little more because of the cost of all forms of pollution control.
3. Determining the cost of not controlling air pollution is much more difficult than determining the costs of pollution control. One estimate of the benefits of cleaner air brought mainly by the Clean Air Act in the late 1970s placed the value of prevented damage to health, vegetation, materials, and property values in the range of $4.6 billion to $51.2 billion, with a best guess of $21.4 billion.
4. If all things had a dollar value, our air pollution control task would simply be to balance or match the costs of air pollution control with the cost of not controlling air pollution. Not all effects of air pollution have universally agreed dollar values, however.
5. U.S. National ambient air quality standards (NAAQS) have been set for particulate matter, sulfur dioxide, nitrogen dioxide, ozone, carbon monoxide, photochemical oxidants, and lead.
6. Emission standards are the maximum rates set by the EPA at which relatively new pollution sources are allowed to emit pollutants. States may set more stringent emission limitations for new plants, and they set limitations for old plants for which there are no federal standards. According to a state's plan to achieve the NAAQS, the EPA also sets emission standards for especially hazardous substances such as asbestos and mercury and for automobiles.
7. A bubble policy permits the EPA to consider a plant's emission sources as if they were all emitted through a common vent.
8. The pollutant standard index rates each pollutant on a common scale based on a value of 100 when a pollutant is at its NAAQS level.
9. Although staying indoors does in fact lessen exposure to bad outside air, the indoors has characteristic air quality hazards of its own, notably particulate matter, asbestos, nitrogen dioxide, radon, and formaldehyde.
10. Although there was a federal Air Pollution Control Act in 1955 and local air pollution laws that go back to the 1890s, modern clean air

legislation in the United States started with the Motor Vehicle Act of 1960 and began to assume its present form in the Clean Air Act of 1970.

11. The Clean Air Act has resulted in improvements in air quality, but the U.S. still has a long way to go. Air pollution is a serious problem throughout the world, in both developed and developing nations.

12. Clean air strategies range from scrubbing stack gases to energy conservation; conservation helps to solve energy, land use, and water resource problems as well as the air pollution problem.

Environmental Career Profile

Bonnie Jean Smith

Bonnie Jean Smith is Coordinator for environmental learning in the U.S. Environmental Protection Agency's Region III, headquartered in Philadelphia. The EPA is in the executive branch of the U.S. federal government and is responsible for setting and enforcing most federal environmental regulations. In particular, the EPA is responsible for implementing and enforcing some 14 major U.S. laws protecting the environment. An important aspect of the work of the EPA is to see to it that citizens recognize the value of a clean environment and of pollution prevention. Through a recently enhanced environmental education thrust directed at schools and other community institutions, the EPA seeks to promote environmental stewardship as an inherent part of life on the part of U.S. citizens. Ms. Smith's work in the Center for Environmental Learning is to promote environmental education, to improve the public's understanding of current and emerging policy issues, and to increase opportunities for the public to have input into the work of the EPA. To do this, Ms. Smith helps to organize and stage lectures, forums, and seminars on such topics as waste minimization, air toxins, risk analysis, indoor air pollution, and dispute resolution. She manages EPA's Region III awards program that recognizes exemplary environmental education. She works with educators, leaders of environmental organizations, officials of local and state government agencies, and corporate leaders. She also works with many capable, dedicated people within the EPA. Ms. Smith finds her work rewarding because she sees environmental education as the key to the EPA's long-term success.

Ms. Smith has a bachelor's degree in communications. Before coming to the EPA she had 15 years' experience working in various public-private partnership programs in both Minneapolis–St. Paul and New York City. Among the skills and qualifications she finds most useful in her work are speaking and writing skills. She says that it is also important to have a basic understanding of contemporary environmental issues as well as some experience in working with partnership programs and in environmental education. It also helps, she says, to have a good sense of humor.

References and Further Reading

References marked with an asterisk are cited in the chapter.

*Barrett, L. B., and Waddell, T. E., 1973. *The Cost of Air Pollution Damage: A Status Report* (Publication No. AP-85). Research Triangle Park, N.C.: National Environmental Research Center, U.S. Environmental Protection Agency.

*Basala, A. C., 1981. "The Office of Air Quality Planning and Standards Benefits Analysis Program," paper presented at the 74th annual meeting of the Air Pollution Control Association, Philadelphia.

*Cohen, B. L., 1989. "Expected Indoor ^{222}Rn Levels in Counties with Very High and Very Low Lung Cancer Rates," *Health Physics* **57**:897–907.

*Cone, M., 1987. "L.A. to E.P.A.: Don't Hold Your Breath," *Sierra* **32**(6):27–32.

Council on Environmental Quality, 1989. *Environmental Trends.* Washington, D.C.: U.S. Government Printing Office.

*Council on Environmental Quality, 1991. *Environmental Quality: 21st Annual Report of the Council on Environmental Quality.* Washington, D.C.

*Freeman, A. M., 1982. *Air and Water Pollution Control: A Benefit-Cost Assessment.* Wiley, New York.

*Gold, M., 1980. "Indoor Air Pollution," *Science '80,* March–April, pp. 30–33.

*Hall, J. V.; Winer, A. M.; Kleinman, M. T.; Lurmann, F. W.; Brajer, V.; and Colome, S. D., 1992. "Valuing the Health Benefits of Clean Air," *Science* **255**:812–816.

*Heintz, H.; Hershaft, A.; and Horak, G., 1976. *National Damages of Air and Water Pollution Control.* Report submitted to the U.S. Environmental Protection Agency.

*Horst, R. L.; Manuel, E. H.; Bentley, J. T.; and Basala, A. C., 1984. "Quantitative Information in the NAAQS Review Process: A Summary of a Benefit-Cost Analysis for Particulate Matter," paper presented at the 77th annual meeting of the Air Pollution Control Association, San Francisco.

*Kinnear, J. W., 1991. "Clean Air at a Reasonable Price," *Issues in Science and Technology,* Winter 1991–92: 28–31.

*Krupnick, A. J., and Portney, P. R., 1991. "Controlling Urban Air Pollution: A Benefit-Cost Assessment," *Science* **252**:522–528.

*Management Information Services, 1986. *Economic and Employment Benefits of Investments in Environmental Protection.* Washington, D.C.: MIS.

*Maugh, T. H., 1976. "Photochemical Smog: Is It Safe to Treat the Air?" *Science* **193**:871–873.

*Pytte, A., 1990. "Clean Air Act Amendments," *Congressional Quarterly Weekly Report,* Nov. 24, pp. 3934–3963.

*Ridker, R. G., 1967. *Economic Costs of Air Pollution.* New York: Praeger.

*Seabrook, C., 1987. "Air Pollution Costing Farmers up to $5 Billion in Lost Crops," *Atlanta Constitution,* Oct. 27, p. 1A.

*Stone, R., 1992. "NRC Faults Science Behind Ozone Regs," *Science* **255**:26.

*Uhling, H. H., 1950. "The Cost of Corrosion in the United States," *Corrosion* **51**:29–33.

*U.S. Environmental Protection Agency, 1979. *The Cost of Clean Air and Water: Report to Congress* (EPA 230/3-79-001). Washington, D.C.: U.S. Government Printing Office.

*U.S. Environmental Protection Agency, 1984. *National Air Quality, Monitoring, and Emission Trends Report, 1982* (EPA 450/4-84-002). Washington, D.C.: U.S. Government Printing Office.

*U.S. Environmental Protection Agency, 1987. *Radon Reduction Methods: A Homeowner's Guide* (OPA 86-005). Washington, D.C.: U.S. Government Printing Office.

U.S. Environmental Protection Agency, 1988a. *National Air Quality and Emissions Trends Report, 1986.* Research Triangle Park, N.C.: EPA.

*U.S. Environmental Protection Agency, 1988b. *Regulatory Impact Analysis on the National Ambient Air Quality Standards for Sulfur Oxides.* Research Triangle Park, N.C.: EPA.

U.S. Environmental Protection Agency and U.S. Department of Health and Human Services, Centers for Disease Control, 1986. *A Citizen's Guide to Radon* (OPA 86-004). Washington, D.C.: U.S. Government Printing Office.

*U.S. Environmental Protection Agency, 1990. *Environmental Investments: The Cost of a Clean Environment* (EPA-230-11-90-083). Washington, D.C., November.

U.S. Environmental Protection Agency, 1991. *National Air Quality and Emission Trends Report, 1990,* EPA 450/4-91-023, Office of Air Quality Planning and Standards, Research Triangle Park, N.C.

12

Water Pollution

Because many kinds of chemicals dissolve in it, water is sometimes called the universal solvent. This remarkable property of water makes its contamination inevitable in the technical sense. After water is purified by evaporation and as it condenses and begins to fall back to the earth, it immediately begins picking up dissolved gases and particulates. Stretching the point a bit, even hundreds of feet above the earth's surface this water would be contaminated from the point of view of a biologist needing ultrapure water for a study of the effects on an organism of the absence of trace elements. Once a raindrop strikes the earth, it picks up materials like calcium, magnesium, iron, and zinc and becomes contaminated at a more rapid rate.

One way to look at water and how living things use it is to note that water serves to carry away waste. Single-celled aquatic organisms discharge wastes directly into surrounding water. More complicated organisms have circulatory systems (carrying mostly water) that bring nutrients to cells and carry away waste. This role of water at the cellular level is closely analogous to the uses of water at the "civilization" level of organization (see Figures 12.1 and 12.2). Major cities are typically located on important waterways, which serve to carry nutrients and other needs in and to carry manufactured products and wastes out.

Years ago, when populations were less dense, water pollution was much less widespread. Though this may be obvious, it is important to note that water pollution has accelerated more quickly than population densities; this is apparently a result of modern ways of doing things.

It has been estimated that every American now uses about four times as much water as Americans living at the turn of the century (see Chapter 7). A typical person in the United States uses about 100 gallons (almost 4 liters) of water each day, well over half of it for flushing away wastes and for showers and baths. Twenty-five gallons or more pass through the average washing machine or dishwasher. Another significant use of water is for miscellaneous things like hosing down driveways and watering lawns.

As our factories became larger, they required increasing amounts of water to remove both chemical wastes and waste heat. According to the U.S. Department of the Interior, it takes 82,000 gallons of water to make a ton of medium-quality paper and 184,000 gallons to make the same amount of fine book paper. Thus about 1 gallon was used in the production of the paper that makes up this page.

Agriculture also makes a significant contribution to water pollution. About 40% of the water supply in

337

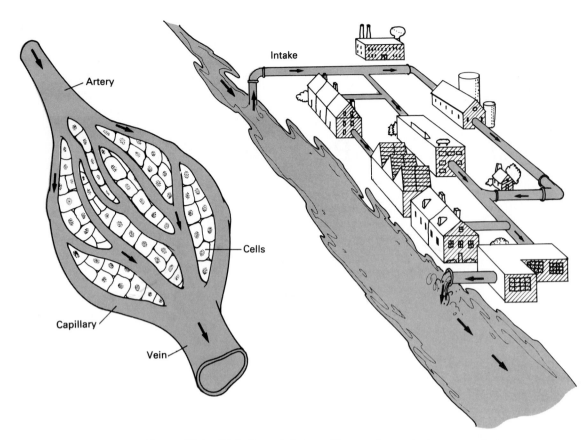

Figure 12.1 Of Cells and Cities. The use of water by humans to flush away or dilute waste parallels one of the major functions of blood at the cellular level.

the United States goes to agriculture, and some of this water ends up as polluted water carrying various types of agricultural wastes and contaminants.

Water pollution comes, then, from industrial, agricultural, and domestic sources and is the result of both the way things are done (technology and lifestyle) and the number of people doing these things in a given area (population density).

In this chapter we look at types of water pollutants, including biological agents, salt, chemicals that enrich, chemicals that poison, sediment, heat, and radioactive waste. We will see that nearly all bodies of water suffer from multiple pollutants. We also look at water pollution from the perspective of sources because control strategies and laws are tailored to sources. We will see that industry, agriculture, and municipal sources all make significant characteristic contributions. Because energy is so important to us, we focus later in the chapter on the problems of oil pollution and acid mine waste. Thereafter, we examine the overall magnitude of the water pollution problem and try to get at some of the reasons why one out of every two or three miles of streams in the United States is significantly polluted. We will focus on one river (the Ohio), one lake (Lake Erie), and one sea (the Mediter-

ranean). We also consider the problem of groundwater pollution in the United States. We examine various technological and nontechnological alternatives for the control or reduction of water pollution. We review some economic factors involved in water pollution and water pollution control, and we conclude with a review of water pollution and the law.

Water Pollution: A Definition

We need a working definition of **water pollution** for the reading that lies ahead. Precise definitions will be developed as we move along, but for the time being, let our definition be this: any change in water caused by *Homo sapiens* and considered by *Homo sapiens* to be unfavorable to *Homo sapiens* or to other forms of life. Basically, water pollution is any human action that impairs the use of water as a resource. It should become clear by the time we finish this chapter that the term *water pollution* is a relative one. Pollution is relative to the intended use of the water. Nearly pure water may be unsuitable for making beer. Water of lesser

Figure 12.2 How Water Gets Polluted. Water gets polluted because it is used
to wash away wastes, including heat.

quality may satisfactorily be used for recreation, fishing, navigation, or irrigation. The problem is that there is much water in the world that might be used for some purpose but is not used because it is polluted.

Water pollution is the red, lifeless, acid mine drainage water of Blacklick Creek, which runs through Nanty Glo, Twin Rocks, and Vintondale, Pennsylvania; it is the sometimes raw sewage–laden waters of the Ohio River; it is the detergent-derived phosphate in Lake Erie, the asbestos fibers in Lake Superior off Duluth, Minnesota, the mercury in the coastal waters of the harbors of Japan, and the heated water from the cooling towers of the large electric-generating plant in Seward, Pennsylvania. Water pollution is the pesticide-laden silt in the Missouri River; it is the oil on the beaches of Brittany, Texas, Chile, and

Massachusetts. Water pollution is a major global environmental problem.

Types of Water Pollutants

Although there might well be dozens of ways to subdivide the pollutants of water, all water pollutants fall into one of four *general* categories. Water can be polluted by (1) biological agents, (2) dissolved chemicals, (3) nondissolved chemicals and sediment, and (4) heat.

Small amounts of certain chemicals that sometimes cause pollution can actually be neutral or even *beneficial* to water quality and aquatic ecosystems (curve A in Figure 12.3). Other chemical agents are harmful in practically any amount (curve B in Figure

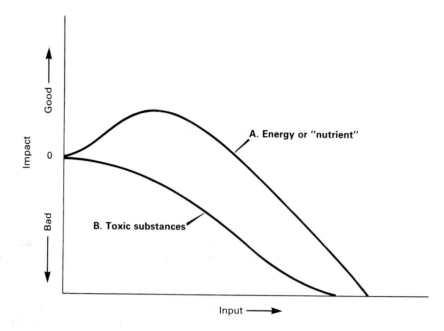

Figure 12.3   Too Much of a Good Thing Is as Bad as Being Poisoned.   Some pollutants are actually beneficial to aquatic ecosystems in small amounts (curve A). Other types of pollutants are harmful at almost any level (curve B).

12.3). Because of the importance of this concept, the following more specific classification will be used from this point on:

1. Biological agents
2. Chemicals that enrich and overenrich (both organic and inorganic chemicals)
3. Chemical poisons
4. Physical agents (including heat and suspended solids)
5. Radioactive substances
6. Inorganic salts, acids, and alkalis

Biological Agents

Many ancient civilizations had rules about water sanitation, so we know that the relationship between human disease and water has been recognized for centuries. A severe outbreak of cholera that occurred in the mid-nineteenth century in London was traced to a contaminated well on Broad Street. The pump handle was removed and the epidemic ended, even though it was not known at the time that a particular microorganism, *Vibrio cholerae,* was the causative agent. This was an early triumph of epidemiology (Chapter 10). Other epidemics of this very dangerous waterborne disease have occurred over the centuries in every part of the world. In 1991 Peru struggled with a cholera epidemic and the problem spread to other Latin American countries as well, including Brazil, Ecuador, Chile, and Colombia.

Epidemics of another waterborne disease, typhoid fever, have also occurred worldwide throughout recent centuries. Notable epidemics occurred in Pennsylvania in 1885, New York State in 1890, and Massa-

chusetts in 1890. In each of these epidemics, the disease was eventually determined to be spread by drinking water.

Near the end of the nineteenth century, after it was demonstrated that cholera could be prevented by water purification and sanitation—keeping *Vibrio cholerae* out of drinking water—there was an abrupt disappearance of this and other waterborne diseases. The curve has flattened out now to the point at which there are only about 4,000 known or documented cases of waterborne disease in the United States every year. In countries with less well developed public health programs and water purification systems, however, current rates are much higher; typhoid fever, dysentery, cholera, and viral hepatitis continue to be serious problems. The World Health Organization estimates that well over 900 million people in developing countries lack access to safe water supplies.

The incredible thing about many countries today and about all countries in the days before good sanitation is that there are not and were not *more* epidemics. Waters contaminated by human waste are and were often directly connected to drinking water supplies, but the relative scarcity of epidemics can probably be explained in part by immunity. People who are chronically exposed to certain microorganisms build up some degree of immunity to them. This is the basis of the gastrointestinal problems people often have when visiting another country; it is also the reason why explorers of long ago sometimes caused epidemics in the countries they visited. In both of these examples, people are exposed to microbes to which their immune systems are not acclimated.

In any case, the problem of waterborne disease can be simply and somewhat grossly stated as the contamination of humans' drinking water with humans' disease organism–bearing bodily waste. There is a long list of human **pathogens** (organisms that cause disease) that can remain viable in raw wastewater (see Table 12.1).

Hepatitis is a viral disease; another potentially important waterborne virus is the poliovirus. Chlorination, which kills most of the bacteria that make their way into water, has little effect on viruses. More than 100 different kinds of viruses have been identified in human wastes, and many of these can remain viable for some time in domestic sewage.

By far the most common disease from waterborne pathogens is acute gastrointestinal illness (AGI). It is caused by a variety of different agents, including bacteria of the genera *Shigella* and *Salmonella*, protozoans, and viruses.

Another very common waterborne disease is giardiasis, also known as "backpacker's disease." It is caused by a protozoan, *Giardia lamblia*. Although its symptoms are similar to AGI, giardiasis is much more severe and can last for several months. The protozoan is found in surface water. When water is not properly filtered or disinfected, the protozoan can survive and raise havoc with the human gastrointestinal tract.

A rather infamous bacterium associated with water is *Legionella*. In July 1976, a pneumonia-type disease affected individuals attending a conference of members of the American Legion at a Philadelphia hotel. It was eventually determined that a bacterium was the culprit. The pathogen was able to survive in the moisture in the hotel's ventilation system. Legionnaire's disease and a milder form called Potomac fever

are caused not by consumption of contaminated water but by inhaling the organism.

Microorganisms other than those found in human waste can also be a problem. This extends to microorganisms that make their way into water supplies as by-products of the processing of fruit and vegetables and as discharges from canneries and slaughterhouses. However, most of the bacterial contaminants that get into water from these sources are relatively harmless and, in fact, may help to degrade organic matter discharged along with the bacteria. Potentially **pathogenic** (disease-causing) microorganisms might also be transmitted from animals in this way. *Anthrax* bacilli or parasites originating from diseased animals in canneries and slaughterhouses, for example, can enter water supplies and infect animals and people downstream.

Chemicals That Enrich and Overenrich

Energy-rich Organic Chemicals. Certain chemicals in the right concentrations can distort and disrupt aquatic ecosystems by overfeeding certain components of such systems. Overfeeding (**eutrophication**) of aquatic ecosystems can occur in two basic ways via two kinds of chemical pollutants. One way is through the addition of inorganic nutrients that are normally limiting for plants (Chapter 3). Another way, the one we will discuss first, is through the addition of organic chemicals that serve as food for decomposers.

The addition of dissolved organic matter to an aquatic ecosystem gives a boost to the decomposers, organisms that use organic materials as sources of

Table 12.1 Examples of Biological Agents of Human Disease in Polluted Water

Types of Organisms	Agent	Disease
Bacteria	*Vibrio cholerae*	Cholera
	Salmonella typhi	Typhoid fever
	Other salmonellas	Enteritis (intestinal inflammation)
	E. coli of pathogenic types	Enteritis
	Shigella dysenteriae	Dysentery
	Legionella	Pneumonialike pulmonary disease
Protozoans	*Entamoeba histolytica*	Amoebic dysentery
	Giardia lamblia	Giardiasis
Helminths	*Schistosoma*	Schistosomiasis
(multicellular parasites)	(trematodes) (flukes)	(Bilharziasis)
	Ascaris (roundworm)	Ascariasis
	Anchylostoma (hookworm)	Anchylostomiasis
Viruses	Poliovirus	Fever, headaches, nausea, diarrhea, muscular pains, meningitis, paralytic polio
	Adenovirus	Fever, acute upper and lower respiratory tract infections, inflammations of the eyes
	Reovirus	Common cold and other upper and lower respiratory tract infections, diarrhea, hepatitis (especially in children)
	Hepatitis virus A ("infectious hepatitis" virus)	Fever, nausea, diarrhea, hepatitis (acute or chronic)

energy and nutrients. The problem comes when this activity increases to the point at which the decomposers use up all of the available oxygen as they oxidize organic matter (see Chapter 1).

The degree to which pollution by organic matter will remove oxygen from water depends on a number of factors, including the amount of oxygen in the water and the amount of water receiving the waste discharge. It would be better from the point of view of a stream community if organic matter were discharged (if it had to be discharged at all) into a very large, rapidly running, cool stream that is saturated with oxygen. Being cool helps in two ways: Decomposition is slower, and oxygen dissolves better in cool water. The degree to which oxygen will be depleted by sewage also depends on how much sewage is discharged in a given time. The oxygen-depleting strength of organic matter is a rather precise parameter called biochemical oxygen demand.

Biochemical oxygen demand (BOD), sometimes referred to as *biological oxygen demand,* is a quantitative expression of the oxygen-depleting impact (via the action of decomposers) of a given amount of organic matter. It is an expression of how much oxygen is needed for microbes to oxidize that organic matter. (Organic matter will undergo chemical oxidation even in the absence of decomposers; there is also a straight *chemical oxygen demand,* or COD.)

In the extreme, large amounts of organic matter could result in a near-absolute depletion of oxygen in a given body or stretch of water. This would clearly make life impossible for the species that need oxygen. Fish and zooplankton die under such circumstances, and even among the bacteria themselves there is a rise in **anaerobic species**—species that can live in the absence of oxygen. This in turn leads to the production of foul-smelling toxic end products of anaerobic respiration such as those listed in Table 12.2.

Short of absolute oxygen depletion, the reduction of oxygen in water can still seriously disrupt natural systems. For most aquatic systems, dissolved oxygen should never be lower than 3 ppm at any time and should actually be above 5 ppm for the greater part of every day. (Since respiration is carried out by both plants and animals at night, there is a nocturnal drop in oxygen content, which is restored in the morning when plants begin photosynthesizing.) Changes in species composition in aquatic ecosystems vary with average oxygen concentrations because different species can tolerate different oxygen limits. Trout require at least 5 ppm of oxygen, whereas certain scavenger fish like carp can survive in water having as little as 1 ppm.

Inorganic Chemicals That Enrich. A second way in which aquatic ecosystems can be overenriched (and thus polluted) is through the addition of *inorganic* matter, such as phosphates and nitrates. These substances can be added to aquatic ecosystems indirectly in the form of phosphorus- and nitrogen-containing *organic* pollutants, and they can also be added as pollutants directly. Some detergents contain large amounts of tripolyphosphates, and as much as 25% of the nitrate and phosphate fertilizer used in agriculture makes its way into water, contributing to eutrophication. Phosphates and nitrates can get into water supplies from other sources; we will consider some of these shortly.

Phosphate pollution is an especially serious problem because in many if not most parts of the world, including the United States, this form of the element phosphorus is often the plant-growth-limiting nutrient in aquatic environments; that is, phosphorus (as phosphate) is the controlling nutrient in terms of Liebig's law, discussed in Chapter 3. What this means is that when phosphate is added to water supplies, it triggers a rapid growth of plants. Somewhat ironically, perhaps, while this plant growth results in an increase in oxygenation due to increased photosynthesis, plant respiration and decomposition of dead plant material create a problem similar to the one we just described for organic pollution of water.

The addition of inorganic or organic matter as sewage can have a devastating effect on a pond or a lake. Although these materials tend to have the same kind of oxygen-consuming effect on rivers, the impact there can be quite variable. Rivers vary considerably in flow, and consequently they have different rates of reoxygenation, mixing, and sediment load, which limits penetration by sunlight (and consequently plant growth). These factors all determine the impact of **eutrophicants** (agents that lead to overenrichment) (see Figure 12.4).

Table 12.2 Chemistry of Aerobic and Anaerobic Decomposition

Chemical Element in an Organic Compound	Compounds in Which Each Element Ends Up	
	In Aerobic Decomposition (where oxygen is present)	In Anaerobic Decomposition (where oxygen is absent)
Carbon	Carbon dioxide (CO_2)	Methane (CH_4)
Sulfur	Sulfate salts (SO_4^-)	Hydrogen sulfide (H_2S) (stinks, is poisonous)
Nitrogen	Nitrate salts (NO_3^-)	Ammonia (NH_3) (stinks, is poisonous)

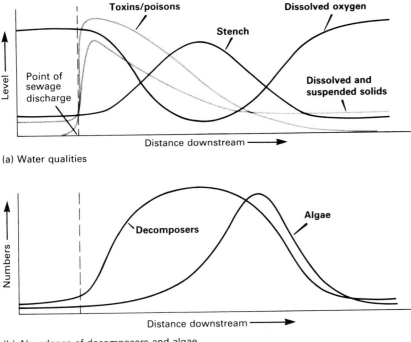

(a) Water qualities

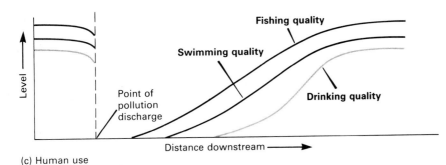

(b) Abundance of decomposers and algae

(c) Human use

Figure 12.4 The Impact of Sewage. These graphs show the impact of sewage on some of the important qualities of a stream of water and how this impact varies from the point of discharge to various points downstream. No distance units are given because the impact at a given distance varies according to stream flow and amount of discharge.

We point out again that what happens when water becomes polluted by overenrichment is not so much a complete wiping out of life forms as it is a change in the forms of life and the qualities of water that is perceived to be detrimental and harmful. Sewage-polluted streams are often described by the uninitiated as devoid of life, dead, when in fact they may support much larger populations of living things than clean water. The problem is that they tend to support microscopic organisms to the exclusion of other aesthetically or economically desirable organisms. In terms of the concepts introduced in Chapters 1–4, *biomass* may stay the same or increase with eutrophication; *species diversity* is usually decreased significantly. Protozoans often exist in great numbers in sewage-polluted waters because food chains become shortened (Chapter 2) by the elimination of upper trophic levels.

Among the most common organisms of polluted water is the *Tubifex* worm. The presence of *Tubifex*

worms in water is almost a certain indication that the water is polluted to some degree. The worms are good sources of food for fish and other aquatic life; they are found in large numbers in polluted water because conditions are unsuitable for the consumers that feed on them.

In summary, small amounts of either organic matter or certain inorganic chemical compounds can actually be beneficial to an aquatic ecosystem in the same way that fertilizers are beneficial in small amounts to gardens, crops, and terrestrial ecosystems. The problem comes when too much is added too soon in too small a space, pushing the ecosystem beyond the point of resilience.

Chemical Poisons

Many of the chemicals that get into water with an assist from *Homo sapiens* are poisons. Poisons are poisons, of course, no matter in what subsphere of the ecosystem they happen to be located. We have already reviewed the properties of some of the poisons found in the atmosphere. Many of these poisons are found in water as well. In Chapter 16 we will review the properties and health effects of pesticides and other long-lived poisons. Here we will simply summarize the poisons found in water and discuss mercury and PCBs briefly.

Among the inorganic toxic chemicals found in water supplies are arsenic (which comes from many types of insecticides), cadmium (from electroplating operations), chromate (from various industrial processes), cyanide, lead, selenium, and mercury. Others are copper, chromium, and zinc. These substances are toxic to fish and other aquatic organisms as well as to human beings because they interfere with the action of enzymes and other biochemicals as we discussed in Chapter 10. Metal toxicity problems are often compounded by the biological magnification of biometallic compounds (metal-organic complexes) in aquatic ecosystems (see Chapter 16).

The Minimata Bay incident in Japan has become the classic example of the impact of water contamination by mercury. During one period in the 1950s there were more than 50 deaths in the seacoast village of Minimata, more than three times that many cases of brain damage, and two dozen children born with neurological problems, all linked to the ingestion of mercury introduced into the food chain. The source of the mercury was effluent from a local factory. Nearly all of the victims, it was found, ate fish from the bay three times a day. Cats and other household pets also showed signs of mercury poisoning. This incident established mercury as a bona fide environmental problem and set off a mercury scare that eventually led people to avoid eating swordfish and tuna.

The **methylation** of inorganic mercury—the incorporation of mercury into an organic methyl (CH_3) compound by microorganisms—in the sediment of lakes, rivers, and waters of other types is apparently the key to the transport of mercury in aquatic food chains leading to human beings. Because inorganic mercury is relatively insoluble in water, the organic variety is much more important. Its organic character makes it lipid-soluble and thus soluble in body tissues; it is therefore subject to biological magnification (see Chapter 16). Methyl mercury accumulates in organisms in aquatic environments according to their placement in the trophic level, the highest concentrations being found in top carnivores. As we discussed in Chapter 9, the greatest effect of inorganic mercury compounds seems to be on the kidney and digestive organs; organic mercury compounds, by contrast, affect neural tissue, principally the brain, and also cause digestive system problems and birth defects.

Many types of organic toxins also get into water supplies as pollutants. Examples of these are pesticides such as DDT and chlordane, chemicals that are actually made to harm living things (see Chapter 16); hydrocarbons such as benzene that cause cancer (see Chapter 18); and a host of chemicals that cause genetic damage and birth defects. Certain organic chemicals—phenols, for instance—give an off taste to water.

A particularly interesting class of organic water pollutants, one that has generated considerable discussion in recent years, is the family of **polychlorinated biphenyls (PCBs)**. PCBs are chlorinated compounds similar to DDT in their extreme stability. Because highly chlorinated PCBs are very stable and are usually present in only trace amounts, they are not a biochemical oxygen demand problem for aquatic ecosystems; they are extremely toxic, however. PCBs originate in a variety of industrial processes including the manufacture of brake linings, grinding wheels, glass, ceramics, various types of coatings, flame-proofing paint, varnishes, sealants, electrical equipment, tires, plastic coatings, soap, and (ironically) water treatment chemicals, to name just a few. Over 1.4 billion tons of PCBs were produced in the United States before production was banned in 1976. Because of their stability, PCBs that escape the manufacturing process or that escape from the products into which they have been incorporated, such as electrical transformers, make their way into aquatic ecosystems.

The maximum PCB level accepted by the FDA in fish to be eaten by human beings is just 2 ppm. Ingestion of large concentrations of PCBs causes death due to various physiological disturbances. We do not yet know the long-term effects of chronic ingestion of low levels of these compounds.

Another category of chemical contaminants that is of increasing interest are volatile contaminants, primarily chlorinated hydrocarbons. The most notable is TCE, which is a common contaminant of groundwater. TCE was used in the past for degreasing, and improper disposal has led to widespread appearance in groundwater.

Physical Agents

Heat. Water has a high heat capacity; that is, it is able to absorb large amounts of heat with relatively small increases in its own temperature while remaining liquid, making it an ideal cooling medium. For this reason, many industrial plants are located on rivers, where water is available to carry away waste heat. As heat-laden water is discharged back into the main water supply, it raises the temperature of the aquatic environment. As much as 80% of the water used in industry in the United States is used for cooling purposes. Electric utilities alone use billions of gallons of water for cooling each day, trillions of gallons each year.

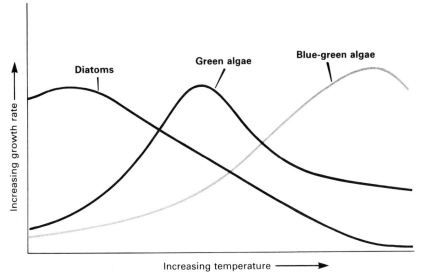

Figure 12.5 Temperature Preferences.
Different species have different temperature optima and tolerance ranges. *Diatoms* are single-celled algae characterized by transparent silica shells or cases and a golden brown pigment in addition to chlorophyll. Blue-green algae are distinguished from green algae by a much more primitive cell type and the presence of bluish green and other pigments in addition to chlorophyll.

Heat affects the life found in water in several ways. First, every life form has a temperature tolerance range, as we discussed in Chapter 3. At some point in the life cycle of every organism there is a most temperature-sensitive stage—hatching eggs, for example. All individual organisms have ranges of temperature within which they can survive and specific temperatures at which they function best (Figure 12.5). Changing water temperatures can therefore result in change in species composition. Waste heat can also adversely affect aquatic ecosystems by increasing temperature *variability*, keeping the natural system off balance. The most important direct effect of heat on organisms that do not have a constant body temperature is the acceleration of metabolic reactions; the warmer it is, the faster the metabolic reactions. Because heat accelerates some kinds of reactions more than others, temperature changes tend to disrupt the chemistry of living things.

Heat also disrupts and changes the chemistry of the *abiotic* environment. Heat increases the solubility of certain chemicals and generally decreases the solubility of gases.

Heat can actually affect the amount of oxygen and other gases dissolved in water in several ways. First, gases dissolve more readily in cold water than in warm water (Figure 12.6). Second, because higher temperatures accelerate metabolism, heat can accelerate decomposition, making BOD more of an immediate and acute problem while at the same time increasing the need for oxygen by fish and other organisms. A third way in which higher water temperatures can affect dissolved oxygen and other gases is in the stratification that can occur when hot water is

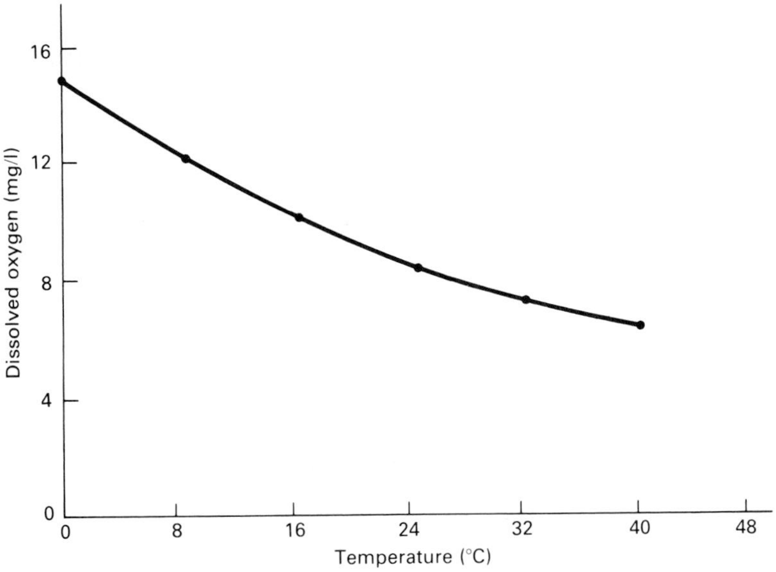

Figure 12.6 Oxygen Solubility and Water Temperature. The quantity of oxygen that can dissolve in water is inversely related to temperature. The lower the temperature, the more oxygen can be held in solution. Thermal pollution thus influences the oxygen content of water.

discharged into a cooler body of water. The effects of stratification on chemical mixing were discussed in Chapter 3.

Getting back to the *direct* effects of heat, temperature changes of only 2 or 3 degrees under certain circumstances can have a very serious effect on fish and aquatic life. Trout eggs, for example, take 165 days to hatch in cool water (3°C, or 37°F) but will hatch in only a month if kept at 12°C (54°F). They will not hatch if temperatures reach 15°C (60°F). Since temperature is a major key to spawning and reproduction for many species (Table 12.3), increased heating can result in such important problems as the hatching of fish eggs out of synchrony with normal peaks in food supply for the hungry hatchlings. Another problem is that blue-green algae tend to do better in warmer water than other types of algae; population explosions in these types of algae are often accompanied by significant levels of the toxins they produce. We will return to pollution by heat later.

Floating Solids and Liquids.
Oil, grease, and a number of other materials that float on the surface of water are another kind of pollutant. The effects of these substances can be aesthetic, of course, since they occur where they can be seen, but certain types of organic materials that float are also toxic. Some of the materials in oil, for example, are flammable, creating the kind of fire hazard that resulted in Cleveland's Cuyahoga River catching on fire from oil on its surface in 1969. Floating materials also decrease light penetration and can retard the diffusion of gases such as oxygen. Floating materials may also contribute to biochemical oxygen demand. If floating materials break up and become suspended in water, they can concentrate toxins (for example, oil droplets can accumulate fat-soluble toxins such as DDT) in water and deliver these to filter-feeding organisms such as clams and mussels.

Suspended or Sedimentary Solids.
Among the things that affect water physically are undissolved solids, some of which dissolve over long periods of time and some of which practically never go into solution. Both types of solids decrease water quality in a physical way; silt and other insoluble materials clog waterways, fill up dams, and make water cloudy or muddy. Such solids can also be physical problems for gill breathers (fish) and filter feeders (such as clams). By adsorption and in other physical and chemical ways, suspended organic and mineral solids can also concentrate metals and other toxins and then deliver them to various organisms in the food chain. Pesticides can be concentrated in and on suspended solids, for example.

Sediment, which gets into streams naturally through runoff, is also contributed by agriculture and industries like the china (clay) industry, construction, and steel.

Color.
Some pollutants change the color of water. Perhaps the most outstanding example of this is the red and yellow colors in acid mine discharges derived from iron oxide and sulfate. Although these discharges are harmful because of the chemicals, which are only incidentally related to color, coloring agents are among the most objectionable types of pollutants as far as the public is concerned. People seem to be raised to indignation most by things they can see. The problem is that funding has a way of going to things that raise public indignation, often before it goes to truly serious problems.

Chemicals That Make Foam.
Some of the chemicals that get into water make it more apt to foam. This was an extremely serious problem in the

Pollutants that cause foaming in water supplies may be less dangerous than other forms of pollution, but they tend to attract more public attention.

Table 12.3 **Recommended Maximum Water Temperature for Life Events in Various Species of Fish**

Temperature	Fish Species and Events
32.2°C (90°F)	Growth of largemouth bass, drum, bluegill, and crappie
28.9°C (84°F)	Growth of pike, perch, walleye, smallmouth bass, and sauger
26.7°C (80°F)	Spawning and egg development of largemouth bass, white and yellow bass, and spotted bass
20.0°C (68°F)	Growth of salmon and egg development of perch and smallmouth bass
12.8°C (55°F)	Spawning and egg development of salmon and trout (other than lake trout)
8.9°C (48°F)	Spawning and egg development of lake trout, walleye, northern pike, and sauger

1960s and earlier, when detergents were **non-biodegradable** (not broken down by microorganisms), but is less of a problem now that non-biodegradable detergents are no longer made. There are still other pollutants that cause foaming in water supplies, and as with color, this visible form of pollution discourages recreational use and is often a more politically important form of pollution than many more serious pollutants are.

Radioactive Substances

Discussions of the problems of radioactive materials in water supplies can be found in Chapters 6 and 13. Chapter 18 contains a detailed description of the nature of radionuclides. Suffice it to say here that radioactive materials make their way into water supplies

as a result of the following activities:

1. The processing of uranium ore
2. The laundering of contaminated clothing from laboratories where radioisotopes are used
3. Wastes released from research laboratories
4. Wastes released from hospitals using isotopes in diagnostic and therapeutic procedures
5. The processing of fuel elements from uranium ore
6. Water released from nuclear power plants
7. Fallout generated by nuclear weapons testing

Low-level waste from hospitals and other laboratories can get into water by leaching from what may have been thought to be secure low-level waste depositories. Some is simply dumped illegally.

Enrichment Box 12-1

The Relationship between pH and Hydrogen Ion Concentration, Including Some Common Expressions of Acidity, and Some Common Acidic and Alkaline Liquids

The pH of a solution is defined as the effective hydrogen ion concentration of that solution expressed as a negative logarithm. The equation looks like this:

$$pH = -\log_{10}[H^+]$$

A 1-molar concentration of H^+ ions would have a pH of 0 because the logarithm of 1 is 0 ($10^0 = 1$). A solution with a pH of 7 would have a 10^{-7} molar concentration because the logarithm of 10^{-7} is -7, and $-(-7) = +7$.

Quantitative Description	pH	Moles of Hydrogen Ions per Liter		Example
Extremely alkaline	14	10^{-14}		Household lye
	13	10^{-13}		Bleach
	12	10^{-12}		Ammonia
Strongly alkaline	11	10^{-11}		
	10	10^{-10}		
	9	10^{-9}		
Slightly alkaline	8	10^{-8}		Baking soda
				Seawater
Neutral	7	10^{-7}	Common range for most natural waters	Blood
				Distilled water
				Milk
Slightly acidic	6	10^{-6}		
	5	10^{-5}		
Strongly acidic	4	10^{-4}		Orange juice
	3	10^{-3}		Vinegar
	2	10^{-2}		Acid mine water
	1	10^{-1}		Lemon juice
				Concentrated lab. acids
Extremely acidic	0	10^{-0}		Battery acid

Acidity is measured in terms of pH units, which inversely reflect hydrogen ion concentrations; pH values go down as the acidity or hydrogen ion concentration goes up (gets stronger).

Note that pH is really the negative exponent of the effective hydrogen ion concentration. Another way of saying this is that pH is the negative logarithm of the effective hydrogen ion concentration. A logarithm of any number is the power to which 10 must be raised to yield that number. Hydrogen ions (H^+) are the "active" components of acids. These positive ions influence chemistry by reacting with negatively charged groups on organic and inorganic chemicals.

Inorganic Salts, Acids, and Alkalis

Some inorganic salts dissolved as ions cause overenrichment; others are chemical agents of hardness—calcium salts, for example. The **hardness** of water is the degree to which certain dissolved salts make it difficult to produce a soapy lather in that water. These salts are present naturally in water supplies, and they are present in high concentrations in certain types of wastes. They can also come out of solution under certain circumstances and become deposited within pipes and water-handling equipment, causing industrial maintenance problems. Agents of hardness also interfere with dyes in the textile industry and cause special problems in the beer-brewing industry. Still other inorganic salts in water pose human health problems. For example, magnesium sulfate, which cannot be absorbed from the digestive tract, has a **cathartic** effect, causing a chronic chemical type of diarrhea when it is present in significant amounts.

Acids and alkalis represent another kind of chemical pollution problem that has to do with ranges of tolerance for particular associations of living things in aquatic ecosystems. Although only a few organisms can survive very high acidity and only a few can withstand very high alkalinity, the acidity of water supplies varies over a broad range; this has an effect on aquatic life and limits the uses of water by humans. Chemical neutrality is pH 7; pH levels below 7 reflect increasing acidity, and pH levels above 7 reflect increasing alkalinity (see Enrichment Box 12.1). Fish can tolerate a pH range from about 4.5 to 9.5. Discharges from industrial sources have been known to range from as low as 2 to as high as 11. Acids are produced as a by-product of many kinds of industries including mining and those that contribute the starting materials for acid rain (see Chapter 10). Alkalis such as sodium hydroxide are produced as a by-product of soap manufacturing, textile manufacturing, and the tanning of leather.

Acids and alkalis influence the chemistry of organisms directly by influencing the shape and function of key cellular molecules. Indirect effects include changes in water chemistry via influences on solubility of abiotic chemicals.

Water Pollution Sources

We have just considered the characteristics of major water pollutants. Ahead, we are going to consider strategies for controlling these pollutants. Before we discuss control strategies, we should know a bit more about the sources of water pollution.

The sources can be generalized as follows. First, U.S. industrial plants and electrical utilities discharge more than 100 billion gallons of used water each day. Much of this water is not adequately treated. Some is not treated at all. Imperfect American public sewage systems contribute another 40 billion gallons a day. There are still several thousand fair-sized communities and towns in the United States that treat sewage very

---

---

little if at all. Fifty billion gallons of water containing pollutants come from U.S. agricultural runoff. These wastes include pesticides and fertilizers as well as the organic waste that comes from hog and cattle feedlots. Accidental oil spills and acid mine drainage also contribute to the water pollution problem.

Electrical utilities use billions of gallons of water every day for cooling, contributing to thermal pollution. Construction sites and farms contribute millions of tons of sediment through runoff. Millions of tons of garbage, sludge, chemicals, miscellaneous dirt, and debris are dumped each year into the ocean. Still another important source of water pollution is leaching from chemical and solid waste dumps (Figure 12.7) (see also Chapters 13 and 17).

The EPA has found it convenient to distinguish point sources from nonpoint sources of pollution, mostly for purposes of control and regulation. A **point source** is a discernible port or channel through which wastes are discharged, including pipes, ditches, channels, sewers, tunnels, and even floating barges from which pollutants are discharged. **Nonpoint sources** of pollution include the runoff from urban areas, agricultural runoff, and the like.

It should be noted that there are *natural sources* of water contaminants. These include silt and sediment from natural erosion, natural oil seeps, and organic material (e.g., leaves) flushed into streams by heavy rains and flooding.

Municipal Sewage

Municipal sewage constitutes one of the largest sources of BOD and suspended solids. Despite significant progress in municipal sewage treatment, municipal discharges are still one of the major sources of water pollution. It seems that the problem here is the existence of combined storm sewers and wastewater sewers in many American cities. When there are heavy rains, the combined flow exceeds treatment plant capacity, and bypass mechanisms operate to allow the sewage to pass directly into the receiving stream *without* treatment. Another part of the problem is that many towns and cities still have inadequate sewage treatment systems. The result is that several billion pounds of BOD and several billion pounds of suspended solids annually pass from the nation's city sewer pipes into waterways.

An extension of the domestic waste problem is the problem of waste discharge by pleasure boats and other craft. There are more than 10 million pleasure boats in the United States, and the number is growing by almost a quarter of a million boats each year. Until the Water Quality Improvement Act was passed in 1970, boats simply discharged wastes directly into the waters on which they floated. Now standards of performance and marine sanitation devices are supposed to prevent the discharges of inadequately treated sewage from boats. However, the standards are often ignored, and there is little enforcement.

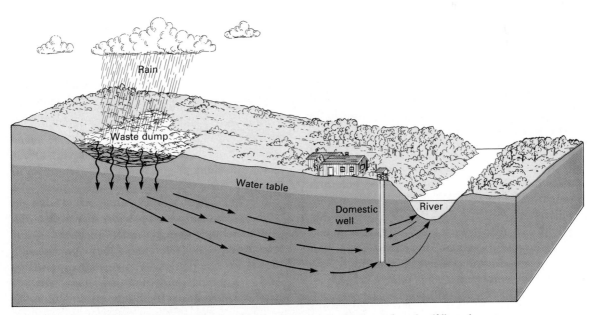

Figure 12.7 Landfill Leaching. Groundwater that seeps from landfills and dumps can reach wells and rivers to become sources of contamination. Especially important in this regard are metals, other persistent toxic chemicals, and radioisotopes.

Table 12.4 The Human Population Equivalent of Fecal Production by Animals

Biotype	Fecal Production (gallons per capita)	Relative BOD per Unit of Waste	Population Equivalent
Human	150	1.000	1.00
Horse	16,100	0.105	11.30
Cow	23,600	0.105	16.40
Sheep	1,130	0.325	2.45
Hog	2,700	0.105	1.90
Hen	182	0.115	0.14

Source: U.S. Department of Agriculture.

Water Pollutants from Farms

Agricultural sources of pollution include the body wastes of farm animals, the runoff of inorganic fertilizers and pesticides, sediment, and salt. Fertilizers and animal wastes both contribute to eutrophication.

The total farm animal population in the United States exceeds 700 million. The waste these animals generate is equivalent to that of about 2 billion people (Table 12.4). Within the last generation or so it has been found that rather than allowing cattle to roam the open range up to marketing time, it is economically beneficial to concentrate them during the last stages of their growing in feedlots, where they can be fattened before being sent to market. American feedlots now hold more than 500 million cattle and more than 1 million hogs. These concentrated sources alone generate a biochemical oxygen demand equivalent to that of almost 100 million people.

The use of commercial fertilizers has grown considerably since the dawn of the modern age of chemistry in the early 1940s. Unheard of before World War II, nitrate fertilizers are now applied at 50 pounds per planted acre in the United States. According to some authorities, these fertilizers are the main reason why this country is able to produce enough to feed all Americans and a considerable number of people in the rest of the world. A problem with this is that such fertilizer is sometimes washed into water, where it con-

tributes to eutrophication; some of it leaches into groundwater, polluting wells.

Nitrate in drinking water is a serious problem. As early as 1945 it was found that high levels of nitrates in well water caused methemoglobinemia, or "blue babies." The condition is the result of changes in the iron in hemoglobin that prevent it from carrying oxygen. There are some indications that nitrates may also be associated with nervous system impairments. Studies are under way to determine linkages with some forms of cancer and birth defects. Nitrate levels in groundwater in agricultural areas are on the rise.

Our review of the available data indicates that although intensifying economic pressures may well take care of farm-related pollution (fertilizer costs much more than it once did), the potential and actual seriousness of the environmental problem warrants significantly more attention to better timing and improved methods of fertilizer application. We have very little good data on the fate of fertilizer applied to farm fields, and we need more. In the meantime, since doing away with inorganic fertilizers is not very likely, a solution would be to control the runoff by using and applying fertilizers in ways that would minimize this problem.

Things are interconnected, and simple solutions are rarely sufficient. It has been argued that banning commercial fertilizer could conceivably cause more water pollution than it would prevent. If the alternative selected was to bring additional marginal acreage into production, this would add to runoff, silting, and other types of pollution. According to some authorities, if inorganic fertilizers were banned, productivity would drop to half of what it is now, and food prices would skyrocket.

Industry

Water is used in four basic ways by industry: for cooling, processing, boiler feed, and sanitary service. The problems of using water for sanitary service are identical to the problems of municipal sewage. Boiler feedwater is a relatively minor problem. Here we will

The concentrated wastes of cattle kept in feedlots have more of an impact on water supplies than the wastes of the same number of cattle scattered across rangelands would.

discuss a few specific examples of the industrial use of water for food processing and the processing of paper.

Fruit and Vegetable Processing.
Water is used in the fruit and vegetable industry in cleaning, sorting, cooking, and other forms of processing. In some industries, lye solutions are used to dissolve vegetable skin; and in some cases, vegetables have to be dipped into hot water to loosen the skins or to denature the enzymes that would later produce off colors and off tastes.

Pulp and Paper Industry.
The paper industry produces pollutants in a number of ways, beginning with forestry practices (including the use of pesticides and the practice of clear-cutting, which increases erosion and siltation). Water is used in the processing of wood to remove the fibers that go into the manufacturing of paper. These processes yield biochemical oxygen-demanding sulfide liquors. Even the processing of recycled paper has potential for pollution in the deinking and bleaching processes.

Electric Utilities

Electric utilities generate most of the waste heat discharged into water supplies in the United States; however, other industries make significant contributions. As was discussed in Chapter 6, after steam has pushed the blades of turbines, causing them to spin, it must be cooled and condensed before it can be returned to the boiler for another cycle. For this to happen, the steam must give up its heat energy, and water is almost always the medium used to carry such heat away. A single 1,000-megawatt plant requires 2 to 3 million gallons of cooling water *every minute*.

Water Pollution from the Production of Coal

Coal production presents water pollution problems at two different stages. The first is during mining, and the second is in the processing of coal once it is mined.

The first coal mine in the United States opened in Pennsylvania a decade or two before the Declaration of Independence was signed. Since then, streams throughout the coal-producing regions in the United States have become polluted from acid drainage to the extent that many are absolutely devoid of normal aquatic life. Acid mine wastes originate as water passes through mines and various iron compounds (particularly iron sulfide compounds) are oxidized. The agent of oxidation is the oxygen dissolved in water or in the air pumped through the coal mines. The acidification process is accelerated by **chemosynthetic bacteria,** microorganisms that oxidize inorganic chemicals (iron sulfides in this case) as chemical energy sources and use the energy in the same way that photosynthetic organisms use sunlight. The equation for the reaction is as follows:

$$2FeS_2 + 7O_2 \xrightarrow{enzymes} 2FeSO_4 + 2H_2SO_4$$
$$+ 2H_2O \qquad\qquad + energy$$

As can be seen from the equation, the active ingredients are sulfides of iron, oxygen, and water. Iron sulfide ($FeSO_4$) can be further oxidized to yield still more sulfuric acid (H_2SO_4). The overall effect of these reactions is the conversion of sulfur compounds into sulfuric acid, one of the most powerful acids known (Chapter 9). The most important offending microorganism is the bacterium *Thiobacillus thiooxidans*. Coal mining can be thought of as the creation or expansion of habitat for this microorganism.

Although acid is the most important constituent in acid mine drainage, other chemicals are present in mine effluents as well, and the composition varies considerably from one location to another. A chemical analysis of an acid mine–polluted stream in comparison to natural waters is presented in Table 12.5.

Table 12.5 Chemical Analysis of a Stream in Pennsylvania Containing Acid Mine Drainage and the Chemical Properties of Natural U.S. Waters. Data for mine drainage (in which pH averages between 3 and 4) can also be compared to U.S. Public Health Service and World Health Organization drinking water standards of pH 7.0–9.0, 0.3 ppm for iron, 200 ppm for calcium, 150 ppm for magnesium, 250 ppm for sulfate, and 500 ppm for dissolved solids. All numbers are in parts per million.

	Average Value for 5 to 11 Sample Sites	Maximum Content of 5% of Natural Waters	Maximum Content of 95% of Natural Waters
Total dissolved solids	1,444	72.0	400.0
Bicarbonate	0	40.0	180.0
Sulfate	1,012	11.0	90.0
Calcium and magnesium	200	18.5	66.0
Iron	1.0	0.1	0.7

Sources: Data from Jones, O. C., 1951. "Acid Mine Water: Its Control Reduces Stream Pollution," *Mechanization,* Part I, Vol. **15** (Oct) 10 and Part II, Vol. **15** (November) 11; M. Lippmann and R. B. Schlesinger, *Chemical Contamination in the Human Environment* (New York: Oxford University Press, 1979) p. 41.

Acid drainage can also come from piles of waste removed from coal mines. Leachates coming from such gob piles can be very similar to the drainage that comes from coal mines directly.

Control of Acid Mine Drainage.

Acid mine drainage presents a unique problem of control. There are a variety of methods for preventing the formation of acid mine water and neutralizing the acid in these waters. Strategies that have been suggested include these:

1. Using landscaping to control the amount of water that flows into mines
2. Sealing coal mines so that water cannot flow in or out
3. Flooding the mines with water and holding it there to keep oxygen out
4. Using lagoons to impound acid mine water, then feeding this into streams gradually in proportion to the amount of flow in the receiving stream
5. Using the acid mine water for washing coal (this, according to some proponents, both neutralizes the acidity and cleans the coal)
6. Using limestone or lime to neutralize the acid
7. Covering strip mines with earth
8. Disposing of waste rock from coal mines in layers sandwiched between layers of earth

It is difficult, though possible, to clean up a stream after it is polluted by acid drainage. One case in point is the Youghiogheny River. Pennsylvania embarked on a ten-year program to clean up this river in the late 1960s. The task was to cost an estimated $500 million, $200 million of which was to go toward stopping mine drainage into the river system. Coal veins and mine shafts were sealed, and the river began to become noticeably cleaner in the early 1970s.

Waste from Coal Preparation.

After coal is mined, it is usually brought to a coal-cleaning plant, where it is processed by first breaking it into small pieces and then washing out impurities. Many coal-cleaning operations produce significant amounts of sediment and suspended solids containing calcium, magnesium, sulfates, iron, and other minerals found in coal and shale. These problems can be solved in part by treating and reusing water used in the coal-cleaning process.

Water Pollution from the Production, Transport, and Use of Oil

The problem of oil pollution of the ocean is not new. Conferences were held as early as 1926 on the pollution of navigable waters by oil, and another such conference was held at the League of Nations in 1935.

The history of oil pollution in most minds begins with the wreck of the *Torrey Canyon*. This infamous tanker, a small one by today's standards, ran aground on what were supposed to have been well-charted rocks in broad daylight in 1967. The ship released over 100,000 metric tons of oil (more than 700,000 barrels; one barrel contains 42 U.S. gallons or 159 liters), most of which eventually washed ashore to pollute the beaches of southern England and northern France. There have been a host of similar incidents involving larger ships since 1967.

A 1990 report issued by the Natural Resources Defense Council revealed that oil spills are common events. Over 80 million gallons of oil were spilled in U.S. waters between 1980 and 1986. In those same years, U.S. tankers were involved in over 400 groundings, nearly 500 collisions or rammings, and 50 or so fires and explosions. There were 95 deaths. Due to technological limitations, typically less than 15% of the oil lost in these types of incidents is recovered.

There have been two notable oil pollution events in the recent past. One was the deliberate dumping of 250 million gallons of oil into the Persian Gulf during the Gulf War in January, 1991. The other was the grounding of the oil tanker *Exxon Valdez*. On March 24, 1989, almost 11 million gallons of oil were spilled in Prince William Sound. It was at the time the worst oil spill in U.S. waters. According to the Coast Guard almost all of the oil was released in the first 12 hours. The oil then spread to cover over 900 square miles of ocean. As much as 4,000 miles of coastline may have been affected. By the end of 1989, the following direct damage had been documented: almost 37,000 dead birds, over 1,000 dead sea otters, and 144 dead eagles. This is estimated to be only a small percentage of the total deaths. The Fish and Wildlife Service estimates that 65% of affected eagle nests produced no young in 1989. Due to concern about catching fish contaminated by oil, fisheries were closed, at a significant loss of revenues.

The world's oil production is approximately 3 trillion gallons per year, well over half of which is moved over the ocean from wellhead to consumer. Although less than 1% of this oil ends up as a pollutant, mostly in the oceans, this amounts to considerable contamination. Thousands of oil spills contaminate coastal waters every year. Figure 12.8 shows the amounts of oil reaching the oceans from the various sources.

Oil spills also occur in inland waters. In January 1988 the collapse of an oil tank near Pittsburgh, Pennsylvania, sent almost a million gallons of oil into the environment, creating a 20-mile-long oil slick that flowed down the Monongahela River and into the Ohio. Water shortages occurred in many of the cities along the way that routinely use the Ohio River as

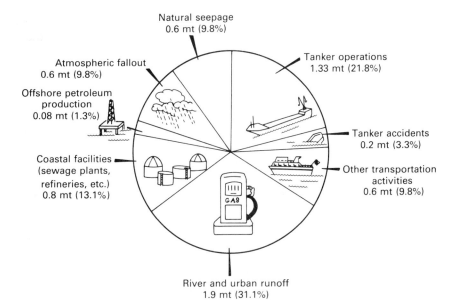

Natural seepage
0.6 mt (9.8%)

Tanker operations
1.33 mt (21.8%)

Atmospheric fallout
0.6 mt (9.8%)

Offshore petroleum
production
0.08 mt (1.3%)

Tanker accidents
0.2 mt (3.3%)

Coastal facilities
(sewage plants,
refineries, etc.)
0.8 mt (13.1%)

Other transportation
activities
0.6 mt (9.8%)

River and urban runoff
1.9 mt (31.1%)

Figure 12.8 Oil Pollution. Sources of the oil that ends up in the world's oceans in millions of metric tons (mt) per year.

their primary drinking water source, since their water intakes were shut off as the slick moved past. Water supplies for over a million people in four states were threatened.

As dramatic as major spills are, much of the oil pollution in the world's waters results from routine operations, in which oil wastes are mixed with seawater that had been taken on as ballast and dumped from oil tankers when the ships return to the sources of oil for refills. Although technology is available for recovering oil from the ballast tanks, sometimes the crews of outmoded oil tankers ignore international law and simply dump this mixture of seawater and oil overboard as soon as they are out of danger of detection. Some 3 to 6 million tons of oil are dumped in the oceans annually as residues of petrochemical opera-

tions, at-sea dumping, and slipshod practices (Smart et al., 1987). A U.S. Forest Service marine bird specialist claims that chronic oil pollution kills half a million marine birds annually at drill sites in the North Atlantic (Turner, 1990).

Two other sources of oil pollution are the occasional accident in offshore wells and inadvertent leakage from oil deposits through cracks developed in the ocean floor offshore as side effects of the drilling and capping processes. There is also a considerable amount of natural oil pollution in the oceans.

Effects. What happens when oil gets dumped into the oceans depends on wind, temperature, ocean currents, geology, and location (open waters or a confined bay or sound). Generally, a predictable sequence

More oil pollution comes from the routine operation of tankers than from highly publicized oil spills. Many tankers flush oil wastes directly into the sea, year after year.

of events takes place. Following the spill, the oil slick is spread out by the wind, waves, and ocean currents. Oil constituents with a boiling point below 200°C evaporate. Other constituents are oxidized by sunlight. Some are degraded by microorganisms. Other components are very stable and can remain in the aquatic environment for many years. Some of these effectively disappear by dispersion. About 24 hours after the spill, **emulsification** of oil and seawater occurs. This is an especially significant event because a sticky, viscous material is formed that adheres to everything with which it comes in contact—fish, birds, mammals, shorelines. It has been found that as little as 0.1 ppm of oil (1 part oil to 10 million parts water) can seriously affect fish, crustaceans, and plankton (Lenssen, 1989).

One episode that was studied quite thoroughly was the Santa Barbara incident of the early 1970s. A leaking offshore well polluted miles of beaches and had dramatic effects on shorebirds, coating their feathers with oil, preventing them from feeding or flying, and killing them. Despite this tragedy, the oil spill apparently had only minimal long-range effects and very few important short-range effects on the marine environment. The earlier *Torrey Canyon* incident was also studied, and it was determined that some of the detergents used to try to disperse the oil may have actually done more harm to marine life than the oil itself.

When oil coats their feathers, birds are unable to fly, and their feathers lose the capacity to retain body heat. Most affected birds die of exposure.

In a 1985 report, the National Research Council confirmed the fact that the biological effects of oil spills vary considerably. They also found that remnants of effects can still be measured many years later in some cases.

On an evening in March 1978 the supertanker *Amoco Cadiz* lost its steering and ran aground near Portsall on the Brittany coast of France. Nearly a quarter of a million metric tons of oil were lost, contaminating 300 km of seacoast. An accounting of the oil after the first few months showed that 13% was incorporated into the **water column** (water between the surface and the bottom), another 4.5% degraded quickly in the water, 8% ended up in tidal sediments, 28% washed into the intertidal zone, 30% evaporated, and the rest was not accounted for (Gundlach et al., 1983). Oil that came ashore ended up in sediments or in salt marshes. Rocky beaches fared better than sandy beaches, which still held oil deposits several years after the spill. Marshes showed no recovery two years later, and it was predicted that it would likely be decades before they would reattain their condition prior to the spill.

The impact of the *Exxon Valdez* spill on the environment is being studied thoroughly by the U.S. government and Exxon. Some estimates are that the full cost may run as high as $2 billion (Turner, 1990). One study estimated that 100,000 to 300,000 marine birds were killed. This would exceed any other oil-related spill to date. A later study indicated that wildlife and marine life losses were greater than originally predicted. Populations of sea otters, harbor seals, bald eagles, clams, and snails all declined. Reproductive irregularities were found in salmon and herring fry.

For the first year of study, $35 million from the Clean Water Act and other federal acts has been authorized for determining spill impacts on coastal habitats, air and water, fish and shellfish, mammals (both marine and terrestrial), and birds. In addition, the funds will be used to determine the economic value of damaged natural resources and restoration efforts (Turner, 1990). Some experts think that the *Exxon Valdez* spill will have significant long-term effects due to the subarctic conditions, which may slow photochemical and microbial degradation; the quantity of the spill; and the enclosed area in which it occurred (also see Enrichment Box 12.2).

Oil floats, and it coats things. Thus it can kill quickly by coating marine and freshwater life, interfering with gas exchanges necessary for life. Some of the longest-lasting impacts, however, come from contamination of fine bottom sediments. An accumulation of oil there can cause conditions to become anaerobic. Benthic organisms are particularly susceptible. Aerobic species may be replaced, thus affecting the food

The Valdez Principles

At the time of the *Exxon Valdez* oil spill in March 1989, the Coalition for Environmentally Responsible Economies (CERES) was working on an ecological code of business. Consequently, they named their code the Valdez Principles. Basically, the code sets out criteria for evaluating how a company operates from an ecological perspective. It is the hope of the group that organizations and individuals will use these principles to guide them into sound ecological investments, putting money into companies that abide by the principles. The public will be apprised of companies that have signed on and have, through some audit process, been found to be operating in compliance with the principles. The idea is based on the same concept as the Sullivan Principles, which were used to limit business involvement in South Africa to oppose apartheid. CERES is a coalition of environmentalists and social investors that was initiated by the Social Investment Forum, a Boston-based group of financial professionals interested in socially responsible investing. The Valdez Principles are endorsed by a variety of environmental organizations.

The Valdez Principles

1. *Protection of the Biosphere.* We will minimize and strive to eliminate the release of any pollutant that may cause environmental damage to the air, water, or earth or its inhabitants. We will safeguard habitats in rivers, lakes, wetlands, coastal zones and oceans and will minimize contributing to the greenhouse effect, depletion of the ozone layer, acid rain, or smog.
2. *Sustainable Use of Natural Resources.* We will make sustainable use of renewable natural resources, such as water, soils and forests. We will conserve non-renewable resources through efficient use and careful planning. We will protect wildlife habitat, open spaces and wilderness, while preserving biodiversity.
3. *Reduction and Disposal of Waste.* We will minimize the creation of waste, especially hazardous waste, and wherever possible recycle materials. We will dispose of all waste through safe and responsible methods.
4. *Wise Use of Energy.* We will make every effort to use environmentally safe and sustainable energy sources to meet our needs. We will invest in improved energy efficiency and conservation in our operations. We will maximize the energy efficiency of products we produce or sell.
5. *Risk Reduction.* We will minimize the environmental health and safety risks to our employees and the communities in which we operate by employing safe technologies and operating procedures and by being constantly prepared for emergencies.
6. *Marketing of Safe Products and Services.* We will sell products or services that minimize adverse environmental impacts and that are safe as consumers commonly use them. We will inform consumers of the environmental impacts of our products or services.
7. *Damage Compensation.* We will take responsibility for any harm we cause to the environment by making every effort to fully restore the environment and to compensate those persons who are adversely affected.
8. *Disclosure.* We will disclose to our employees and to the public incidents relating to our operations that cause environmental harm or pose health or safety hazards. We will disclose potential environmental, health or safety hazards posed by our operations, and we will not take any action against employees who report any condition that creates a danger to the environment or poses health and safety hazards.
9. *Environmental Directors and Managers.* At least one member of the Board of Directors will be a person qualified to represent environmental interests. We will commit management resources to implement these principles, including the funding of an office of vice president for environmental affairs or an equivalent executive position, reporting directly to the CEO, to monitor and report upon our implementation efforts.
10. *Assessment and Annual Audit.* We will conduct and make public an annual self-evaluation of our progress in implementing these principles and in complying with all applicable laws and regulations throughout our worldwide operations. We will work toward the timely creation of independent environmental audit procedures which we will complete annually and make available to the public.

---

chain. Also, fish populations, such as herring, which deposit their eggs in the sediments, are affected.

Much damage to the benthic community—the organisms in the tidal zone and just below—was caused by the oil pollution that occurred when the supertanker *Metula* went aground off the coast of Chile in August 1974, releasing 54,000 tons of oil. Studies showed that the benthic community began to recover within several months. None of this is to say that the overall effects of oil on aquatic systems are not significant. The full story on chronic effects is not yet known, and the acute effects can be significant. A February 1976 barge accident dumped a million gallons of heating oil into Chesapeake Bay and resulted in the deaths of some 20,000 birds, a notable effect indeed. Some of the chemicals in crude oil and in partially refined products are known to be carcinogens; we will discuss these in some detail in Chapter 18.

As yet, there is practically no data on the effect of oil pollution on human health. Oil pollution seems to have little long-range effect on fish, and there does not yet seem to be evidence for food chain magnification of the hydrocarbons in oil. Although many microscopic marine animals appear to take them up, the general fate of these hydrocarbons is to be rather quickly converted (photochemically and metabolically) to oxygenated metabolites (some of which may be more toxic than the parent compounds) and then to simple, harmless compounds. Long-term effects on aquatic communities have yet to be fully documented.

Dealing with Spills. Strategies for dealing with oil pollution once it has occurred have not been very effective and some such as hot water washing may do more harm than good. The failures include using detergents to disperse oil that can do more harm than the oil and using absorbants like straw to hold the oil, creating a new disposal problem. Burning the oil is usually not very effective because flammable materials escape quickly and leave the more fire-resistant chemicals behind. Various new containment methods are being tried and tested, for example, deploying barriers that contain the oil so that it can be siphoned up, skimmed off, or recovered in some other way. In general, all of these approaches have proved ineffective.

Some of the latest methods being considered include the use of microorganisms that can degrade oil. Some natural microorganisms degrade oil, and others have been genetically engineered to do so. The addition of fertilizer containing the limiting nutrients nitrogen and phosphorus is effective in accelerating the growth of the naturally occurring microorganisms and hence the degradation of oil. This use of fertilizer succeeded in cleaning some of the shoreline after the *Exxon Valdez* spill. Further investigation is needed to see if the bacteria actually consumed the oil or simply loosened it to be washed back into the water. Preliminary data showed no adverse impact on the aquatic environment due to overfertilization.

It is clear, however, that the best way to deal with this problem is to prevent it from happening in the first place. This can be done by better enforcement of regulations governing bilge-pumping and tanker-cleaning operations, better control of loading and unloading operations, maintenance and controls in the construction of supertankers, better training of tanker crews, better training and methods for managing offshore drilling rigs, and better regulation of aboveground storage tanks.

In August 1990, new federal oil spill legislation was signed into law to address some of these points. The law attempts to resolve the issue of prevention by mandating double-hulled construction for new vessels and requiring retrofitting of existing vessels. A study of the use of liners and secondary containment for facilities and aboveground storage tanks is to be conducted and its recommendations are to be implemented. The national contingency plan for oil spills is to be updated with duties specified at the federal, state, and local levels, as well as strategies for responding to oil spills.

The billion-dollar Oil Spill Liability Trust Fund was established to pay for oil spill cleanups. A 5¢-per-barrel excise tax on domestic and imported oil will finance the fund.

Water Pollution as a By-product of Refining. Many millions of gallons of oil wastewater are generated every day in petroleum refining. Water is used in many ways in petroleum refining, chiefly to remove salts and other impurities from oil. According to the U.S. Department of the Interior, 10 liters of water are used to refine a liter of gasoline.

Polluted Water: Some Notable Examples

Having considered the types of pollutants and some of the more important sources of water pollution, we would like to consider the magnitude of the water pollution problem by looking at what it has done to a river, a lake, and a sea.

We have chosen a lake, a river, and a sea as representatives of the major bodies of water affected by pollution; there are different considerations for each. Because the flow in rivers tends to increase oxygenation, for example, rivers can handle biological oxygen demand more readily than lakes. The differences between rivers and lakes in this regard are becoming less now that many major rivers are being converted into chains of long lakes through the construction of flood control, recreational, and navigational dams. The Ohio River offers a good example of this trend. Coastal problems are the main problems in oceans and seas; the open oceans tend to minimize the impact of pollution by their sheer volume. That is why we picked the relatively compact Mediterranean. Lakes as a group, though they too vary considerably in size, are especially prone to damage from pollutants because of their low flow and stratification. Lake Erie in North America is an outstanding example of the problems of lake pollution.

The Trouble with Lake Erie

North America's Great Lakes are the largest reservoir of fresh water on earth, containing one-fifth of the world's supply. The natural resources of these lakes include water for drinking, navigation, recreation, and industrial uses; fish as sources of food; and the kinetic energy of water as hydroelectric power as the water spills from one lake into another on its way to the sea.

The Great Lakes are like tubs connected by very small pipes to one another and eventually to the ocean. Things dumped into them tend to settle out or accumulate in other ways rather than wash into the Atlantic Ocean. This presents a problem for all five Great Lakes, and it is most serious in Lake Erie. Lake Erie is both the shallowest of the five and the lake serving the greatest number of people and industrial centers.

In the 12,000 years since it was born as a gouge left by a receding glacier, Lake Erie has suffered many abuses. In addition to natural crises, Lake Erie has long been receiving pollutants created as by-products of the industrial age. Lake Erie drains some 30,000 square miles of North America; 70% of the drainage comes from the American side. More than 11.5 million Americans dump their sewage (after various degrees of treatment) into Lake Erie, and the wastes from a million Canadians come in from the other side. Even on normal days the Detroit River alone brings nearly 5,000 cubic yards of mostly agriculturally derived sediment into the lake. On stormy days the same river delivers 50,000 cubic yards of soil from Michigan and Canada. A hodgepodge of industries dump hundreds of millions of gallons of spent water into the Cuyahoga River each day.

Pollution has changed the biological character of the lake, and detergents and other eutrophicants have apparently caused significant oxygen depletion. Sturgeon and other commercially important fish began to disappear from the lake nearly a century ago. Contrary to much that has been said about Lake Erie's death, the eutrophication has not brought about a decrease in the *amount* of life in the lake; it has brought about a change in the *character* of that life.

What has happened to Lake Erie has been described as accelerated aging. By some estimates, the lake looks twice its age. We know from the early chapters of this book that lakes do not last forever. They gradually become shallower and shallower as sediment falls into them and begins filling the lake in from the edges. Eventually, lakes become marshes and advance toward some kind of terrestrial community (Chapter 4). Through overenrichment from phosphates, nitrates, organic matter, and other by-products of civilization and through erosion-derived sediment, Lake Erie has become shallower and contains more organic matter and less oxygen. It is not uniformly polluted. The shallow, heavily industrial western basin is in worse shape than the eastern basin.

There is no question that the impact of water pollution on Lake Erie has been considerable or that the many and varied users and potential users of the lake have a problem. The hope is that the problem is not irreversible or at least that the rate of aging can be slowed. There is cause for some optimism.

By early 1980, three-fourths of the more than 100 industrial polluters on the American side of Lake Erie cited for violations in the 1960s were meeting minimum treatment requirements. Municipal sewage treatment has also improved. The United States and Canadian governments have invested more than $7.5 billion since 1972 to limit the amount of phosphorus in municipal discharges into the lake. This resulted in an almost 85% reduction in annual phosphorus discharge by 1985. Improvements in discharges into the Detroit River which empties into Lake Erie were also significant (Makarewicz and Bertram, 1991). Restocking and fish management programs have started to bring the fish back. There is still room for improvement.

No sooner had Lake Erie begun to show signs of recovery from its eutrophication problems than its pollution by persistent toxic chemicals became increasingly apparent. Significant levels of toxaphene, DDT, dioxin, and PCBs have been reported in fish throughout the Great Lakes system. These toxic chemicals are apparently making their way into the lakes from farm fields and from leaky disposal sites in the drainage system.

A 1985 report issued jointly by the National Research Council and the Royal Society of Canada assessed the progress that has been made in cleaning up the Great Lakes. Phosphate loads and levels of heavy metals, pesticides, and some organic chemical contaminants from point sources have been decreased. However, airborne contaminants and contaminants from nonpoint sources such as leaching landfills are unchanged or possibly on the increase. The report called for increased efforts to control nonpoint sources of pollutants and development of a comprehensive toxic substances management strategy.

Atmospheric deposition is emerging as a significant factor in water pollution. It may be that DDT is being transported through the atmosphere to the Great Lakes from as far away as Mexico and Central America (Eisenreich, 1987). Toxic pollutants from incinerators, from evaporation from agricultural land, and from coal combustion can be airborne over long distances before they wash out as precipitation or settle out as dry deposition onto land or surface water. Processes such as these are affecting the quality of water in the Great Lakes and, without question, elsewhere as well.

The Ohio River

In 1936 a Kentucky congressman called the Ohio River a "cesspool." Everything was dumped into the river; pollution control was decades in the future. The Ohio River is a good example of what water pollution has done and can do to a river.

The Ohio River Basin includes nearly 200,000 square miles of Indiana, Kentucky, Ohio, West Virginia, Illinois, Virginia, North Carolina, Tennessee, Maryland, Pennsylvania, and New York. About 20 million people live and work in this basin, and it has over 2,000 major industrial operations. The valley produces about 75% of the nation's coal and more than one-third of its steel. The Ohio River Basin has over 94 million acres of cropland. Obviously, a proportional share of the water pollution related to these activities ends up in the Ohio River. The river is already polluted at its source, where it is formed by the confluence of the far-from-pure Allegheny and Monongahela rivers at the Golden Triangle in Pittsburgh, Pennsylvania. Things get worse as the Ohio flows some 981 miles to the southwest, where it joins the Mississippi at Cairo, Illinois. The urban centers along the Ohio include Paducah, Covington, Newport, Owensboro, Louisville, and Ashland, Kentucky; Evansville, Indiana; Cincinnati, Portsmouth, Ironton, Marietta, and Steubenville, Ohio; Huntington, Parkersburg, Weirton, and Wheeling, West Virginia; and Pittsburgh, Pennsylvania.

According to some sources, many industries along the Ohio still do not meet current emission standards. Numerous reports have identified hundreds of errant chemical compounds in Ohio River water; many of these chemicals are known to produce cancer in humans and other animals.

Some of the Ohio River pollution problem stems from the fact that the Ohio has been converted over the years into a series of long lakes created by dams built by the Army Corps of Engineers to ensure navigation depths along the entire length of the river. A related problem is that the percentage of oxygen saturation during the summer drops considerably from Pittsburgh to Cairo.

According to the Environmental Protection Agency's Office of Water Planning and Standards, pollution problems of the Ohio River include (1) low alkalinity (high acidity)—a result of the fact that the Ohio drains many coal-mining areas in Pennsylvania, West Virginia, and other states; (2) a high total fecal **coliform** count*—indicating that sewage treatment in a number of cities along the Ohio is inadequate; (3) a high iron and manganese concentration—reflecting the character of the industry along the Ohio; and (4) some toxic substances—DDT and chlordane, for example.

A 1987 report released by ORSANCO, an organization of the states bordering the Ohio River, summarized the results of testing for toxics in the Ohio River from 1976 to 1985. Toxic substances were found throughout the length of the river, varying in types and amounts depending on the use of the watershed and the types of industrial discharges. The most frequently found toxics were zinc and copper. Others commonly

detected were chloroform, lead, phenols, nickel, and chromium. The toxic that most often exceeded the EPA's cancer risk criteria was chloroform. Arsenic also exceeded the cancer risk criteria, but not as frequently. Lead was the toxic that most frequently exceeded the EPA's aquatic life criteria and, at times, the EPA's human health criteria. Only on rare occasions were the safe drinking water standards exceeded for arsenic, cadmium, chromium, copper, lead, or mercury. Although pesticide and PCB residues were not found often in water samples, they were routinely detected in fish. Health officials cautioned against eating, more than once a week, fish caught in the Ohio River. The commission concluded that the river overall, however, is cleaner than is generally perceived.

Some of the things that get into the Ohio River cancel some other things out. For instance, the Ohio has high levels of suspended solids, apparently the result of sediment being washed into the stream during high flows along its length. Although the Ohio also has a great potential for algae growth, the cloudiness of water, caused by suspended solids, blocks out sunlight, inhibiting algal growth.

As late as 1948, still 99% of the raw sewage from toilets, slaughterhouses, and similar sources went directly into the Ohio—untreated. Although there was some very slow progress in the 1950s, significant improvement did not begin until the environmental movement in the late 1960s. Because of the Clean Water Act and other legislation, much of the raw sewage and odor of the Ohio are gone now. Many of the cities along the river are fast approaching the EPA standard of reduction of sewage by 85%. Many of the coal mines have been closed and sealed; operational mines are better controlled. Also on the comeback trail are sauger, crappie, white bass, and freshwater drum and paddlefish. According to ORSANCO, Ohio River fish now carry less heavy metal.

Largely as a result of clean water legislation of the early 1970s, the 700 industrial plants along the Ohio have sharply reduced the amount of waste they dump into the river. But plants are relatively easy to monitor and to regulate. The greatest unresolved problems now come from farms, construction sites, septic tank leakage, and the like—nonpoint sources and sources otherwise difficult to regulate.

A report issued in May 1990 by ORSANCO made an assessment of nonpoint source pollution on the Ohio. It found that agricultural runoff was the most widespread nonpoint source, but mining was the most severe. Impacts from mine runoff were most severe in the northern third of the river. Acid mine drainage carries sulfates and heavy metals into the river. Agriculture is the predominant source of nonpoint pollutants in the bottom third of the river. Sediment, pesticide, and fertilizer residues are all carried into the river.

---

* Coliforms are bacteria that inhabit the digestive tract of *Homo sapiens* and other animals. Though not usually pathogenic in themselves, their presence indicates contamination by fecal waste and warns of potential infection by less common but more virulent inhabitants of the gastrointestinal tract, such as the causative agents of cholera and hepatitis.

PCBs were found to be a problem along the entire length of the river.

Even the point sources have considerable room for improvement. Because of combinations of technical problems, design problems related to the connection of storm drains to sewage systems, and other problems, some Ohio River towns are removing less than half the contaminants from their sewage.

There has been dramatic though incomplete success with other of the world's great rivers. The Potomac River, which runs through Washington, D.C., used to smell of human excrement; today, the cities that affect the Potomac remove more than 85% of pollutants from the sewage they generate before the Potomac gets it. It was 0% not very long ago, but Congress has apparently acted to get its own house in order. Other notable success stories include the Willamette River in Oregon and the Thames in England—both of which, like the Ohio and the Potomac, were once virtual cesspools and are now relatively clean.

A study of water quality trends in U.S. rivers from 1974 to 1981 (Smith, Alexander, and Wolman, 1987) made the following observations:

1. Because of nonpoint source pollution, national water quality goals may not be met, even if all point source pollution controls are implemented.
2. Contaminants in the atmosphere are having an impact on water quality.

Regarding specific contaminants, the report found:

1. Decreases in fecal colifor were widespread, owing primarily to improvements in municipal wastewater treatment.
2. Overall, *decreases* in dissolved oxygen–deficient conditions exceeded *increases* by 3 to 2. Improvements in dissolved oxygen occurred in New England, the Middle Atlantic states, and the Ohio and Mississippi basins; the greatest frequency of increased dissolved oxygen–deficient conditions occurred in the Southeast. There was no strong correlation between decreased BOD loads from point source controls (especially municipal treatment facilities) and improved levels of dissolved oxygen, even though municipal BOD had decreased by 46% and industrial BOD by 71%. There was some evidence that the increasing nonpoint source BOD diminished the benefits derived from decreasing point source BOD. During this time period, the population increased by 11% and the GNP by 25%; some observers suggest that just remaining stable during this period can be considered as improvement. The location

of sampling monitors was cited as another possible factor in the failure to find a significant relationship between municipal point source controls and improved oxygen conditions.
3. Nationally, suspended sediments showed increasing concentrations in river basins with large amounts of cropland within their watersheds. Decreases in suspended sediments were noted in other watersheds.
4. Total phosphorus concentrations showed decreases in areas with point source control of phosphates but increases where nonpoint sources dominate.
5. Increases in total nitrate were frequent and widespread and could be correlated with agricultural activities such as increased use of fertilizer and livestock density. Atmospheric nitrate deposition was also a significant source of water pollution, especially in the Ohio, Middle Atlantic, Great Lakes, and Upper Mississippi basins.
6. The salinity of the nation's rivers has increased significantly. A likely culprit is the use of salt on highways, increasing the sodium and chloride content of water, especially in the Ohio, Tennessee, Lower Missouri, and Arkansas-Red basins. Increased chloride content is also correlated with increased population and hence increased human waste laden with chloride. In the western states, irrigated agricultural land has an influence on river water salinity. In Missouri, Arkansas, and Tennessee, an increase in sulfate from surface coal production is a factor. Salinity decreased in the Upper Colorado River Basin, where salinity control measures had been implemented.
7. Records show an increase in arsenic and cadmium, especially in the northern Midwest. Atmospheric deposition from fossil fuel combustion and runoff from fly ash storage are considered the primary sources of these metals.
8. Decreases in dissolved lead concentrations occurred along the East and West coasts and the Missouri and the Mississippi river basins, most likely because of the decrease in consumption of leaded fuels. However, there were no significant declines in lead concentration in the Ohio and Great Lakes basins, where other as yet undetermined factors must be affecting lead levels.

The Mediterranean Sea

The Mediterranean Sea is of interest because it has enjoyed a delicate balance with humankind for thousands of years. Areas along the Mediterranean are highly populated and have been for some time; there were reportedly 50 million people in the Mediterranean basin when the Roman Empire was at its peak. Today, more than 200 million people live in the countries that border the Mediterranean Sea; some 44 million of these live adjacent to the sea and interact with it directly.

The Mediterranean is of special interest because of these questions: Can humankind pollute the oceans? Has it done so already? The Mediterranean is a small sea, it is relatively shallow, and for all practical purposes it is isolated. If humans have truly had an impact on oceans, nowhere should this impact be more evident than in the Mediterranean Sea.

Many of the nations ringing the Mediterranean and many others whose rivers empty into it are now highly industrial. Helmer (1977) describes an inventory made by the UN Environment Programme in which pollution of the Mediterranean was cataloged according to the broad categories of (1) domestic sewage, (2) industrial waste (including pollution by the petroleum industry), (3) agricultural runoff, (4) radioactive waste, and (5) river discharge (rivers coming into the Mediterranean treated as point sources). A summary of the findings of this study follow. (This study was republished intact in 1984; see United Nations Environment Programme, 1984).

Domestic sewage comes into the Mediterranean in great volumes both directly and in the rivers that empty into it. Although this situation has improved since the early 1970s, virtually none of the communities along the Mediterranean bothered to treat sewage as recently as the mid-1970s. In 1987, fully 90% of the sewage was still dumped untreated (Smart et al., 1987). An Italian cholera epidemic in 1973 has been blamed on sewage contamination of shellfish, but relatively few other sewage-related health problems have been documented along the Mediterranean coast. Domestic sewage is a source of organic matter, bacteria, and nutrients such as phosphate from detergents and some metals. The main incentive for cleaning things up even today seems to be the potential health impact and the aesthetic impact on the tourist trade in places like the French Riviera.

Industrial waste comes from many types of Mediterranean industry, and it comes in great volumes. The Mediterranean is actually one of the most highly industrialized areas in the entire world; its industrial spectrum includes iron and steel, food processing, pulp and paper, leather, textiles, and petroleum-based chemical operations. Industrially derived metals including mercury, cadmium, copper, lead, and zinc have been measured in the open Mediterranean and have been found to be present in amounts no greater than the background levels in the world's major oceans. As in the oceans, however, there are high concentrations at various localized spots along the coast. High concentrations of copper have been measured in the coastal waters off Marseille, for example. Industrial waste also adds organic matter and suspended solids to the Mediterranean.

Oil pollution is a problem of yet uncertain dimension in the Mediterranean. Much oil is spilled there, as would be expected from its role as one of the major highways linking the oil-rich Middle East with the rest of the world (nearly 400 million tons of petroleum are transported over the Mediterranean each year). Most spilled oil comes from the routine tanker petroleum refining and drilling operations that ring the Mediterranean.

The impact of spilled petroleum on the marine environment is similar to that reported along other shipping lanes of the world. The most visible problem is that of tarry lumps that wash up on beaches. Tainting of commercial species of fish and even decimation of populations of some species have been blamed on oil. Spain has reported the tainting of gray mullet and mussels in oil-polluted harbors. Both France and Italy have reported polluted shellfish and fish that taste bad to the point of being inedible. Turkey has reported that the reproduction of bonita and mackerel has been negatively affected by oil pollution. Nearly all of the oil problems are in harbors and coastal areas rather than in open areas.

The Mediterranean also receives considerable *farm-related pollution*. Rivers carrying nutrient runoff from the farming areas of southern Europe contribute most to this problem. In addition, these rivers carry metals, organic matter, suspended solids, pesticide residues, and other pollutants found in domestic, industrial, and agricultural discharges. Suspended solids and pesticides run off into the Mediterranean from the surrounding watershed, but the contribution made by this runoff is less significant than that carried into the Mediterranean by rivers. Pollutants from atmospheric transport were not included in the study.

Although there are nuclear reactors in Spain, France, and Italy and although there are medical and research installations all along the Mediterranean, *radioactive waste* does not appear to be a problem at present. However, considerable expansion of the atomic power industry in the Mediterranean basin is projected over the next few decades, and the situation may change.

If marine pollution is defined as the introduction into the oceans by *Homo sapiens* of materials and energy that have the effect of harming living resources, harming human health, hindering marine activities such as fishing, limiting any use of seawater (including recreation), then the Mediterranean Sea is polluted—in some places more than in others. Most of what we know about so far is localized and coastal in nature. The data collected to date shows little if any pollution of the open Mediterranean Sea. Despite its long-term, intense insult by humans (relatively speaking) and despite its being much smaller in volume than the earth's oceans, there is little difference between the levels of contamination in the Mediterranean and in the oceans. But coastal pollution remains a serious problem. One ship found the waters around the coast of Italy polluted with sewage and industrial waste and the beaches littered with garbage (Smart et al., 1987).

The Oceans in General

It is almost incomprehensible to think we could pollute the oceans. They make up 70% of the earth's surface and are seemingly inexhaustible sinks for waste. Perhaps this is why we continue to dump thousands of tons of garbage into them every year. Although the point can be argued at great length, it is perhaps impossible to pollute the oceans to a uniformly dangerous level. Although the dynamics of biomagnification may indeed lead to the contamination of high-trophic-level organisms over wide expanses of the world's oceans, the most serious problem for the moment is localized pollution.

Already the pollution of coastal areas of the oceans is very serious. Coastal pollution affects some of the most important natural systems on earth—the estuaries—and extends over much of the adjacent continental shelves in many locations. Estuaries are important because they serve as breeding areas for species that hold key positions in many oceanic food chains and marine ecosystems. We saw in earlier chapters that estuaries are among the most productive ecosystems on earth; they constitute an optimal life support interface between the atmosphere, lithosphere, and hydrosphere.

Overenrichment of localized areas of the ocean cause algae to proliferate dramatically in what is termed an *algae bloom*. As this abundance of algae dies and decays, it depletes the area of oxygen and thus causes the death of numerous other aquatic organisms. Recently, toxic algae blooms called *red tides* have been found to be associated with overenriched waters (Lenssen, 1989). Previously, they had been thought to be purely natural events. Algae blooms have caused problems in the Baltic and Adriatic seas and along the coasts of Denmark, Germany, and the eastern United States.

In the United States, one-third of all coastal shellfish beds are closed at any one time due to pollution. The $6 billion commercial and recreational fishing industry is threatened by polluted waters (Smart et al., 1987). A 1987 report by the Office of Technology Assessment predicts that the quality of coastal waters and estuaries will diminish over the next decade if additional efforts to control pollution are not made. Specifically, control of runoff (nonpoint sources of pollution), dumping, and disposal of dredged materials are cited. High levels of organic chemicals, disease-causing organisms, and nutrients and low oxygen levels were found to be characteristic of coastal waters.

Chemical pollutants in the ocean are of growing concern. Ash from incineration at sea, runoff containing pesticides and other chemicals, industrial discharges that flow into coastal waters, and airborne pollutants are just some of the sources of toxic chemicals and heavy metals that have been found in excessive amounts, especially in the upper surface waters. This is especially significant because this is the area inhabited by the phytoplankton and zooplankton that form the basis of the ocean food chain.

In the open oceans, about 6 million metric tons of litter annually is thrown overboard from ships, including those of the U.S. military (Smart et al., 1987). Other materials are discarded by offshore oil rigs.

People have lived in towns and cities in the Mediterranean basin for thousands of years, but the pollution problem in the open parts of the small, shallow Mediterranean Sea appears to be no worse than that in the open oceans of the world.

Fishers lose about 135,000 tons of plastic nets and fishing apparatus annually. Plastic and synthetic materials that do not biodegrade can entangle or be ingested by marine life, resulting in death. This has affected marine mammals, seabirds, and endangered species such as the sea turtle. The International Convention for the Prevention of Pollution from Ships prohibits disposal of any plastics in the sea. This prohibition, ratified by 39 nations, became effective in January 1989.

More than 60 nations are party to the Convention on the Prevention of Marine Pollution by Dumping of Wastes and Other Matters (known as the London Dumping Convention), which regulates wastes loaded on ships for the specific purpose of dumping. It prohibits the dumping of plastics and other persistent materials and bans ocean incineration of toxic substances by 1994 (Lenssen, 1989). Although parties to the convention have called for greater efforts to prevent dumping of persistent substances, enforcement remains a problem.

Sewage, pesticides, oil, and plastic have had effects on oceans as far removed as the Arctic and Antarctica. Gradual degradation by pollution may be having a much greater impact on marine wildlife in the long term than outright habitat destruction. Pollution of the oceans is indeed a problem, the seriousness of which is only beginning to be recognized.

The Pollution of Groundwater

Groundwater (see Chapter 7) is also subject to pollution. Since a great deal of this water source is tapped for various uses, groundwater pollution has important implications. Consider the following:

__ Ninety-five percent of rural households and one-third of the United States' 100 largest cities use groundwater as their main source of drinking water and irrigation.
__ Around 1990 the United States used about 100 billion gallons of groundwater per day.
__ Groundwater is the source of fully half the irrigation water used in the American West.
__ Groundwater is the next largest reservoir of water on earth after the oceans.
__ Once contaminated, groundwater may remain contaminated for hundreds of years.

Obviously, groundwater is important. Even so, an understandable first reaction to the notion of groundwater pollution might well be, How can groundwater become polluted? It's underground! Figure 12.7 illustrates how contaminants can get into surface water from waste dumps by being carried by groundwater. Any such groundwater would itself be polluted,

as would groundwater that receives other contaminants from above. It is not difficult to imagine that the 10,000 or more toxic waste dumps identified by the EPA (see Chapters 13 and 17) are an enormous source of potential groundwater contamination. In even more quantitative terms, government statistics indicate that 28 to 54 million tons (EPA) or as much as 275 million tons (Office of Technology Assessment) of federally regulated hazardous waste is released into the environment each year. Much of this has ended up in leaky landfills and ordinary garbage dumps from which there has been seepage into groundwater. Agriculture and household wastes and deliberate deep-well disposal are additional sources of groundwater contamination (Figure 12.9). Septic tanks alone are the source of hundreds of billions of gallons of waste discharged into the ground each year. Leaking underground storage tanks are another major source. Many of these steel tanks buried years ago are beginning to corrode and leak. The Steel Tank Institute has estimated that 350,000 tanks containing gasoline would leak in a period of just five years (Sun, 1986). This did not include tanks holding other wastes and chemicals.

There is considerable evidence that wastes do contaminate groundwater. In recent years, as many as one-third of all cities relying on groundwater for drinking have experienced some form of contamination, mainly from leachate from toxic waste dumps. Groundwater contamination has been reported in every state, and the EPA has reported contamination of drinking water wells by toxic organic chemicals in virtually every state. Although less than 2% of underground aquifers are badly polluted at present, many of these are in large population centers.

Four of the most common pollutants of groundwater are chloride, nitrate, heavy metals, and hydrocarbons. Information is needed on which pesticides have the greatest potential for leaching into groundwater and under what conditions, as well as on how they move and persist in aquifers. Such data is lacking for most groundwater contaminants. The specific health effects of some of these groundwater contaminants are described in Chapters 9, 10, and 18.

The EPA undertook an assessment of pesticide and nitrate contamination of groundwater. In a report issued in late 1990, the agency found contamination by at least one pesticide in 10% of the community drinking water wells in the United States and about 4% of rural domestic wells. Nitrates were found in more than half of the wells tested. The levels of pesticides that were found were considered low. Some of the nitrate levels were high. This was the first assessment of its kind by the EPA. It was intended as a first step in accumulating groundwater data in the United States. Such data is necessary before a federal groundwater law can be written.

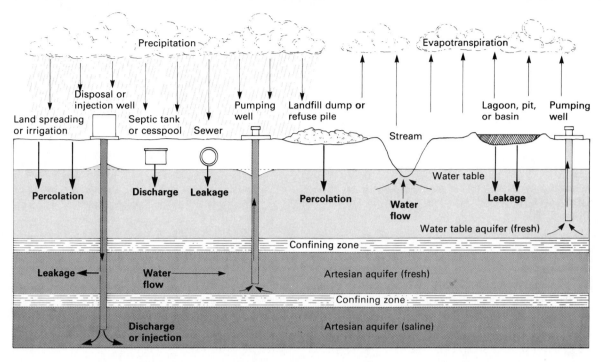

Figure 12.9 Sources of Groundwater Contamination. As this figure shows, there are a variety of sources that can contribute to contamination of groundwater.

The problem of groundwater contamination is especially difficult because of the out-of-sight nature of the problem; we know comparatively little about what goes on "down there." We know that the groundwater flows relatively slowly, but we know little about the pathways. The assumption that water will be filtered of contaminants before it reaches major underground aquifers has proved false. Nature can provide some cleansing in various ways—surface adsorption, dilution, mechanical filtration, precipitation following chemical interaction, buffering, chemical neutralization, microbial degradation, and plant uptake—but we do not know to what extent these occur in different locations. Very few groundwater sources have ever been tested. Nor have many, if not most, of the chemicals contaminating groundwater today been adequately tested, particularly at low levels, for health effects; no standards exist for their allowable concentrations.

Because of the site-specific problems associated with groundwater and the fragmented authority for handling the problem, little progress has been made in the United States on establishing a national groundwater protection policy or on defining local, state, and federal roles in the process. At least 16 pieces of federal legislation and 11 federal agencies are involved in groundwater protection.

According to the U.S. Environmental Protection Agency (1987), efforts to protect groundwater generally fall into eight categories:

1. Development of statewide groundwater plans and strategies
2. Aquifer classification laws
3. Standards setting limits on contaminants
4. Land use management laws to control activities on the land that may contaminate groundwater
5. Groundwater funds to be used for replacing drinking water supplies or for monitoring or cleanup
6. Laws regulating agricultural chemicals
7. Laws regulating underground storage tanks (see Chapter 17)
8. Water use laws relating to water withdrawals from underground supplies

The EPA concluded, however, that overall, few comprehensive groundwater protection programs exist.

Legislation to accomplish a national groundwater protection program has been introduced several times in Congress. To date, nothing has been enacted into law. Areas of debate include what roles the federal and state governments should assume. There is also controversy about the exact strategies that should be included. Does classifying aquifers for certain uses relegate some aquifers to further degradation? Should any aquifer be written off for future uses because of current contamination? As the awareness of the seriousness of groundwater pollution continues to emerge, public outcry may cause the piecemeal approach of regulating via the Safe Drinking Water Act, the Clean

Activated sludge process: a process that removes organic matter from sewage by saturating it with air and adding biologically active sludge.

Adsorption: an advanced way of treating wastes in which activated carbon removes organic matter from wastewater.

Aeration tank: a chamber for injecting air into water.

Algae: plants that grow in sunlit waters. They are a food for fish and small aquatic animals and, like all plants, put oxygen in the water.

Bacteria: small living organisms that often consume the organic constituents of sewage.

Biochemical oxygen demand (BOD): the dissolved oxygen required by organisms for the aerobic decomposition of organic matter present in water.

Coagulation: the clumping together of solids to make them settle out faster. Coagulation of solids is brought about with the use of certain chemicals such as lime, alum, and iron salts.

Combined sewer: a sewer that carries both sewage and stormwater runoff.

Comminutor: a device for the catching and shredding of heavy solid matter in the primary stage of waste treatment.

Diffused air technique: a technique whereby air under pressure is forced into sewage in an aeration tank. The air is pumped down into the sewage through a pipe and escapes out through holes in the side of the pipe.

Electrodialysis: a process that uses direct current and membranes to separate soluble minerals from water.

Flocculation: the process by which chemicals dissolved in water are made to precipitate by chemical, physical, or biological action.

Organic matter: the carbonaceous waste contained in plant or animal matter and originating from domestic or industrial sources.

Oxidation pond: an artificial lake or body of water in which wastes are consumed by bacteria. It is used most frequently with other waste treatment processes. An oxidation pond is basically the same as a sewage lagoon.

Primary treatment: treatment that removes material that floats or will settle in sewage. It is accomplished by using screens to catch the floating objects and tanks for the heavy matter to settle in.

Sand filters: filters that remove some suspended solids from sewage. Air and bacteria decompose additional wastes filtering through the sand. Cleaner water drains from the bed. The sludge accumulating at the surface must be removed from the bed periodically.

Secondary treatment: the second step in most waste treatment systems in which bacteria consume the organic parts of the wastes. It is accomplished by bringing the sewage and bacteria together in trickle filters or in the activated sludge process.

Sedimentation tanks: tanks that help to remove solids from sewage. The wastewater is pumped to the tanks, where the solids settle to the bottom or float on the top as scum. The scum is skimmed off the top, and solids on the bottom are pumped off for incineration, digestion, filtration, or other means of final disposal.

Septic tanks: cisterns used for domestic wastes when a sewer line is not available to carry them to a treatment plant. The wastes are piped to underground tanks directly from the home. The bacteria in the wastes decompose the organic waste, and the sludge settles on the bottom of the tank. The effluent flows out of the tank into the ground through drains. The sludge is pumped out of the tanks, usually by commercial firms, at regular intervals.

Sewers: pipes that collect and deliver waste water to treatment plants or receiving streams.

Sludge: the solid matter that settles to the bottom, floats, or becomes suspended in the sedimentation tanks and must be disposed of by filtration and incineration or by transport to appropriate disposal sites.

Sterilization: the destruction of all living organisms. In contrast, disinfection is the destruction of most of the living organisms.

Suspended solids: the small particles of solid pollutants that are present in sewage and resist separation from the water by conventional means.

Trickle filter: a support medium for bacterial growth, usually a bed of rocks or stones. The sewage is trickled over the bed so that the bacteria can break down the organic wastes. The bacteria collect on the stones through repeated use of the filter.

Water Act, and waste management acts to give way to a more coordinated and consolidated approach. Much public debate is needed, however, so that the implications of long-term strategies are fully explored.

Wastewater Treatment and Other Types of Water Pollution Control

One way to control pollution is not to generate pollutants or to generate less. This is more applicable to some sources than to others, however. Except for long-range population control, there is really no way to reduce the amount of human and animal bodily wastes generated. These must be dealt with by using some kind of strategy to keep them from becoming water pollution problems. Industry is another story. There *are* ways to reduce industrial waste, and we will discuss some specifics shortly.

Self-purification of Water

As indicated in Table 12.6, there are natural mechanisms of water purification; some of these are physical, some are chemical, and some are biological. Physical purification includes obvious things such as dilution and some less obvious ones like the absorption of chemical contaminants into suspended clay parti-

Table 12.6 Sinks for Water Pollutants in the Hydrosphere and in Soil

Chemical Group	Some Sources	Major Sinks	Examples
Pesticides	Agricultural operations; public health programs	Photo-oxidation on surface of soil or water	Dieldrin; 2,4-D
		Hydrolysis in waterways	DDT
		Oxidation and reduction catalyzed by organic and mineral fractions in soils and sediments	Organophosphate pesticides
		Adsorption on particles in soil or suspended in water	DDT
		Microbially mediated degradation in soils and sediments	Organophosphate pesticides; carbamate insecticides
Hydrocarbons	Industrial operations; petroleum spills	Photo-oxidation on surface of soil or water	Some petroleum components
		Adsorption on particles in soil or suspended in waterways	Polycyclic aromatic hydrocarbons
		Evaporation	Low-boiling-point materials
		Microbially mediated degradation	Polycyclic aromatic hydrocarbons; naphthalene
Halogenated hydrocarbons	Industrial operations	Adsorption onto particles suspended in waterways	Polychlorinated biphenyls (PCBs)
Synthetic polymers	Industrial operations	Autooxidation on surface of soil or water	Butyl rubber
		Microbially mediated degradation in sediments and soils.	Cellulose nitrate, cellulose acetate, styrene polymers
Fertilizers	Agricultural operations		
Nitrogen		Chemodenitrification	Nitrate
		Volatization from topsoil to air	Ammonium
		Microbially mediated reactions; production of nitrous oxide and nitrogen gas, which may diffuse into atmosphere	Nitrate, ammonium
Phosphorus		Microbially mediated reduction of phosphate; precipitation from solution in ground and surface waters	Phosphate

Source: R. B. Schlesinger, "Natural Removal Mechanisms for Chemical Pollutants in the Environment," *BioScience 29* (1979): 97, 99.

cles that eventually settle out. As a result of absorption and sedimentation, some pollutants end up as deep-ocean sediment and are more or less permanently sequestered from the biosphere. Chemical purification also occurs naturally; organic and inorganic acids may be neutralized by reaction with lime (calcium carbonate) in limestone creekbeds, for example. Biologically, many naturally occurring microorganisms are able to degrade the organic matter that reaches waterways.

All of the natural sinks described in Table 12.6 can be overloaded. They are variable in how well they work, and many circumstances affect them. The amount of suspended clay available for absorption varies, as does the amount of calcium carbonate dissolved in water; microorganisms work better at some temperatures than others; and so on. Even the most optimally functional natural sinks can become overloaded by the sheer volume of pollutants.

We have tried to develop water treatment systems to solve this problem. At first our efforts were largely engineering-intensive and ecologically inap-

propriate. Lately, we seem to be discovering that dealing with this problem by gently helping nature along (Figure 12.10) may be the best answer.

Primary Sewage Treatment

Primary sewage treatment is the simple removal of the solids from wastewater. As illustrated in Figure 12.11, the basic primary treatment system includes some combination of screens, filters, grit chambers, and sedimentation tanks connected in series. Some primary treatment systems also use grinders, which shred solid materials to be collected later in a sediment tank. Primary treatment systems often include a chlorinator, which is designed to kill all or most of the bacteria present in wastewater. It has been suggested that chlorination may be a double-edged sword. It may kill bacteria and some, though not all, viruses, but it also causes the chlorination of certain organic chemicals, perhaps creating compounds, such as trihalomethanes, that are carcinogenic (see Chapter 18).

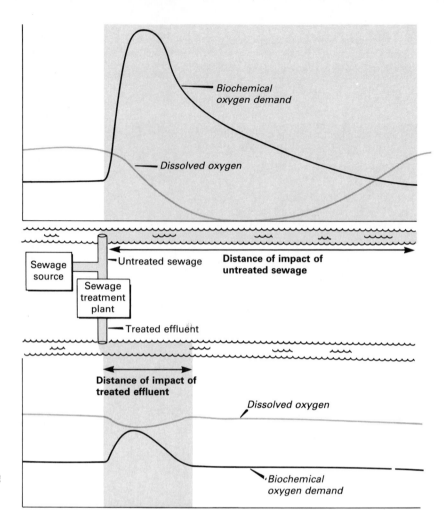

Figure 12.10 The Impact of Sewage Treatment. Sewage treatment plants compress into a very small space the cleansing action that would normally take place over much longer distances in a flowing river. The ideal treatment plant reduces the impact to near zero.

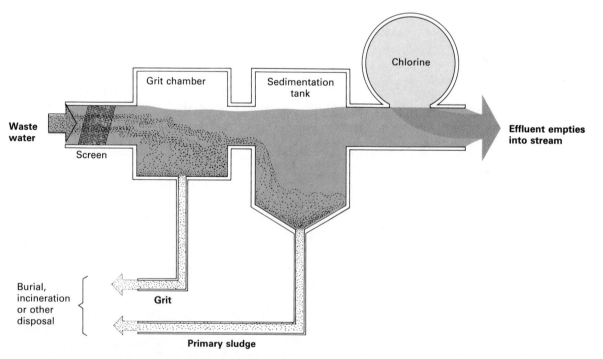

Figure 12.11 Primary Treatment of Sewage. Primary treatment involves physical screening and physical removal of grit and sedimental solids. Chlorine is usually added at the end of this short, ineffective process to kill potentially harmful bacteria.

The final product of primary treatment is wastewater containing assorted toxic substances and a large amount of dissolved organic matter and salts. These can constitute a considerable BOD and potential for eutrophication. Thus primary treatment alone is generally inadequate.

The captured products of primary treatment, the sludge and solids, must be used or disposed of by some means. Common methods include allowing the sludge to dry and then spreading it in a landfill or incinerating it. Primary treatment does little more than help to slow down the aging of waterways by keeping out grit and solids that would tend to fill them up. The problems that remain in wastewater after primary treatment can be reduced considerably by secondary treatment.

Secondary Sewage Treatment

Secondary sewage treatment employs the kinds of decomposers discussed in Chapter 1 to break down the organic matter in sewage before it gets to a river or lake. The target for this stage of treatment is the dissolved organic matter that constitutes the BOD in sewage. There are several types of secondary treatment processes, two of which are the activated sludge process and the trickle filter.

In the **activated sludge process,** illustrated in Figure 12.12, aerobic microorganisms decompose the organic matter in wastewater; as they do so, they increase in numbers. Some of the enriched or biologically activated material thus generated is used to reseed the incoming wastewater. In a **trickle filter,** which is usually a big vat filled with crushed stones on which aerobic decomposers become established, wastewater is trickled downward over the stones while air (oxygen) is being forced upward, creating an enormous oxygen-rich surface on which organic material is reduced to carbon dioxide, water, and mineral salts. Both of these processes work on the liquid portion of sewage from which the solids have been removed. Secondary treatment of solids is accomplished by allowing it to be digested in an **anaerobic digester** (Figure 12.12) at relatively high temperatures. Here microorganisms that do not use oxygen break down the organic matter in sludge, producing gases like methane, which have a potential to be used as a source of energy to heat sewage treatment plants. Sludge residence time in a sludge digester is about 15 days, and the product that emerges is not markedly different from organic humus. In some places it is even packaged and sold as a soil conditioner.

Whereas primary sewage treatment is basically *physical*, secondary treatment is *biological*. The biological nature of secondary treatment makes it susceptible to poisons that can kill decomposers. Pollution by chromate compounds has been known to shut down secondary treatment systems for long periods of time.

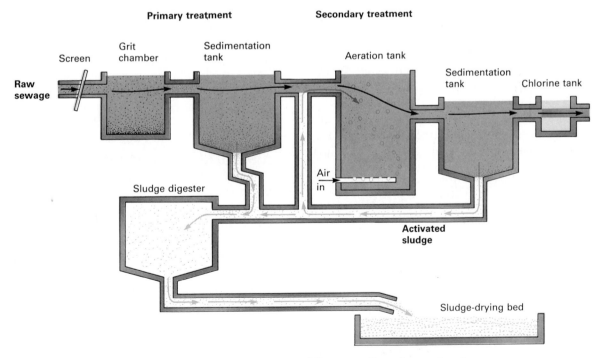

Figure 12.12 Secondary Treatment of Sewage. Secondary treatment systems involve biological oxidation of wastes using, for example, an activated sludge system (shown here). In other secondary systems, sludge is subjected to anaerobic biological digestion; the treated sludge is dried and disposed of by incineration or land disposal.

Table 12.7 Performance of Primary and Secondary Stages of Sewage Treatment

Component Removed	Pollutants Removed (percent)	
	Primary Treatment	Secondary Treatment
Biological oxygen demand (BOD)	30	90
Suspended solids	60	90
Nitrogen compounds (total)	20	50
Phosphorus compounds (total)	10	30

Source: American Chemical Society.

A comparison of primary and secondary treatment in the removal of wastewater components is made in Table 12.7. Note that secondary treatment removes nearly all the biochemical oxygen demand.

Although secondary treatment systems are very effective in reducing organic matter, they are not effective in removing the inorganic salts that can also cause eutrophication. This is where tertiary treatment comes in.

Tertiary Sewage Treatment

Primary treatment is physical, secondary treatment is biological, and tertiary treatment is chemical. The nature of **tertiary sewage treatment** is complicated because it varies from location to location, but it is basically the use of chemical methods to remove some of the chemicals remaining in sewage wastewater after primary and secondary treatment. An example is the precipitation of phosphorus compounds through the addition of iron or aluminum salts. Nitrogen compounds including ammonia can also be removed by chemical processes.

It goes without saying that as we improve the quality of water by increased degrees of treatment, the treatment becomes more and more expensive. Now even more expensive ways of treating sewage are emerging.

Modern Physicochemical Treatment Systems

New physicochemical treatment systems do not rely on any type of biological treatment. The basic components are pretreatment, clarification, filtration, absorption, and disinfection. Pretreatment is actually done by industries which discharge their effluents to public sewers. They are required to remove toxic chemicals that could disrupt the sewage treatment process, cause damage to the treatment plant, or contaminate the sludge resulting from treatment. Clarification is a process by which chemicals such as alum or lime are used to precipitate wastes into particulate masses that settle out. Filtration is a step that removes any remaining suspended solids. Absorption removes dissolved organic matter by chemical charge attraction using such materials as activated carbon. Some physicochemical treatment systems use **electrodialysis,** a system in which charged poles attract negatively and positively charged ions from wastewater, causing them to pass through tiny pores in membranes and then allowing the water to move on relatively free of dissolved ions.

Physicochemical systems are not affected by toxins that occasionally neutralize or disrupt secondary treatment plants, and they are relatively unaffected by cold weather. Physicochemical treatment systems are in operation in Niagara Falls, New York; Garland, Texas; Pittsburgh, Pennsylvania; Fitchburg, Massachusetts; and Cleveland and Plainsville, Ohio. These systems are expensive to build and operate because they are material- and energy-intensive. In the perspective of the total environment, some of them may actually cause as much environmental disruption as they prevent.

Package Treatment Plants

In wide use these days in many parts of the suburban United States are so-called package treatment plants. These are small wastewater treatment systems, usually less than a few hundred square feet in size, that treat the waste from small subdivisions and small communities. When present in large numbers in urban areas, package plants are a problem because they require considerable maintenance and attention. Regulation and monitoring of the effectiveness of these plants are difficult. Many of them discharge partially treated waste into low-capacity streams and have been known to present health hazards in heavily populated areas.

Industrial Water Pollution Control

Some of the especially toxic substances produced by industry—metals and toxic organic chemicals, for example—can be removed from effluents chemically. A number of strategies other than directly removing waste from the effluent are also possible. These include volume reduction, strength reduction, neutralization, and equalization of discharge. The first three are self-explanatory and have to do with changing processes and procedures within plants. **Equalization of discharge** means using a holding system for pollutants that allows them to be discharged in proportion to the amount of flow or capacity of the receiving waters to absorb and neutralize them.

Figure 12.13 illustrates two common methods for dissipating waste heat. The wet tower and dry tower

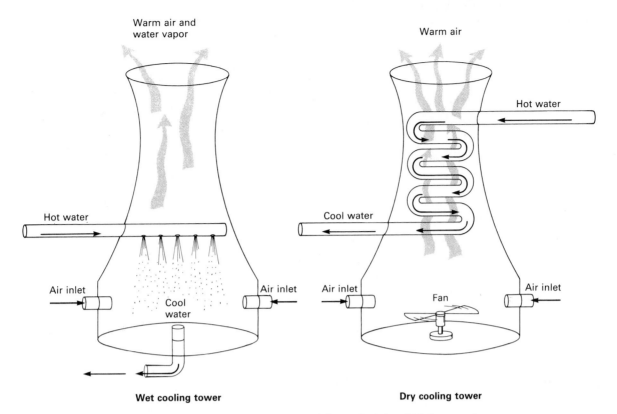

Figure 12.13 Cooling Towers. Two of the methods by which heat can be removed from water used in industrial cooling are illustrated. The wet cooling tower is a device in which hot water is exposed to air, which takes the heat away via evaporation. The dry tower uses the principle of the automobile radiator; heat is given off by convection to air forced over coils containing the hot water.

work by dissipating heat into the atmosphere instead of into the hydrosphere. The *wet tower* works by removing the most kinetically active water molecules by evaporation. The natural draft up and out of such towers carries away heat as evaporated water. There are two problems with this method: Water is lost in this process, and the addition of water vapor to the air in the vicinity of such cooling towers makes the area foggier than normal; under certain conditions this vapor can also lead to highway icing and altered patterns of precipitation.

The *dry tower* works on the same principle as the automobile radiator. The kinetic energy of hot water passing through coils is transferred to air molecules surrounding the coils, which are removed rapidly by a fan. Because this system requires power to run the fan, it is slightly more expensive to operate than the wet tower.

Another solution to the thermopollution problem is the holding or cooling pond. In this system, warm water is simply held in a pond for a certain time before it is released into a stream. Natural cooling by evaporation and radiation dissipates the heat before it gets into the natural aquatic ecosystem. Ponds take up space, of course, and this can be a problem in areas

where land is at a premium. The most ecologically sensible way to reduce thermopollution is to adopt technologies and practices that reduce the need for energy.

Deep-well Injection: Waste Under the Rug

One way of dealing with hazardous waste is to inject it into limestone or sandstone strata. Ideally, this is done where such permeable strata are sandwiched between relatively impermeable layers (Figure 12.14) well below the water table so that contamination of the water table is prevented. Some experts say that perhaps half of the surface of the United States may be suitable for deep-well injection disposal, but others contend that the dangers of leakage and contamination are too great and hence that this method should not be used.

Near Denver, Colorado, it has been shown that there is a high correlation between deep-well injections and earthquakes, presumably a result of the transmission of the pressure of injected liquids into subsurface strata.

There are already several hundred deep-well injection operations in the United States, and the num-

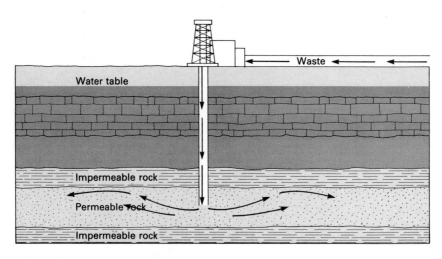

Figure 12.14 Deep-well Injection. One method of disposing of hazardous waste is the deep-well injection system. Ideally, waste fluids are deposited well below the water table into permeable layers of rock sandwiched between impermeable layers of rock.

ber is increasing each year. Millions of barrels of wastes are being injected into such wells each day.

Recycling Water and Waterborne Wastes: An Ecologically Sensible Water Pollution Control Strategy

More and more companies, particularly those that need large quantities of water, are adopting systems in which water is used over and over. A steel-manufacturing operation employing recycling methods can cut the need for water from 65,000 gallons per ton of steel produced to about 1,000 gallons per ton. This is certainly a big difference. Such recycling methods will make ever-greater sense as clean water becomes increasingly scarce and more expensive.

Another dimension to water recycling is the fact that pollutants are resources out of place. At least some industrial pollutants can be recovered profitably. The recovery of zinc from wastewater generated in the production of rayon is one such example. Zinc is a dwindling resource (Chapter 7), and it is harmful to fish, even in concentrations of 1 ppm. Processes have now been developed whereby much of the 50 million pounds of zinc sulfate used annually as a catalyst in the rayon industry can be recovered rather than lost as a water pollutant.

Still another dimension of water conservation strategy is using water for more than one thing in the same use cycle or use stream. You may have had the experience of cleaning something in the kitchen sink while looking out the window at a rather dry garden and wondering why there couldn't be a system whereby the water being used in the sink could somehow be diverted to the garden, where it would serve a good purpose and reduce the burden of water to be treated at the local sewage treatment plant. There are many other such examples of possible multiple uses of water. For instance, water from the kitchen sink could be diverted for flushing toilets. We need not use drinking-quality water to flush away wastes. It has been estimated that such multiple uses of water could reduce the water to be treated in sewage treatment plants by 40% or more.

Beneficial Use of Sewage Sludge

What about carrying this a step further? What about directing the water carrying human wastes in shorter loops? Although this is a somewhat less aesthetic concept, it has already been demonstrated to be practicable.

Sewage enrichment farming has been practiced in many countries of the world for centuries. Human waste, called "nightsoil" in Asian countries, is collected and used as the primary source of fertilizer. The system has been updated in some places by the use of vacuum trucks for collecting the waste. However, with urbanization and the increasing availability of commercial fertilizers and livestock waste, nightsoil is being used less and less. Officials in China are concerned; they know that nightsoil is a soil conditioner as well as fertilizer, and they fear the long-term consequences of abandoning this approach (Lowe, 1989).

Sophisticated systems for using wastewater both as a source of water and as a source of organic and inorganic nutrients are on their way to being a practical reality in the United States. As early as 1962, scientists at Pennsylvania State University began evaluating the use of sewage for irrigation and enrichment of forests and other types of land. They studied the application of partially treated wastewater—water free of suspended solids—to fields, forests, and pastures at rates that allowed the water to be absorbed by these terrestrial systems. They found that soil functioned as a perfect filter and that clean water percolated through to the water table. Some of the water and most of the nutrients dissolved in the wastewater were picked up by vegetation, increasing its yield.

This practice has been used to solve the problem of revegetation of strip-mined areas. Even when returned to contour, the spoil piles that result from strip mining remain nonfertile and acidic. Spraying partly treated wastewater onto such surfaces leaches the acid out of the topsoil rather quickly, provides a wet environment, and adds organic matter. The effect is to accelerate ecological succession (Chapter 4). Strip-mined land in the Shawnee National Forest in southern Illinois, which once resembled the surface of the moon, has been restored to grass and brush with the aid of Chicago sewage.

A major problem with this approach is the risk of chemical contamination and infection with water-borne human pathogens. Wastewater and solid residuals carry variable chemical loads and may carry a full complement of pathogenic organisms (see Table 12.1). The degree of risk depends on the treatment process employed before the sewage is applied and the survival of the pathogenic viruses, parasite eggs, and bacteria in soil, crops, groundwater, and runoff. The problem is a very real one since, for example, shellfish can accumulate human viruses.

Although the risks of contamination and infection are serious indeed, they can be avoided. Tertiary treatment of municipal sewage, control of industrial water pollution, and the strict control of persistent toxic chemicals could eliminate or greatly diminish chemical contamination. New technologies for disin-

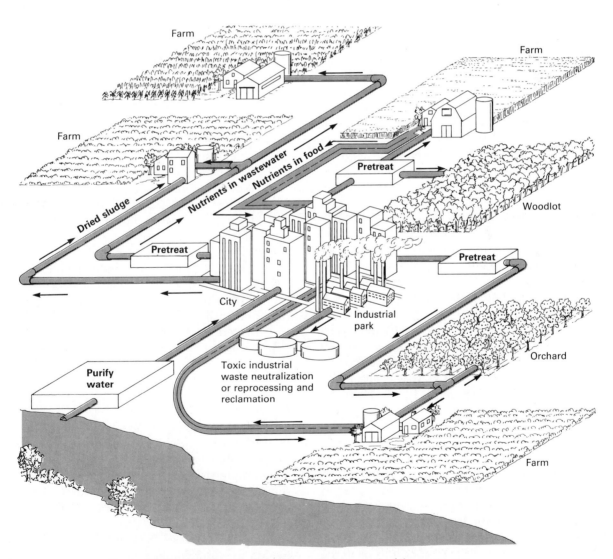

Figure 12.15 Wastewater Recycling. Perhaps much of the wastewater problem could be solved by an improved system of recycling and reuse. Such systems could also help to solve other problems, e.g., the need for irrigation water and the need for energy in the manufacture of chemical fertilizer.

fecting wastewater should all but eliminate the risk of disease.

The EPA estimates that sludge generated from U.S. sewage treatment plants has nutrients equivalent to 10% of the fertilizers farmers use—at a value of more than $1 billion annually (Lowe, 1989). The EPA is working on regulations to promote the beneficial uses of sludge and to protect human health. Worldwide, wastewater is used for irrigation. Unlike chemical fertilizers, organic materials in wastewater decompose gradually and release nutrients slowly. Although the nutrient content of sludge or wastewater is not consistent, in Thailand and India the use of wastewater has increased yields of rice, wheat, cotton, and potatoes 25% to 50% over commercial fertilizers (Lowe, 1989). The World Health Organization is working on guidelines for safe use of municipal sewage and is updating its wastewater irrigation guidelines for the Third World.

Why not plan large-scale systems for multiple use of water such as that depicted in Figure 12.15? Apparently, they are on the way. In Muskegon County, Michigan, a 5,500-acre farm is used as a tertiary treatment plant for the effluent from the county sewage treatment plant, about 35 million gallons per day. This effluent, containing nutrients not removed by secondary treatment, is used to irrigate and fertilize corn fields, producing a crop of corn that brings in about $1 million annually. The income from the corn crop offsets some of the operating expenses. This farmland advanced treatment process to remove nutrients costs about 45¢ per 1,000 gallons of water, compared to $1.23 per 1,000 gallons in Lansing and $1.17 per 1,000 gallons in Ann Arbor, both of which have in-plant advanced treatment. There is an underground drainage system that permits system operators to check the effluent before it goes into the Muskegon River and eventually into Lake Michigan.

Wetlands have proved to be effective natural sewage treatment systems. In the United States, artificial wetlands associated with a settling pond have proved effective and cost less than half of what conventional treatment plants do (Goldstein, 1988). However, proper system design is crucial.

Let us summarize by saying that the deficiencies in nearly all the methods used for treating the effluents, the energy that it takes to treat or dispose of sewage, our dwindling water supplies (Chapter 7), and the increasing value of some of the materials discarded as wastes seem to suggest that the real solution to the wastewater problem in the future is the tightening of water use loops, reduction of the volume of the waste stream, and recycling. Sewage treatment through the tertiary stage is expensive. It is expensive precisely because this type of solution is linear rather than cy-clic. The cycling of wastewater makes profound ecological sense because it conforms best to two major ecological principles, namely, that all parts of the environment are interrelated anyway and that since the system is closed, nothing can really be disposed of, and everything must go somewhere. We expect to see increasing use of recycling systems and procedures that are more in harmony with natural systems than the technologically heavy systems we have relied on in the past.

The Purification of Drinking Water

The treatment of water for drinking is much like the treatment of wastewater, but the process is more intense and more complete. Whereas purer sources of water require very little treatment or monitoring of quality, river water such as that used in many cities throughout the world for drinking receives a variety of more energy-intense treatments. These procedures involve some combination of the following:

1. Spraying the water into the air to release dissolved gases like hydrogen sulfide that give water a bad taste
2. Treating the water chemically, using coagulants like aluminum sulfate (alum), a compound that clumps together finely suspended particles, protein, and other material having electrical charges
3. Allowing the suspended solids and the coagulated solids to settle in sedimentation basins
4. Passing the water through a system of filters such as sand filters and adsorbants that take out previously unsedimented materials as well as bacteria and even viruses
5. Using chlorination to kill any remaining microorganisms.

Such systems serve nearly every community in the United States.

We have come to rely on the safety of water treated by these systems. But in 1987 the Center for Responsible Law, an affiliate of Ralph Nader's consumer protection organization, issued a report saying that one out of every five public water systems was contaminated by chemicals, many of them toxic. More than 2,000 contaminants were found, mostly organic compounds, and 190 were known to be harmful to health. Research into the safety of drinking water has been intensified recently because of the discovery of carcinogens in drinking water supplies in many communities in the United States and other parts of the world (see Chapter 18). The U.S. federal government has attempted to improve the situation through the Safe Drinking Water Act. This law directs the EPA to establish minimum drinking water standards for the nation.

The Economics of Water Pollution and Water Pollution Control

Proposals to control water pollution have included abolishing the free enterprise system, apparently because in the minds of some people, water pollution is an inevitable consequence of free enterprise capitalism. It is true that water pollution, like all environmental problems, has an interesting economic aspect, but the relationship between water pollution and enterprise has nothing directly to do with whether or not the enterprise is free. Water pollution has been a problem in countries like China, the Soviet Union, all of Europe, and indeed all of the countries of the world from at least the beginning of recorded history.

There is water pollution in the republics that made up the former Soviet Union for many of the same economic reasons that there is water pollution in the United States. The fact is that water has rather uniformly been considered by all societies as a free medium that can be used to carry waste away. In economic terms, this means that the pollution of water is and has been treated as an externality (Chapter 19). In other words, it does not cost anything to dispose of waste in water because, in the attitude of most human beings, water is inexhaustible; therefore, it does not need to be paid for or taken into account in the cost of doing business. The roots of the water pollution problem go back to times when very few people were affected in any economically determinable way. As population densities have increased, the negative economic impact of pollution has correspondingly increased. More people are now concerned about the cost of *not* controlling pollution.

Action must now be based on a consideration of both the cost of water pollution control and the cost of *not* controlling pollution. Let us begin by looking at some of the costs of water pollution control.

The Cost of Water Pollution Control

Though it is difficult to establish the economic impact of water pollution in all of its various aspects, determining the cost of water pollution control is relatively simple and straightforward.

The cost of treating sewage varies with the sophistication of the equipment and the energy required to operate it. At one end of the spectrum are relatively cheap means of treating sewage, such as the waste stabilization pond. At the other end of the spectrum are the very highly sophisticated tertiary treatment systems we described earlier. Secondary treatment costs two or three times more than primary treatment; tertiary treatment can cost up to six times as much. Sophisticated tertiary treatment systems that produce drinkable water push the cost even higher.

There has been a great increase in sewage treatment plant construction in the United States since the mid-1950s. Much of the increase has come since the early 1970s and the passage of the Federal Water Pollution Control Act, which had a provision for partially financing the construction of sewage treatment facilities.

The Cost of Not Controlling Water Pollution

In Chapter 11 we discussed in considerable detail the difficulties in assigning dollar values to the negative impact of air pollution. The problem is nearly as difficult for water pollution. We could say that the economic impact of uncontrolled pollution can be estimated from the degree to which we are now willing to pay for controlling pollution. We could also deduce the negative economic impact from the benefits of minimizing the wastewater problem.

The benefits of industrial waste treatment given by Nemerow (1978) range from the value of the resources recovered to the value of the water that is recycled or reused. Nemerow gives a whole range of secondary benefits, including increased economic growth of the area served by a particular industry or municipal sewage treatment plant. There is also a host of intangible benefits such as the public relations factor for industry and a potential for development of land areas that could not be developed if there was polluted water.

Using his own set of assumptions and parameters, Freeman (1979) derived the annual value of water pollution control benefits as shown in Figure 12.16. Freeman's estimates were based on 19 individual reports by others published between 1966 and 1979. In Figure 12.16, the ranges are the estimates from different sources using different assumptions; the single values above the ranges are best estimates. Freeman indicated that there were many problems with his summary report. He cited difficulties related to the fact that water quality varies considerably on a national scale, making it nearly impossible to generalize about water quality for the nation as a whole or about the degree of cleanup that has occurred. Freeman also cited difficulty in coming to economic conclusions

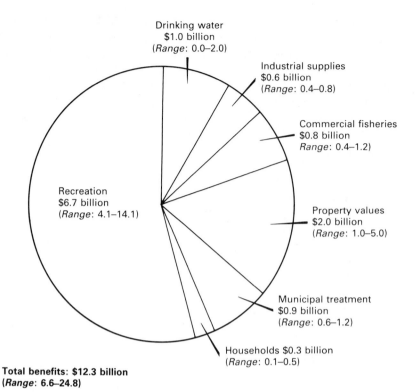

Drinking water
$1.0 billion
(*Range*: 0.0–2.0)

Industrial supplies
$0.6 billion
(*Range*: 0.4–0.8)

Commercial fisheries
$0.8 billion
Range: 0.4–1.2)

Property values
$2.0 billion
(*Range*: 1.0–5.0)

Recreation
$6.7 billion
(*Range*: 4.1–14.1)

Municipal treatment
$0.9 billion
(*Range*: 0.6–1.2)

Households $0.3 billion
(*Range*: 0.1–0.5)

Figure 12.16 Benefits of Water Pollution Control. An estimate of the benefits derived in 1985 from the removal of conventional water pollutants (in billions of 1978 dollars). Some of the limitations of these estimates are given in the text.

Total benefits: $12.3 billion
(*Range*: 6.6–24.8)

about the effect of legislation because the principal legislative tools have tended to focus on point sources and because most of the data he had to work from was based on the benefits to be enjoyed after 1985—after all of the stages of the implementation of the best available technology were completed as specified in the Clean Water Act.

Finally, Freeman indicated that his conclusions risked grossly underestimating true water pollution control benefits because early estimates tended to focus on such items as fecal coliform bacteria and BOD, largely failing to take into account the fact that the 1977 amendments (to be discussed shortly) redirected attention to toxic substances.

The Balance Sheet

It is tempting to note that $28.2 billion was projected to be spent on water pollution control each year from 1978 to 1987 and that Freeman's high estimate of the annual benefit to be enjoyed during 1985 was $24.8 billion (Figure 12.16), but there are many reasons, technical and otherwise, why these figures cannot be directly compared.

Despite the problems of doing an exacting cost-benefit analysis, it is obvious that economic factors in the solution to the water pollution control problem will force industry and municipalities to treat the cost of dealing with pollutants as a direct cost of doing business—as every bit as much of a direct cost as raw materials and labor.

Water Pollution and the Law

Water pollution has a worldwide legislative history going back to prehistoric times. Basically, clean water laws were designed to clean up water resources and to keep them clean. In the United States, the laws specifically did the following:

1. Directed the states to establish criteria for meaningful water quality standards based on the intended use of the water
2. Gave the federal government (the EPA) the job of developing guidelines for setting water quality standards and the authority to set technology-based upper limits on what could be dumped into water by municipal and industrial sources, without regard to the intended use of the receiving water
3. Gave the EPA the job of setting performance standards for new water pollution sources, mainly industrial and sewage treatment plants
4. Authorized the regulation or control of water pollution via a permit system until pollution of navigable water ceases altogether

The implementation strategy in all clean water legislation was, in essence, to define polluted water (or, conversely, to set water quality goals) and then to encourage the prevention, reduction, and eventual elimination of water pollution. Encompassed in this strategy of implementation were programs of research into pollution control and monitoring technology, grants to states and municipalities for the construction of wastewater treatment facilities, and grants to states to help them carry out their water resource management responsibilities over the full spectrum, from setting standards to enforcing them.

In the United States, water pollution control legislation goes back to the Refuse Act of 1899, which was intended to address only problems of navigation, disease, and oil discharges. Although Water Pollution Control Acts were passed in 1948, 1956, 1961, 1965, 1966, and 1970, landmark legislation did not come until the 1972 Federal Water Pollution Control Act, a response to the deep public concern about the environment born in the late 1960s. This act called for the establishment of effluent limitations, that is, limits on what can be discharged into waters from factories, sewage treatment plants, and other point sources of pollution; established the National Pollutant Discharge Elimination System (NPDES), aimed at establishing schedules for reducing pollutants and complying with effluent limitations through permits; and called for performance standards for new plants and industries. Other provisions set up a national system of water quality surveillance and areawide planning. Amendments to the act in 1977 adjusted some compliance timetables and further clarified some provisions of the 1972 law.

The last major change in U.S. water pollution control laws was the Water Quality Act of 1987. It made several significant changes. The sewage treatment construction grant program was phased out, and a mechanism was established for states to set up revolving loan funds to pay for future water pollution control facility needs. States are now coping with how they intend to fund water pollution control facilities in the future. This act also set up new programs of grants to the states to assess which segments of streams are not likely to meet water quality standards because of nonpoint source pollution or toxic pollutants. Management strategies to address these problems and to bring these stream segments into compliance are to be established. This act also directed the EPA to set standards for contaminants in sludge and to set out best management practices to meet these standards. The National Estuaries program, with grants for development of management plans, and the Clean Lakes program were also established.

At the time of this writing, Congress is again considering a revision to the Clean Water Act. However, prospects for passage of a bill in 1992 are not good. Major issues include funding for the state revolving loan funds for sewage treatment facility construc-

tion. In addition, issues relating to wetlands protection, groundwater, and toxics are under discussion either as part of the Clean Water Act reauthorization or as separate bills.

The National Environmental Policy Act of 1970

Because it contained some legislation relevant to water pollution, we will include a brief discussion of the National Environmental Policy Act here. As its name implies, this act actually encompasses all kinds of pollution. The National Environmental Policy Act (NEPA), which was signed into law on January 1, 1970, established the **Council on Environmental Quality** and charged this group with the study of the condition of the nation's environment on a regular basis. The act also required every federal agency to take environmental factors into account in its decision making and to show how this was done by publishing an **environmental impact statement** in advance of each major activity that uses federal money and potentially affects the environment. The Council on Environmental Quality is the main force in the federal government behind the development of every impact statement. The EPA, in contrast, is charged with reviewing and commenting in writing on the environmental impact statements made by all other federal agencies.

The environmental impact statement clause in the NEPA had an enormous impact on the environmental scene. It has been called the most important piece of environmental legislation ever passed. *When done properly,* environmental impact statements must assess the potential impact any project will have on the environment. They must also analyze in straightforward terms the cost-benefit ratios of all legitimate alternatives to proposed actions. If statements are not done properly, the agency can be sued and the project halted. A flood of suits have been filed since the NEPA was passed, raising charges that it is too easy to sue and that all progress has become mired in red tape. But many people feel that it is about time that a lot of the reckless disregard for the environment, formerly mislabeled as progress, is systematically dragged into the light and more carefully considered from all angles.

The Safe Drinking Water Acts

The first Safe Drinking Water Act in the United States was enacted in 1974. It provided for the establishment of regulations covering the taste, odor, and appearance of drinking water and the development of measures to protect underground sources of drinking water. The law was intended to standardize the purity of water throughout the United States. The states have primary responsibility for enforcement (**pri-**

macy), provided that they have acceptable standards, regulations, surveillance and enforcement procedures, provisions for variances, and reporting and record-keeping systems.

In 1986, because of public concern over contamination of drinking water and underground supplies, Congress reauthorized the Safe Drinking Water Act and strengthened it significantly. A schedule in the act speeded up the rate at which the EPA sets national primary drinking water standards. These standards set maximum contaminant levels (MCLs) and outline treatment technologies for specific contaminants.

In addition, the act obliges the EPA to issue regulations requiring that public water systems monitor at least once every five years for contaminants that are as yet unregulated. The EPA will list which contaminants are to be evaluated.

The act also orders the EPA to issue regulations requiring that operators of injection wells monitor for migration of contaminants if wastes are injected below a drinking water aquifer and bans the use of lead in plumbing in or connected to public water systems.

Two new programs were established in the 1986 amendments aimed at protecting groundwater. They are the Wellhead Protection Program and the Sole Source Aquifer Demonstration Program. The act requires states to establish plans to protect the surface and subsurface areas around wells (wellheads) used as public drinking water sources from contamination that could affect health. States must develop plans for protecting these areas from contamination by such things as pesticide runoff, hazardous waste leachate, and leaking underground gasoline storage tanks. The Sole Source Aquifer Demonstration Program was established to encourage localities to develop programs to protect critical aquifers that serve as the sole or principal water supply for an area. States, local governments, or planning agencies may prepare plans to protect these aquifers from contamination and submit them to the EPA for approval. Approved plans are eligible for dollar-for-dollar matching money for implementation.

The Impact of Water Pollution Control Legislation

How well has clean water legislation worked? Success has been mixed. We should qualify this answer by pointing out again that there are literally dozens of ways of assessing water quality and that any given degree of pollution has varying importance depending on how the water is to be used and the relative scarcity of water in a particular location.

By some standards—for example, the removal of total contaminants from sewage—there has been con-siderable progress; as a result of this and other gains, the quality of water in some lakes, rivers, and estuaries has been improved considerably. However, most lakes, rivers, and estuaries have about the same quality of water now as they did in 1972; some have actually deteriorated.

Considering the overall negative trend throughout most of this century, the fact that some water pollution problems have stopped getting worse is an element of considerable progress. The 1980 annual report of the Council on Environmental Quality summed up progress in the 1970s by saying that water quality *did not get any worse* throughout that decade and that water pollution was still widespread coming into the 1980s. Recent water quality inventories by the EPA and assessments made by environmental organizations indicate that this state of affairs persisted through the mid-1980s. Serious problems remain with treatment plants that have overflows of combined sewer and rainwater runoff (Figure 12.17), and new problems with toxic substances in water seem to appear daily. Eutrophication and toxic chemicals are still problems in the Great Lakes, and groundwater contamination has become one of the most important environmental issues of the day.

Much remains to be done about toxic chemicals. We need to know much more than we do about the threshold levels (assuming that there are such) of most of the compounds that are now found as water contaminants. (Chapter 18 discusses the problems of determining the threshold of carcinogenicity and the broad philosophical questions associated with acceptable risk.) The setting of specific effluent standards has been a major problem of implementing clean water laws for years. The 1972 Water Pollution Control Act specified that the federal government set specific effluent standards for toxic substances, standards based not on the intended use of the water but on toxicity. Economic, political, and legal pressures have generated considerable drag on this part of the law. As late as 1977, only six effluent standards had been set, and a coalition of citizens took the EPA to court to order the agency to get busy setting guidelines. This contributed to Congress's amending the legislation to set up three categories of pollutants with respect to effluent standards: conventional pollutants (bacteria, BOD, etc.), toxic pollutants, and nonconventional pollutants (nitrates, phosphates, and other chemicals not falling into either of the other two categories). Toxic pollutants now number well over 100 different chemicals. The best available technology for limiting effluents of the substances was to be implemented by 1984; the deadline was extended to 1989 by the 1987 amendments. The issue of toxics is currently under review as part of the reauthorization discussions on the Clean Water Act.

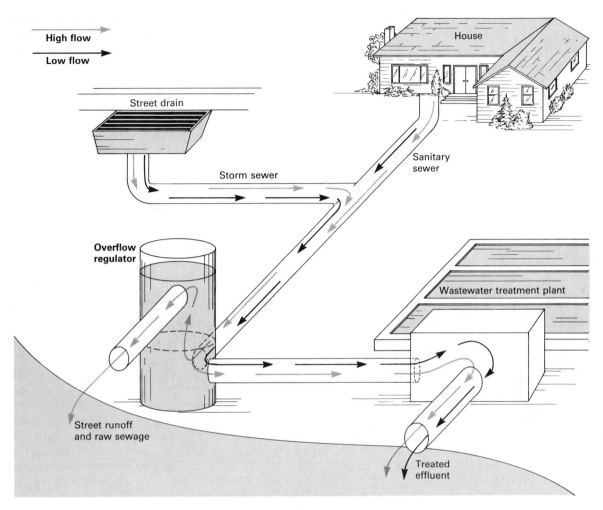

High flow

Low flow

House

Street drain

Sanitary
sewer

Storm sewer

Overflow
regulator

Wastewater treatment plant

Street runoff
and raw sewage

Treated
effluent

Figure 12.17 Storm Sewers and Sanitary Sewers. The sewer systems in U.S. municipalities are of two types. The combined type carries both rain runoff and waste. In other systems these two functions are separate. Combined systems like the one shown here present a problem at times of heavy rainfall, the common solution being to allow the flow to bypass the treatment plant—rainwater, wastewater, and all. Separate systems are much more amenable to good water pollution treatment.

What Needs to Be Done: The Role of the Citizen

Though some progress has been made in controlling certain point sources of water pollution, much remains to be done. Legislation now on the books seems adequate; major programs for implementation and conforming with the laws are now needed. Water quality standards and water pollution laws, like any laws or standards, are useless unless put into practice and enforced. Implementation plans are just so much paper until they are carried out. This is where citizens will play a major role in the future.

We can all do something about water pollution. Individually, we can obviously stop dumping garbage into water supplies and change habits at home that generate wastewater. In a complex society, however, the effectiveness of each one of us is limited by the degree to which we are able to work with other members of our communities to bring about adequate treatment systems for wastewater and to establish new patterns that minimize water pollution problems. The key is education. To be effective, a citizen must be well informed. Armed with reliable information and a solid understanding of the problem of water pollution, citizens and citizens' groups can then effectively make their interest in clean water known to elected officials and administrators. Such individuals and groups can exert pressures by notifying candidates for public office that their votes will go to candidates who support clean water.

Any citizen awed by feeling like a little fish in a big pond should remember that the pond water is being polluted.

How Environmentally Sound Are Your Water Habits?

Remember that water is precious, that the water you use came from upstream and, even during a single cycle, may have been used possibly many times—and that it will be used by other people downstream.

Do you have water-saving fixtures on showers and faucets?

Do you take long showers?

Do you have a water-saving toilet, or have you put a brick or some other device in the toilet tank to decrease the volume of water used in each flush?

Do you wash full loads or use washing machines that allow the amount of water used to vary?

Do you use natural alternatives to chemical cleaners?

Do you "wash organic waste away" in the garbage disposal or put it in a compost heap?

Do you use lawn chemicals? If you do, do you know anything about them?

Are you careful about what you send to the landfill?

What do you think about the importance of what individuals do about water pollution?

Concepts to Remember

1. The properties of water make it an ideal cleaning agent and heat remover. The problem is that when water is used in these ways, it gets dirty and warm.

2. Water pollution is any change in water quality brought about by human activity that impairs the use of water as a resource or that adversely affects other forms of life.

3. Water pollutants come principally from domestic, agricultural, and industrial sources. The four most general kinds of water pollutants are biological agents, dissolved chemicals, nondissolved chemicals and sediment, and heat.

4. Many kinds of diseases can be transmitted by drinking water, and such diseases are still major world health problems.

5. Organic chemicals added to aquatic ecosystems can lead to oxygen depletion of the water as the decomposers oxidize the organic matter. Inorganic chemicals that serve as nutrients for aquatic vegetation can indirectly cause oxygen depletion by causing the overgrowth of plant material, which is then later oxidized by decomposers.

6. Many toxic substances are soluble in water, and many of those that are not can be carried

on the surface of particles suspended in water.

7. Heat disrupts aquatic ecosystems in several ways. Heat makes gases less soluble in water and drives certain solids into solution; both of these things amount to changes in the chemistry of water. Heat accelerates the metabolism of aquatic organisms at the same time that it makes oxygen less available for metabolism. Accelerated metabolic activity can carry certain organisms beyond their tolerance limits; short of that, accelerated metabolism can disrupt balances in aquatic systems, for example, by causing egg hatching to speed up.

8. The effects of certain contaminants of water can be magnified as they pass through food chains.

9. A point source of water pollution is any discernible port or channel through which pollutants enter the water. Point sources are generally easier to control than nonpoint sources such as farm fields and construction sites.

10. Oil spills in coastal areas can be great ecological disasters. They have profound effects in the short term; long-range effects have not been fully documented. Strategies for dealing with oil spills after the fact have not been very effective.

11. By far the most serious water pollution problems in the oceans occur in coastal areas.

12. Groundwater pollution is a major environmental issue today partly because about half of the people of the United States now use groundwater for drinking.

13. In sewage treatment plants, water pollutants are removed by physical (sedimentation), biological (oxidation), and chemical (stripping) methods.

14. The most ecologically effective water pollution control methods are volume reduction, short-loop recycling, and using water more than once in the same use stream (e.g., irrigating forestland with dirty water).

15. It costs a lot to control water pollution, but it also costs money not to control water pollution.

16. There are some very good laws on the books for water pollution control and safe drinking water; citizens can help the clean water effort by insisting that these laws be enforced.

Environmental Career Profile

Thomas D. Forsythe

Dr. Tom Forsythe is an environmental planning specialist with the Tennessee Valley Authority's Land Between The Lakes. The Land Between The Lakes (LBL) is a national recreation and environmental education area framed by Lake Barkley and Kentucky Lake in western Kentucky and northwestern Tennessee. One of the principal purposes of LBL is to demonstrate "integrated resource management" and sustainable, multiple use of natural resources. Dr. Forsythe's job is to help to keep LBL's natural resources management plan fine-tuned. He prepares environmental impact statements for developing or expanding campgrounds, establishing trails, and other LBL projects. He also heads the Land Between The Lakes Biosphere Reserve Program—a United Nations-U.S. Department of State program designed to identify and provide for the protection of different types of biotic communities throughout the world. In addition, Dr. Forsythe serves as liaison between the TVA and the many colleges and universities that make use of the LBL for teaching and research. Although Dr. Forsythe does quite a bit of desk work, he also spends time out and around the LBL's 170,000 acres working on water-monitoring projects, supervising the work of interns, assisting with educational programs for school groups, and sometimes even working in his specialty area of aquatic and fisheries science. Dr. Forsythe likes his job very much (he has worked at LBL since 1978) because of the opportunity it gives him to learn new things. In addition to aquatics and fisheries work, Dr. Forsythe is involved in forestry, terrestrial ecology, and geographic information systems. He likes the variety.

Dr. Forsythe has a bachelor's degree in biology, a master's degree in water pollution biology, and a doctorate in fisheries science. According to Dr. Forsythe, the basic qualifications for the work he does would include an in-depth knowledge of chemistry, ecology, and microbiology; some experience linking science and practical management; and a knowledge of academic and government agencies.

References and Further Reading

References marked with an asterisk are cited in the chapter.

* Eisenreich, S. J., 1987. "Toxic Fallout in the Great Lakes," *Issues in Science and Technology* **4**:71–75.

* Freeman, A. M., 1979. *The Benefits of Air and Water Pollution Control: A Review and Synthesis of Recent Estimates.* Brunswick, Maine: Bowdoin College. Report prepared for the Council on Environmental Quality.

Freeman, A. M., 1983. *Air and Water Pollution Control: A Benefit-Cost Assessment.* New York: Wiley.

* Goldstein, B. E., 1988. "Sewage Treatment, Naturally," *Worldwatch*, July-August, pp. 5–6.

* Gundlach, E.; Boehm, P.; Marchand, M.; Atlas, R.; Ward, D.; and Wolfe, D., 1983. "The Fate of Amoco-Cadiz Oil," *Science* **221**:122–129.

* Helmer, R., 1977. "Pollutants from Land-based Sources in the Mediterranean," *Ambio* **6**:312–316.

Holloway, M., 1991. "Soiled Shores," *Scientific American* **264**(4):101–116.

* Lenssen, N., 1989. "The Ocean Blues," *Worldwatch*, July-August, pp. 26–35.

* Lowe, M. D., 1989. "Down the Tubes," *Worldwatch*, March-April, pp. 22–29.

* Nemerow, N. L., 1978. *Industrial Water Pollution: Origins, Characteristics, and Treatment.* Reading, Mass.: Addison-Wesley.

* Makarewicz, J. C. and Bertram, P., 1991. "Evidence for the Restoration of the Lake Erie Ecosystem," *BioScience* **41**(4):216–223.

* Office of Technology Assessment, 1987. *Wastes in Marine Environments.* Washington, D.C.: U.S. Government Printing Office.

Okun, D. A., 1991. "A Water and Sanitation Strategy for the Developing World," *Environment* **33**(8):16–20ff.

* Smart, T.; Smith, E.T.; Vogel, T.; Brown, C.; and Wolman, K., 1987. "Troubled Waters," *Business Week*, October 12, pp. 89–104.

* Smith, R. A.; Alexander, R. B.; and Wolman, M. G., 1987. "Water Quality Trends in the Nation's Rivers," *Science* **235**:1607–1615.

* Sun, M., 1986. "Ground Water Ills: Many Diagnoses, Few Remedies," *Science* **232**:1490–1493.

* Turner, M. H., 1990. "Oil Spill: Legal Strategies Block Ecology Communications," *BioScience* **40**:238–242.

* United Nations Environment Programme, 1984. *Pollutants from Land-based Sources in the Mediterranean* (UNEP Regional Seas Report and Studies No. 32). Geneva, Switzerland: UNEP.

* U.S. Environmental Protection Agency, 1987. *Survey of State Ground Water Quality Protection Legislation, 1985.* Washington, D.C.: Office of Ground Water Protection.

13

Land Use and Land Misuse

What is good for the land and the biotic communities supported by it is good for humankind. This was expressed well by Aldo Leopold in *The Sand County Almanac,* published in the 1940s. Leopold's book called attention to the need for a rational land ethic. (An **ethic** is a system of values by which decisions are made and on which actions are based.) It was an appeal that land decisions be based on the ecological importance of the land. Historically, our use of land has been mostly haphazard and exploitive. Land is still taken for granted in modern U.S. society. Our present ethic treats land as a commodity to be bought and sold to the highest bidder, to be used at its highest economic value for the landowner. The problem is that this may not always coincide with what is best over the long run. Land is a finite resource; we must make wise decisions about its use that satisfy human needs for eons to come.

In Chapters 9–12 we looked at how air and water are harmed by human activity. In this chapter we will focus on land, the third great sphere (see Chapter 3). We will see how land is harmed by things we put into it (waste disposal), the process of taking things out of it (mining), and simply using it (agriculture, forestry, development).

Soil: The Lithosphere Connection

To appreciate the human impact on land, it is necessary to have a basic understanding of the nature of soil. Such an understanding will enable us to get at such questions as, What soils are best for agriculture? Why are clay soils often used for land disposal of waste? How can land disposal of waste threaten underground water supplies? Why does clear-cutting of forests cause a loss and degradation of the soil? How is it that soil in the tropics can support lush natural vegetation and yet is poor for farming?

Soil Defined

Soil is the part of the lithosphere with which living things interact most directly. It is the part of the lithosphere that overlaps with the atmosphere, the hydrosphere, and the biosphere most intensively. As such, it provides many of the basic needs of all terrestrial species. Soil provides the nutrients that allow us to obtain food, clothing, and shelter. We grow food in soil, we grow crops to produce fabrics, we graze livestock on it, and we grow trees on it. All this, and yet this vital resource averages only about 6 inches deep worldwide.

The soil situation around the world varies from large expanses of thick, good soil to no soil at all. In many places erosion runs ahead of soil formation, and in other places there is insufficient water to support the plant growth that allows soil to develop. Shifting sand is not really soil; bare rock surfaces in mountains and deserts and fresh lava surfaces have no soil layer. The ability of the soil to support food crops is the major determinant of whether or not a particular place can support civilization. Transportation notwithstanding, nearly all of the people on earth live in places where soil can furnish food.

The appearance of soil and land is deceiving. Land looks stable and even permanent at first glance. Yet processes that change the lithosphere and cause soil formation are going on constantly, but they are processes that bring about change very slowly. The problem is that we can damage soil—and even lose it—rapidly.

How Soil Is Formed

We saw in Chapter 3 how solid bedrock is broken down by weathering with time. We saw that water plays an important part in this process as it dissolves certain minerals and as it causes cracks and fragmentation by freezing and thawing. We described how living things help soil formation along, by growing or tunneling in and among broken particles, releasing organic acids causing the particles to break down even further. As vegetation dies and decays, it mixes with the mineral particles, bringing an organic dimension to soil, conferring properties on soil that better enable it to support vegetation and animal life. All of this has gone on for eons, and the soil we have today represents a kind of net gain for these processes against the processes of soil erosion. The fact that soil making is still going on makes soil a renewable resource. However, the time frame for producing an inch of soil from bedrock may be centuries or, in environments lacking warmth or water, tens of centuries. Whenever soil is eroded away from the surface faster than it is regenerated, trouble becomes inevitable.

The soil formation process ultimately defines three regions: bedrock or **parent material,** the region of fragmented bedrock called the **regolith,** and the region of plant growth (to the depth of root penetration) called **soil.**

Soil formation is a complex process involving the parent material, climate, topography, animal and plant life, and time (see Figure 13.1). Initially, the type

Figure 13.1   The Process of Soil Formation.   Soil is formed from the interactions of climate, topography and plant and animal life with bedrock.

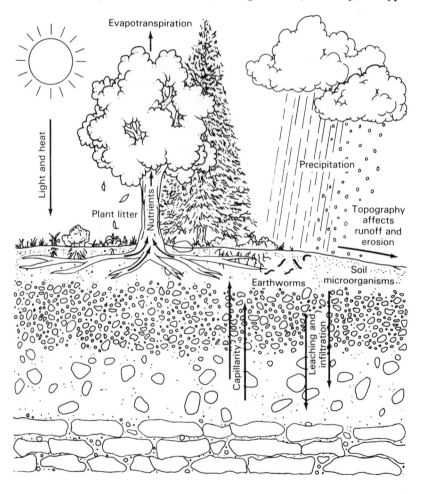

of parent material is significant. However, with time, its importance diminishes, as is shown by the fact that soil profiles are more closely aligned with major vegetative regions (biomes) than with geologic formations. Moisture and temperature are the primary climatic factors that influence soil formation. We will discuss the role of moisture in detail later. Temperature is important because of its impact on water, causing freezing and thawing and contributing to rock fragmentation. Temperature also influences the activity of soil microorganisms in the decomposition process. Warmer temperatures increase the rate of activity, speeding up decomposition. In addition, the temperature of rainwater tends to affect what minerals will be leached by water. Cold water tends to leach iron and leave silica; warm rains of the tropics tend to leach silica and leave iron. A reddish color in soil usually indicates the presence of iron. Topography (slope) affects drainage and the tendency of soil to erode.

Animal and plant life affect the organic content of the soil as biological material within the soil or on the surface dies and decays or as waste products are deposited. Microorganisms in the soil also play a vital role in the decomposition process. Decaying organic matter in the soil, known as **humus,** provides nutrients and gives soil spongelike properties, better enabling it to hold water. Soils with abundant organic matter tend to be black. Plants and tunneling animals, especially earthworms, also affect the structure of the soil as their roots or travel paths provide channels for

air and water. Earthworms eat their way through the soil, and their wastes, called casts, leave a mixture of organic and inorganic nutrients behind. They bring organic materials down from the surface and move materials upward. Under certain conditions, hundreds of thousands of earthworms may live in an acre of soil.

Soil in Profile

Because of the way soil is formed, it has a cross-sectional pattern, or **soil profile,** as can be seen along the side of a trench or gully. There are generally three main layers, or **horizons.** Starting from the top is the **A horizon,** which is composed of fresh and decaying plant litter on the surface and topsoil. Topsoil contains humus. The surface of the A horizon is sometimes covered by a layer of fresh, decaying litter called the A_o or O horizon. The next layer, the **B horizon,** is sometimes called the *subsoil.* Minerals leached from the topsoil tend to be deposited here. The third layer is the **C horizon,** which is the parent material. Generally, when we talk of soil, we are talking about the A and B horizons. The depth of each of these layers varies and depends on weathering and other aspects of the environment.

To understand the significance of various soil types and their vulnerability to human activities, let us compare the soil profiles found in prairie, rain forest, and evergreen (coniferous) forest (see Figure 13.2) (also see Enrichment Box 13.1).

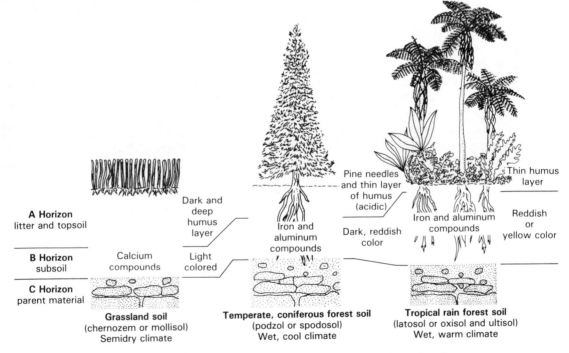

Figure 13.2 Soil Profiles. By looking at a cross-section of soil, three layers can be distinguished—topsoil, subsoil, and parent material. Soil layers typical for various biomes are shown here.

Soil Classification Schemes

As is the case for many components of the ecosphere, soils have been classified to distinguish one soil type from another. A classification system developed by Russian soil scientists in the late 1800s was the basis for soil classification worldwide. Soils were classified on the basis of climate and vegetation interactions. The classification system is known as the Great Soil Groups. In 1960 the U.S. Soil Conservation Service developed a new classification scheme. It is called the U.S. Com-

prehensive Soil Classification System, or the Seventh Approximation. The latter name comes from the fact that the system is the seventh version; six earlier versions were tested and changed. This system is based on observable soil characteristics. The major unit of classification is the soil order. The Great Soil Groups and the soil orders do not correspond exactly. The table below relates some of the soil groups to the soil orders.

Name	Derivation	Classic Great Soil Group
Entisol	Nonsense symbol	Azonal [simple profile or purely local as waterlogged (gley) soils]
Inceptisol	L. *inceptum,* beginning	Brown forest soils, some poorly drained [gley] soils
Aridisol	L. *aridus,* dry	Desert soils
Mollisol	L. *mollis,* soft	Chernozems and prairie soils
Spodosol	Gk. *spodos,* wood ash	Podzols
Alfisol	Nonsense symbol	Gray-brown podzolic soils
Ultisol	L. *ultimus,* last	Red-yellow podzolic soils, latosols and laterites
Oxisol	F. *oxide,* oxide	Latosols and laterites
Histisol	Gk. *histos,* tissue	Bog soils

Source: Adapted from P. Colinvaux, *Ecology.* © 1986 John Wiley & Sons, used with permission.
Note: For an in-depth description of the U.S. soil classification system, see McKnight (1987).

A typical prairie soil (**chernozem** or **mollisol**) is shown in Figure 13.2. The deep A horizon is due to the decaying grasses and the action of worms and other organisms that mix organic matter deeper into the soil. The moderate rainfall prevents significant leaching of minerals. Generally, rainwater will penetrate to some depth, then be pulled back to the surface by evaporation and capillary action. The water will often deposit its mineral content in the subsoil, forming a layer of calcium at the line of deepest water penetration. This leaves the horizon white, characteristic of the deposition of mineral salts. Cropping of prairielands or grazing must somehow provide for the return of decaying matter to the A horizon, or the humus is soon depleted. Humus is formed by decaying grasses in undisturbed prairie.

The typical soil for coniferous forests is known as **podzol** or **spodosol.** Here the surface of the A horizon is composed of a layer of pine needles. Precipitation that flows through this litter layer becomes acidic from the decomposing pine needles. As it travels through the A horizon, the acidic water leaches many minerals, especially iron and aluminum, and deposits them in the B horizon. This causes the lower part of the A horizon to be gray or whitish from the remaining silica, while mineral deposits in the B horizon make it red and brown. The acidic nature of the soil also

prevents the deeper mixing that is provided by earthworms or burrowing animals that live in less acidic soils, and so there tends to be a rather sharp distinction between the A and B horizons. The term *podzol* is Russian for "ash-earth" or "ashes," named for the white appearance of the A horizon. Clear-cutting of conifers or forest fires leaves the podzol soils very susceptible to erosion because of the lack of undergrowth in the evergreen forest and the friable structure of the A horizon. The shallow A horizon can easily be lost.

The profile of a tropical rain forest soil (**latosol** or **oxisol** and **ultisol**) illustrates why there are problems farming soils from tropical forest biomes. The lateritic A horizon is very shallow because organic material decays quickly under tropical conditions. Constant rainfall carries nutrients into the soil, and they are taken up by the roots of the lush tropical trees. Constant warm rainfall leaches silica from the soil but leaves aluminum and iron compounds. This produces a reddish subsoil that, if exposed to heat, will harden. When tropical rain forests are cleared for agriculture, the thin topsoil is quickly lost and the unproductive subsoil is left exposed. The roots of tropical trees no longer absorb the nutrients and return them to the nutrient cycle. What was once covered by lush vegetation becomes hardened, nutrient-poor land (laterite).

Physical and Chemical Properties of Soil

Soil is variably porous, containing spaces or **interstices** filled with air or water. Well-aerated soils are important for plant growth. Oxygen and carbon dioxide are interchanged in the soil through the root system of a plant. If the soil is not well aerated, pore spaces can become filled with carbon dioxide. If drainage is poor, spaces may become filled with water or become **waterlogged,** causing the death of the plant. Many a houseplant has met its end as a result of too much watering.

If there are a lot of connecting spaces between soil particles for water or air, the soil is said to have high **porosity.** Porosity is different from permeability. **Permeability** is the ease with which water can move through soil. The types and sizes of particles in the soil affect porosity and permeability (see Figure 13.3). The largest and coarsest particles found in soil are sand. Silt particles are somewhat smaller, and clay particles are the smallest of all. The percentage of each type of particle in soil gives soil certain characteristics. Sandy soils, because of the large particles, have large air pockets, and water drains through them easily. As it does, it carries soluble minerals with it. Sandy soil has high porosity and high permeability. Clay soils have high porosity but low permeability. Although there are lots of interstices, clay particles tend to soak up and hold water and expand like a sponge instead of letting

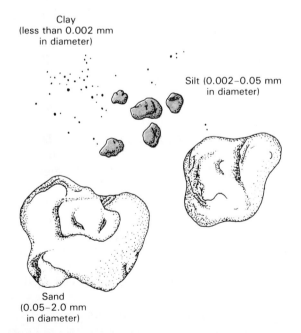

Figure 13.3 Relative Sizes of Sand, Silt, and Clay Particles. This figure shows the relative size of clay, silt and sand particles. The size of the particles influences the porosity and permeability of each material.

it percolate through. Clay is often used to line or cover landfills because it does not allow water to infiltrate as much as soils with less clay and more sand or silt particles. The soil type most amenable to plant growth contains a mixture of all three types of particles (sand, silt, and clay) and is called **loam.**

Because the interaction of soil and water is so important in understanding the leaching problems associated with waste sites (see Chapter 12) and surface pollutants, we will go into some detail on this soil-water relationship. Water infiltrates the soil from rainfall and melted snow. As it percolates through the soil, it picks up minerals and any other fine particles or materials that are soluble in water and carries them some distance before depositing them deeper in the soil or even in the groundwater supply if permeability is very high. Some important nutrients susceptible to leaching are nitrates, sulfates, potassium, and calcium.

In addition to the water percolating through the soil via gravity, some water called *capillary water* remains in the soil. It is this water that dissolves nutrients and keeps them available for use by plants. Capillary water moves from wet areas to drier areas and can move up, down, or laterally in the soil. Clay and humus play significant roles in retaining capillary water. They also influence the chemistry of soil in other ways.

Soil particles smaller than one-tenth of one-millionth of a meter (0.1 μm) are called **colloids.** Soil has both mineral and organic colloids. Because colloids have electric charges on their surfaces, they play an important role in the dynamics of mineral ions dissolved by the water in soil. The tendency of the negative charges on colloidal particles to hold positively charged ions such as calcium, sodium, magnesium, and potassium offsets the tendency of water to leach these ions out of the soil. Since the ions bound to colloidal particles can be given up to plants, the colloids act as nutrient storage depots, greatly retarding the loss of important mineral nutrients from soil.

There is a complex equilibrium set up between the nutrients in capillary water and the nutrients bound to clay particles or humus. As nutrients are used up in capillary water, more are released from the colloidal particles. Changes in soil acidity can disturb this equilibrium.

As described in Chapter 12, acidity is really a measure of the concentration of hydrogen ions in water. Hydrogen ions are positive ions, and as such they have the ability to displace other positive ions on the binding sites (negative charges) on colloids. Displaced ions are then leached out of the soil, and it becomes less fertile. Lime is used by farmers and gardeners to deacidify soil. In effect this recharges soil colloids with minerals needed by plants. This phenomenon has relevance to the problem of acid precipitation. Acid deposition may act to leach minerals from the soil, rendering them less fertile. Under certain conditions, harmful amounts of positive ions may be washed from the soil into aquatic ecosystems (see Chapter 10).

The movement of minerals vertically in soil has relevance to both the law of the minimum and Shelford's law of tolerance (see Chapter 3). Soil can have too much or too little in the way of dissolved minerals as a consequence of the upward and downward movement of minerals. In wet climates, where soils tend toward the acidic, the downward movement of water carries important minerals with it out of the root zone. Minerals such as calcium carbonate can be carried out of the A horizon into the B horizon or, under certain climatic conditions, to the C horizon, where it is deposited in whitish crystalline form. Generally speaking, the drier the climate, the higher in the soil dissolved minerals tend to be deposited. In extremely dry climates with high evaporation rates, minerals tend to be drawn upward to be deposited on the surface. In such climates, water is drawn upward for the same reason a wick draws oil upward in an oil lamp. As water moves up, it brings dissolved minerals with it. When the water evaporates, it leaves minerals behind as a whitish deposit. This process, called **salinization,** can eventually make the soil toxic to many kinds of plants. Irrigation of desert soils is made problematic because of the tendency of desert soils toward salinization resulting from the progressive accumulation of salts and alkalies present in small quantities in the irrigation water.

Not only are plants affected by the vertical movement of minerals in the soil, but they play an active part in it. Plants take up minerals as they grow, in effect drawing minerals up. When plant material dies, the minerals they release are carried back into and through the soil, to be taken up again by vegetation. The organic colloids added to soil as humus act to hold nutrient ions in the soil. The net effect of all of this is a short-loop biogeochemical cycle.

Soil Conservation

Good soil produces good vegetation, which provides food and habitat for animals. A major concern of good land management is keeping soil in place and maintaining its fertility. This is especially true of agricultural land. If we are to sustain food production over the long term, we must succeed in keeping soil on croplands. All of the fertilizer available cannot change bedrock into soil. Soil erosion and the loss of soil fertility as a result of poor farming practices are serious problems worldwide.

Soil Erosion

Problems with soil erosion in America were probably noted following the first clearing made by the first American settlers. Because land seemed unlimited in those days, erosion and the loss of soil fertility caused little concern. Folks simply moved on after a farm started to wear out. Because this frontier attitude had a long life—and indeed persists to this day in many people—soil conservation has been an uphill battle throughout the past two centuries. One of the pioneers in noting the connection among agricultural practices, soil erosion, and the loss of productivity was Jared Eliot (1685–1763). Although he lived and died a long time ago in New England, he advocated practices to minimize runoff and even the use of cover crops such as clovers (known in Europe to restore fertility even

---

Mini-Glossary

Assimilative capacity: the ability of the air, water, and land to purify themselves of pollutants; problems appear when pollutants occur in concentrations that exceed the assimilative capacity of the system.

Contour plowing: plowing at right angles to the slope; furrows follow the contour of the hill rather than the slope to help to decrease erosion.

Cover crops: the planting of crops to provide cover for bare fields to prevent erosion and to restore nitrogen and other nutrients to the soil; leguminous plants that fix nitrogen are often used (see Chapter 3).

Crop rotation: the practice of growing different crops in succession on the same land to help to maintain productivity; cover crops that fix nitrogen are often used in alternating sequence with other crops.

Eminent domain: the right of a government to take private property for public use.

Multiple use: harmonious use of land for more than one purpose.

No-till farming: a system that consists of planting a narrow slit trench without tillage and with the use of herbicides to control weeds.

Open or green space: a relatively undeveloped green or wooded area usually left within a developed area.

Prime farmland: land capable of producing consistently high yields because of its soil quality, topography, moisture availability, and long growing season.

Silviculture: management of forest land for timber.

Sustained yield: renewable resources managed so as not to be depleted over time.

Zoning: the process of establishing specific areas for specific types of uses and activities and prohibiting other types of uses and activities; for example, areas are commonly zoned agricultural, residential, commercial, or industrial.

---

then) to protect and restore the land. Despite his efforts and those of even more notable early Americans like Thomas Jefferson, ignorance and apathy about soil conservation predominated throughout the nineteenth century and beyond, and much good American soil now lies at the bottom of our coastal waters. It is astounding that this once nearly totally agrarian nation really did not get serious about soil conservation until the 1920s. The Soil Conservation Act was passed in 1935, but even then it took a while for the Soil Conservation Service (SCS) created by the act to sort its place out with other federal agencies and with the states. Soil conservation districts were eventually established, and the SCS began its efforts to control erosion throughout America.

Soil Conservation Practices

We have known for some time that such things as contour plowing, the use of cover crops (especially "green manure" crops), no-till farming, crop rotation, and other means are available to retard both soil loss and diminished fertility. However, in establishing such practices, our progress has been slow. The American farm economy has been such that short-run return is the paramount consideration. A lot of agricultural economics can be summed up by saying that farmers who know how to conserve soil and who would even like to do what it takes simply cannot afford to spend the money on soil management programs in a system where others are free *not* to. Part of the problem is that much soil erosion occurs so gradually that it is almost imperceptible. Fifteen tons of soil lost from a hectare of land in a single storm will diminish soil depth by 1 mm (Pimentel et al., 1987). Erosion, then, may not, in the short term, appear nearly as serious to the farmer as making ends meet.

All of this makes soil conservation far more an econosociopolitical problem than a technical one. We know how to conserve soil; we just haven't figured out how to see that what we know is practiced. While we struggle to solve the problem, nearly 1,000 tons of soil is ejected by the Mississippi River into the Gulf of Mexico every minute; 3 billion tons of sediment enters U.S. waterways from runoff annually. According to the U.S. Department of Agriculture (USDA), soil loss from erosion threatens the productivity of one-third of the nation's cropland.

It is certainly worth noting here that erosion affects more than soil depth. Erosion reduces productivity of the land by depleting it of organic matter and nutrients, reducing its water-holding capability, and limiting rooting depth as the A horizon is lost. Pimentel and colleagues (1987) cite studies that indicate a 15% to 30% additional loss in productivity when these total effects of erosion, rather than just the effects of the reduced soil depth, are considered. Some studies estimate that as much as 4 billion tons of topsoil is lost annually from wind and water erosion. This translates into 8.2 million tons of nitrogen, 0.06 million tons of phosphorus, and 2.0 million tons of potassium lost from U.S. cropland; the cost to restore these nutrients would run at least $5 billion (Pimentel et al., 1987).

Others estimate the value of this nutrient loss at $7 to $18 billion annually (Pimentel et al., 1987). Since soil cannot be truly lost from the ecosphere, it must go somewhere. Soil out of place can cause air pollution, water pollution, and land pollution. Pimentel and colleagues (1987) set the short-term cost of soil erosion in the United States at $24 billion annually (value of nutrients lost, reduced crop yields, and sediment and flood effects on water resources). Certain agricultural

Contour plowing and strip cropping are good land management practices.

practices also diminish soil fertility. Soil that is continually cropped or grazed will eventually lose its fertility because the normal cycle that returns mineral-rich dead plant matter to the soil is disrupted. This can be overcome by adding back nutrients as animal manure, by practicing crop rotation—with nitrogen-fixing cover crops, for example (see Chapter 3)—and by adding chemical fertilizers. There are discussions elsewhere in the text about problems associated with the use of synthetic fertilizers (high cost of production in energy and dollars) (Chapter 6) and groundwater contamination (Chapter 12). Agricultural practices also degrade soil in other ways. Soil compaction that results from use of heavy farm machinery can destroy the structure of the soil, restrict water infiltration, and lead to increased erosion. Use of irrigation in dry areas can result in salinization, as described earlier (also see Chapters 8 and 14).

Soil Erosion in Other Countries

The United States is not the only country with soil conservation problems. Pimentel and colleagues (1987) state that soil loss rates ranging from 10 to 100 tons per hectare per year worldwide are ten times higher than soil formation rates (see Table 13.1). Soil formation occurs in the range of 0.3 to 2 tons per hectare per year, the rates generally varying inversely with latitude.

Degradation of agricultural lands by erosion, salinization, and waterlogging (from poor drainage) results in an irretrievable loss of 15 million hectares annually. Pimentel and colleagues (1987) report a study that concludes that degradation of arable land

worldwide will depress food production 15% to 30% based on actual soil loss and projections for the period 1975–2000. It is indeed ironic that the primary cause of erosion in the Third World is the clearing of land to grow food. Third World countries in particular, trying to feed themselves or grow crops for export, have created major soil erosion problems by clearing vegetation and forest cover and cropping land that is sloped or has poor soils. In Latin America, soil erosion from forest removal, cultivation on slopes, and salinization are all problems. Erosion and salinization are major problems in Africa.

Degradation of land in India is a serious problem. Around 35% of the land in India that is potentially productive is affected by water and wind erosion and salinization (Postel, 1989). The United Nations Environment Programme estimates that 11 billion acres— 35% of the earth's land—is threatened by desertification (see Chapter 8).

Many problems in Third World countries make good land management difficult. Because most of the land is in the hands of a few landowners, poor workers must farm less desirable land and crop it intensely. Population growth puts stress on the need for increased agricultural production as well as clearing of land for cropping and firewood supplies.

Despite the problems, some efforts to improve erosion control have been successful. In Ethiopia, *bunds* or earthen and rock walls are built across hillsides to catch eroding soil. This helps to form natural terraces to lessen erosion further (Postel, 1989). In China, pilot efforts are under way with consortia of local farmers and politicians and international agen-

Table 13.1 Erosion Rates in Selected Geographic Regions

Country	Erosion Rate (tons per hectare per year)	Comments
United States	18*	Average, all cropland
Midwest deep loess hills (Iowa, Missouri)	36*	MLRA† No. 107, 2.2 million hectares
Southern High Plains (Kansas, Oklahoma, Texas, New Mexico)	52*	MLRA† No. 77, 6.2 million hectares
China	43	Average, all cultivated land
Yellow River Basin	100	Middle reaches, cultivated rolling loess soils
Java	43	Brantas River Basin
Belgium	10–25	Central Belgium, agricultural loess soils
Ethiopia	20	Simien Mountains, Gondor region
Madagascar	25–40	Nationwide average
Nigeria	14	Imo region, includes uncultivated land
Salvador	19–190	Acelhuate Basin, land under basic grains production
Guatemala	200–3,600	Corn production in mountain region
Thailand	21	Chao River Basin
Burma	139	Irrawaddy River Basin
Venezuela and Colombia	18	Orinoco River Basin

* Combined wind and water erosion; all others are water only.

† Major Land Resource Areas.

Source: Adapted from D. Pimentel et al., "World Agriculture and Soil Erosion," *BioScience* 37:278. Copyright © 1987 by the American Institute of Biological Sciences.

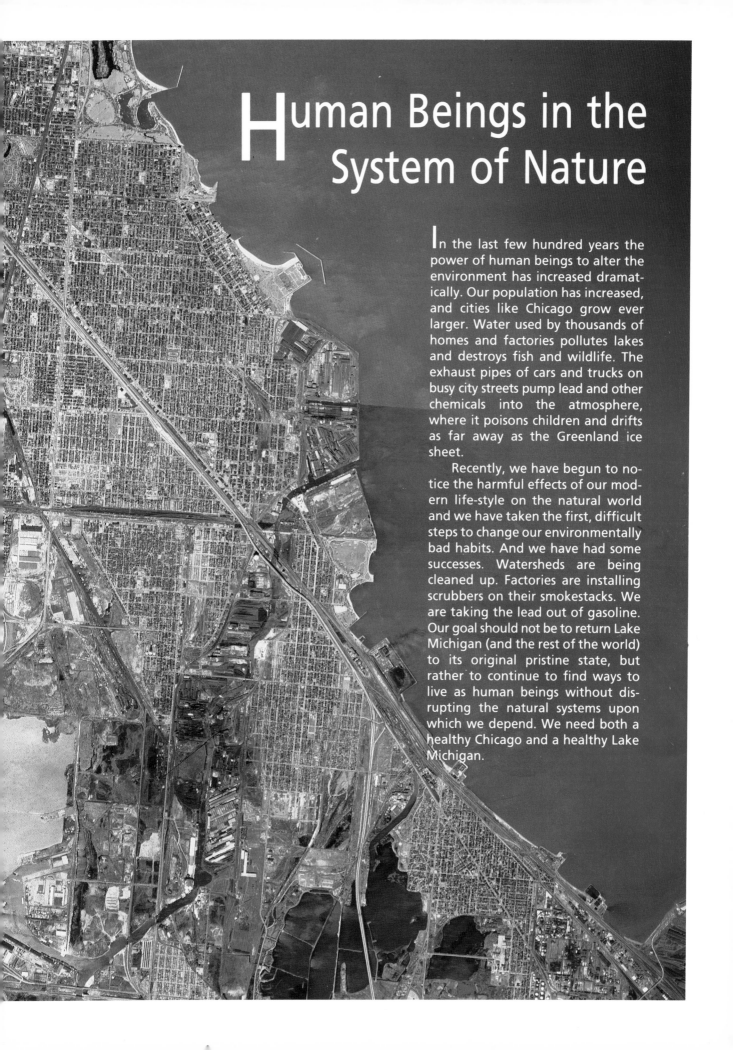

Human Beings in the System of Nature

In the last few hundred years the power of human beings to alter the environment has increased dramatically. Our population has increased, and cities like Chicago grow ever larger. Water used by thousands of homes and factories pollutes lakes and destroys fish and wildlife. The exhaust pipes of cars and trucks on busy city streets pump lead and other chemicals into the atmosphere, where it poisons children and drifts as far away as the Greenland ice sheet.

Recently, we have begun to notice the harmful effects of our modern life-style on the natural world and we have taken the first, difficult steps to change our environmentally bad habits. And we have had some successes. Watersheds are being cleaned up. Factories are installing scrubbers on their smokestacks. We are taking the lead out of gasoline. Our goal should not be to return Lake Michigan (and the rest of the world) to its original pristine state, but rather to continue to find ways to live as human beings without disrupting the natural systems upon which we depend. We need both a healthy Chicago and a healthy Lake Michigan.

The Midwestern United States provides homes for millions of people, hosts heavy and light industry, and supplies food for tables around the world. People in the Midwest must meet the needs of their modern life-style without compromising nature.

The industries that we depend upon to manufacture the toys and tools of our modern life have had a tremendous impact upon our mineral, air, and water resources. Factories like this one in Indiana consume enormous amounts of mineral resources and fossil fuels. Through pollution they change the composition of the air and water. The people who work in and live near factories have begun to demand a cleaner environment, but the cost will continue to be high.

Agriculture is at the heart of the human relationship to the environment. In Ohio, Amish farmers practice small-scale farming using animals and avoiding fossil fuels. In Nebraska, fallow farming, irrigation, and mechanical harvesting allow for efficiencies of scale without waste. Each approach has both benefits and drawbacks.

Cities, like factories, require huge amounts of energy and generate tons of waste. Plastics and metals decompose very slowly. We must find ways to safely move the resources trapped in our trash back into the resource stream. With care and planning, our cities can be clean, comfortable places to live.

Planning should also allow for some wilderness to remain. We have an obligation to preserve some of the habitat of all of the living things with which we share this planet. Not only does this enrich our world and the world of those to come after, but we may well depend on the creatures who depend on the habitats we preserve.

What use are satellite images? Besides being fascinating pictures, satellite images can be used to pinpoint different kinds of terrain and to help analyze patterns of use and development in an area. This ERTS photo shows mining operations in Belmont County Ohio in September 1973. The light blue parts are water, green is forestland, red is strip-mined land, brown is reclaimed land, and gold is rangeland.

cies to change land uses and to help to improve erosion control techniques and productivity. (Other strategies are described in Chapter 8.)

Another aspect of the soil conservation problem is the question not of maintaining the fertility of existing croplands, but of protecting agricultural lands from irreversible conversion to other uses.

Land as a Resource

There are just under 2.3 billion acres of land in the United States. Of this, almost one-third is owned by the federal government. The rest is privately owned (59%) or owned by state or local governments (7%) or Indian tribes (3%). These facts of ownership make regulating land very different from regulating air and water. Air and water are largely perceived as belonging to everyone; land, however, is owned by individuals and groups, who have deeds to prove it. How individuals and groups decide to use their land is a function of ever-changing population and economic shifts and cultural trends.

Although there are a variety of possible uses for any one parcel of land, all land is not suitable for all uses because of topography, location, soils, or climatic factors. It is also a fact that some land uses are, for all practical purposes, irreversible.

Land for Farming

The SCS has developed a relatively simple system for classifying the suitability of land for farming (Figure 13.4). There are eight classes, each divided further on the basis of characteristics such as erosion potential, wetness, shallowness of root zones, and climatic limitations. Class I lands are best for farming; Class VIII lands are not suitable for farming.

At first glance, it appears that farmers are doing a good job of cropping the most suitable lands. Most Class I land is being used to grow crops. According to the results of the 1987 National Resources Inventory (NRI) conducted by the SCS, 70% of prime farmland is being used for crops (George and Choate, 1989). Even more Class II and Class III land is available, and so most of our food is grown on these types of lands. According to the SCS, the quality of land used for cropping has improved since the 1960s. In addition, a fair amount of good land now in pasture, range, or forest could be converted to crop use. According to the 1987 NRI, 2.7 million acres of rangeland and pastureland in Classes I, II, and III were converted to cropland between 1982 and 1987.

Figure 13.4   The Soil Conservation Service Land Classification System.
There are eight groups based on suitability for farming. Class I land is the best for growing crops; Class VIII is not suitable for cropping owing to its soil type, topography, and wetness. A sample of each class type is shown here.

Table 13.2 Rural Nonfederal Land (in millions of acres)

Cropland	422.4
Rangeland	401.6
Forest	393.9
Pastureland	129.0
Minor land uses	59.8
Total	1,406.7

Source: 1987 Natural Resources Inventory, U.S. Soil Conservation Service.

The problem is that although the land being farmed is generally good land, the total number of acres of good land available for farming is diminishing. Good land in Classes I–III is being used for places to live, places to work, and roads. Once land is given over to these uses, it cannot realistically be converted back to farmland. Strong economic forces come into play in such land use conversions. Farmers in rural areas on urban fringes find it increasingly more profitable to sell their land to developers than to farm it. Between 1982 and 1987, developed land increased by 4 million acres. Overall, during this same time period, cropland increased by 1.4 million acres due to conversion from pastureland and rangeland. Prime farmland, however, decreased by the same 1.4 million acres. Some of the prime land was developed, and some was reclassified due to a loss of water available for irrigation (George and Choate, 1989).

Table 13.2 shows how nonfederal rural land is being used. Figure 13.5 summarizes major land use changes in the United States from 1982 to 1987.

In response to the need for a more useful tool to measure suitability of land for agriculture, in 1981 the SCS created the agricultural Land Evaluation and Site Assessment (LESA) system. The rating produced from this system is based on both physical and economic factors, which include the land suitability classification, soil productivity, soil potential, location of the land, distance to market, uses of adjacent land, zoning, and the availability of water and sewer lines.

Figure 13.5 Land Use Changes in the United States, 1982–1987. Although there was a net increase of 1.4 million acres in cropland from 1982 to 1987, that same amount of acreage of prime farmland was lost to development and degradation.

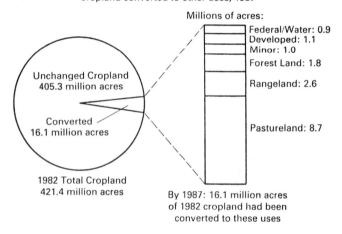

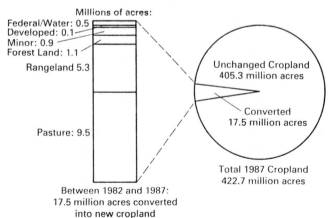

The 1985 and 1990 Farm Bills*:
A New Beginning

Because of the seriousness of the problems relating to erosion and misuse of lands by converting highly erodible lands or wetlands to farmland, Congress in December 1985 passed the Farm Security Act, also known as the 1985 Farm Bill. Several conservation provisions were included in the act to tie federal farm benefits such as price and income supports, commodity-related payments, crop insurance, Farmers Home Administration loans, farm storage facility loans, and Commodity Credit Corporation benefits to the use of soil conservation practices. In addition, a payment system for *not* cropping highly erodible land was established. The four major conservation provisions of the act are as follows.

1. *Conservation Reserve.* The Agricultural Stabilization and Conservation Service (ASCS) will pay up to half the cost of retiring highly erodible land by making permanent plantings of grasses, legumes, trees, or wildlife cover or establishing windbreaks. A farmer signs a ten-year contract to keep the land as a conservation reserve, and the ASCS makes annual rental payments to the farmer for this land as long as the conditions of the contract are met. This provision is to encourage farmers to stop cropping highly erodible land.

2. *Conservation Compliance.* If a farmer continues to crop highly erodible land, the farmer must have an approved conservation plan for those lands fully implemented by January 1, 1995, to continue to be eligible for federal farm benefit programs. This provision is to *require* the use of conservation practices if highly erodible soil is to continue to be cropped.

3. *Sodbuster.* If highly erodible lands not used for crop production from 1981 to 1985 are cropped, there must be an approved conservation plan in place for the farmer to remain eligible for federal farm benefit programs. This provision is geared toward preventing the conversion of highly erodible land to cropland.

4. *Swampbuster.* If natural wetlands are converted to cropland after December 23, 1985 (the date the farm bill was signed), the farmer can lose eligibility for federal farm benefits for all land that is cropped (not just the converted wetland). This provision is obviously to protect wetlands, which have been disappearing rapidly due to conversion to cropland.

* Technically the Farm Bills should be called Acts after they are passed. But they are still commonly referred to as Farm Bills.

Results from the first five years of these programs indicate that **excessive soil erosion** has been cut by one-third (Weber, 1990). Excessive soil erosion is erosion that occurs above and beyond what is replaced by new soil formation.

The 1990 Food, Agriculture, Conservation and Trade Act. In 1990 Congress enacted another farm bill with several provisions concerned with promoting environmental protection. The 1990 farm bill established the Agricultural Resources Conservation (ARC) Program, consisting of three separate initiatives. The first was the continuation of the Conservation Reserve Program. By 1995 between 40 and 45 million acres were to be enrolled in the program. Second, a new wetlands reserve program was established. Paid easements would be provided for up to 1 million acres of wetlands. The third initiative was the Water Quality Incentives Program. Government payments would be made to farmers who implement approved plans to reduce water pollution. The goal was to enroll 10 million acres in the program.

The new agricultural policy established in the 1985 and 1990 farm bills provides economic incentives to farmers to use soil conservation practices. The law is noteworthy in that it also recognizes the general ecological importance of land.

The Ecological Importance of Land

With increasing demand for diminishing plant, animal, and mineral resources, we have developed recovery techniques—such as clear-cutting of timber and surface mining of minerals—that provide for a large return but are harmful to land and thus ecologically unsound. We have harvested living resources of the land in a way that has brought many plant and animal species to extinction or near-extinction. By destroying wetlands and forests in the name of agriculture and other uses, we have brought some species to the point of extinction indirectly by destroying their habitats (see Chapter 14).

In the early pages of this text we discussed the interrelationships between the great spheres—land, air, water—and the biosphere. When an area that was once forest or farmland is converted to an industrial park, each of the great spheres and the relationships between them are affected in some way. Among the questions that should be asked are these: How will this change affect the quality of the air? Will the industries release pollutants? What effect will the change have on the watershed in which it lies? What wastes will be produced by the industrial plants, and where will they go? How will drainage and runoff change?

In a somewhat less drastic change—for example, converting a farm woodlot to cropland—the questions might be these: Once the trees are gone, what precautions will be taken to prevent erosion? What wildlife species might be eliminated? What new species might be attracted? How will air temperatures and humidity be affected? How will drainage patterns be altered? The point is that any change in land use will have direct effects on the air and water and on living systems as well.

Thus it is important for land use planners and managers to have some familiarity with ecological interactions and with the concept of **assimilative capacity.** By this we mean that land, air, and water have the ability to cleanse themselves through physical, chemical, and biological reactions. They can assimilate or neutralize pollutants. For example, bacteria can break down organic matter that might enter waterways or be deposited on the land. However, self-cleansing is limited and is subject to overload. When an area is developed, bringing concentrations of people, animals, industry, and associated wastes, it is important that the assimilative capacity of the great spheres be taken into account as part of the planning process.

The length of time needed for a succession from high-rise apartment house to woodlot is many hundreds of times more than the succession time from woodlot to high-rise apartment. When a land use decision will result in changes that are, for all practical purposes, irreversible, it must be made in full knowledge that future generations will be bound and limited by that decision. Good land use decisions require long-range vision—something our species has yet to master. Perhaps there is an inherent incompatibility or irresolvable conflict between economics and ecology when it comes to land use.

Land for Other Things: The Economics of Land Use

Economics plays a big part in the way land is used. From an economic perspective, the value of land depends on many factors, including physical characteristics, location, climate, topography, institutional factors, public opinion, and technological ability to use the land. All of these affect the price of a given piece of real estate. When we talk about land use capacity from an economic perspective, we are referring to the ability of that parcel of land to return a net profit. The highest and best use in economic terms is the use that will

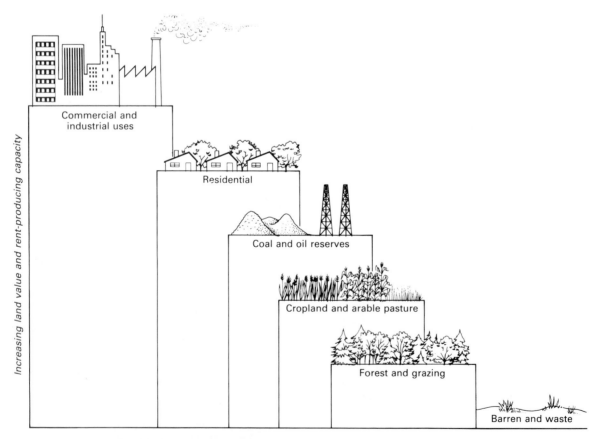

Figure 13.6 Valuing Land Uses. Economically, the highest and best use of land is the use that is the most profitable to the landowner. The problem is that what is most economically profitable is not always the most ecologically sound and may not always coincide with the long-term general welfare.

provide the landowner with the greatest economic return. All other uses are lower uses, economically speaking. The economics of land and land use are based on the concepts of supply and demand. A unique economic characteristic of land is that the supply of land is ultimately inelastic. There is only so much land area on the surface of the earth; land is located in a given spot, and there is no relocating it. From an economic perspective, the supply of land is elastic only through variation in the ways in which it is used.

It may well be that in most cases the highest and best immediate use of land for the landowner coincides also with what is best for society. Theoretically, as the needs of society change, this is matched by changes in the uses of land that return the highest profit (see Figure 13.6). The problem is that certain land use decisions made out of immediate considerations may be irreversible. What then?

Just how sacred is the right of landowners to do anything with their land? Should someone be permitted to sell good farmland for an apartment complex when land nearby that is unsuitable for farming would serve just as well for the apartments? Should a builder be permitted to build new houses in an area that is likely to flood every five years—perhaps creating a general tax burden later? Should the owners of relatively expensive houses be able to block the construction of low-cost houses nearby? Should a landowner be permitted to drain a swamp to bring extra acres into production during a time of high soybean prices?

Should someone be permitted to build a tavern in a residential neighborhood? Should a farmer be permitted to bring extremely hilly land into production even though the topsoil will almost certainly erode away within a few years? Some of these questions have been dealt with by governments and the courts, and the answer to many of these questions is no. The question of whether landowners should be permitted to do whatever they please with their land has been resolved: They are not. The question now seems to be, How should decisions related to the best uses of land best be made, and who should be involved in making them?

Land Use and the Law

There are several important historical concepts at the heart of the land use issue. An appreciation of these is important to an understanding of where we have been and where we might be going.

The Fifth Amendment to the U.S. Constitution states, "No person shall . . . be deprived of life, liberty, or property without due process of law; nor shall private property be taken for public use without just compensation." Most of the original European immigrants who came to the United States were not landowners; they were peasants, debtors, and others looking for a better life. Their right to own land and use it as they saw fit was of prime importance.

As times changed and neighbors lived closer and closer to one another, it became necessary to place some restrictions on land use by zoning in order to protect the public interest. **Zoning** is a tool that can be used by a community to limit the development or use of land to ensure compatibility with surrounding uses. For example, areas that are zoned residential cannot be used for industrial development. Thus public interest was for the most part wrapped up in protecting land values; a secondary application was to protect people from dust, smoke, and noise.

The essence of the controversy over placing restrictions on land use was the issue of *taking*. Governments have the **right of eminent domain**, that is, the right to take private property for public use. In exercising this power, the government must determine what is fair compensation to the landowner and pay that amount to the landowner (although landowners often feel that the compensation is not sufficient and they have recourse to contest the amount of compensation through the courts). This occurs, for example, when a dam is built and land will be flooded by the reservoir formed by the dam. But the question of taking without just compensation also arises when restrictions are placed on land use. How many restrictions can be placed on a landowner before it is necessary to compensate the person for the land because the uses left are very limited? Specific cases bearing on this question come occasionally before the courts. The

most recent ruling from the U.S. Supreme Court was issued in 1992. This is a topic of continuing controversy.

Most current laws let land use decision-making powers rest at the local level, where the people have the most control. This concept has become firmly entrenched as land use restrictions become increasingly inevitable. From an environmental perspective, the limitations of the local approach lie in the fact that land resources, like air and water resources, are not always best managed within arbitrary political boundaries.

Federal Involvement in Land Use Management

Although land use decision making has traditionally been done at the local level, both the federal government and the states have had some influence on land use planning. The Northwest Ordinances of 1785 and 1787 instituted a rectangular survey system for land and a system for land records. As is readily apparent by their rectangular configurations, most of the states west of the Mississippi were carved out after these ordinances were established. In the early 1800s the federal government began a system of land grants to states for educational purposes and for the development of railroads, roads, and lands. By the end of the nineteenth century, few grants of land were available; federal cash grants followed in their place.

For Further Thought . . . 13-2

Competing Uses of Public Lands

We have limited land resources. Some uses of land are simply not compatible with others, and even under a policy of multiple use, choices must be made. For example, what criteria should be considered in designating land as wilderness? What criteria should be considered in reversing a wilderness designation? Some people say that we need a certain amount of unspoiled wilderness in our world. Others argue that wilderness serves only wealthy environmentalists. What do you think?

Perhaps the most wide-ranging federal land use planning effort came in 1934, when President Franklin D. Roosevelt created the National Resources Planning Board. This was a data-gathering group concerned with the development of natural resources, cities, and growth. This board was to work with all levels of government. However, concern grew over the power it seemed to wield, and it was abolished in 1943. Emphasis shifted again to the state level.

Today we can best summarize federal involvement in land use management and planning in two ways: First, the federal government owns approximately one-third of the land area of the United States; federal lands amount to about 760 million acres. Second, the federal government continues to use its traditional carrot-and-stick approach with regard to land use planning. That is, it provides incentives to get states to move in certain directions.

Numerous federal legislative acts affect land use; for example, in 1962 the Federal Highway Act required regional transportation planning as a prerequisite to granting federal highway money. In 1966 the Demonstration Cities and Metropolitan Development Act instituted the Budget Bureau Circular No. A-95 process. A so-called A-95 review means that projects using federal funds must go through a regional or state clearinghouse for review and comment before submittal to the appropriate federal agency. This was intended to align federal grants with overall regional and state plans and goals. This A-95 review process was essentially eliminated under the Reagan administration. Other examples include the National Flood Insurance Program, which requires participating communities to issue flood plain zoning ordinances.

Federal environmental laws also tend to force land use planning. Section 208 of the Federal Water Pollution Control Act requires areawide water management planning. Noise control by land use control is part of the Noise Control Act of 1973.

Perhaps one of the most controversial of major federal laws governing land use is the law regulating surface mining.

Surface Mining Regulation. Surface extraction is a method used for the mining of rocks, phosphates, and copper as well as coal. There are several methods of surface mining (see Figure 13.7); topography, depth and size of the coal seam, and other factors determine which will be used. When a coal seam is very deep, underground mining is the most practical method of extraction. If a coal seam is located near the surface, it is easier and cheaper to remove the earth above the seam (the *overburden*) and extract the coal from the surface. Surface mining of

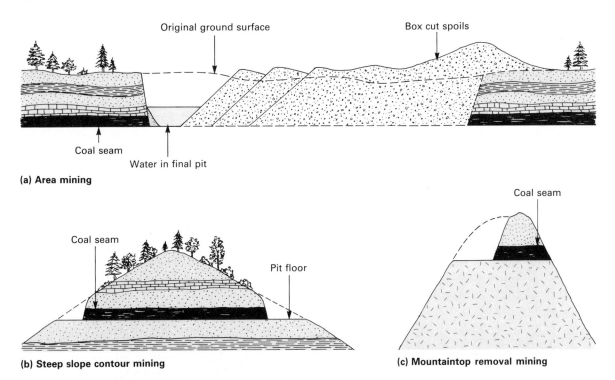

Figure 13.7 Three Types of Surface Mining. (a) Area mining is used to get at shallow seams in gently rolling terrain. (b) Steep-slope contour mining is used to mine relatively deep-lying seams in steep-slope areas. It recovers only the outer edge of the seam. (c) Mountaintop removal is used to mine high-lying seams by completely shearing off the mountaintop. Federal law requires careful placement of the overburden (not shown here).

coal is common in the eastern, central, and western United States.

Because of distinct regional differences, the environmental problems associated with surface mining in each region vary. In the eastern coal region, erosion and acid mine drainage are the ecological problems of greatest importance. Degradation of the landscape has been another consideration; it is addressed in current federal law, which requires backfilling to eliminate high walls left after mining is completed. Chemical pollution of surface areas is the primary ecological problem resulting from surface mining in the central coal fields. Water—or rather the lack of it—dominates the ecological problems associated with surface mining in the West. Lack of rainfall hinders revegetation and permits wind erosion.

Potential strip mine land in the western region is now primarily in pasture and rangeland. In the central region, many potential surface mining areas are currently used as timberland, pasture, and cropland. In the eastern region, most of the land area overlying coal deposits is mountainous.

Because of its highly visible ecological effects on the landscape and on water quality, strip mining is an old environmental issue. Perspectives on surface mining range from a call for a total, unqualified ban to the feeling that because of energy needs, coal resources should be developed at all costs.

On August 3, 1977, the federal **Surface Mining Control and Reclamation Act,** Public Law 95-87, was passed. In PL 95-87, Congress states that it is the purpose of the act to "strike a balance between protection of the environment and agricultural productivity and the nation's need for coal as an essential source of energy."

The aspects of the law that affect land use include some of its most controversial provisions:

1. Postmining land use must be described in the application for a permit to mine and is the basis for determining some variances. (**Variances** grant permission to vary, with justification, from the letter of the law or regulation.)
2. Prime farmland may not be mined unless the operator can restore the land to an equivalent or higher level of yield. Also, the secretary of the USDA is directed to issue specific requirements for soil removal, storage, and replacement.
3. Return to approximate original contour will be the general rule. Land must be restored to a condition capable of supporting the premining land use or a higher or better use. A variance from return to original contour may be requested by the surface landowner if the postmining use is industrial, commercial, residential, or public or if the variance would improve the watershed.

4. Each state, in order to assume primacy (authority) for the program, must establish a mechanism for declaring lands unsuitable for mining. Lands may be considered unsuitable if their reclamation is not technologically or economically feasible, if it is contrary to local land use plans, or if the mining affects fragile or historic lands. Under this provision other lands, the loss of which would affect water, food, or fiber supply or endanger life or property by disturbing flood-prone or geologically active areas, could also be considered unsuitable.
5. Mountaintop removal of coal is permitted if the postmining land use is for industrial, commercial, agricultural, residential, or public facilities use. Mountaintop removal is used in areas with steep slopes where it is easiest to shear off the top of the mountain as the coal is being removed, leaving a flat-topped mountain. It must be shown that the new use is an equal or better economic or public use.
6. The act places a severance tax on coal, a large percentage of which will go into the Abandoned Mine Reclamation Fund for the reclamation of **orphan lands,** that is, lands that were mined in the past and improperly reclaimed. This may turn some productivity to now spoiled land. It will also eliminate some of the environmental quality problems such as acid drainage and erosion.

Challenges to the strip mine law have been made in the courts and are aimed to a large extent at land use issues. They are based on the premise that the law amounts to an unjust taking of land without compensation, especially in the area of declaring land unsuitable for mining and in the requirement that land be returned to its approximate original contour even if the landowner would prefer a different postmining contour. According to the advocates of this position, mining operations in steep-slope areas could produce a bench of flat land for housing and development if the slope did not have to be returned to the approximate original contour. The U.S. Supreme Court has upheld the constitutionality of PL 95-87. Other individual court challenges to the law in specific circumstances will no doubt be made.

Protecting Critical Land Areas: The Coasts and Other Wetlands. Another example of federal involvement in land use management has been the management of coastal lands and other wetlands. These are considered critical natural areas for shore protection, flood protection, recreational opportunities, and spawning grounds for marine life. The **Coastal Zone Management Act of 1972** was designed to encourage states to regulate development of coastal areas in order to protect them from degradation. The law was set up to provide planning and implementation grants to states with coastal areas.

Wetlands have been lost at an amazing rate since the 1950s. Estimates of the original U.S. wetland acreage range from 185 to 215 million acres. Around 50% of the original wetlands in the 48 coterminous states may have already been irreversibly altered or destroyed. From the mid-1950s to the mid-1970s, net annual wetland losses averaged around 500,000 acres. Since then, losses have averaged around 300,000 acres a year. About 95% of all wetland losses are attributable to human activity and 5% to the natural process of succession. The primary human activity resulting in inland drainage of wetlands is agriculture.

Inland freshwater marshes, swamps, bogs, and ponds make up 95% of U.S. mainland wetlands; coastal marshes comprise the rest. Because wetlands have no visible use and because dry lands do, protection of wetlands has been difficult, and filling in of wetlands has been a common practice. (See Chapter 14 for a discussion of the value of the wetland habitat.)

Other federal policies have been promoted to protect wetlands. In 1977, President Carter issued an executive order making it the policy of the nation and all federal agencies to conserve wetlands and to assist in construction projects on wetlands only as a last resort. President Bush established a policy of "no net loss of wetlands," though there has been controversy surrounding its implementation. Several federal agencies are involved in wetlands. The U.S. Army Corps of Engineers is charged with issuing permits for placing dredge or fill material in wetlands. In its decision to issue or not issue a permit, the Corps consults with the Fish and Wildlife Service, the Environmental Protection Agency, and the National Marine Fisheries Service. Generally, however, the Corps does not consider agricultural expansion into wetlands to be within its jurisdiction.

Federal monies for wetlands protection have come from the Department of Housing and Urban Development for purchase of wetlands through revenue sharing, the Department of the Interior through purchase of outdoor recreational areas, and the Fish and Wildlife Service for fish and wildlife restoration projects. However, pressures to develop wetlands continue.

Thus land use is being addressed and encouraged in bits and pieces, rather than in an overall, comprehensive land use plan at the federal level. Federal grants are also influential in the construction of housing, highways, airports, mass transit, and sewers—the so-called infrastructure of urban development. However, little has been done with these as tools in limiting urban sprawl and in land use planning.

A federal land use bill has been debated in Congress several times. Most versions called for states to define their roles in land use decision making, to plan for the regulation of projects with regional impacts, and to protect the agricultural use of land. To date, no comprehensive land use bill has been enacted.

State Involvement in Land Use Planning

For many years, almost the only role of the states in land use planning was the enactment of legislation permitting local communities to plan and zone. The U.S. Department of Commerce developed model laws in 1924 and 1928 called the **Standard Zoning Enabling Act** and the **Standard City Planning Enabling Act,** respectively. These model laws were prototypes for the states to use in drafting their own specific legislation authorizing planning and zoning.

In time, the states became more involved in land use management. A few states, beginning with Hawaii in 1962, passed state land use plans. Several approaches to statewide plans have emerged. In some states—Hawaii, Connecticut, and New Jersey, for example—the state plan is only advisory. Local governments are encouraged to make planning decisions consistent with the state plan, which may go so far as to label areas in the state as suitable or not suitable for development. However, the state has no enforcement authority. Another approach is to set up statewide standards that local governments are required to meet in their planning and regulation of development. For example, there may be specific standards set for subdivision plans, and the local planning body is required to disapprove a plan that does not meet these standards. Colorado requires a developer to show that adequate water and sewer service is available to support the proposed development; a local authority would have to disapprove any plan not containing such information.

There are also states that use a combination of zoning maps, goals, and mandatory standards. In Oregon, state goals and guidelines were set up in 1974. Local governments are required to develop land use plans consistent with these guidelines. This comprehensive approach remains the exception rather than the rule, however. Instead of dealing with the overall uses of land, many states have narrowly focused on the management of specific land resources. Generally, these are areas in which impacts are felt beyond local jurisdictions. Examples of how states have acted with regard to critical areas include siting legislation establishing guidelines for the locating of power plants. Control of state-owned lands and parks is a direct means by which states affect land management.

Regional Planning

Regional planning within states and between states has also emerged. Some states—Kentucky, for example—have divided the state into development districts. Each multicounty district is involved in acquiring data and planning for growth and services in that district.

Perhaps one of the earliest, and still ongoing, attempts at regional land use planning within a state has been the agency set up to control land use in Adirondack Park in New York State. In 1971, as a result of concern over development in the park and the potential for degradation of a unique recreational resource, the New York legislature established a commission to develop a land use plan for the park. In 1973 a plan was approved by the New York State legislature. Zoning classifications were set up for private and public lands within the park based primarily on the intensity of use. A wilderness classification on state lands would prohibit high-intensity use such as off-road vehicles (ORVs); the intensity of use of private lands is controlled by restricting the number of buildings in a given area.

The success of this regional land use program has been mixed. There has been local opposition; permits have at times been granted more on what some claim to be political grounds than according to sound land use policy. Although private and municipal development is regulated, the Adirondack Park Agency is only advisory when it comes to projects of state agencies such as highway development. This double standard further aggravates some local residents. Development around the 2,300 lakes and ponds is not well controlled; staff is lacking to follow up on permit holders to be sure they are in compliance with the conditions of their permits. Recently, concern over the park led the governor of New York to appoint the Commission on the Adirondacks in the Twenty-first Century. Almost 250 recommendations were made concerning the future of the park and its management. Basically, the commission called for making environmental concerns the first priority over development and all other concerns. So even after all these years, the same tensions are present. But in a complicated area such as land use, a prototype of state involvement in resource protection has proved that it can at least provide an opportunity to direct development, if not control it. The lessons learned from the record of this regional prototype may prove helpful to other states as natural resources become increasingly important to the general citizenry.

Land Use Management at the Local Level

The concept of land use planning includes zoning in both urban and rural areas. Land use planning on a regional, nationwide, or global scale can be thought of as an extension of the concept of local zoning.

Traditionally, urban land use planning has been left to the local communities. Several mechanisms are used generally by localities for such planning. The **planning board** or planning commission is an advisory group that analyzes available data and develops a comprehensive plan for directing growth in the community on the basis of this data. The plan is primarily a policy statement. To make certain that the master plan is implemented, a locality has other tools. The most common of these are zoning ordinances and subdivision regulations.

Zoning and Subdivision Regulation. In 1926 the Supreme Court held that zoning was a legitimate use of a government's police power, that is, its right to protect the health, safety, and welfare of its citizens as specified in the Tenth Amendment to the U.S. Constitution (*Village of Euclid* v. *Ambler Realty Co.*). Zoning is implemented by zoning ordinances that consist of a zoning map, a narrative to accompany the map, and a listing of regulations that apply in each district such as building size and parking requirements. Basically, zoning ordinances set forth what uses are permissible and where. The districts or zones are usually classified as commercial, industrial, and residential (see Table 13.3).

In recent years, other special zoning districts have emerged, such as agricultural, flood plain, and open-space districts. Such designations are meant to protect an area and to keep it relatively undeveloped. Agricultural districts will limit nonagricultural development; open-space districts will prohibit or restrict development, leaving a natural, open, green space in an area. Open-space districts may be useful in fast-growing urban areas. An even broader concept has emerged in the form of special districts called environmental quality districts. Such a designation requires any development to meet certain environmental quality standards such as erosion control or protection of natural or mature vegetation. Cincinnati has such a program, which permits the consideration of factors that probably should be, but are not usually, considered in zoning regulation procedures and guidelines (Merriam et al., 1986).

In general, zoning ordinances are developed by planning boards or commissions but are approved and passed by city councils or other local elected governing bodies. Zoning laws allow for zoning appeals in cases in which zoning ordinances are alleged to cause undue hardship. This and the fact that planning boards are purely advisory often allow political decisions to override zoning logic.

Related to zoning is the regulation of subdivisions through the specification of certain physical parame-

Table 13.3 Some Zoning Categories and Restrictions for Use

Zoning District	Principal Permitted Use
Residential	
R-2	Low density; not to exceed two dwellings per acre
R-4	Medium low density; single- and two-family dwellings not to exceed four per acre
R-8	Medium density; single- and two-family dwellings; mobile home parks; not to exceed eight dwellings per acre
RMF-16	Medium high density, multifamily; not to exceed 16 dwellings per acre
HRR-32	High density, high-rise; not to exceed 32 dwellings per acre
Business or Commercial	
PRD (or C-1)	Professional, research, office
LBD	Local business district (convenience businesses serving an immediate neighborhood)
GBD	General business district (shopping centers and other large-space uses)
Industrial	
I-1	Light industry
I-2	Heavy industry
I-3	Extractive industry (mining activities)
Special	
A	Agricultural districts
FP	Flood plain districts
OS	Open-space districts (recreation and conservation)

ters such as minimum lot size, street design, open space, and service lines. The developer is usually required to follow a step-by-step procedure, and approval by a planning board is required before the lots are recorded with the county.

Is Zoning Good or Bad? Not everyone agrees that zoning works as well as it should. Some critics of zoning think that it restricts innovative development. Zoning does not compensate a landowner for the restrictions placed on the use of the land. However, the courts have upheld the legality and constitutionality of zoning. Land development is not a right, they say, and therefore compensation for restricting development is not always necessary. Zoning is also described as too much of a negative tool; it *prohibits* certain uses in certain areas but is not geared to encouraging specific kinds of development.

A sociological criticism of zoning is that it may be used to preserve the status quo and to prevent the integration of neighborhoods. Minimum lot sizes and prohibitions on multiple dwellings put certain areas out of reach of middle- and low-income families and prevent development of low- and moderate-cost housing in certain areas. Critics claim that zoning has not halted urban sprawl. Another problem is that governments depend on property taxes for revenue, and this influences zoning decisions.

Even with all its limitations, zoning is one of the most widely used land use planning tools.

Other Land Use Planning Tools. Additional standard tools at the local level for influencing land use include so called "official maps" (maps that designate future rights-of-way for streets, drainage systems, parks, and other public lands) and building and housing codes that serve to regulate building size, frontage, and the like.

Localities can provide positive encouragement for development via tax incentives, low-interest loans, aids in land acquisition, and many other stimuli to

Zoning laws can help communities to regulate population density, preserve green space, and encourage the best use of land.

businesses locating in a specific area. For example, a local government may establish special **enterprise zones** such as in a center city area where industrial development is needed but not likely to occur and offer various economic incentives such as tax breaks to encourage business to locate there.

Taxation.

With the need to encourage the keeping of good agricultural land in agricultural production, the concept of **differential assessment** for land use was developed. Under this system, agricultural land is taxed (assessed) on the basis of its use as agricultural land rather than at its fair **market value,** that is, rather than at its economically highest potential land use. This offers some relief to farmers with land near suburban areas. If such land were taxed at its market value as subdivision lots, it would become too great a financial burden on the farmer. Differential assessments for agricultural lands give farmers a tax break as long as the land is in production. Differential assessments may be absolute, or they can have rollback provisions. **Rollback** means that the full tax payment based on market value is only deferred and becomes due when the land use is changed from agriculture to some other use. Such a provision is included to discourage speculators from purchasing and holding agricultural land. Land is more likely to be developed immediately if the speculator is required to pay deferred back taxes. Under rollback, the farmer is still protected indefinitely as long as the land stays in agricultural use.

A problem with the rollback concept is that it tends to encourage leapfrogging development. Since land farther from the city would tend to have a smaller deferred tax than that closer to the city, developers would be likely to jump over some agricultural areas with high deferred taxes to areas a little farther from the city. This, of course, means increasing urban sprawl. Further applications of the differential assessment are under consideration by various government bodies as a tool for encouraging land use other than agricultural—such as providing for open green space. Differential assessments in those cases would be based on a landowner's agreement to restrict development. As green space and open space become more and more desirable commodities, use of taxes to encourage their maintenance will be more likely. At least 48 states have some type of differential tax assessment of farmland.

Limiting Growth and Development.

In 1636, Boston passed one of the first ordinances limiting population growth. It required that any resident obtain permission for any stranger or house guest who intended to stay more than two weeks. Over the years, other cities have tried similar actions. In recent times, cities have restricted the number of new residences that can be built in an area, thus restricting population density. Courts have given various interpretations of such ordinances. There is great controversy over whether or not such laws accomplish their intended purpose. Some people seem to think that these laws bear the old zoning stigmas associated with restricting lot size—that is, they result in increasing land prices because of the simple laws of supply and demand, thus stifling poor and moderate-income families. A side effect is the encouragement of growth outside the restricted area—people must go somewhere.

Development Rights.

One of the most innovative approaches to land use control is that of transferable development rights. Under this system, a unit or large parcel of land is assigned a certain degree of **development right,** the right to change the use of the land from a natural area or an agricultural one to a commercial, residential, or industrial use. Such rights go with the land, just like mineral rights and air rights (the right to use the air and space above one's land). Just as mineral rights can be sold, so can development rights. The system involves the sale of development rights by owners of land that will not or cannot be developed to owners who wish to develop their own land. This is done by requiring developers to buy a certain number of rights from other landowners in the area before a given type of development can take place. Under a development rights program, regardless of other requirements, you need a certain number of rights. The number of "rights" that come with the land are not sufficient for major developments. You have to purchase more. For example, if you owned land and wanted to build condominiums you would have to purchase development rights from landowners who will not be using their land for development. In this way an owner of undevelopable land is compensated for the fact that the land cannot be developed, and development is limited.

In other development rights programs, state or local governments may purchase development rights. Money for these purchases might come from a bond issue supported by the community as occurred in King County, Washington. In other cases, such as the state of Maryland, state appropriations and county funds are used to purchase development rights. Problems with the development rights system lie primarily in lack of experience with it. There are no simple answers to how many development rights "units" should be assigned to a particular use—for example, how many rights are needed for an apartment complex versus an industrial park—or what the going price for a development right should be. However, the concept is promising and will probably become more common.

Easements. Another legal means of providing for the balanced use of land is the acquisition by a public group or agency of easements. **Easements** are a legal means by which permission is obtained to use land for a specific purpose when that land is owned by another person. Easements are required for installing sewer or gas lines by a public utility. Easements for public purposes usually give the landowner no option on whether to grant the easement or not. The utility offers the landowner monetary compensation, although landowners often question its adequacy. Ultimately, the land in question can be taken by invoking eminent domain powers. Scenic and conservation easements are tools used to keep a land area from being developed from its natural state. The landowner keeps the title to the land but is often given a tax break rather than any direct monetary compensation for agreeing to keep the land in an undeveloped state through a scenic or conservation easement. Organizations such as the Nature Conservancy use this as a tool when purchase of the land is not economically feasible.

Let the Market Do It. There is another perspective at the opposite end of the spectrum of ideas on land use regulation. This view holds that regulation is not necessary and that economic forces alone will lead to a balanced use of land. The city of Houston, Texas, has used such a nonzoning approach. The concept underlying it is that if certain uses are economically advantageous, landowners will enter into "restrictive agreements or covenants" to preclude an uneconomic use. For example, under a nonzoning approach, a developer interested in changing a large tract of land from a residential to a business development would not have to apply to a zoning board for a zoning change and be subject to a public hearing and formal public input. Under the nonzoning approach, a developer who can acquire a tract of land can then proceed with the project. Residents who are concerned about the potential for commercial development in their residential area could formally covenant or contract with each other not to sell for a use that may be incompatible with the single-family dwelling.

Studies of the Houston nonzoning approach show that a separation of land uses occurs without formal zoning. Preservation of large areas of single-family residences was less likely than under zoning; however, this could have been the result of poorly developed restrictive covenants. Multiple-family developments under the nonzoning approach were found to be easier to locate than under the zoning process. However, planning for sewer and other utility services to commercial areas was much more difficult and in some cases more costly to the taxpayers initially. A nonzoning approach is less political and more responsive to economics, since there is no board to approach for a decision. If and when a particular development can be justified economically, it is likely to occur. Supply and demand dictate land use. The danger is that longer-term environmental considerations and social impacts may take a back seat.

As an example of the abuse of restricted covenants, some covenants still barred the lease or sale of real estate to nonwhites as late as December 1984. Although such provisions became illegal in 1968 when the Fair Housing Law was passed, lawsuits had not yet been filed to overturn all existing racially restrictive real estate deed covenants.

Federal Lands and Land Use Management

So far we have looked primarily at how the use of *private* lands can be directed toward the use that is the best for the common good, and we have talked primarily in terms of economics. However, one-third of the land in the United States is part of the public domain, that is, land owned and managed by the federal government. Because the federal government is primarily concerned with the general welfare of the public, values other than purely monetary ones come into consideration in managing public lands. Let us now look at the federal policy on managing public lands, alternative uses, and competing interests and values.

About 8 billion acres of land in the United States is publicly owned. This land is rich in resources. About one-third of U.S. timber production occurs on federal lands. The federal government may lease minerals for extraction by the private sector, provide for the grazing of livestock on the land, and use the land for public recreation and protection of fish and wildlife resources.

Public lands are managed by several federal agencies. The Bureau of Land Management in the Department of the Interior has control over the greatest portion of the land—almost 50%. The Forest Service in the Department of Agriculture manages around 25%. The National Park Service and the Fish and Wildlife Service each manage about 10%. The Department of Defense manages 4% to 5%.

In looking at land use on private lands, we focused on techniques for providing incentives for the landowner to use land for the general welfare and not necessarily its highest economic return. In looking at land use on public lands, we find that federal agencies are directed to manage the land in a way that will provide the greatest good to the greatest number of people in the long run. The issue that arises here is that as resources on private lands become scarce, the pressure for development and use of resources on public lands increases. Finding the balance among the conflicting uses that best serves the general welfare is obviously no simple matter.

In the states west of the Rockies, the United States manages about 60% of the land. Consequently, in the western part of the nation, land management by the U.S. government is coming under increasing scrutiny. Typical of such reaction to federal land ownership is the so-called **sagebrush rebellion,** the challenge western states such as Nevada, Arizona, Utah, New Mexico, and Wyoming have made to the federal title to millions of acres in these states. The federal government has provided payments to the state for federal land in lieu of taxes, since federal land is not taxable by the states and consequently deprives the states of revenue. This is becoming less and less satisfactory to the states because they want the land and its resources for development as they see fit.

Valid uses of public land include grazing, timber production, fish and wildlife management, watershed protection, outdoor recreation, mining, research, wilderness protection, education, and protection of historical, cultural, and archaeological artifacts. All of these purposes are to be served by our federal lands. The question is, How much for each?

The policy for public lands set down by statute in 1960 is that they be managed under the principles of **multiple use and sustained yield.** Public lands are to be used in a variety of noncompeting fashions (multiple use). For example, timber harvesting can coincide with recreational activities and can be done in a way that protects the watershed. Renewable resources are to be managed in such a way that future generations can rely on these same resources (sustained yield); management of fish, wildlife, and timber is to be done in a manner that does not deplete the resource over time.

To ensure that land and land resources are conserved as well as used, Congress has set up four programs that are aimed at keeping the land in its natural state. These four programs are the National Park System—set up to protect lands for recreation; the National Wildlife Refuge System—set up to protect wildlife species; the Wild and Scenic Rivers System—set up to protect these resources in their natural flowing state; and the National Forest System—set up to maintain forest resources. Then, consistent with the concept of multiple use, Congress has specifically allowed certain other subclassifications within each system and permitted activities to occur in lands dedicated to these systems as long as these activities are consistent with the primary purpose of the land designation. The national programs and the activities permitted within each classification are shown in Table 13.4. For example, under the National Park System a land area may be designated a park, a monument, a preserve, or a recreation area. Hunting is not allowed in a park or monument but is allowed in a preserve or recreation area.

Realizing the importance of keeping some federal lands in a state relatively untouched by human activities, Congress has called for another designation to be added to all of the four major federal land protection systems just described. This is the designation of **wilderness.** If an area within any of these major programs is designated as wilderness, all development and consumptive uses of the land and its resources are prohibited. Wilderness areas are to be used in a manner that results in no, or very little, human impact. Because use of wilderness lands is so restricted, wilderness designations are strongly fought by advocates of resource development.

Table 13.4 Federal Land Systems and Restrictions on Land Use

| | National Park System | | | | National Wildlife Refuge System | | | Wild and Scenic Rivers System | | National Forest System |
	Park	Monument	Preserve	Recreation Area	Refuge	Range	Monument	Wild Rivers	Scenic or Recreational Rivers	
Sport or trophy hunting	no	no	yes	yes	yes*	yes*	yes*	yes	yes	yes
New mining claims	no	no	no	yes	no†	no†	no	no	yes	yes
New oil and gas leasing	no	no	no	yes	yes*	yes*	no	no	no	yes
Commercial timber cutting	no	no	no	no	yes*	yes*	yes*	no	no	yes
Commercial cultivation	no	no	no	no	yes*	yes*	yes*	no	no	yes
Sport fishing	yes	yes	yes	yes	yes	yes	yes	yes	yes	yes
Commercial fishing	yes	yes	yes	yes	yes	yes	yes	yes	yes	yes

* Permitted as long as it is found by the secretary to be compatible with the purposes of the unit.

† Some existing units are open to new mining claims to the extent allowed in the secretarial or presidential order establishing the unit.

Note: Wilderness classification, by law, is supplemental to the purposes for which National Parks, National Wildlife Refuges, and National Forests are established and administered and does not change those purposes. Wilderness designation does preclude developments and consumptive uses, such as commercial timber harvest, which would infringe on the wilderness character of an area.

Source: M. K. Ritter, *National Resources Defense Council Newsletter,* March-April 1979, p. 25.

Mining on Federal Lands. The leasing of federal lands for mining is administered by the Department of the Interior. The original legislation regulating this leasing was the **Minerals Licensing Act of 1920.** This law was enacted at a time when western coal resources were not in demand, and from 1920 to 1960, relatively few leases were granted. In the 1960s, demand for western coal increased, and the provisions of the 1920 act were found to be inadequate. A moratorium on the leasing of federal coal reserves was formally adopted in 1973, although it was informally in effect as early as 1971. In 1976, Congress passed the **Federal Coal Leasing Act,** which was designed to discourage speculative acquisition of coal resources on public land. The same Congress enacted the Federal Land Policy and Management Act mentioned earlier, which placed land use planning requirements on the leasing process.

In December 1987, final regulations were issued requiring the Bureau of Land Management to give more attention to environmental effects of mining on federal lands. This was the result of a 1984 study by the Office of Technology Assessment and a decision by the secretary of the Interior that environmental concerns deserved more consideration.

Land Disposal of Waste

Land is affected not only by how we use and treat its surface but also by what we put into it. Land has always been the ultimate garbage can. We dump waste on it, and we bury waste in it. Even when waste is treated or incinerated, this residue finds a final resting place within the earth. This is not likely to change with existing feasible technologies. Because of this, and because we are having major problems finding land for waste disposal, we will look at land disposal of waste (solid, hazardous, and radioactive) in this chapter on land.

Solid Waste

As of 1975, the most common way to dispose of solid waste in the United States was still the open dump. It was not until October 1976 that Congress passed the Resource Conservation and Recovery Act, which, among other things, prohibited open dumps and required a national inventory of dumps and a compliance schedule for converting open dumps into sanitary landfills (see Figure 13.8).

In 1987, landfilling was still the most common form of waste disposal; there are about 9,300 operating municipal sanitary landfills in the United States (Kelly and Kubetin, 1987).

Sanitary Landfills. Very noticeable to travelers to Virginia Beach on the Norfolk Expressway is a recreational area known as Mount Trashmore. In 1969 the U.S. Department of Health, Education and Welfare funded the building of this hill, which is 72 feet high, 800 feet long, and 100 feet wide and is composed of 85% solid waste. It is perhaps one of the most publicized of all sanitary landfills. It was to serve as an innovative approach to useful solid waste disposal. Once completed and topped off, properly managed landfills may be used for parks, playgrounds, golf

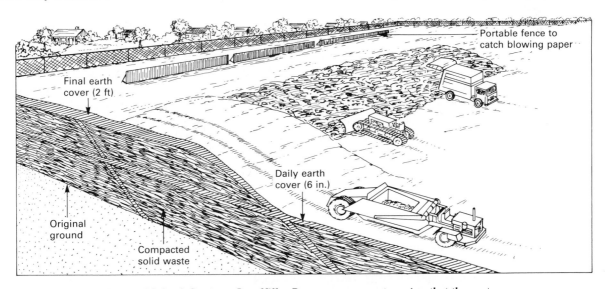

Figure 13.8 A Sanitary Landfill. Proper management requires that the waste be compacted and that a cover of earth be placed over the waste daily. If done properly, this eliminates rodents and blowing trash and reduces odors. If other mechanisms are used for solid waste disposal, including incineration and resource recovery, some residue remains. Sanitary landfills are needed for burial of such residues.

courses, and other purposes that do not require heavy construction.

Figure 13.8 illustrates the most common procedures for operating sanitary landfills. Some states are now setting more stringent requirements for operation of solid waste landfills, including liners, leachate and gas collection systems, and cover seals. The cost of disposal by landfill is low in comparison to the cost of incineration. Only open dumping is cheaper. However, there are problems with landfills that will continue to increase costs. Federal regulations require landfills to be upgraded to meet stricter standards to protect groundwater aquifers and to address other environmental and health factors. Land costs are constantly rising. Land in and near metropolitan areas is expensive, and urban areas are the most concentrated sources of wastes. As one moves away from the city, the cost of land decreases, but the cost of transportation increases.

A major social problem with landfills is the siting of new ones. Communities selected as sites for landfills often oppose them because of concerns about safety, health, property values, truck traffic, odors, litter, rodents, and insects. Sites that are suitable geologically may be passed over for political reasons; sites that are not suitable geologically may be chosen for political reasons.

The technical problems of landfills include improper management and improper selection of soil types; both of these may lead to health and pollution problems. Leaching of materials from the infiltration of rain and surface water may lead to pollution of groundwater. The accumulation of methane gas from the anaerobic decomposition of waste may lead to a fire or explosion. (See Chapters 7 and 17 for more information on solid waste disposal.)

Composting. **Composting** is a method of land disposal of waste that sets up ideal conditions for decomposition and for the return of organic materials to the soil as conditioner and fertilizer. Composting has been used much more extensively in European countries than in the United States largely because cheap commercial fertilizer has kept the market for compost very limited here. Figure 13.9 illustrates composting. The basic principle involves a mix of organic material, bacteria, and oxygen to allow rapid decomposition. Unfortunately, composting is not suitable for industrial refuse, and changes in the composition of municipal wastes to include more and more non-biodegradable waste and less organic waste also make composting less efficient. Bacteria do not work on plastics and synthetics, and they are very slow to break down rubber and cellulose wastes such as lumber and pressed boards.

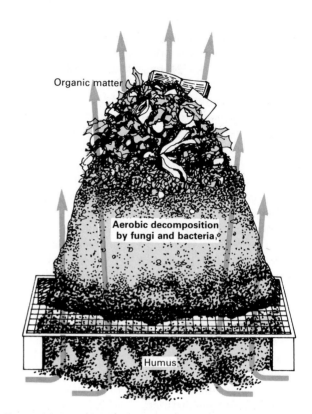

Organic matter

Aerobic decomposition by fungi and bacteria

Humus

Figure 13.9 A Compost Heap. Compost heaps may be large-scale operations or backyard projects. The critical mix includes organic matter, oxygen, and decomposers. Decomposers feast on the organic matter and break it down in the presence of oxygen into simpler, partly degraded organic and mineral components. Compost may then be used as a soil conditioner.

Hazardous Waste Disposal

The last resort for hazardous wastes is to store them underground in deep wells or in landfills. It is the last resort because the waste remains toxic and the potential remains for premature or unexpected release into the environment; thus continuous long-term monitoring and maintenance are required. Eventually, the waste will surely leach or make its way through landfill barriers. A basic premise in this strategy is that release will occur at such a slow rate as to be negligible in its effects or will occur after the waste itself has had time to be degraded. It is an undesirable approach because it may postpone the handling of hazardous wastes until future generations. Although restrictions on landfilling of hazardous waste have been put in place, the potential for disaster is always there. The problems associated with landfilling of hazardous waste were brought to the attention of the nation in the case of the Love Canal.

The Love Canal. In August 1978, President Carter declared the Love Canal in Niagara Falls, New York, a federal emergency disaster area. Federal assistance became available to evacuate some of the families in the area. This disaster, the result of improper disposal of hazardous waste, was the first national emergency ever to be declared for such a reason.

For several years until 1952, the Hooker Chemical Company disposed of some of its wastes by burying them in 55-gallon drums in a ditch known as the Love Canal. When the burial area was filled, it was capped with clay. The land was sold for $1 to the Board of Education of Niagara Falls, and a school and some residences were built on the site. Twenty-five years later the wastes began to leach. Construction caused the clay caps to crack; water infiltrated the trenches containing the waste. Leachate penetrated basement walls in the area, surfaced in backyards, and volatilized in the air. Over 300 different chemicals were identified, many of them known or suspected carcinogens.

The federal government and the state of New York spent around $250 million to clean up the former waste site. (Recovery of these costs from the responsible party was pursued in the courts.) The state of New York has declared parts of the Love Canal area suitable again for human habitation.

The Love Canal experience was not an isolated one. A 1985 report by the Office of Technology Assessment estimated that as many as 10,000 sites in the United States might need federal cleanup assistance over the ensuing 50 years at a cost of $100 billion. This estimate takes into account surface impoundments, closed solid waste landfills, and leaking licensed hazardous waste facilities.

Hazardous Waste Landfills. The **hazardous waste landfill** is similar to a solid waste landfill except that greater precautions must be taken to prevent leaching, to separate incompatible wastes, and to check for migration of the hazardous waste from the trench. Wells are dug at strategic depths to check for leaching or migration of the waste. Knowing exactly where and when to monitor is crucial and requires information about subsurface geology and water movement.

Figure 13.10 shows the general scheme for a hazardous waste landfill. Usually several barriers are used to ensure that the waste will be contained as long as possible. These barriers include impermeable clay liners and synthetic liners. Both may be used to line the trench and to cap it to retard infiltration of rainwater. A leachate collection system is constructed to collect any water that may infiltrate the trench. Sometimes two leachate systems are used as a safeguard, one in the bottom of the trench and one beneath the liner. As the trenches are completed, they are capped with clay and contoured to allow for rapid runoff of rainwater and to decrease infiltration.

The technology associated with liners for hazardous waste landfills is only now emerging. No research projects have been under way for long enough to determine the life span of such liners under field conditions.

The EPA restricts the disposal of liquids in landfills. Liquids in trenches pose a problem not only because they can migrate but also because quantities can build up like water in a bathtub. This accumulation of liquid can then produce pressure that pushes the liquid downward and outward. There are ways to solidify liquid waste. These include cement- or lime-based solidification; use of thermoplastic binders such as bitumen, asphalt, paraffin, and polyethylene; and use of organic polymers. Cement- or lime-based binders can be used with wet waste, but they are subject to leaching by acidic solutions. Thermoplastic binders may be susceptible to breakdown by organic solvents. They are, however, resistant to aqueous solutions, and they do reduce migration caused by rainfall infiltration.

The EPA regulations require that landfills be insured against damages that could result from leaching

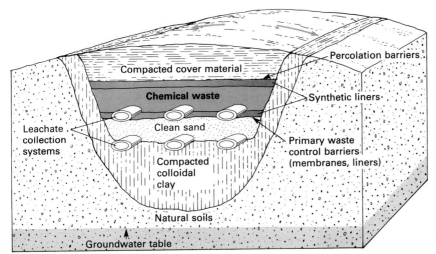

Figure 13.10 Hazardous Waste Disposal. Because of the serious health implications of groundwater contamination from hazardous waste landfills, numerous checkpoints are required in their design. The fill must be in an area of tight clay soils, or such clays must be imported to line the trench. Pipes and conduits to collect any leachate are required. The leachate can be pumped out and its composition tested. Synthetic liners can be another barrier between the waste and the environment or part of the cap to prevent water infiltration. The problem is to design a system that will work as long as the wastes remain hazardous—an indefinite period for some of these wastes.

and that a fund be established to pay for cleanup or closure and to cover monitoring and maintenance of the site for a given period of years after closure. Implementation of these requirements should make use of other treatment and disposal technologies more attractive economically (see Chapter 17).

Deep-Well Injection. Deep-well injection has been used for years in the oil fields to dispose of brine, and it has also been used for toxic waste disposal. Essentially, it involves pumping wastes into sandstone or limestone formations 1,000–3,000 meters deep (see Chapter 12). Opponents of this technology claim that it has caused earthquakes because of the increased underground pressures and that it has a potential for contaminating groundwater. Proponents claim that such episodes have been the result of poor engineering. Compatibility of wastes with well material and with the geology of the injection zone and the confining layers is also a major concern.

A study of hazardous waste injection wells in operation in six states by the U.S. General Accounting

Office (1987) found only two cases of contamination of drinking water but cited difficulty in detecting contamination. This was due to the fact that monitoring wells assess only a small portion of the large underground areas involved in deep-well injection. The EPA does not require monitoring wells; mechanical integrity tests of the injection wells are required. There are over 175 hazardous waste injection wells operating in the United States.

Radioactive Waste Disposal

The history of radioactive waste disposal has been a history of disposal in the land. Most radioactive wastes are not covered by the federal definition of hazardous waste, since they are regulated under the Atomic Energy Act. They are a subset of "solid wastes" and certainly constitute a special environmental problem. As we noted in Chapter 6, one of the problems of the nuclear industry has been the lack of a sound radioactive waste disposal program.

Radioactive wastes are classified as low-level and high-level wastes. **High-level wastes** are basi-

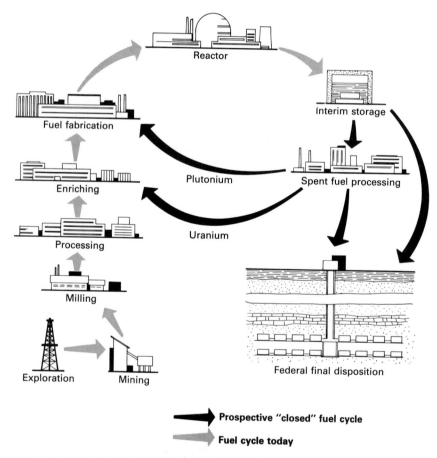

Figure 13.11 The Nuclear Fuel Cycle. Radioactive wastes are generated at each step in the cycle from uranium mining to processing to use as reactor fuel to reprocessing or disposal. Spent fuel is called high-level waste. Any other wastes or contaminated material have historically been classified as low-level. There are no permanent disposal facilities for high-level wastes in operation.

cally the spent fuel (fuel already used) from power reactors. Should reprocessing of commercial waste, which was halted under a presidential directive, be started again, high-level waste would be produced from fuel reprocessing (see Figure 13.11). The status of disposal technologies for high-level radioactive waste is discussed in Chapter 6.

Low-level wastes are any radioactive wastes that are not, by definition, high-level wastes. The terms *low-level* and *high-level* refer to how the waste is generated and not to what it is or how radioactive it is. Low-level wastes include a broad spectrum of radioisotopes and are produced by the mining and processing of uranium, by nuclear power facilities, and by medical and research activities. They include radioactive isotopes used for cancer treatment and contaminated glass, pipes, clothes, animal carcasses, dry trash, and other laboratory equipment.

Low-Level Radioactive Waste Disposal. Low-level wastes have largely been disposed of by **shallow land burial.** Originally, the intent was to have the waste permanently contained in the trenches by the impermeability of the rock in which it was buried. Consequently, artificial liners were not required; wastes were buried in cardboard boxes, and even liquid wastes were buried. These practices caused many problems, including leakage. Criteria for siting and managing of such low-level facilities have been tightened over the years.

Six such commercial low-level nuclear waste disposal facilities have been located in the United States.

Low-level radioactive waste was buried in unlined trenches as late as 1981 in Los Alamos, New Mexico. Because this method proved inadequate for safety, liners are now required.

Most of these facilities are currently not accepting commercial wastes or have been temporarily closed at one time or another because of water management problems, mismanagement by site operators, or inadequate enforcement of packaging and transportation requirements. The federal government operates several such facilities, but these have not been open to burial of commercial wastes. Experiences with low-level commercial sites have not been good overall. The impact this has had on the public credibility in the area of waste disposal technology is apparent in the growing opposition to the siting and operating of such facilities.

Federal Policies on Low-Level Radioactive Waste Disposal. In December 1980, Congress enacted Public Law 96-573, the **Low-Level Radioactive Waste Policy Act,** which established the policy that states shall be responsible for providing the disposal capacity for low-level nuclear waste. The legislation authorized states to establish multistate compacts. These would have a board or commission composed of representatives of the states with powers set out in the compact relating to siting of low-level nuclear waste disposal facilities and responsibilities of member states. Compacts are subject to ratification by the state legislatures and by Congress. The law also provided that after 1986, compacts could prohibit use of their facilities by noncompact states. This was to give impetus to the states to find alternatives for handling their own wastes or to join compacts by 1986. The 1986 deadline was extended to 1992; however, states not in compliance were subject to increasing surcharges.

U.S. Supreme Court rulings in 1992 raised questions about several provisions in the Act. Also, some questions are beginning to emerge about the economic viability of an increasing number of low-level radioactive waste disposal sites. The volume of low-level radioactive waste generated in the United States dropped by almost half from 1980 to 1990. With the potential for new sites in each of the compact regions, there is concern that the cost per cubic foot for low-level radioactive waste disposal could increase dramatically to cover the cost of developing these new facilities. The potential for proliferation of land disposal sites cannot be taken lightly in view of past problems.

Saving Land from Waste

In Chapter 17 we will look at alternatives for disposing of waste so that what is put in the ground will result in as few and as minimal problems as possible in the long run. We will also look at the real solution to the problem of waste, and that is generating less of it.

Concepts to Remember

1. Soil provides the basis for food and shelter for all terrestrial species.

2. Soil is, in a sense, a renewable resource, since soil formation is ongoing; it is, however, a very slow process. The problem is that soil is allowed to erode at a much faster rate than it is regenerated.

3. Soil formation is a complex process involving the parent material (bedrock), climate, topography, animal and plant life, and time. All of these factors result in many different types of soils. Not all soils are good for all uses.

4. Humus and certain small clay particles are important constituents of soil because they hold water and act as storage sites for plant nutrients.

5. Water can leach minerals from the soil. Soil acidity can increase the leaching of minerals, resulting in a loss of soil fertility and the potential of contaminating aquatic systems with harmful levels of some minerals.

6. Soil conservation is aimed at preventing erosion and maintaining soil fertility.

7. Soil erosion and loss of soil fertility from failure to return nutrients to the soil, salinization, and waterlogging are serious problems worldwide.

8. In attempting to increase food production by farming marginal lands, soil erosion is increased; this in turn results in rapidly declining productivity.

9. Land is a finite resource.

10. Land possesses some elastic qualities; its use can and does change over time with the demands of society.

11. Not all land is suitable for all uses because of its location, soil type, topography, or climatic conditions.

12. Some changes in land use are, for all practical purposes, irreversible.

13. Loss of prime farmland to development is a serious problem for everyone.

14. Land use is to a great extent governed by economics.

15. The best use of land from society's long-term perspective may not be the use that will provide the greatest immediate economic benefit to the landowner.

16. Certain government programs are designed to give a landowner incentive to use land in the best way for the public good over the long run.

17. The notion that any person should be able to do anything with land owned personally is outdated.

18. By being familiar with the tools of land use planning in urban and suburban areas and by being familiar with legal alternatives for preserving farmland in rural areas, individuals can become involved in effective land use management in their own communities.

19. Federal policy requires that federal lands be used and managed for multiple uses and sustained yields to ensure resources for future generations. The designation as wilderness prevents many developmental uses and so is often opposed by developmental interests.

20. Sound methods for managing the resources of the land are likely to be appreciated and practiced best by people who have an understanding of ecological relationships.

21. If we fail as a society to develop an ecologically sound land ethic, land will continue to be abused and will eventually no longer produce what humans need.

22. We all have responsibility as stewards of the land.

23. Land disposal of wastes in landfills or dumps has been the traditional method of waste disposal in the United States. With land disposal there is the likelihood of leaching from the landfill or dump site. From a long-term perspective, it is the least desirable disposal method.

24. As long as land disposal is the most economical method of waste disposal, it is likely to remain the most commonly used method. Factors that will raise the cost of landfills include the high cost of land, the enforcement of long-term liability, the requirements for upgrading landfill operations by requiring liners and leachate collection systems, and groundwater monitoring.

25. Pollution of groundwater from leaching of landfills is a major health concern. Because of lack of control over the types of materials placed in sanitary landfills and open dumps over the years, many wastes currently classified as hazardous have been buried in sanitary landfills. Safeguards required for hazardous waste landfills to control leaching have generally not been applied to sanitary landfills.

26. The primary method of disposal of low-level radioactive waste has been shallow land burial. There is still no repository for high-level radioactive waste, and none is likely before the year 2000.

Sherry W. Klosiewski

Sherry Klosiewski is a district environmental educator and naturalist with the Wisconsin Department of Natural Resources. She lives in Rhinelander, Wisconsin. As its name implies, the Wisconsin Department of Natural Resources (DNR) manages and protects natural resources of Wisconsin, ensuring the continued quality of the environment for future generations. Ms. Klosiewski plans and coordinates all facets of the education and interpretive program for a ten-county district. She works with teachers and youth group leaders, helping them to be better environmental educators. She consults with other DNR staff to help to make the agency more effective in educating and reaching out to the public. A major part of her job is to coordinate the nature-interpretive programs for the district's five state parks and forests. Day to day, Ms. Klosiewski trains and evaluates the summer naturalists and is involved in the development of interpretive trails and exhibits. Ms. Klosiewski sees her work as dynamic and exciting. She takes great satisfaction in working with educators and other concerned citizens, finding new and better ways to help them to become better educators and more effective environmental caretakers.

    Ms. Klosiewski has a Bachelor of Science degree with majors in wildlife management and English. Before taking up her current work, she served as a nature center naturalist and as assistant coordinator for Wisconsin Project WILD and Project Learning Tree. Ms. Klosiewski counts as her most valuable assets a knowledge of environmental education principles and her verbal and written communication skills.

References and Further Reading

References marked with an asterisk are cited in the chapter.

Collins, R. C., 1991. "Land Use Ethics and Property Rights," *Journal of Soil and Water Conservation* **46**(6):417–418.

* George, T. A., and Choate, J., 1989. "A First Look at the 1987 National Resources Inventory," *Journal of Soil and Water Conservation* **44**(6):555–556.

* Kelly, W. J., and Kubetin, W. R., 1987. "Barge Carrying Unwanted Garbage from Long Island Becomes Symbol for Larger Problem of Solid Waste Disposal," *Environment Reporter* **18**(3):332–337.

* Leopold, A., 1949. *A Sand County Almanac.* New York: Oxford University Press.

* McKnight, T. L., 1987. *Physical Geography: A Landscape Appreciation,* 2d ed. Englewood Cliffs, N.J.: Prentice Hall.

* Merriam, D. H.; Andrew, C. I.; Pagini, J. D.; Bressler, M. S.; and Dexter, A. D., 1986. *Environmental Regulations: Techniques to Protect the Natural and Built Environments.* Presented at the Sixth Annual Zoning Institute of the American Planning Association, November.

* Pimentel, D.; Allen, J.; Beers, A.; Guinand, L.; Linder, R.; McLaughlin, P.; Meer, B.; Musonda, D.; Perdue, D.; Poisson, S.; Siebert, S.; Stoner, K.; Salazar, R.; and Hawkins, A., 1987. "World Agriculture and Soil Erosion," *BioScience* **37**:277–283.

* Postel, S., 1989. "Land's End," *Worldwatch,* May-June, pp. 12–20.

Tanner, T. (ed.), 1989. *Aldo Leopold: The Man and His Legacy.* Ankeny, Iowa: Soil and Water Conservation Society.

* U.S. General Accounting Office, 1987. *Hazardous Waste: Controls over Injection Well Disposal Operations* (GAO/RCED 87-170). Gaithersburg, Md.: U.S. General Accounting Office.

* Weber, P., 1990. "U.S. Farmers Cut Soil Erosion by One-Third," *Worldwatch,* July-August, pp. 5–6.

14 Wildlife, Wilderness, and Other Biological Resources

Human beings have a significant impact on other species. This should be clear now that we have considered the many ways we indirectly affect other species through our impacts on air, water, and land resources. But these are not the only ways in which we affect other species. Indeed, we have many significant *direct* effects on the plants and animals that share this planet with us. Through overexploitation we have played a major part in rendering many species extinct and in bringing many others to near-extinction. The passenger pigeon and the buffalo are perhaps the most familiar examples of the former and the latter, respectively. By changing forests, prairies, and other kinds of land into farmland and cities, we have eliminated or threatened many plant species directly and many animal species indirectly—by taking away their habitat. Likewise, by filling in wetlands, we have eliminated or threatened still other species. Of all the environmental insults we inflict on the earth because of our numbers and lifestyle, the impact on other biological resources is one of the most irreversible. It is also one for which we have little scientific data to determine how significant in the long run our impact is. All of this deserves, and will get, some special consideration here as we examine the direct and indirect effects of human activities on our biological resources.

Biological Resources Defined

That we depend on other species for food and fiber is obvious. But far more species are important to us than the ones we eat or convert into clothing and shelter. In fact, since the overwhelming majority of the species on earth have not yet been described and are essentially unknown to us—and since we are still exploring the usefulness of the nonfood and nonfiber species we know very well, we must assume that *all* species are ecologically important. We must also assume that *all* other species are potentially useful to us and are thus, by definition, potential **biological resources.**

Cause for Concern, Rationale for Action

Extinct, Threatened, and Endangered Species

By some estimates, there are 5 to 10 million species on the earth. Our ignorance of the biology of our planet is such that there may be as many as 30 million species; we are not sure (see Table 14.1). Only

410

about 1.7 million species have been adequately described by biologists; the rest remain to be discovered. In Chapter 5 we pointed out that most of the species that have ever existed on the earth are now extinct. With so many species, and with extinction so much a part of the natural order of things, why worry about extinction? So what if the likes of the red wolf, Kirtland's warbler, and the Devil's Hole pupfish disappear from the face of the earth? Can we really do anything more than delay the extinction of most nearly extinct species, even if we wanted to? Even if the answer to this question is no, should we not at least avoid being the active cause of extinctions? Certainly we humans can and should take into account the potential threat to endangered species in the things we do. We should be concerned that the combined direct effects on individual species through exploitation and indirect effects through habitat destruction threaten many species. By some estimates, human activities have helped to bring rates of extinction today close to the rates of the mass extinctions of the past (see Chapter 5). A 1989 report by the World Resources Institute estimates that 25% of the species of the world could face extinction in the next 25 years at current rates of loss (Reid and Miller, 1989). There are some indications that biological diversity is currently being reduced to its lowest level since the end of the Cretaceous period (Wilson, 1989) (see Figure 14.1).

The complete rationale for concern about endangered species has elements that are ecological, technological, and philosophical.

Humans are part of the natural system; thus what is good for the system is good for humankind. As we have already said, the stability of the ecosphere is related at least in part to diversity (Chapter 4). We sometimes use the expression *biological richness* to convey the high regard we have for diversity. From the standpoint of the ecosphere as a whole, it makes sense to view the richness of the earth's species composition as an "index of flexibility" for the biosphere. Large numbers of organisms adapted to a wide variety of conditions give the biosphere a kind of "ready for any-

Table 14-1 Known and Estimated Diversity of Life on Earth

Form of Life	Known Species	Estimated Total Species
Insects and other arthropods	874,161	30 million insect species, extrapolated from surveys in the forest canopy in Panama; most are believed unique to tropical forests.
Higher plants	248,400	Estimates of total plant species range from 275,000 to as many as 400,000; 10% to 15% of all plants are believed undiscovered.
Invertebrates*	116,873	True invertebrate species may number in the millions; nematodes, eelworms, and roundworms each may comprise more than 1 million species.
Lower plants†	73,900	Not available.
Microorganisms	36,600	Not available.
Fish	19,056	21,000, assuming that 10% of fish remain undiscovered; the Amazon and Orinoco rivers alone may account for 2,000 additional species.
Birds	9,040	Known species probably account for 98% of all birds.
Reptiles and amphibians	8,962	Known species of reptiles, amphibians, and mammals probably comprise over 95% of total diversity.
Mammals	4,000	
Total	1,390,992	10 million species is a conservative estimate; if insect estimates are accurate, the total exceeds 30 million.

* Excludes arthropods; includes 1,273 miscellaneous chordates.

† Fungi and algae.

Source: Based on E. C. Wolf, *On the Brink of Extinction* (Washington, D.C.: Worldwatch Institute, 1987).

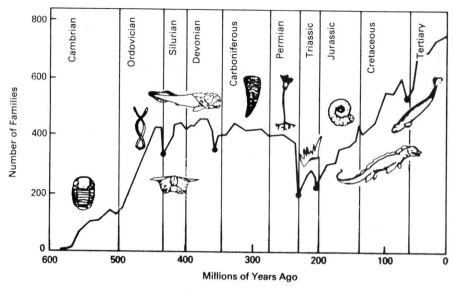

Figure 14.1 Mass Extinctions in Geologic Time. Five major mass extinctions
have occurred in geologic time. Biological diversity is currently declining at a
dramatic rate as a result of human activity. Some observers think we are nearing the
mass extinctions of the past.

thing" character. The richer the biosphere is—in
other words, the more adaptable the biosphere is to
whatever comes along—the more stable it is. Losing
species is like losing parts of finely tuned machines.

How much less stable does a specific ecosystem
become when a species is lost? It depends; the specific
species, the circumstances, and other factors are important. Some species do play more significant roles in
an ecosystem than others. In fact, they may control the
relative abundance of other species in that ecosystem
(Westman, 1990). These are termed **keystone species.** The problem is that we do not have much data
about which species are keystone species. Certainly,
the importance of a species to a system is *not* a function of whether or not a species can walk on two legs or
be seen without a microscope. Without decomposers
like fungi and bacteria, the nutrient cycles in the ecosystem would soon come to a halt. Clearly, we cannot
prioritize the species to be protected on the basis of
size, cuteness, or complexity.

Species preservation can also be argued from a
human-centered technological perspective. If a species becomes extinct, we have lost the chance of using
that species for improvement of the human condition.
Plants, for example, may contain undiscovered antibi-
otics, anticancer drugs, and other useful substances.
Digitalis, a drug used by patients with heart disease, is
made from the foxglove plant. The evening primrose
has been found to contain a compound previously
thought to be found naturally only in human milk. It
has potential benefits in treating hardening of the ar-
teries and arthritis. The periwinkle produces a substance used in treating Hodgkin's disease, a cancer of

the lymphatic system. The medicinal potential of yet-
undiscovered substances in discovered species and
yet-to-be-discovered species could be immense. About
50% of the drugs sold in the United States today are
derived from wild plants and animals. This takes on
even more importance since the development of **genetic engineering,** our ability to transfer genetic
traits from one species to another, creating new combinations of genes.

Our ability to transfer genes for resistance to
pests and disease could transform the entire agricultural system and end its dependence on synthetic pesticides (see Chapter 8). There is a variety of wild corn
found in Mexico that is a perennial; because it comes
up every year, it does not have to be replanted. The
genes in this variety offer some intriguing possibilities
for corn production. Ironically, the only known patch of
this wild corn was found on a hillside that was just
about to be plowed under.

In Chapter 8 we discussed the need for a gene
pool to develop new strains of crops that could grow in
less favorable climates and could use plant nutrients
more efficiently, as well as other such plant crop improvements. The gene pool formed by all the earth's
species is the raw material that contains the potential
for "engineered"—and natural—adaptation to a
changing world. To lose species, then, is to diminish
the size and richness of the gene pool. For this reason
alone it is clearly in our best interest to avoid causing
species to be lost and to preserve species in danger.

Some people argue for species preservation from
the position that all components of the ecosphere have
a right to be and that they do not exist solely for

purposes useful to humankind. Do trees have rights, perhaps? Can any one element really have special standing in a complex system in which all elements are interdependent?

Species Introductions as a Special Problem

There is no question that humans can alter habitats and squeeze out other species. Two-thirds of the birds on the Hawaiian Islands became extinct over a 200-year period following settlement of the islands by the Polynesians (Lewin, 1987). Studies also show that the immigration or introduction of species other than humans can threaten indigenous species. Hawaii has more endemic species than any other place on earth: 100% of its invertebrate species, 98% of its birds, and 93% of its flowering plants occur nowhere else on earth (Holing, 1987). Its location 2,500 miles from any other landmass does not make this surprising. There are at least 150 different types of natural communities on the Hawaiian Islands; however, since the coming of *Homo sapiens,* about 4,000 new plant species have been introduced. New insect species continue to enter at the rate of a couple of dozen a year. At least 70 types of birds have become established. All of this provides competition for—and otherwise threatens—native species. About 800 endemic plants are categorized as endangered today, and 30 bird species are on the Fish and Wildlife Service's threatened and endangered list. Introduced species have been most successful in areas where native vegetation has been affected by disturbances such as trampling by livestock and fire.

Introduced species also cause other kinds of problems. A prime example is the tamarisk tree. The tamarisk, known for its hardiness, was brought to the United States from Eurasia in the mid-1800s as an ornamental windbreak and means of erosion control (Johnson, 1986). Indeed, it did prove hardy, particularly in arid regions. The tamarisk is a **phreatophyte,** a plant that can survive in dry climates because of its long root system. The problem with the tamarisk is that it has an almost insatiable thirst: A large tree can absorb 200 gallons of water per day. The tamarisk is also a great competitor. Known also as the salt cedar, it secretes salt, which drops to the ground, making the environment much less suitable for native species. Its tiny seeds are easily dispersed by wind, water, and wildlife. The tamarisk is found today on 1 million acres of U.S. land, primarily along rivers and streams in 15 states. Its high rate of transpiration and its need for water has affected springs, pools, and fragile oases in the West. To date, the only successful elimination tactic has been to attack the tamarisk tree plant by plant. Pools in a Death Valley National Monument, for example, were restored by cutting individual trees and applying herbicide to the stumps. Exotic species can become economic pests or artificial keystones that change the relative abundance of species in an area (Westman, 1990). Removal of exotic species has become an issue in national parks.

Elimination of Habitat

Clearly, the destruction of a particular kind of habitat leads to the elimination of any species dependent on that habitat. Less obvious, perhaps, is that even the fragmentation or chopping up of large expanses of similar habitat also has negative impacts on dependent species. When islands of original habitat are left as an area is "developed," the fragmentation of habitat that results can lead to **faunal collapse,** the loss of animal species. Some species simply require large expanses of habitat in order to flourish. Even in expanses of territory as large as our national parks there have been significant losses of species (see Table 14.2), especially among mammals (Wolf, 1987). Ecologists speak of **biological impoverishment** as the decline in species diversity that stems from various stresses put on ecosystems. The survivors in such situations tend to be ecological opportunists. We call many of these *pests* because great numbers of them are in places we wish they were not.

Pine Barrens and Blue Butterflies. The relationship between the Karner blue butterfly and the Pine Bush in New York State is a prime example of the way that loss of habitat results in loss of species (Stewart and Ricci, 1988). The Karner blue butterfly is found in the Pine Bush, primarily associated with the lupine plant. The Pine Bush area is an inland sand dune area. It is an arid region characterized by periodic fires that prevent the successful intrusion of other plant and tree species. The common scrub oak and pitch

Table 14-2 Habitat Area and Loss of Large Animal Species in North American National Parks, Assessed in 1986

Park	Area (square kilometers)	Original Species Lost (percent)
Bryce Canyon	144	36
Lassen Volcano	426	43
Zion	588	36
Crater Lake	641	31
Mount Rainier	976	32
Rocky Mountain	1,049	31
Yosemite	2,083	25
Sequoia–Kings Canyon	3,389	23
Glacier-Waterton	4,627	7
Grand Teton–Yellowstone	10,328	4
Kootenay-Banff-Jasper-Yoho	20,736	0

Source: E. C. Wolf, *On the Brink of Extinction* (Washington, D.C.: Worldwatch Institute, 1987).

pine are adapted to these periodic fires. Without them, other tree species such as aspen, maple, white pine, and black locust would tend to invade and take over.

The lupine plant, with its small blue flowers, is well adapted to the dry pine barrens. It thrives in sandy soil and can survive periods of drought. The larvae of the Karner blue butterfly feed on the leaves of wild lupines as their only food source. The butterfly has only about an inch of wingspan. It has been on the New York State endangered species list since 1977. The butterfly lives only three to five days and cannot travel long distances. Consequently, if a subdivision or broad highway must be traversed to reach pine barren habitat and more lupine plants, the Karner blue is not likely to make it.

Once covering a 40-square-mile area, the Pine Bush is now segmented by major highways and surrounded by suburbs. Fires in the Pine Bush area are being controlled by fire departments from adjacent suburbs. The highways that traverse the area serve as fire walls to keep the fires from spreading. The sandy soil is easy to move, and since 1950, the development of residential, commercial, and business buildings has escalated.

There are some hopeful signs. In July 1987 the New York Supreme Court halted development in the Pine Bush, based on the fact that the environmental impact statement did not determine the minimum acreage necessary to protect the Pine Bush ecosystem. Both the local and the state government and the Nature Conservancy are working to set aside and manage 2,000 acres of the Pine Bush.

The Karner blue is only one species. Others that call the Pine Bush home are the spadefoot toad, the worm snake, the hognose snake, Fowler's toad, and the box and spotted turtles. Often seen as a wasteland by unknowing persons, the pine barrens, like every other ecosystem on earth, is home to a variety of plant and animal species. As the Pine Bush area goes, so goes the lupine and so goes the Karner blue butterfly and numerous other species as well. The same story is repeating itself, but with different characters, in your area as well as worldwide.

Wetlands: A Special Case

Wetland is a general term referring to any area where land and water come together. In such areas, the edge effect (see Chapter 4) is readily apparent in the wide variety of plant and animal species found there. Wetlands are characterized by soils saturated with water and vegetation adapted to periodic flooding or submersion. More common names for wetlands include estuaries, marshes, bogs, swamps, and peatlands. In North America, coastal wetlands include tidal marshes (salt water and fresh water) and mangrove systems. Inland wetlands include freshwater

marshes, peatlands and bogs, swamps, and **riparian lands** (lands that border rivers and streams). Some wetlands are forested, and the Fish and Wildlife Service has established five classes of these forested wetlands: broad-leaved deciduous (in which the tree species include maple, ash, gum, and oak), needle-leaved deciduous (cypress and tamarack), broad-leaved evergreen (mangroves, loblolly, and sweet bay), needle-leaved evergreen (black spruce, white cedar, and pond pine), and a fifth class of wetland characterized by dead woody vegetation caused by prolonged inundation.

Wetlands have a variety of forms that range from cypress domes and mangrove swamps to bottomland hardwoods to prairie potholes to mudflats—wherever in the world land and water meet.

Wetlands are among the most biologically important and productive ecosystems on earth (see Figure 2.8). Table 14.3 lists some of the important functions of wetlands.

Wetlands are important as stopping-off points and wintering and nesting areas for migratory waterfowl and other birds and as spawning grounds for fish and shellfish. It has been estimated that two-thirds of the commercially important fish and shellfish along the Atlantic Coast and the Gulf of Mexico and 50% along the Pacific Coast depend on wetlands for their food sources and spawning grounds. On their long journeys, migrating shorebirds need stopping-off areas where they can feed and restore fat reserves for the next leg of the journey. These migrating birds tend to use the same areas every year. If wetland areas are lost, new stopping-off places must be found, and others may become overpopulated.

Figure 14.2 shows a disturbing trend in the numbers of ducks migrating each fall. The estimated duck population has declined from 92 million birds in the

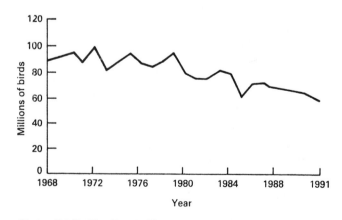

Figure 14.2   Decline in Migrating Waterfowl in the United States.   The number of ducks migrating each fall in the United States has been decreasing. Some species have declined more than others. Loss of wetland habitat is a significant factor in this decline.

Table 14-3 Wetland Functions and Conservation Concerns

Wetland Function	How Wetlands Perform Function	Conservation Concern
Flood storage	Some wetlands store and slowly release floodwaters.	Fill or dredging of wetlands reduces their flood storage capacity.
Flood conveyance	Some wetlands (particularly those immediately adjacent to rivers and streams) serve as floodway areas by conveying flood flows from upstream to downstream points.	If flood flows are blocked by fills, dikes or other structures, increased flood heights and velocities result, causing damage to adjacent upstream and downstream areas.
Erosion control/wave barriers	Wetland vegetation, with massive roots and rhizome systems, binds and protects soils. Vegetation also acts as wave barriers.	Removal of vegetation increases erosion and reduces capacity to moderate wave intensity.
Sediment control	Wetland vegetation binds soil particles and retards the movement of sediment in slowly flowing water.	Destruction of wetland topographic contours or vegetation decreases wetland capacity to filter surface runoff and act as sediment traps. This increases water turbidity and siltation of downstream reservoirs, storm drains, and stream channels.
Pollution control	Wetlands act as settling ponds and remove nutrients and other pollutants by filtering and causing chemical breakdown of pollutants.	Destruction of wetland contours or vegetation decreases natural pollution control capability, resulting in lowered water quality of downstream lakes, streams, and other waters.
Fish and wildlife habitat	Wetlands provide water, food supply, and nesting and resting areas. Coastal wetlands contribute nutrients needed by fish and shellfish to nearby estuarine and marine waters.	Fills, dredging, damming and other alterations destroy and damage flora and fauna and decrease productivity. Dam construction is an impediment to fish movement.
Recreation	Wetlands provide scenery, wild areas, habitat, wildlife and water for recreational use.	Fill, dredging or other interference with wetlands causes loss of area for boating, swimming, bird watching, hunting and fishing.
Water supply (surface)	Some wetlands store flood waters, reducing the timing and amount of surface runoff. They also filter pollutants. Some serve as sources of domestic water supply.	Fills or dredging cause accelerated runoff and increase pollution.
Aquifer recharge	Some wetlands store water and release it slowly to groundwater deposits. However, many other wetlands are discharge areas for a portion or all of the year.	Fills or drainage may destroy wetland aquifer recharge capability, thereby reducing base flows to streams and groundwater supplies for domestic, commercial, or other uses.

Source: Council on Environmental Quality, *Environmental Quality, 1985: The Sixteenth Annual Report* (Washington, D.C.: Council on Environmental Quality, 1987), p. 280.

1970s to around 70 million in recent years. Certain species have declined even faster: pintails by 54% and mallards by 20% over their averages for the last 30 years (Lenssen, 1989). A major factor in this decline is loss of adequate habitat, i.e., wetlands. One of the major contributors to wetland loss is agriculture. Although the swampbuster provisions of the 1985 Farm Bill (see Chapter 13) were intended to curtail drainage of wetlands for agricultural purposes, enforcement has been lagging in some cases (Lenssen, 1989). There are indications, too, that contamination of wetlands by toxic chemicals and by acid deposition have had an impact on duck populations (Lenssen, 1989). Other contributors to wetland loss are urban development, mining, and highway development.

Wetlands are important economically also for such cash crops as timber, marsh hay, wild rice, cranberries, blueberries, and peat moss. The areas are used for hunting, fishing, and boating. Wetlands also play a role in protecting shorelines and banks from erosion.

In some cases they may store floodwater and thus moderate floodwater levels. By purchasing the wetlands along the Charles River in Massachusetts, the local government made sure that the lands would be preserved as wetlands with their ability to retain flood waters and that they would not be converted to other uses.

Wetlands are important in the maintenance of groundwater supplies and in the purification of water. Runoff containing nutrients feeds into wetlands, which can then hold and use the nutrients. Bulrushes and cattails, common wetland plants, utilize nutrients from sewage and absorb some toxic chemicals. Researchers in Louisiana have estimated the value of phosphorus removed by wetlands at $500 to $1,500 per acre annually (Hodges-Copple, 1989).

Wetlands help to buffer temperature changes—a valuable service in Florida and other citrus-growing areas that are susceptible to drastic temperature fluctuations.

Wetlands are not the vacant, useless areas they might seem. Besides serving as spawning and nesting grounds, wetlands are buffer zones that minimize flooding and erosion and moderate temperature changes.

It has been estimated that the functions provided by an acre of wetland are worth $10,000 to $30,000 a year (Hodges-Copple, 1989). Unfortunately, the market value for these lands is much below this amount. This makes such wetlands vulnerable to drainage or fill in order to increase the economic value of the land by converting its use to farming or development.

In colonial days, coastal wetlands were valued for use as sources of salt for livestock, rushes for household use such as making brooms, and habitat for ducks, geese, and other shorebirds and for shellfish (Platt, 1987). However, when Congress passed the Swamp Land Acts in 1849 and 1850, the general common use of wetlands was supplanted by a national policy that encouraged draining or filling wetlands for agricultural use. There were 215 million acres of wetlands in colonial times. Today there remain fewer than 100 million acres, 90 million of them inland and the rest coastal estuaries. About half of the inland wetlands are forested. In the United States, wetlands continue to be lost at the rate of 300,000 to 500,000 acres a year. Forested wetlands have been lost at a rate perhaps five times that of nonwetland forests (Abernethy and Turner, 1987).

Loss of wetlands is especially severe in the South. Of the 9 million acres of wetlands estimated to have been lost in the United States between 1955 and 1975, fully 8 million were in the South (Hodges-Copple, 1989). In 1989, Congress enacted the North American Wetlands Conservation Act. The law provides funding for the government to purchase and manage areas identified as critical wetland habitat for migrating waterfowl. Canada, the United States, and Mexico are parties to a 1986 treaty to establish the North American Waterfowl Management Plan. As many as 6 million acres in the three countries may fall under the protection of this treaty.

Loss of wetlands is a worldwide problem. Population centers have grown up in places all over the world where water was available, and this has often meant wetland "development." Because the ecological function of wetlands has not always been apparent, wetlands have been drained and filled to provide solid land for expansion. Trees have been cut for construction and firewood. Mangrove systems have been especially vulnerable in developing countries; the Philippines has already lost almost 50% of its mangrove swamps. In some developing countries, **mariculture** (saltwater aquaculture) projects have been encouraged. However, because they are not always properly managed, the waters become contaminated and acidified. When this happens, the system is abandoned and the operator moves to another area. Called "slash-and-burn aquaculture," the potential is for severe, long-term damage to wetlands. Major water development projects in developing countries also affect wetlands, especially those involving the construction of dams that flood some wetland areas and cause others to dry up. Certain international development agencies (such as the World Bank) are devoting more effort to preserving biological diversity in the projects they support. This should help in these areas.

Pollution of wetlands and of coastal areas carries beyond into nearshore areas. In great danger from pollution of these waters are the highly productive coral reef communities. Coral reefs are in decline worldwide, especially in the Caribbean, the Pacific, and the Indian Ocean. Although several factors come into play, excessive nutrients and sediments are thought to be primary contributors to the decline. These materials dumped into the Caribbean from the Orinoco River in Venezuela affect the World's End Reef 80 miles away near the Grenadine Islands (Kohn, 1990). Nutrients result in the growth of algae that compete with coral for space and usually win. Sediment blocks out light. Nutrients come from coastal runoff and sewage from coastal developments. Sediments come from construction, forest cutting, and agriculture.

Tropical Forests: Another Special Case

Perhaps the most significant worldwide threat to biological diversity is the destruction of tropical forests. There are varying estimates of how fast tropical rain forests are disappearing and how many species are being affected. Norman Myers (1987), a British environmental consultant, concluded that 20% of the earth's species could be lost by the year 2000. He reasoned that if there are 5 million species on earth (a conservative estimate), and two-thirds of them live in the tropics, then 3.3 million species live in the tropics. If at least 50% of these are found in tropical rain forests, and if one-third to one-half of all tropical rain forests will be gone or seriously disturbed by the year 2000, then two-thirds or three-fourths of a million species are in danger of extinction. If one adds other species, and rich habitats such as woodlands, wetlands, and coral reefs, the earth is in danger of losing 1 million species by the year 2000, or 20% of the species now in existence (one out of five).

Remote sensing studies indicate that tropical forests are disappearing at a rate 50% faster than 1980 estimates. Some 40 to 50 million acres are lost annually, compared to the 1980 estimate of 28 million acres annually (Associated Press, 1990a).

Daniel Simberloff of Florida State University estimates that if 50% of the Latin American forests are disturbed by the turn of the century, 15% of plant species in the forest will be lost. If the current rate of destruction and degradation increases, loss of plant species could rise to 66%. Although the numbers vary, the conclusion is the same: Massive loss of tropical forest species is likely in our lifetimes.

Biological Diversity in the Tropical Forests.

The wealth of species in tropical rain forests is almost beyond imagination. Ten acres of tropical forest contain 300 species of trees, compared to 20 or so species in the same area of temperate forest. In 1986 a survey by the Smithsonian Institution of insects in the canopy of a rain forest in Peru found 41,000 species (including 12,000 different kinds of beetles) in an area no larger than a hectare (Lewin, 1986). This led to a rethinking of the total number of extant species. Prior to this finding it was commonly thought that there were 5 to 10 million species on earth. The implication of the Smithsonian data is that there may be as many as 30 million species on earth.

The Tropical Forest as a Resource.

Tropical forests have been shown to play a significant role in limiting erosion, protecting soil quality, moderating seasonal flooding, and protecting waterways from silt buildup. Clearing of rain forests in Panama is resulting in a rapid buildup of sedimentation in the reservoirs that feed the Panama Canal (Whelan, 1988). Rivers in Brazil now flood and at other times run dry; their even year-round flow has been lost with the watershed protection provided by the forests.

In addition to wood products, tropical forests produce oils, spices, and rattan used in worldwide commerce. Tropical forests provide habitats for many native and migratory species. Last but not least, the diverse plants hold untold treasures in their gene pools for medicinal, agricultural, and other benefits for humankind.

The rate at which tropical forests are disappearing is of significant environmental concern, since some of the impact, including loss of species, is irrevocable. On a global scale it is not known what impact deforestation will have on atmospheric conditions, in particular on the hydrologic cycle or on climate (see Chapter 10). According to some research, deforestation could cause rainfall to decline by more than 25% in the Amazon River Basin, reducing annual rainfall from 97 to 72 inches a year. This climate change would doom the rest of the forest area. In addition, the impact of reduced rainfall could be global, affecting runoff into the Atlantic Ocean (Associated Press, 1990b). Also, because tropical forests provide a sink for CO_2 in the atmosphere, their destruction contributes to the buildup of carbon dioxide and thus to global warming.

Pressures leading to increased deforestation include agricultural needs, areas for livestock grazing, and need for timber and fuel. Less than one-tenth of the land that is cleared each year is reforested.

Some techniques for lessening the problem of massive deforestation include the encouragement of agroforestry systems, that is, combining farming and forestry; the development of forest management plans; and new approaches to park design and management. However, the solutions also involve social, political, and economic considerations.

According to one report, government policies play a major role in logging and the conversion of rain forests to cattle raising and agriculture (Repetto, 1990). Many of these policies are the result of pressure from growing populations and increasing national debt. The policies of developed countries, financial institutions, and international agencies also play a part. For example, the financing of roads by the World Bank may actually encourage development in forested areas.

If the forests are to be saved, efforts to ensure sustained utilization of resources must become an essential component of all development policies. Illegal trade in plant and animal species indigenous to the tropical forests is also a problem. We will discuss worldwide efforts to protect tropical forests shortly.

We hear a lot about loss of tropical rain forests. But loss of rain forests in temperate zones is also a problem (see Table 14.4). Temperate rain forests are found in many countries, including Australia, Chile, Japan, New Zealand, and Norway, and in the United States from southeastern Alaska to northern California (Ryan, 1989).

Table 14-4 The Disappearing Temperate Rain Forest

- An estimated 15% of the original old-growth forests of the U.S. Pacific Northwest remains.
- Less than 10% of the tall eucalyptus forests that existed in Tasmania 200 years ago remains.
- Only 2% of British Columbia's coastal rain forest is protected from logging.
- The U.S. Forest Service lost $41 million selling timber in Alaska's Tongass National Forest in 1988.
- At current rates of logging, the native coastal forests of Chile's Maule region will disappear in 30 years.
- Unfragmented old-growth in Washington's Gifford Pinchot National Forest may disappear within three years.
- The Douglas fir rain forest ecosystem has been eliminated from British Columbia.

Source: J. C. Ryan, "Plight of the Other Rain Forest," *Worldwatch,* March-April 1989, p. 41.

Biological Resource Management Strategies

Biological resource management takes many forms, and agreement is far from universal, even among biologists, as to the effectiveness of various strategies. Still, taking steps to save a particular endangered species, passing laws protecting threatened species, controlling the harvest of wildlife through hunting regulations, using optimal methods of harvesting forest products, managing federal lands for multiple use (e.g., hunting, fishing, hiking, and timber harvest), and establishing systems of state or provincial and national parks and wilderness areas are all forms of biological resource management. Boiled down, all biological resource management strategies amount to just two kinds of action: action aimed at particular species—for example, captive breeding or species-specific hunting licenses—and actions that amount to manipulation of habitat so as to favor and sometimes limit species dependent on that habitat.

Good management of forests or other plant life and of wildlife begins with good management of the land. In its most basic form, good management of land means keeping land in place. This is an obvious part of good farmland management, but good land management is just as important to a continued good harvest of trees year after year. Management by the people involved in construction of highways, buildings, or homes is also important to protect streams from sediment runoff and to leave some soil for landscaping purposes.

Good land produces rich plant life, which in turn provides food, cover, and habitat for wildlife. The key to good land management is disturbing the land as little as necessary and, when it must be disturbed, taking steps to minimize erosion and repair the disturbance as quickly as possible.

But this is not to say that good biological resource management always means leaving nature alone or trying to hold it constant.

Habitat Management

As we saw in Chapter 4, areas undergo succession from one type of plant life to another over time. The pace of succession can be altered by various techniques. Mowing, grazing, and plowing all slow succession by maintaining a given transitional stage. One tool of nature for slowing succession has been fire. This same tool can be used by humankind.

Although we usually think of fire as a destroyer of plant and animal life, it can actually play a very positive role in the maintenance of certain habitats and consequently in the support of certain wildlife species. The Pine Bush area in New York State, discussed earlier in this chapter, is a good example. Fire is a natural occurrence in some habitats and can, when properly controlled, be used by humans for habitat management.

Forest fires have been around much longer than humankind. Fires caused by lightning are natural phenomena that serve to clear out dead forest debris. At the same time, fire provides a mechanism for a rapid return of nutrients to the soil. As material burns, it releases carbon dioxide and water, and its nutrients remain as ash. Destruction by fire results in a surge of nutrients for the next generation. Fire also allows for reestablishment of some sun-tolerant species. Certain pines even have cones that do not open to disperse the seeds unless certain fire temperatures are reached. The open area from a controlled burn also provides an edge where a variety of transitional species may grow (see Chapter 4).

Burning of prairie land is another example of fire helping to maintain an ecosystem. Because these areas are grazed so closely today, such fires are less likely to occur. In some preserved prairie areas, fires may be set intentionally to maintain the system.

The chaparral areas in the southwestern United States are composed of dense, scrubby brush and low vegetation. Characteristically, chaparrals ignite and burn every eight to ten years when the vegetation becomes extremely dense. This prevents succession to taller tree species and thus perpetuates what would otherwise be a transitional stage.

Wildlife Management

Species are managed indirectly by good land and habitat management; populations can also be managed directly. Good wildlife management is concerned with the production and harvesting of game species, the maintenance and protection of nongame species, and the control of certain species in certain locations. The purpose of such management is to prevent conflict with human health and human activities such as crop and livestock production. Two species population management tools are particularly controversial. Let us look in turn at the roles of predators and hunting in wildlife species management.

Predator Control. Perhaps the ecological role that is most misunderstood by the public is that of the predator (see Chapter 4). Predators have traditionally been seen as villains and their prey as innocent victims. Humans have spent much time and energy attempting to control or eliminate predators. We have at times put bounties on predators or promoted large-scale poisonings.

A prime example of modern-day attitudes toward predators is the controversy over coyotes in the West. Many ranchers are convinced that coyotes are costing them millions of dollars by attacking their herds. Others believe that coyotes only cull the old and sickly from herds or that only a few individual coyotes in certain areas are to blame and not the whole species. Nevertheless, large poisoning campaigns were conducted in the 1960s. They succeeded in killing not only coyotes but also prairie dogs, eagles, and vultures. Poisoned bait was eaten indiscriminately by many species, not just the target species or the individual animal causing the problem. Some of the poisons, such as **"1080" (sodium fluoroacetate),** were distributed in dead carcasses. Not only the animal that ate the carcass died but also any animal that fed on the animal that ate the carcass.

Public outrage resulted in the establishment of a committee to look into the problem; the committee was headed by Aldo Leopold's son, Starker. This committee adopted two principles regarding predator control. It concluded that all native animals should be considered resources and that control programs should be local and aimed at individual animals causing the problem. However, it was not until 1972 that the coyote-poisoning campaigns were prohibited on public lands. Even now, the attitudes expressed by Leopold's group have not fully taken hold in predator management and in the minds of the public. In late 1983 the ban on "1080" was modified to allow its use in single lethal doses.

There are alternatives to poisoning for managing predator-prey relationships. Managing for wildlife may involve the provision of good cover habitat for the prey species. For management of game species it might involve the enhancement of nongame buffer species— species that are not hunted by humans but that a given predator would find satisfactory. Rodents often serve as buffer species. Care must be taken, however, to ensure that the buffer species do not compete with the game species for its food or habitat, carry disease, or interfere in any other way with the life cycle of the desired species. Care must also be exercised so that the buffer species do not proliferate to the extent that they draw more predators. In addition, the buffer species must not have a hibernating stage, at which time it would leave the desired species even more vulnerable for a period of time.

Hunting. Although some people deplore hunting as a sport, if it is managed properly, hunting not only does not adversely affect natural systems but may in fact help. Any healthy animal population produces a surplus. The taking of this surplus will have no adverse effect on the population.

Hunting regulations determine what can be hunted (even what sex), when, and for how long. Bag limits are set to ensure that a population is not over-hunted. Seasons ensure that hunting does not occur at times when a species is carrying its young or is unable to escape—for example, at times when ducks' plumage hinders their flight.

Fall is a prime hunting season because populations largely contain adults at a maximum in size and numbers before the winter. The length of the hunting season may vary on the basis of updated population surveys. The amount of the take is recorded annually for large game species, and age and sex are determined when possible. Every effort is made to gauge the condition of the species so that the take can be properly controlled or prohibited for the next season. Some species such as the dodo and the passenger pigeon were overhunted, and this contributed to their extinction. It should be pointed out that 33% of all mammals and 42% of all birds that have become extinct since 1600 have disappeared because of overkill.

Problems arise when hunting goes unregulated, as it did before 1900. Animals were shot strictly for sport, for their plumage, or for their taste. The American bison and the passenger pigeon are two examples. In the late 1800s and early 1900s, however, legislation was enacted to control some of these abuses.

Ecologically, hunters assume the role of predators, which are now missing in many cases. Hunting of deer may prevent overpopulation and later starvation within a herd. Thus hunting does have a place in the scheme of things.

Species Protection and the Law

One way to protect species is through government policies. The United States is one country that has passed a species protection law. In the United States and elsewhere, pressures come to bear on the implementation of such laws.

In 1973, Congress passed the **Endangered Species Act.** This act sets out procedures for having species listed as endangered or threatened. The Endangered Species Act defines as **endangered** any species that is in danger of extinction throughout all or a significant portion of its range and as **threatened** any species that is likely to become an endangered species within the foreseeable future throughout all or a significant portion of its range. The act also provides that federal agencies may not proceed with or initiate projects that "jeopardize the continued existence" of endangered or threatened species. This includes private projects using federal monies or requiring federal permits. A serious challenge to the effectiveness of the act occurred early on. It pitted a fish against the Tennessee Valley Authority.

The snail darter, a small fish belonging to the perch family, was discovered in the Tennessee River in an area to be impounded by the Tellico Dam, a multimillion-dollar TVA project in Tennessee. In 1975 the snail darter became listed as an endangered species. Soon lawsuits were filed to stop the completion of the Tellico Dam because impounding the water would allegedly lead to the disappearance of the snail darter, which is adapted to running stream water. Attempts were made at that time to transplant the snail darter to the Hiwassee River. In June 1978 the Supreme Court ruled in favor of the snail darter and against the TVA. As a result of this and other controversies, Congress amended the Endangered Species Act to establish a special committee to resolve such conflicts. In January 1979 this committee also ruled in favor of the snail darter and against completion of the dam, primarily on *economic* grounds—the costs of the dam did not justify the benefit.

After the Supreme Court decision, after the amendment to the law, and after the committee finding, Congress passed legislation exempting the Tellico Dam from the Endangered Species Act and funded the dam's completion. The provision was attached as an unobtrusive amendment to an appropriations bill for energy and water projects. In late 1979 the gates of the Tellico Dam were closed, and the reservoir began to fill.

The transplant of the snail darter to the Hiwassee River was a success. In 1984 the snail darter was also found to be thriving in four branches of the Tennessee River and in streams in Alabama and Georgia. The U.S. Fish and Wildlife Service has since upgraded the status of the darter from endangered to threatened.

The Endangered Species Act was reauthorized in 1982. Opponents of the act wanted only "higher life forms" protected and organisms such as mosses, molds, algae, and invertebrates excluded from protection. Others wanted economic and not just biological factors considered in whether or not a species should be protected. Others proposed weakening of the restriction on federal projects that jeopardize protected species. These proposals threatened the very essence and spirit of the act and would have made it much less valuable in protecting species.

Congress did not weaken the act. The 1982 reauthorization streamlined the process for listing species. It allowed greater flexibility in regulating incidental taking of species when continued existence of species was not at stake. It provided for protection of endan-

gered plants on public lands. And it provided for citizen lawsuits in defense of endangered species.

In 1987, Congress extended the act through 1992 and strengthened it in several ways. It increases funding significantly, provides greater protection for plant species, and raises fines for violations of the law. Species waiting to be listed will be monitored more closely in case extinction becomes imminent.

A 1988 assessment of the implementation of the Endangered Species Act by the U.S. General Accounting Office found a need for a more centralized data system. It also found that recovery plans were lacking for several species and that plans were not regularly monitored or updated. Additional studies are under way (see Enrichment Box 14.1).

There is no doubt that much is at stake and that the debate on the future of the Endangered Species Act will involve many interests. A major area of discussion will be the balance between economic interests and endangered species preservation. As we discuss later, the habitat of the northern spotted owl conflicts with logging operations in the Pacific Northwest. The sockeye salmon was declared endangered in late 1991. Because the region in which the salmon is found is dependent on hydropower as its source of electricity, any requirements to vary water levels in reservoirs to assist the spawning runs of the salmon could lead to increased electricity rates. The Fish and Wildlife Service is studying the need to list the gnatcatcher, a small bird that inhabits the California coast, as threatened. This could significantly affect construction along the coast—concerning new homes and jobs. The ability of Congress to maintain a strong national policy toward species protection will face a major test.

Regulation of endangered species is also conducted at the international level. In 1973 the **Convention on International Trade in Endangered Species of Wild Flora and Fauna (CITES)** was written to prohibit commercial trade in endangered species, restrict trade in threatened species, and regulate trade in protected species. Applicable species are listed as appendixes in the agreement. Currently, 60 countries, including the United States, have ratified the agreement. It is this agreement that banned trade in elephant ivory as of January 1990.

Species Preservation: The Condor and the Red Wolf

Intense efforts have gone into saving individual species. The California condor and the red wolf are two such examples.

The condor, a huge bird with a wingspan of almost 10 feet, has been around since the last ice age. It once ranged throughout the western United States. Loss of habitat, hunting, lead shot, egg collectors, utility power lines, oil sumps, and pesticides all have contributed to its demise (Nielson, 1987). In the late 1930s there were about 100 condors in existence. In 1985 there were 26 condors in captivity and one in the wild known as AC–9 (Adult Condor–9). Condors are now extinct in the wild. The story of attempts to save the condor has been a story of controversy. There have been two views on how to proceed. One was a hands-off approach designed to save the condor solely by protecting its habitat. The other was a hands-on approach to learn more about the species and orchestrate its survival. The hands-on approach involved tagging the birds, providing them with a tracking device, and visiting their nests. As the number of condors in the wild continued to fall, the hands-on approach became the only viable alternative. Whether or not it will save the species remains to be seen. Eggs were taken from nests, and hatchlings have been raised successfully in zoos. The tracking devices have provided information on the range and habitat of the condor. There has been some success in placing closely related Andean condors, raised in captivity, in flocks living in the wild.

The red wolf is another species preservation story. It is the story of the return of a near-extinct species to its natural habitat. The red wolf once roamed wild throughout the southeastern United States. It is smaller than the gray wolf but larger than the coyote, both closely related species. Red wolves are not all red in color; they may vary from black to cinnamon to gray. The decline of the red wolf was the result of a variety of factors, including hunting for bounties, poisoning by U.S. government–supported programs, and loss of forest habitat. In Texas, where the range of the red wolf overlapped with that of coyotes, interbreeding occurred. In 1967 the red wolf

---

Enrichment Box 14.1

Good News for the Bald Eagle

At the time of this writing, the Fish and Wildlife Service was considering upgrading the status of the bald eagle from endangered to threatened. Since the early 1970s, about $25 million has been spent to try to save the national symbol of the United States. The bald eagle population increased from fewer than 1,500 birds in the lower 48 states in 1974 to over 11,500 in 1989. Only about 20 other species, including the American alligator, have ever been reclassified upward, from endangered to threatened status.

Zoos: Are They Humane?

Some people see zoos as animal prisons that cause neurotic behavior and suffering. Others see zoos as important to the long-range protection of endangered species through education and in other ways. Do zoos have a place in maintaining species diversity and educating the public? What do you think?

was declared an endangered species, and in 1977 the Fish and Wildlife Service established a recovery plan. Red wolves were captured in the wild and bred in captivity. To ensure a pure strain of red wolf, individuals were carefully studied prior to breeding. Offspring were checked to be sure there were no coyote characteristics present, indicating a hybrid wolf. The breeding program took several years. When pairs were ready to be introduced into the wild, a problem arose as to where to put them. Local opposition caused a change in plans several times. The stigma of the red wolf as livestock predator remains. Finally, in 1987, four pairs of red wolves bred in captivity were released in the Alligator River National Wildlife Refuge in North Carolina. The wolves lived in pens at first to acclimate them to their new surroundings. A dry dog food diet was replaced by red meat and eventually by live prey. Releases to refuges in North and South Carolina and Louisiana were made by late 1989. The red wolf population increased to 25; about half were born in the wild. The success of returning the red wolf to the wild will be closely monitored by the Fish and Wildlife Service as well as by other agencies. Some observers are quite skeptical that an animal bred in captivity can be successfully returned to the wild.

Zoos. Zoos play an important role in the survival of endangered species. Many have captive breeding programs to increase the populations of endangered species and, by preventing too much in-

breeding, to keep a variable gene pool. Some, like the National Zoo in Washington, D.C., are looking at high-tech reproductive biotechnology to improve the long-term survival chances of some species. This includes in vitro fertilization, freezing of sperm and eggs (cryopreservation), and use of surrogate females from closely related species to carry an implanted embryo to term. Several other species have been reintroduced into the wild from captive breeding programs. These include the Arabian oryx, an antelope reintroduced into Oman and Jordan, and the golden lion tamarins, released into the rain forests of Brazil. About 100 zoos in North America, Europe, and Australia conduct captive breeding programs, and plans are to expand these efforts. But the capacity of zoos and their resources for this purpose are limited. Captive breeding programs have not always been successful. The ability of a species raised in captivity to be successfully introduced into the wild is mixed. Biotechnical approaches are relatively new and will require much additional research and experimentation. Although zoos provide one important part of the effort to save endangered species, they are not the ultimate solution.

Habitat Protection

Some General Considerations
Some people argue that all of the money, time, and effort being put into saving a single species could

be better spent saving large expanses of habitat and, in that process, large numbers of species. The problem of biological diversity can be addressed on many levels from ecosystem and community to population and species. Some observers believe that "emergency room conservation" puts large monetary resources into a few species whose chances of survival are minimal anyway (Scott et al., 1987). Perhaps we should concentrate on preserving and restoring habitats rather than individual species.

As we have already mentioned, loss of tropical rain forests threatens two-thirds or more of the species on earth. We cannot explore and discover all of the species in the tropical rain forest as fast as they are disappearing. The immediate solution is to identify the undisturbed areas of greatest species diversity and work to protect them. Computer maps and models, along with data from remote sensing satellites, can be applied toward finding areas most in need of protection now. For example, fairly good data is available on the distribution of known vertebrates. Distribution maps can be overlaid with maps of existing habitat preserve areas and potential preserve areas (see Scott et al., 1987). Those areas with the greatest variety of species can be identified and earmarked for further conservation efforts, as can those areas in greatest danger of encroachment by development. This type of analysis is

already being done, and it has shown that more efforts of this sort can be fruitful. An ecosystem approach to the problem of loss of biological diversity can protect common species and save endangered species in an efficient use of available resources. Information, however, is sorely needed on how much area is required for a population of a certain size to be maintained. More research in this area could provide conservationists with extremely important data from which to develop strategies to protect natural areas.

Restoration of Habitat

While efforts to protect existing habitat are being made worldwide, other efforts are under way to attempt to restore native habitat that has already been lost. Restoration of habitat is not an easy undertaking. In some cases, natural vegetation must be planted in the normal sequence of successional stages to prepare the area for the climax vegetation. In some habitats, fire is encouraged—in prairie areas, for example— while in others the habitat must be protected from fire or other human-made disturbances.

In Costa Rica, University of Pennsylvania biologist Daniel Janzen is attempting to reestablish dry tropical forest in the Santa Rosa National Park. This effort will involve not only replanting and reintroducing

For Further Thought . . . 14-3

Species versus Habitat Protection

It has been argued that we put too much effort and money into saving individual species and that we should concentrate more on habitat protection. What do you think?

animal species but controlling other factors—such as fires, livestock grazing, and hunting—as well. The hope is that in time, natural regeneration will diminish the need for active human management. However, some experts believe that active human management will of necessity be a continuous part of restored habitats. In the United States a major prairie restoration project is ongoing at the Fermi National Laboratory (Fermilab) in Batavia, Illinois. Already 180 hectares have been restored, and native animal species are being reintroduced.

Habitat Preservation: Wetlands

As the significance of wetlands has come to be realized, efforts have been made to protect them. There are several U.S. laws that bear on wetland protection.

The **Coastal Zone Management Act** encourages protection of coastal wetlands by providing grants to states for developing and implementing coastal management plans. The act also established the Estuarine Sanctuaries System. Several sanctuaries have been established with 50% matching grants available to states from the federal government.

It is expected that by the year 2000, fully 80% of the population of the United States will live within 50 miles of an ocean or the Great Lakes (Bohlen, 1990). Some observers question the adequacy of the Coastal Zone Management Act under such population pressure. Changes were made in the act by Congress. These changes put all activities, including federal, under the state management plans unless the president exempts activities because they are in the "paramount interest" of the nation. States are also required to develop programs to control pollution runoff that affects coastal areas.

The 1986 **Emergency Wetlands Resources Act** provided $40 million a year for wetland purchase. The Fish and Wildlife Service is to establish wetland acquisition priorities. Acquisitions from public and private sources are not expected to be able to protect nearly the amount of wetlands that should be protected.

There are several other acts that have some impact on wetlands, but the primary legislative tool available for wetland protection in the United States is Section 404 of the Clean Water Act. Section 404 requires that any person must have a permit from the U.S. Army Corps of Engineers to discharge dredged or fill material into the waters of the United States. Although use of Section 404 has been somewhat successful in protecting coastal wetlands, it has not been so useful in protecting inland wetlands, which make up the bulk of remaining wetlands in the United States. There are several reasons for this. There is some controversy over what extent Section 404 can be used on inland wetlands. New development and construction activities are definitely covered by the section. Normal farming and timbering activities are not covered unless they include discharges that will restrict the flow of U.S. waters. Eighty percent of wetland loss has been related to conversion of wetland to farmland according to the U.S. Fish and Wildlife Service (Tripp, 1986). Draining, excavating, or flooding of wetlands is not explicitly addressed in Section 404.

To issue a Section 404 permit, the Corps of Engineers is bound by guidelines issued by the U.S. Environmental Protection Agency. The guidelines provide that no permit shall be issued if there is a practicable alternative that will not affect the aquatic environment. Consequently, permits to fill wetlands for projects that are just as suitable on nonwetlands should not be issued in most cases. If a permit is issued, the project is to be done in a manner that will minimize the impact on the wetlands, or else some compensation for loss of the wetlands is required. However, all of this was open to interpretation. Often the interpretation of the EPA and that of the Corps of Engineers did not coincide. The EPA can veto the issuance of a permit if it finds that the activity would have an adverse impact on water supplies, fish and wildlife, or recreation. In 1986 the Office of Wetlands was created in the EPA to carry out responsibilities in regard to Section 404. In 1989 the Corps and EPA signed a memorandum of agreement (MOA) setting out uniform procedures for applying this wetland destruction mitigation policy. It was revised in 1990 because of outcries from the regulated interests. However, it remains an important instrument for consistent issuance of Section 404 permits.

There does seem to be general agreement that this case-by-case permit issuance activity under Section 404 is not adequate in the long term to protect wetlands. Some authorities for *regional* planning efforts already exist. They include advanced identification by the EPA and the Corps of Engineers of lands suitable and unsuitable for Section 404 permits. The Coastal Zone Management Act provides for the development of Special Area Management Plans (SAMPs) to protect critical coastal wetlands. Additional provisions for protecting wetlands were included in the 1990 farm bill. However, none of these tools is being used to its full potential.

Recently, once again wetlands have become a major focus of Congress. In 1989, federal agencies (Army Corps of Engineers, EPA, Fish and Wildlife Service, and the Soil Conservation Service) adopted a manual that established a uniform method for identifying wetlands subject to Section 404 based on hydrology, soils and vegetation. Varying interpretations of the manual caused much concern that the methodology had significantly expanded the wetlands subject to

Section 404 permitting. The furor resulted in a movement to revise the manual and also reopened the issue in Congress. Wetlands policy is the subject of several bills in Congress. The Bush administration established a goal of "no net loss of wetlands." How this will be carried out, especially in light of its position in the current debate, remains to be seen.

Protecting Tropical Forests

The seriousness of the problem of tropical deforestation is appreciated by many of the countries involved. Many have made efforts to address the problem. Some of the factors that hinder their activities include lack of data, lack of trained conservationists, poor economies, and lack of a coordinated long-term strategy. It is interesting that in relative terms, 25% more land is protected in South America than in the United States (three times as much in actual total acreage) (Mares, 1986). Some countries have been more inclined to take action than others. Chile, Colombia, Ecuador, Peru, and Venezuela have all set aside much more land proportionately than has the United States; Brazil, Argentina, Uruguay, French Guiana, and Guyana have set aside proportionately less. All of the countries have laws to protect habitat, plants, and animals. But the ability to use these tools to prevent deforestation is restrained due to poverty, growing populations, poor economies, and huge debts. Forests are cleared for slash-and-burn farming, either to eke out a subsistence or to grow crops for export. As we saw in Chapter 13, the soils in most tropical rain forests are poor for farming. They soon lose their fertility, and the farmer moves on to another area. Forests are also being cut for firewood, a major energy source for the people. Some forests are being cleared for cattle grazing. Ghellean T. Prance of the New York Botanical Gardens states that 5 acres of Amazonian jungle can support 1,200 individual trees representing 200 different species—or one cow (Myers, 1987). With poor economies and large debts, governments that set aside land for parks and reserves lack the funds to manage them properly. In addition, the lack of data on a system as rich and significant as tropical rain forests is appalling. Again, poor countries lack trained researchers and conservationists to acquire the data needed. It takes money for research and for the development and implementation of plans to manage the rich resource. Development and implementation of a plan require political stability, also a problem in many Central and South American countries.

Costa Rica is one of the model countries in Central America. Its government has been stable, and it has set aside 27% of the country's land as protected national park and reserve areas. But Costa Rica has one of the highest deforestation rates in Latin America. Most of the land outside the reserves has been cleared or altered. The nation's problems lie in rapid population growth, unemployment, decreased demand for its exports, increasing national debt, and exploitation of its natural resources (Tangley, 1986).

Aware of this problem, the government has established a national biodiversity institute. The purpose of the institute is to inventory the rich diversity of the country, with emphasis on plants and insects. Birds and mammals have been fairly well inventoried already (Tangley, 1990). Ultimately, it is hoped that such an inventory will open the way for research into products useful to society. Such uses make protection much more likely in the long run.

Loss of tropical rain forests is a global problem. As such, the countries with this rich resource cannot be expected to bear the monetary cost of protecting rain forests on their own.

Addressing a Global Problem. In 1985 an international plan sponsored by the World Bank, the World Resources Institute, and the U.N. Development Programme estimated a cost of $8 billion over five years to stem tropical forest deforestation. The report, *Tropical Forests: A Call for Action,* set out specific recommendations for action in 56 countries. Five issues were included in the report: fuelwood and agroforestry, land use on upland watersheds, forest management for industrial uses, conservation of tropical forest ecosystems, and strengthening institutions for research, training, and extension. It is a promising sign that the world community has begun to realize its responsibilities in regard to addressing tropical reforestation.

In recent years the discipline of **conservation biology** has emerged with an understanding that preservation and restoration of habitat will indeed require active management; this discipline combines genetics, ecology, natural resource management, and several other disciplines to be applied to the primary goal of conserving biological diversity. Some experts have pointed out the need not only to send expert assistance to the tropics but also to train people in tropical countries in conservation biology and its techniques. There is an awareness of the need for research that will look at conservation of ecosystems, not just individual species.

As far back as 1963, the University of Costa Rica established the Organization of Tropical Studies (OTS) to join with other universities throughout the world to train field biologists at the tropical rain forest preserve in Costa Rica. More than 44 universities now operate within the OTS, and increasing numbers of biologists from countries suffering major destruction of their tropical forests are being trained.

In 1979 the Minimum Critical Size project was implemented in Brazil in coordination with the World

Wildlife Fund and the National Institute for Amazonian Research to study the effects of cattle ranches on species extinction due to the isolation of segments of forests (Wolf, 1987). Studies of tropical forest regeneration in southern Venezuela are being undertaken to see how tropical forests return to abandoned agricultural lands. Indications are that the full recovery of tropical forests after slash-and-burn farming will take 150 years. Such studies, though not always optimistic, are essential and provide baseline data on which to make management decisions.

What is even more promising is an awareness in the political and economic realm that preservation of natural habitats and species is important. A new policy established by the World Bank in 1986 focused on a commitment to protect wildlands in the projects it selects to receive World Bank financial assistance. The policy states that the World Bank will not finance projects in most cases that convert lands of exceptional biological diversity or environmental significance; where possible, projects will be sited on already converted land; where this is not possible, sites of lesser wildland value will be selected. Where significant natural areas are converted, an ecologically similar area will be preserved *elsewhere*. The World Bank cites the value of retaining wildlands as economically beneficial in regard to drought and flood control, agriculture, tourism, fisheries, and conservation of biological diversity.

In 1990 the World Bank reported that 50% of the new loans approved in the year before had an environmental component. Although some observers are skeptical of the World Bank efforts, the fact that the Bank felt the need to make such a statement indicates a growing environmental awareness. Good environmental management is good economic management as well. The significance of an economic and development-oriented group like the World Bank issuing such a policy certainly helps drive this point home.

Another interesting economic phenomenon that has emerged is the trading of debt for conservation. Third World countries have run up enormous debts. Banks in the United States know that repayment of these debts is unlikely because of the continuing poor economies. Private organizations have attempted to have debtor nation debt reduced in return for a guarantee by the debtor nation to protect certain habitats, parks, or regions in the country. Conservation International in Washington, D.C., was the first to do this, for Bolivia. With money from a private foundation, Conservation International reduced Bolivia's debt by $650,000 by purchasing a $650,000 note at an 85% discount. In exchange, Bolivia will protect 3.7 million acres in the Amazonian forest and set up a fund to manage it.

Costa Rica established the Guanacaste National Park through this same process. Although the area has been cleared for cattle ranching, attempts will be made to regenerate the dry tropical forest ecosystem in the area. Since the loss of tropical forests has to some extent been because of the need to improve the economy, the debt-for-conservation approach tends to interrupt the vicious circle of habitat destruction for short-term economic gains.

This debt-for-nature swap is especially applicable in Africa, where 80% of the debt is publicly held; it is owed to governments and international financial institutions, not to private banks. Canada has converted over $500,000 of development assistance loans to sub-Saharan Africa to grants, thus eliminating the debt on what were initially low-cost loans.

General international organizations are already involved in attempting to identify and protect critical natural areas worldwide. The International Union for Conservation of Nature and Natural Resources (IUCN) is heading an international effort to establish a world system of reserves and to improve ecosystem management. Nations that are party to the Convention Concerning the Protection of the World Cultural and Natural Heritage have designated 57 natural sites as World Heritage Sites.

The **Man and the Biosphere Program (MAB)** is an intergovernmental organization established through the U.N. Educational, Scientific and Cultural Organization (UNESCO) to focus research, technical training, and public education on the need to know more about the intricacies of the whole-world biosphere. In its most basic form, the purpose of MAB is to study the earth's ecosystems and how we relate to them. The work of MAB is coordinated by an international council composed of representatives from 30 nations elected by the UNESCO General Conference.

One of the projects of MAB is the Biosphere Reserve Project. This project recognizes that each of the earth's ecosystem types is different and that each must be preserved and studied if we are to understand the whole. Through this project, reserves are being established and recognized throughout the world. There are currently almost 300 biosphere reserves in more than 60 countries; 41 are in the United States.

Many countries, including Brazil, Peru, and Indonesia, have established national parks and reserves. Private organizations similar to the Nature Conservancy in the United States are also emerging in other countries. The **Nature Conservancy,** founded in 1951, is a private, nonprofit international membership organization dedicated to the protection of critical natural areas. It employs a variety of legal tools to protect such lands from development. These may involve outright purchase of land, trading of lands, or acquisition of conservation easements. Often the Conservancy attempts to find local entities, public or private, to manage the areas it acquires. The organization is responsible for the protection of over 3 million acres in the United States, Canada, Latin America, and the Caribbean.

Regardless of the approach taken, there is increasing agreement that attention must be given to the needs of the local population bordering reserve areas. In the long run, reserves will not be safe if growing populations demand food and fuel. People's short-term needs must also be met. Thomas Eisner (1989) has proposed that efforts be made to "prospect for nature's chemicals" in these tropical areas. By this he refers to the need to develop uses for some of the chemicals found in plants in tropical forest habitats. The results can be monetary returns for the country, the development of research facilities in those areas, the targeting of research to problems in those areas, and ultimately to greater support for conservation.

Federal Management of Biological Resources

Government policies do have a major impact on biological resources. In addition, governments themselves are stewards and managers of public lands. We will focus on the U.S. government as we further explore the issues of biological resources on public lands.

As the U.S. government has attempted to manage the nation's biological resources, much controversy over strategy and implementation has been generated. We would now like to examine the involvement of the federal government in biological resource management, looking in turn at rangeland management, management of national forests, national parks, and wilderness.

Rangeland

The largest U.S. public landowner, the Bureau of Land Management (BLM), has managed public lands for decades. Its management has been controversial. Generally, the BLM was directed to follow the policy of multiple use and sustained yield (see Chapter 13). Many people question its success in doing this.

A prime example of controversy over BLM has been its management of federal rangelands.

The problem of overgrazing of public lands is not new. Even in the late 1800s and early 1900s, reports of overgrazing and destruction of rangeland were prominent. In 1934, frustrated by the inability to control grazing on public land, the **Taylor Grazing Act** was passed. It closed the public domain to further settlement and set up grazing districts with local advisory boards to manage public-domain rangelands. The local boards were strongly influenced by the stockmen. The districts were organized under the Federal Grazing Service, which became the BLM in 1946.

BLM rangelands have been open to grazing by privately owned livestock. It has been estimated that about 27,000 ranchers graze 8 million head of livestock on public lands. It has also been estimated that one-quarter of the grazing is done by livestock owned by fewer than 2% of the ranchers. BLM does charge these ranchers a nominal grazing fee, but this fee has traditionally been about one-fifth the fee charged for grazing on private lands. It has provided a few large ranchers with very inexpensive grazing opportunities. This has led to deterioration and overgrazing of public rangelands. Lawsuits requiring the BLM to assess the environmental impact of grazing on public lands led to the passage of the **Public Rangelands Improvement Act of 1978.** This act directed that grazing fees be increased and that a study be done to arrive at an equitable formula for assessing grazing fees after 1986. Half of all monies collected from the fees were to go into rangelands improvement. However, the Act was not effective as later studies show.

A 1988 assessment by the U.S. General Accounting Office found that although most public rangelands are stable or improving, 50% of rangelands are categorized as being in poor or fair condition, and emphasis is not being placed on protecting the acreage that is declining and overstocked. The study found that 19% of the grazing allotments permit grazing at levels higher than the land can support over the long term. A 1989 report issued by the National Wildlife Federation and the Natural Resources Defense Council concluded that there had been no significant improvement in rangeland habitat since 1985 and that none was likely under present policies and funding.

A movement is under way again in Congress to raise grazing fees to market levels.

National Forests

Whereas the Federal Land Policy and Management Act addressed land use on federal lands controlled by the BLM, lands under the Forest Service are governed by other laws. The Forest Service regulates both forest and grasslands. In recent years the policy for managing forest lands has been set by the **Forest and Rangeland Renewable Resources Planning Act (RPA) of 1974** and the **National Forest Management Act (NFMA) of 1976.** These acts direct that a long-range plan for the management of U.S. Forest Service lands, including some rangelands, be developed so that these resources will be available on a sustained basis. The acts also set up a process for evaluating the status of forest and range resources, that is, a method for assessing how well the plan is working. Before this act, plans for the forest were localized, and no comprehensive overall program was in operation. According to the act, comprehensive multiple-use plans were to have been developed for all forests and grasslands by 1985.

More than 50 of the final plans have been appealed. Appeals have cited emphasis in the plans on increased commercial activities such as timber cutting and grazing, increased road building, and a passive approach to other Forest Service mandates, including watershed protection, recreation, and wildlife resources. One such plan for the Ouachita National Forest in Arkansas is especially controversial. Ouachita is the oldest and largest national forest in the South, comprising about 1.6 million acres. The plan opens up two-thirds of the timber base to cutting and permits clear-cutting on 844,000 acres (Danforth, 1987).

The U.S. Forest Service has been criticized over the years primarily because it has been perceived as being overly concerned with timber production and less concerned with its other charges, including watershed protection. The issue of clear-cutting of national forests epitomizes this controversy.

Clear-cutting is a forest-harvesting technique whereby all of the trees in a given area are leveled. The land is scraped, and trees are replanted, resulting in an "even-age" stand of timber (a stand where all of the trees are about the same age) for the next cutting cycle. This technique provides an environment for the regeneration of sun-tolerant species, especially evergreens. Timber harvesters like it because it is relatively quick and easy and accommodating to machinery in comparison to **selective cutting,** whereby the stand of timber must be surveyed to mark mature trees of economic value, which then are cut, leaving a mix of tree sizes and types for the next cutting cycle.

Clear-cutting mars the landscape aesthetically; this is what brought it to the attention of the public. It has been used inappropriately for hardwood species, resulting in a shift from mixed hardwood (deciduous) to softwood (evergreen) species. Clear-cutting over large expanses causes destruction of habitat, erosion, and sedimentation in streams. Although the even-age stand resulting from the replanting of clear-cut areas is beneficial in terms of the next harvest, the benefits can be short-lived. Monocultures lack diversity and are susceptible to pest invasions. This fact and soil loss will eventually catch up with the timber owner of a patch of woodland that is repeatedly clear-cut.

It should be noted that on a small scale (30 acres or less), when employed in a staggered or checkerboard fashion and where erosion is not a major problem, clear-cutting may not result in significant environmental damage. Clear-cutting may even provide

When done carefully, on a small scale, clear-cutting may be a useful forest-harvesting technique. However, it can lead to erosion, sedimentation, and habitat destruction.

habitat, expanding edge effects within large wooded expanses (see Chapter 4). However, use of this technique in inappropriate locations and other large-scale misuses have resulted in major environmental damage and calls for the banning of all clear-cutting. In the long run, better management practices at the expense of convenient harvesting will allow for a longer-term timber operation—sustained yield as Congress has mandated.

The U.S. Forest Service policy on logging on public lands has come to a head in the Pacific Northwest with the controversy over logging old-growth forests and maintaining habitat for the northern spotted owl. We will briefly discuss this controversy because it is symptomatic of the issues at stake on both sides. The issue is one of national policy. Do we want to protect old-growth stands of timber and the species they house on public lands, or do we want to harvest them?

Old-Growth Forests. As defined by the U.S. Forest Service, **old-growth forests** are characterized by trees that are at least 200 years old. Such forests have other characteristics as well. They are a complex ecosystem and house a great diversity of species. Some experts believe that like tropical rain forests, they hold as yet unknown sources of natural pesticides and medicines that may someday prove useful to human society. In addition, they also house the data bank on forest ecosystems over time. Once they are gone, we have no laboratory to simulate the same natural relationships. About 87% of old-growth forest in the United States is gone. Concern is also that the stands that do exist are becoming so fragmented that the stand in any one area is not sufficient to accommodate the needs of the species that live there. One of the many species that calls old-growth forests home and that cannot likely survive without it is the northern spotted owl.

The Spotted Owl Controversy.
The Forest Service studied the northern spotted owl because it is considered an indicator species of a healthy forest. (See Enrichment Box 14.2) The agency determined that the owl was facing extinction. Lawsuits were filed using the Endangered Species Act to curtail logging in areas inhabited by the spotted owl. The lawsuits resulted in injunctions on logging. The industry claimed that thousands of jobs would be lost.

Because of the alleged severe economic implications and the power of the forest industry lobby, Congress passed legislation to declare that the courts should assume that the logging practices of the federal government (in this case, logging allowed by the Forest Service) complied with environmental laws. This legislation was found to be unconstitutional. The courts said that Congress could modify or waive environmental laws but could not interfere with a court's ability to review compliance with the law.

Prior to that ruling, the Fish and Wildlife Service officially declared the northern spotted owl a threatened species. Without further congressional action, then, the owl and its habitat fell under the protection of the provisions of the Endangered Species Act.

The spotted owl is not the only species dependent on old-growth forests—it gained the limelight because of its formal threatened status. The marbled murrelet, Vaux's swift, the northern goshawk, the pine martin, the Olympic salamander, the tailed frog, and many other animals are all in the same predicament (Corn, 1990).

The Timber Impact.
Even before the spotted owl controversy, jobs in the timber industry were being lost in the Northwest. According to one study, more than half of the job losses were the result of automation, about 20% were due to reduced logging from increased protection of forest areas, and 25% were jobs that were transferred overseas (Stiak, 1990). Even though jobs decreased, the amount of timber harvested almost doubled in the 1980s. The industry sees old-growth forests as decadent environments that can be harvested to provide jobs and goods. Furthermore, the timber in old-growth forests is some of the best available. A complicating factor is that the Forest Service and the Bureau of Land Management obtain significant revenue from timber sales in the Pacific Northwest. This causes a conflict of interest when it comes to curtailing timber sales.

In 1991, the Fish and Wildlife Service halted proposed timber sales on 44 of 175 tracts of federal lands proposed by the Bureau of Land Management because they affected the habitat of the northern spotted owl. The Bureau of Land Management requested the Secretary of the Interior to convene the Endangered Species Committee composed of Cabinet officials which can authorize exemptions to the Endangered Species Act, and permit the proposed timber sales to proceed. The

---

Indicator Organisms

Although we have designed and developed sophisticated technological means of monitoring pollution, one of the most effective ways of monitoring the condition of an ecosystem is studying the status of certain species in that system. Because these biological monitors can catch synergistic impacts and bioaccumulation effects, they are in some respects superior to machines. The concept is not new. Many years ago, miners carried canaries into the mines with them; if the canaries died, it meant that "bad air" was around and it was time to leave.

In some cases, certain species can be monitored for specific chemical content. Bees are used to sample for a wide variety of pollutants that might be picked up in the air and in dust and be returned to the hive. Earthworms are studied to determine soil contamination. They pick up numerous contaminants as they eat their way through the soil. Likewise, shellfish continuously pull water through their systems and filter out food and with it pollutants. They can be analyzed for contaminants.

Some organisms are monitors by their presence (or absence) in an ecosystem. Amphibians, for example, are good overall indicators. Frogs and their kin can absorb contaminants from soil and water through their skin; they eat insects that may contain contaminants. They have complex life histories involving the egg, larva, and adult forms. It is therefore alarming that amphibians have been disappearing at a rather rapid rate worldwide since the late 1970s. Is it an indication of overall environmental deterioration?

---

committee met and decided to suspend the Act to allow logging on 13 of the 44 tracts.

In the meantime, several proposals have been debated in Congress to balance the interests and needs of the timber industry and the spotted owl. The crux of the issue is U.S. policy on harvesting timber on public lands. The issue is similar to that faced by developing countries with tropical rain forests. The definitive answer has not yet been given.

New Forestry. The controversy over harvesting national forests has led to innovative approaches to addressing the issue. One such approach has been called **new forestry.** Its goal is to allow for harvesting while maintaining diversity and ecological integrity. The means by which this is to be done is by an ecosystem approach. The new forestry tries to maintain the ecosystem by providing for less disturbance of the area. Under new forestry, snags and woody debris, which have been shown to be important to invertebrates and fungi, would be left and not burned. Partial cutting is espoused over clear-cutting or selective cutting. In partial cutting, 85% to 90% of the trees are harvested. One drawback is that since not all the trees are harvested, larger areas might need to be partially harvested to arrive at the same number of board feet. The techniques of new forestry have not been definitively established. There is consensus, however, that forest policy and management must become more ecologically based for a sustained harvest.

National Parks

There are almost 300 million recreational visits to lands administered by the National Park Service each year. The lands administered by the U.S. Forest Service actually receive more recreational visits; U.S. Army Corps of Engineers facilities run second, and National Park lands are third. Americans need and use outdoor recreational opportunities.

Deterioration of national parks is a major concern. The National Park Service oversees 354 sites comprising more than 75 million acres.

Some of the factors leading to deterioration come from within the parks and some from without. A 1980 report by the National Park Service, called the *State of the Parks Report,* cites acid rain, air pollution, urban encroachment, and industrial and commercial development on adjacent areas and roads as external components of deterioration in the parks. There are also some internal components.

The purpose of national parks has been to preserve natural habitat and provide for its enjoyable use. In some ways this is a conflict of directives, and the Park Service has attempted to reconcile both over the years in several ways. From 1964 to 1977, parks were designated as historical, recreational, and natural. Each designation had its own set of priorities and management guidelines. Emphasis in natural areas focused on preservation, in historic areas on maintaining historical features, and in recreational areas on outdoor recreation. In 1977 this separate designation was replaced by a zoning scheme within each park whereby various priorities can be emphasized in each zone. Zoning is based on natural and cultural features. Guidelines for managing for the various types of units within the parks have been established.

But there are indications that the role of the parks in preserving natural habitat is not being met. Fourteen national parks in the western United States have lost 42 populations of mammals (Chase, 1987). This is believed to be related to a decline in natural habitats in the parks, which in turn is related to over-

---

For Further Thought . . . 14-4

What Should Be the Primary Purpose of National Parks?

This is becoming a major public policy question in the United States as many of its national parks are suffering environmental deterioration from overuse. Should the parks be preserves, protecting wildernesslike areas, or should they be used to meet human needs for interaction with the natural world? If use of park lands is to be restricted, on what basis and in what ways should use be restricted? Should limited financial resources be put into improving and maintaining the existing park system or in acquiring new park lands? How should pollution and other impacts from areas adjacent to national parks be handled? What are your thoughts?

---

use, development near park borders, and air and water pollution. In a 1990 report, the GAO found that the EPA's program to prevent significant deterioration of air quality in park areas was inadequate; 99% of the sources and 90% of the pollution were not regulated.

Chase (1987) cites the Park Service itself as a threat because its personnel, especially at supervisory levels, tend to be largely those with backgrounds in law enforcement, not resource management. The Park Service budget does not emphasize research. Chase maintains that parks are considered as complete ecosystems, with the management strategy being to do nothing. He cites that historically these park areas have changed dramatically since the time prior to the coming of settlers. Predators have been eliminated, new species of plants and animals have been introduced, and fire control practices have been imposed. Chase maintains that what national parks need is not protection but restoration. To do this would require the collection of baseline data (what were the park areas like in pre-Columbian times), an inventory of current species, the removal of exotic species, and the development of strategies to model conditions existing in earlier times in reference to population and species management. Chase bases his scenario on the report of the Advisory Board on Wildlife Management, set up in 1962 to look at wildlife problems in national parks. Chaired by A. Starker Leopold, the board concluded that a reasonable illusion of primitive America should be re-created in the parks. The conclusions of the board were interpreted and served as the basis of the "green book" issued in 1968 that contains National Park Service policies for managing natural areas. To date, restoration has not emerged as a priority of the Park Service.

Recreation is also low on the list of budget priorities, and this has led to a deterioration of facilities at national parks. The use of off-road vehicles (ORVs) has also contributed to the deterioration of federal recreational lands. ORVs include snowmobiles, motorcycles, and four-wheel-drive vehicles. ORVs tend to destroy vegetative cover, leading to erosion and other environmental damage. By executive order in 1977, President Carter directed federal agencies to control the use of ORVs to reduce environmental damage. Implementation has been and will continue to be difficult. There may be over 10 million ORVs in operation in the United States, and the expanses and types of land involved are not easily patrolled. To help in addressing some of the problems, Congress passed the National Park Access Act in 1978 to improve access to the parks consistent with preservation and good energy conservation. The act calls for the development of plans for mass transit within the parks and discouragement of private vehicles.

With the need for recreational lands increasing, Congress in 1978 also passed the National Parks and Recreation Act, which added to the lands designated as national parks and wild and scenic rivers. However, policy changes in the early 1980s focused on upgrading and maintaining existing federal parks and recreation areas rather than designating new areas. Proponents of new land designations cautioned that such a policy leaves potential future parklands open to development and aggravates the existing problem of park land overuse.

Interest in the issue of the availability of outdoor recreation areas for U.S. citizens caused President Reagan to establish the President's Commission on Americans Outdoors in 1985. The commission was made up of members of Congress, state and local governments, park officials, recreation industry representatives, and environmental representatives. The commission made its report in 1987. Among its recommendations were the following: The Land and Water Conservation Fund should be converted to a billion-dollar trust fund to pay for federal, state, and local land acquisition and the development and rehabilitation of recreational facilities; environmental laws must be strictly enforced at all levels of government; greenways should be established near metropolitan areas where most people live so that open space is readily accessible; congressionally authorized land acquisitions should be expedited. With an emphasis on action at the local level, the report proposed the creation of a new quasi-governmental institution to distribute grants to communities and local organizations to initiate local action. The report has been controversial. In fact, a lawsuit was filed *before* the report was published, to prevent its publication and distribution. The report was released by a private publisher, and the lawsuit was dismissed. Critics of the report fear that it endorses massive federal land acquisition and land use control. We are back to the old idea that anyone should be able to do anything with his or her own land (see Chapter 13).

Wilderness
Congress has set forth the policy governing wilderness areas in the United States.

> In order to assure that an increasing population, accompanied by expanding settlement and growing mechanization, does not occupy and modify all areas within the United States and its possessions, leaving no lands designated for preservation and protection in their natural condition, it is hereby declared to be the policy of the Congress to secure for the American people of present and future generations the benefits of an enduring resource of wilderness. For this purpose there is hereby established a National Wilderness Preservation System to be composed of federally owned areas designated by Congress as "wilderness areas,"

and these shall be administered for the use and enjoyment of the American people in such manner as will leave them unimpaired for future use and enjoyment as wilderness, and so as to provide for the protection of these areas, the preservation of their wilderness character, and for the gathering and dissemination of information regarding their use and enjoyment as wilderness; and no federal lands shall be designated as "wilderness areas" except as provided for in this Act. [PL 88-577, enacted September 3, 1964, Section 2(a)]

By its designation as a wilderness area, a given parcel of land is automatically eliminated from some land uses such as timber cutting and mining. Land is effectively taken out of production (in the common sense) and used entirely intact for educational, research, aesthetic, and limited recreational purposes. Some people feel that in an age of shortages, too much area is being designated as wilderness and that the concept of multiple use, which calls for federal land to accommodate many uses at one time rather than a single use, is a more realistic and sensible approach. Others see the use of true wilderness lands for mining, timber harvesting, or developed recreation as an irreversible land use decision. By statutory definition, wilderness is any area "where the earth and its community of life are untrammeled by man, where man himself is a visitor who does not remain."

The need for wilderness is difficult to document or to quantify. Consequently, challenges to wilderness designations have been loud and long.

A number of monumental battles over the issue of wilderness versus development erupted during the 1970s and continued through the 1980s. We will examine two: the Alaska lands issue and the designation of wilderness areas in the national forests.

The Alaska Lands and Arctic National Wildlife Refuge Issues.
Perhaps the major battle of the 1970s over an imminent wilderness designation was the battle of Alaska. Alaska encompasses an area of 375 million acres, an area so large that at one time it spanned four time zones. (In 1983, Alaska consolidated these into two zones.) Alaska's coastline is longer than that of the continental United States, and it has 365,000 miles of streams and rivers. The only U.S. populations of Dall sheep, polar bear, and musk oxen are found in Alaska. The state has numerous mineral deposits and expansive areas of forests. It has a population of 550,000 persons. These features all combined to put Alaska in the middle of a national controversy involving conservationists and developers and raised questions extending to states' rights and national security.

The controversy began after the **Alaska Native Claims Settlement Act (ANCSA)** became law in 1971. This act had come about because the purchase of Alaska by the United States from Russia did not provide for the land rights of native Alaskans. When oil was discovered in the arctic slope in 1968, the natives formed a coalition and filed land claim suits. ANCSA was enacted to deal with the issues raised. Under the provisions of the act, native Alaskans, including Aleuts, Indians, and Eskimos, were given the authority to select 44 million acres of land from the public domain.

Two especially controversial sections of that act are Sections 7(d)(1) and 7(d)(2), known popularly as D-1 and D-2. Under Section D-1 the secretary of the interior is to review all public lands in Alaska to see whether in the public interest any of them should be withdrawn from the native selections and from state selections that were authorized when Alaska became a state in 1959. (Alaska was given 25 years to select 104 million acres of land from the federal public domain for state lands. These selections were not complete when ANCSA was enacted.) Under Section D-2 the secretary was to designate 80 million acres for inclusion in the national park, national forest, national wildlife refuge, and national wild and scenic rivers systems.

Numerous ways of designating the Alaska lands were debated in Congress. The Alaska lands bill was finally signed into law in December 1980. Facing a new president to be inaugurated in January 1981, Congress settled for a compromise. As enacted, the bill set aside 56.4 million acres as wilderness and 49 million acres for various other designations including national parks, wildlife refuges, and wild river areas. The act also completed distribution of federal lands to the state of Alaska and the 44 million acres to Alaskan native groups. Although neither side—those in favor of protecting Alaska wilderness or those concerned with developing its natural resources—was completely satisfied, all concerned considered it a landmark after years of debate. The final decisions will lead to irrevocable changes in land areas heretofore substantially untouched.

As part of the compromise on the legislation, the Arctic National Wildlife Refuge (ANWR) was doubled in size to 18.1 million acres; 8 million acres were designated wilderness. The 1.5 million acres comprising the coastal plain were given no designation; rather, the Department of the Interior was directed to study the area for its oil and gas potential and to make a recommendation to Congress.

In 1987 the Department of the Interior made its recommendation that drilling for oil and gas in the coastal plain of the ANWR be allowed. Interior estimates that 600 million (95% chance) to 9.2 billion (5% chance) recoverable barrels of oil lie beneath this area. The mean estimate is 3.2 billion barrels. Needless to say, the recommendation has been extremely controversial. Oil companies support it, contending that they can drill and still protect the environment because of their experience in Prudhoe Bay and that oil operations would touch less than 1% of the coastal plain. They cite U.S. Fish and Wildlife data to show that calving areas used by the porcupine caribou are extensive, and they claim that drilling will not, overall, affect calving as caribou can use alternative areas without any effect. They say that caribou herds have prospered on the North Slope oil fields. In addition, drilling will decrease dependence on foreign sources of oil, improve U.S. balance of payments, and create jobs.

Opponents of allowing development of oil fields in the ANWR claim that Prudhoe has not been without its environmental problems. They cite thawing of the permafrost and flooding from road and building construction as affecting wetlands; air pollution; release of wastewater containing hydrocarbons, chemical additives, lead, and arsenic directly into the environment; and oil spills. They claim that the amount of oil available may be no more than enough to supply U.S. needs for six months. The EPA noted that the porcupine caribou herd has some significant differences from the central arctic herd. In the case of Prudhoe, calving grounds were largely outside the development area, the herd has a lower population density, and predation by wolves and bears has been minimized. The EPA criticized the environmental impact statement done by the Department of the Interior. They indicated that

---

Do Endangered Species Need More Protection?

If it is true that most species that have ever inhabited the earth are now extinct—and were in fact extinct long before we humans made our presence felt on the earth—extinction must be every bit as common as the origin of new species on this planet. In this light, some people argue that it is no big deal that humankind occasionally hastens the departure of species that were likely soon to be gone anyway. We cannot, they say, support the human population without destroying certain species-sustaining habitats. If the species that depended on these habitats have got to go, so what? they say. Others argue that we have no right to force other species out, and besides, they help to sustain the ecosystems that sustain us. Still others argue from a more human-centered point of view that we shouldn't force the demise of species because we may need them for some yet unknown reasons. What do you think our approach to the issue of endangered species should be?

---

issues relating to air pollution, noise, cumulative effects, marine transportation facilities, and water and gravel supply needs were not adequately addressed. The EPA criticized Interior's definition of "major" effects. The department defined an effect as major only if the disturbance lasted more than 30 years. The EPA cites the potential effect on 12,650 acres of wetlands as major, whereas Interior classifies it as minor. The EPA cites some inconsistencies where effects are called moderate in the text of the report and minor in the conclusions. Congress has not voted to allow oil exploration in the ANWR. In the wake of problems in the Persian Gulf in 1990, the governor of Alaska called for Congress to take action to open the ANWR for oil exploration.

Drilling in the ANWR is not the only issue of economic development versus environmental protection Alaska faces. With its huge expanse and resources, numerous confrontations occur at the state and local levels. In 1991, natives were allowed to sell their shares in the corporations that were set up under ANCSA. Any person with one-quarter native blood became a shareholder in a village and regional corporation. The 44 million acres of land were distributed to these corporations to manage the lands and cash settlement that were part of the law. Many of these lands are within the national parks and refuges. Many of the corporations are having financial problems. Native holdings were subject to state and local taxes in 1991. If many shares are sold to commercial interests, timber cutting, oil exploration, and other developments could threaten park and refuge lands.

The tension in Alaska is likely to continue for some time. There is much at stake.

Wilderness and the National Forest System.

There are 187 million acres of land in the U.S. Forest System. Under the 1964 Wilderness Preservation Act, the U.S. Forest Service was directed to inventory the 60 million acres of roadless areas and to make recommendations for designation of these lands as part of the wilderness system. Finally, in January 1973 the recommendations for RARE I (Roadless Area Review and Evaluation) were announced. They included 12.3 million acres to be designated wilderness, only 45,000 acres of which were east of the Rockies. Environmentalists challenged the findings in court. A second study, RARE II, was undertaken. More liberal criteria for "human impact" were used in eastern forests so that areas with some human impact could still be designated wilderness; the act provides only that humankind's imprint be "substantially" unnoticed. RARE II recommended that 15.4 million acres (577,000 east of the Mississippi) be declared wilderness areas, 36 million acres be nonwilderness and open for timber production, and 11 million acres remain under study. Generally, Congress has dealt with the wilderness designations on a state-by-state basis.

Wilderness and the Fish and Wildlife Service.

The Fish and Wildlife Service is responsible for managing over 450 refuges involving almost 90 million acres in the United States. Refuges were established to be managed primarily for the benefit of wildlife. A 1989 report by the GAO found that secondary uses of refuges such as recreation, mining, and grazing are overshadowing the primary purpose of the refuges. They are in some cases allowed, even though they are harmful to the wildlife. The report stated that the reasons for this situation include agency response to local pressure or pressure from economic interests. In some cases the agency does not have complete ownership and hence control of the lands.

In Closing

The "bottom line" message of this chapter is that we humans need to be careful to preserve the biological diversity of Planet Earth. We must assume that we are part of a seamless web of life, and we must behave as if the inadvertent elimination of any one species could be the beginning of an unraveling of the fabric of the biosphere. Not only must we be concerned about the overexploitation of individual species, but we must also be careful not to wipe out species by wiping out habitat. We must leave some expanses of land undisturbed, as wilderness; we may even have to restore some. This is obviously a matter of broad public policy, and there is reason for hope.

In March 1987 the U.S. Office of Technology Assessment stated the need for a comprehensive coordinated strategy to preserve biological diversity in the United States. In particular, the report stated, there is a need to conserve ecosystem types—from one-quarter to one-half of the ecosystem types in the country are not currently protected. The report went on to recommend that Congress establish a national policy, mandate a conservation strategy, and amend existing authorities to provide a coordinated effort to preserve biological diversity.

In the past several years, Congress has considered legislation relating to biological diversity in the United States, including many of the factors noted in the OTA report. The legislation that has been considered has included a national policy statement endorsing the conservation of biological diversity as a national goal and establishing a coordinated federal strategy for maintaining and restoring biodiversity in the United States. At the time of this writing, none of the proposals had been passed.

There has also been some enlightened movement at the international level. We described the International Man and Biosphere Program and the World Bank's interest in biological diversity earlier as just two excellent examples. The United Nations Environment Programme held an international conference on the environment in 1992 in Brazil, to examine the issue of biodiversity and protection of biological systems and to develop a biodiversity conservation strategy for the world, among other things (see Chapter 19). A treaty was signed by 153 nations in which they agreed to inventory and protect endangered species. As part of the agreement, *developed* nations commit to helping *developing* countries fund their programs and to share technology and profit when developed nations gain from using the genetic resources of developing countries. The U.S. did not sign the biodiversity treaty citing concern about its impact on the biotechnology industry in the United States. The humans of the world have apparently begun to recognize the importance of a global ecology and the seriousness of threats to biological diversity in the ecosphere. Recognition is the first step toward solving problems.

Concepts to Remember

1. A wide diversity of species provides an index of flexibility for the biosphere, i.e., the ability to adapt to changing conditions.

2. The potential benefits of genetic engineering are enhanced by the existence of a large gene pool.

3. By some estimates, human activities have helped to bring rates of species extinction close to the rates of the mass extinctions of the past.

4. Species preservation is important. Species diversity helps to protect the integrity of the ecosphere and serves as a resource for improving the human condition.

5. Species that are introduced into new environments by accident or for specific reasons often have no local predators or controls and threaten the survival of native species.

6. Zoos provide a resource for protecting individual members of threatened or endangered species by captive breeding programs. In some cases this is done with the purpose of introducing offspring to the wild; success has been mixed.

7. The destruction of habitat leads to the elimination of species. Even the fragmentation of habitat into smaller parcels leads to elimination of species.

8. Wetlands are among the most biologically productive ecosystems on earth. They are important in flood protection, erosion control, pollution control, surface water and groundwater supply, and fish and wildlife habitat. They are important to the fishing and shellfish industries. Wetlands are being lost at an enormous rate worldwide because of development and pollution.

9. The destruction of tropical forests is a significant worldwide threat to biological diversity and perhaps climate. Two-thirds of all species are thought to exist in the tropics.

10. There are ways to manage biological resources to provide for sustained yields, to protect the diversity of species, and to meet other human needs all at the same time.

11. Habitat protection and restoration is a way of protecting endangered or threatened species and common species simultaneously. Studies are needed to find undisturbed areas with the greatest species diversity for protection. Research is needed to identify the size of an area required for survival of particular species.

12. The international community is taking an interest in protecting tropical forests. Research, cooperative restoration and preservation projects, and training of local people in habitat management are all under way. Economic incentives to encourage habitat protection in developing nations are being implemented by international development agencies and private organizations.

13. Federal lands in the United States are to be managed according to the policy of multiple use and sustained yield. There is controversy over how well this policy is carried out on federal lands by the Bureau of Land Management, the Forest Service, and the National Park Service. Overgrazing on federal lands, clear-cutting in national forests, and development and lack of habitat management in parks are some of the practices in question.

14. Conservation biology is emerging as a discipline that combines genetics, ecology, and natural resource management to conserve biological diversity.

15. Preservation of wilderness areas is an ongoing struggle as development interests seek access to potential oil, gas, mineral, and timber resources in wilderness-designated areas.

16. One-quarter to one-half of the ecosystem types in the United States are not protected. A national policy is needed to provide a coordinated effort to preserve biological diversity.

17. The loss of a species is irreversible.

James R. Aldrich is director of the Kentucky chapter of The Nature Conservancy. The Nature Conservancy is an international, private nonprofit organization with a single mission: to preserve rare and endangered species and natural communities through protection of the ecosystems that sustain them. The Nature Conservancy takes pride in its nonconfrontational, businesslike approach to conservation; it protects land by buying it and then by looking after it. The work of The Nature Conservancy is international in scope ranging from Brazilian rain forests to the Kentucky River Palisades in Mr. Aldrich's adopted state. Mr. Aldrich's job is to develop and implement a comprehensive program in Kentucky to protect important natural areas. Simply put, his job is to develop a plan of just which areas in Kentucky need protection and protect them. In carrying out this responsibility, Mr. Aldrich is assisted by a board of trustees, The Nature Conservancy's national headquarters staff, the regional director of The Nature Conservancy, and a small office staff that includes a development director (in charge of fund-raising) and a biologist whose job it is to inventory and document Kentucky's biological resources. Day to day, Mr. Aldrich deals with landowners, lawyers, county clerks, and various other people as he implements his chapter's conservation plan by buying and looking after particular parcels of land. He works in close cooperation with agencies such as the Kentucky Nature Preserves Commission and the U.S. Forest Service. A big part of his work is raising the money from private and government sources to purchase land; he makes fund-raising presentations to individuals and groups throughout Kentucky. The Nature Conservancy clearly has important work to do in Kentucky. More than 80% of the wetlands Kentucky once had are already gone; only a few hundred acres of the original prairie and a few hundred acres of virgin forest remain—but over half of Kentucky's original forestland is still forestland. Among the things that Mr. Aldrich likes about his job is the knowledge that he has something very directly to do with preserving some of our natural heritage for future generations to learn from and enjoy. "How many people can come to work every day knowing that they have done and will likely do something more to help protect the environment?"

Mr. Aldrich has a double major in biology and environmental geography. Previous work experience that has helped prepare him for the job he does now includes work as botanist and environmental specialist and as coordinator of the Indiana Natural Heritage Program, both with the Indiana Department of Natural Resources. Mr. Aldrich says that the basic qualifications for the kind of work he does include a bachelor's or higher degree in a related field, management experience, an understanding of land protection techniques, the ability to write well and to speak well in public forums, the ability to deal with a variety of people (landowners, corporate officials, and government personnel), and a tolerance of frequent travel, often on short notice and frequently on weekends.

References and Further Reading

References marked with an asterisk are cited in the chapter.

*Abernethy, T., and Turner, R. E., 1987. "U.S. Forested Wetlands, 1940–1980," *BioScience* **37**:721–727.

Arundel, J., 1990. "The Eagle's Return," *New York Times*, March 26.

*Associated Press, 1990a. "Study Boosts Deforestation Estimates by 50 Percent," June 8.

*Associated Press, 1990b. "Study Warns against Amazon Deforestation," March 16.

*Bohlen, C. C., 1990. "Protecting the Coast," *BioScience* **40**:243.

*Chase, A., 1987. "How to Save Our National Parks," *Atlantic Monthly*, July, pp. 35–44.

*Corn, M. L., 1990. "The Endangered Species Act: The Storm's Eye," *BioScience* **40**:637.

Cowen, R., 1990. "Vanishing Amphibians: Why They're Croaking," *Science News* **137**:116.

*Danforth, D. K., 1987. "Battle over Clear-cutting Flares Anew with Plan for Forest in Arkansas," *National Geographic News Service*, May 25.

Davis, P. A., 1992. "Economy, Politics Threaten Species Act Renewal," *Congressional Quarterly*, January 4, pp. 16–18.

Davis, P. A., 1991. "Cry for Preservation, Recreation, Changing Public Land Policy," *Congressional Quarterly*, August 3, pp. 2145–2151.

Durning, A. B. and Brough, H., 1991. *Taking Stock: Animal Farm-*

ing and the Environment. Washington, D.C.: Worldwatch Institute.

*Eisner, T., 1989. "Prospecting for Nature's Chemical Riches," *Issues in Science and Technology* **VI**(2):31–34.

*Hodges-Copple, J., 1989. *Beyond Making Trades and Buying Time: Wetlands Conservation in the 1990s*. Research Triangle Park, N.C.: Southern Growth Policies Board.

*Holing, D., 1987. "Hawaii: The Eden of Endemism," *Nature Conservancy News*, February–March, pp. 6–13.

*Johnson, S., 1986. "Alien Plants Drain Western Waters," *Nature Conservancy News*, October–November, pp. 24–25.

*Kohn, H., 1990. "Reefs on the Rocks," *Greenpeace*, July–August, pp. 20–21.

*Lenssen, N., 1989. "Where Have All the Ducks Gone?" *Worldwatch*, January–February, pp. 8–9.

*Lewin, R., 1986. "Damage to Tropical Forests, or Why Were There So Many Kinds of Animals?" *Science* **234**:149–150.

*Lewin, R., 1987. "Hand of Man Seen in Birds," *Science* **236**:1522.

*Mares, M. A., 1986. "Conservation in South America: Problems, Consequences, and Solutions," *Science* **233**:734–739.

Moyer, S. and Feierabend, J. S., 1991. Statement of the National Wildlife Federation before the Subcommittee on Environmental Protection of the Senate Environment and Public Works Committee on Wetlands Protection and Federal Wetlands Legislation, 55pp.

*Myers, N., 1987. "The End of the Line on Natural Diversity," *Technology Review* **90**:76–77.

*Nielson, J., 1987. "Last Chance for the Condor," *Sports Illustrated*, March 23, pp. 62–66.

*Platt, R. J., 1987. "Coastal Wetland Management: The Advance Designation Approach," *Environment* **29**:17–20ff.

*Reid, W. V., and Miller, K. R., 1989. *Keeping Options Alive*. Washington, D.C.: World Resources Institute.

*Repetto, R., 1990. "Deforestation in the Tropics," *Scientific American* **262**(4):36–42.

Root, M., 1990. "Biological Monitors of Pollution," *BioScience* **40**:83–86.

*Ryan, J. C., 1989. "Plight of the Other Rain Forest," *Worldwatch*, May–June, pp. 10–11ff.

Ryan, J. C., 1992. *Life Support: Conserving Biological Diversity*. Washington, D.C. Worldwatch Institute.

*Scott, M. J.; Csuti, B.; Jacobi, J. D.; and Estes, J. E., 1987. "Species Richness: A Geographic Approach to Protecting Future Biological Diversity," *BioScience* **37**:782–788.

*Simberloff, D. "Are We on the Verge of a Mass Extinction in Tropical Rain Forests?" in David K. Elliott, ed., *Dynamics of Extinction*. N.Y.: John Wiley & Sons, 1986.

*Stewart, M. M., and Ricci, C., 1988. "Dearth of the Blues," *Natural History*, May, pp. 64–70.

*Stiak, J., 1990. "Old Growth," *Amicus Journal* **12**(1):35–45.

*Tangley, L., 1986. "Costa Rica: Test Case for the Neotropics," *BioScience* **36**:296–300.

*Tangley, L., 1990. "Cataloging Costa Rica's Diversity," *BioScience* **40**:633–636.

*Tripp, J. T. B., 1986. "The Status of Wetlands Regulations," *Environment* **28**:44–45.

*U.S. Office of Technology Assessment, 1987. *Technologies to Maintain Biological Diversity*. Washington, D.C.: U.S. Government Printing Office (see also 1992 update).

Wald, J., and Albersworth, D., 1989. *Our Ailing Public Rangelands: Condition Report*. New York: Natural Resources Defense Council.

*Westman, W. E., 1990. "Managing for Biodiversity," *BioScience* **40**:26–33.

*Whelan, T., 1988. "When the Forests Go," *Environmental Action*, January–February, p. 7.

*Wilson, E. O., 1989. "Threats to Biodiversity," *Scientific American* **261**(9):108–116.

*Wolf, E. C., 1987. *On the Brink of Extinction: Conserving the Diversity of Life*. Washington, D.C.: Worldwatch Institute.

15

Noise as an Air Pollutant

Julius Caesar reportedly barred chariots from certain parts of Rome during certain hours of the day and most of the night. Chariot wheels on cobblestones must have made quite a racket. Although there are other such references to noise throughout ancient literature, noise, like other pollution problems, is by and large a modern problem. In our stampede for bigger and better we have literally set the environment vibrating with misspent energy.

Although estimates vary, some experts say that background noise levels have been increasing in industrialized societies for decades by as much as 1 decibel per year. There is general agreement among the experts that background noise long ago passed the point of causing significant harm. We have jackhammers, power mowers, demolition equipment, air compressors, generators, motorcycles, cars, trucks, aircraft, washing machines, dishwashers, food processors, stereos, meat grinders, chain saws, and much, much more. Most of these are associated with urban life, but even in what used to be secluded, quiet areas we have off-road vehicles and appliances that disturb the serenity of the outdoors and damage the hearing of the operators.

Noise is a pollutant under any definition of the term. Noise hurts, and it is a by-product of human activity. Noise can be diminished, but only at some cost. As with the effects of other pollutants, the effects of noise are difficult to measure.

We do know that very loud sounds and unwanted sounds affect humans in three general negative ways. First and foremost, loud sound can impair hearing and thus interfere with the functioning of our bodies and in this general respect is somewhat like ozone. Second, very loud sounds and unwanted sounds can also have other negative physiological effects—for example, change in blood vessel diameter, heart rate, and blood pressure. Third, unwanted or startling sounds can affect humans psychologically.

When Is Sound Noise?

The word *noise* has the same Latin root as the word *nausea*. This leaves little doubt as to what the originators of the term *noise* had in mind, and it sets the stage for our definition.

The American National Standards Institute defines noise as any "undesired" sound. Noise has also been described at times as sound without value. As such, these terms and expressions do not define noise adequately; they fail to take into account the fact that harmful sounds are not always *perceived* as harmful. Any definition of noise should include words like *hurt* or *harm*. Our definition of **noise** is unwanted sound or sounds of a duration, intensity, or other quality that cause physiological or psychological harm to humans or other living things. Under this definition, *noise* would apply to any loud sound above 90 decibels (see Figure 15.1), no matter how "beautiful" it happened to be.

While loudness is relatively nonsubjective—at least in terms of the limits of physical damage—many factors are involved in the individual determination of other characteristics of sound. These include, in addition to frequency and intensity, the complexity or irregularity of the sound, the rate at which sound gets louder or quieter, and even more complex psychological factors such as whether sound is perceived by a listener as being necessary. Still other factors include

Figure 15.1 The Decibel Scale. The locations of familiar sound sources are shown on the scale. Note that many common sound sources fall within the range of psychological and physical harm.

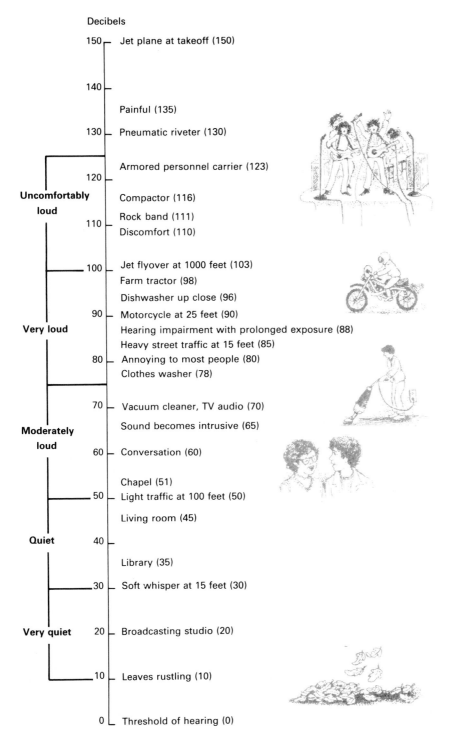

Decibels

150 — Jet plane at takeoff (150)

140 —

Painful (135)

130 — Pneumatic riveter (130)

Armored personnel carrier (123)

120 —

Uncomfortably loud

Compactor (116)

Rock band (111)

110 — Discomfort (110)

100 — Jet flyover at 1000 feet (103)
Farm tractor (98)
Dishwasher up close (96)

90 — Motorcycle at 25 feet (90)

Very loud

Hearing impairment with prolonged exposure (88)
Heavy street traffic at 15 feet (85)

80 — Annoying to most people (80)
Clothes washer (78)

70 — Vacuum cleaner, TV audio (70)

Sound becomes intrusive (65)

Moderately loud

60 — Conversation (60)

Chapel (51)

50 — Light traffic at 100 feet (50)

Living room (45)

Quiet 40 —

Library (35)

30 — Soft whisper at 15 feet (30)

Very quiet 20 — Broadcasting studio (20)

10 — Leaves rustling (10)

0 — Threshold of hearing (0)

the relationship of a sound to sounds people are accustomed to hearing and—believe it or not—the listener's age, degree of training and education, and socioeconomic status. More on this later.

The Effects of Noise

The Auditory Effects of Noise: Hearing Loss

The magnitude of noise-induced hearing loss in the United States can be estimated from occupational hearing loss, worker compensation claims, and the number of Americans with impaired hearing. Hundreds of thousands of workers between the ages of 50 and 59 are eligible for worker's compensation because of hearing impairment caused by their jobs (see Figure 15.2). Studies have shown that factory workers have double the rate of hearing loss that white-collar workers have at comparable ages. It has been estimated that perhaps half of the machinery used in industry operates at levels that can impair hearing. But workplaces are by no means the only source of hearing impairment. A 1977 EPA study showed that about one in ten people in the United States is exposed to noises of duration and intensity sufficient to cause hearing impairment.

The problem of noise-induced hearing impairment begins to occur somewhere between 80 and 90 decibels (see Enrichment Box 15.1 and Figure 15.1). Regulatory agencies continue to struggle with where to set limits for noise levels in industry. Some studies

Job-related noise is a major cause of hearing impairment. Federal regulations now require companies to protect their employees from the effects of sustained high-decibel noise.

have shown that sound at 80 decibels (on the A-scale) can produce a temporary elevation in the threshold of hearing and that this may become permanent with repeated exposure. Studies in both industrial and mili-

Figure 15.2 Hearing Loss in the Workplace. Occupational hearing loss over a range of frequencies is shown as a function of both frequency and duration. These data clearly show that hearing loss is a function of noise level and frequency.

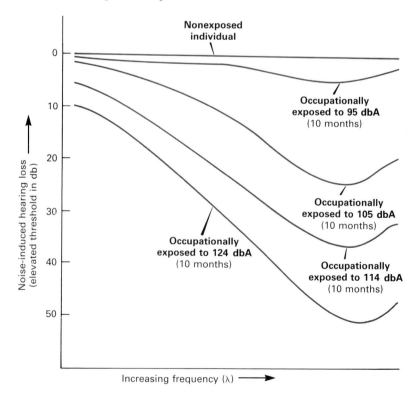

Enrichment Box 15.1

The Decibel Scale

Deci- comes from the Latin word for "ten," and a *bel* is the logarithm of a ratio of any two acoustical (sound) or electrical intensities. In terms of sound, a **decibel** is 10 times the logarithm of the ratio of two sound intensities, one being the intensity of any sound of interest and the other being a reference sound. In the United States, the reference sound intensity is 10^{-12} watts per square meter; this is the intensity of a sound barely audible to a human being. A sound 100,000 times louder (more intense) than the reference level would be called a 50-decibel sound. Why? Because the logarithm of 100,000 is 5, and 10 times 5 is 50. Sound with 10 times the intensity of the reference level would be a 10-decibel sound because the logarithm of 10 is 1, and 10 times 1 is 10. A jet plane at takeoff generates—up close—a sound intensity 1 quadrillion times that of a barely audible sound. The logarithm of 1 quadrillion is 15, and 10 times 15 is 150; a jet plane at takeoff—up close—produces a 150-decibel sound (see Figure 15.1).

The really important thing to remember about the decibel scale is that it is a reference scale, meaning that it is a logarithmic way of expressing how much louder a particular sound is than a sound level chosen arbitrarily. This scale is very appropriate to human hearing because we also happen to hear in a logarithmic way. That is, we actually perceive an increase from 10 to 20 decibels as about a doubling of sound intensity, even though it is really a tenfold increase in power intensity.

The A-Scale

Sometimes sound levels are expressed according to the decibel-A(dbA)-scale. In this scale the frequencies to which humans are most sensitive are given more weight or more importance in the assessment of effects on human hearing.

---

tary situations have shown that progressive noise-induced hearing impairment can be caused by exposure to sound levels slightly above 80 decibels if exposure occurs repeatedly over an eight-hour day.

Whereas we once accepted hearing loss as an inevitable consequence of old age, it is now apparent that it might instead be due to the cumulative effects of noise over one's life span. It may in fact *not* be natural to lose one's hearing—at least not at the rate we lose it in modern society. Significant percentages of American teenage and pre-teenage children have measurable hearing loss. Studies have shown that people living in environments that are relatively free of noise do not have hearing loss even at advanced ages. Very old Mabaans, for example, members of an African tribe in the southeastern part of the Sudan, have about the same hearing acuity as American children (Rosen et al., 1962).

Studies of young people listening to rock music have shown that such music, generated at more than 92 decibels throughout the 500- to 8,000-Hz range and sustained over one hour or so, can produce a 40-decibel *threshold shift* in about 10% of the listeners—and somewhere between a 20% and 30% threshold elevation in the remainder. Though most of this threshold shift is usually temporary, repeated exposure can make it permanent. A **threshold shift** is an elevation in the threshold of hearing—that is, the quietest sound that can be heard becomes a louder sound.

Where Does the Auditory Damage Occur?

A number of years ago, researchers found that when guinea pigs were exposed to loud noises—rock music, to be exact—the hair cells of the inner ear (Figure 15.3), cells responsible for the conversion of

---

Mini-Glossary

Decibel (db): a relative measure of sound intensity; ten times the logarithm of the ratio of two sound intensities, one being a reference sound level.

Frequency: rate; sound frequency is the rate at which compression waves arrive at or pass a fixed point. Frequency is perceived as pitch.

Hearing threshold shift: a change in the loudness of the quietest sound that can be heard. Exposure to loud sound can cause the threshold to rise.

Hertz (Hz): cycles per second; a measure of sound frequency.

Infrasound: sound too low in frequency to be heard by humans (below 20 Hz).

Intensity: loudness; acoustical power (the energy that sound delivers) per unit of area.

Loudness: the human perception of the intensity of sound. Loudness is actually determined by both intensity and frequency; a 20-decibel sound will not necessarily sound equally loud at different frequencies.

Noise: sound with a duration, intensity, or other quality that causes physical, physiological, or psychological harm or stress to a human being.

Pitch: the human perception of sound frequency (and to some extent intensity).

Sonic boom: human perception of the shock wave generated by an object traveling faster than the speed of sound.

Ultrasound: sound too high in frequency to be heard by humans (above 20,000 Hz).

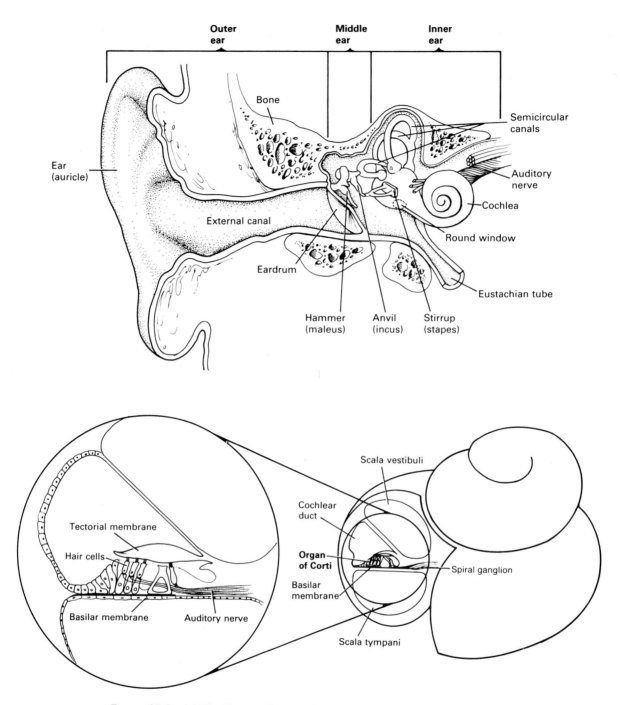

Figure 15.3 (a) The Human Hearing Apparatus. The outer middle, and inner ears. **(b) The Organ of Corti.** If the cochlea illustrated in (a) were cut away, the organ of Corti would be as shown here. Within the inner ear the organ of Corti runs along the entire length of the basilar membrane, which is wound within the spiral cochlea. The basilar membrane vibrates in particular patterns with particular sound frequencies, thus causing distortion of particular groups of hair cells to different degrees. This is in turn responsible for the generation of characteristic patterns of nerve impulses that travel along auditory nerve fibers to the brain.

mechanical energy into nerve impulses, collapsed and shriveled. It is believed that loud sounds destroy these hair cells. Once destroyed, hair cells are not replaced. The hair cells that enable us to perceive the character-istic frequencies of human speech are often the first to be destroyed. Thus as hearing loss proceeds, the first sounds to go are those by which human beings communicate.

Noise and Health

The people in the apartment next to you play their music so loud that you can hear it. It is not your kind of music; it makes you irritable. What rights do you have? What rights should you have?

Other Physiological Effects of Noise

Anyone who has experienced sound near the threshold of pain would not be surprised to learn that sound in the range of 120–150 decibels can affect the respiratory system and affect balance to the extent of dizziness, disorientation, nausea, and vomiting. But even sounds as quiet as 70 decibels can have measurable physiological effects. Such effects may not result in any immediate impairment, but they do emphasize the remarkable magnitude and variety of effects noise has on human beings. Let us examine some reasons why a relationship between noise and physiology might exist.

Nerve fibers that leave the inner ear carry impulses elicited by sirens, trumpets, or Madonna to the medulla of the brain stem, where they meet other fibers going to other parts of the brain. Nerve pathways permit both ears to communicate with numerous parts of both sides of the brain, including the centers of consciousness and the control centers that regulate breathing, blood pressure, and other bodily functions below the level of consciousness. Figure 15.4 illustrates the pathways connecting the ear and key glands and organs of the body.

As part of the body's "early warning system," ears never really sleep; they are connected to the brain's arousal center and can wake us up. They are also able to muster glands like the thyroid and the adrenal gland to secrete hormones that prepare the body for fight or flight in the so-called startle reaction. Loud noises and explosive sounds can cause what physiologists call the *sympathetic reaction* because most of the characteristic physiological changes are controlled by the **sympathetic nervous system.** This is a division of the nervous system that controls many bodily functions below the level of consciousness, for example, heart rate, blood pressure, glandular secretions, and digestive tract motility. Generally speaking, when the sympathetic nervous system is activated, the things that support a flight-or-fight reaction are turned on; unessential activities, such as digestion, are shut down.

Overall, loud sounds can cause increased production of most hormones of the pituitary gland; among the most important of these is the adrenocorticotropic hormone (ACTH). ACTH in turn stimulates the adrenal gland, which secretes several different hormones. Through a variety of influences, these hormones in turn (1) enhance the body's sensitivity to adrenaline, (2) increase blood sugar levels, (3) suppress the immune system, and (4) decrease the liver's ability to detoxify blood (Moller, 1975).

In the evolutionary period during which the ears and their relationships with various organs and glands of the body were emerging, very loud sounds were rare; when loud sounds occurred, they almost always indicated danger. This connection could well have served as a selective pressure (see Chapter 5) leading to the "wiring" that now prepares human beings to react automatically to loud sounds. Somewhat surprising perhaps is the fact that noise can elicit a number of automatic responses even at rather low levels. It has been shown that constriction of blood vessels in the skin can occur at only 70 decibels on the A-scale, the level of sound found on some residential streets.

The physiological responses and problems we have with noise and sound, then, may be the result of the fact that we have dragged a body shaped for a quiet, primitive environment into the bustling modern world. As we are besieged by noises, part of each of us wants to run and hide, while another part is saying, "Don't worry about it!" Perhaps this conflict is responsible for some of the psychological effects of noise.

Psychological Effects of Noise

The German philosopher Arthur Schopenhauer said, "I have long held the opinion that the amount of noise which anyone can bear . . . stands in inverse

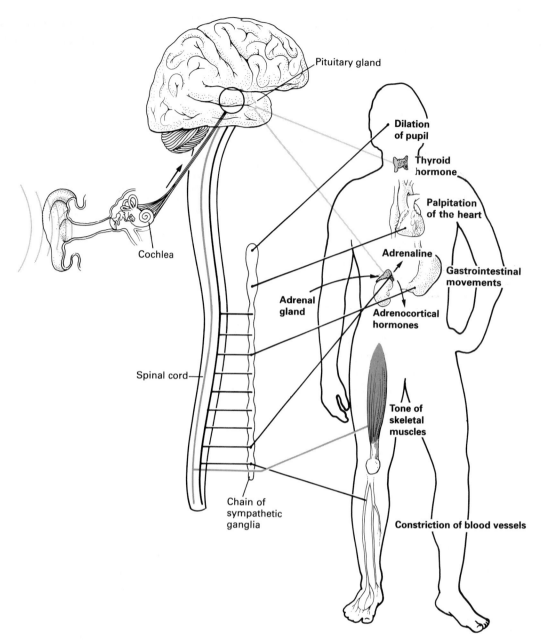

Figure 15.4 The Auditory Connection. This illustrates how noise might affect various important body functions via the sympathetic nervous system and hormones from the pituitary gland.

proportion of his mental capability. . . . Noise is a torture to all intellectual people." Perhaps Schopenhauer was emphasizing the fact that noise interferes with high-order mental function. Certainly, to some extent, noise is a torture to all people.

All of us are aware that noise can interfere with speech, thinking, and studying and that it can be a source of stress. Noise can have psychological effects on humans ranging from mild distress to complete unhinging. Noise has been used as a form of torture for centuries.

Noise and Speech. The ability of noise to interfere with speech may be its most important psychological effect. Noise interferes with communication in our species, and since communication is one of our biological specialties, such interference is a rather widespread source of stress. Perhaps largely because of this impact on communication, noise might have something to do with various forms of social conflict as well as with indigestion, nervous breakdown, heart failure, high blood pressure, and insanity. Zuremstehung (1959) found that steelworkers assigned to the noisiest parts of steel mills had significantly more social difficulty, both at the plant and at home, than workers in less noisy parts.

Noise, Sleep, Rest, Relaxation, and Mental Stability.
Because sleep is important to emotional stability, it is worth noting that a study in Philadelphia some years ago showed that city noises interfered with sleep both by prohibiting a deep sleep and by interrupting sleep, especially among elderly people.

It seems plausible that through some combination of its effects on sleep, its interference with communication, and other forms of noise-induced distress, noise may contribute to emotional disturbances. A study done in France (see Smith, 1970) suggested that noise may cause as much as 70% of the neuroses found in major cities like Paris. A study reported in *The Lancet* (Abey-Wickrama et al., 1969) showed that for certain categories of mental illness, admissions to psychiatric hospitals in London occurred at significantly higher rates for people living in areas of maximum noise levels near Heathrow Airport. The authors suggested that even if noise might not actually be the cause of mental illness, it might aggravate conditions leading to mental illness.

Despite such reports and conclusions based on personal observation and common sense, occasionally a writer will suggest that humans are able to adapt to noise. We suspect that the truth is that although humans may well be able to adapt to noise to some extent, the importance of noise as an alarm signal precludes extensive adaptation. Studies have shown that the *physiological* effects of noise are independent of culture and adaptation, and we suspect that the same is true of its psychological effects.

Psychological Impact and the Ear of the Beholder.
Below the levels of sound intensities that are known to be physically harmful, human beings seem to be able to tolerate certain kinds of noise more than others. As we stated earlier, human beings are least able to tolerate sounds they consider unnecessary. An example is the unnerving tapping of a foot. Obviously, sounds that are considered necessary are less likely to be bothersome. Almost any kind of sound could variously be described at the same time by different people as pleasant, irritating, annoying, beautiful, terrible, disturbing, cruel, or unbelievable. A sports car climbing through the gears to a high rate of speed might be one such example. Subjectivity is what makes it so difficult to quantify the annoyance or the psychological impact of noise on human beings. There is simply too much variation from one human being to another. We are pretty much stuck with having to

make inferences from what we see and observe. This puts advocates of reducing noise levels in our environment in the position of having to prove a point that is at best extremely difficult to prove.

The Prospects for Noise Control

One of the ways in which noise is unique as a pollutant is that we have the technology by which to control nearly every kind of noise. We know more about noise control than we know about the effects of noise. This suggests that the main problem with noise is that people are not aware that it *is* a problem.

Noise may be unique as a pollutant not only in that it is far less generally recognized as a pollutant than ozone and the like but also in that many human beings actually seek it out. Motivational studies suggest that human beings equate noise with power. Manufacturers of appliances designed to do powerful things have found that the public develops an impression of how powerful and therefore how good an appliance is partly by how much noise it makes (Brody, 1979). This means that even though noise may be the most controllable pollutant, social motivation for its control actually runs in the wrong direction. Little wonder that there has not been more progress with noise control.

Engineering Quiet
Basically there are four ways to control noise:

1. Modify the ways things are done so as to generate less noise
2. Shield the noise-generating devices or processes at the source
3. Shield the receiver
4. Move noisy things away from people

Examples of the first approach would be reducing automobile traffic, outlawing sirens, making it impossible for sound to come out of a stereo except through headsets, and using glue instead of rivets. An example of the second approach would be the use of vibration-damping or -absorbing materials in dishwashers and automobiles. Earplugs and control booths are examples of the third approach. Isolating new jetports from people would be an example of the fourth. There is a fifth approach, called "antinoise," now on the horizon. This approach quenches compression waves by detecting them and generating waves in the opposite direction (see Alper, 1991).

We have already identified a major problem associated with controlling sound in appliances, tools, and equipment by direct engineering—namely, that

Jet traffic is a noise problem mainly because most large airports are located near residential areas.

the public seems to equate noise with power. Perhaps this could be changed somewhat by education, but it could certainly be changed by laws requiring noise sources to meet certain standards. Possible improvements in machinery and appliances would be sound-absorbing motor mountings, better installation, motor enclosures, and other improvements in design. Sound-absorbing materials have long been used to cut down on the noise produced by electric motors, engines, and machinery; these just need to be applied more generally. There is hardly a machine that could not be better designed with noise reduction in mind. Though some people say that building codes must be changed to permit more flexible use of noise-absorbing materials, there is not a house, office building, restaurant, terminal, or public building that could not now be constructed so as to reduce noise.

Legislating Quiet

As with other environmental pollutants, the control of noise pollution can conveniently be divided into two categories, occupational and nonoccupational. We will consider laws covering the general environment first.

The number of local ordinances and local noise control laws have increased in the United States almost as fast as the seriousness of the noise problem. Many of these laws are ineffective because they are based on subjective criteria and are very difficult to enforce. Because many of the most notable sources of noise (automobiles, trucks, trains, machines) are made or imported for distribution throughout the United States, noise control logically falls under the general welfare responsibility of the federal government. Federal laws tend to be more effective because they can deal more effectively with noise at its sources.

The present law of the land regarding noise control in the general environment is the Noise Control Act of 1972 as amended by the Quiet Communities Act of 1978. As a result of this legislation, the U.S. Environmental Protection Agency developed noise regulations for interstate motor carriers, other medium and heavy trucks, portable air compressors, motorcycles and their exhaust systems, and interstate rail carriers. General labeling of some products was begun in 1979, as was a labeling program for products designed to reduce noise. Somewhat problematic is the fact that the EPA's Office of Noise Abatement Control was abolished in 1981 (see Raloff, 1991).

The very first U.S. standards for occupational noise levels were established in 1969. In 1974 the Occupational Safety and Health Administration (OSHA) set the current standards at 90 decibels over an eight-hour day with the duration of exposure to be cut in half for every 5 decibels over 90. The National Institute for Occupational Safety and Health (NIOSH) and the EPA have proposed an 85-decibel limit in the workplace over eight hours. Economic and technical difficulties involved in achieving an 85-decibel limit have apparently kept OSHA, the agency charged with establishing regulations, from lowering the limit, although the Hearing Conservation Amendment to the Occupational Safety and Health Act now requires companies to have a hearing conservation program for employees exposed to 85 decibels or more. Such programs are to include annual hearing tests, provision of personal protective equipment, and area monitoring.

For Further Thought . . . 15-2

Noise: Is There Really a Problem?

Obtain a decibel meter. After you have learned how to use it and have planned where to go and when, visit various locations in your community and make measurements at various distances from sound sources. Make a presentation on your thinking about the noise issue after you have made the measurements.

The Economics of Noise Control and the Impact of Noise

How does one weigh the relative costs and benefits—for example, a sonic boom versus getting passengers from California to New York as fast as technologically possible? How does one compare a poorly understood, slightly negative effect on thousands of people in relationship to a benefit—which may also be slight—derived by 100 passengers? How does one measure how much a worker's productivity will decrease in a noisy environment? How would one bring into a cost-benefit analysis such things as days off due to an illness that may never be perceived to be connected to noise? How would one go about determining the cost of a sleepless night? Of a fight in a family? Of indigestion? You may have heard questions such as these before;

they are only partly rhetorical. They are a real challenge to people who would make the case that money must be spent to control noise. With noise, as with other pollutants, environmental cost-benefit analysis is largely an art form.

Clearly, the cost of noise pollution is significant. The World Health Organization estimates that industrial noise costs billions of dollars annually worldwide in absenteeism, inefficiency, and accidents as well as in direct compensation. Studies in England have shown that workers exposed to continuous noise in the 90-decibel range, not far beyond the limits now currently set for U.S. industry, made more errors and were generally less productive. Only when we have a better grip on these kinds of costs can they be measured against the costs associated with restricting jet traffic in airports during certain times of day, the costs of modifying jets to make them less noisy, the costs of changes in automobile design and maybe even in highway design, and in general the costs of making environments less noisy.

Concepts to Remember

1. Noise—unwanted, loud, or otherwise irritating sound—is a genuine pollutant. It is a human-generated feature of the environment that causes human health problems.

2. Noise harms people in three basic ways: (a) Loud sounds can impair hearing; (b) loud and otherwise irritating sounds can elicit physiological changes, e.g., in blood pressure, heart rate, and digestive function; and (c) irritating sounds can also serve as a source of psychological stress.

3. The fact that noise has physiological and psychological effects is related to the importance to survival of sound and hearing.

4. The decibel is a relative measure of sound

   intensity based on the reference sound pressure of the quietest sound that can be heard by a normal human.

5. Hearing damage associated with loud sound occurs in the inner ear and is permanent.

6. The psychological impact of sounds below the threshold of physical damage is almost entirely dependent on the listener. One listener's symphony is another listener's noise.

7. Noise control technology is well developed, but motivation to apply this technology is lacking or even negative—people equate loud machinery with power and may actually seek out loudness.

Janet A. Marvin

Janet A. Marvin is an outdoor recreation planner with the U.S. Fish and Wildlife Service's Erie (Pennsylvania) National Wildlife Refuge. The Fish and Wildlife Service's mission is to conserve, protect, and enhance fish and wildlife by protecting places where they live, for the continuing benefit of the American people. In her work at the Erie Wildlife Refuge, Ms. Marvin plans and implements programs for the public and evaluates public use of the refuge. She gives presentations on the refuge to groups of visitors and shows them around the refuge describing how waterfowl and other wildlife live and interact. Outside the refuge, she gives conservation lectures and slide presentations and helps local schools to develop and implement environmental education projects. Part of her job is to put together pamphlets and nature guides for refuge visitors. In the public relations aspects of her work, she initiates and maintains contacts with the news media, federal, state, and local governments, and various other groups to enhance their relationships with the Wildlife Service. Part of this aspect of her work is to prepare and distribute timely news releases to keep the public informed of refuge programs and activities. Her supervisory duties cover groups of volunteers and seasonal staff. She also oversees the operation and maintenance of public-use facilities, including the refuge's buildings, grounds, equipment, exhibits, and trails. What appeals to Ms. Marvin most about her work are its variety and the fact that dealing with a diverse public provides an interesting daily challenge. She also likes the feeling that she is making a positive difference, influencing the attitudes of people, from

preschoolers to senior citizens. It is exciting, she says, to teach people to value fish and wildlife resources so that they can act more responsibly on environmental issues and promote fish and wildlife conservation.

Ms. Marvin has a bachelor's degree in biology and has taken additional professional development courses in environmental education, interpretation, business supervision, and first aid.

References and Further Reading

References marked with an asterisk are cited in the chapter.

*Abey-Wickrama, I.; A'Brook, M. F.; Gattoni, F. E.; and Herridge, C. F., 1969. "Mental-Hospital Admissions and Aircraft Noise," *Lancet* **2**:1275–1277.

*Alpher, J., 1991. "Antinoise Creates the Sounds of Silence," *Science* **252**:508–509.

Aniansson, G., and Peterson, Y., 1983. "Speech Intelligibility of Normal Listeners and Persons with Impaired Hearing in Traffic Noise," *Journal of Sound and Vibration* **90**:341–360.

*Brody, J. E., 1979. "A Quiet Protest against Noise," *New York Times*, January 28.

Crocker, M., and Price, A., 1975. *Noise and Noise Control* (2 vols.). Boca Raton, Fla.: CRC Press.

Environmental Education Group, 1973. *Noise Pollution and Solutions for Silencing the Problem.* Los Angeles: Environmental Education Group.

Fahy, F., 1990. "Tracking Down the Noise Polluters," *New Scientist* **10**:43–46.

Harris, C. M., ed., 1979. *Handbook of Noise Control*, 2d ed. New York: McGraw-Hill.

Holding, D.; Loeb, M.; and Baker, M., 1983. "Effects and Aftereffects

of Continuous Noise and Computation Work on Risk and Effort Choices," *Motiv. Emotion* **7**:331–344.

*Moller, A. R., 1975. "Noise as a Health Hazard," *Ambio* **4**(1):6–13.

National Academy of Sciences, 1981. *Effects on Human Health from Long-Term Exposures to Noise.* Washington, D.C.: National Academy of Sciences.

*Raloff, J., 1991. "Dormant Noise Program's Silent Reverberations," *Science News* **140**:100.

*Rosen, S.; Bergman, M.; Plestor, D.; El-Mofti, A.; and Hamad-Falti, M., 1962. "Presbycusis Study of a Relatively Noise-free Population in the Sudan," *Annals of Otology, Rhinology, and Laryngology* **71**:727–743.

Rosenberg, J., 1991. "Jets over Labrador and Quebec: Noise Effects on Human Health," *Canadian Medical Association Journal* **144**:869–875.

Ryals, B. M., and Rubel, E. W., 1988. "Hair Cell Regeneration after Acoustic Trauma in Adult *Coturnix* Quail," *Science* **240**:1774–1776.

*Smith, L. K., 1970. "Noise as a Pollutant," *Canadian Journal of Public Health* **61**:475–480.

*Zuremstehung, J. G., 1959. "Vegitative Funktionsstörungen durch Lärmeinwirkung," *Archiv für Gewehrpathologie und Hygiene* **17**:238–261.

16

The Problem
of Persistent Hazardous
Materials
in the Ecosphere

Up to this point we have considered environmental problems that have been more or less limited to a single "great sphere." We have looked in turn at air pollution, water pollution, land misuse, and abuse of our biological resources. In this chapter and the next two, we are going to take a more focused look at environmental pollutants that get into the ecosphere in a much more general way, sometimes contaminating all of the great spheres simultaneously. Some get into the ecosphere when we try to dispose of them as wastes. Others—pesticides, for example—escape into the general environment as a consequence of their overuse, persistence, or misapplication. In some cases, we learned after the fact that pesticides have had unwanted side effects, even when used properly. We will consider the environmental and health effects of pesticides in this chapter.

Hazardous materials in the ecosphere cause both ecological problems and human health problems. In Chapter 17 we will examine the problem of solid and hazardous wastes, reviewing the environmental and health consequences of various waste disposal alternatives. Because of the increasing concern over human illness resulting from environmental contamination from persistent hazardous chemicals, we

will focus on cancer and some other illnesses and their relationships to the environment in Chapter 18.

The Ecological Connection

The Problem of Persistence

Rachel Carson, in her classic 1962 book *Silent Spring*, alerted the public to the possibility of environmental disruption by pesticides and other persistent synthetic chemicals. Evidence accumulated since then confirms that the ecosphere is indeed contaminated with human-made chemicals and that these do have a negative impact. Metals, pesticides, and other synthetic chemicals that are degraded very slowly, if at all, are caught up in the flow of materials in the ecosphere. They end up in organisms that are very distant and very different from their original location or target.

Because by definition **synthetic chemicals** have not evolved in the ecosphere—they are created in laboratories (see Table 16.1)—biological systems do not have the enzymes to break them down. The breakdown of such chemicals thus often depends on physicochemical processes that may take a long time to bring about complete degradation. The persistence

449

Table 16.1 U.S. Production of Intermediate and Finished Synthetic Organic Chemical Products, 1967–1988

Product	Quantity*	1967	1982	1984	1986	1988
Crude coal tar	m gal	NA	316	342	169	162
Petroleum†	b lb	NA	93	109	114	112
Cyclic intermediates	b lb	21	38	47	50	58
Dyes	m lb	206	222	233	236	280
Organic pigments	m lb	53	71	86	89	116
Medicinal chemicals	m lb	180	227	279	264	258
Flavor and perfume	m lb	112	156	179	138	162
Plastic and resin	b lb	14	38	48	52	63
Rubber processing	m lb	264	232	288	324	353
Elasticizers	b lb	4	4	5	4	5
Plasticizers	b lb	1	1	2	2	2
Surface-active agents	b lb	4	4	6	6	7
Pesticides	b lb	1	1	1	1	1
Miscellaneous	b lb	60	115	116	119	134

* m gal = million gallons; m lb = million pounds; b lb = billion pounds.

† Primary products from petroleum and natural gas for chemical conversion.

Source: U.S. International Trade Commission.

of certain chemicals increases the chance that they will inadvertently enter organisms through food intake, water use, or gaseous exchanges with the atmosphere.

Living organisms are likewise unable to handle many naturally occurring substances, such as metals, in some of the concentrations and chemical forms that are now common because of human activities. Lead poisoning and mercury poisoning are familiar examples in humans and other species (see Chapter 10).

The persistence of radioactive material is predictable. The half-lives of various radioactive isotopes are known. Half-lives range from a few seconds to tens of thousands of years. Human activities that may cause radioactive materials to be released into the environment include weapons testing, uranium mining, burial of radioactive materials in poorly suited disposal sites, and nuclear power plant accidents such as the one at Chernobyl (see Chapter 6).

Biomagnification

Perhaps the most significant impact of metals, persistent chemicals, and radioactive materials on ecosystems and human health stems from their magnification in food chains. As living things break down the complex molecules of carbohydrate, fat, and protein in their food, they assimilate some components and excrete others. Still other components move into storage compartments such as fat and bone. Chemicals that are stored in this way accumulate over time and are passed up the food chain in higher and higher concentrations.

At each link in the chain, the amount of pesticide passed on to an individual organism is increased. For example, earthworms may ingest a pesticide by eating contaminated organic matter. A single earthworm will consume many times its own weight in organic matter and accumulate much of the pesticide residue that the organic matter contains. The more it eats, the more pesticide it accumulates. At the next level, a robin eats numerous earthworms. The accumulated pesticide in all the earthworms ingested by a robin may reach a level at which there are negative, even fatal, effects. This is exactly what happened with the pesticide DDT in 1963 in Hanover, New Hampshire. Seventy percent of the robin population died after a spraying of DDT for Dutch elm disease. Events like this were the basis for Rachel Carson's concern about a silent spring when no birds would be left to sing. Logically, top carnivores such as *Homo sapiens* are particularly vulnerable to biomagnification.

A classic example of biomagnification was reported in California's Clear Lake (see Figure 16.1). DDD, a derivative of DDT, was used to spray for gnats around the 46,000-acre lake in 1949, 1954, and 1957. Concentrations in water reached 0.02 ppm. In 1960, studies of the biota in the lake indicated that the plankton contained 250 times the amount of DDD in the water; frogs contained 2,000 times more; sunfish contained 12,000 times more; and grebes, the top carnivores, contained up to 80,000 times more. This resulted in numerous deaths among the grebes in 1954

Mini-Glossary

Biomagnification or **bioaccumulation:** the process whereby persistent chemicals can be passed along in the food chain and concentrated in larger amounts at each level in the food chain.

Chlorinated hydrocarbons: a subgroup of insecticides containing carbon, hydrogen, and chlorine. They are fat soluble and persistent.

Integrated Pest Management (IPM): an approach to pest control that involves the minimal use of chemicals, good agricultural practices, and use of knowledge of the life cycle of the pest.

Persistent: degraded only slowly in the environment

Organophosphates: a subgroup of insecticides containing carbon, hydrogen, and phosphorus. They tend to be water soluble and break down rapidly in the environment. They are highly toxic and care must be taken to avoid exposure when they are used.

Pesticides or **biocides:** chemicals that are toxic to a wide variety of living things; they may be *herbicides* (toxic to plants), *insecticides* (toxic to insects), *fungicides* (toxic to fungi).

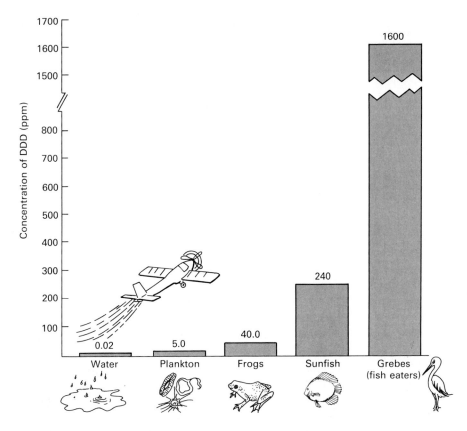

1700
1600
1500

800
700
600
500
400
300
200
100

Concentration of DDD (ppm)

1600

240

40.0

5.0

0.02

Water

Plankton

Frogs

Sunfish

Grebes
(fish eaters)

**Figure
16.1 Biomagnification.** This graph shows how various species studied in California's Clear Lake, north of San Francisco, accumulated DDD, a metabolite of DDT, through the food chain.

and 1957. Edible fish caught ten years later were found to have concentrations of DDD above FDA standards.

A similar occurrence was reported in the Long Island tidal marshes (Woodwell, Wurster, and Isaacson, 1967). DDT, used here for mosquito control for 20 years, was found in the following concentrations:

___ Plankton: 0.04 ppm
___ Shrimp (which feed on plankton): 0.16 ppm
___ Minnows (which feed on plankton): <1 ppm
___ Predatory fish (which feed on minnows): 1–2 ppm
___ Fish-eating birds: 13–26 ppm

DDT has been shown to be lethal to shrimp at levels as low as 0.0055 ppm. Concentrations of DDT in the brains of birds at levels of 30 ppm or greater present a lethal threat. DDT is stored in fat in living things. Under cold conditions or poor diet conditions, such as during hibernation, when energy is needed, fat reserves are mobilized; as this happens, DDT residue reenters the bloodstream. DDT residues of 10 ppm in fatty tissues of birds can concentrate under these conditions to as much as 20 ppm in the bloodstream.

These examples show biomagnification to be a straightforward, real environmental problem, the basis of which lies in the way organisms interact ecologically. This phenomenon of biomagnification or bioac-

cumulation is especially significant in attempting to set standards for contamination in air, soil, or water.

Air, water, and soil are all vehicles that transport pesticides, stable toxic chemicals, and radioactive materials. The rate and distance of transport are influenced by climatic factors, the characteristics of the specific chemical, and how the chemical is applied or released.

Little is known about the impact of long-term low-level exposure to many of these substances. What we do know is that any stable substance released in the environment can and probably will make its way into the nutrient cycles that include *Homo sapiens* and other organisms. Pesticides make up the bulk of persistent synthetic chemicals deliberately released into the environment. We will start with them as we scrutinize the overall environmental toxin problem and the choices we face.

Pesticide Ecology

Defining Terms
A **pest** is any organism that is unwanted by a given species for one reason or another. It may be an annoying gnat, a dandelion in an otherwise homogeneously grassy yard, or a borer on a corn plant. In some cases—the dandelion, for example—the pest is a social or cultural one. Corn borers are more important

Agricultural pests like corn borers are not simply nuisances; they can affect the world's supply of food. Still, the environmental side effects of pesticides must be carefully considered in dealing with pests.

The chemistry of the two major groups of insecticides causes them to behave very differently in the ecosphere. Organophosphates are water-soluble and usually break down rather rapidly in the ecosphere. They are often less persistent but may be more toxic than chlorinated hydrocarbons to humans and other mammals. They represent much more of an acute health concern and require special precautions when they are applied. Chlorinated hydrocarbons are the more persistent of the two; they do not break down rapidly in the environment, and they tend to accumulate. For example, DDT in soil may persist ten years or longer under certain circumstances. Chlorinated hydrocarbons are fat-soluble and not water-soluble; they accumulate in the fatty tissue of animals.

Pesticide Pathways

Hazardous materials may enter the ecosphere by way of any of the great spheres—air, water, or soil. Once in the ecosphere, hazardous materials may cycle through the spheres and into living organisms. We saw earlier in Part Three how substances can enter and pollute air, water, and soil. Here we will look at pesticides to show the paths by which they enter and contaminate all three spheres. These pathways hold true in a general sense for all other persistent, stable, toxic chemicals. Their specific use or method of application will affect the degree to which each sphere is an entry to the biosphere.

Synthetic organic pesticides are produced and sold in a concentrated form that must be diluted for application. Dilution not only decreases pesticide strength but also facilitates an even distribution of the chemical over a large area. Different materials that are used as carrier materials for pesticides include dusts, impregnated granules, solutions, emulsions, and suspensions of wettable powders. Diluting agents include clays, talc, water, and nonaqueous solvents. These are dusted or sprayed onto plants or worked into the soil. The pesticides then follow various transport routes elsewhere (see Figure 16.2).

Air. Spraying and evaporation are the two major ways for pesticides to enter the atmosphere; smaller amounts may enter from improper incineration of pesticide containers. (Incineration may provide a more significant pathway into the ecosphere for other types of persistent chemicals.) In the spraying of pesticides, the climatic factors of wind, temperature, and humidity are extremely important, as is the method of application. Naturally, spraying when winds are light will reduce drifting. For efficiency in getting pesticides to the target, spraying is preferable to dusting. Dust particles are smaller than spray droplets. Their lesser gravitational pull allows dust particles to

pests. They and their kind have great economic impact and even affect human survival in a hungry world. Cultural and socioeconomic factors are important in any discussion of the necessity of pest control.

There are several categories of pesticides based on the type of target organism that a particular pesticide has been developed to eliminate. For example, **herbicides** are plant killers, **insecticides** are insect killers, and **fungicides** kill fungi. We may refer to all of these as **biocides.**

Insecticides: A Closer Look

The best-known pesticides are the insecticides. There are three principal chemical subgroups of these: chlorinated hydrocarbons, organophosphates, and carbamates.

Chlorinated hydrocarbons contain specific arrangements of atoms of hydrogen, carbon, and chlorine. Insecticides in this class include aldrin, dieldrin, endrin, heptachlor, DDT, DDE, DDD, methoxychlor, lindane, toxaphene, and mirex.

As their name implies, all **organophosphates** contain carbon, hydrogen, and phosphorus. Representatives of this group are diazinon, malathion, methylparathion, parathion, and phorate.

The carbamate insecticides contain arrangements of carbon, hydrogen, and nitrogen. **Carbamates** are not so numerous, and the most common one is carbaryl. Here we will focus primarily on the chlorinated hydrocarbons and the organophosphates.

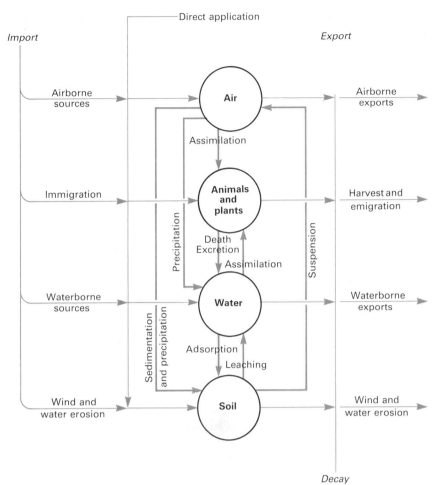

Figure 16.2 Pesticide Pathways.
The routes by which pesticides spread
through the ecosphere are many and
varied. They may be applied directly or
imported from (and exported to) other
ecosystems. The chemical characteristics
of the pesticide and the means of
application determine the pathways taken.

stay in the air longer, increasing the chances of drifting.

Ground spraying is more efficient than spraying by aircraft. In ground spraying, as much as 90% of the pesticide will remain in the targeted area under ideal conditions. In aircraft spraying, only about half is likely to stay in the targeted area. Pesticides worked into the soil may also get into the air by volatilization (evaporation). Pesticides in the air may be broken down by oxidation and photolysis by sunlight. They may also be washed out by rain or fall by gravitational pull and land in places they were not intended to reach.

The helicopter is spraying a fungicide on an orange grove in Florida. A helicopter can spray small and irregularly shaped fields more efficiently than an airplane can.

Water. Water contamination may result directly from pesticide application for insect or weed control. In other cases, runoff may carry soil particles laden with pesticides into bodies of water. Pesticides in the air may also drift over and be deposited on the water surface. Leaching of pesticides from soil occurs primarily with water-soluble pesticides (organophosphates) and may lead to contamination of groundwater. Improper disposal of used pesticide containers may contribute residues that are carried off in runoff. One of the major problems of pesticides in waterways is their presence in the bottom sediments of surface waters. Chlorinated hydrocarbons tend to adhere to soil particles; in deep lakes that stratify, fall and spring overturns (Chapter 3) can cause resurgences of previously sedimented residues. As long ago as 1969, the American Chemical Society reported that most surface waters in the United States contain chlorinated hydrocarbon insecticides, especially DDT, dieldrin, and the herbicide 2,4-D. The EPA has confirmed the presence of more than 45 pesticides in the groundwater of 26 states.

Soil. Adsorption of pesticides into the soil and their persistence in soil varies with the composition, pH, and temperature of the soil. Soils that have more humus tend to hold pesticides more strongly. Sandy soils tend to lose pesticides to crop absorption. At cooler soil temperatures, less pesticide volatilizes, and hence more is absorbed. Soil moisture influences the type of microorganisms present; this in turn influences breakdown of the pesticide by bacteria. Pesticides such as aldrin and heptachlor are displaced from soil particles by moisture and evaporate. What about pesticides in the soil taken up by crops? Most organophosphates can be degraded by plants, but chlorinated hydrocarbons are not; plants can change chlorinated

hydrocarbons into more toxic substances, however. Aldrin and heptachlor, for example, can be converted within plant cells to forms having even greater persistence. As indicated above, pesticides adsorbed onto soil particles may be carried into surface waters; pesticides may also leach from the soil into groundwater.

Pesticide Use
There *were* pesticides before the so-called age of chemistry, which began shortly before World War II; however, they were few, and nearly all of them were natural derivatives of plants. Pyrethrins are produced from chrysanthemums, nicotine sulphate from tobacco, and rotenone from a tropical plant. After World War II, the use of synthetic chlorinated hydrocarbons for pest control increased dramatically.

Today almost a billion pounds of synthetic organic pesticides are being used annually in the United States alone. Of that billion pounds, 69% is herbicides, 19% insecticides, and 12% fungicides (Pimentel, 1991).

The actual amount of pesticides used varies regionally, with greater use in more pest-prone regions such as the warm and humid South. Overall, agricultural applications account for 60% of the pesticides used in the United States; the remaining 40% is used in public health, industry, commercial pest control, and home applications.

The Impact of Pesticides on Pests: A Clouded Record
At first glance, pesticides appear to have had, in the short term, a significant positive impact on agriculture and health. Some scientists say that pesticides may have been directly responsible for a doubling of food productivity in the first half of the twentieth cen-

---

For Further Thought . . . 16-1

DDT: A Persistent Problem

Although some resistance is beginning to show in mosquitoes, DDT is still an effective method for controlling mosquitoes that transmit malaria. In developing countries that can't afford major drainage projects and public health systems, malaria is still a serious problem. DDT is still used in these countries to help to stop the spread of malaria. The use of DDT has been banned in the United States and elsewhere because of its impact on wildlife and its persistence. Traces of DDT continue to show up in the soils and sediments and even in the tissues of humans in the United States. This can be partly attributed to DDT used in other countries. Should developing countries curtail their use of DDT for malaria control? Should chemicals banned in the United States be made available to other countries by U.S. manufacturers?

---

tury. DDT has been successful in controlling mosquitoes that transmit malaria. DDT is also directly credited with controlling such other insectborne diseases as yellow fever, viral encephalitis, typhus, plague, cholera, and various tickborne diseases. For developing countries, even now, disease control is a major problem favoring the continued use of chemical insecticides.

However, the record is not really so cut and dried. The emergence of resistant strains of insects threatens the progress made by the use of synthetic pesticides. We may find ourselves back where we started. Outbreaks of insectborne diseases are occurring again. The amount and toxicity of pesticides used is ten times greater now than it was in 1945, yet losses of crops attributable to insects have nearly doubled (Pimentel, 1991).

Pests are indeed a significant human problem. Insecticides have helped us win an occasional battle, but the war is far from over, and the enemy may be getting stronger.

The Emergence of Resistant Strains

Insect resistance to insecticides is a real problem, not just a theoretical one. As we saw in Chapter 5, certain insects in a population may have greater resistance to a pesticide than other members of that population because of genetic variation. The resistant insects tend to survive spraying and reproduce, passing on their resistance. Thus over time the proportion of relatively resistant individuals increases as less resistant individuals are killed off.

About 450 species of insects, mites, and ticks worldwide have genetic strains that are resistant to one or more pesticides (Dreistadt and Dahlsten, 1987). A resurgence of malaria has been reported in more than a dozen countries, attributable to mosquito resistance to pesticides. A similar phenomenon has occurred in a number of other pest species. The EPA reports that 75% of the most serious agricultural insect and mite pests in California are resistant to one or more insecticides. In a few cases the development of resistance has had an unusual twist. It was found, for example, that spraying a pesticide for pests on cotton in Central America caused the emergence of larger numbers of mosquitoes that were resistant to the insecticide.

Even with all we know about the impact of DDT on the environment, we keep repeating our short-sighted mistakes. In the summer of 1990, Brazil sprayed DDT as part of a program to address an epidemic of malaria. The World Health Organization still recommends the use of DDT, and the World Bank funds its use to fight malaria. It is effective—in the short term. But in the long term, one female mosquito resistant to DDT can, in theory, produce 20 million

offspring in a couple of months (McCoy-Thompson, 1990). DDT is not a solution; history shows it is a stopgap measure with long-term negative effects that are not necessarily localized. DDT used in Brazil, once it volatilizes, may travel in the upper atmosphere to the Great Lakes and beyond (McCoy-Thompson, 1990).

The cotton-growing industry in northeastern Mexico and southern Texas was destroyed by the development of resistance to all registered insecticides in the tobacco budworm, a pest that infests cotton and tobacco (Dover and Croft, 1986). Various weeds have also been reported to have developed *herbicide* resistance. Food crops currently threatened by pest resistance include cabbage and rice in Southeast Asia, corn and potatoes in the United States, potatoes in Europe, and sugar beets in the United Kingdom (Dover and Croft, 1986).

There are also indications that replacing one type of pesticide by another type in the same group, one organophosphate for another, for example, preserves and extends previously selected resistance. Thus the use of chemicals to improve public health and control agricultural pests must be carefully considered; there is real potential for disease epidemics caused by resistant strains and crop losses from resistant pests.

Nonspecificity

Being nonspecific, insecticides kill beneficial insects along with pests. In a very real way, spraying DDT on a forest to kill bark beetles is like dropping a bomb on a city to stop urban crime. Competitors and predators of the target pest may be diminished in numbers along with the pest. There is the possibility that the pest will recover long before the predators that helped to keep it in check, creating a problem worse than the original one. Insects that may have held still another pest in control (through predation, competition, etc.) may be eliminated, and a secondary pest invasion may result. An example of a combination of such problems occurred in the Canete Valley in the late 1940s.

The Canete Valley is a cotton-producing area in Peru. Before 1949, older pesticides such as nicotine sulphate and arsenic compounds were used. Soon thereafter, chlorinated hydrocarbons were introduced to the area, and yields increased tremendously for a while. Then the increase slowed somewhat, and problems began to arise as resistant pest strains and new pest types appeared. This resulted in a disastrous crop failure in the mid-1950s. A 1977 UN report on the consequences of pesticide use in cotton-growing areas of Central America indicates similar problems. After several seasons of pesticide use, the primary pests were controlled, but previously minor insect pests sometimes increased in numbers, resulting in severe crop damage.

A study (Dreistadt and Dahlsten, 1986) of the effects of spraying of malathion to eradicate the Mediterranean fruit fly in California showed negative secondary impacts: increased garden whitefly, aphid, and mite outbreaks; scale outbreaks; and increased honeybee mortality. Some positive side effects included the suppression of other scale diseases.

The message in all of this is that the indiscriminate use of nonselective pesticides causes natural selection of resistant strains (see Chapter 5) and disturbs natural predator-prey relationships (see Chapter 4). This in turn sets the stage for an increase in resistant strains and invasions by strains normally held in check by predators. The result is that more intense and more frequent applications of pesticides become necessary, and these result in increased costs, increased pressure for selection and resistant strains, and increased opportunity for contamination and disruption in the ecosphere.

Pesticides and Health

Exposure Routes
Pesticides may enter the bodies of animals, including humans, by ingestion, inhalation, or absorption through exposed surface areas. Fish can absorb pesticides through their gills or body surfaces or by ingesting contaminated food. Land-dwelling wildlife ingests pesticides primarily in food but may ingest some in contaminated water and may pick up some from the air (for example, in licking their fur, rabbits may ingest pesticide adsorbed onto the fur from the air). Low-level exposure to pesticides over a long period of time can result in selective accumulation in living organisms.

Storage itself is not the problem. Problems come (1) when accumulated doses are activated, moving out of storage in high concentrations and causing one kind of health effect or another, or (2) when accumulated pesticide is further magnified as it passes up food chains. Lethal concentrations and doses (see Table 16.2) may be reached, and short of these there may be important sublethal effects.

Chlorinated Hydrocarbons in Animals Other Than Humans
Lethal effects of chlorinated hydrocarbons on fish, birds, and mammals usually involve the central nervous system, and the ultimate effect—death—may be preceded by symptoms such as respiratory difficulty, sluggishness, and neurological complications.

The sublethal effects of chlorinated hydrocarbons on fish include lowered reproductive potential, which is partly the result of the accumulation of pesticides in the yolk sac of fish eggs. As the developing fry absorb the yolk sac for nourishment, the pesticide is

Table 16.2 Insecticide Toxicity. The amount of specific pesticides lethal to half the test organisms exposed to that dose are shown here. LD_{50} is the lethal dose for 50% of the test organisms; LC_{50} is the lethal environmental concentration for 50%. LD_{50} and LC_{50} are standard toxicological terms reflecting the fact that not all of the organisms in a test group will be equally susceptible.

	LD_{50} (mg/kg in white rats)	LC_{50} at 11°C (mg/l in fish)
Chlorinated hydrocarbons		
Aldrin	40	0.0082
Dieldrin	46	0.0055
DDT	250	0.005
Endrin	12	0.0044
Heptachlor	90	—
Lindane	125	no effect at 0.03
Toxaphene	69	0.0022
Organophosphates		
Malathion	1,500	0.55
Parathion	8	0.065
Methyl parathion	15	irritation at 1.0
Carbamates		
Carbaryl	540	—
Zectran	15-36	no effect at 1.0

Source: Scientists Institute for Public Information.

released and may kill the fry. Other effects have been documented as well, including liver and kidney damage, damaged gills, and modified (usually increased) metabolism. In addition, behavioral changes such as seeking warmer waters have been noted. Other studies have shown that in the presence of endrin, minnows tended to abandon their protective behavior of staying under rocks during daylight hours.

The sublethal effects of chlorinated hydrocarbons on birds include lowered reproductive potential resulting from the failure of eggs to survive to maturity. Thin shells have been identified as the cause of this problem, and they apparently result from abnormal calcium metabolism in some birds affected with pesticides, notably birds of prey such as the bald eagle, the osprey, and hawks (see Figure 16.3). Other biochemical health effects include changes in the metabolism of certain substances and changes in liver functioning. Behavioral changes include delayed nesting due to alterations in hormone balances, which may cause delayed ovulation.

The effects of chronic low levels of persistent insecticides on land mammals are not well known. More work is clearly needed in this area.

Thus chlorinated hydrocarbons can in fact alter an ecosystem by negatively affecting its components. Even the base of the food chain may be affected. One study demonstrated that photosynthesis in marine phytoplankton was diminished in the presence of DDT. Since such phytoplankton are the base of all aquatic

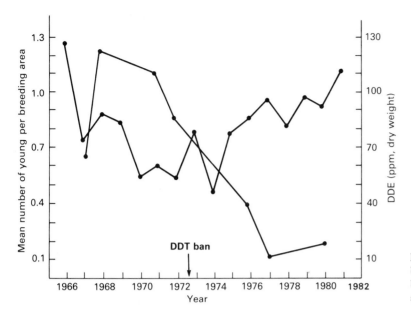

Figure 16.3 Eagle Eggs and DDT. The DDT ban was successful in improving the survival rate of bald eagles. DDT residues in their eggs declined significantly.

food chains and are a major source of oxygen in the ecosphere, such effects can be very important.

Organophosphates in Animals Other Than Humans

Organophosphates are also toxic. However, because they break down rather quickly, they are less likely to become chronic problems for fish and wildlife. Strict safety precautions are nevertheless necessary to guard against inadvertent acute exposure of animals during pesticide application. Should a wandering animal be exposed to organophosphates or carbamates, these pesticides may affect the animal's nervous system by inhibiting an enzyme called cholinesterase, which regulates the transmission of nerve impulses at nerve endings.

At many nerve-nerve junctions, nerve impulses are transmitted by the release of a chemical called **acetylcholine.** Acetylcholine released by the stimulating neuron stimulates the next nerve fiber in line. Once acetylcholine triggers the impulse in the receiving nerve cell, the acetylcholine is usually destroyed quickly by the enzyme **cholinesterase.** Without cholinesterase, a single stimulus would be greatly amplified and would be sustained for much longer. Organophosphates and carbamates inhibit the activity of cholinesterase, resulting in tremors or even death. Enzymes that are important in metabolizing foods may also be affected by these pesticides.

How severe the effects of organophosphates are depends on many factors, including *synergistic* effects. The effects of two or more pesticides acting at the same time may be different from the effect of one of the pesticides acting independently. For example, the organophosphate malathion is normally detoxified by enzymes from the liver and so may not be considered extremely harmful to mammals. However, if other organophosphates inhibit the action of the liver enzyme system, the malathion's toxic effects will be greater. Experiments on fish and wildlife have been geared primarily to exposure to one specific pesticide. Because species are likely to be exposed to several pesticides at a time in nature, the synergistic effects of pesticides need further investigation.

Human Health and Pesticides

The validity of comparing pesticide effects on animals to effects on humans might be questionable. However, there *have* been direct studies of the effects of pesticides on people. Case reports of overexposure in the workplace or accidental ingestion have provided opportunities for direct study. In addition, epidemiological studies of groups of people who are chronically exposed to pesticides at work have provided some data, as have studies of pesticide residues in the general population.

General Effects. The fate of a pesticide inside the human body varies with the pesticide. Chlorinated hydrocarbons tend to be stored in body fat and in the fat portions of the blood and human milk. Normally, organophosphates and carbamates are metabolized and excreted rather quickly. Certain herbicides pass through mammalian systems intact.

The immediate effects of organophosphate poisoning on humans include stomach cramps, dizziness, vomiting, and heavy sweating. Direct exposure of humans to organophosphates is also reported to cause symptoms of mental derangement, memory loss, sleepwalking, speech difficulties, and depression. Generally, organophosphates affect the activity of certain enzymes. As we mentioned earlier, cholinesterase is inhibited, and this affects the transmission of nerve impulses. Other enzymes that metabolize foods, food additives, drugs, and other chemicals are also inhibited, as are enzymes that detoxify other organophosphate compounds.

There are complications with the breakdown products of some organophosphates. Some research indicates that in the presence of high ozone concentrations, parathion produces 30 times as much paraoxon as under normal conditions. Paraoxon, a compound in which the sulfur atom in parathion is replaced by an oxygen atom, is 10 to 100 times more toxic to human red blood cells than parathion. Poisonings of farm workers in central California may have resulted from this process.

Food intake is the major route by which chlorinated hydrocarbons enter human systems. Some studies indicate that perhaps half of a person's food intake may contain measurable residues of persistent insecticides. Generally, the chlorinated hydrocarbons affect the central and peripheral nervous systems, perhaps by interfering with the transmission of nerve impulses. The symptoms of acute DDT poisoning include facial numbness, **malaise** (feeling unwell), headache, vomiting, dizziness, confusion, and tremor. Many chlorinated hydrocarbons are suspected carcinogens; these include DDT, chlordane, aldrin, dieldrin, and heptachlor.

Accidental poisoning by chlorinated hydrocarbon pesticides is not rare. Perhaps one of the most noted occurred in Hopewell, Virginia, where in 1976 a chemical company was found to be discharging wastes into the James River. These wastes contained residues of kepone, a persistent water-insoluble insecticide. The James River and Chesapeake Bay were substantially polluted by these discharges. Plant workers began to exhibit symptoms of poisoning, including tumors, skin discoloration, blurred vision, memory loss, coordination problems, and joint and chest pain. Kepone is also suspected to be a human carcinogen; it causes tumors in animals.

Body Burdens. We all carry the residues of pesticides in our bodies. Table 16.3 shows for some common pesticides (1) the important residue-containing foods, (2) the **tolerance level** (the concentration in parts per million that may be present in food considered safe for human consumption), and (3) the **acceptable daily intake (ADI),** the amount considered to be safe for human consumption on a daily basis. Tolerance levels and ADIs are set by the World Health Organization and the Food and Agriculture Organization on the basis of the best available data from manufacturers and the scientific literature. Table 16.4 compares the estimated daily per capita intake of insecticides for a 50-kg person with FAO/WHO standards for various countries.

The EPA runs the National Human Monitoring Program for Pesticides to determine environmental levels and trends in human contamination. Samples of adipose (fatty) tissue, blood, and urine are collected and analyzed. A number of interesting observations have been made as a result:

1. DDT was stored in larger amounts than any other organochloride found in the survey; it was in virtually all tissue examined.
2. Although consistently present, DDT residues have tended to be reduced in recent years. (Agricultural use of DDT was banned in the United States in 1972; only limited use by permit is still allowed.)
3. Dieldrin also occurred in low levels in almost every tissue examined, and its levels did not

Table 16.3   Selected Pesticides Whose Residues May Be Present in Foods

Pesticide	Major Food Source	Typical Tolerance (ppm)	FAO/WHO Acceptable Daily Intake (mg/kg body weight)
Organochlorines	Many agricultural products; concentrate in fat: fish, animal tissue, milk, eggs		
Toxaphene		7.0	—
Lindane		1.0	0–0.01
DDT		1.0	0–0.02
Dieldrin		0.1	0–0.0001
Heptachlor		0	0–0.0005
Organophosphates	Fruits, vegetables, grains		
Malathion		8.0	0–0.02
Parathion		1.0	0–0.005
Diazinon		0.75	0–0.002
Carbamates	Fruits and vegetables		
Carbaryl		10.0	0–0.01
Carbofuran		0.1	0–0.01

Sources: U.S. Food and Drug Administration; World Health Organization.

Table 16.4 Estimated Daily per Capita Intake of Insecticides for a 50-kg Person Compared with FAO/WHO ADI Values for Various Countries

Insecticide	Acceptable Daily Intake (μg)	Country	Total Intake (μg)
Total DDT	250	United States	55.0
		England	44.0
		Spain	78.4
Lindane	625	United States	3.0
		England	6.6
		Spain	13.8
Dieldrin	5	United States	3.4
		England	4.7
		Italy	1.9

Source: Adapted from S. McKerchner and F. W. Plapp, Jr., "Measuring the Residue," *Environment* 22:10.

change greatly over five years. Dieldrin also reflects exposure to aldrin, since the body rapidly converts aldrin to dieldrin.

4. Metabolites of heptachlor and chlordane were found in low levels in nearly all tissues.

5. Residues of lindane, mirex, and other pesticides were also found, but at low frequencies.

Results of urine sample analyses by the EPA also indicate the presence of pesticides including organophosphates, chlorinated hydrocarbons, and carbamates. A few of the samples contained carbaryl, parathion, and the herbicide 2,4,5-T.

The presence of pesticide residues in human milk is of particular concern. Several types of residues are found in human milk; DDT and its derivatives are found in higher concentrations than any of the other pesticides.

A report on pesticide residues in food for human consumption was conducted and published in 1984 by the Natural Resources Defense Council (NRDC). The NRDC conducted a survey of California-grown fruits and vegetables sold in San Francisco. Of the 71 samples taken, 44% contained detectable residues of 19 different pesticides, 18% contained residues of more than one pesticide, and three samples contained four pesticides. Four of the pesticides detected were DDT, dicofol, endosulfan, and trifluralin. There are several hypotheses as to where the DDT comes from. Some experts conjecture that the DDT is simply residue still in the soil from use before the ban. Others hypothesize that DDT is obtained illegally from Mexico and is used actively on crops today. In addition, dicofol (kelthane) contains DDT as a contaminant. A study by the U.S. General Accounting Office (1986) found that sampling pesticide residues in imported foods by the FDA was inadequate. According to the GAO, the rate of pesticide contamination in imported foods is twice the rate in domestic foods.

Pesticides around the Home.

Although several studies indicate that people in lower socioeconomic classes seem to be more exposed to pesticides because of greater prevalence of pest problems, bug-conscious Americans in all classes are exposed to home pesticides. Pest strips, bugproof shelf paper, termite control, rodent poisons, lawn treatments, and garden applications all mean considerable human exposure.

Lawn care is a $1.5 billion business in the United States, involving over 7 million households and 40 different pesticides. Of the 40 pesticides used by lawn care companies, Public Citizen, a nonprofit group formed by consumer activist Ralph Nader, (Weiss, 1989) classifies 12 as suspected human carcinogens, 21 as causing long-term health hazards, and 20 as causing short-term nervous system damage. The EPA reports that there is very little information available on the impact of these chemicals on humans. State regulation of lawn care companies is minimal. Public Citizen found that eight states require warning signs in the yard, eight permit consumers to request advance notice of spraying on their or adjacent lawns, three states require that customers receive a consumer information sheet, one state requires written contracts, and two states require every pesticide applicator to be trained and tested.

Because of public pressures, lawn care companies are starting to provide options such as fertilizer only or pesticide application only if pests are found. Some organic lawn care companies are also emerging.

A study of soils from 1968 to 1974 showed that urban soils actually tend to have a higher level of pesticide residue than agricultural soils (see Table 16.5). The only exceptions were in some areas of the South where DDT was used heavily on cotton crops. Lawns in the United States may receive 5–10 pounds of pesticide per acre—an amount significantly higher than most farms.

Consumers Union did a supermarket study of home-use pesticides and found 50 common "active" ingredients in 250 products ranging from flea collars to weed and bug killers and insect repellants. The EPA has little or no data on chronic toxic effects for over

Table 16.5 Occurrence of DDT and Chlordane in Urban and Agricultural Soils (percent)

Locality	DDT and Analogs		Chlordane	
	Urban	Agricultural	Urban	Agricultural
Pittsfield, Mass.	55.6	26.3	11.1	5.3
Washington, D.C.	59.1	20.0	33.3	0
Greenville, S.C.	61.6	75.0	9.3	0
Tacoma, Wash.	34.7	30.2	14.7	0

Source: U.S. Environmental Protection Agency.

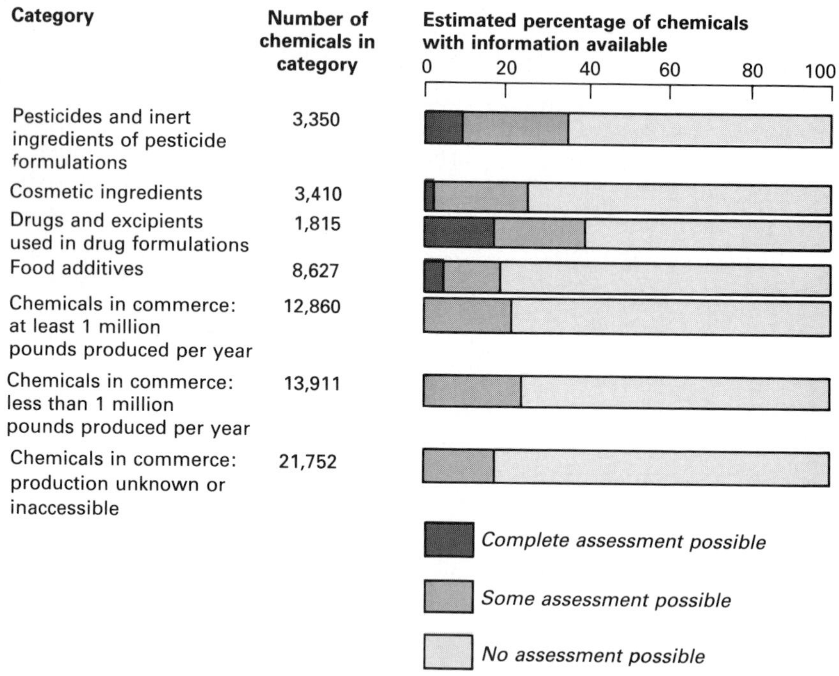

Category	Number of chemicals in category	Estimated percentage of chemicals with information available
Pesticides and inert ingredients of pesticide formulations	3,350	
Cosmetic ingredients	3,410	
Drugs and excipients used in drug formulations	1,815	
Food additives	8,627	
Chemicals in commerce: at least 1 million pounds produced per year	12,860	
Chemicals in commerce: less than 1 million pounds produced per year	13,911	
Chemicals in commerce: production unknown or inaccessible	21,752	

Complete assessment possible

Some assessment possible

No assessment possible

Figure 16.4 Adequacy of Toxicity Information for Assessing Human Health Hazards. A National Academy of Sciences committee estimates that humans may be exposed to almost 66,000 chemicals; we know very little about the effects of most of them on human health.

half of these active ingredients ("Pests at Home," 1988). In particular, Consumers Union looked for data concerning carcinogenesis, mutations, reproductive effects, birth defects, and neurobehavioral effects. Problems are even emerging with the so-called inert ingredients found in such products. There are about 1,200 inerts used in 50,000 pesticides; the EPA states that 100 have known or suspected negative health effects, 800 are relatively untested, and 300 appear safe for use (Kistner and Porterfield, 1987).

So how safe are we? No one really knows. We continue to pour chemicals into our environment and make them part of ourselves with little knowledge of their effects (see Figure 16.4).

Pesticide Bans and Restrictions

An Overview

Questions and concerns about the long-range effects of pesticides on humans and the rest of the biosphere have led to bans or restrictions on the use of specific pesticides. Such bans and restrictions have been hard fought because agriculture, forestry, and other interests depend so much on pesticides and because there is little solid scientific data on the health and environmental effects of pesticide use. (See Enrichment Box 16.1)

The following case studies illustrate many of the real problems associated with pesticide regulation. It

---

Enrichment Box 16.1

An Apple a Day . . .

Until the spring of 1989, an apple a day was considered a healthy habit. In the spring of 1989, people weren't so sure. It all had to do with Alar, a chemical used by apple growers to keep the apple on the tree longer, promote color, and increase storage time. In February 1989, the EPA announced that some studies showed that Alar had some carcinogenic effects on mice. The EPA concluded that continued use of Alar could result in 45 cancers per million exposed humans annually. The EPA decided to ban Alar effective July 31, 1990. The Natural Resources Defense Council and the media proposed even more serious implications from Alar use based on the fact that children are large consumers of apples and apple products. The carcinogen in Alar is UDMH (unsymmetrical dimethylhydrazine). Alar contains about 1% UDMH; the human body converts an additional 1% of Alar to UDMH. When Alar is

heated, as it is in the production of applesauce and apple juice, the conversion of Alar to UDMH is accelerated so that 5% of the Alar ends up as UDMH.

With increasing consumer concern, the company producing Alar stopped its production effective June 2, 1989.

There has been some controversy over the validity of the tests on the carcinogenicity of Alar and whether or not the media presented a balanced picture. Some people think that the Alar episode shows the need for greater understanding by the public and the media on risk assessment associated with chemicals. Others think that the episode shows the need for better methods of determining chemical toxicity on humans. Still others simply point to the power of the consumer to cause the early elimination of an alleged carcinogen from the marketplace.

---

takes years for pesticides to be reviewed. Even after they are shown to be carcinogenic in the laboratory, it might be years before any restrictive action is taken. Once action is initiated, it can be held up for years in the courts, and emergency exemptions can be granted to allow some uses. Once a substance is restricted or canceled, it may be found that substitutes are equally hazardous or have unknown effects.

Agent Orange

The most infamous of the herbicides is 2,4,5-T, or **Agent Orange,** used as a defoliant in Vietnam. (Stripes on the side of the containers holding the compound were orange; hence the name.) Agent Orange contains contaminants called dioxins. Although dioxins are extremely toxic, their long-term effects are currently only suspected. Dioxins are carcinogenic and cause birth defects in laboratory animals. A 1991 study by the National Institute for Occupational Safety and Health indicated only a slightly increased cancer risk for workers in the chemical industry who had exposure rates to dioxin 500 times greater than the norm (see Chapter 18).

In 1966, dioxin residues were found in fish, shellfish, and mother's milk in the Vietnam target areas. In 1978 and 1979, Vietnam veterans in the United States began complaining of a variety of symptoms such as numbness, skin rashes, liver problems, and birth defects in their children, all allegedly linked to exposure to Agent Orange. Several damage suits were filed by alleged victims. At least one major class action suit was settled out of court. Congress passed legislation mandating some compensation to veterans exposed to Agent Orange. The law cited some evidence linking exposure to chloracne (a skin disease similar to acne), liver dysfunction, and soft-tissue sarcoma.

In the 1970s, the U.S. Department of Agriculture restricted use of 2,4,5-T near waterways and homes and on food crops. However, its use continued in forest areas, along highway and power line rights-of-way, and on lawns and golf courses. In March 1979 the EPA suspended most uses of 2,4,5-T when a study indicated a possible link between dioxin and human miscarriages in Oregon. Miscarriages and infant deformities were found to occur with frequencies more than three times greater than expected in Alsea, Oregon, a mountain community located near a national forest that had been heavily sprayed with 2,4,5-T. Although the U.S. Forest Service and the timber industry supported the continued use of 2,4,5-T, its suspension was upheld by the courts. EPA hearings on the ban on 2,4,5-T continued. In October 1983, Dow Chemical withdrew its protest of the EPA's plan to cancel registration of 2,4,5-T and silvex (another toxic herbicide) for use. Dow cited the emotional and political concern about dioxin contamination as the reason for withdrawing its protest. Four days later, the EPA issued the cancellation notice.

Mirex and the Fire Ants

Another controversial ban has been on the use of the pesticide mirex, which was used throughout the South from 1958 to 1978 to combat the fire ant. The fire ant does not attack crops but is a pest because it creates large mounds for its homes in the fields. The ant does have a sting and is considered a general nuisance. The federal government has contributed close to $1 billion in a 20-year campaign against the fire ant, yet the problem remains.

Mirex is a persistent chlorinated hydrocarbon. When it decays, it degrades into kepone, a substance that is neurotoxic in humans. Mirex and kepone have been associated with cancer in laboratory animals. An EPA survey conducted in 1976 showed that 23% of all samples of human tissues contained mirex.

In 1977 the EPA ordered the phasing out of mirex by the end of June 1978 and approved the use of ferriamicide in its place. Although ferriamicide breaks down more quickly in sunlight than mirex, it contains some mirex and may have health effects of its own. In 1982 the EPA issued a temporary restraining order on the sale and use of ferriamicide pending further investigation. Several requests for emergency exemptions to use ferriamicide have been denied on the grounds that there is no evidence of significant economic or health problems arising from *not* using it on cropland.

EDB

Ethylene dibromide (EDB) is a halogenated hydrocarbon pesticide in the same family as DDT, chlordane, heptachlor, aldrin, and dieldrin. It has been used since the 1940s for fumigating soils before planting to kill such things as fruit fly larvae and roundworms and for fumigating stored grains, milling machinery, and logs. EDB is also used to fumigate citrus, tropical fruits, and vegetables.

In the 1970s, federal government studies indicated that EDB is a carcinogen. It also causes gene mutations and reproductive damage. In 1977 the EPA first moved to control the use of EDB, but strong industry lobbies and lack of initiative under the Reagan administration resulted in no formal action.

In the summer of 1983, reports from California, Georgia, Florida, and Hawaii indicated groundwater contamination by EDB. In September 1983 the EPA ordered the immediate suspension of the use and sale of EDB for soil fumigation. In March 1984 the EPA announced that as of September 1, 1984, citrus fruit and papayas with EDB residues could not be sold in the United States.

The most recent substitute for EDB is methyl bromide. Methyl bromide has also been around since the 1940s, but it has only recently been tested. Tests now indicate that methyl bromide is also a carcinogen.

Wood Preservatives

Pesticides are used not only in agriculture and forestry but also in manufacturing. In July 1984, after six years of special review, the EPA placed restrictions on three pesticides used as wood preservatives on items such as railroad ties, telephone poles, picnic tables, and desks. The restricted pesticides are creosote, pentachlorophenol ("penta"), and inorganic arsenicals; they account for 97% of all wood preservatives used. These pesticides are used to deter fungi, insects, bacteria, and marine borers in wood; they extend the service life of wood by up to five times. However, creosote was found to cause cancer in laboratory animals and skin cancer in workers regularly exposed to it. "Penta" has a dioxin impurity shown to cause cancer in animals. Epidemiological studies have shown arsenic to be associated with cancer in humans. "Penta" and the arsenicals caused birth defects in test animals; creosote and the arsenicals caused mutations in bacteria and lab animals. The restrictions are basically designed to protect workers but will require the wood industry to initiate a consumer awareness program on the handling and disposal of treated wood. Limits are placed on the use of treated wood indoors. Manufacturers were to significantly reduce the dioxin contaminant in "penta."

Laws Regulating the Manufacture and Use of Pesticides

Pesticide regulation in the United States began in 1947 when pesticides were lumped together with other chemicals in legislation dealing with "economic poisons." The **Federal Insecticide, Fungicide and Rodenticide Act (FIFRA)** of 1947 addressed itself to pesticides, but only those being shipped across state lines, all of which were to be registered and appropriately labeled. The *use* of pesticides was not specifically addressed. When attention was drawn to the dangers of pesticides in the 1960s, the public pressed for more comprehensive laws. In 1972 and again in 1975 a number of far-reaching amendments to FIFRA were enacted. These amendments extended the regulatory aspects of the 1947 law to all pesticides whether they were shipped across state lines or not. The amendments required that the EPA classify all pesticides as either general or restricted. **Restricted-use pesticides** can be used only under the supervision of a certified applicator, and there are specific testing re-

quirements for certification. States administer these programs if the state certification program is approved by the EPA. The intent is for pesticides that are potentially harmful to the biosphere to be applied only under supervision by individuals with some sense of the potential for harm as well as for good.

The 1972 and 1975 amendments also require registering of all pesticides intended for sale in the United States regardless of place of production. In addition, all pesticide producers must now register and keep detailed production and distribution records.

Under the amendments to FIFRA it is illegal to use pesticides improperly. The EPA has the authority to inspect establishments and private property with the owner's consent or, if there is some basis for believing that there is wrongdoing, with a warrant. The EPA may recall pesticide products, stop their sale or use, seize illegally used products, or seek court injunctions. Both civil and criminal penalties are specified in the amendments.

Because pesticides fall under the jurisdiction of numerous federal agencies, the federal act now also calls for cooperation among the agencies involved, including the EPA, the Food and Drug Administration, the Occupational Safety and Health Administration, the Federal Aviation Administration, and the Fish and Wildlife Service.

Amendments to FIFRA passed in 1988 require the EPA to complete its reregistration of the 600 active ingredients in pesticides within nine years. The amendments also transfer responsibility for the cost of storing and disposing of suspended or canceled pesticides from the EPA to the pesticide registrant. As of 1987, the EPA had banned four pesticides on an emergency basis—EDB, dinoseb, 2,4,5-T, and silvex. Under the old law, the EPA had to compensate each manufacturer for the remaining inventory and for storage and disposal—a cost estimated at $200 million. This change alleviated the need for the EPA to consider such costs before suspending or banning a pesticide.

There are other federal laws that affect pesticide regulation. The Federal Water Pollution Control Act and its amendments set effluent limits for toxic pollutants, including aldrin, dieldrin, endrin, and DDT. These laws are described in Chapter 12, where we also discuss pesticides as sources of water pollution and the Safe Drinking Water Act. The Federal Food, Drug and Cosmetic Act regulates the types and amounts of pesticide residues allowed on raw agricultural commodities and specifies labeling requirements. Many of the provisions of the Toxic Substances Control Act of 1976 also cover pesticides, since it provides for broad control of the production, distribution, and use of all potentially hazardous chemicals. All of these acts address individual facets of the production, transportation, use, and disposal of pesticides and other chemicals that are

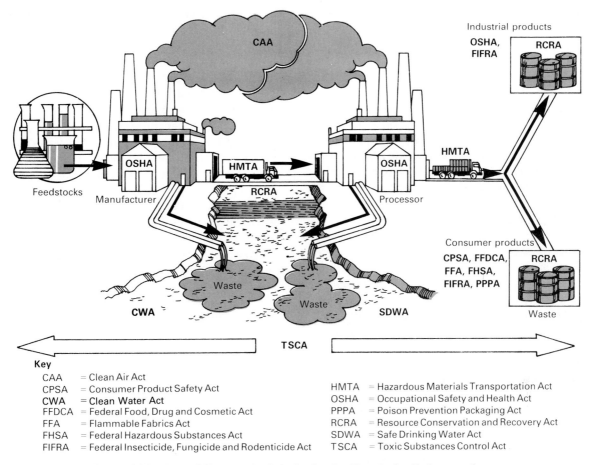

Figure 16.5 Laws Affecting the Life Cycle of a Chemical. Each piece of legislation is directed toward a different aspect of the problem from generation to consumption to disposal. There is obviously some overlap.

Key

CAA	= Clean Air Act		HMTA	= Hazardous Materials Transportation Act
CPSA	= Consumer Product Safety Act		OSHA	= Occupational Safety and Health Act
CWA	= Clean Water Act		PPPA	= Poison Prevention Packaging Act
FFDCA	= Federal Food, Drug and Cosmetic Act		RCRA	= Resource Conservation and Recovery Act
FFA	= Flammable Fabrics Act		SDWA	= Safe Drinking Water Act
FHSA	= Federal Hazardous Substances Act		TSCA	= Toxic Substances Control Act
FIFRA	= Federal Insecticide, Fungicide and Rodenticide Act			

potentially harmful to humans and to ecosystems (see Figure 16.5).

Integrated Pest Management and Other Alternatives to Insecticide Dependence

The University of California Biological Control Division releases millions of natural enemies of pests every year. University of California scientists claim that the program has probably saved more than $300 million in the past 50 years.

In 1944, Klamath weed, which had been imported from Europe inadvertently, was taking over prime rangeland in California. After the strategic release of natural enemies, leaf beetles from Australia, the weed all but disappeared. The resultant savings to the livestock industry are estimated to be $3.5 million a year.

In 1981 the cotton crop in Nicaragua cost more to produce than was returned for selling it. A good 30%

of the production cost was attributed to pesticides. In 1982 the government initiated a program to reduce pesticide use by using cotton stubble in the field baited with a sex attractant for the boll weevil. Only the baited areas were sprayed with pesticide. In 1983 data indicated a healthy cotton crop, increased acreage in the program, and a 60% decline in pesticide use, comprising 10% to 15% of production costs.

Cotton growers in Arkansas claim to be cutting pest control costs by $480,000 a year by using fewer pesticides and filling in with natural predators. Biological control agents used in management strategies also include viruses, bacteria, and fungi. There is an accelerating trend toward using combinations of some of the above strategies, the particular combination being based on the characteristics of the pest species in question. This combination approach is called **integrated pest management (IPM).** Given the economic and environmental effects of large-scale pesticide use, IPM is catching on fast. IPM is pest *control,* not pest *eradication.* It tries to keep pests at an economically tolerable level by using a combination of good agricultural practices and chemical and biological control agents. In IPM the most effective use of pesticide application is related to such things as the life cycle of the target organism, size of the target pest population, and weather. IPM takes into account as many strategic factors as possible.

The key to successful use of IPM is knowledge of the ecology of a pest species, including the micro-ecosystem of which it is a part. The basic idea is to make the system unfavorable for successful reproduction of the pest species. An IPM program may be developed as a combination of physical, chemical, and biological techniques used at strategic times in the life cycle of the pest. These techniques range from the simple to the highly sophisticated.

Some Elements of Integrated Pest Management

IPM may use simple agricultural practices like crop rotation to prevent a buildup of pests year after year or tilling a field at an appropriate time to expose insect eggs to drying by the air or to crush eggs and immature stages of the pest. It may involve strip-cropping to break up a large expanse of a single crop. For example, some pests that may attack cotton prefer alfalfa as a food source. If alfalfa is planted in strips between rows of cotton, the pest will tend to stay on the alfalfa and feed on it. Some pests, of course, will spread to the cotton, but the infestation and the damage to the cotton will not be as severe. As long as alfalfa is available, the pest will tend to leave the cotton alone. Experiments with corn, beans, and squash in Costa Rica have also demonstrated the effectiveness of this type of strategy.

The development of resistant strains of crops continues. One strategy is to breed strains with better natural defenses. Another is to create strains with shorter growing seasons so that the crop can be harvested before pests reach the most harmful point in their life cycles (Figure 16.6).

Sterile-Male Technique. A direct method aimed at decreasing the reproduction of pests is the sterile-male technique. One of the earliest uses of this technique was in the 1950s against the screwworm fly, which attacks cattle. Releasing radiation-sterilized males to mate with the female flies results in infertile eggs. This plan works because females mate just once, whether or not they are fertilized. Initial attempts were successful, and even now the screwworm fly is kept under control in many places by this technique. Control costs are estimated to be about one-fifteenth of the cost of potential annual losses due to fly infestations. In the summer of 1990, a release of sterile male screwworm flies in Libya was undertaken to stem the spreading of the pest in Africa. The fly is thought to have been carried to Libya three years ago in a livestock shipment from South America (Palca, 1990).

Pheromones and Juvenile Hormones. Many animals, including insects, communicate with members of their own species by secreting chemical substances called **pheromones.** Sex-attractant pheromones can be used to draw insects into traps, disorient insects, or otherwise decrease reproduction. This technique has been successful against the gypsy moth. Pheromones are most effective in controlling low-density populations. They are also useful in monitoring infected areas; the extent of infestation can be determined on the basis of the number of insects trapped over a specific period of time. **Juvenile hormones** are substances produced by insects that maintain them in an immature state. Chemicals that have similar effects can prevent insects from reaching sexual maturity and reproducing. The chief advantage of the hormone approach is its species specificity.

Biological and Natural Enemies. We have already described some uses of natural insect predators or parasites to control a pest population. In the

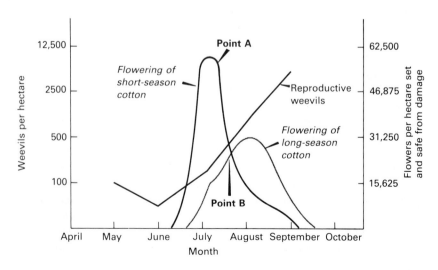

Figure 16.6 Outfoxing the Boll Weevil.
This graph shows how a variety of cotton that has a short growing season can be used to prevent substantial crop loss from the boll weevil. Point A shows when second-generation weevils begin to emerge in short-season cotton; by this time, most flowers have set and are safe from damage. At point B, second-generation weevils begin to emerge in long-season cotton; only a small percentage of flowers are safe by then.

early 1970s, a wasp parasite was imported into Connecticut to help to control elm spanworms, which were destroying elm trees. The parasites lay their eggs in the spanworm eggs. The wasp eggs then develop at the expense of the spanworm eggs. The wasp does not sting humans.

In a similar success story, a small fish introduced into California's Clear Lake appears to be helping to control the gnat problem. Both bacteria and viruses have also been used successfully to control specific pests. Bacteria have been used to control Japanese beetles, for example. Programs are testing the effectiveness of these biological pesticides and improving the efficiency of use and production so that their costs and control effects are comparable to those of their synthetic chemical counterparts. Like chemical pesticides, these alternatives are monitored and controlled by the EPA.

Resistance Management

Because it is a multifaceted approach to the control of pests, IPM should diminish the development of resistance in pest species. Resistance management refers to any attempts to prevent, delay, or reverse the evolution of resistance through various chemical and ecological techniques (Dover and Croft, 1986). The development of resistance is costly in terms of crop loss, increasing volumes of pesticide applied, and the use of new, more expensive pesticides to replace those no longer usable. More research is needed to determine how insect populations develop pesticide resistance and to find ways to block or reverse it. Monitoring resistance in insects is being done by various researchers, but there are no national systems for consolidating data. Manufacturers do test for the potential for resistance to their pesticide products, but no general test is available for pesticide groups. Biotechnology is expected to play a significant role in resistance management, since genes for resistance or susceptibility can be altered. The use of nonchemical control methods should certainly be considered as a promising alternative in pesticide resistance management.

Low-input Sustainable Agriculture

Whereas IPM is concerned primarily with controlling pests, a broader concept of agriculture, termed **low-input sustainable agriculture (LISA)**, is developing. This approach to farming minimizes the use of synthetic chemical fertilizers and pesticides and focuses on practices that will maintain the ability of the land to grow crops for an unlimited period of time. Thus erosion control and good irrigation practices are part of the program. The actual methods used depend on the characteristics of the specific farm and the crops to be grown but include such practices as crop rotation, use of leguminous crops and animal manure for fertilizer, use of cover crops, crop diversification, and mechanical cultivation to control weeds (also see Chapter 8). Field monitoring to determine soil needs and pest management strategies is important.

A report issued by the National Academy of Sciences called *Alternative Agriculture* (National Research Council, 1989) asserted that farmers who use little or no chemicals are usually as productive as those who do. The report called for changes in U.S. farm policy that promote use of chemicals and continued planting of the same crops year after year on the same land. Basically, the study found that "wider adoption of proven alternative systems would result in ever greater economic benefits to farmers and environmental gains for the nation." As we have seen before, sound environmental practices make good economics, especially in the long term.

Pesticides in the Third World

Pesticides banned in the United States may be sold in other parts of the world. The World Health Organization indicates that 500,000 people in the world each year are poisoned by pesticides and 10,000 of them may die; the numbers may actually be much higher. Third World countries eager to increase food production for domestic use or for export are often willing buyers. The United Nations Food and Agriculture Organization orchestrated a treaty to provide for "prior informed consent" by the receiving country before the exportation of specified pesticides. Exporter countries are to inform Third World consumers of restrictions on their products in Western countries. Studies by the U.S. GAO and the EPA indicate that "prior informed consent" does not always work well. Notification may be incomplete, may not be written in the language of the receiving country, and may not go to the right person. The process is burdensome, and receiving countries may not have the bureaucracies to handle it.

In addition, the ability to export pesticides that have been banned for use in the United States leads to what is called the "circle of poison." Pesticides that are banned for use in the United States can still be manufactured in the country. The manufacturer can export these pesticides that are then used in other countries and return to the United States in and on imported produce. Although a very small percentage of imported food is sampled, about 5% of what is sampled has been found to be contaminated by pesticides that are illegal in this country.

Indonesia has taken a major step to diminish pesticide use and to encourage IPM. Pilot projects in Indonesia showed that farmers using IPM had increased profit and slightly higher crop yields. Training of farmers nationwide in Indonesia is being undertaken, and the use of certain pesticides has been banned. The action was in response to the ineffectiveness of pesticides due to the development of resistance in the pests that attack Indonesian rice crops.

Sociological and Cultural Aspects of Pests

Monoculture: Invitation to a Feast

As we mentioned early in the chapter, social and cultural factors determine what is and what is not a pest. Many of our traditional practices ignore ecological realities and actually invite major pest problems. For example, the large farm with its huge expanse of one or two crops favors the invasion of pests and the development of diseases. Where there are large expanses of corn, large numbers of corn eaters will tend to move in, and corn disease can spread quickly. In 1970 some 15% of the U.S. corn crop was wiped out by southern corn leaf blight because of the lack of resistance in the varieties of corn used and the habitat provided by large, unbroken fields. Lawns of pure bluegrass or other grasses may be more susceptible to damage by sod webworms or other pests.

Beautiful Tomatoes

In recent years, pesticide use has become more important because of a demand for aesthetically pleasing raw foods. Some pesticide use is for producing cosmetically flawless fruits and vegetables that chain stores will select and shoppers will purchase. Our culture has put so much emphasis on looking good that we forget that invisible chemicals can be much more harmful than a slightly blemished piece of fruit.

The FDA sets **defect action levels (DAL)** for foods to limit the number of insects or insect parts in various kinds of food. Here are some examples:

___ Shelled peanuts: 5% insect-infested
___ Tomato paste and pizza: 30 fruit fly eggs per 100 grams or 15 fruit fly eggs and one larva per 100 grams or 2 larvae per 100 grams
___ Frozen broccoli: 60 aphids, thrips, and/or mites per 100 grams

DAL tolerances have become stricter in recent years mainly because of cosmetic considerations. Meeting the standards generally means additional pesticide use. The agricultural community is being pushed for quality control while at the same time being squeezed by increased pest control costs and environmental interests. However, there are indications that consumers are willing to sacrifice a few blemishes for organically grown food. They are even willing to pay a little more for such food. The number of certified organic farmers is increasing, as well as the demand for produce grown without pesticides.

What would happen if pesticide use on 40 crops in the U.S. were cut by 50% and nonchemical controls were used? David Pimentel (1991), a noted agricultural expert from Cornell University, concludes that the new controls would cost about a billion dollars and food costs would rise 0.6%. He estimates the social and environmental costs of pesticide use in the United States alone at $2.2 billion annually and believes that it may be even twice that. In 1985 Denmark approved a plan to reduce pesticide use by 50% before 1997. In 1988 Sweden set a goal of 50% reduction in pesticide use by 1993. The Netherlands is developing a program for 50% reduction in pesticide use by the turn of the century (Pimentel, 1991).

Global Metal Pollution

As discussed in Chapter 7, we humans rely on minerals in many of our daily activities. Minerals, specifically metals, are used in industrialized societies to manufacture appliances, automobiles, and other goods; to produce pipelines to carry water and oil; to produce farm machinery; and for many other purposes. Metals are released in the burning of fossil fuels and in mining, smelting, and metal refining. Metals are not merely persistent; they don't degrade at all. Once they enter the ecosphere, they will stay somewhere. There are indications that toxic metals have the potential to become a serious problem worldwide.

The issue of global metal pollution has been studied by Jerome Nriagu (1990), a research scientist for Environment Canada. Some of his findings are summarized here.

1. The industrial consumption and discharge of toxic metals have increased significantly since the early 1900s. A comparison of the emissions of trace metals into the air from natural and synthetic sources shows clearly that industrial emissions are the dominant source of trace metals in the air (see Table 16.6).
2. Transport through the air is a significant source of trace metals found in water.
3. Other major sources of trace metals in aquatic systems are wastewater and sewage from industrial and domestic sources and urban runoff. These cause localized problems, but such problems have been documented worldwide. Freshwater systems appear to be extremely susceptible to trace metal pollution. Bioaccumulation in the food chain can be significant.
4. Most of the contamination of soils from heavy

Table 16.6 Industrial and Natural Mobilizations of Trace Metals in the Biosphere (in thousands of tons per year)

Element	Production from Mines*	Total Industrial Discharges	Weathering Mobilization
Antimony	55	41	15
Arsenic	45	105	90
Cadmium	19	24	4.5
Chromium	≈6,800	1,010	810
Cobalt	36		120
Copper	8,114	1,048	375
Lead	3,077	565	180
Manganese	≈16,000	1,894†	4,800
Mercury	6.8	11	0.9
Molybdenum	96	96	15
Nickel	778	356	255
Selenium	1.6	76	4.5
Vanadium	34	75†	855
Zinc	6,040	1,427	540

* Only a fraction of each metal mined each year is released into the environment in the same year.

† Geochemistry and industrial discharges of these metals are largely unknown.

Source: J. O. Nriagu, "Global Metal Pollution," *Environment* 32 (1990):28.

metals comes from the disposal of metal-containing wastes from industry, agriculture, wood ash, and sewage sludge.

5. Urban areas are especially susceptible to trace metal pollution because they are closer to the synthetic sources. Airborne trace metals in urban areas may be 5 to 10 times higher than in rural areas and 100 times greater than in remote areas.

6. Trace metal pollution has been documented in antarctic ice layers and in the Arctic. Even in an environment as expansive as the ocean, the profile of lead follows the atmospheric flux of lead.

7. Since trace metals are deposited on the land surface and in topsoil, they are readily available to some crops and may enter the human food chain in that manner. Likewise, fertilizers and pesticides contain metals. In Japan, almost 10% of rice paddy soils are no longer suitable for growing food for human consumption because of excessive metal pollution.

8. There is great potential for metal pollution to become an even more serious problem in developing countries. Increased industrial development, lack of environmental regulation, and expanding needs of growing populations make increased metal pollution highly likely. Already airborne lead is higher in some cities in developing countries, such as Guatemala City and Buenos Aires, than in urban developed areas.

9. The health implications of global metal pollution

are difficult to know explicitly. There is little data on long-term, low-level exposure to most metals for humans and other living things (see Chapter 9).

10. Humans have become the key agents in the global redistribution of trace metals in the biosphere.

Persistent Chemicals Other Than Pesticides

We have focused on pesticides in this chapter. The problems we have discussed that surround pesticides apply to other persistent chemicals as well. These chemicals enter the biosphere when they are purposely or inadvertently released into the air, land, or water. Once in the biosphere they enter biogeochemical cycles. Those chemicals that are fat soluble have the danger of being biomagnified in the food chain and stored in fat tissue in animals, including humans. The toxicity of some chemicals may change as they are broken down within the environment or within living things. We discuss toxic chemicals in more detail elsewhere (see Chapters 12, 17, and 18). The major federal law in the United States enacted to address the issue of toxic chemicals is the Toxic Substances Control Act.

The Toxic Substances Control Act

Nearly everybody knows at least one toxic substance horror story. Chemically notable train derailments, evacuated neighborhoods, chemical dumps like Love Canal, vinyl chlorides, kepones, and asbestos have all had their share of the headlines in the United States. The U.S. Congress, in a characteristic zigzag approach to the problem, passed a number of laws designed to control toxic substances between 1970 and 1977. (These were the Clean Air Act, 1970; the Occupational Safety and Health Act, 1970; the Federal Insecticide, Fungicide and Rodenticide Act, 1972; the Federal Water Pollution Control Act, 1972; the Safe Drinking Water Act, 1974; the Resource Conservation and Recovery Act, 1976; and the Toxic Substances Control Act, 1977.) Of these, the Toxic Substances Control Act (TOSCA) (PL 94-469), which became law on January 1, 1977, is the most significant. As is suggested by the fact that six years of debate spanning three Congresses preceded the law, TOSCA is surely the most complex and the most far-reaching.

TOSCA is unique because it for the first time gave a regulatory agency, the EPA, the power to insist that new chemicals be considered guilty until proved innocent. The reverse had been true until TOSCA.

TOSCA recognizes two broad categories of chemicals—old ones and new ones. Old chemicals are those already in use when TOSCA became law; new chemicals are those that have or will come into use after January 1, 1977. The EPA was given the primary burden of assessing the risk associated with the old chemicals. The law intended that the burden for evaluating new ones should fall to industry.

During the first few years, two sections of TOSCA were the keys. One of these required that an inventory of chemicals be drawn up. The other was the establishment and definition of procedures to be used for pre-manufacturing notification.

It turned out to be an unexpectedly large problem that an inventory of existing chemicals had to be completed *first*. Since the law deals with new chemicals and with significant new uses of old ones, a list of the old ones had to be compiled first. As many as 5 million chemicals were already in existence, 63,000 of them in common use in the United States. By some estimates, as many as three or four *new* chemicals would have to be reviewed by an understaffed EPA every day!

As if this were not enough, when asked to provide lists of the chemicals they used, some companies claimed that such information was a trade secret and that providing it would tip off their competitors.

In the language of the law itself, testing is *not required*. Industry can decide when and when not to test. This was apparently built into the law because of our traditional American reverence for free enterprise. The way the law works, however, is that anyone intending to introduce a new chemical (or to use an old one in some new way) must file notice with the EPA at least 90 days in advance. Upon notification the EPA can do nothing, ban the chemical, limit its use, or ask for more information.

A second part of this problem is that the methods available for testing chemicals are far from perfect. For cancer-causing chemicals, for example, standard testing protocols are designed to establish any cancer-causing *potential*—not to predict the frequency with which cancer will appear in humans for a given exposure. The fact that we have decided to assume that there is no threshold for carcinogens does not mean that all carcinogens are equally potent. Some carcinogens may be a million times more potent than others. This obviously has profound regulatory implications.

TOSCA acknowledges the importance of economic considerations and in doing so adds another dimension to the complexities of implementation.

A clause in TOSCA that brings much litigation is one qualifying the authority of the EPA: "This authority should be exercised . . . as not to impede . . . or create . . . barriers to technological innovation while . . . [ensuring] that such innovation . . . [does] not present an unreasonable risk." Neither *impede* nor *unreasonable risk* is defined in the act. You can see at once how this clause might be subject to legal interpretation.

The effectiveness of the law has been mixed. Even given a rigorous application of the act, it is far from perfect. There will be chemicals that screens will miss, and humans will continue to be at risk.

Persistent Chemicals and the Biosphere

The problems associated with pesticides, other persistent chemicals, and metals show the connections between land, air, water, and living things. We are part of a single, interactive system. We must act with full knowledge that what we put into that system, what we disturb by our activities, may come back to us and stay with us a very long time.

Concepts to Remember

1. Materials cycle in the biosphere. Any stable substance that enters the biosphere can and probably will make its way into the nutrient cycles of humans.
2. Synthetic chemicals did not evolve in the ecosphere but have been created in the laboratory; biological systems have not generally developed enzymes to break down many synthetic chemicals. Physicochemical breakdown is much slower; hence some of these chemicals may remain in the ecosphere for a relatively long time.
3. Organisms cannot handle many naturally occurring substances such as metals in some of the high concentrations and chemical forms in which humans stir them up.
4. Chemicals may inadvertently enter organisms through food intake, water use, or gaseous exchanges with the atmosphere.
5. Persistent chemicals may affect organisms in ways and amounts not anticipated because of biomagnification, the fact that small amounts are passed up the food chain and concentrate in ever-higher amounts in individuals at each trophic level.
6. Pesticides are classified into the major groups of chlorinated hydrocarbons and organophosphates. Chlorinated hydrocarbons are

generally more persistent than organophosphates and are fat-soluble and thus may accumulate in the fatty tissue of organisms.

7. One danger of sustained pesticide use is the development of pests that are resistant to the pesticide. The emergence of resistant strains of insects jeopardizes the progress made by use of synthetic pesticides.

8. Most pesticides are nonspecific and kill organisms other than the target pest. They are biocides.

9. Pesticide residues are present in the fatty tissues, blood, urine, and milk of humans.

10. Integrated pest management (IPM) is an approach to pest control designed not to eradicate pests but to keep damage minimal. IPM minimizes the use of pesticides and gears pest control to the life cycle and habits of the pest species. It consists of good agricultural practices used in concert with chemical and biological control agents.

11. Alternatives to using synthetic pesticides for pest control include timely planting and harvesting, crop interspersal, use of biological and natural enemies, use of pheromones and juvenile hormones, and genetic breeding for pest resistance.

12. Many farm practices such as monoculture ignore ecological realities and invite major pest problems.

13. Low-input sustainable agriculture (LISA) minimizes synthetic chemical inputs and promotes the use of IPM and other practices in an attempt to maintain the land resource over time.

14. Control of persistent chemicals is a global problem, since air transports contaminants around the earth.

15. Further data must be obtained on the health effects of toxic chemicals, and governmental regulatory programs must be improved to control toxic chemicals in the ecosphere.

Environmental Career Profile

Laura A. Mahoney

Laura Mahoney is senior manager for risk assessment services with Eckenfelder, Inc., in Nashville, Tennessee. Eckenfelder is an environmental consulting firm specializing in finding solutions to hazardous waste problems. Ms. Mahoney's firm works throughout the United States, usually dealing with problems covered by the Resource Conservation and Recovery Act and the Comprehensive Environmental Response, Compensation and Liability Act. Ms. Mahoney's work in particular deals with state and federal superfund sites and other hazardous material storage and waste sites. After field investigations have been performed and before engineering solutions are described, Ms. Mahoney and her co-workers evaluate environmental data, sometimes employing models, to determine the fate of hazardous materials in the environment. She evaluates potential migration and exposure pathways and calculates potential human health risks from a particular site. For each project, her work culminates in a (usually lengthy) technical report. In addition to her own fieldwork, Ms. Mahoney also directs the work of others and participates in meetings with clients and with representatives of regulatory agencies. Ms. Mahoney finds her work both challenging and rewarding. She says that it calls on most areas of her academic background (she has a bachelor's degree in biology and a master's degree in physics and biophysics). The work is challenging, she says, because every problem she investigates is different—different chemicals, different geologic setting, different regulatory environment (laws are constantly changing), and different people. It is stimulating and satisfying to work in an emotionally charged area with a high degree of public awareness—to help to find solutions to problems that

people care very much about. For anyone interested in doing the kind of work she does, Ms. Mahoney recommends a technical degree in an environmental or related field plus five to ten years of related work experience. Strong writing and speaking skills are essential.

References and Further Reading

References marked with an asterisk are cited in the chapter.

Battista, G. S., 1973. "The Conviction of DDT," *Environment Reporter* **3**(39):1–21. (Monograph 14)

*Carson, R., 1962. *Silent Spring.* Boston: Houghton Mifflin.

"Dioxin Risks Exaggerated, Study Concludes," 1991. *Washington Post,* January 24.

*Dover, M. J., and Croft, B. A., 1986. "Pesticide Resistance and Public Policy," *BioScience* **36**:78–92.

*Dreistadt, S. H., and Dahlsten, D. L., 1986. "Medfly Eradication in California: Lessons from the Field," *Environment* **28**:18–20ff.

*Dreistadt, S. H., and Dahlsten, D. L., 1987. "Pesticide Resistance: Strategies and Tactics for Management," *Environment* **29**:25–27.

Foran, J. A. and Glenn, B. S., 1991. "Reducing the Health Risks of Sport Fish," *Issues in Science and Technology* **VIII**(2):73–77.

*Kistner, W., and Porterfield, A., 1987. "Bug-Spray Alert," *Technology Review,* May–June, pp. 10–12.

Lappé, M. 1991. *Chemical Deception: The Toxic Threat to Health and the Environment.* San Francisco: Sierra Club Books.

*Mattes, K., 1989. "Kicking the Pesticide Habit," *Amicus Journal* **11**(4):10–17.

*McCoy-Thompson, M., 1990. "Brazil Enlists DDT against Malaria Outbreak," *Worldwatch,* July–August, pp. 10–11.

*National Research Council, 1989. *Alternative Agriculture.* Washington, D.C.: National Academy Press.

*Nriagu, J. O., 1990. "Global Metal Pollution," *Environment* **32**:7–11ff.

*Palca, J., 1990. "Libya Gets Unwelcome Visitor from the West," *Science* **248**:117–118.

*"Pests at Home," 1988. *Environmental Action,* January–February, pp. 8–9.

*Pimentel, D.; McLaughlin, L.; Zepp, A.; Lakitan, B.; Kraus, T.; Kleinman, P.; Vancini, F.; Roach, W. J.; Graap, E.; Keeton, W. S.; and Selig, G. 1991. "Environmental and Economic Effects of Reducing U.S. Agricultural Pesticide Use," *BioScience* **41**(6):402–409.

Powledge, F. 1991. "Toxic Shame," *The Amicus Journal* **13**(1):38–44.

*Rosenblum, M., 1987. "World's Poorer Nations Spreading Pesticide Poison," Associated Press, November 11.

Stigliani, W. M.; Doelman, P.; Salomons, W.; Schulin, R.; Smidt, G. R. B.; Van der Zee, S., 1991. "Chemical Time Bombs," *Environment* **33**(4):4–9ff.

*U.S. General Accounting Office, 1986. *Pesticides: Better Sampling and Enforcement Needed on Imported Food* (GAO/RCED 86-219). Gaithersburg, Md.: U.S. General Accounting Office.

*Weiss, L., 1989. *Keep off the Grass.* Washington, D.C.: Public Citizen's Congress Watch.

*Woodwell, G. M.; Wurster, C. F.; and Isaacson, P. A., 1967. "DDT Residues in an East Coast Estuary: A Case of Biological Concentration of a Persistent Insecticide," *Science* **156**:821–824.

Yett, S. P., 1989. "What Is Agent Orange?" *Congressional Quarterly Weekly Report,* December 16, p. 3430.

Zilberman, D.; Schmitz, A.; Casterline, G.; Lichtenberg, E.; Siebert, J. B., 1991. "The Economics of Pesticide Use and Regulation," *Science* **253**:518–522.

17

Waste Disposal in the Ecosphere

Every year around the world, billions of tons of waste are generated. This waste is the result of activities in our own homes, businesses, and industries. Disposal of this unwanted material is a monumental environmental problem with many dimensions.

Waste amounts to lost mineral resources; we covered this aspect of the problem in Chapter 7. Waste hastens the exhaustion of valuable energy resources because it generally takes more energy to make things from scratch than to make them from recycled material, and it takes more energy and dollars to process lower-grade ores as the higher-grade ores are exhausted. These dimensions of the problem were explored in Chapters 6 and 7.

If waste is incinerated, it contributes to the air pollution problem; if it is buried, it contributes to the water pollution problem and takes up land and space; if it is dumped at sea, it diminishes marine resources and may wash up on beaches. Regardless of what is done with waste—short of not generating it in the first place—valuable energy resources are consumed in collecting, transporting, and disposing of it. These things are all well known, and yet the problem keeps getting worse.

As we review problems of disposal and solutions to those problems, it should quickly become apparent that we have waste problems because our socioeconomic system fails to take into account some of the most basic ecological principles governing material cycles. Although some recycling of waste is done in manufacturing places (as for cullet in the glass industry) and some municipalities separate and recycle paper, aluminum cans, and glass, by and large, industry and consumers generate a one-way flow of materials (see Figure 17.1). This flow ultimately piles up and causes problems somewhere in the ecosphere.

The Origin and Nature of Solid Waste

What Is Solid Waste?

Solid waste may not be a solid at all. As defined by U.S. federal law, **solid waste** may be a liquid, a semiliquid, or even a liquefied gas. It may be generated from industries, municipalities, businesses, or homes. It may be a discarded item that is worn out, broken, or otherwise no longer of use; it may be packaging or organic food waste; it may be a toxic substance produced as a by-product in some industrial process.

From location to location, the composition of solid waste differs significantly. For example, rural

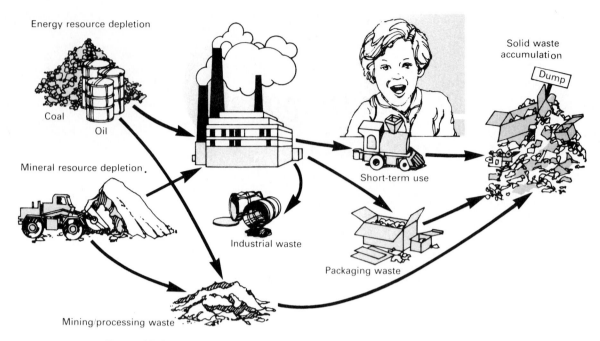

Figure 17.1 **Typical One-Way Flow of Materials.** Most of our packaging and manufactured products move in a one-way flow from resource extraction to disposal. This results in resource scarcity at one end and disposal problems at the other.

communities may have more food and other organic waste, whereas cities tend to have more paper waste. Figure 17.2 gives a general breakdown of municipal waste in the United States by type of material and type of product. On the average, 70% to 80% of residential and commercial solid waste is combustible. It is worth noting that combustible solid waste yields about 9 million BTUs per ton, compared to 24 million BTUs per ton for coal.

Where Does Waste Come From?

Of the more than 4 billion tons of waste produced annually in the United States, about 75% comes from mining, 12% from agriculture, 9% from industry, 3%

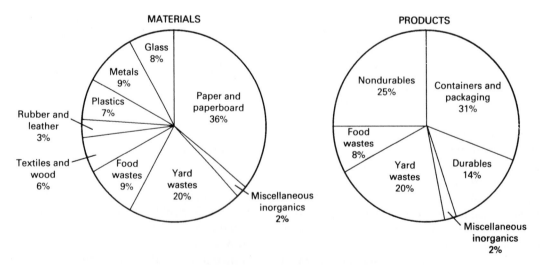

Figure 17.2 **Content of Municipal Solid Waste, by Weight.** This figure breaks down the content of municipal solid waste in the United States by type of material and type of product. Amounts vary from time to time and place to place. For example, paperboard may vary from 15% to 40% of the total by weight. Notice that packaging makes up a significant percentage of the waste stream. It is used once and thrown away. Is there a better way to use our limited resources?

Hazardous waste: useless, unwanted, or discarded material that may pose a threat to human health or the environment.

High-level radioactive waste: spent fuel or solidified waste from reprocessing spent fuel.

Leaching: the process by which nutrient chemicals or contaminants are dissolved and carried away by water or are moved into a lower layer of soil. The contaminated liquid is called *leachate.*

Residue: material that remains after gases, liquids, or solids have been removed. Incineration residue refers to solid material collected after an incineration process is completed.

Resource recovery: the process of obtaining material or energy from waste. Several processes are used to convert solid waste into solid, liquid, or gaseous fuels.

Sewage sludge: the concentration of solids removed from sewage during wastewater treatment.

Solid waste: useless, unwanted, or discarded material. *Agricultural waste* results from the rearing and slaughtering of animals and the processing of animal products and orchard and field crops. *Commercial waste* is generated by stores, offices, and other activities that do not actually turn out a product. *Industrial waste* results from manufacturing and industrial processes. *Municipal waste* is residential and commercial waste generated by a community. *Residential waste* is household solid waste.

Solid waste management: the purposeful, systematic control of the generation, storage, collection, transport, separation, processing, recycling, recovery, and disposal of solid waste.

Source reduction: the alteration of processes, practices, and policies to reduce the amount of waste generated.

---

from municipalities, and 1% from sewage sludge. Much of this waste could be classified as hazardous.

The Role of Culture and the Economy

Some estimates say that each person in the United States produces over half a ton of domestic solid waste per year—almost 4 pounds a day. Domestic solid waste includes residential household waste and waste from commercial businesses. Per capita industrial and domestic waste loads vary considerably around the world. Among industrialized nations, for example, U.S. and Canadian residents generate almost twice the amount of waste per person as western Europeans and the Japanese. On a per capita basis, the United States is the top generator of waste in the world (Young, 1991). Perhaps the legacy of the seemingly unlimited space and resources of the frontier sets the United States apart from most other industrial nations. Being younger, with lower population densities, and endowed with a lot of space and an immense material resource base, the United States has been able to prosper even with an inefficient materials system. Rapidly escalating costs of mineral imports, energy imports, and solid waste disposal should change this rather abruptly.

Although the amount of waste per capita generated in developed nations varies, one trend is clear: More and more municipal solid waste is being generated (see Table 17.1). Although a comfortable standard of living does not necessarily have to result in a high production of solid waste, there does seem to be a relationship between affluence and waste generation. The differences between developed and developing nations bear this out. The average person in Hong Kong or in Hamburg generates twice the waste each day as the average person in Calcutta or in Kano, Nigeria. The average person in New York City generates five times as much!

During the recession of 1974–1975, the volume of solid waste in the United States decreased (see Figure 17.3). Paper and board waste seem to be the best indicators of the relationship between affluence and waste. Nonfood waste decreased from 84.8 million tons in 1974 to 77.5 million tons in 1975. Likewise, in 1975, container and packaging waste volumes decreased to 1971 levels, and the volume of discarded newspapers, books, and magazines dropped even lower. Some combination of the production of fewer

Table 17.1 Percent Change in Municipal Solid Waste Generation in Selected Industrialized Countries, 1980–1985

Country	Total	Per Person
Ireland (1980–1984)	+72	+65
Spain (1978–1985)	+32	+28
Canada	+27	+21
Norway	+16	+14
England and Wales (1980–1987)	+12	+11
Switzerland	+12	+9
Denmark	+6	+6
Sweden	+6	+5
France	+7	+5
Italy	+7	+4
Portugal	+13	+4
United States	+8	+3
Austria (1979–1983)	+3	+3
Luxembourg	+2	+2
Japan	0	−3
West Germany (1980–1984)	−10	−9

Source: Adapted from J. E. Young, *Discarding the Throwaway Society* (Washington, D.C.: Worldwatch Institute, 1991), p. 14. Data from the Organization for Economic Cooperation and Development and the U.S. Environmental Protection Agency.

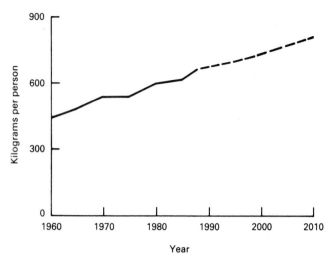

Figure 17.3 Estimated Average Individual Domestic Waste Generated, 1960–1988, with Projections to 2000. Although the number of pounds of solid waste generated daily per person cannot be determined exactly, estimates indicate that per capita generation of solid waste has increased over the years. The dip in the mid-1970s reflects a recession, which apparently resulted in more conservative use of materials.

items, corner cutting, increased scavenging, and frugality apparently accounted for these trends.

Composition Trends

The composition of solid waste has changed over the years. An increase in plastic waste and paper used as packaging and a decrease in food wastes have characterized the change in the United States. Plastic con-

Does an affluent society have to generate so much waste?

tinues to grow as a component of the solid waste stream. It increased from less than 5% by weight in 1980 to 9% in 1990. (See Enrichment Box 17.1) This amounts to about 20% of the waste stream by volume (Young, 1991). Since plastics and paper comprise most of the combustible components of solid waste, today's waste would seem to be more burnable than waste in the past. However, concern over toxics released from incineration of plastics (and waste in general) is growing. Some of the toxic chemicals used in plastic packaging are benzene, cadmium compounds, carbon tetrachloride, lead compounds, styrene, and vinyl chloride. The increase in plastic waste also means that the waste stream is less biodegradable than it was in the past. Degradable plastics are available, but there has been little experience with their degradation products.

Solid Waste Disposal Practices

If materials are not reused directly or recycled, there are only three places for the waste to go: into or onto the land, into the air, or into the water. There are no other options. (See Enrichment Box 17.2). We considered landfills in Chapter 13; here we will focus on ocean dumping and incineration.

Ocean Dumping

Solid waste that does not end up in or on the land or in the air ends up in the water. Water can be contaminated by waste through runoff and leaching. For many years the dumping of wastes into surface waters, in particular ocean dumping, was practiced as an inexpensive and convenient method of waste disposal. More stringent and expensive on-land requirements for incineration or other disposal further encourage ocean dumping. Today approximately 50 million tons of waste is discharged into the oceans worldwide each year. This waste consists of 80% dredge spoils (materials scraped from the bottoms of bodies of water to improve channel flows), 10% industrial wastes, 9% sewage sludges, and 1% miscellaneous. Since much of the discharging takes place near the shore, the dilution benefits of the large, open sea are not available.

Field studies at ocean sewage outfalls in California show high concentrations of some metals in sediments and in marine fish and shellfish, eutrophication, and alteration of ocean bottom habitats. Some organisms seem to flourish near the discharge areas; others disappear, and still others exhibit infections. Studies on the East Coast show that in one acid dump site off New York, used since 1948 for so-called nontoxic chemical wastes, sediments are contaminated, and marine organisms show concentrations of contaminants. Infections and fin rot are common.

Plastics

Plastics make up 9% of the waste stream by weight and 20% by volume. But it is estimated that their contribution to the waste stream will grow. The EPA estimates that it could be 40% of the waste stream by weight in 2000. Due to consumer concern about plastics, industry has made an effort to address the role of plastic in the waste stream in two ways. One is by experimenting with degradable plastics. The other is by initiating plastic recycling programs.

Degradables

Plastics are petroleum-based materials composed of long chains of molecules made up of smaller molecules. They do not degrade easily. In fact, degradation may take 500 years or more. Public concern has increased over the lack of degradability because of the volume that plastics occupy in landfills, the litter problem, and the danger to wildlife from ingesting or becoming entangled in plastic wastes on land and in the water.

Plastic that would degrade seemed like a dream come true. But now, there is general consensus that degradable plastics have limited value and may, in fact, be detrimental to environmental concerns.

Degradable plastics are formed in two basic ways. The plastic is made in a way that is attractive to some microbes and thus truly degrades. This type of plastic is very expensive and limited to such uses as sutures. The second method is to use filler in the production of the plastic that breaks down by microbial action, such as cornstarch (biodegradable), or reacts with light (photodegradable).

There are several problems with the so-called degradables. No one really knows how long they will take to degrade or what the final product will be. Theoretically, these plastics can be broken into small enough bits to be fully degraded, but studies show slow progress to this end. Degradables do not address the landfill space problem. Photodegradables will not have access to light. Recent landfill studies show that under current landfilling methods, even naturally degrading materials do not degrade (see Rathje, 1989). Conditions are not right. Even newspapers 40 to 50 years old that have been landfilled are still readable. Degradability to save landfill space is a moot point.

The value of degradable plastics in addressing the litter and wildlife problems is a function of how fast they degrade. This must be balanced with an appropriate shelf life. Degradables do, in fact, have very limited use, except perhaps as sheet plastic or plant pots for home garden use.

What is of even more concern is that the degradable particles may prove to be themselves a nuisance and perhaps even toxic. Promoting degradable plastics may deter the more environmentally sound approaches of source reduction, reuse, and recycling. Fillers in degradable plastics contaminate the recycling process. Manufacturers say that in producing degradables, they are responding to public demand.

Recycling

Efforts by industry to recycle plastics are increasing. Actions by states to ban plastics and public concern have provided the impetus for industry recycling efforts. Generally, the plastic is recycled into new products, not more of the original product. The two types of plastic most frequently recycled today are PET (polyethylene terephthalate), used in soft drink bottles, and HDPE (high-density polyethylene), used for milk jugs.

There are several problems to be overcome. Collection is a problem. Many curbside collection programs do not include plastic because of the volume involved. Areas with container deposit programs provide one of the best means of obtaining a high return. Once the plastic is collected, separation can be a problem. There are five common types of plastic. Pure melted resins from a single type of plastic are the most useful, since each type has different strength and flexibility characteristics. Some states are requiring that codes be placed on plastic containers to identify their resin type. Europeans have found uses for melted mixed plastic waste, and the technology is slowly finding its way to the United States.

All indications are that the amount of plastic we use will increase. Currently, half of our plastic waste is packaging. A fundamental question must be addressed: Do we want to continue to use our finite petroleum resources to produce products that we often use once and throw away? Does recycling of plastic into new products save additional petroleum inputs? Are there adequate substitutes?

Floating Dumps

In March 1987 the problem of where to dispose of solid waste received international attention when a barge carrying 3,186 tons of solid waste left Islip, New York, for a landfill in Morehead, North Carolina. The waste was refused in North Carolina. The barge traveled 6,000 miles, and operators conducted negotiations to dispose of the waste with six states (North Carolina, Alabama, Mississippi, Louisiana, Texas, and Florida), Mexico, Belize, and the Bahamas. Finally, in May, the barge made its way back to New York. In September the waste was incinerated in Brooklyn and the ashes returned to the Islip landfill on Long Island. The Islip landfill was being closed because it overlies a drinking water aquifer. Many communities are facing the problem of diminishing landfill space. The barge *MOBRO 4000,* the tug *Break of Dawn,* and the journey they made with the Islip waste are just the tip of the trash heap.

In some places, routing sewage into the ocean is still practiced. But the oceans cannot flush away our wastes indefinitely.

The dump site used by Philadelphia from 1973 to 1980 was located off the coast of Maryland. Studies show effects on sediments and bottom-dwelling organisms but no long-term effects on the water column. A 106-mile dump site off the New York coast used since 1961 for chemical wastes is characterized by depths of 1,700–2,750 meters and shifting water masses. Field studies have shown little impact on marine life here; elevated concentrations of pollutants have been temporary and limited to the dumping area. The volume and dynamic exchange of water in this area is thought to be an important factor in apparently diminishing long-term environmental impacts.

Changes in waste content and volumes, including increased amounts of nonbiodegradable and hazardous chemical wastes, have made ocean dumping ever less satisfactory. Coastal seafood operations are affected by chemicals and by potential pathogens from sewage sludge that get into the food chain. Floating debris from ocean dumping tends to wash back onto beaches, causing both health and aesthetic problems. During the summer of 1988, episodes of garbage washing up on beaches, including medical wastes such as hypodermic needles, vials of blood, and other hospital garbage, were reported from Maine to Texas.

The 1972 Marine Protection and Sanctuaries Act requires that a permit be obtained for dumping in the ocean. This legislation also prohibited any further dumping of sewage sludge in the oceans after December 31, 1981. **Sewage sludge** is the solid residue from wastewater settling tanks. Today it is often composed of toxic heavy metals and synthetic organic compounds as well as other organic materials, pathogens, nutrients, and water. However, as of 1988, sewage sludge was being dumped in ever-increasing amounts in the oceans, and the prospect was for a continuation of that trend.

What happened to the deadline? New York City, having run into numerous problems—one of which was cost—in attempting to convert to land-based sewage sludge disposal, challenged the deadline in court. The court held the ban to be invalid and said that New York City had to be given the opportunity to show that its dumping would not "unreasonably degrade" the marine ecosystem.

Because of the low cost of ocean dumping, other municipalities considered it, including Philadelphia, Baltimore, Boston, Washington, D.C., and Jacksonville, Florida. Consideration was also being given to using more deep-water sites. Although that involves higher transport costs, such sites might provide for more rapid dilution and less bottom accumulation of materials. However, continuing problems with contamination of beach areas in 1988 provided increased pressure for Congress to act, and legislation was passed to ban all U.S. ocean dumping by 1992.

Table 17.2 compares the effects of using various types of disposal technologies including ocean disposal for sewage and industrial wastes containing certain contaminants.

Lahey and Connor (1983) attempted to compare the health risks associated with ocean disposal with those associated with land disposal. For residents of Nassau County, Long Island, they compared the cancer risk associated with soluble chemical compounds that are likely to leach from land disposal sites and contaminate groundwater with the cancer risk of compounds that are likely to be absorbed and stored by fish eaten by humans (Figure 17.4). They calculated that the additional cancer risk over a lifetime for a 155-

Contaminants	Effects of Ocean Disposal	Effects of Land Disposal	Effects of Incineration
Pathogens (bacteria, viruses, and parasites)	Contamination of shellfish	No problem if sludge is stabilized; should not apply sludge on crops grown for raw consumption	Destroyed by incineration
Metals	Mostly nontoxic as a result of chemical changes after dumping	Absorption of cadmium, zinc, nickel, and copper potentially toxic to plant growth; potential hazard to human health from eating crops contaminated with cadmium	Metals in air emissions and ash
Organic chemicals	Bioconcentration by fish, a potential health hazard	Potential absorption by cattle grazing on land where sludge has been spread; potential groundwater contamination by smaller, more soluble compounds	Largely degraded by incineration; possible formation of suspected carcinogens

pound person who drinks 2 liters of water per day is 34 in a million. If the same individual eats 6.5 grams of fish per day, the lifetime cancer risk is 65 in a million. Both levels are greater than the risk threshold considered acceptable by the EPA. The authors concluded that there should be as much concern about disposal of pollutants such as PCBs in shallow ocean water as there is about dumping them on the land.

Incineration

Solid waste that does not end up on or in the land or in the ocean may end up in the air. Incineration is a relatively common method of solid waste disposal in areas with concentrated populations. In these areas, scarce land and transportation costs make incineration an economical alternative. Combustion of organic matter produces carbon dioxide, water, particulate matter, various gases, and ash. Particulate matter and gases that enter the air cause both health and economic problems (see Chapters 9–11). Ash causes a residual solid waste disposal problem.

The coupling of waste incineration with energy recovery is gaining increasing attention as a compelling solution to two environmental problems. We will

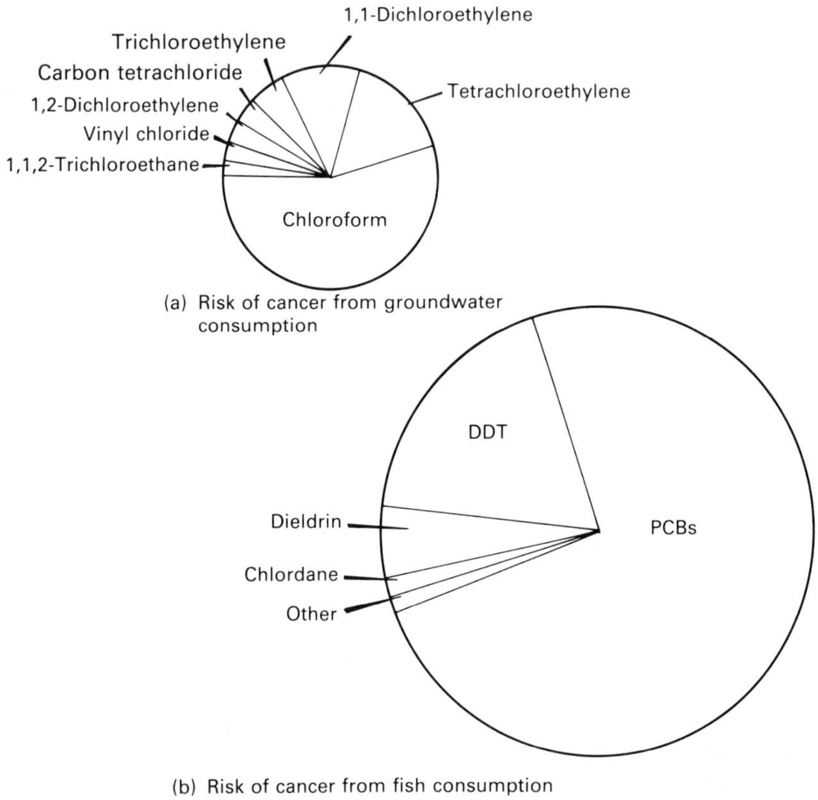

(a) Risk of cancer from groundwater consumption

(b) Risk of cancer from fish consumption

Figure 17.4 Risks of Getting Cancer from Drinking Tap Water versus Eating Locally Caught Fish. (a) The risk from drinking water is about 34 in a million; more than half of the risk comes from chloroform contamination. (b) The risk from consuming fish is about 65 in a million; more than 75% of the risk comes from PCBs.

discuss this, as well as the relationship between recycling and incineration, in subsequent sections.

Waste as an Energy and Material Resource

When Is a Waste Not a Waste?

One way to diminish the problem of disposal is to reuse the waste. Very simply, **resource recovery** is a method for turning wastes into resources by recovering usable products—both materials and energy. Resource recovery may involve simple facilities such as aluminum and paper sorting and recycling centers or complex automated materials recovery facilities. In any case, as the cost of disposal continues to rise because of land prices and pollution controls, resource recovery is likely to become more and more common (see also Chapter 7).

Resource recovery can begin at home. Sorting paper, glass, and aluminum waste to be delivered to the nearest recycling center primarily involves human energy. Placing organic wastes in a backyard compost heap reduces the volume of waste put into trash cans and saves valuable nutrients. The sorting of materials is the preliminary step in most resource recovery operations.

Resource recovery can involve high-level technology, as shown in Figure 17.5. This is often referred to as *mixed waste recovery*. Separation at the source is the most common separation technique for glass; mechanical recovery of glass is possible, but it is not economical if glass is the only recovery product. Table 17.3 lists the useful products that can be made from a variety of reclaimed materials.

The economics of high-technology facilities increasingly requires that energy recovery be an inherent part of these systems. Because of the quantity of waste required for economic feasibility, energy recovery facilities are generally located near high concentrations of wastes and industrial markets to purchase the recovered products, which may include heat (steam), electricity, synfuels, methane gas, methanol, or ammonia.

The energy potential in solid waste is reported to equal at least 28% of that expected from the oil in the Alaska Pipeline. Generally, about 70% to 80% by weight of municipal solid waste (domestic and commercial, not industrial, agricultural, construction, or sewage) is combustible. Table 17.4 compares the energy values of solid waste and other fuel sources. One ton of municipal solid waste is approximately equal to 9 million BTUs, 65 gallons of No. 1 (kerosene) fuel oil, or 9,000 cubic feet of natural gas. Figure 17.6 shows that although resource recovery from municipal waste in the United States is increasing, the percentage is still quite small. Other developed countries exceed the United States in their rates of recycling. Recycling rates are difficult to determine and compare due to the many variables to be considered and the varying definitions of solid waste from country to country. But studies indicate that Japan recycles between 26% and

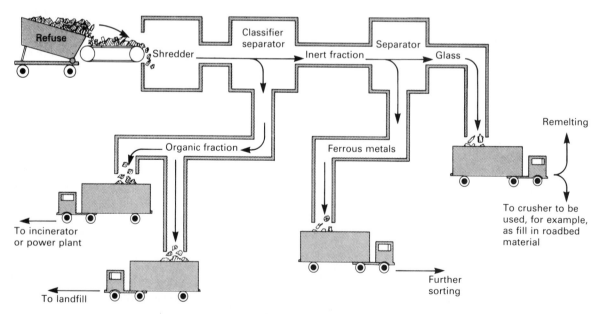

Figure 17.5 Generalized Resource Recovery. In general, all high-technology facilities involve the same basic processes. The refuse is shredded; metals and glass are separated and recovered for sale or use. The organic fraction may be incinerated, buried, or used in an energy recovery facility.

Table 17.3 Products Made from Recycled Waste

Waste Material	Product
Paper	Printing paper
	Writing paper
	Sanitary paper
	Packaging
	Insulation
	Construction paper
	Hardboard
Sludge	Compost
	Roadbeds
Rubber (tires)	Pavement
	Retreads
	Reefs
Plastic	Pipes
Glass	Ceramic bricks
	Glasphalt
	Concrete
Iron and steel	Bars
	Cast-iron pipes
	Structural shapes
Aluminum	Siding
Slag	Cement
Fly and bottom ash	Roadfill
	Roadbase stabilizer
	Asphalt
	Concrete additive
	Cement
	Aggregate
Sulfur	Asphalt cement
Oil	Oil
Refuse-derived fuel	Energy
Various wastes	Automobiles
	Hand tools
Various chemicals	Paint
	Soap and wax
Wood, metal, textiles	Office furniture
Kiln, lime, and gypsum dust	Chemical waste neutralizer
	Fertilizer

Source: U.S. General Accounting Office. 1980. *Federal Industrial Targets and Procurement Guidelines Programs are not Encouraging Recycling and Have Contract Problems.* Washington, D.C.: U.S. Government Printing Office, p. 13.

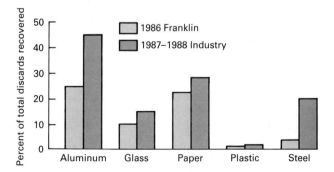

Figure 17.6 Municipal Solid Waste Recycling Rates in the United States. Two estimates of the recycling rates for aluminum, glass, paper, plastic, and steel are shown here. Generally, industry estimates are somewhat higher than those arrived at by Franklin Associates, a consulting group.

Table 17.4 Approximate Energy Content of Wastes in Comparison with Standard Fuels

Fuel	Heating Value (percent)
Solid	
Coal (standard)	100
Bark	37
Wood waste (general)	46
Sawdust	54
Coffee grounds	46
Corncobs	59
Liquid	
Oil (standard)	100
Oil waste	97
Paints/resins	54
Dirty (used) solvents	86
Old grease	76
Gaseous	
Natural gas (standard)	100
Coke oven gas	78
Refinery waste gas	271

Source: D. R. Nichols, Jr., "Are We Ready to Convert Solid Waste to Energy—Profitably?" *Resource Recovery and Energy Review* **4** (1977):9.

42% of its waste; just before reunification, West Germany recycled 15% to 33% of its waste (Young, 1991; Thurner and Ashley, 1990).

Energy Recovery Through Incineration

Many communities incinerate solid waste. Usually, the heat produced by the burning of waste simply dissipates and is lost. However, the heat produced from burning wastes can be used to convert water into steam (Figure 17.7), and the steam can then be used for heating or for running a turbine to produce electricity. This has been done since World War II in Europe, where land and resource scarcity made this an economically feasible approach. Frankfurt, Germany, and Amsterdam produce some of their electricity by this means.

As of late 1990, about 128 waste-to-energy incineration facilities were in operation in the United States. They produce energy equivalent to nearly 30 million barrels of oil annually ("1990 Predicted," 1990).

A heat recovery incinerator plant was completed in Nashville, Tennessee, in 1974. By 1977 this plant was burning an average of 400 tons of municipal solid waste daily. The plant, which is designed to handle 720 tons per day, reduced landfill space requirements by 90%. Steam produced from the burning of the waste is used for district heating and cooling.

A plant in Saugus, Massachusetts, is processing about 1,300 tons of solid waste per day. In its first seven years of operation (1975–1982), the facility processed well over 2 million tons of waste. Until 1985, steam generated by the process was piped to a local

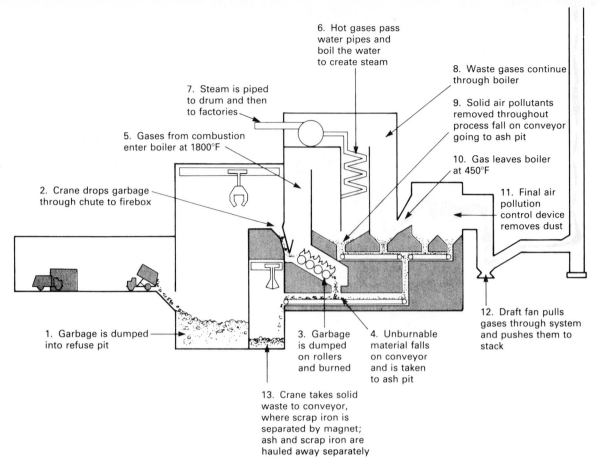

6. Hot gases pass water pipes and boil the water to create steam

8. Waste gases continue through boiler

7. Steam is piped to drum and then to factories

9. Solid air pollutants removed throughout process fall on conveyor going to ash pit

5. Gases from combustion enter boiler at 1800°F

10. Gas leaves boiler at 450°F

2. Crane drops garbage through chute to firebox

11. Final air pollution control device removes dust

1. Garbage is dumped into refuse pit

3. Garbage is dumped on rollers and burned

4. Unburnable material falls on conveyor and is taken to ash pit

12. Draft fan pulls gases through system and pushes them to stack

13. Crane takes solid waste to conveyor, where scrap iron is separated by magnet; ash and scrap iron are hauled away separately

Figure 17.7 Operation of a Refractory-Wall Incinerator. This incinerator recovers the heat energy from burning municipal solid waste by converting it to a more usable form as steam.

industry. In 1985 the facility converted to power generation from steam.

Codisposal. An emerging technology called thermal **codisposal** involves the simultaneous disposal of solid waste and sewage sludge. Because of its volume and the presence of persistent chemicals such as heavy metals, use of sewage sludge as a soil fertilizer has been questioned. However, because of its high water content, it does not burn readily. Thermal codisposal involves the use of energy from burning solid waste to dry the sludge to a point at which it will burn without supplemental fuel or in combination with more refuse. Rising land costs, which make land disposal of sludge more and more expensive, and rising fossil fuel costs, which make incineration of sludge more expensive, combine to make this technology look promising. Codisposal has been used in Europe for well over a decade.

Fuel from Waste

Refuse-derived Fuel Systems. In some municipal solid waste processing facilities, combustibles are separated from noncombustibles. The combustibles are shredded and burned in utility and industrial boilers as a primary fuel or as a supplement to fossil fuels. This type of solid waste processing operation is called a **refuse-derived fuel (RDF)** system.

From 1972 to 1976 an RDF demonstration plant was operated in St. Louis, Missouri. The city, Union Electric Company, and the EPA cooperated in the venture. The plant converted 300 tons of municipal solid waste to RDF daily, and this was used in boilers initially designed for the combustion of pulverized coal. The RDF provided approximately 10% of the fuel for the boilers. An RDF system in Ames, Iowa, processes about 180 tons of waste per day. RDF is used to supplement other fuel sources in a ratio of 20% RDF to 80% fossil fuel.

Methane Recovery. As organic wastes decompose in landfills, they produce gases, predominantly methane and carbon dioxide. A buildup of methane can be dangerous, since it can cause fire or explosion. It has been found, however, that such methane is a commercially recoverable fuel. From April 1974 to February 1985 the city of Los Angeles used gases vented from a landfill in Sun Valley to produce enough electricity for 350 homes. A landfill in Menlo Park, California, has provided sufficient gas to provide electricity to 1,000 homes. The Fresh Kills landfill on Staten Island sells methane to fuel 50,000 homes—about 5 million cubic feet daily. The EPA estimates that over 38 billion cubic feet of methane may be recoverable annually from landfills located near large metropolitan areas in the United States.

Pyrolysis. **Pyrolysis** is the process of heating refuse in a nearly oxygen-free environment to produce oil, gas, or a char as an end product. A ton of solid waste processed in this manner yields the energy equivalent of about one barrel of oil, a little more or less depending on the recovery efficiency. The city of Baltimore began operating a commercial-scale facility in 1975 to produce gas from municipal solid waste. The plant was to produce 4.8 million pounds of steam daily from the low-BTU gas produced from pyrolysis. This would accomplish a savings of 357,000 barrels of oil annually. Other revenue would be available from the sale of ferrous metals that had been sorted out and the sale of glassy aggregate used in concrete manufacturing and street paving. However, after numerous modifications, the process was replaced by a mass-burning system.

Energy recovery efficiencies range from 37% to 62% for pyrolysis, compared to 62% to 71% for direct combustion. The advantage of pyrolysis is that a more generally useful and transportable form of energy is produced.

A Special Problem: Old Rubber Tires. Rubber tires continue to pose disposal problems. As many as 240 million tires are discarded annually in the United States alone, and 2 to 3 billion more are stockpiled or resting in landfills and dumps around the country (Mahtesian, 1991). Incineration of rubber must be done in specially equipped facilities to prevent air pollution and to accommodate the intense heat released by the burning rubber. Tires tend to cause problems in landfills because they spring back into shape, tend to rise to the surface after burial, and do not decompose well. About 60% of the scrap tires in the United States are simply stockpiled aboveground. They are eyesores, breeding grounds for disease-carrying insects, and fire hazards. Some systems for burning rubber for fuel have been successful; they are generally small-scale operations. However, the BTU content of burning rubber is approximately equal to that of coal. Pyrolysis has been used successfully on a small scale to produce fuels from rubber. Several states have placed taxes on retail tire sales to generate funds for cleaning up old tires.

Bioconversion. Although solid wastes decompose to methane and carbon dioxide under landfill conditions, this process can be accelerated artificially by means of an anaerobic digester. Sewage sludge containing decomposers is mixed under anaerobic conditions with shredded combustible components of municipal solid waste. The mixture is heated in a digester for about a week. A demonstration bioconversion plant has been in operation in Pompano Beach, Florida, since 1978. Methane is being produced from a mixture of RDF and sewage sludge.

The Economics of Resource Recovery

Many factors are involved in successfully initiating, constructing, and operating a high-technology resource recovery facility. More research and development are clearly needed for resource recovery prototypes both large and small. Once there is public acceptance of a resource recovery project, the most crucial factor is securing a market for recovered materials in business and industry.

A stable and dependable market is an economic must in resource recovery. Creating markets for recycled materials will be a function of economics and federal encouragement and a shift in federal policies that have an impact on recycled products. Virgin materials have traditionally been given economic advantages over recycled goods in transportation costs. Such barriers must be removed. As we shall see, legislation passed in 1976 does require the federal government to increase its use of products containing recycled materials. Many states have followed suit to require recycled content in state procurements.

Other policy changes and incentives may accelerate the growth of the recycling industry. Dwindling resources may provide additional pressures to make these adjustments occur more rapidly.

Resource recovery facilities must be tailored to the characteristics of particular locations, that is, to the volume of waste available and to the needs of the industries that will buy the products. To ensure that the necessary volume of waste from the community actually goes to the facility, mandatory participation may be necessary. Mandatory universal collection is a major issue, especially in rural areas, where rubbish is burned and sinkholes have served as free disposal sites. Collection is a greater problem too where population is more widely scattered. Individuals, businesses, and waste collection services may be required to take their waste to a particular waste disposal facility. Recycling programs need to be established prior to determining the needed capacity for the facility because recycling programs will take materials from the waste stream and thus diminish the amount and change the mix of waste to be handled by the resource recovery facility.

Another problem is that there is an 18- to 40-month lead time in getting high-technology resource recovery facilities operational. As demand for such facilities increases and as research and development make the technology more reliable, the lag time should decrease. However, the need for site-specific design will continue to make the design and planning phase extremely important.

There are other institutional and financial constraints to resource recovery. Generally, solid waste has been considered a local problem. Disposal is paid for from general funds or from specific fees. Local governments may not have the expertise to examine and administer resource recovery facilities or to evaluate small-scale alternatives. Small municipalities may not generate enough waste to support a large resource recovery facility. Although regional approaches to resource recovery may be needed to produce a sufficient volume of waste, there are few, if any, defined channels for such cooperation between governments. In fact, the importation of waste from other jurisdictions has become a major issue. Also, cities must become better able and willing to enter into long-term agreements in order to justify investments in high-technology resource recovery facilities.

Serious questions are emerging about incineration in general as a disposal option. Incineration at high temperatures can produce toxic air pollutants including dioxins, acidic gases, and heavy metals. The ash from incineration may be toxic. The classification and handling of this waste remain controversial. There is a need for highly trained individuals to run incinerators. Mass burning of waste without preprocessing creates serious environmental problems ("Institute Says," 1986). *Preprocessing* refers to the separation of recyclable materials prior to incineration.

Waste Collection

One cost that is similar for all waste management systems is the cost of collection—getting the waste from its source to the point of disposal or processing. The average cost of collecting and disposing of 1 ton of municipal waste runs from $30 to $100. Greater transportation distances increase disposal costs at the rate of around 50¢ to $1 per mile per ton of garbage disposed. Collection accounts for 50% to 80% of the total spent annually in the United States on solid waste management.

Waste and Health

Leaching
Improper disposal of solid waste can generate health hazards (see Figure 17.8). Groundwater pollution from leaching from landfills is a major health concern. About 50% of the domestic water supply in the United States comes from underground sources. Trace metals and improperly disposed toxic waste could contribute greatly to groundwater contamination problems. Identifying leaching-related health problems is difficult, and correcting such problems once they arise is extremely difficult and costly—if possible at all.

Leachate is formed from the infiltration of rain through the soil covering. As the water seeps through the buried waste, water-soluble chemicals may be picked up and carried along (see Figure 12.7). This leachate then follows the normal hydrologic flow and may enter the groundwater system. The constituents of the leachate and their concentrations vary with local conditions. However, there are certain common problems.

Carbon dioxide is produced from the decomposition of the solid waste. In solution, carbon dioxide forms carbonic acid. This acid may increase the dissolution of the rock that forms the groundwater aquifer and thus increase the amount of minerals in the groundwater.

An even more dangerous situation arises because of lack of control over the types of materials placed in sanitary landfills and open dumps. Many wastes that are now classified as hazardous have been buried in dumps and landfills over the years. More than 20% of the hazardous waste sites on the list for cleanup under U.S. federal legislation are municipal landfills. Safeguards currently required for hazardous-waste landfills, such as the use of water-impermeable liners and leachate collection systems, are now being applied to sanitary landfills. EPA regulations require, among other things, remedial action to clean up landfills leaking into aquifers; the placement of waterproof covers over landfills when they are closed; in some cases, the installation of liners and leachate collection systems in new landfills; and monitoring to prevent the disposal of hazardous wastes and to detect the presence of methane.

Other Health Considerations
Gases that are generated in and escape from landfills include methane, carbon dioxide, hydrogen sulfide, and hydrogen. As these gases migrate laterally, they may kill vegetation. Methane mixed with air is explosive. Hydrogen sulfide is toxic (Chapter 10), and it smells bad. Decaying organic matter may serve as food or as a breeding habitat for rats and flies. Less degradable solid waste may shelter rodents. In addition, runoff may carry litter and dissolved waste into surface waters, adding limiting nutrients that overenrich. Stray solid waste can cause navigational and flooding hazards (by clogging drainage ways). Wood and paper wastes are fire hazards.

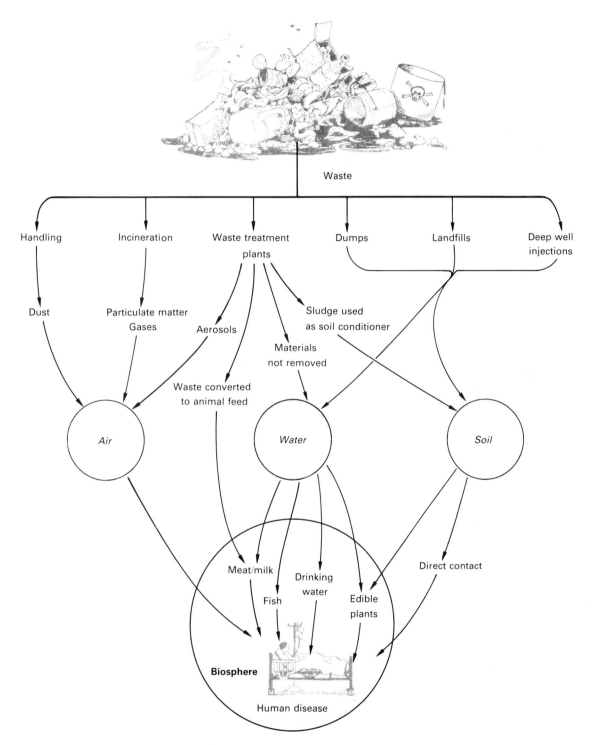

Waste

Handling Incineration Waste treatment Dumps Landfills Deep well
plants injections

Dust Particulate matter Sludge used
Gases Aerosols as soil conditioner

Materials
not removed

Waste converted
to animal feed

Air *Water* *Soil*

Meat/milk Direct contact

Drinking
Fish water Edible
plants

Biosphere

Human disease

Figure 17.8 Waste and Disease. The impact of waste on human health may
be direct or indirect. Everything must eventually go somewhere, and some pathways
eventually lead to people. This flowchart shows how the trash you dispose of today
may come back to you later in an even less desirable form.

Solid wastes along roadsides can cause accidents from falls or cuts; blowing paper has on occasion blocked a driver's vision. Roadside mowers can turn waste into lethal projectiles. Livestock and other animals may be harmed if they swallow litter.

Solid Waste Management: Comparing the Options

Generally, there is an environmental hierarchy for handling solid waste that has been adopted by most industrial nations. It establishes the following sequence of management practices for solid waste:

1. Source reduction (changing processes and products to eliminate waste)
2. Reuse for the same purpose (e.g., refilling glass bottles)
3. Recycling—preferably back into the same product or, if unfeasible, into a new use or product (the criterion of importance here is whether or not substituting the recovered material for a virgin one actually reduces overall raw material extraction; see Young, 1991)
4. Incineration—with energy recovery, if possible
5. Landfill, as a last resort

Although many nations have adopted this hierarchy in principle, they have not followed it. Without question, the last two options, incineration and landfilling, are the most common means of waste disposal. Government policies focus on waste management rather than waste prevention. In the United States, municipal and state governments are beginning to take recycling seriously, but it is estimated that eight to ten times as much money will go into increasing incineration capacity as will go into recycling in the first five years of the 1990s (Young, 1991).

Currently in the United States, about 75% to 83% of municipal solid waste is landfilled, 6% to 12% is burned, and 10% to 11% is recycled. In countries such as Denmark, France, Sweden, and Switzerland, half or less of the waste is landfilled (Young, 1991).

Although incineration does diminish the volume of waste by 80% to 90%, from an environmental perspective incineration is inferior to recycling programs. Incinerators are not useful in conserving materials. Although they may be designed for energy recovery, in most cases, more energy would be saved by recycling the material and reducing the need for new virgin extraction than by burning it. For example, recycling paper can save five times as much energy as could be recovered from burning it; recycling plastic milk jugs saves about twice as much energy (Young, 1991).

Incineration is an expensive alternative compared to recycling and composting. According to the Institute for Local Self-reliance in Washington, D.C., $8 billion could build incineration capacity to burn 25% of the estimated solid waste to be generated in the United States in the year 2000 or build enough recycling and composting facilities to handle 75% of the waste that year (Young, 1991).

The management options are not all compatible but do compete for the same items in the waste stream. According to the Center for the Biology of Natural Systems at Queens College in New York, source reduction, reuse, and recycling compete directly with incineration for about 80% of the waste stream. The center also estimates that 85% to 95% of the current waste stream could be recovered (Young, 1991).

Promotion of composting of yard waste is a practice whose time has come in the United States (see Chapter 7). Europe has been doing it for some time. Sweden composts about one-fourth of its solid waste. Yard wastes make up about one-fifth of all solid waste. In the fall, leaves can comprise a walloping 75% of municipal solid waste. With a national goal of 25% waste reduction in the United States and many state and local reduction goals, composting of yard wastes is likely to increase.

Table 17.5 summarizes some of the impacts of various approaches to solid waste management.

Hazardous Chemical Waste

With the advance of technology and the development of new industrial processes has come the rise of chemical wastes, some of which are toxic or poisonous. As we have already seen, these can cause serious air, water, and land pollution problems.

What Is Hazardous Waste?

According to the definition found in U.S. federal law, **hazardous waste** is any waste that because of its quantity, concentration, or physical, chemical, or biological nature may cause mortality or irreversible or incapacitating illness or pose a threat to either human health or the environment. Practically speaking, federal regulations allow a generator of wastes to use two different approaches to tell whether the waste produced is hazardous. A generator may test according to EPA standards for ignitability, corrosiveness, reactivity, or toxicity. The law itself requires that persistence, degradability, and potential for bioaccumulation also be considered, but these properties are not as easily determined. Or the generator may check to see whether the waste under consideration is on the federal list of hazardous wastes. Estimates of the amount of hazardous waste generated in the United States

Table 17.5 Managing Garbage from Homes: Options and Impacts

Option	Number of Employees		Landfill Needs Per Year (cubic yards)	Net Cost (dollars per year) (includes sale of any energy produced)		Amount of Energy (gallons of gas equivalent)		Environmental Issues	Citizen Convenience
(a) **Landfill** everything (landfill 15 miles away)	Collection Landfill Total	40 2 42	52,000	Collection Landfill Total	1,300,000 520,000 1,820,000	Collection Landfill Total used	30,000 13,000 43,000	• is unattractive • uses land • can pollute water and air • can create hazardous gases (methane) • bury/lose natural resources	• just put waste at curb
(b) **Voluntary Recycling** Curbside pickup of glass, newsprint, plastic, aluminum. Landfill remainder.	Collection Recycling center Landfill Total	44 8 2 54	47,000	Collection Recycling (profit) Landfill Total	1,400,000 10,000 470,000 1,860,000	Collection Recycling (saves) Landfill Total saved	33,000 300,000 12,000 255,000	• reduces impacts at landfill • reduces pollution from manufacturing • reuses natural resources	• need to separate recyclables • builds good habits
(c) **Mandatory Recycling** (as in "b" above)	Collection Recycling center Landfill Total	48 15 2 65	42,000	Collection Recycling (profit) Landfill Total	1,500,000 60,000 420,000 1,860,000	Collection Recycling (saves) Landfill Total saved	36,000 600,000 9,000 555,000	same as "b"	• need to separate recyclables • requires enforcement for noncompliance • builds good habits
(d) **Mandatory Composting** of yard waste. Landfill remainder. (Figures assume 1/2 yard waste is composted at home.)	Collection Composting Landfill Total	42 1 2 45	45,000	Collection Composting Landfill Total	1,350,000 50,000 450,000 1,850,000	Collection Composting Landfill Total used	33,000 1,000 10,000 44,000	• reduces need for landfill • reduces methane gas pollution • reduces strength of leachate • produces fertile humus • reuses natural resources	• need to separate yard waste • builds good habits
(e) **Incinerate** for energy recovery. Landfill ash and nonburnables. (incinerator in town)	Collection Incinerator Landfill Total	38 12 1 51	10,000	Collection Incineration Landfill Total	1,250,000 750,000 200,000 2,200,000	Collection Incinerator (produces) Landfill Total produced	28,000 840,000 2,000 810,000	• reduces need for landfill • produces fly ash high in heavy metals that requires special handling • produces air pollutants • consumes natural resources	• just put waste at curb

Note: Example compares costs for a community producing 100 tons/day, 5 days/week.

Numbers presented are realistic but not specific to any one community.

Other options and combinations of options exist.

Source: Wisconsin Department for Natural Resources, *Recycling Study Guide* (Madison, 1991), p. 25.

---

alone range from 264 million to as high as 1 billion tons.

Where Does It Come From?

Most hazardous waste is generated by industry. The EPA lists more than 125 industrial processes that may generate it. Generally, about 70% of waste characterized as hazardous is generated by the chemical and petroleum industries, about 20% by metal-related industries, and 10% by all other industries. Certain products and the types of hazardous waste generated by their production are shown in Table 17.6.

There are two very distinct problems associated with the management of hazardous waste. One is the management of waste generated now and in the future; the other is how to deal with waste that was improperly disposed of in the past.

Treatment and Disposal Alternatives

In the United States, most hazardous waste (96%) is handled on-site, that is, where it is generated. The rest is disposed of by commercial storage, treatment, or disposal companies. Figure 17.9 shows how the commercial facilities handle this waste. Large gen-

Table 17.6   Sources of Hazardous Waste

Product	Hazardous Waste
Plastics	Organic chlorine compounds
Pesticides	Organic chlorine compounds, organic phosphate compounds
Medicines	Organic solvents and residues, heavy metals (e.g., mercury and zinc)
Paints	Heavy metals, pigments, solvents, organic residues
Oil, gasoline, and other petroleum products	Oil, phenols, and other organic compounds, heavy metals, ammonia salts, acids, caustics
Metals	Heavy metals, fluorides, cyanides, acid and alkaline cleaners, solvents, pigments, abrasives, plating salts, oils, phenols
Leather	Heavy metals, organic solvents
Textiles	Heavy metals, dyes, organic chlorine compounds, solvents

Source: U.S. Environmental Protection Agency.

erators find it economically feasible to treat on-site; small generators tend to use commercial facilities. More than 80% of all generators ship at least some waste to off-site facilities.

For many years, most hazardous waste was disposed of in unlined lagoons or nonsecure landfills and in miscellaneous ways such as by dumping into sewers or deep wells and by incineration.

Although there is disagreement on the best ways to deal with hazardous waste, it can clearly be handled better than in the past. Table 17.7 compares some of the advantages and disadvantages of various management technologies. The technology and policies associated with the storage, treatment, and disposal of hazardous waste are similar to those for solid waste but provide for detoxification or more stringent containment requirements.

Source Reduction, Reuse, and Exchange. As with other wastes, the best method of managing hazardous waste is simply to produce less of it, that is, waste minimization. Figure 17.10 demonstrates various facets of waste minimization. **Waste minimization** refers to reducing the amount of waste actually generated, recycling of waste, or treatment of waste to render it nonhazardous. All three processes result in less hazardous waste. However, a 1986 report by the Office of Technology Assessment makes a clear distinction between source reduction and the other two waste minimization processes. The report points out that source reduction involves changes in the production and operation process itself; it is not concerned with the waste stream. Source reduction is prevention; recycling and treatment are waste management techniques. In *Pollution Prevention Pays* (1979), Michael Royston presented a collection of case studies that showed how changes in production reduced waste and saved money. Several U.S. states have initiated programs to offer technical assistance to industry to evaluate its production processes to reduce waste. Basically, an audit of processes and practices is done to identify opportunities to reduce waste. Solutions may involve improved housekeeping practices so that hazardous wastes don't mix with nonhazardous wastes and other such measures, substituting materials with less toxic residues, or redesigning equipment, operations, or products. Often, however, implementation of the findings results in monetary savings to the company through less costs for waste disposal and more efficient use of raw materials. North Carolina's Pollution Prevention Pays Program is a model.

Congress passed legislation in 1990 establishing a national policy of promoting the reduction or elimination of hazardous waste at its source of production. The EPA is charged with developing a strategy for

Figure 17.9   Waste Management Methods Used by Commercial Facilities.   Only 12% of these wastes were recovered or incinerated.

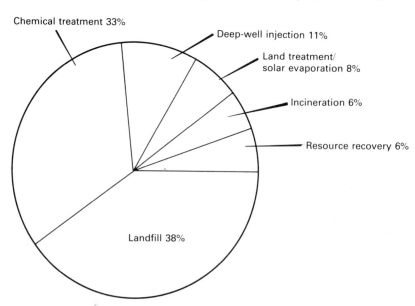

Chemical treatment 33%

Deep-well injection 11%

Land treatment/ solar evaporation 8%

Incineration 6%

Resource recovery 6%

Landfill 38%

Table 17.7 Comparison of Some Hazardous Waste Reduction Technologies

	Disposal		Treatment		
	Landfills and Impoundments	Injection Wells	Incineration and Other Thermal Destruction	Emerging High-Temperature Decomposition*	Chemical Stabilization
Effectiveness: How well it contains or destroys hazardous characteristics	Low for volatiles, questionable for liquids; based on lab and field tests	High, based on theory, but limited field data available	High, based on field tests, except little data on specific constituents	Very high, commercial-scale tests	High for many metals, based on lab tests
Reliability issues	Siting, construction, and operation Uncertainties: long-term integrity of cells and cover, liner life less than life of toxic waste	Site history and geology; well depth, construction, and operation	Monitoring uncertainties with respect to high degree of DRE; surrogate measures, PICs, incinerability†	Limited experience Mobile units; on-site treatment avoids hauling risks Operational simplicity	Some inorganics still soluble Uncertain leachate test, surrogate for weathering
Environmental media most affected	Surface water and groundwater	Surface water and groundwater	Air	Air	Groundwater
Least compatible wastes‡	Liner reactive; highly toxic, mobile, persistent, and bioaccumulative	Reactive; corrosive; highly toxic, mobile, and persistent	Highly toxic and refractory organics, high heavy metals concentration	Some inorganics	Organics
Costs (Low, Moderate, High)	L–M	L	M–H (Coincineration = L)	M–H	M
Resource recovery potential	None	None	Energy and some acids	Energy and some metals	Possible building material

* Molten salt, high-temperature fluid wall, and plasma arc treatments.

† DRE = destruction and removal efficiency; PIC = product of incomplete combustion.

‡ Wastes for which this method may be less effective for reducing exposure, relative to other technologies. Wastes listed do not necessarily denote common usage.

Source: U.S. Office of Technology Assessment.

reducing the generation of hazardous waste. The EPA estimates that source control techniques could reduce hazardous waste by one-third by 2010. As the requirements for proper disposal of hazardous waste are en-

forced, some experts believe that industry will find it more economical to change its processes to decrease the volume of hazardous waste it produces or to find mechanisms for reusing it. The Chemical Manufac-

The best way to manage hazardous waste is to produce less of it.

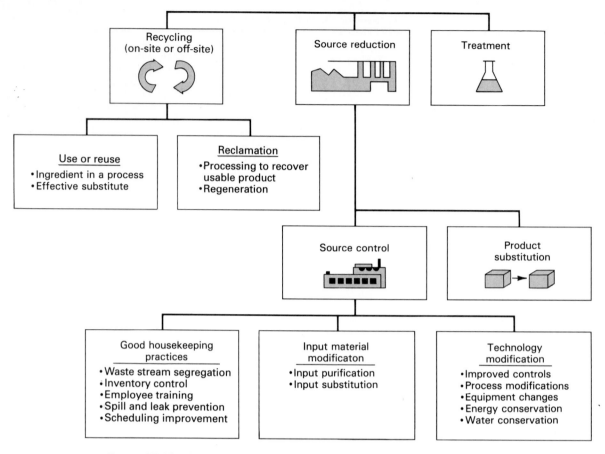

Figure 17.10 Waste Minimization Techniques. The best solution to our waste problem is to produce less waste. This diagram shows several ways to accomplish this.

turers Association claims that hazardous waste generated by the chemical industry decreased by more than 20% between 1981 and 1985 ("Chemical Industry," 1987). The Office of Technology Assessment report, however, suggests that additional economic incentives will be necessary for industry to put its limited resources into waste reduction instead of waste management (see Table 17.8).

The EPA estimates that perhaps 20% of the hazardous waste currently generated could be reused or recycled. In other cases, when one industry cannot reuse its own waste, another industry can. A number of waste exchange programs are in operation in the United States under the direction of trade associations, chambers of commerce, universities, and state or local governments. Industries provide information on the wastes they have available for reuse; this information is advertised, and inquiries from industries wanting to use the waste are returned to the generator for follow-up. The exchange program seems most appropriate for solvents, oils, and surplus chemicals.

Simple volume reduction methodologies are also useful and will become more important as disposal costs increase. Evaporation of water from holding ponds, for example, is a simple means of concentrating and thus reducing the volume of hazardous chemicals that happen to be in aqueous solution.

To date, however, source reduction or waste prevention has not been seriously undertaken by the marketplace. There are many reasons for this. Changing processes or equipment can be expensive and can interrupt production for a period of time. Products are not designed with ease of disposal in mind, and consumers do not often consider this in their purchases. A report by the Office of Technology Assessment recommends that economic incentives be considered at the federal level, including a grant fund for industrial research and development on waste reduction, for covering some of the capital cost of plant and process modification, and for public education. It also recommends improvements in information flow to the consumer, including product labeling to indicate goods that in their production help to achieve the goal of reducing waste. The report also suggests that outright bans or the threat of bans on certain types of products, especially packaging products like polystyrene foam containers and some types of plastic, have caused industry to take action.

Table 17.8 Incentives and Disincentives for Various Waste Minimization Techniques

	Source Reduction	Off-Site Recycling	On-Site Recycling
Incentives			
Increased cost of waste management	X	X	X
Difficulties in siting new hazardous waste management facilities	X		X
Permitting burdens and corrective action requirements	X	X	X
Financial liability of hazardous waste generators	X		X
Shortages of liability insurance	X		X
Public perception	X	X	X
Disincentives			
Economic barriers			
Lack of capital	X		X
Financial liability		X	
Technical barriers			
Attitudes toward unfamiliar methods	X	X	X
Batch processes			X
Lack of information	X	X	X
Technical limits of process	X		
Technical quality concerns	X		
Regulatory barriers			
Need to obtain treatment, storage, or disposal permits	X	X	X
Perceived stigma of managing hazardous waste		X	X
Revisions to other environmental permits	X	X	X

Source: U.S. Environmental Protection Agency.

Chemical Treatment. If hazardous waste cannot be reused, the next preferred treatment is detoxification before disposal. **Detoxification** is the conversion of a toxic substance into something that is not hazardous. There are numerous ways to detoxify wastes; many of them are expensive. A common detoxification process is the neutralization of wastes by adjusting the pH. Acidic wastes can be neutralized by the addition of lime. This has the side benefit of precipitating out heavy metals. The iron and steel industry, the electroplating industry, and other metal-finishing industries use this process. Some wastes can be detoxified by oxidation-reduction reactions. For example, cyanides oxidized with sodium hypochlorite produce carbon dioxide and nitrogen. Carbon absorption can be used to remove organics from wastes. It has been used, for example, in the textile industry for removing dyes from effluents.

Biological Treatment. Most organic hazardous waste can decompose and hence is amenable to biological treatment such as composting. Precautions must be taken, however, to prevent the disposal of heavy metals or nonbiodegradable wastes in this manner, since they could accumulate in the soil and be taken up by plants, be carried into streams by runoff, or leach into groundwater. **Land farming** is a particular kind of composting involving the incorporation of the waste into the soil surface and mixing to provide aeration; decomposition is ideally taken care of by microorganisms. Land farming is suitable for more voluminous hazardous wastes such as sewage sludge and petroleum refinery sludge. Composting is a more efficient method of biological treatment because decomposition occurs at higher temperatures. Composting requires protection from precipitation and usually some kind of bulking agent to provide a texture that is porous to ensure aerobic decomposition.

Incineration. Incineration of hazardous waste is a treatment that may or may not result in a hazardous residue. If the residue is hazardous, it must be properly disposed of by some other means. Hazardous waste may be burned alone or with supplemental fuel. Traps for particulates, air pollution equipment such as scrubbers for toxic gases like hydrogen chloride, and high burning efficiencies are required. Several hazardous waste incinerators operate in the United States.

Several companies have developed systems for the incineration of hazardous waste on ships at sea. Two such incineration ships are the *Vulcanus I* and *Vulcanus II,* which are owned by a U.S. waste management company. Incineration ships have proved to be efficient burners if properly operated. In 1977, *Vulcanus I* burned surplus quantities of the herbicide Agent Orange. The EPA monitored the air during the incineration and found that hazardous by-products were not released in significant amounts. The environmental impact of incineration at sea has yet to be fully determined.

Integrated Hazardous Waste Management, European Style

European nations have relatively advanced technologies for managing hazardous waste. Because of the scarcity of land in Europe, detoxification and destruction of hazardous waste are favored over land disposal. In Denmark, nearly all hazardous waste is destroyed or detoxified; in Germany, well over half is detoxified. Although some destruction and detoxification technologies yield residues that must be placed in secure land disposal sites, the volume is much reduced.

A good example of an integrated system for hazardous waste management is the Kommunekemi or "community chemical" plant in Denmark located near Nyborg. Its hazardous waste management system begins with industrial waste collection stations operated by municipalities and dropoff stations for household hazardous wastes. Wastes collected at both types of stations are funneled into the Nyborg plant. The plant itself consists of three rotary kiln incinerators. The heat from the incineration process is used to generate steam to provide more than 60% of the heating demand for the 12,000 inhabitants of Nyborg. The chemical and physical processes on-site include destruction of cyanides by oxidation-reduction reaction using sodium hypochlorite, neutralization of acids and bases, chemical reduction of chromium, and precipitation of heavy metals. Solid residues from incineration and treatment are disposed of in a landfill near the plant.

Unless industries have special permission for on-site management, they are required by law to take their wastes to the municipal transfer and collection stations. Industries are charged for disposal of their waste, costs being dependent on the technology used.

Similar integrated arrangements are found in other European countries. In Germany, Bavaria has seven collection stations that separate oil and water, neutralize acids and bases, and thicken sludges. The wastes are then sent on to one of three destruction facilities that include incineration and other treatment technologies. Bavaria's facilities are owned by local governments in some cases and are government-private cooperative ventures in other cases.

Waste Facility Siting

A major problem in waste management is deciding where to put the treatment and disposal facilities. For many reasons, no one wants to live near waste facilities. Locating facilities in remote areas means higher transportation costs and increased danger of a spill, derailment, or other accident. Policies for siting such facilities are being sought at all levels of government.

The United States leaves questions of siting to state and local governments. States have used different approaches to the siting of waste facilities. Some have given the authority for siting decisions to local governments; some have required local governments to follow certain state-level regulations in their decision making; some have prohibited local siting and have made it exclusively a state function. Several states have set up siting boards composed of government, expert, and citizen members (in some cases representing the local community) to make the final decision. Some states have made the siting board an appeals board; some use it as a board for final arbitration if the negotiation process between the industry and the local community becomes stalled.

Incentives to local communities to allow such facilities in their areas have been tried with some success (Table 17.9). Special taxes on such facilities to be returned as revenue to the county are used in some states. By law, Kentucky has proposed linking waste facilities with industrial parks, to provide jobs, a buffer

Table 17.9 Compensations and Incentives for Allowing Hazardous Waste Facilities in a Community

Impact Issue	Compensation	Incentive
Truck traffic	Improve or partly maintain roads; provide traffic lights	Completely maintain roadways
Aesthetic impact	Offer direct cash payments to affected individuals or groups	Build aesthetically pleasing park
Groundwater pollution risk	Provide liability insurance	Develop additional water supplies
Loss of wildlife area	Provide fund for endangered wildlife	Build additional recreation area
Property value decline	Provide land value guarantees and direct payments	Buy and provide additional property to affected residents
Uncertainty about potential damages	Provide performance bond liability insurance, emergency response fund; provide tipping fees to community	Purchase or provide guarantees or backing of municipal bonds
		Donate to local charitable organizations
		Provide free disposal service to local industry
		Clean up existing waste site

Source: U.S. Environmental Protection Agency.

zone between a facility and the community, and an increased tax base. Public education, public involvement from the initial phases of site planning, and taking steps to establish credibility with the public are among the measures being used to resolve hazardous waste siting problems.

Public relations problems associated with siting today are the result of poor waste disposal practices in the past, increased public awareness, and lack of public trust that industry will act responsibly in handling waste or that government can properly monitor and enforce hazardous waste regulations.

The siting of new facilities slowed in the late 1970s and early 1980s for other reasons as well. Industry was waiting to see what the federal requirements would be for certain types of hazardous waste facilities. As landfills started to become more expensive and their long-term liabilities made them less desirable, industry began to examine its waste stream for more economic handling methods.

The Legislative History of Solid and Hazardous Waste in the United States

In 1965 the first national solid waste legislation was enacted. The **Solid Waste Disposal Act of 1965** (PL 89-272) provided money to the states to develop solid waste management plans and to survey current disposal practices. In 1970 the **Resource Recovery Act** (PL 91-512) marked a significant policy change from focusing on *disposal* problems to examining *recovery* processes for materials and energy. However, the act was basically nonregulatory. In October 1976 the **Resource Conservation and Recovery Act (RCRA)** was enacted. This was an omnibus bill with regulatory implications. It was an important addition to federal comprehensive air and water laws passed earlier in the 1970s (see Chapters 9–12).

The Resource Conservation and Recovery Act (RCRA)

The Resource Conservation and Recovery Act addresses the problems of both solid and hazardous waste. The law requires states to develop solid waste management plans and prohibits open dumps. All dumps were to be closed or upgraded to landfills by the mid-1980s. In addition, federal procurement agencies were directed to begin purchasing materials that use the greatest amounts of recycled materials. Recognizing the need for the development of markets for recycled material, the law directed the Department of Commerce to assist in developing specifications for recovered materials to ensure good-quality products. The department was also directed to look at developing markets for recycled materials and to help to promote them. The Department of Mines and Minerals was directed to focus on mineral waste problems, including recovery from industrial wastes and junk car processing. Freight rates and their implications for recycled materials were to be examined by the Interstate Commerce Commission. The Department of Energy was directed to promote research and development for production of energy from solid waste.

In the area of hazardous waste, the Act provides for the development of criteria for identifying and listing hazardous waste and of standards for generation, storage, treatment, and disposal of such waste by the EPA. The law also calls for the development of a system to track wastes from point of generation to point of disposal by means of a *manifest*.

The manifest is a multicopy form (similar to a bill of lading or shipping voucher) that identifies, among other things, the origin, quantity, composition, and destination of the hazardous waste. The form is completed and signed by the waste generator, the transporter, and the disposer on delivery of the waste. The disposer returns the original manifest to the generator. Such a system is intended to track the waste from "cradle to grave." All parties are required to report periodically to the regulatory agency and to notify it if manifests are not properly received.

To assist the states in developing hazardous waste policies, the EPA has set this order of priorities for disposing of hazardous waste:

1. Reduce the generation of hazardous waste.
2. Separate out and isolate hazardous waste from other industrial waste to keep the volume small.
3. Use the waste through exchange and recovery.
4. Incinerate the waste; detoxify or neutralize it, where possible.
5. Dispose of the waste in secure landfills.

The hierarchy is very similar to that outlined earlier in this chapter for solid waste.

The Resource Conservation and Recovery Act was reauthorized by Congress in 1984. The revised legislation brought small generators of hazardous waste (100–1,000 kg/month) under regulation and placed substantially greater restrictions on land disposal of hazardous waste. The legislation banned land disposal for certain types of waste and established stricter standards for landfills and surface impoundments. It further established a program for the regulation of underground storage tanks. There may be 100,000 such tanks leaking gasoline, pesticides, or industrial solvents below the ground.

Congress hopes to have another revision and re-authorization of RCRA by the end of 1992. Major issues currently under discussion are the rights of states to regulate the importation of out-of-state waste, recycling mandates, and the regulation of municipal incinerator ash.

Many aspects of waste management have been left to state and local governments. Waste, however, according to the courts, is an item of interstate commerce. Thus states have no right under the U.S. Constitution to restrict its importation. This has become a controversial issue. States in the Northeast can dispose of their wastes more cheaply by sending them to other states in the South and the Midwest. This is because of the dwindling landfill space in the Northeast and the higher costs of landfilling there because of stringent requirements. Congress is considering proposals to allow states with EPA approved solid waste plans to regulate out-of-state waste imports in some manner, such as charging significant surcharges to make the importation economically unattractive.

Recycling issues revolve around setting recycling goals and establishing markets for recycled goods. One option under consideration for increasing the use of recycled materials is mandating manufacturers use a minimum content of recycled material in their products and packaging. The issue on municipal incinerator ash is whether or not it should be regulated as a hazardous waste. It is not now and such a designation would significantly increase the cost of its disposal. The ash does often contain heavy metals and toxic residues.

Superfund

The **Comprehensive Environmental Response, Compensation and Liability Act (CERCLA)** of 1980, the so-called **Superfund** legislation, was designed primarily to address the problem of financing cleanup of abandoned or illegal hazardous waste sites. In October 1986, Congress reauthorized the Superfund and increased it from $1.6 billion to $8.5 billion. The **Superfund Amendments and Reauthorization Act (SARA)** included several new provisions. The EPA was put on a timetable for cleaning up sites, and some guidance was provided for cleanup standards. A national community "right to know" program was established, requiring industries to report the chemicals they handle to local officials so that proper emergency response programs can be in place. States are required to ensure the availability of adequate disposal facilities for hazardous waste anticipated to be generated within the state over the next 20 years. Effective October 1989, the EPA can withhold Superfund money from any state that cannot provide this assurance.

Other Solutions

The ideal solutions to our waste problem in the long run appear to be reduction in the actual amount of waste produced and maximum reuse and recycling of what is generated. Both approaches conserve resources and energy. Given that waste cannot be eliminated altogether, resource recovery at least helps to diminish what has been a one-way flow of materials. Resource recovery is the complement to source reduction, and both are essential to a more efficient materials system—ecologically speaking. More and more economists, business executives, and citizens are beginning to agree.

Changing the materials system is basically a problem of changing consumer patterns. For the most part, this is a question of economics. How does one convince a person to separate waste? How can people be motivated to use products that are most easily disposed of or recycled or simply to use less? How can they be encouraged to use recycled materials? How does one encourage manufacturers, processors, packagers, and distributors to change what they do so as to produce less waste? Here are some alternatives that have been proposed or attempted.

Container Deposit Laws

One of the most controversial approaches to the problem of solid waste is the so-called container deposit law or **bottle bill.** A bottle bill simply imposes a deposit of several cents on most types of beverage containers. The deposit is returned to the purchaser when the empty beverage container is returned to the place of purchase. Bottle bills have the effect of returning potential solid waste to the source; they also tend to shift beverage packaging from nonreturnable containers to returnable ones (see Figure 17.11).

Several countries have mandatory beverage container deposit programs, including Sweden, Denmark, Norway, and the Netherlands. Several provinces in Canada have deposit programs. Bottle bills have also been enacted in various states and areas in the United States. Federal military installations have initiated deposit requirements.

Pros and Cons. Bottle bill proponents claim that such laws reduce litter, reduce the total volume of solid waste, and conserve energy by promoting a shift to refillable containers and forcing recycling (see Figure 17.12). Bottles can be refilled ten times or more; cans can be recycled, eliminating the energy-intensive process needed to produce aluminum from bauxite. Both practices reduce air and water pollution and help to conserve natural resources.

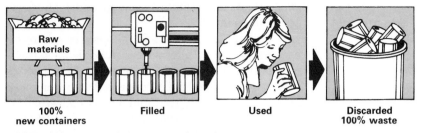

100% new containers → **Filled** → **Used** → **Discarded 100% waste**

(a) Simplified present beverage container system

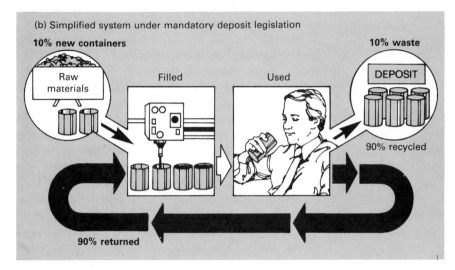

(b) Simplified system under mandatory deposit legislation

10% new containers

Raw materials → Filled → Used → **DEPOSIT** **10% waste**

90% recycled

90% returned

Figure 17.11 Managing Used Beverage Containers. Proponents of bottle bills claim that such legislation significantly reduces litter. Opponents claim that bottle bills are ineffective and discriminatory, since beverage containers make up only 25% of all litter.

Opponents claim that bottle bills are discriminatory, since beverage containers constitute only 25% of all litter. They further object to increased labor costs from the additional handling of containers and possible sanitary problems. Opponents say that bottle bills that do not put an equal deposit on all beverage containers influence competition. Such bills, they state, result in undue hardships on beverage industries that produce nonreturnables or throwaways, since their products cannot be refiled or recycled and thus, when returned, are of little economic benefit to the company. However, plastic recycling is on the increase. Container deposit laws boost return rates considerably.

Proponents counter that the impact of throwaway containers on small business was such that the number of breweries decreased from 300 in the 1950s to fewer than 100 in the 1970s. Soft drink bottling plants decreased from 4,500 in the 1960s to 2,300 in the 1970s. Because throwaways did not need to be returned to the bottling plant, bottlers could ship beverages over longer distances. Large national brewers and bottlers took advantage of this potential by building larger bottling facilities and expanding their markets. Smaller bottlers could not compete with the economies of scale of the larger operators.

The GAO evaluated studies done on the costs and benefits of state bottle bill laws. It found that container deposit laws are generally compatible with curbside recycling programs. State deposit laws have reduced beverage container litter by 79% to 83% and total solid waste by 6% by weight and 8% by volume. However, studies do not compare these benefits to the cost of implementing the system. Higher costs result from increased handling, transport, and storage of containers. It is interesting to note that a brewery with locations in Oregon and Washington changed from disposable bottles back to refillables in 1990 (Young, 1991).

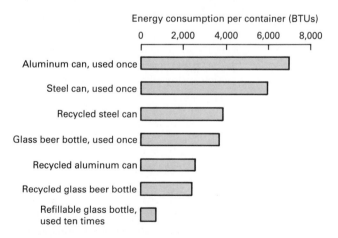

Energy consumption per container (BTUs)

Aluminum can, used once
Steel can, used once
Recycled steel can
Glass beer bottle, used once
Recycled aluminum can
Recycled glass beer bottle
Refillable glass bottle, used ten times

Figure 17.12 Energy Consumption per Use for 12-Ounce Beverage Containers. Without question, the most energy-efficient container for beverages is the refillable glass bottle.

Washington State has been active in promoting litter cleanup and recycling. Among its programs has been the hiring of summer Job Corps workers to pick up litter along highways.

plastics, and throwaways in general and not just beverage containers.

Although litter bills and recycling programs have been promoted as alternatives to bottle bills, the two are not mutually exclusive and may in fact be complementary. The bottle bill is concerned with stopping the production of litter; the litter bill is concerned with cleaning up litter. Some states are now considering enactment of both types of bills. The litter bill approach does seek a remedy for the aesthetic and economic issues related to solid waste collection and disposal. However, it does not address the ecological issues associated with the conversion of resources and energy, as the bottle bill does.

Mandatory Recycling Programs

Another emerging management tool at the state level is mandatory recycling programs. Oregon started the process in 1983 by requiring any communities over 4,000 in population to establish recycling programs at their disposal sites. In 1986 all citizens were to have recycling stations accessible or have curbside pickup of recyclables. In 1987, New Jersey passed a statewide mandatory recycling program aimed at reducing its solid waste volume by 25%. Residents must separate at least three "marketable waste materials." At least half of the states have mandatory waste reduction laws or goals and are moving toward recycling as a tool to reduce solid waste.

Individual cities are taking the same approach. In 1988, instead of building an incinerator, Seattle began an intensive recycling program with the goal of reducing the amount of waste disposed of by 60% by 1998. In 1989 the city had achieved a recycling rate of 37%. Heidelberg, Germany, a city of about 135,000, has achieved a similar 37% recycling rate.

Mandatory recycling must be coupled with incentives to stimulate market demands for goods, or the collected materials have no outlet (see Table 17.10).

Litter Bills

The beverage industry proposes community beautification and litter pickup programs in lieu of beverage container legislation. Along with this the industry supported the enactment of a "litter law." The **litter bill** approach first took form in Washington State. An assessment is placed on industry in the state to cover the cost of litter cleanup. The assessment is levied on industries that generate the products that often end up as litter. The Washington approach is supported by the beverage industry; they see it as more equitable than bottle bills, since it deals with paper,

Table 17.10 Overview of the Recycling System for Various Materials: Limiting Factors and Incentive Points

| | Limiting Factors | | Appropriate Incentive Points | | |
| | | | Collection and | | Industrial and |
Materials	Supply	Demand	Processing	Mills	Commercial Consumers
Old newspapers		x		x	x
Old corrugated containers	x		x	x	
Office papers	x		x		
Mixed papers		x		x	x
Plastics	x	x	x	x	x
Glass	x		x		
Tin cans	x		x		
Aluminum	x		x		
Used oil		x		x	x
Tires		x		x	x

Source: U.S. Office of Technology Assessment.

Economic Incentives

Many types of economic incentives and disincentives have been proposed to push the marketplace toward use of the most desirable waste management options. These include product fees that would vary with the recycled content of the product or the product's ability to be reused or recycled; grants for product and process modification to reduce waste; and the elimination of subsidies for virgin production, such as depletion allowances for mineral extraction, which run counter to the goal of maximum reuse and recycling. Some areas are considering assessments for the use of virgin materials. Others provide tax breaks for transporting secondary goods within a given jurisdiction. Charging residential waste generators for each garbage can or bag collected has proved successful in promoting recycling and reducing waste. Since this program was initiated in 1981 in Seattle, the number of full garbage cans collected per household each week has declined from 3 to 1.

Washington State has established a sales tax on hazardous substances sold in the state, including solvents, petroleum products, pesticides, and fertilizers. The tax is placed on the first sale of the product in the state as a type of **advance disposal fee.** The consumer will be the ultimate payer of the tax. The rationale is that higher-cost products should cause the consumer to seek less costly, less toxic products and provide an incentive to the manufacturer to produce less toxic products. Money from the sales tax is used to fund cleanup of hazardous waste sites and to support public education programs on recycling, disposal of hazardous waste by small generators of waste, and other waste-related programs ("Poisons Tax," 1988).

Forcing the Market

There are other tactics that have been proposed for forcing a market response. Minimum warranty requirements could be used to improve the durability of products. Bans on certain types of waste, such as yard waste, in landfills is another approach that has been used. Bans on certain types of packaging, such as polystyrene foam containers and plastic six-pack holders, have been undertaken in some areas. Bans or the threat of bans have proved to be an impetus for change. Another potent factor has been consumer education. McDonald's agreed to phase out its use of polystyrene foam containers because it found good alternatives that addressed its customers' concerns.

Education and dissemination of information to consumers is important. Surveys show that many consumers would pay more for products that are less environmentally harmful, if they knew how to select them. Canada and Japan have "ecolabeling" programs. A couple of nonprofit groups in the United States are working on establishing green label programs.

Regardless of the options selected, the bottom line will require that each of us reassess our role in waste generation—what we do, how we do it, and what we purchase—at home and on the job. Sound ecologically based choices at all levels will be fundamental to achieving waste reduction goals.

Risk Assessment and the Future

How society chooses to handle hazardous materials and waste will depend to a large extent on how it perceives the risk (see Enrichment Box 17.3).

There are many hazards in our world—some are significant; others are not. Because it is not always easy to tell the difference, people have trouble putting risks into perspective. We see bizarre behavior such as smokers refusing to use artificial sweeteners, and we see people in states of confusion throw up their hands and disregard all risk. Confusion about risk also shows up as outrageous public policy, unbalanced legal decisions, and ineffective risk regulation. We need better ways of dealing with risk than approaching all risk as if it must be reduced to zero.

Consider the following: We smoke billions of cigarettes per year, but we ban saccharin. We eat fatty diets while worrying about pesticide residues in food. We worry about dental X-rays, yet live in houses full of radon. We avoid the oceans because of sharks. Some of us refuse to fly in airplanes, but we drive at high rates of speed in automobiles. People are likely to overestimate the effect of synthetic things yet tend to downplay natural hazards—believing, perhaps, that nature knows best. We need to remember that nature is far from benign.

Why do we find it difficult to put risk into perspective and so act accordingly? The answer is partly obvious, but it has many dimensions.

First of all, there are many ways of expressing, defining, and perceiving risks. The debate over the safety of nuclear power has raged for several decades, and experts on both sides often cite and emphasize *different* statistics—to prove either the safety or the danger of nuclear power. Consider the dilemma of a nonexpert listening to an antinuclear spokesperson emphasizing what *could* happen and a nuclear power advocate who points out that few people are killed by nuclear power each year relative to the hundreds killed by natural gas. Consider the listener's confusion, after hearing about the disaster at Chernobyl, in having the *actuarial* risk of nuclear power compared to crossing the street a few extra times in a lifetime or to raising the national speed limit from 55 to 55.003 miles per hour. The listener also plays a role in coloring risk. Allman (1985) points out that when the risk is for gain, people tend to be conservative, greatly favoring a small cer-

Ranking Environmental Risks

In 1986 a special task force of senior staff at the U.S. Environmental Protection Agency was asked to assess and compare the risks associated with a wide spectrum of environmental problems. When their results were compared against rankings by the public, quite a difference was found. It also appeared that funding priorities to address the problems were more closely aligned with public perception than with the agency analysis.

How EPA Experts Rank Environmental Risks (Highlights)

Overall High/Medium Risk	High Health; Low Ecological and Welfare Risk	Low Health; High Ecological and Welfare Risk	Overall Medium/Low Risk Groundwater-related Problems	Mixed and/or Medium/Low Risk
• "Criteria" air pollution from mobile and stationary sources (includes acid precipitation) • Stratospheric ozone depletion • Pesticide residues in or on foods • Runoff and air deposition of pesticides	• Hazardous/toxic air pollutants • Indoor radon • Indoor air pollution other than radon • Drinking water as it arrives at the tap • Exposure to consumer products • Worker exposure to chemicals	• Point and nonpoint sources of surface water pollution • Physical alteration of aquatic habitat (including estuaries and wetlands) and mining waste	• Hazardous waste sites—active (Resource Conservation Recovery Act) • Hazardous waste sites—inactive (Superfund) • Other municipal and industrial waste sites • Underground storage tanks	• Contaminated sludge • Accidental releases of toxic chemicals • Accidental oil spills • Biotechnology (environmental releases of genetically altered materials)

How the Public Ranks Selected Environmental Risks

High Risk	Medium Risk		Low Risk	
1. Chemical waste disposal 2. Water pollution 3. Chemical plant accidents 4. Outdoor air pollution	5. Oil tanker spills 6. Exposure to pollutants on the job 7. Eating pesticide-treated food	8. Other pesticide risks 9. Contaminated drinking water	10. Indoor air pollution 11. Exposure to chemicals in consumer products 12. Genetic engineering (biotechnology) 13. Waste from strip mining	14. Nonnuclear radiation 15. "Greenhouse effect" (carbon dioxide and global warming)

Source: EPA Journal, November 1987, p. 11.

tain gain over a large probable one. Yet when risk involves loss, people tend *not* to be conservative.

It is not the record or projected number of deaths but rather a *perception* that one has some personal control over risk that seems to diminish its seriousness in people's minds. In assessing riskiness, we tend to accept the risk that comes with smoking, skiing, and hang-gliding, and we have less tolerance for hazards associated with food additives, pesticides, and nuclear power. It is interesting that this holds true even when the perception of control is wrong. For example, we

tend to think that we ourselves are better drinker-drivers than those who kill themselves or others in automobile accidents.

Another reason for the confusion about risk is that some risks get mixed up with benefits. An X-ray adds to one's risk of cancer, but it can bring far more benefit in diagnosis of disease.

Still another problem is that the media distort risk. The distortion comes in decisions about newsworthiness. Most of us will die of heart disease or cancer, but such deaths rarely make the news. The

Toxics, Wastes, and You

How do you do when it comes to sound waste management practices and toxic chemical use?

1. Are you careful about what you send to the landfill?
2. Are you careful about following directions if you use biocides in your garden?
3. Do you have any idea what your lawn treatment service is doing?
4. What do you do with used motor oil?
5. Do you separate your garbage into recyclables and compostables?
6. Do you choose products with minimal packaging and reusable or recyclable packaging?
7. Do you buy food products in bulk?
8. Do you take your own reusable bags to the grocery?
9. Do you always choose reusables over disposables?

What do you think about the impact individuals can have in dealing with our solid and hazardous waste problems?

media would have us believe that automobile accidents and serial killers are among the most common causes of death. Many Americans stayed away from Europe in droves a few years ago because of terrorists' attacks on airports, yet many of them suffered far greater risk just by living in large cities.

We can express risk by equating and comparing things that we do know something about—for example, things that would increase one's risk of dying by 1 in a million. Allman (1985) points out that this way of comparing risks would equate smoking 1.4 cigarettes, drinking a half liter of wine, and living five years at the boundary of a nuclear power plant. Another way of equating or comparing risks is to look at the estimated average number of days certain habits or actions would take off of the risk taker's life. Allman cites the following: being a coal miner, 1,100 days; being 20% overweight, 900 days; being an unmarried male, 3,500 days.

But not all risks can be so neatly placed on such a scale. Some risks are unknown, and the quantification of some may even be beyond the reach of science. A committee that studied the biological effects of ionizing radiation for the Nuclear Regulatory Commission once said, "The Committee does not know whether those doses of gamma or X-rays of about 100 millirads per year are detrimental or not." Straightforward, truthful expressions of this type are not common enough. More often we talk about what *might* be true, given lots of qualification; others try to make sense out of it, usually by dropping or misstating the qualifiers and thus making it sound as if someone *does* know.

Some observers think that we have to find better ways in which to focus on *significant* risks in our society, ignoring the insignificant ones. Byrd and Lave (1987) suggest that zero-risk goals, such as the one expressed in the Delaney amendment to the Pure Food, Drug and Cosmetic Act, are unrealistic and, in a world with limited resources to apply to safeguards, impede efforts to deal with the risks that are significant (see Chapter 18).

Byrd and Lave propose working definitions of both significant and *de minimis* (trivial) risk. They define significant risk as risk large enough to be potentially observable, with some adjustment for rare diseases (because rare diseases tend to be overemphasized by observation). They say that epidemiologically observable increases in death rates occurring somewhere between 2 deaths per 100 and 3 per 10,000 constitute significant risk. Until someone comes up with a better number, they suggest that a functional lower limit of "significant" might be 1 in 1,000. They suggest the upper limit of *de minimis* risk—the limit below which we would ignore risk—is somewhere just under 1 in 100,000. For risks that fall between significant and *de minimis*, Byrd and Lave suggest that cost-benefit analyses would be helpful in determining acceptability.

Byrd and Lave's definitions can be debated. Moreover, they leave a lot of doubt about what to do about hazards whose range of uncertainty spans both definitions. Often the public disagrees with the experts. Despite all this, some experts propose that, given limited time and resources, regulatory agencies must adopt approaches such as the ones that Byrd and Lave suggest, "perfecting" them as they are implemented. It is impossible to achieve zero risk, and trivial risk is hardly worth worrying about in a world so full of serious environmental risks in need of immediate attention.

Concepts to Remember

1. Once waste has been generated, it must go somewhere. It goes into either the land, the air, or the water.

2. For regulatory purposes, waste is usually classified as solid, hazardous, or radioactive.

3. The amount of waste generated per capita by a nation is not necessarily related to affluence or a certain standard of living. The United States generates more waste per capita than any other industrialized nation.

4. The composition of waste has changed over the years. In the United States there has been an increase in plastic wastes and packaging wastes and a decrease in organic wastes.

5. European nations are far ahead of the United States in developing waste disposal technologies that do not center on land disposal. This has been necessary because of the density of population and scarcity of land in Europe.

6. Ocean dumping as a disposal option must be examined carefully, and attempts are being made in the United States to phase it out.

7. Resource recovery is a method of turning wastes into resources by recovering usable energy and material products.

8. One ton of municipal solid waste is equivalent to 9 million BTUs, 65 gallons of No. 1 (kerosene) fuel oil, or 9,000 cubic feet of natural gas.

9. For a resource recovery facility to be successful, a community must plan it well. Management options are not all compatible but may compete for the same items in the waste stream. Recycling programs should be developed prior to designing a high-tech resource recovery facility. It is crucial that markets be identified for the recovered materials or energy.

10. Governments and individuals have a role to play in developing markets by purchasing products containing recycled materials.

11. There is a generally accepted hierarchy for preferred waste management alternatives, in this order: source reduction, reuse, recycling into the same product, recycling into a new product, incineration with energy recovery, incineration without energy recovery, and, as a last resort, landfill. Although source reduction is cited as a major goal by many governments, few government resources go into achieving that goal. Resources have gone toward incineration and landfilling. Policies and programs have focused on waste management rather than on waste prevention.

12. Recycling fulfills its ecological goal when it reduces the amount of virgin raw material that must be extracted or used. The value of a proposed recycling effort should be weighed against this criterion.

13. Estimates of the volume of hazardous waste generated increase yearly. Improper disposal of hazardous waste in the past has resulted in many costly cleanup problems and will continue to haunt our world for years to come.

14. The siting of new facilities for waste management has become difficult because of improper management practices in the past and lack of government enforcement. All levels of government and industry are seeking new and equitable methods for facility siting that involve compensation to and negotiation with the host community.

15. Changing a throwaway society is basically a problem of changing consumer patterns. Some methods of changing consumer patterns include full-cost pricing (include the cost of disposal in the cost of the product), deposit requirements on beverage containers to encourage their return for reuse or recycling, and incremental user charges (the more waste you produce, the more you pay for collection).

16. Waste minimization includes reuse and recycling of waste. It also involves changing manufacturing and production processes to produce less waste; this is called source reduction. Case studies of process changes show that source reduction can also be more economical; in short, "pollution prevention pays."

17. Waste reduction is one part of the solution to the waste problem. Reuse and recycling are other options. Detoxification by chemical and biological treatment is also desirable. Incineration in most cases reduces the combustible components of the waste to ash. Land disposal via landfills, lagoons, and ponds and deep-well injection may cause more

serious environmental problems for future generations; these practices should be the disposal methods of last resort.

18. Some serious questions are being raised about the environmental impact of incinerating waste, including toxic air pollutants and possible toxic ash generated. Highly trained individuals are needed to run incinerators properly. The environmental impact of ocean-based incineration has still not been fully evaluated.

19. Any solutions to the waste problem should be grounded solidly in the ecological principles that materials cycle in the ecosphere and everything must go somewhere. To continue to use resources linearly is to continue to burden ourselves and future generations with the specter of scarce resources, waste disposal problems, and long-term monitoring of waste sites.

20. With the many chemicals that are entering the biosphere daily, there must be some method to assess the risk in order to make informed decisions and to focus research and regulatory efforts on the most serious problems.

Environmental Career Profile

Kenneth Scott Harris

Scott Harris is an Environmental Health and Safety Manager with Tremco Inc. in Barbourville, Kentucky. Tremco is a manufacturer of industrial sealants, caulks, and adhesives. Mr. Harris is responsible for all reporting, monitoring, and control of air emissions and hazardous waste generation and disposal. He evaluates new chemicals that may be used in manufacturing processes and is also responsible for waste-water treatment and employee chemical exposure. Mr. Harris's safety duties include general safety and ergonomics. Mr. Harris's job involves both desk work and work out in the plant. He interacts with state and federal regulatory agencies in developing for his plant the assurances required by law. What he likes best about his job is the opportunity it gives him to see how individual decisions and actions affect the entire plant, from the products themselves to the employees and the budget, and extending even to the greater community outside the plant. He says that every day, he gets to see that everything is indeed connected to everything else.

Mr. Harris obtained a Bachelor of Science degree in geology with a minor in chemistry; he has a master's degree in environmental health. Previous work that helped to prepare him for his present job includes positions with the Kentucky Department of Surface Mining, the Kentucky Division of Water, the Weyerhaeuser Company and the Lord Corporation. Mr. Harris lists knowledge of wastewater treatment, aptitude

for process analysis, a background in applied chemistry, knowledge of environmental administrative regulations, and curiosity as the basic requisites for the work he does. He says that environmental engineering, environmental health, and industrial hygiene are growing fields with good pay and a good long-term job outlook.

References and Further Reading

References marked with an asterisk are cited in the chapter.

Arrandale, T., 1991. "Recycling: Getting Down to Business," *Governing* **4**(11):35–44ff.

*Allman, W. F., 1985. "Staying Alive in the 20th Century," *Science 85* **6**(10):31–41.

*Byrd, D., and Lave, L. B., 1987. "Narrowing the Range: A Framework for Risk Regulators," *Issues in Science and Technology* **III**(4):92–100.

*"Chemical Industry Cuts Hazardous Waste by More than 20 Percent, CMA Study Says," 1987. *Environment Reporter*, June 12, pp. 517–572.

Frankel, C., 1992. "Blueprint for Green Marketing," *American Demographics* **14**(4):34–38.

*"Institute Says Incineration of Trash Is Uneconomical, Environmentally Harmful," 1986. *Environment Reporter*, May 30, pp. 122–123.

*Lahey, W., and Connor, M., 1983. "The Case for Ocean Waste Disposal," *Technology Review*, August-September, pp. 60–70.

*Mahtesian, C., 1991. "Five States Join Battle against Tires," *Governing*, January, pp. 15–16.

*"1990 Predicted to Become Record Year for New Waste-to-Energy Facilities in U.S.," 1990. *Environment Reporter*, November 30, pp. 1487–1488.

---

*"Poisons Tax Will Fund Cleanup Efforts in Washington," 1988. *State Legislatures*, February, p. 6.

*Pollock, C., 1987. *Mining Urban Wastes: The Potential for Recycling*. Washington, D.C.: Worldwatch Institute.

Powell, J. 1992. "Federal Disincentives to Recycling," *Resource Recycling* **XI**(6):45–46.

*Rathje, W. L., 1989. "Rubbish!" *Atlantic Monthly*, December, pp. 1–10.

*Royston, M. G., 1979. *Pollution Prevention Pays*. Oxford: Pergamon Press.

*Thurner, C., and Ashley, D., 1990. *Developing Recycling Markets and Industries*. Denver: National Conference of State Legislatures.

U.S. General Accounting Office, 1990. *Solid Waste: Trade-offs Involved in Beverage Deposit Legislation*. Washington, D.C.: U.S. Government Printing Office.

U.S. Office of Technology Assessment, 1989. *Facing America's Trash: What Next for Municipal Solid Waste?* Washington, D.C.: U.S. Government Printing Office.

*Young, J. E., 1991. *Discarding the Throwaway Society*. Washington, D.C.: Worldwatch Institute.

18

Cancer
and the Environment

There are many kinds of diseases associated with environmental degradation and hazardous materials to be sure. We have described environmental problems associated with cancer, infectious diseases, birth defects, neurological problems, and many kinds of skin, liver, respiratory, and kidney diseases throughout the book. Because cancer is largely an environmental disease, and because it is the disease most feared and most often associated with hazardous materials, we have chosen to feature the cancer-environment connection in this chapter. As we examine cancer here, we will also use the cancer-environment connection in a more general way to illustrate (1) the broadness of the term *environment,* (2) the significance of the connection between environmental problems and serious disease, and (3) the difficulty of linking particular diseases to specific environmental causes. Many of the things we will have to say about cancer, in other words, will be applicable to other kinds of environmental diseases as well.

The World Health Organization and other sources estimate that up to 95% of all cancer is caused by environmental factors. This may seem to imply that pollutants such as agricultural and industrial chemicals are the culprits, but other environmental factors—radiation, diet, tobacco use, and sunlight in-cluded—are largely responsible for the fact that depending on where we live in the world, as many as one in three of us in some countries will develop cancer and as many as one in five of us will die of it. Nearly 73 million people now living in the United States will get cancer in their lifetimes; hardly a family in the developed world will be untouched by the disease. Over a million Americans will be diagnosed as having cancer this year. If current trends continue, nearly 50 million Americans now alive will die of cancer; nearly 500,000 die of it each year. The direct cost of cancer care in the United States alone exceeds $500 million annually, and it has been estimated that total direct and indirect costs, including such things as lost earnings, exceed $25 billion each year. Obviously, cancer is a serious environmental problem. And it is getting worse.

The incidence of cancer in the United States has been increasing by about 1% per year for several decades (see Figure 18.1). Some of this can be attributed to gains made against other diseases and the resultant increase in the number of people living long enough to develop cancer; some of the increase can also be attributed to improved detection and diagnosis. But there is evidence that much of the increase in cancer is the result of a greater number of environmental causes of the disease (Figure 18.2). After adjusting for age, the

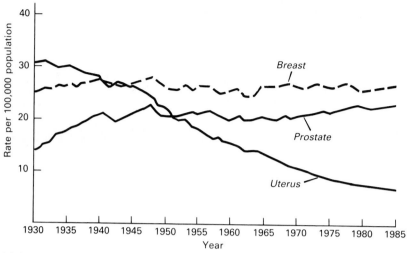

(a) Age-adjusted death rate for cancer of the breast, prostate, and uterus

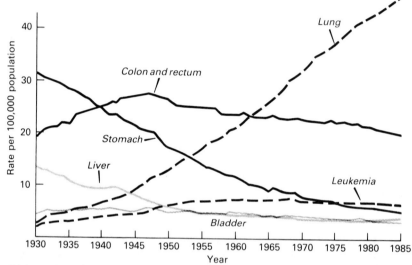

Figure 18.1 Trends in Cancer Mortality Rates by Site. Much of the increase in cancer mortality in males since 1930 has been due to lung cancer; this trend did not occur in females in the same way with time. However, lung cancer mortality rates for females have recently begun to rise dramatically. Deaths due to stomach cancer and cancer of the uterine cervix have declined significantly since the 1930s.

(b) Age-adjusted death rate for leukemia and cancer of the lung, stomach, liver, bladder, colon, and rectum

number of new cases of cancer rose 14.5% from 1973 to 1987.

Cancer Defined

Cancer cells differ from normal cells in that they divide in an out-of-control fashion, producing tissue disorder; they tend to leave their sites of origin and spread to other parts of the body; and they produce effects on the host that include lowering of immune defenses, fever, disruption of blood coagulation, and weight loss.

A major clue to the nature of cancer is that when cancer cells divide, the daughter cells are also cancer cells. The daughter cells in turn pass the transformed traits along to their daughter cells, and so on. Thus

cancer must have something to do with genes (Chapter 5). Cancer must be a result of some kind of permanent alteration in the genes or in factors that control the expression of genes.

Recently discovered cancer genes, called **oncogenes,** are now believed to be at the root of cancer. Oncogenes are thought to be present in all cells, where they serve the normal function of keeping cell division turned on during early stages of development of the organism, after which they are permanently shut off or turned way down. At least some oncogenes keep cells dividing by means of oncogene products that in various ways mimic the action of natural growth factors that stimulate cell division during processes such as wound repair. According to the **oncogene theory** (see Kupchella, 1987; Bishop, 1982; Weinberg, 1983), the in-

Part Three   The Impact of Human Activities on Health and the Environment

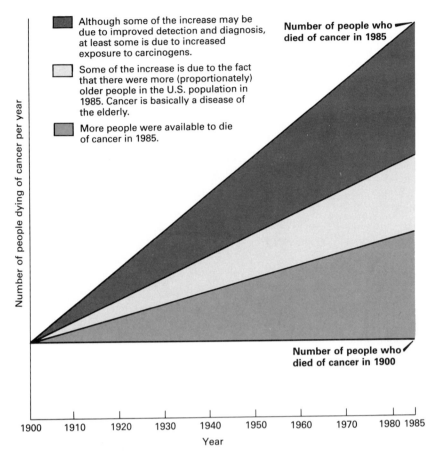

Figure 18.2 Increase of Cancer with Time. This shows cancer mortality in absolute numbers. Some of the increase in cancer that we see today can be accounted for by the fact that there are more people alive today to die of cancer. Also, because cancer is a decrease of the old, some of the increase is due to increased life expectancy. Although some of the rest of the increase may be due to improved detection and diagnosis, much of the excessive increase most likely results from an increase in the barrage of environmental insults that cause cancer.

The legend within the figure reads:

- Although some of the increase may be due to improved detection and diagnosis, at least some is due to increased exposure to carcinogens.
- Some of the increase is due to the fact that there were more (proportionately) older people in the U.S. population in 1985. Cancer is basically a disease of the elderly.
- More people were available to die of cancer in 1985.

Axis labels: Number of people dying of cancer per year (vertical); Year (horizontal, 1900–1985). Number of people who died of cancer in 1985; Number of people who died of cancer in 1900.

Mini-Glossary

Alpha particle: a positively charged particle emitted by certain radioactive substances, consisting of two protons and two neutrons. Because of their mass, alpha particles are not very penetrating.

Benign neoplasm: neoplasm that does not spread to other sites and that will cause death only rarely.

Beta particle: a particle, equal in mass to an electron, emitted by the nuclei of certain radioactive substances.

Burkitt's lymphoma: a lymphoma of the jaw endemic to Central Africa, first described by Dennis Burkitt.

Cancer: synonym for malignant neoplasm.

Carcinogen: an agent that can initiate the development of a malignant neoplasm.

Carcinoma: a malignant neoplasm originating in any epithelial cell.

Gamma ray: a ray similar to an X-ray but with a shorter wavelength; these very penetrating rays are emitted in the decay of certain radioactive substances.

Host: the organism infected by a disease-causing organism.

Initiator: an agent that can convert a normal cell into a cancer cell; *initiation* is the first step of a process of cancer induction. All carcinogens are initiators (see *promoter*).

Lymphoma: malignant neoplasm of certain kinds of cells of the lymphatic system.

Malignant neoplasm: a neoplasm of uncontrolled character that has some possibility of spreading and that will eventually kill the host.

Mesothelioma: a malignant neoplasm arising from cells of mesodermal origin in the chest or abdominal cavity.

Neoplasm: literally, new growth; a mass of cells produced by abnormal cell division.

Promoter: an agent that cannot initiate cancer but can push it along by inducing cell division once cancer has been initiated. Promotion is the second step in the process that leads to the expression of some cancers (see *initiator*).

Proximate carcinogen: a chemical that can be converted metabolically into an active carcinogen.

Sarcoma: a malignant neoplasm originating in any mesodermal cell.

Tumor: any abnormal swelling or lump; this nonspecific term is *not* synonymous with cancer or neoplasm.

Ultimate carcinogen: a carcinogenic agent derived by metabolic activation from a proximate carcinogen through one or more steps.

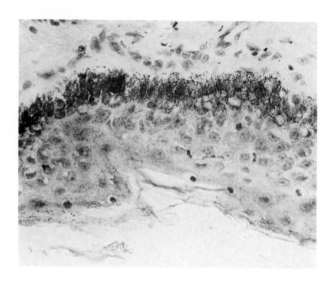

At left is normal heavily pigmented skin (magnified 290×). At right is malignant skin cancer (magnified 180×), showing the disruption of normal patterns by the cancer cells.

duction of cancer amounts to inappropriate reactivation or enhanced expression of oncogenes in one or more of several possible ways, for example:

1. Mutations (Chapter 5) in regulator genes that can then no longer hold oncogenes in check
2. Mutations in oncogenes that enable them to escape control and produce more products
3. Mutations in oncogenes that specify more powerful or possibly more stable oncogene products
4. The introduction by viruses of uncontrollable oncogenes into what had been normal cells
5. Derangement of control of oncogenes in ways not involving mutations, e.g., via chemically or physically induced chromosome breaks that separate oncogenes from genes that control them

Cancer-causing agents are all things that can cause mutations in oncogenes or related genes or can cause other gene control problems. There are many such things in the natural environment, and human beings have added many more.

Present theories of *carcinogenesis* (the initiation of cancer) are based on the idea that a number of changes must occur in a cell before it becomes a cancer cell. The increase of cancer with age in human beings probably reflects the fact that the conversion of a normal cell to a cancer cell requires a number of permanent alterations in a cell's genetic makeup. The more time goes by, the greater the likelihood that all the events will have occurred. Similarly, the more carcinogens in the environment, the more likely that the required number of carcinogen-mediated events will occur in a given amount of time.

The Causes of Cancer

Three classes of agents can cause cancer and are present in the environment—chemicals, radiation, and viruses. Each of these can interfere with genes by damaging or otherwise altering DNA or by interfering with the control of DNA expression.

Viruses and Cancer

One of the most intriguing aspects of cancer being studied today is the involvement of viruses in human cancer. It has been known for nearly a century that viruses can cause cancer in animals. For a number of years the search has gone on for a viral agent in human cancer. Investigators have looked with their electron microscopes and other tools for viruses that might be associated with certain kinds of cancer. There have been searches for outbreaks of cancer within families and for other cluster patterns that would indicate an infectious agent. These studies have turned up only limited evidence linking viruses to human cancer.

Viruses cause literally hundreds of cancers in animals, and since humans are animals, it is highly likely that there are also human cancer viruses. Human cancer-virus connections implicated thus far include warts (benign papillomas of the skin), Burkitt's lymphoma (Epstein-Barr virus) (see Figure 18.3), cer-

Figure 18.3 Burkitt's Lymphoma. The fact that Burkitt's lymphoma is common to low parts of central Africa is one of the bits of evidence suggesting that a virus transmitted by mosquitoes inhabiting low marshy areas may be important in causing this type of cancer.

vical cancer (herpes type II and certain papilloma viruses), hepatocellular cancer (hepatitis B virus), and some types of leukemia and lymphoma.

Radiation and Cancer

Before Wilhelm Roentgen discovered X-rays in 1895, human beings were exposed only to the radiation that came from natural sources. Radiation from human activities has increased since 1900, most precipitously since the advent of the atomic bomb and the harnessing of nuclear energy. Human activity now contributes significantly to the radiation pollution of the atmosphere, the hydrosphere, the lithosphere, and—unavoidably—the biosphere.

There are many kinds of ionizing radiation. Different sources produce radiation with different energies and different biological effects. For the discussion that follows, it is important to know the following:

___ **Ionizing radiation** is radiation of sufficient energy to convert uncharged chemicals into charged ion pairs (one negative part and one positive part).
___ Ionizing radiation comes from the radioactive decay of the nuclei of atoms. There are more than 1,000 different **nuclides** (atoms with different total numbers of protons and neutrons in the nucleus); some are stable, and some spontaneously

split into parts and give off radiation in the process (see Figure 6.29). The latter are called radioactive nuclides or **radionuclides.** The spontaneous disintegration of the nucleus is termed **radioactive decay.**
___ All nuclides that have the same number of protons (and are therefore forms of the same element) but different numbers of neutrons are called **isotopes** of one another. Isotopes are not equally stable; those that undergo spontaneous fission are called radioactive isotopes or **radioisotopes.** For example, iodine 131 is a radioisotope of nonradioactive iodine 127.

On the sun, atomic fission, atomic fusion (Chapter 6), and other processes give off all types of electromagnetic radiation including both ionizing and nonionizing forms (Chapter 2). Fortunately, little harmful radiation reaches the earth's surface; most of what does reach the earth is absorbed in the upper atmosphere (Chapters 2 and 9). Two exceptions are gamma radiation, which can be a problem for crews of high-altitude, supersonic jets, and ultraviolet radiation. The relationship between ultraviolet rays and skin cancer is now well documented; it is an important part of the natural cancer-environment problem.

Radioisotopes like radium 228 and uranium 238 occur naturally and can be purified from ores. Other natural sources of radiation (e.g., radon 222 and thorium 220) can be released from soil, from rocks, and from coal when it is burned. Certain kinds of useful radioisotopes can be made in **cyclotrons** (devices in which a nuclide can be made from larger nuclides by bombarding them with high-speed charged particles). As was discussed in Chapter 10, depletion of the ozone layer threatens to make radiation an even bigger problem.

Most of the radiation received by the general population comes from medical uses of radiation and from natural background radiation (see Table 6.9). Most radioactive pollution comes from atomic testing. The source has added 15% to natural radiation worldwide since the 1940s. Among the nuclides produced and released into the atmosphere in atomic explosions are cesium 137, iodine 131, and strontium 90 (see Table 18.1).

Although the greatest potential for pollution from nuclear reactors is the accident (see Chapter 6), perhaps the greatest practical problem related to atomic power is the reprocessing of spent reactor fuel and the disposal of low-level and high-level radioactive wastes (Chapters 6, 13, and 17). Normally, negligible amounts of radioactivity are produced by activation of nonradioactive substances present in the cooling water in nuclear reactors.

Table 18.1 Some Important Radionuclides and Their Half-Lives

Radionuclide	Target Tissue	Half-life
Calcium 45	bone	165 days
Carbon 14	whole body	5,760 years
Cesium 137	soft tissues, genital organs	27 years
Iodine 129	thyroid	17 million years
Iodine 131	thyroid	8 days
Plutonium 239	bone, liver, spleen	24,400 years
Radium 226	bone	1,620 years
Strontium 90	bone	28 years
Tritium (^{3}H)	whole body	12.3 years

Some of the artificial (human-made) radionuclides used diagnostically in medicine and as tracers in research end up in sewage and in water supplies (because of illegal disposal and because of leaching from supposedly secure landfills). From water they can be transmitted to humans directly or through the food chain.

The Biochemistry of Radionuclides. Radioactive isotopes of carbon, hydrogen, iodine, phosphorus, and other elements behave chemically exactly like their stable counterparts. Radioactive nuclides end up in tissue wherever their nonradioactive isotopes would end up. Carbon is a constituent of all organic molecules by definition and is present in all tissues and organs. Carbon 14, a radioactive isotope of carbon 12, also distributes itself relatively uniformly throughout any living organism. The same would be true of hydrogen and its radioisotopes. Radioactive iodine 131 is concentrated in the thyroid gland because the thyroid normally concentrates iodine. Strontium 90 concentrates in bone because strontium is chemically very similar to calcium and mimics that element. Some of the more important radionuclides and their half-lives are listed in Table 18.1

The Cancer Connection. The experience of painters of watch dials early in this century established the relationship between radium and bone cancer in humans. So that our grandparents could see their watches in the dark, painters were employed in watch factories to dab radium-containing, glow-in-the-dark pigments on the hands of watches. Imagine them using their lips to bring the tips of their tiny, radium-contaminated brushes to fine points. In this and other ways the painters absorbed radium, which lodged in their bones. Years later, a significant number of watch dial painters—many of them long since retired from watch painting—developed **osteogenic sarcoma,** a form of bone cancer.

The story of the watch dial painters raises the important point that there are two distinctly different ways of getting radiated. One is to be hit by rays that come from some *external* source. The other is to be hit by rays emitted from some *internal* source—a radioactive isotope that is breathed in, eaten, or absorbed through the skin.

Other radiation-related tumors have been reported in humans. These include cancers that have resulted from the radiation produced by the atomic bombs dropped on Hiroshima and Nagasaki in 1945 and some that have resulted from occupational exposure. In 1957 a fire burned for 15 hours in the uranium core of a reactor at Windscale, England. By one estimate, more than 20,000 curies of iodine was released—hundreds of times more than was released at Three Mile Island in 1979. A threefold increase in leukemia experienced in that area of England may be a result of that incident.

Low-Level, Long-Term Exposure. In 1943, Hanford Works, an atomic plant in Richland, Washington, began to monitor its workers. In a 1977 report, Mancuso, Stewart, and Kneale compared the long-term radiation exposure of the workers with the causes of death reported on their death certificates. The result indicated that plant workers receiving exposure rates *below* the annual dosage established as safe by the federal government had more than double the normal expected rate of certain types of cancer.

However, this study and its methodologies have been the subject of some controversy (see U.S. General Accounting Office, 1981). Until now, there really has been no generally accepted definitive study of the impact on humans of long-term, low-level exposure to radiation. Data gathering continues as the U.S. government tries to prescribe standards for exposure well within an acceptable risk level. Summaries of 52 studies involving humans are presented in the GAO report just cited.

Biological Magnification of Radionuclides. Generally, when radionuclides are released into the environment, they are dispersed. We say generally because there are radioactive waste disposal sites where high concentrations are accumulated. Biologically, as we have seen with iodine 131 and the bone-seeking radionuclides, radioactive substances can become concentrated in particular organs, reversing the tendency toward dilution. Radionuclides can also become concentrated, as we discussed in Chapter 17, in food chain transfers.

Chemicals and Cancer

Table 18.2 lists some of the chemical carcinogens, chemical mixtures, and classes of chemical carcinogens that have been linked to cancer in humans or are strongly suspected of being linked. The first purified chemicals identified as carcinogens were *po-*

Table 18.2 Some of the Major Known or Suspected Human Chemical Carcinogens

Chemical or Industrial Process	Main Type of Exposure*	Target Organs in Humans	Main Source of Exposure†
Arsenic compounds	Occupational, medicinal, environmental	Skin, lung, liver‡	Inhalation, skin, oral
Asbestos	Occupational	Lung, pleural cavity, gastrointestinal tract	Inhalation, oral
Benzene	Occupational	Blood-cell-forming system	Inhalation, skin
Benzidine	Occupational	Bladder	Inhalation, skin, oral
Bis(chloromethyl)ether	Occupational	Lung	Inhalation
Cadmium	Occupational	Prostate, lung	Inhalation, oral
Chromate-producing industries	Occupational	Lung, nasal cavities‡	Inhalation
Cyclophosphamide	Medicinal	Bladder	Oral, injection
Diethylstilbestrol (DES)	Medicinal	Uterus, vagina	Oral
Hematite mining	Occupational	Nasal cavity, larynx	Inhalation
Isopropyl oil	Occupational	Nasal cavity, larynx	Inhalation
Melphalan	Medicinal	Blood-cell-forming system	Oral, injection
Mustard gas	Occupational	Lung, larynx	Inhalation
Nickel refining	Occupational	Nasal cavity, lung	Inhalation
Phenacetin	Medicinal	Kidney	Oral
Soot, tars, and oils	Occupational, environmental	Lung, skin, scrotum	Inhalation, skin
Vinyl chloride	Occupational	Liver, brain‡, lung‡	Inhalation, skin

* Association demonstrated.

† May not be the only routes by which such effects could occur.

‡ Indicative evidence.

Source: L. Tomatis et al., "Evaluation of Carcinogenicity of Chemicals," *Cancer Research 38* (1978):877–885.

lycyclic hydrocarbons. It has since been determined that most polycyclic hydrocarbons are noncarcinogenic. Those that are carcinogens include derivatives of benzene and anthracene and various products of fossil fuel combustion. Unfortunate experiences with bladder cancer in the dye industry established that a number of azo dyes and other *aromatic amines* were carcinogenic. *Nitrosamines* are well-established carcinogens, having been shown to produce cancer in dogs, monkeys, parakeets, rats, mice, hamsters, guinea pigs, and rainbow trout (Shapley, 1976). A number of miscellaneous organic compounds, including the pesticides DDT, dieldrin, chlordane, aldrin, endrin, heptachlor, and chlorodecone (kepone) (see Chapter 16) and the artificial sweeteners saccharin and cyclamates, are capable of causing tumors in animals. Vinyl chloride, estrogens, and (ironically) certain anticancer drugs are all strongly implicated in

Table 18.3 Natural Products Associated with Cancers in Humans

	Source	Target Organ
Aflatoxins	Aspergillus flavus (a mold)	liver, kidney
Alkaloids	plants	liver
Cycasin	plants	lung, bladder,
Saffrole	plants (sassafras)	mouth, esophagus,
Tobacco smoke products	cigarettes, pipe, cigars	pharynx, larynx

Source: L. Tomatis et al., "Evaluation of Carcinogenicity of Chemicals," *Cancer Research 38* (1978):877–885.

human cancer. A number of metals—some of which are essential nutrients in small amounts—and other inorganic compounds have also been implicated in human lung and skin cancer. These include arsenic, chromates, asbestos, beryllium, nickel compounds, and cobalt. Rounding out the list of kinds of carcinogens are a number of *natural products* including the aflatoxins produced by the mold *Aspergillus flavus*, certain plant alkaloids, and saffrole, an extract of sassafras bark (Table 18.3).

A quick glance at the kinds of tumors produced in humans by the agents listed in Tables 18.2 and 18.3 reveals that the skin and lungs—the places where humans are in direct contact with the environment—and the organs of excretion and detoxification are common sites of environmental cancers.

Epidemiology of Human Cancer

Much—maybe most—of what we know about the relationship between environmental factors and cancer has come from epidemiology (Chapter 10). In humans, long-term exposure to pure carcinogens rarely occurs. Epidemiology can help to sort out the major factors related to carcinogenesis from a complex combination of factors.

Patterns of Cancer Death Worldwide

The estimate that 90% to 95% of all cancers are environmentally related is based on epidemiologic data for cancer worldwide. There are great differences

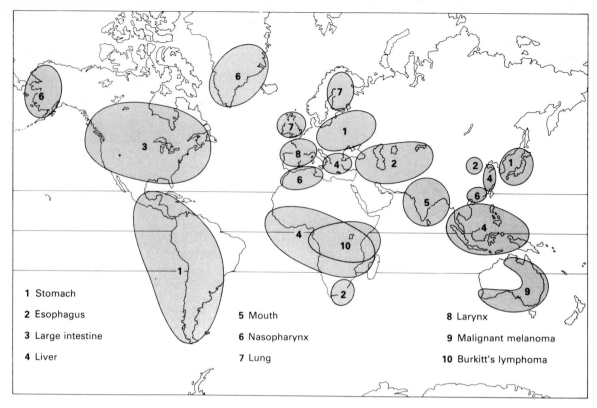

Figure 18.4 Cancer Hotspots around the World.

1 Stomach

2 Esophagus

3 Large intestine

4 Liver

5 Mouth

6 Nasopharynx

7 Lung

8 Larynx

9 Malignant melanoma

10 Burkitt's lymphoma

in the kinds of cancer that predominate from country to country (see Figure 18.4). There is good evidence (from observations of immigrants) that these patterns are related to environmental differences rather than genetic differences (see Tables 18.4 and 18.5). Theoretically, if the mortality rate for each specific cancer could be reduced worldwide to the level found in the country in which it is presently *lowest*—by making some appropriate environmental change—cancer mortality could be reduced to one-tenth its current prevalence.

Japan has the highest death rate for stomach cancer and one of the lowest for lung cancer. Nearly

the reverse is true for U.S. whites. There are fourfold differences in mortality due to uterine cancer worldwide; U.S. whites and nonwhites are almost at opposite ends of the spectrum. These and other observations indicate that America is apparently *not* the land of equal opportunity when it comes to cancer.

Patterns of Cancer Death in the United States

The importance of the environment is also suggested by the distribution of cancers of various types throughout the United States. The U.S. Department of Health and Human Services published an extensive

Table 18.4 Comparison of Cancer Incidence Rates for Japanese in Japan and for Japanese and Caucasians in Hawaii (annual incidence per million people)

Primary Site of Cancer	Patient's Sex	Japan		Hawaii, 1968–1972	
		Miyagi Prefecture, 1968–1971	Osaka Prefecture, 1970–1971	Japanese	Caucasians
Esophagus	M	150	112	46	75
Stomach	M	1,331	1,291	397	217
Colon	M	78	87	371	368
Rectum	M	95	90	297	204
Lung	M	237	299	379	962
Breast	F	335	295	1,221	1,869
Uterus (cervix)	F	329	398	149	243
Ovary	F	51	55	160	274

Note: Ages 35–64, standardized for age.

Source: International Agency for Research on Cancer.

Table 18.5 Comparison of Cancer Incidence Rates for Nigerians in Ibadan, Nigeria, and for Blacks and Whites in Two Parts of the United States (annual incidence per million people)

| Primary Site of Cancer | Patient's Sex | Ibadan, Nigeria, 1960–1969 | United States | | | |
| | | | Blacks | | Whites | |
			San Francisco, 1969–1973	Detroit, 1969–1971	San Francisco, 1969–1973	Detroit, 1969–1971
Colon	M	34	349	353	294	335
Liver	M	272	67	86	39	32
Pancreas	M	55	200	250	126	122
Lung	M	27	1,546	1,517	983	979
Breast	F	337	1,268	1,105	1,828	1,472
Uterus (cervix)	F	559	507	631	249	302
Lymphosarcoma at age 15*	M	133	10	5	4	3

Note: Ages 35–64, standardized for age.

* Including Burkitt's lymphoma. The cited rates are the average of the age-specific rates at ages 0–4, 5–9, and 10–14.

Source: International Agency for Research on Cancer.

series of maps (Pickle et al., 1987, and Figures 18.5 and 18.6) in which the mortality due to various types of cancer was presented by county or economic subunits of the states. Among the notable observations that can be made from these maps are the following:

1. The mortality due to cancer of all types is extremely high in the Northeast—the most heavily industrialized and most heavily populated part of the country (Figure 18.5).

2. From this same map it is evident that cities present an especially high risk of cancer; nearly all of the major old cities, particularly those of the eastern United States, stand out as having cancer mortality in the highest decile.

3. The Northeast has a high death rate for cancer of

Figure 18.5 Areas of the United States in Which Death Rates for All Types of Cancer in White Males Were Significantly High during the Period 1970–1980.

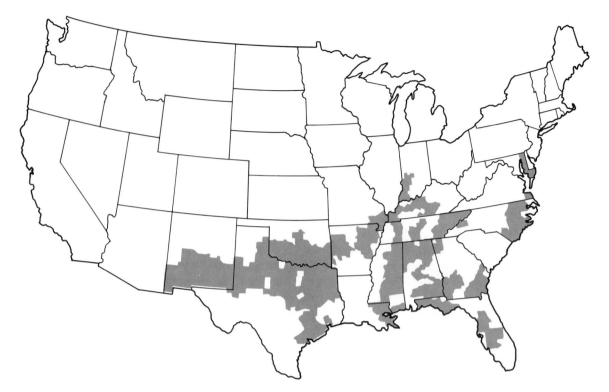

Figure 18.6 Areas of the United States in Which Death Rates for Skin Cancer (Other than Melanoma) in White Males Were Significantly High during the Decade 1950–1959.

the female breast, but the difference between North and South is diminishing.

4. Cancer of the cervix shows a particularly high mortality pattern in Appalachia.

5. Figure 18.6 illustrates the effect of a natural environmental carcinogenic agent—sunlight. Skin cancer (including malignant melanoma) mortality is particularly high in the Sun Belt states.

6. Stomach cancer mortality is high in certain states of the Upper Midwest. It has been suggested that mortality for stomach cancer there closely matches the rate found in the countries from which the people of the Upper Midwest migrated. Cultural habits that have persisted—such as diet—might account for the high rate of stomach cancer there.

The fact that there are unusually strong patterns for some types of cancer is highlighted by the observation that for some other cancers—pancreatic cancer, for example—there appears to be no unusual distribution pattern. An average amount of pancreatic cancer is found just about everywhere in the United States.

Heredity or Environment?

Cancer is obviously a product of the interaction of the host and the environment. However, epidemiologic observations of the sort we have just considered indi-cate that in general the environment is a much more important contributor to cancer than heredity. Considerable support for this comes from cancer data for immigrants and their children. The children of immigrants have death rates from cancer that match those prevailing in the country to which their parents have immigrated. The immigrants themselves, however, get cancer at rates prevailing in their homeland.

In one study it was shown that people born in Israel, whether they were Jewish or not, had about the same *cancer mortality* (cancer deaths per 100,000 population). Jewish and non-Jewish Israelis living in Israel who were born in Europe or the United States had stomach cancer mortality equivalent to that in their place of origin. These immigrants to Israel apparently carried with them the effects of the environmental exposures in the countries from which they emigrated.

A study of immigrants to California from Japan revealed that over successive generations, the incidence of several different kinds of cancer gradually approached the incidence of those kinds of cancer in California (Figure 18.7). The transition required several generations, perhaps because it takes that long for a family of immigrants to fall completely under the influence of a new environment. It would obviously take a few years for Japanese immigrants to be influenced by the parts of the environment that are cultural

Environmental Cancer: Whose Fault Is It?

Epidemiologists say that as much as 90% of all human cancer is caused by environmental factors. This is based on observations concerning the differences in cancer incidences throughout the world and changes in cancer risk that occur in people who move from place to place. When people hear that cancer is caused by environmental factors, they assume that this is because of things that industry or other people do to the environment. Yet it is estimated that about 30% of all cancers in the United States are linked to the use of tobacco, another third are related to diet (the overall nature of the diet, not necessarily food additives and contaminants), and that another large fraction of all cancers are caused by exposure to sunlight. This seems to implicate lifestyle—things that we do

to ourselves—as the most important aspect of cancer causation. Environmental cancer is thus one serious environmental problem over which the individual has the most control. Does this take industry and the EPA off the hook? Cancer will strike one out of every three or four of us. Where should government's effort to reduce cancer be directed?

in origin—diet, for instance. These studies, plus studies of Japanese immigrants to Hawaii (Table 18.4) and of blacks in Africa versus blacks in the United States (Table 18.5), all show the same thing: *Place* is more important than *race* when it comes to cancer risk.

None of this is to say, however, that heredity is not important. Many kinds of cancer in experimental animals show differential susceptibilities for different varieties of the same species. We should expect that

this same sort of thing will be found in human beings; there may be certain physiological types that have a higher risk than the population at large of developing certain kinds of cancer. There is evidence of hereditary risk of developing cancer of the breast, stomach, large intestine, lining of the uterus, prostate, lung, and possibly ovary. And of course, it is not possible to completely separate the genetic factors from environmental factors. Families tend to be exposed to the same

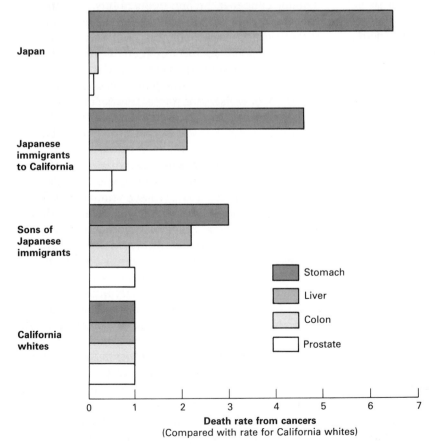

Figure 18.7 Incidence of Various Cancers in Japanese, Californians, and Immigrants from Japan to California. The death rates for stomach, liver, colon, and prostate cancer in California white males are normalized here to 1 so that the rest of the figure can be read as a multiple or a fraction of 1. The gradual shift to California rates of death from cancer following immigration suggests both that the environment, not heredity, is important here and that persistent cultural habits may be important in cancer.

environment as well as to carry common genetic characters. So the question of the relative importance of genetics in cancer is still somewhat open.

Urban Cancer: The Mystery of Increased Cancer Among City Dwellers

Cancer is an urban disease in many respects. Figure 18.5 shows that many major U.S. cities have had significantly high mortality associated with all types of cancer combined. The differential between cities and rural areas seems to be especially striking for certain kinds of cancer. For instance, the incidence of cancer of the lower digestive tract in urban areas of the United States is nearly twice as high as it is in rural areas; the overall incidence for U.S. cities is about ten times greater than it is in Uganda, a relatively rural country. Lung cancer is another example. People in cities tend to have higher lung cancer rates than those who live in rural areas. The presence and concentrations of certain pollutants in cities is undoubtedly at the heart of this urban-rural differential, most likely because environmental contaminants act synergistically with carcinogens such as cigarette smoke.

In a study reported in the *British Journal of Cancer* more than three decades ago (Stocks, 1960), the concentration of hydrocarbons in the air over certain cities in England was highly correlated with lung cancer, even after social factors were considered. In the same study, lung cancer mortality was found to be correlated with atmospheric smoke density in more than 25 areas in northern England and Wales. Cancer of the stomach was also related to smoke in various boroughs. The hydrocarbon benzopyrene was implicated as a prime suspect.

Despite these and other more recent studies showing, for example, that ozone is associated with cancer (see Chapter 10 and Hassett et al., 1985), Page and Asire (1985) suggest that, relatively speaking, there is little evidence that air pollution poses a serious cancer risk. Serious risk may be present despite the scarcity of evidence, however. The connection may be obscured by other influences such as indoor air and smoking.

Nitrosamines are also prime suspects in urban cancer. Nitrosamines come from a great variety of sources and can be found almost everywhere in the urban environment. Nitrous oxide and nitrogen dioxide can combine with water to form nitrous acid, which can then combine with amines under certain conditions to form nitrosamines. Such a reaction might account for the statistical correlation between nitrogen dioxide and cancer in urban areas.

The causes of urban cancer probably extend to greater uses of tobacco and alcohol among city dwellers. Stress, which has been shown to be an important immune system suppressor, may also be a factor here.

Overall, there are many potentially cancer-related differences between city dwellers and rural residents. There are industrial pollutants in the air, and they occur in pockets of exotic and intense combinations. City dwellers take in between 500 and 1,000 liters of city air every day. City dwellers might also be drinking water that is less pure or that in some cases is contaminated from various sources like cities upstream. Urban cancer is even more complex a problem than occupational cancer, which we will take up next.

Occupational Cancer

The first connection between an occupational exposure and a cancer was made more than 200 years ago when Percival Pott described the occurrence of testicular cancer in chimney sweeps. Since then, many epidemiologic studies have indicated that occupational factors are important in the *etiology* (cause) of cancer (Table 18.2).

Although relatively few specific chemical agents have been implicated in occupational cancer (see Shottenfeld and Haas, 1979), various estimates indicate that a significant percentage of cancer deaths in males may, in the future, prove to be occupational in origin. Occupational cancers now include lung cancer and pleural mesothelioma (a cancer of the lining of the thoracic cavity between the lungs and the chest wall) in insulation workers and others exposed to asbestos; bladder cancer in aniline dye workers; a number of cancers in rubber industry workers exposed to such chemicals as 2-naphthylamine, benzidine, and certain nitrosamines; lung cancer in coke oven workers; skin cancer in cutting oil workers; nasal sinus cancer in woodworkers; cancer of the pancreas and lymphomas in organic chemists; and angiosarcoma of the liver and perhaps brain cancer in workers in the polyvinyl chloride manufacturing industry (Shottenfeld and Haas, 1979).

Reflecting the complexity of carcinogenesis and the importance of cofactors is the fact that the specific risk of *lung* cancer has *not* been shown to be significantly increased among nonsmoking asbestos workers. A study by Selikoff, Hammond, and Churg (1968) showed that asbestos workers who smoked cigarettes had a far greater risk of developing lung cancer than people who smoked but did not work with asbestos. Asbestos workers who smoke, according to the study, have nearly 100 times more risk of death from lung cancer than an otherwise similar individual who neither smokes nor works with asbestos. Asbestos workers who do not smoke had no greater risk than the normal population, a finding that indicates a strong synergistic effect of tobacco smoking and exposure to asbestos.

Another kind of cancer associated with exposure to asbestos—mesothelioma—exhibits no such rela-

Asbestos fibers may be trapped in the lungs of people who work with asbestos. There is a strong correlation between asbestos and cancer, particularly in people who smoke.

tionship; that is, the increased risk of developing this rare cancer is equally great in smokers and non-smokers and may be a risk even to members of the families of asbestos workers (Selikoff, 1975; Nicholson, 1977).

Pesticide Workers and Cancer. In 1986 the National Cancer Institute reported that farmers and other workers exposed to herbicides face a higher risk of lymphatic cancer (non-Hodgkin's lymphoma).* The study, of a group of Kansas farmers, showed that those exposed to herbicides 20 days or more each year were 600% more likely to develop lymphatic cancer. Overall, farmers who used herbicides were found to have a 1.6-fold increased risk for developing non-Hodgkin's lymphoma than nonfarmers (Hoar et al., 1986). The report specified that **2,4-D** (2,4-dichlorophenoxyacetic acid), a widely used ingredient in many brands of herbicides, was particularly implicated.

In a case-control interview study in Iowa and Minnesota designed to evaluate the role of pesticides in leukemia and non-Hodgkin's lymphoma, no increase was found in risk for non-Hodgkin's lymphoma for farmers. For leukemia, no excess risk was found in association with raising a specific crop or animal or

* In the same year the EPA banned the herbicide dinoseb after it was found by Hoechst AG, a chemical and pharmaceutical company in Germany, to be associated with irreversible neurologic and skeletal malformations in the offspring of rabbits exposed during pregnancy. The EPA also had data indicating that the herbicide affected fertility in male rats and mice. At the time of the ban, some 7 to 11 million pounds of dinoseb-containing herbicides were used annually in the United States. The EPA also placed restrictions on the most widely used herbicide in the United States, alachlor, after test results showed that it caused tumors in mice and rats.

with exposure to major classes of pesticides (see Blair et al., 1985; Cantor, 1982; and Hoar et al., 1986).

Recently, the U.S. firms that manufacture pesticides that had been widely used for termite control—chlordane and heptachlor—agreed to stop selling them until ways are found to apply them that eliminate long-term human exposure. Both pesticides have been classified by the EPA as probable carcinogens.

Industrial and other occupational settings are obviously important in linking environment to cancer in humans because these settings provide relatively controlled, though inadvertent, exposure of human beings to potential carcinogens. Workers are intimately exposed to relatively high concentrations of the very same agents to which the general population is exposed at lower concentrations. Although the overall problem of occupational cancer may be relatively small, evaluation of the experiences of occupational groups should provide valuable information and allow us to identify suspected cancer agents in the general environment.

Diet, Food Additives, and Drugs

Another important part of our environment is what we eat and drink, and cancer has been linked to the overall quality of diet. Of the one-third of all cancers believed to be associated with diet in the United States, nearly all of the risk comes from traditional, everyday foods. There is a strong association between some cancers (e.g., of the breast and the colon) and fat intake and obesity (Cohen, 1987). This is probably related to fat's ability to influence the levels of certain hormones that may be cancer **promoters** (substances that are known to be able to accelerate the development of cancer). Worldwide, overall cancer mortality by country correlates directly with red meat consumption (Figure 18.8). Countries that consume more meat have more cancer. It is not known, however, whether this is due to the meat itself, to something in the meat, or to the fact that when one eats meat, one eats less cereal or plant material. There may be other factors that parallel meat and cereal consumption patterns that are ultimately responsible for this correlation.

Food additives also have an environmental cancer connection. Food additives are chemicals added to food to enhance color or to preserve it. In the United States, such items fall under a clause of the 1958 Pure Food, Drug, and Cosmetic Act. The clause, usually referred to as the **Delaney Amendment** (because it was sponsored by Representative James J. Delaney of New York), specifically bans the use of any food additive found to cause cancer when eaten by humans or animals or when subjected to any other test appropriate for evaluating the safety of food additives. The clause has been used in the banning of such substances

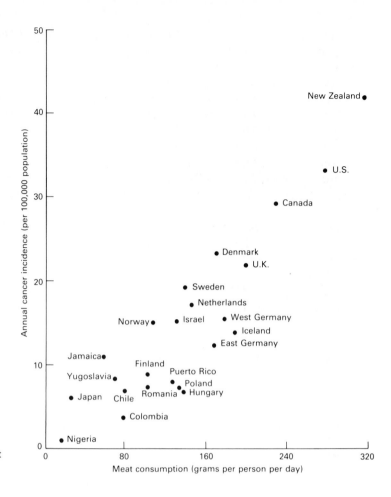

Figure 18.8 Meat Is a Carcinogen—or Is It?   What do data like these, showing only correlation, prove?

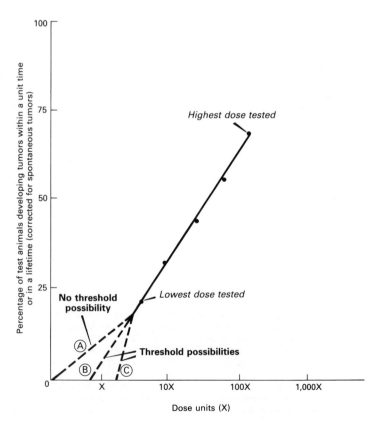

Figure 18.9 Low-Dose Wobble.   Animal tests for carcinogenic potential are usually done by using several relatively high-dose regimens. Testing at very low doses becomes increasingly less feasible economically because as zero dose is approached, extremely large number of animals must be exposed to produce even a few tumors for counting.

as cyclamates and saccharin, saffrole (root beer flavoring), diethylstilbestrol (DES, an estrogenlike dietary substance fed to livestock to accelerate meat production), and red dye No. 2. All of these produced excess numbers of cancers in experimental animals.

The Delaney Amendment is controversial because it admits no threshold level (Figure 18.9). It is based on the assumption that there is no low level of any carcinogenic substance that is completely harmless in food or drink. The study on which the banning of saccharin was originally based, for instance, involved feeding about 100 rats a diet of 5% pure saccharin from the time they were born until they died. Fourteen of these rats developed bladder cancer, compared with only two such animals in a group of 100 animals given no saccharin. Although a human being would have to drink several hundred 12-ounce diet sodas a day for life to accumulate an equivalent dose, strict application of

the Delaney Amendment called for the removal of saccharin from the consumer market in the United States.

The banning of saccharin (and its subsequent exemption from the Delaney Amendment by Congress) offers a particularly poignant example of the difficulty of regulating carcinogens in the environment. If there is indeed a relationship between obesity and cancer, if the banning of saccharin results in a return to natural sweeteners like sugar, and if this produces in some people a weight increase that predisposes them to certain fat-related cancers, the banning of saccharin might result in more cancers than if saccharin were left on the market—if, of course, saccharin would indeed cause tumors in human beings (see Enrichment Boxes 18.1 and 18.2).

The difficulty associated with risk assessment was illustrated in the United States over the regulation of the use of Alar, a chemical used to hold apples on

Enrichment Box 18.2

How Much Is a Human Life Worth?

In several places in this book we have glossed over the question of how much a human life is worth. It bothers most people even to consider the notion that human lives might fall short of invaluable. Yet we tolerate many kinds of activities that clearly *will* take a small number of randomly selected lives prematurely. Coal mining, automobile racing, war, hunting, shipping, driving, and football are but a few examples of essential and nonessential activities in this category. With environmental chemicals the question somehow seems more important.

In the case of benzene, for example, OSHA attempted to set new low limits on exposure in the workplace on the basis of a study showing that of 748 workers exposed to benzene in two chemical plants in Ohio between 1940 and 1949, nine died of leukemia. This was roughly eight more than would be expected in such a group. Of course, nobody knows exactly what levels of benzene were involved for any of the workers. Although even the orders of magnitude are in dispute, exposure could have been as high as several hundred parts per million (Carter, 1979). This is a central point in the debate over a proposed reduction from 10 ppm to 1 ppm, which is reported to cost somewhere between $500 million and $1 billion.

Even if we knew for a fact that a certain number of lives would be saved by lowering exposure limits to benzene—or to anything, for that matter—we could still face having to make decisions like this one: *Just how many lives saved would justify the expenditure of even $1 million?* Keeping in mind that everyone will die eventually of something (though it seems heartless to bring this up) and that a chemically induced cancer death really means a shortened life span, how much is ten years of a randomly selected life worth?

Perhaps some consensus could be derived from what we actually do pay to save otherwise randomly taken lives. The U.S. government, for example, is apparently willing to spend about $140,000 to save one life in highway construction costs (Fischhoff, Slovic, and Lichtenstein, 1979). But even this seems to leave us hanging.

---

the tree longer and to promote color development and increase storage life. The worst-case estimate in studies of the carcinogenicity of Alar was that there would be some 45 cancer cases per million people; however, one group estimated that the probable risk was more on the order of three or four cancers per trillion people (see Rosen, 1990).

It must be pointed out that there is no scientific evidence to suggest that there *is* a carcinogenic threshold—that is, a dose of any carcinogenic substance that will not produce tumors if given to enough people in a large population.

A study of the effectiveness of the Delaney Amendment by the U.S. National Academy of Sciences (1987) found some shortcomings. The clause applies only to pesticides in processed foods and not in raw foods. Moreover, the U.S. EPA has been enforcing the clause only on pesticides registered since 1978, and the report says 90% of the cancer risk comes from pesticides approved prior to 1978. In place of the current Delaney Amendment, the report supports the concept of a "negligible risk" standard for suspected carcinogens. Such a standard would cancel use of any pesticide that had a combined cancer risk from residues on raw and processed food of greater than one in a million. This standard, applied to 28 pesticides regarded as the most dangerous, would reduce cancer risk by 98% but cause cancellation of only 32% of the uses. Continued application of Delaney as is reduces total risk by only 55%.

Perhaps the biggest problem with the zero-risk approach to regulation exemplified by the Delaney Amendment is that it can miss many carcinogens. The food additives sodium nitrate and sodium nitrite are not themselves carcinogenic, but there is evidence indicating that they may be converted into nitrosamines under acidic conditions in the stomach or perhaps by bacteria in the digestive tract. Nitrosamines, as we indicated earlier, have carcinogenic potential in many species of animals. It has been suggested that a high incidence of stomach cancer in Japan is due to nitrosamines that are present in fish and other smoked foods common in the Japanese diet.

Most of the cancer risk associated with diet apparently comes from natural components of food or from dietary imbalances. In a report by the Committee on Diet, Nutrition and Cancer of the National Academy of Sciences (1982), a number of specific recommendations were made about what to eat and what to avoid in order to minimize the risk of cancer:

— Reduce fat consumption from the 40% typical of the American diet to 30%. Epidemiological studies have shown colon, breast, and prostate cancers to be linked to fat intake.

— Reduce consumption of salt-cured, pickled, and smoked foods. Diets heavy in these items have been linked to cancer of the stomach and of the esophagus. The smoking of foods is known to produce nitrosamines and other proven carcinogens.

— Include whole-grain cereals, fruits, and vegetables, especially those rich in vitamin C and beta carotene (precursor to vitamin A), in the diet.

The recommendations specifically mention the vegetable family *Cruciferae*—cabbage, brussels sprouts, broccoli, and cauliflower—as being especially rich in vitamins A and C and other substances believed to inhibit the formation of the active forms of cancer-causing chemicals.

__ Avoid excess alcohol consumption especially if you smoke. Cancers of the mouth, larynx, esophagus, and respiratory tract have been linked to smoking and drinking together; heavy drinking even without smoking has been linked to colorectal cancer.

Japan's National Cancer Center offers the following list of cancer prevention dos and don'ts, most of which have to do with food or drink: (1) have a nutritionally balanced diet; (2) have a variety of types of food; (3) avoid excess calories, especially fat; (4) avoid excessive drinking of alcohol; (5) smoke as little as possible; (6) take vitamins in appropriate amounts; eat fiber and green and yellow vegetables rich in carotene; (7) avoid drinking fluids that are too hot and eating foods that are too salty; (8) avoid the charred parts of cooked food; (9) avoid food with possible contamination by fungal toxins; (10) avoid overexposure to sunlight; (11) have an exercise program matched to your condition; and (12) keep the body clean (see Sugimura, 1986).

Water and Other Drinks. If you like bourbon and water or scotch and water, you may be in double jeopardy. As we have just pointed out, alcohol predisposes one to oropharyngeal cancer in smokers, to cancer of the esophagus, and maybe even cancer of the liver. Of possibly more interest to all of us, carcinogens can be carried in drinking water.

A curious pattern in America and in other countries is that we take our drinking water from the same rivers that we use as dumps for human and chemical waste. To be sure, we purify drinking water; but just as surely, we do not get absolutely everything out during the purification process. Drinking water supplies are subject to contamination by carcinogens from industrial plants, parking lots, accidental spills, and pesticides in agricultural runoff. Petroleum refinery wastes also find their way into our lakes and rivers. Other miscellaneous chemical carcinogens get into our water supplies from gas plants, coke ovens, distilleries, and wood processing plants. Tars, pitches, and creosote are washed into our water supply from a variety of sources. Interestingly and ironically, water *purification* may lead to the formation of chlorinated hydrocarbons that have carcinogenic potential. In December 1980 the U.S. Council on Environmental Quality issued a report linking rectal, colon, and bladder cancer to the chlorination of drinking water (see Chapter 12).

The waterways that are contaminated by our wastes may also be sources of our drinking water.

In an EPA study of 80 American cities in which water supplies were sampled for halogenated compounds, chloroform was found in all of them (Harris, 1975). In light of these things, it is perhaps not surprising that a significant relationship was found between cancer death rates in Louisiana and drinking water obtained from the Mississippi River. The population whose drinking water came from the Mississippi had higher total cancers, as well as more cancers of the urinary organs and cancer of the gastrointestinal tract (Page, Harris, and Epstein, 1976).

Drugs. Perhaps the chemicals we consume as drugs are even more important than food additives; the doses of drugs we take are generally higher. The most important cancer-related drugs are the estrogens. Millions of American women today take some form of the hormone estrogen. Men and women alike are exposed to some extent—at least they were in the past—to a derivative of the same hormone called diethylstilbestrol (DES), once widely used in animal feed for increasing meat production. At one time, DES was also given to women during pregnancy for certain disorders. Now a strong association has been found between pregnant women's use of DES and vaginal cancers in their daughters. To date, several hundred such daughters have been found to have vaginal cancer. The Food and Drug Administration now requires labeling on all estrogen preparations, warning of increased risk of certain cancers. Although recent reports seem to have cleared birth control pills as a factor in breast cancer, cancer of the uterine lining, and ovarian cancer, there is still believed to be an excess risk of cancer in the uterine cervix in women who use the pills regularly. Birth control pills have also been linked to certain benign neoplasms of the liver. Male hormones used therapeutically have also been related to liver cancer (Antunes and Stolley, 1977). Another class of drugs that apparently has some carcinogenic potential is, ironically, anticancer drugs. Some of these drugs act by binding to DNA, which normally kills cancer cells. They have been shown to cause cancer in experimental animals. Actinomycin D, for instance, is carcinogenic in some doses and under some circumstances (Curtis, 1976).

Radon and Cancer

We considered this topic in the context of indoor air pollution in Chapter 11. Suffice it to say here that radon may well turn out to be among the most significant natural environmental causes of cancer. The EPA and other sources indicate that a significant fraction, maybe 10%, of the 136,000 lung cancer deaths each year in the United States may be attributable to the gaseous decomposition products of radon.

Linking Specific Environmental Agents to Human Tumors: The Difficulty

Although we can conclude that there are numerous agents in our environment that can cause cancer, we are *certain* about a few dozen chemical-human tumor relationships and only a few radionuclide-human tumor relationships. Many problems stand in the way of conclusively demonstrating whether any particular chemical agent is or is not carcinogenic in humans.

First of all, the ultimate experiments cannot be done. One cannot inject suspect chemical agents into humans to see directly whether or not cancer results. Second, cancer takes a very long time to develop. Even if the above experiments could be done, 20, 30, or even more years might pass between the exposure to a carcinogen and the appearance of a tumor in humans (Figure 18.10). Even experiments with animals take a long time, and they are very expensive. Third, the results of studies on laboratory animals are not uniformly and unequivocally applicable to humans. Fourth, it is difficult to generalize about carcinogens because they

Figure 18.10 The Correlation of Cigarette Smoking and Lung Cancer. A considerable time lag apparently occurs between exposure to tobacco smoke and the onset of cancer.

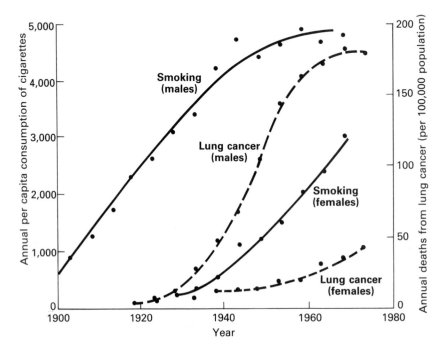

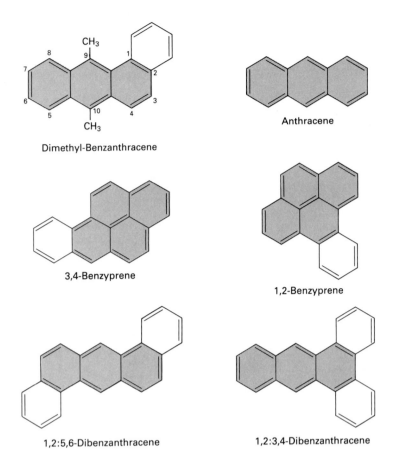

Figure 18.11 Polycyclic Hydrocarbons. The chemical structures on the left are moderate or strong carcinogens; those on the right are weakly carcinogenic or have no carcinogenic properties at all. Apparently even very small changes in chemical structure can mean the difference between carcinogenicity and noncarcinogenicity.

do not share many general reactive or structural features (Figure 18.11). Fifth, many chemical agents are moving targets; they must be converted to carcinogens by the enzymes in human tissues or by the bacteria in the human digestive tract; others may be converted to carcinogens in the air by sunlight or by interaction with other pollutants. It is even possible for two individually noncarcinogenic agents to be carcinogenic together or in the right sequence (Figure 18.12) on page 520.

The Many Faces of Lifestyle

We must point out once again that when we speak of chemicals, we are using the term in the broadest possible sense. We certainly do not mean to include only synthetic chemicals. We certainly *do* mean to include chemical hormones such as estrogen that are normal human secretions. We even mean to implicate any yet unknown balances or states in body chemistry brought about by particular cultural and social customs or lifestyle. Lifestyle is a far more important cause of environmental cancer than all synthetic chemicals combined. This is all complicated by the fact that lifestyle is undoubtedly highly correlated with chemical exposure. In a study that compared Mormons and non-Mormons (and thus Mormon and non-Mormon lifestyles) in Utah (Lyon et al., 1976), it was found that cancer incidence rates were much higher for the non-Mormons (see Figure 18.13). For cancers known to be associated with tobacco and alcohol the incidence rates were 30% to 800% higher for the non-Mormons.

The point is that it is ultimately one's total lifestyle that specifies one's risk of cancer. This means that defining this risk, the factors involved, and the degree of involvement will be at least as difficult as describing all the many aspects of lifestyle.

A summary of all cancer-associated environmental factors with estimates of the cancer fraction attributable to each factor is given in Table 18.6 on page 521.

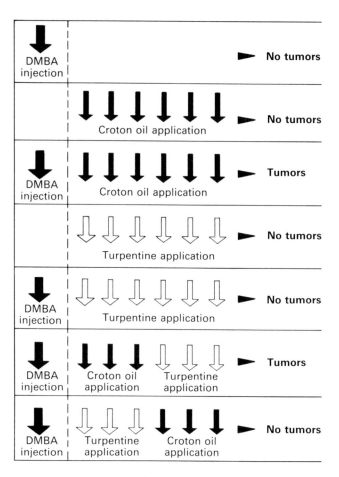

Figure 18.12 Carcinogenic Synergism. Although neither a subthreshold dose of dimethyl benzanthracine (DMBA) or repeated applications of croton oil individually will produce tumors, they do produce tumors when applied together in a certain sequence. Though turpentine, another kind of irritant, does not have the same effect when substituted for croton oil, it does when substituted for late applications. This suggests that carcinogenesis may be a highly complicated process with a number of different stages.

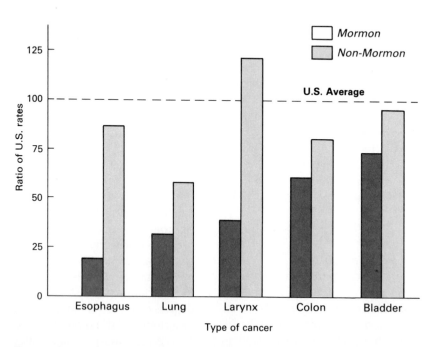

Figure 18.13 Cancer in Mormons and Non-Mormons in the State of Utah, 1966–1970. Note that although incidence rates are generally higher for non-Mormons, they are much higher for cancers (esophageal and laryngeal) associated with alcohol and tobacco. Mormons do not use either of these substances.

Table 18.6 Summary of Cancer-associated Environmental Factors

Factor	Sites Considered in Drawing the Estimates	Range of Estimates Associated with Factor* (percent)
Diet	Digestive tract, breast, endometrium, ovary	35–50 (35)
Tobacco	Upper respiratory tract, bladder, esophagus, kidney, pancreas	22–30 (30)
Occupation, all exposures	Upper respiratory tract, others	4–38 (4–10)
Alcohol	Upper digestive tract, larynx, liver	3–5 (3)
Infection	Uterus (cervix), prostate, and other sites	1–15 (10)
Sexual development, reproductive patterns, and sexual practices	Breast, endometrium, ovary, cervix, testis	1–13 (7)
Pollution	Lung, bladder, rectum	1–5 (2)
Medical drugs and radiation	Breast, endometrium, ovary, thyroid, bone, lung, blood (leukemia)	1–4 (1)
Natural radiation	Skin, breast, thyroid, lung, bone, blood (leukemia)	1–3 (1)
Consumer products	Possibly all sites	<1
Unknown associations	All sites	1–10 percent

* Most commonly cited "best" estimate given in parentheses.

Source: C. E. Kupchella, "Environmental Factors in Cancer Etiology," *Seminars in Nursing Oncology* 2(1986):162. Reprinted by permission of Grune & Stratton, Inc.

Linking Environmental Factors and Cancer: Prospects for the Present and the Future

The point was made in an earlier section that if 90% to 95% of most cancers are of environmental origin, we ought to be able to identify the environmental factors and eliminate them, thereby bringing about an enormous decrease in mortality due to cancer. This, of course, presumes that we can actually test a substance and show whether or not it has carcinogenic potential. Evidently, an even bigger presumption is that we can in fact do something about the carcinogens we *do* identify.

Testing for Carcinogenic Potential: Where Do We Stand, and What Is Happening?

Of the millions of known chemical compounds, only a few thousand have been adequately tested for carcinogenicity, and there are only about 1,000 chemicals that have been shown to produce tumors in experimental animals. Many times this number are under suspicion for one reason or another. Not all of the millions of chemicals already in existence have the same priority for testing, of course, because many of them are not in common use; however, in the 1970s the U.S. was producing about 120 billion pounds of 9,000 or so synthetic chemicals in large-scale com-

Lifestyle can lead to environmental cancer.
In seeking the perfect tan, sunbathers increase their risk of skin cancer.

mercial manufacturing. In addition, the number of chemicals to which workers and the population as a whole are exposed is growing rapidly. Hundreds of new compounds are mass-produced each year.

A carcinogenesis bioassay program has been under way for some time at the U.S. National Cancer Institute. The protocol for the testing there requires lots of animals, lots of time, and lots of money. Testing a single chemical may take up to three years and as much as several hundred thousand dollars. Although many chemicals are being tested, there are not enough qualified testing specialists or testing facilities in the United States to test all chemicals fully. New chemicals are created faster than the old ones can be tested. Various factors such as the presumed likelihood that a chemical is carcinogenic and the use to be made of the chemical are considered in deciding which chemicals to test first.

During recent years, chemical screening systems that are relatively fast, very cheap, and fairly reliable have been developed. Bruce Ames (1979) developed a test that assesses carcinogenic potential via a mutagenic assay using microbes. One version of the **Ames test** makes use of the bacterial species *Salmonella typhimurium*. The parent, or natural, strain of this species is able to make the amino acid histidine, but a mutant tester strain is unable to make this essential amino acid. Billions of mutant organisms are exposed to a test chemical; if the chemical is able to change DNA, some of the exposed mutants experience a reversal of the very mutation that made them unable to make histidine. They signify that this back-mutation has occurred by growing on a medium that lacks histidine—something they could not do if the back-mutation had not occurred. The beauty of Ames-type tests is that they are quick (overnight) and cheap. But such tests miss some known carcinogens, presumably because they fail to replicate certain features of what can happen in a mammalian organism.

For example, this system fails in its basic form to take into account any species-specific metabolic activation of carcinogens. Humans and other mammals have metabolic enzymes that bacteria do not have. In view of this limitation, testing protocols have been used in which test chemicals are first exposed to mammalian liver enzymes before the chemicals come into contact with bacterial DNA. Mammalian cell culture systems that will do the same sorts of things that the Ames bacterial systems will do are being developed, but these are more expensive and more time-consuming. Despite their limitations, Ames-type tests serve as valuable prescreening tests for chemicals, identifying chemicals that are mutagenic and thus should be subjected to more expensive, time-consuming, whole-animal carcinogenicity tests.

Cancer, the Environment, and Personal Choices

One of the purposes of this chapter was to reinforce a point made in some of the preceding chapters, namely, that some environmental contaminants occur in air, water, land, or in the biosphere—all at the same time. The fact that carcinogens and other toxic materials move between and among the great spheres makes them especially difficult to deal with once they are set loose in the ecosphere.

A more important purpose of this chapter was to make two other points. First of all, from the perspective of an individual organism—a human, for instance—the environment includes everything from the skin out and from the lining of the digestive tract in; we must think of the environment in the broadest possible terms—to include even such nonspecific things as lifestyle—when considering the effects of the environment on human health. Second, qualities of the environment other than technology-derived air pollution and water pollution may be far more important to personal health. Human-made chemicals and radiation, though certainly important and in need of attention, account for a rather small part of the cancer problem.

Cancer is the most feared environmental disease, and rightly so. A great majority of human cancer has been linked to the environment, and far less than half of the people in whom cancer is diagnosed are being cured, even in countries with advanced health care. The good news is that the most important environmental causes of cancer are things over which we already have control as individuals. The bad news is that there is no equivalent of the EPA able to regulate what we do to ourselves. If skin cancer is counted, sunlight and tobacco account for over half of the cancer problem. If the effects of natural components such as fats in the diet are added to this, we have probably accounted for more than 80% of all cancers in the developed world. Infectious agents and reproductive practices account for much of the remainder (see Table 18.6). As the realization that the enemy is indeed ourselves continues to sink in, we must obviously continue to clean up the environment and keep carcinogens and other toxic materials from entering the ecosphere—even as we contemplate changes that we should make in our personal choices.

Marvin D. Mills

Dr. Marvin D. Mills is concerned about the environment inside industrial plants. He is an educational specialist with the National Institute for Occupational Safety and Health (NIOSH) in Cincinnati. NIOSH is a federal agency charged with conducting research and otherwise finding information that will serve as a basis for setting standards for occupational safety. The agency is also responsible for safety education and development and maintenance of a cadre of occupational safety and health care workers. This is where Dr. Mills comes in. NIOSH conducts courses for industrial hygienists, safety professionals, environmental scientists, occupational health nurses, and management personnel. Dr. Mills develops and improves teaching materials for these instructional programs, and he evaluates the programs themselves, making recommendations as to how they might be made more effective. Another important part of his work is to plan, organize, and coordinate the professional internship program at NIOSH for college juniors and seniors.

Dr. Mills has a bachelor's degree in biology, a master's degree in health education, and a doctorate in safety and health management. Previous work experience that has helped to prepare him for his current job includes many years'

experience as university professor, teaching safety and occupational health.

Concepts to Remember

1. Epidemiological evidence indicates that most human cancer is caused by factors in the environment. These environmental factors include sunlight, tobacco smoke, naturally occurring radiation, and dietary factors, as well as synthetic chemicals, radiation, and other environmental pollutants.

2. Cancer is a family of diseases in which the regulation of cell division is deranged by gene mutation or other events that can affect genetic control of cell division.

3. There are three classes of agents that can cause cancer: chemicals, radiation, and viruses.

4. Evidence linking viruses and a few human cancers is strong but circumstantial; hundreds of different kinds of cancer in animals and plants are known to be caused by viruses.

5. Most of the radiation received by the general public comes from medical uses of radiation (X-rays, etc.) and from natural background radiation.

6. Radioactive isotopes behave chemically like their nonradioactive counterparts.

7. Chemicals, both natural and synthetic, are the most important class of environmental carcinogens.

8. The basis of the statements attributing up to 95% of cancers to environmental factors are the dramatic differences in cancer patterns from country to country and from place to place throughout the world. The fact that the descendants of immigrants exhibit the patterns of cancer death of the countries to which their parents move indicates that the differences are more geographic than genetic.

9. In general, we know more about the connections between environmental cancers related to occupation than we do about any other environmental source. This is because occupational exposures have tended to be relatively easier to document and because patterns within small, distinct groups of people are easier to identify.

10. We ingest carcinogens in even totally natural foods; food additives, food processing, and drugs add still more.

11. The drinking water of many cities contains measurable amounts of carcinogens.

12. It is difficult to link cancers with specific environmental causes because (a) suspected agents cannot be given to people to see what happens; (b) cancer takes a very long time to develop—20, 30, even 40 years in some cases;

(c) the results of animal studies are not necessarily applicable to humans; (d) it is difficult to generalize about chemicals because carcinogens do not share general reactive features; (e) chemical agents can be moving targets—some substances must be converted through several steps before the ultimate carcinogen appears; (f) combinations of chemicals may be important; and (g) there are literally millions of chemicals already in existence, very few of which can be tested because of the amount of work and expense involved.

References and Further Reading

References marked with an asterisk are cited in the chapter.

Ableson, P., 1992. "Diet and Cancer in Humans and Rodents," *Science* **255**:141.

American Cancer Society, 1983. "The Geography of Cancer," *Cancer News,* Autumn.

American Cancer Society, 1991. "American Cancer Society Guidelines on Diet, Nutrition and Cancer," *Ca-A Cancer Journal for Clinicians* **41**(6)334–338.

American Cancer Society, 1991. *Cancer Facts and Figures, 1991.* Atlanta: American Cancer Society.

*Ames, B. N., 1979. "Identifying Environmental Chemicals Causing Mutations and Cancer," *Science* **204**:587–594.

Ames, B. N.; MaGaw, R.; and Gold, L. S., 1987. "Ranking Possible Carcinogenic Hazards," *Science* **236**:271–280.

*Antunes, C. M. F., and Stolley, P. D., 1977. "Cancer Induction by Exogenous Hormones," *Cancer* **39**:1896–1898.

Beebe, G. W., 1982. "Ionizing Radiation and Health," *American Scientist* **70**:35–44.

*Bishop, J., 1982. "Oncogenes," *Scientific American* **246**(3): 80–92.

*Blair, A.; Malker, H.; Cantor, K. P.; Burmeister, L.; Wiklund, K., 1985. "Cancer among Farmers: A Review," *Scandinavian Journal of Work, Environment and Health* **11**(6):397–407.

*Cantor, K. P., 1982. "Farming and Mortality from Non-Hodgkin's Lymphoma: A Case-Control Study," *International Journal of Cancer* **29**:239–247.

*Carter, L. J., 1979. "Dispute over Cancer Risk Quantification," *Science* **203**:1324–1325.

*Cohen, L. A., 1987. "Diet and Cancer," *Scientific American* **257**(5):42–48.

*Curtis, C., 1976. "The Carcinogenicity of Anticancer Drugs in Man," *Cancer* **37**:1014–1023.

Epstein, S. S., 1976. "Cancer and the Environment: A Scientific Perspective, Facts and Analysis," *Occupational Health and Safety, Industrial Union Department AFL-CIO* **25**(2):1–13.

*Fischhoff, B.; Slovic, D.; and Lichtenstein, S., 1979. "Weighing the Risks," *Environment* **21**(4):17–20.

Gough, M., 1987. "Environmental Epidemiology: Separating Politics and Science," *Issues in Science and Technology* **3**(4): 20–31.

*Harris, R. H., 1975. *The Implications of Cancer-causing Substances in Mississippi River Water.* Washington, D.C.: Environmental Defense Fund.

*Hassett, C.; Mustafa, M. G.; Coulson, W. F.; and Elashoff, R. M., 1985. "Murine Lung Carcinogenesis Following Exposure to Ambient Ozone Concentrations," *JNCI* **75**:771–777.

Henderson, B. E.; Ross, R. K.; and Pike, M. C., 1991. "Toward the Primary Prevention of Cancer," *Science* **254**:1131–1138.

*Hoar, S. K.; Blair, A.; Holmes, F. F.; Boysen, C.D.; Robel, R.J.; Hoover, R.; Fraumeni, J.F., Jr., 1986. "Agricultural Herbicide Use and Risk of Lymphoma and Soft-Tissue Sarcoma," *JAMA* **256**:1141–1147.

Kritchevsky, D., 1991. "Diet and Nutrition," *Ca-A Cancer Journal for Clinicians* **41**(6):328–333.

*Kupchella, C. E., 1987. *Dimensions of Cancer.* Belmont, Calif.: Wadsworth.

*Lyon, J.; Klauber, M.; Gardner, J.; and Smart, C., 1976. "Cancer Incidence in Mormons and Non-Mormons in Utah, 1966–1970," *New England Journal of Medicine* **294**:129–133.

*Mancuso, R. F., and Brennan, J. J., 1970. "Epidemiological Considerations of Cancer of the Gall Bladder, Bile Ducts and Salivary Glands in the Rubber Industry," *Journal of Occupational Medicine* **12**:333–341.

*Mancuso, T. F.; Steward, A.; and Kneale, G., 1977. "Radiation Exposures of Hanford Workers Dying from Cancer and Other Causes," *Health Physics* **33**:369–384.

*Marx, J. L., 1976. "Estrogen Drugs: Do They Increase the Risk of Cancer?" *Science* **191**:838.

*National Academy of Sciences, 1982. *Diet, Nutrition, and Cancer.* Washington, D.C.: National Academy Press.

*National Academy of Sciences, 1987. *Regulating Pesticides in Food: The Delaney Paradox.* Washington, D.C.: National Academy Press.

*Nicholson, W. J., 1977. "Cancer Following Occupational Exposure to Asbestos and Vinyl Chloride," *Cancer* **39**:1792–1801.

*Page, H. S., and Asire, A. J., 1985. *Cancer Rates and Risks* (NIH 85-691), 3d ed. Bethesda, Md.: National Institutes of Health.

*Page, T.; Harris, R. H.; and Epstein, S. S., 1976. "Drinking Water and Cancer Mortality in Louisiana," *Science* **193**:55–57.

*Pickle, L. W.; Mason, T. J.; Howard, N.; Hoover, R.; and Fraumeni, J. F., 1987. *Atlas of U.S. Cancer Mortality among Whites, 1950–1980.* Washington, D.C.: U.S. Department of Health and Human Services.

*Rosen, R. D., 1990. "Much Ado about Alar," *Issues in Science and Technology* **7**(1):85–90.

*Schottenfeld, D., and Haas, J. F., 1979. *Carcinogens in the Workplace.* New York: American Cancer Society.

*Seilikoff, I. J., 1975. "Recent Perspectives in Occupational Cancer," *Ambio* **4**(1):14–17.

*Seilikoff, I. J.; Hammond, E. C.; and Churg, J., 1968. "Asbestos Exposure, Smoking, and Neoplasia," *JAMA* **204**:106–112.

*Shapley, D., 1976. "Nitrosamines: Scientists on the Trail of Prime Suspect in Urban Cancer," *Science* **191**:268–270.

Stern, J., 1988. "British Nuclear Accident in 1957 Seen Linked to Leukemia Rise," *Oncology Times,* March 1, pp. 1, 15.

*Stocks, P., 1960. "Atmospheric Pollution and Mortality," *British Journal of Cancer* **14**:397–418.

*Sugimura, T., 1986. "Studies on Environmental Chemical Carcinogenesis in Japan," *Science* **233**:312–317.

*U.S. General Accounting Office, 1981. *Problems in Assessing the Cancer Risks of Low-Level Ionizing Radiation Exposure.* 2 volumes. Gaithersburg, Md.: U.S. General Accounting Office.

*Weinberg, R., 1983. "A Molecular Basis of Cancer," *Scientific American* **249**(5):126–143.

In Chapters 19 and 20 we summarize what we have covered up to this point, focusing on the roots of our environmental problems and our prospects for solving them in the future. Among the questions to be explored are these: How has each of our institutions contributed to our most pressing environmental problems and to our failure to head them off? In what ways will these institutions be likely to help or hinder us in our search for solutions? In summary, just what are our most important environmental problems? How shall we deal with them in the future? What are the prospects for success?

part four

Points of View

19 Human Institutions: Problems and Solutions

Although the word *institution* may conjure up images of hospitals, schools, or big buildings, we use it here in a more general way. We refer to the formal ways in which human beings organize civilization and the things that make it up. We will use the word **institution** here to refer to religion, government, law, economics, education, science, medicine, and cities. We are devoting a chapter to institutions because (1) the way in which our institutions function has much to do with the fact that we have environmental problems, and (2) they will also have a lot to do with our prospects for dealing with these problems effectively.

It is important that we appreciate the general role of institutions in human affairs in order to appreciate their relationship to our environmental problems.

Institutions serve to stabilize and ease human interaction, making it orderly and more civilized. Institutions can also hamper interaction and stand in the way of needed changes. Civilizing human interaction while preserving flexibility is difficult. Some institutions achieve a better balance than others.

Many systems of government have carefully spelled-out, built-in mechanisms that provide for change and at the same time ensure that change will not be capricious, arbitrary, or too sudden. In other types of institutions—education, for example—mechanisms for change are not very well established,

and response to the need for change occurs via less formal and somewhat more incremental mechanisms.

History indicates that in general, institutions typically lack sufficient flexibility to persist for very long without change. Forces that are constantly at odds and partly account for this historical pattern are the inevitability of change and a strong human drive to live by institutions. Change or replacement of existing institutions is nearly always difficult because not all of the elements of the culture or society recognize the need for adaptation at the same time. If the natural resistance to change holds sway too long, pressures build up until needed change can no longer be accommodated by the existing structure. Then new institutions replace the old, sometimes traumatically. Institutions work best when permitted to evolve gradually.

Our ability to cope with the future will be determined in part by how adaptable our institutions are and how readily they can conform to new circumstances, new pressures, new opportunities, and new dangers related to the environment. It is somewhat encouraging, then, that clean, nonpolluted, well-planned environments are starting to become fixed within our institutions and our institutional value systems. The pace is not so encouraging. Some commentators say that the pace is so slow that our only hope is to recast some of our institutions—to bring about an

Citizen suit: a provision in some environmental laws that gives citizens the right to bring suit—that is, gives them *standing*—to require enforcement of the law.

Class action suit: a suit filed on behalf of all members of a given group even though they may not be known individually.

Commons: resources that belong to everyone but to no one in particular.

Economics: the study of how a society and its members choose to use scarce resources to produce goods and services, now and in the future, from all the competing demands (with or without the use of money).

Environmental ethic: a way of looking at a thing and evaluating it in terms of how well it conforms to the laws of ecology—whether the thing is bad or good ecologically.

Externality: a cost not taken into account in figuring the cost of doing business.

Gross national product (GNP): the total value of the goods and services produced by a country annually.

Institution: any of a number of the fundamental parts or aspects of a culture—for example, government, law, education, religion, politics, science, and medicine.

Plaintiff: the person who brings suit in a court; the party who opposes a defendant.

Public trust: the responsibility of government to hold and manage public property for the benefit of the public.

Standing to sue: the right of a party to assert or enforce legal rights and duties in court by virtue of having a sufficient stake in the outcome.

Suburbanization: increase in the proportion of people who live in the suburbs.

Trespass: unlawful interference with one's person, property, or rights.

---

abrupt change to a new way of living and relating to one another and to the environment.

Religion

Early in the environmental movement, the historian Lynn White wrote a widely reproduced and widely quoted essay blaming much of the environmental crisis on elements of the Judeo-Christian ethic. White and other writers have suggested that the notion featured in many religions that the earth is only a place we will use temporarily and then move on accounts for a basic disregard for the natural world and its life support systems. The attitude that we are simply passing through makes the world a disposable item.

Others disagree with White, claiming that Judeo-Christian teaching has had only an indirect influence on how the Western world treats the environment. They claim that religious tradition accounts for only a small part of the environmental problem and that other factors, including rapidly changing technology, urbanization, and especially increased individual wealth, are considerably more important.

Regardless of what has been true in the past, perhaps religious institutions could serve humankind better by putting more emphasis on ecological responsibility. What do you think?

Science and Technology

Is Science a "First Cause"?

Many observers have blamed science for the sad state of our environment. The reasoning is that science begets technology, and technology often causes prob- lems more serious than those it solves. Science does provide the basic information out of which new technologies emerge. It is also generally true that nearly all of our environmental problems can be traced directly to technology—to the ways we have of doing things. But new technologies and ways of doing things have accomplished much good at the same time. This being the case, it would seem that our strategy for the future should be to be more careful in embracing new technologies.

Do We Need a Watchdog over Science and Technology?

A widely held opinion is that all new knowledge is good—that knowledge is better than ignorance. Though this may be true in the purest form, it is not equally true that *all* technology is good. We clearly need better systems for appraising the possible consequence of new technologies well before they are allowed to become established and recognized, by their effects, for what they are.

Obviously, no institution can be relied on to police itself very well with the interests of all society in mind. We may need to find new and better ways of involving all parts of society in evaluating new technologies. An example of such a process is the system of **human studies review** now entrenched in U.S. biomedical research.

By law, any U.S. federally funded study involving human beings in which drugs will be given or that will involve risk must have its aims, objectives, and methods reviewed and certified by a duly constituted committee in each research institution or medical school. The composition of these committees is also specified by law and must include lay people, ministers, and lawyers as well as medical doctors and scientists. Such committees are charged with evaluating the risks and benefits to be derived from human studies. They must also assure themselves that the research subjects will be systematically informed of all the risks and the benefits they face by their participation in the study.

Our reason for bringing this up here is to suggest that perhaps this concept should be extended in some form to deal with all technological activities—all new ventures that amount to ecological experiments with human subjects. How do you see it?

Scientific and Technical Solutions

Science obviously also holds promise with regard to environmental problems. Perhaps the major hurdle here is that not much of the scientific establishment has been set to work on our environmental problems. Over half of the scientific and technical community of the world is usually at work on weapons of war. In the United States, as much as 70% of federal research and development (R&D) money has gone for military projects (see Figure 19.1). Less than 3% of military research is basic science; the bulk is directed toward demonstration projects and prototypes (Chandler, 1988) (see Table 19.1).

Expenditures for environmental research worldwide fall significantly short of what is needed to do the job. In the United States the *increase* in the 1985 defense budget was three times the *total* budget for natural resources and environmental programs in all federal agencies combined (*Outdoor News Bulletin*, 1984).

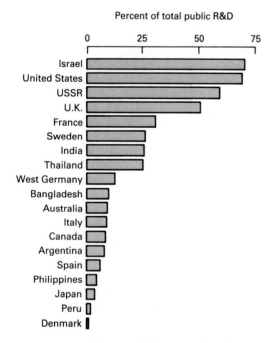

Percent of total public R&D

Figure 19.1 Expenditures for Military Research and Development in 19 Countries in the Mid-1980s.

Table 19.1 Trade-offs between Military and Social or Environmental Priorities in the Mid-1980s

Cost (billions of dollars)	Military Priority	Social or Environmental Priority
100	Trident II submarine and F-16 jet fighter programs	Estimated cleanup cost for the 3,000 worst hazardous waste dumps in the United States
68	Stealth bomber program	Two-thirds of the estimated cost to meet U.S. clean water goals by 2000
39	Requested Strategic Defense Initiative ("Star Wars") funding for fiscal 1988–1992	Disposal of highly nonmilitary radioactive waste in the United States
30	Two weeks of world military expenditures	Annual cost of the proposed UN Water and Sanitation Decade
11	West German outlays for military procurement and R&D, fiscal 1985	Estimated cleanup costs for West German sector of the North Sea
8	Approximately four days of world military expenditures	Five-year action plan to save the world's tropical forests
6	Development cost for the Midgetman intercontinental ballistic missile	Annual cost to cut U.S. sulfur dioxide emissions by 8 to 12 million tons per year to combat acid rain
5	Approximately two days of world military expenditures	Annual cost of proposed UN action plan to halt Third World desertification over 20 years
4	Six months of U.S. outlays for nuclear warheads, fiscal 1986	U.S. government outlays for energy efficiency, fiscal 1980–1987

Source: Adapted from M. Renner, *National Security: The Economic and Environmental Dimensions* (Washington, D.C.: Worldwatch Institute, 1989), p. 48.

As we have seen throughout this book, environmental problems are not limited to the borders of individual nations; they are global in scope and resources, and data are also needed on this level.

Medicine

Individual physicians around the world have been at the forefront of the environmental movement, studying environmental problems and warning societies of the hazards of various forms of pollution. However, the medical establishment is really not set up well to deal with environmental health problems. Physicians have been admonished throughout recorded history to treat the sick and leave the well alone. The problem with this is that for the most part the diseases (most of which have clear environmental relationships) that are killing people in developed countries today cannot be treated effectively after they have emerged. Some cannot be treated at all. Clearly, medicine must adopt more of a prevention format.

Carnow (1971) and many others since have recommended that physicians become knowledgeable about environmental factors in disease and change their way of looking for the cause of disease to include environmental causes. Physicians must not only support but also participate in the search for complex environmental, physical, biological, psychological, and causal factors in disease. They should be prepared to assume leadership roles in the effort to identify environmental problems and deal with them once they have been defined. The philosophical basis of medical practice and medical training must be reconsidered and restructured.

Agriculture

A number of events are occurring that are having an impact on agriculture. They include the growth of interest in organic farming and the public demand for food grown free of chemicals, the growth of interest in regulating and controlling the direction of biotechnology, and interest in low-input sustainable agriculture (LISA).

At least 17 states in the United States have regulations relating to what food can be labeled as organically grown. A 1989 study by the National Academy of Sciences indicated that farmers who use little or no chemicals on their crops are about as productive as those farmers who depend on large amounts of chemical fertilizer and pesticides. The study recommends changing the U.S. farm subsidy program, which has the effect of encouraging the use of chemicals to increase crop yields.

Biotechnology has great potential. Most of the research and development has been done by the private sector. In some cases it has resulted in some rather self-serving developments, such as chemical companies developing crops resistant to herbicides, thus encouraging the sale of more herbicides. There is some movement now for the public sector to become involved in order to promote research in the areas of greatest public need and to regulate the dispersal of biologically engineered species.

Many improvements in agricultural efficiency have been geared more toward agribusiness than the small farmer. Some observers believe that if small farms are to remain viable, public technical assistance will have to target their needs. Providers of information and assistance to farmers must focus on methods of low input rather than high-tech and chemical alterna-

tives. The transfer of agricultural technology to developing countries must likewise focus on the cultural and social factors in which the technology must operate in order to be effective over the long haul. Attention must be paid to the loss of agricultural lands as a result of poor land management leading to erosion and desertification. Sustainable agriculture that makes maintenance of the land resource over time a central focus must be practiced on all agricultural lands throughout the world. Research must be directed toward developing knowledge of plant ecology, soils, and pest life cycles to provide the information necessary to make wise decisions and improve even more the effectiveness of LISA and integrated pest management (see Chapter 8).

Cities

Urbanization is a pattern that has been building momentum for quite some time throughout the world. In 1900, less than 14% of the world's people lived in cities; in 1987, 43% did. In 1950, about 600 million people lived in cities; in 1986, over 2 billion did. The number of people living in cities is expected to double between 1986 and 2010. Most of this increase in city populations will occur in the Third World (see Table 19.2). Already, 65% of Latin Americans live in cities; this is expected to increase to 75% by the year 2000 (Brown and Jacobson, 1987).

What does this mean? In a very real sense, the city concentrates all the impacts humans have on nature. Cities are places in which most goods and services are consumed and in which nearly all of the by-products of civilization are concentrated.

Many U.S. cities are beginning to coalesce into supercities or megapolises such as those in Figure 19.2. This development is a consequence of what we know of in the United States as **suburbanization**—

Table 19.2   Urban Share of Total Population, 1950 and 1986, with Projections to 2000   (in percent)

Region	1950	1986	2000
North America	64	74	78
Europe	56	73	79
Soviet Union	39	71	74
East Asia	43	70	79
Latin America	41	65	77
Oceania	61	65	73
China	12	32	40
Africa	15	30	42
South Asia	15	24	35
World	29	43	48

Sources: Worldwatch Institute; Population Reference Bureau.

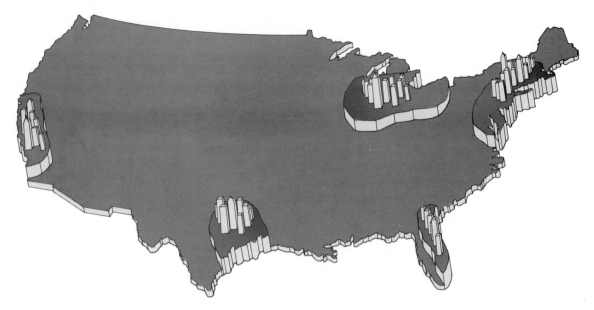

Figure 19.2 Megapolises. Much of the U.S. population is concentrated in five supercities.

moving to the suburbs. In 1950, one in four people lived in suburbs; in 1970, the number was more than one in three. It is still roughly one in three. Suburbanization presents special problems of governance, lack of community, and lack of control over such things as zoning and planning.

Suburbanization presents a political dilemma. Suburbanites are able to benefit from having a city adjacent but, because of tax structures and political boundaries, may contribute little, if anything, to the upkeep of the cities. The erosion of the tax base is the single most important problem facing American cities. It is clearly a problem that stands in the way of city governments' being able to improve the urban environment. Many of today's so-called metropolitan areas are made up of dozens of quasi-independent municipalities in an ungovernable patchwork of law enforcement and zoning.

A second major problem with suburbs is that they are really not communities in the strict sense of the word. Many are just large collections of houses with no community identity—people working in different places in the city and having only a zip code in common. This is a problem because some sense of community must certainly be a factor important to any interest in looking after the community environment.

Cities as Ecological Problems

As incomplete ecosystems, cities are all part of larger wholes encompassing outside areas that are needed to support the city—land "out there" on which grain is grown to feed the chickens that lay the eggs that are brought to the city to be sold for city breakfasts every morning. The unchecked growth of cities, especially the coalescing of many of them into large megapolises, creates the threat of a breakdown in the ecological exchange between cities and their support areas. There are problems getting food to these cities and getting wastes away from them. Air pollution becomes a logarithmically increasing problem as cities get larger.

In general, as cities grow, efficiencies of ecological exchange drop. It has been suggested that the urban environment requires a much larger supply of energy and material per person than the small town or rural environment to satisfy even basic human needs (Brown and Jacobson, 1987). Water supplies (see Figure 19.3) have long been a problem in southern California cities and are now becoming problems in cities of the American Northeast and throughout many of the cities of the world. The larger cities get, the longer and more inefficient the supply and waste removal lines become.

Some of the ways in which cities interact with the environment are summarized in Figure 19.4.

Cities and Artificial Impressions

Cities surely contribute to the artificial and incorrect impression held by many people that humans are independent of the natural environment. How could it be obvious to a person born and raised in the city that he or she has any fundamental connection to the earth? It could be argued that two kinds of influences might balance each other out, that city dwellers see the most concentrated effects of the human impact on the environment and thus might have their consciousness raised by this experience—at the same time that

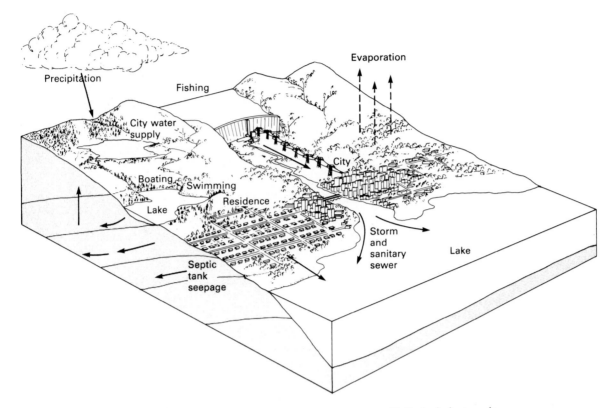

Figure 19.3 City Water. Cities interact in many ways with the hydrologic cycle.

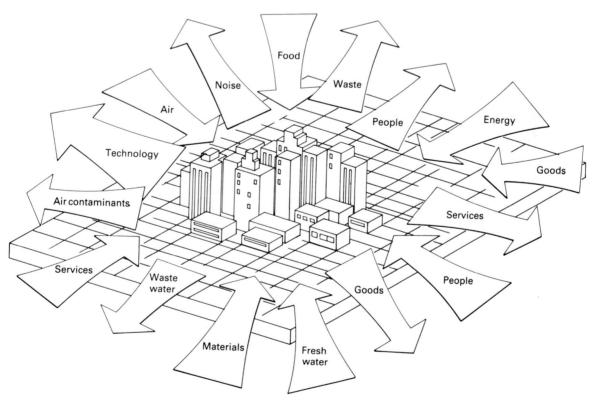

Figure 19.4 Urban Metabolism. High rates of exchange with the surrounding environment are necessary to keep cities viable.

Chapter 19 Human Institutions: Problems and Solutions

When people spend almost all of their time in cities, they may lose sight of their connection to the natural environment.

they tend to lose the perspective of being connected to the land.

It is more likely, we think, that those who experience the concentrated effects of pollution come to accept them as the reality. Perhaps there is a special need for environmental education of city dwellers.

What Can Be Done?

A few things can be done to improve cities. Such activities as redevelopment and urban renewal should incorporate all that we know about our relationship with the environment. New cities that will grow out of smaller entities offer even more promise.

New Cities: Avoiding Old Mistakes. It seems inevitable that new big cities will continue to arise from small cities and small towns. If we assume that there will be new cities, even if they simply rise from the crumbling old ones, it is worthwhile to consider how old ecological mistakes might be avoided.

Developing Cities in Concert with Nature. Developing cities should be shaped as integral parts of larger living systems. Cities, after all, are very much like multicellular organisms requiring energy and the means of excreting wastes. Cities should be designed not as parasites on the land around them but as symbionts with the life systems that support them—cities whose wastes do not represent an insult to the surroundings but are incorporated into the environment in accordance with the principles of nutrient cycling.

Cities might be designed to fit into natural systems without sharp boundaries. There should be more integration along the lines of the greenbelt concept long advocated by urban planners. Perhaps a green checkerboard concept—the green mixed throughout—would be better. Trees and plants add a dimen-

sion to life that goes beyond aesthetics. Trees create cool environments, they dampen noise, they remove certain impurities from the air, and they serve as a reminder that nature is still there and that we are dependent on it.

It should be recognized that there is undoubtedly an optimum size for a city. Perhaps developing cities could have predetermined sizes, enabling more complete planning for all the activities and interrelationships within the city and between the city and the adjacent environment.

Perhaps the most important factor to be taken into account in the development of new cities is that they must be built with resource conservation and resource cycling in mind.

It is exciting to think about all the possibilities. There are different ways of building buildings and laying out residential areas that would conserve natural resources better than we have done in cities today. Buckminster Fuller described the house of the future, for example, as one that could be deposited on any site and would not have any lines for sewage or electricity coming to it because the houses would be "energy-harvesting and growing machines." He envisioned energy-self-sufficient units, drawing energy locally from sources such as the sun and the wind. Many houses could be built underground so that heat could be conserved and maintenance would be less of a problem.

According to the environmental planner Ian McHarg, planning with nature can be sold on the basis that it is economical. By going with the natural "grain," so to speak, by taking ecological principles into account, such problems as land erosion and flood damage can be minimized. Overall, our prospects for extended survival should be enhanced as we build more pleasant environments.

Transportation

Although transportation-associated environmental problems are especially acute in cities, their magnitude and universal importance require that they be considered in a separate section. However, we will focus on the urban transportation problem.

First, as we have discussed in other chapters, the automobile is responsible for much of the air pollution problem and other general problems of the urban environment throughout the world. Second, automobile-dominated transportation systems do not work very well. They are much less efficient than some alternatives (see Chapter 6). Most cities have poor systems of intraurban transit and poorly integrated systems of interurban transit; it can take more time to get to an airport than it does to fly from one airport to another. Beyond the development of new systems of integrated transportation, what we really need—particularly in the United States—is to recast our cities so that the number of automobile passenger miles can be reduced.

The Automobile as a Nonsolution

We emphasize again that the automobile is the greatest single source of dirty air, contributing large fractions of the carbon monoxide, hydrocarbons, and nitrogen dioxide in the air of cities. Because the layout of most modern, recently developed cities has been determined largely by patterns of automobile use, the automobile is also largely, though indirectly, responsible for many other problems, including the dissolution of neighborhoods and the general decline of cities.

Many urban problems in the past half century have been the result of a rapid increase in automobilization. According to Greenwood and Edwards (1973), automobilization increased 20 times faster than population in the United States during the middle portion of the twentieth century (Figure 19.5). Since 1940, the number of auto passenger miles traveled per person in the United States rose from about 1,800 per year to more than 4,000 per year. Most of these miles have been traversed within cities. Today more than 80% of all commuters in the United States use the private automobile as their means of transportation, and over half of commuting cars carry only the driver.

We need to reduce the number of auto miles traveled and reduce the per-mile impact of the automobile. The latter is a problem of law and automotive engineering. The former may be approached by developing integrated mass transit systems, improving alternative modes of interurban transportation, and/or somehow bringing about drastic changes in the patterns of where people live in relation to where they work.

Mass Transit in the United States

The mass transit problem is more than one of building up integrated systems; transit systems have

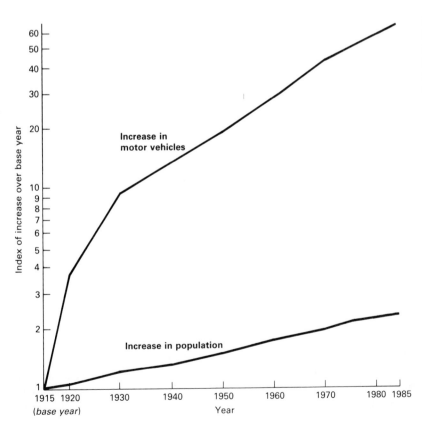

Figure 19.5 Automobilization and Population Growth. Comparison of population and the growth of automobilization in the United States, 1915–1985. During the base year, 1915, there were 100 million Americans and only 2.5 million automobiles.

It takes more than physical components to make a mass transit system successful. People's attitudes are also important. They must believe that the system is convenient, economical, and safe.

actually been in decline for many decades in the United States. More than 200 transit companies have gone out of business in recent years in the United States. Between 1947 and 1970, private automobile travel increased 5.8% per year; during the same period, travel on public transit systems increased less than 1%. Rail travel *decreased* almost 6% during the same period. What we have on our hands in the United States, then, is the problem of reversing a trend that has momentum in the wrong direction.

How might mass transit be restored in America? Some observers believe that this can be accomplished by generating negative pressures on the use of automobiles—making it more expensive for cars to be brought into cities. Mass mobilization will be expensive in any case. Rapid urban mass transit systems will require that 50 miles of track be laid for every million people, and this could be done at present for about $30 million per mile. If such systems were developed in 25 or 30 metropolitan areas with 1 million inhabitants each, the total price tag would approach $50 billion.

Even with alternative long-range improvement and new developments in the future, the world should plan on having the automobile around for the indefinite future. Accordingly, we should continue to direct our attention to ways of improving the automobile and reducing its negative impact on urban life. Elsewhere we have discussed various approaches to this problem, including electric automobiles, new types of engines, restrictions on the size of automobiles, and new kinds of fuel. These should all continue to be pursued with vigor.

Interurban Transit in the United States
Although there have been many years of decline in American rail transportation, it is an efficient mode of transportation and should be reinstituted as a pri-

mary means of getting people and freight from one place to another. European systems, much more highly developed than U.S. systems, could be studied as models.

Recasting an American Institution
Perhaps the American cultural practice of commuting to work and in effect having enormous "home ranges" is so out of tune with ecological constraints that the entire system will have to be scrapped, melted down, and recast over the long term.

As the realities of the laws of ecology become incorporated into our world view, it may well be that the design of cities will change, and some hallowed aspects of the American way of life will change so as to render the automobile, and even the need to get around so much, obsolete.

Education

To change human behavior so that it is aligned better with ecological reality will require a major adjustment of the philosophy underlying education throughout the world. The task in many countries will be to generate more ecologically appropriate values, attitudes, and ethics. Our current ethic or world view—the way we look at things—developed when environmental problems were not recognized as problems. This ethic no longer serves us well.

But an ethic does not change quickly. Adjustment does not immediately follow awareness of a problem. This is because adjustments in ethics run deep. As Aldo Leopold (1949) pointed out, "No important change in ethics was ever accomplished without an internal change in intellectual emphasis, loyalties, affections, and convictions" (pp. 209–210). Much needs to be done in education to accelerate the necessary shift to a more ecologically sound world view or ethical framework and way of life.

An Environmental Ethic

"A thing is right when it tends to preserve the integrity, stability, and beauty of the biotic community. It is wrong when it tends otherwise." This is how Aldo Leopold (1949, pp. 224–225) defined a land ethic or, as we now call it, an **environmental ethic.** Leopold also stated that an ethic is a limitation on freedom of action in the struggle for existence. Any ethic puts some constraints on total freedom. The frontier ethic that carried Americans well into the twentieth century emphasized freedom, including the freedom to bring about environmental degradation. The world has learned that freedom in this extreme is at odds with more basic freedoms or rights to breathe, drink, eat, and enjoy a clean environment. A sound environmental ethic would cause us to go about the business of being human such that everything we did would be measured consciously and even unconsciously by how well it conformed to the laws of ecology. Actions would be judged by whether they were bad or good ecologically.

An environmental ethic is what we need. Environmental education is a means by which we can develop one.

Environmental Education

In the fall of 1975 an international environmental education workshop was held at Belgrade, Yugoslavia. The charter that resulted from that meeting established the goal of environmental education:

> To develop a world population that is aware of and concerned about the environment and its associated problems, and . . . has the knowledge, skills, attitudes, motivations, and commitment to work individually and collectively toward solutions of current problems and the prevention of new ones.

Acknowledging that this is much easier said than done, the conference also set forth some guiding principles for environmental education programs:

1. Environmental education should consider the environment in its totality—natural and human-made, ecological, political, economic, technological, social, legislative, cultural, and aesthetic.
2. Environmental education should be a continuous, lifelong process, both in school and out of school.
3. Environmental education should be interdisciplinary in its approach.
4. Environmental education should emphasize active participation in preventing and solving environmental problems.
5. Environmental education should examine major environmental issues from a world point of view while paying due regard to regional differences.
6. Environmental education should focus on current and future environmental situations.
7. Environmental education should examine all development and growth from an environmental perspective.
8. Environmental education should promote the value and necessity of local, national, and international cooperation in the solution of environmental problems.

One definition of environmental education (established by the U.S. Office of Education) is this:

> Environmental education is the education process dealing with humankind's relationship with the natural and human-made surroundings, and includes the relation of population, pollution, resource allocation and depletion, conservation, transportation, technology, economic impact and urban and rural planning to the total human environment.

Environmental education is *not* simply conservation, outdoor resource management, or nature study (although these areas may be included in an environmental education program). Above all, environmental education is not concerned with indoctrinating students on what to think but rather with helping them to develop

---

For Further Thought . . . 19-1

Changing Paradigms

What do you think our society might look like and how would it function if it had as its basis an *environmental* ethic rather than an *economic* one—if the environment could *not* be treated as an externality? How might transportation, economic development, waste management, lawn care, home building, packaging, consumer purchasing, and land use be different?

---

skills on how to include environmental considerations in their thinking.

Public Education: Reading, 'Riting, 'Rithmetic, and Ecology

Early efforts to clean up U.S. public drinking water supplies late in the nineteenth century and early in the twentieth were stymied for a long time because the public did not appreciate the germ theory of disease. As the story goes, once the public was won over, cleaning up public water supplies proceeded with haste. There is an important general lesson here. Educational institutions worldwide must work harder and more effectively to help the general public to appreciate and understand the natural interrelationships that tell us how environmental problems came to be and how they might best be corrected. The current generally poor understanding of environmental issues by the public came about because in the past there was little in the way of environmental education. Now environmental education must become the keystone of general education; environmental principles must become the paradigm on which *all* education is based. Curricula in schools from preschool through graduate education should be restructured to feature core concepts and fundamental principles having to do with the human relationship with the environment.

Higher Education: The Need for Renaissance People

One would expect universities to be front-runners in the attempt to "environmentalize" the world ethic. One reason why universities have not done more than they have is that environmental education does not fit the university department structure very well. Environmental education touches on all subject disciplines, and universities tend to be organized along traditional disciplinary lines with little crossover.

In recent years the trend has been, particularly at the graduate and doctoral levels, to produce superspecialists in exceedingly narrow disciplines. This trend has reached the point where many people with doctoral degrees have limited views of even the fields in which they work. Spending lifetimes dealing with single enzymes may well produce significant narrow advances. But unless there are also people trained to distill a new, interdisciplinary understanding and apply it to the broad, multifaceted problems, whose domain will be the big picture?

The academic world has always been slow to recognize, formalize, and respect new areas of inquiry as the need for new approaches and new disciplines is fashioned by the times. In this age of rapid changes and new and complex problems, this has become especially problematic. There must be provisions whereby new specialists can be broadly as well as narrowly trained. We need big-picture people who can see ramifications and interconnections and, in general, help society deal with problems involving many disciplines. The artificially sharp boundaries between academic disciplines are out of date.

Informal Environmental Education and Environmentalists

The people who carry the cause of environmentalism forward, at the forefront in the advocacy of a clean environment, have a special responsibility to inform themselves (see Kupchella and Levy, 1975). We

---

For Further Thought . . . 19-2

Generalists versus Specialists

Some people believe that we spend too much time compartmentalizing the way we learn and the way we attack problems and structure government. Some say that we need more integrated solutions and more procedures for addressing multidisciplinary issues. Others say the issues are so complex that we must have specialists who know the technical details to understand the problem fully. Do you think we need more generalists or more specialists?

have already experienced counterecological backlash resulting from announcements of imminent disasters that never come to pass, from overstatement of cases, and from the expending of great amounts of energy for inappropriate or low-priority purposes. To be effective, environmental advocates must first be environmentally educated.

U.S. Federal Involvement in Environmental Education

The implementation of the **Federal Environmental Education Act,** passed in 1970, acknowledged the fact that environmental education is for everyone. Federal funds were made available for elementary and secondary education, curriculum development, teacher training, and community workshops for leaders of business, labor, and government.

In 1974 the Federal Interagency Commission on Education (FICE) was created by executive order to improve communication and coordination at the federal level. A subcommittee on environmental education was established as a part of the FICE. Many federal agencies have formal environmental education programs. Among these are the Tennessee Valley Authority, the National Park Service, the Soil Conservation Service, and the Forest Service.

In 1990, the U.S. Congress renewed a commitment to environmental education initiated in 1970 but rescinded in 1981. The legislation created the Office of Environmental Education within the EPA. Among other things, the office is to support both the development of educational materials and environmental education programs and fellowships for in-service teachers with federal agencies.

In many states, environmental education has been integrated into the basic public school curriculum. Many states have laws requiring that environmental education be taught; some states have laws or policies requiring that textbooks contain environmental perspectives. Some states have policy statements on environmental education.

Government and Environmental Law in the United States

The involvement of national government in environmental protection and the evolution of the environmental laws have developed differently from country to country. We will examine some of the underlying concepts and the constitutional basis of environmental law in the United States. The underlying principles we will consider are found to some degree in all governments and in all constitutions. This section may serve as a point of departure, leading to a detailed consideration of the laws and the governmental approaches to environmental protection in any nation or all nations.

Constitutional Provisions for Environmental Protection in the United States

Several arguments are given for interpreting the U.S. Constitution as providing for protection of the environment. Its "general welfare" clause, for example, can be construed to apply to the environment. The Ninth Amendment to the Constitution states, "The enumeration in the Constitution of certain rights shall not be construed to deny or disparage others retained by the people." The right to a decent environment is clearly one such right.

The Fourteenth Amendment to the U.S. Constitution says that no state shall "deprive any person of life, liberty, or property, without due process of law." To diminish human prospects for survival is to deprive unnamed persons of life—by the extra cancers that result from carcinogen contamination of the environment, for example. To deprive humans of clean air or water is to deprive them of a very basic dimension of freedom. The concept of an environmental bill of rights has been proposed at various times as an amendment to the U.S. Constitution—to explicitly set forth human rights to clean air and water, resource conservation, and environmental protection.

General Grounds for Environmental Litigation

Long before there were specific environmental laws, legal mechanisms for challenging polluters existed through "nuisance," "trespass," and "negligence" laws. In all three categories, the burden of proof is on the **plaintiff,** the one bringing the charges. Often in environmental suits, the plaintiff is seeking **injunctive relief,** that is, the plaintiff wants the source of the harm stopped. To succeed at enjoining a defendant, the plaintiff must identify a legally defensible interest to be protected and demonstrate irreparable harm—which makes long-range or cumulative effects very poor bases for such suits—and show that the damage is greater than the benefits to be derived by continued action by the defendant.

Generally speaking, nuisance, trespass, and negligence are private remedies for special legal wrongs. In some cases, **class action suits** are brought against polluters. Such suits are filed on behalf of all persons affected in the same negative way that the named plaintiff is affected. Such a mechanism makes it possible for small damages to a number of individuals to be added together and thus make the harm larger than any benefit. Also, if found guilty, defendants have greater damage assessments. The value of this approach was restricted in 1973 when the U.S. Supreme Court declared that in order for a *federal* class action suit to be filed, each member of the class must suffer damages greater than $10,000 (*Zahn* v. *International Paper Company*).

Another limitation on the use of common law remedies for environmental purposes has resulted from U.S. Supreme Court decisions relating to the federal Clean Water Act. In 1981, in a case involving the city of Milwaukee and the state of Illinois, the Supreme Court ruled that Illinois could not sue under federal common law on the basis of "public nuisance" to force Milwaukee to meet water quality standards that were stricter than federal standards; according to the Supreme Court, federal statutes displaced federal common law.

The right to bring a case to court is known as **standing.** Generally, to establish standing, a plaintiff must show that he or she is aggrieved, and a specific statute indicating congressional recognition of the plaintiff's cause must be found.

The classic court case that established the precedent for granting standing to environmentalists was *Scenic-Hudson Preservation Conference* v. *Federal Power Commission.* The Consolidated Edison Company of New York wanted to construct a hydroelectric project on Storm King Mountain. When the standing of the environmental group was challenged, the court ruled that an aggrieved or adversely affected party did not have to have "a personal economic interest," that a noneconomic injury was sufficient to establish standing. The precedent was refined further in *Sierra Club* v. *Morton* in 1972 (known as the Mineral King case). The Sierra Club sued to prevent Walt Disney Productions from developing an area in the Sierra Nevadas. The Supreme Court held that members of the club had to use the area in order to have standing. Many states have passed **citizen suit** legislation, which guarantees citizens standing in matters of environmental protection.

The Public Trust

The concept of **public trust** has also been used in environmental protection. This doctrine is based on the premise that the government holds public resources (such as air, water, and land) in trust for the public and that there are limits on the power of government to dispose of these resources. Government is the trustee; the citizenry, both present and future, are the beneficiaries. When the trustee fails to carry out its responsibility, beneficiaries may force the trustee to protect their interest through litigation.

Federal law now spells out a public trust doctrine in the National Environmental Policy Act (NEPA), which states:

It is the continuing responsibility of the Federal Government to use all practical means, consistent with other essential considerations of national policy, to improve and coordinate Federal plans, functions, programs, and resources to the end that the Nation may . . . fulfill the responsibilities of each generation as trustee of the environment for succeeding generations. [Section 101(b)(1)]

By constitutional amendment, Pennsylvania has embraced a public trust doctrine at the state ("commonwealth") level:

The people have a right to clean air, pure water, and to the preservation of the natural, scenic, historic and aesthetic values of the environment. Pennsylvania's public natural resources are the common property of all the people, including generations yet to come. As a trustee of these resources, the Commonwealth shall conserve and maintain them for the benefit of all the people. (Article 1, Section 27)

Michigan, Wisconsin, and other states have also passed environmental public trust laws.

Specific environmental legislation has of course been enacted at the national level in the United States to deal with air, waste, water, and public health. Generally, such federal environmental laws provide broad outlines that are filled in by administrative regulations. The enactment of these laws has been an uphill battle for many reasons:

Public-interest legislation poses many political difficulties because, as a rule, the benefits are often intangible, long-range, and distributed among a very large public, while the costs are often tangible, immediate, and imposed on a specific set of organized interests. (Rosenbaum, 1973, p. 104)

Special-interest groups have traditionally fought such legislation with the objective of blocking or weakening emerging laws. Vocal active public support for environmental concerns has been the key to passage of environmental legislation. Public "watchdogging" and follow-through have also been—and will continue to be—important. Laws can be amended, and laws can go unfunded. A law may contain authorizations for funding, but it must be followed by sustained *appropriations* before it has any real meaning.

The environmental movement has largely been a popular grass-roots movement. It is fitting, then, that environmental laws have explicit provisions by which citizen participation is to be encouraged, solicited, and made part of the administrative process. Citizens' rights to sue government officials for failing to carry out their mandates are spelled out in many environmental laws. A problem that citizens' groups face is that they are usually poorly financed volunteer organizations; this can and does limit their effectiveness.

540

Government

It should be intuitively obvious that environmental improvement can be brought about only at some high level of human organization. A citizen with riverfront property cannot make much headway cleaning up the river by doing anything to the part of the river that sweeps by that property. Environmental improvement can come only through concerted, highly organized efforts by all people along rivers. This is where government comes in.

Government is an institution whereby we spread responsibility over the whole for the whole. It can also be thought of as a device for ensuring that actions take not only existing people into account but also those who will come in the future. Government is the principal institution by which a society can protect the **commons**—those things that belong to everyone and yet belong to no one.

There are different approaches to rationalizing government's role in environmental protection. We touched on one under the heading of environmental law:

> There is a definite ethical basis to a national policy for the environment. This centers upon the responsibility of government as the agent of the people to manage the environment in the role of steward or protective custodian for posterity. It requires the abandonment of government's role as umpire among conflicting and competing resource interests and the adoption of the total environment as a focus for public policy. (Wandesforde-Smith, 1970, p. 208)

According to this view, government is the ultimate steward or trustee for the environment for present and future generations.

No one will deny that environmental issues are fraught with conflicting views and often require a referee. Sometimes the role of referee is at odds with the stewardship role. Government—the courts, lawmakers, and executive officers—must often consider whether or not a straight compromise among those with conflicting views is in the best public interest. "Government's greatest peril arises when the views ought not to be compromised because one view is in error" (Murphy, 1967, p. 290).

Government has and ought to have an important role in environmental protection. The complexity of environmental problems requires personnel, money, authority, and large-scale coordination that can be provided only by government.

Administering Environmental Law

As we indicated earlier, once laws have been enacted, they are further defined and implemented by administrative regulation. Many environmental laws in the United States are under the administrative jurisdiction of the Environmental Protection Agency (EPA).

The EPA was established in December 1970 by executive order of the president. The EPA has ten regional offices, and each state is in one of the ten regions. The agency has broad administrative and coordinating responsibilities for environmental affairs. Its statutory responsibilities cut across those of many other agencies in the executive branch of the federal government and reach down to the local level. Environmental statutes covering air, water, and solid and hazardous waste call for state and local programs to be established but give the EPA the responsibility for guiding, reviewing, and ultimately approving and monitoring programs.

Many other federal agencies also have environmental responsibilities. The Department of the Interior carries out laws dealing with public lands (among other things); within that department, specific offices have particular responsibilities. The Office of Surface Mining, Reclamation and Enforcement (OSMRE), for example, oversees the Federal Surface Mining Act. The Department of Agriculture administers forest and pesticide programs. Even the Department of Commerce and the Department of Defense (Corps of Engineers) administer environmental programs dealing with oceans and dams, respectively.

It has periodically been proposed that a department of natural resources be established at the federal level, combining programs from the Department of the Interior, the U.S. Forest Service, the Department of Agriculture, the National Oceanic and Atmospheric Administration (NOAA), and the Department of Commerce, among others, to help to streamline the administration of natural resource policy. Currently, both the Bureau of Land Management and the Forest Service regulate activities on federal lands, though they are in different departments. The NOAA protects the endangered sea turtle when it is in the ocean; the Fish and Wildlife Service protects it when it comes ashore to lay its eggs. Action to streamline federal agencies and programs is always easier described than done, and little movement in this direction has occurred. Most recent efforts have involved making the EPA a cabinet-level agency.

State Government and the Environment

The role of the states in administering environmental laws has changed over the past few decades. Before 1972, most federal acts established general environmental goals and provided research, technical assistance, and funding to the states to carry out state programs. Federal environmental legislation enacted after 1972 was much more specific and exacting. For example, recent federal environmental legislation tends to establish compulsory compliance deadlines and quantifiable standards.

Federal environmental laws of the past 20 or more years give specific responsibilities to the states. These laws contain "carrot and stick" incentives in the form of significant federal financial assistance, the threat of a general cutoff of federal funds, and clauses that say "if you don't do it, we will."

In some cases, states have modeled their laws after federal laws. Some states require environmental impact statements for projects using state monies, just as NEPA requires them for projects using federal monies. Some states designate wild rivers for protection, just as there are federal designations.

In other cases, states have initiated environmental protection laws of their own as a result of public pressure and lack of action at the federal level. This appears to be a growing trend. Many states have established stringent controls in the area of solid and hazardous waste management. States have also initiated their own programs to authorize the granting of scenic easements, to deal with nonreturnable bottles, to grant citizen standing in environmental matters, to mandate recycling, and to ban chlorofluorocarbons, among other things. States are also joining together on a regional basis to promote specific environmental legislation. A coalition of northeastern governors has supported the same model legislation in each of their states to ban certain toxics in packaging.

Most states have special agencies for environmental matters. Some people think that as emphasis on environmental problems continues to shift to such items as toxic chemicals and the need for waste management facilities, state (and local) governments will need to be more involved in working with local citizens and groups to find locally acceptable solutions to these problems.

Regulatory Approaches

Government has many tools for bringing about the results it desires in environmental control. As we have already discussed, the federal government uses a subsidy approach, providing money for irrigation, drainage, and program administration and implementation. Another type of subsidy is the tax break for pollution control equipment. Such an approach has been criticized for promoting pollution control equipment over changes in processing that might be more effective. Later environmental laws used a more direct regulatory approach whereby the government set standards and maximum discharge levels. This approach has been criticized for its lack of flexibility and for allowing polluters to pollute to a given level without charge. A third approach, which we will take up at some length when we discuss economics, employs disposal charges. Under this approach, a fee or charge for

each unit of pollution is set; the amount is set high enough that it is more economical for the polluter to reduce the discharge than to continue paying the disposal charge.

In all of these approaches, what government does is to cause needed change to take place in *other* institutions, using the time-honored methods of reward and punishment; sometimes government has to make things happen.

Economics can deal easily enough with a straightforward situation such as a person wanting to buy a ticket to a concert. Someone who is unwilling to bear his or her portion of the cost of the concert cannot get through the door. Hueckel (1975) points out that this is in sharp contrast to a situation in which a person wants some air—for breathing. People have never paid for air, and they surely cannot be excluded from using it very easily. This is where government comes in. Government finds a way through taxes, assessments, laws, and the like, to create circumstances in which individuals must pay the cost of keeping air clean. Another example offered by Hueckel (1975) is the situation in which one person believes that airplanes are a nuisance and should not be permitted to exist because of the particulate pollution and noise. The role of government here might be to ensure that all of the costs of air travel, including the cost of abating noise pollution, the cost of minimizing particulate pollution, and other assorted indirect costs be taken into account and divided among the people who travel by air. Government can force the inclusion of various indirect and external factors into cost-benefit equations and cause air travelers to consider them in deciding exactly how necessary air travel is.

Can Government Deal with the Long Range?

Democratic government is supposed to reflect consensus attitudes and feelings of the people it serves. If this is true, governments should be able to deal with long-range problems only to the extent that "the people" are able to perceive the need for such considerations. Herein lies the rub. It seems that most human beings are unable to deal in a very effective way with the long range. The stresses, pressures, and problems people face every day are such that most people cannot deal beyond the immediate. It is easy to say that governments *must* take the long range into account despite this obvious limitation, but that is easier said than done.

First of all, if the government goes very far beyond devoting its attention to immediate problems, it runs into the danger of being called unresponsive. Consider the dilemma. A U.S. congressional representative elected for a two-year term knows that if he or she is not reelected, he or she will not be able to do *any* good for the people. (We will disregard the further complexity caused by the fact that corporations, political action groups, and other organized special-interest groups play a significant role in the election of representatives.) Even though that representative might know that deregulation of gas prices would allow the price to rise, cause a drop in the demand for natural gas, and give the country more time to find new sources of energy, he or she also knows that a vote in favor of deregulation will gain the displeasure of constituents who would have to pay higher prices for gas at a time when all prices are high and getting higher. This, it seems, tends to strap us with a seat-of-the-pants system of management by government at a time when such management is increasingly inappropriate.

However, there are a few examples of attempts to look farther down the road in addressing problems. Although it did not pass, legislation was introduced in Congress to establish an Office of Critical Trends Assessment at the federal level. Programs to identify emerging issues and to consider long-term implications of policies are being initiated in many states. The complex nature of issues and the need to use limited financial resources wisely over the long term are pushing governments to make decisions with a broader picture in mind. Political realities, however, cannot be ignored or underestimated.

International Government?

There are signs of environmental awareness and conscience throughout the world—in western Europe, Japan, throughout the former Soviet Union and various developing countries. It is also becoming increasingly recognized that many environmental problems must be dealt with on a worldwide basis.

In 1972, the United Nations held a conference on the human environment in Stockholm, attended by representatives of 114 nations in the industrialized world and the Third World. The group agreed on a number of principles and promulgated these in its Declaration on the Human Environment. Achieving such agreement was no small task; the declaration had to contain principles assuring Third World nations that environmental concerns would not restrain development. Among the outcomes of this important conference was a proposal that an environmental assessment system, **Earthwatch,** be established to look after water pollution, air pollution, endangered species, and the like, on an international scale.

In May 1982 a ten-year follow-up conference was held in Nairobi. Out of that conference came an effort by representatives of six continents to establish the World Campaign for the Biosphere. The objectives of this campaign are to promote educational programs on the operation of the biosphere, enhanced scientific understanding of the biosphere, and governmental action relative to worldwide environmental problems.

The Earth Summit. In June 1992, the United Nations invited the nations of the world to Rio de Janeiro for a Conference on Environment and Development (UNCED) that came to be known as the **Earth Summit.** The Earth Summit was the most significant gathering of nations ever for the purpose of acting to protect the earth. Delegates from 170 nations negotiated for two years leading up to the conference and produced several agreements addressing sustainable development, global warming, biological diversity, and the relationship between developed and developing nations in environmental matters. Among the key outcomes of the Earth Summit:

1. **The Rio Declaration on Environment and Development:** a non-binding statement setting forth 27 principles that should govern environmental and development policy worldwide. The goal of the Rio Declaration is to ensure that the environment becomes "an integral part of the development process."
2. **Agenda 21:** an 800-page action plan for environmental protection and sustainable development serving as the blueprint for implementing the Rio Declaration. The cost of implementing "Agenda 21" is estimated to be $125 billion, annually. Although several nations did make financial pledges toward meeting the goals of Agenda 21, this amounted to only a small fraction of the amount needed. Nevertheless, a U.N. Sustainable Development Commission was established to monitor progress in implementing the "agenda."
3. **Global Warming Convention:** one of two "binding" agreements adopted at the Summit, this convention was designed to be a legal commitment of the nations of the world to reduce greenhouse gas emissions. All target dates for specific reductions were deleted from the convention prior to its adoption on the insistence of the United States.
4. **Biodiversity Convention:** the second of two binding agreements resulting from the Summit, this convention was designed to address the extinction of species (see Chapter 14). Over 150 nations signed the Biodiversity Convention; the United States did not. This Convention requires signatory nations to inventory species and protect those that are endangered; it also provides for developed nations to help developing nations finance these efforts and requires developed nations to share research, profits, and technology with developing nations from which they

take genetic resources. Both of the foregoing conventions had to be ratified by the appropriate legislative bodies in each of the signatory nations.

5. **Statement on Forest Principles:** this is a non-binding statement setting out 17 points concerning protection of the forests of the world.

Only the years ahead will tell the ultimate success of the Earth Summit. The event itself speaks to the growing concern for the environment worldwide.

Economics

In Chapters 11 and 12 we discussed the significance of economic factors to air and water pollution. Elsewhere we have pointed out that economics is an extremely important factor in all of our environmental problems. Because of the pervasive importance of economics, we devote a section here to economics and business and how these relate to both the state of the environment and our environmental future.

Some Relevant Economic Principles

Perhaps the chief fact relating economics and the environment is that in the past, ecological and environmental factors were not taken into account as part of the cost of doing business, but in the future they must be. This means that there must be a transition from environmental protection's being an external factor to its being fully included in the cost accounting that leads to the determination of the prices of products that are bought and sold in our society.

Among the important principles and concepts in economics that bear on the need for this transition is the so-called law of supply and demand and the concept of margin.

Supply and Demand. It is said that almost every economic system is based on **supply and demand** or some similar market system. In this system, a need is expressed in terms of willingness to pay, which in turn serves as a stimulus to someone else to meet the need—in exchange for money. There are many believers in the market system who claim that it is unmatchable in terms of satisfying countless different, ever-changing tastes and needs at minimum resource costs. Ideally, if consumers buy less, either because they do not want the product or because the prices are too high, demand falls, and producers quit producing. If, by contrast, something is demanded by many people, prices rise, serving as an incentive for the producer to produce more or to find more, and the result is an increased output of the things that people want. The system works to achieve a balance between

supply and demand through feedback. As more and more is produced, supply increases, prices drop, and some stable state is reached wherein what is needed is provided. You may appreciate the fact that, in theory at least, the law of supply and demand operates through feedback much in the same way that feedback operates in ecological systems.

Some observers question the "purity" of the concept of supply and demand, believing that the market system itself often sets up distorted demand through devices such as advertising. Economists like John K. Galbraith argue that consumers tend to buy what is made, advertised, and put before them in shop windows. Producers, according to this view, are not simply providing people with what they want; they are taking an active part in defining what people can or cannot have. The counter point of view is that competition is an extremely strong force operating within systems of supply and demand, and competition brings into consideration all the information necessary for making optimal decisions about what should be produced.

Marginalism. Although the economic concept of **marginalism** is quite complicated and grounded in differential calculus and the mathematics of maximization and minimization, it is in basic outline a straightforward concept and one that is important to our discussion (Figure 19.6). Perhaps it can be illustrated by an example.

Consider a patch of blueberries. If I hired a blueberry picker to pick the blueberries in a certain patch, set this picker to work, and paid an hourly wage, the plentifulness of the blueberries would ensure that for a certain period of time the picker could sustain a high rate of blueberry harvesting—say, 4,000 blueberries per hour. If I could sell 4,000 blueberries for $5 and paid the picker $3 an hour, I would be able to make a $2-per-hour profit (disregarding for the moment any cost associated with growing the blueberries). After a while, as the blueberries became increasingly scarce, the yield per hour would decline, and inevitably, at some point the picker would be hard pressed to gather $3 worth of blueberries in a given hour. It would be at this point that I would have maximized the gain I could derive from my blueberry patch, and any additional time spent by the blueberry picker in that patch should give me more cost than benefit even though some blueberries remained to be picked (see Figure 19.6). If I had to include some of the other costs that sustained growing blueberries, the point at which maximization occurs would be earlier. Obviously, I would do well as a manager to know, to the second, the point at which my costs began to exceed income. To consider environmental problems appropriately, this concept has to be kept well in mind. It does not do in the real world to advocate that all

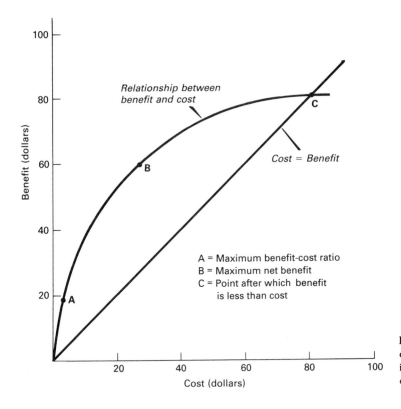

Figure 19.6 The Cost-Benefit Ratio. This ratio can vary significantly with such factors as level or intensity of economic activity. After some point (B), cost increases faster than benefits.

pollutants be removed from the air. Our economic system—any system, in fact—demands that all the costs of removing or controlling pollutants be considered in relationship to the costs of not removing them.

On the Internalization of Externalities

One of the most important adjustments to be made in free-market economic systems is the internalization of social costs or the inclusion of environmental cost in calculating the ratio of cost to benefit.

Economic theory has historically been far more concerned with private transactions and how these lead to the establishment of value than it has with social costs. There are many examples of how various costs of doing business are ignored—treated as **externalities.** Perhaps one example here would help.

A chemical company that had to dispose of a waste product might find that the cheapest way to do it would be to pay somebody to haul it away, no questions asked. By "no questions asked" we mean that any moral responsibility to see to it that the waste is disposed of properly is transferred by the company to the hauler. After the point of transfer to the hauler, any additional cost, to whomever or to whatever, becomes an externality as far as the chemical company is concerned. The hauler might (if able to circumvent the federal tracking system) simply drive the chemical drums to a relatively unpopulated area and just dump

them without regard to what might happen after that. Then the chemicals might leak into a stream or render that particular piece of land and the surrounding area unsuitable for many purposes. In effect, this would transfer the cost of having generated the waste product to people who might move there and have to deal with the waste in the near or distant future.

Externalization has often been accomplished with the complete acquiescence of the citizens affected. Consider a town a generation or two ago with a mine-acid-polluted stream. Although the stream was remembered by old-timers as once good for fishing and most of the town's people were outdoor-oriented, no one ever seemed to mind the loss of the stream—apparently because people saw it as directly related to their livelihood. Pollution was something accepted as necessary to their being able to exist there at all. They went on accepting this as a social cost, driving many miles to fish and paying more than they would have had to otherwise to pipe their drinking water in from another stream. Social costs have often been accepted willingly, and this is precisely why more of them have not been internalized before now.

The divergence between private and social costs of doing business has been a fundamental factor in many types of pollution. The divergence occurred because managers have traditionally had to deal only with private cost to arrive at the profit-maximizing output levels and prices in their business. Dumping wastes into air or into water has, until now, been free.

At first these practices were free because few, if any, people were affected by them, and then, for a long time, those who were affected by them did not care. What must be done now is clear. Private and social costs of doing business must both be considered.

It is obvious that as long as free-market economics neglects the external costs of doing business, it will be a science of illusion, and business will continue to be a game of transferring external costs onto the shoulders of society and the citizens of the future. What we have had in the past could really be called institutionalized inhumanity to humankind. What we must have in the future is **ecological impact costing**—an economic system in which ecological costs are reflected in prices. Many economists have made proposals as to how this might be done. Some have outlined the problems that get in the way of doing it. We would like to take up some of these problems first.

Factors Operating Against the Internalization of Ecological Costs

Among the factors operating against bringing ecological costs into the price-output equations are (1) the inertial resistance built into our present economic and political system, (2) the magnitude of the costs and whether they can be passed along to consumers without any fatal skips in economic heartbeats, and (3) the problem of determining long-term and even short-term social costs (in terms of dollars).

Resistance to Internalization. It seems likely that from now on there will always be a conflict between forces tending to maximize profits and forces that would tend to minimize environmental impact. It has become a rigidly institutionalized business principle that managers will try to externalize as many costs as possible. It is useless to blame managers for this; they would argue that they are simply playing by rules established for some time; the "game" is won by managers who are best at cutting costs.

Predictably, managers of plants and corporation presidents resist every suggested internalization, and it is highly unlikely that the business system will adopt the concept of internalization quietly. The record already indicates that players will warn of every sort of economic disaster when internalization is about to be forced upon them.

The Magnitude of the Cost of Ignoring Social Costs. It is being argued that the magnitude of the cost of protecting the environment as goods and services are produced will be staggering to the point of bringing an end to civilization. The facts simply have not borne out such fears. Although it cost more than $750 billion between 1981 and 1990 in tax money and

corporate money to install required environmental protection devices and to adopt environmental protection practices in the United States, this is expected to amount to only about 2% of the nation's GNP. Our estimates in Chapter 11 of the cost of environmental damage from air pollution alone during the same decade, you may recall, would amount to $250 billion if prevailing practices were allowed to continue. This seems to indicate that the cost of not cleaning up will more than outweigh the cost of keeping the environment clean.

It has also been argued that the cost of pollution control equipment will not add value to the goods being produced and thus the worker hours and dollars that go into pollution control will not show up in increased productivity. It would seem that the best way to counter this argument would be to find ways to assess the value of improved air and water and somehow add this to the GNP, using whatever formulas and methods seem appropriate.

Environmental Blackmail. There is a thing called **environmental blackmail**. Companies argue that if they are forced to clean up their act, they will shut down, and therefore a clean environment will come only at the expense of jobs. There is no doubt that plants have closed and will close as a result of environmental protection measures. But this is, as it should be, a form of natural selection. Environmental quality simply has not been a factor in determining what types of industries and what quality of industries would spring up and thrive in various places. Environmental considerations should obviously have a place in doing business.

When we woke up this morning, we faced a world with an array of devices, practices, and patterns that have been determined by factors other than the social costs of doing business. We have had boisterous lawn mowers, noisy airplanes, and loud trucks because heretofore no pressures existed to select against them. We have dirty factories that cause disease or death in workers and diminish the lives of people in the communities in which these plants are located. Although a plant closing might cause pain and suffering at the local level, society must look at the big picture. Overall, a pollution control industry has grown up around the new environmental laws; this industry could mean more jobs than have been lost as a result of the closing of inefficient polluting plants (see Chapter 11).

The Impact of Environmental Protection Costs on the Poor. Among the problems of economic readjustment to include environmental costs is the idea that relatively poor people can have more of the "better" things in life only if social costs are not passed along to them. Presumably, if various environmental externalities were internalized and cost increases had to be borne by consumers, this would have an especially serious effect on the poor—those only marginally able to accumulate food, shelter, and clothing under present conditions. It is a sobering thought that there may be some people who are alive today only because technology has been able to borrow against nature and the people of the future. Governmental mechanisms can be used to ameliorate the impact of cost internalization on the poor. Continuing to borrow against nature is clearly *not* the solution. It will only lead to greater problems in the future.

The Problem of Determining Social Cost.
Given the nature of our economic system, it would be best if clean air, clean water, unspoiled landscapes, and quiet had physical form and could be subdivided and transferred from one person to another. We would then be able to develop more complete assessments of the value of these commodities. We would be able to see how many dollars changed hands for these commodities. All the rest would fall into place. Our legal and economic systems could handle this kind of transaction in a straightforward manner; we would be much better able to assess damage and the extent of various impairments. As pointed out by Mishan (1972), just because the environment is not divisible in this way is no reason to treat environmental resources as if they were not valuable and as if they were not scarce. The problem is that we have very little basis on which to determine social cost and assign value to things like clean air and water (see Chapters 9 and 10).

How Can Internalization Be Accomplished?
Up to now we have assumed that there are ways of internalizing what have heretofore been externalities. In his essay "Not Peace but Ecology" (1969), Garrett Hardin gives an example of how this might be done. He cites the following case: Before the appearance of the insurance industry, accidents in industry were thought of as the fault of workers. They were unexpected, and they happened as a result of random and miscellaneous causes. They certainly could not be planned for and certainly could not always be blamed on the company. The cost of accidents was therefore an externality.

In more recent years we have come to appreciate that although accidents are somewhat capricious, over large numbers of companies and large numbers of workers they are predictable consequences of people working. Because they happen with a certain frequency, they can be cost-accounted, and the risk can be spread over industry by means of worker's compensation insurance. With worker's compensation the cost of covering accidents became assimilated into the cost of doing business. All companies now either pay insurance premiums or make plans and provisions for insuring themselves. The cost of accidents has thus been internalized. Some economists suggest that there are actually several ways by which environmental and social costs have been, and can be, internalized, including these:

1. Appeals to reason—seeking voluntary internalization of costs
2. Use of the courts, relying on the threat of fines, suits, and judgments to serve as the pressure leading to internalization
3. Government regulation using a combination of the threat of economic sanction and imprisonment to force internalization
4. Taxing pollutants and granting pollution rights
5. Using various subsidies and incentives

Many free-market economists argue that it would be better to put price tags on pollution and to incorporate the environment into standard pricing practices in economics than to muddy the waters with regulations and prohibitions that would be subject to abuse. They see a system of taxing pollution as more compatible with existing economic theory.

Paying for the Privilege of Polluting
This approach would actually allow an industry to pollute, provided that it pays a fee based on volume—so many dollars per ton of sulfur dioxide or per ton of particulate matter, and so on. Analysts who trust our economic system say that even though this does say to the polluter, "Pollute all you want—but it's going to cost you," market forces will eventually cause the environment to be brought into the cost of doing business. Why? Because it would become each manager's responsibility to reduce the amount of money the firm has to pay for pollution, and managers would find ways to cut down on pollution.

Possible variations on this approach make it even more intriguing. The more toxic a substance is, the higher the prices that could be put on it. A very toxic substance could have a prohibitively high price for even a little pollution. There would be problems to be worked out. What about such vagaries as thermal inversions? Would the dollar price vary with the time of the year or the hour of the day? Should it cost more to generate a given amount of pollution close to a population center? This might be effective in causing industries to be located where the cost would be lower. There would be an economic incentive for the polluter to move or to build a plant in places remote from population centers so as to generate less social cost.

A system such as this would require that regulatory agencies be charged with metering the output of all sources of pollution and collecting fees.

The Problem of Uniformity

The major problem with what we have just discussed is that for it to be done right, it would have to be carried out on a global scale. It would not do, after all, to have the steel industry in the United States internalize its social costs if another country can make steel without such internalization and sell the steel in the United States at a much cheaper price, creating an unfair advantage. Such problems could be solved in part within a large country through a system of tariffs, but it would obviously be best if this type of approach were undertaken at the international level.

Is It Good to Have a High Gross National Product?

It is important to appreciate that economics merely keeps score of things that are exchanged for other things and that there is really no place for value judgment in keeping score. The dollar value of goods and services generated in a country—its **gross national product (GNP)**—is a nonjudgmental numerical expression of the total number of dollars paid for goods and services. Good and bad are not distinguished.

As pointed out by Wallace (1972), much of the GNP in the United States consists of junk and of services that are of marginal importance. What this means is that GNP *does not necessarily reflect quality of life;* in other words, GNP is not a measure of personal freedom, natural resources left, happiness, or quality and stability of life in the immediate or long-range future.

The British economist John Maynard Keynes first articulated the theory that led to the economic ideology credited with bringing the world out of the Great Depression, wherein maximum production would be ensured by government spending to stimulate the economy. This ideology, which has dominated Western economic policymaking since before World War II, apparently allowed the concept of production for production's sake to establish some place in our economic thinking. An example actually used in setting forth this theory is that it would be better to pay people to dig holes and pay other people to fill them up than for the diggers and fillers to be out of work. Although such make-work approaches had some value during the depression, many economists believe that the concept actually allowed productivity to become established in our economic system as an end in itself. The fact that we still use an ill-defined concept like GNP in itself suggests that we do indeed have a problem. Think about the fact that the cost of cleaning up

after the *Exxon Valdez* oil spill had a *positive* effect on the United States' GNP. However, some countries, including Norway, France, and Germany, have taken steps to begin to supplement their national economic accounting systems with accounts for nonrenewable, biotic, and environmental resources. Third World nations are also becoming involved. The World Resources Institute is sponsoring pilot projects in India, Costa Rica, and China to develop better resource accounting systems. The United Nations and the Organization for Economic Cooperation and Development (OECD) are looking at ways to revise the standard systems of national accounts to reflect environmental degradation (Repetto, 1989).

Economics as a Descriptive Science

Economics is a science of identifying the factors that have relevance in making maximization decisions; it is, as we said, value-free. Economics generally has nothing to say about whether certain definitions of values are better than others. If the law still recognized indentured servitude, the cost of labor in economics would be no more than that necessary to indenture and maintain a servant. If, by contrast, the law required that workers be looked after and have worker's compensation and other benefits, the cost of labor would include these factors. As another example, suppose we were trying to determine the cost-benefit ratio for producing coal. Being environmentally minded, we might want to include the cost of avoiding contaminating streams and certain costs to minimize the health and safety problems faced by coal miners. Depending on the value assigned to a human life, the coal industry might be required by the unions and the government to establish safety conditions to the point at which it would be absolutely impossible for anyone *ever* to be killed or injured in mining coal. Somewhere short of this, however, the cost of providing this safety would well surpass any economic benefit that might be derived from mining coal. This, of course, would be unacceptable to coal companies, unions, employees, and just about everybody else.

Economic Growth and the Environment

A major question that has much to do with the relationship between economics and ecology is the question of whether or not sustained economic growth is a requisite for a strong, healthy economy. Many economists are growth-oriented and feel that continued economic growth is necessary for the survival of any economic system. Increasing numbers of economists feel otherwise, however. Ecologists know that anything—any good or any type of service that is based in some way on production by ecosystems—will have an upper limit to growth.

To Grow or Not to Grow

When Growth Is Not Really Growth. When the advocates of sustained economic growth speak of growth, they are referring to real increases in GNP, that is, growth over and above the illusion due to inflation. But there is another relativistic quality about GNP worth mentioning here. Human beings do not seem to be satisfied as much by the absolute value of the goods and services they command as by their relatively high standing in relationship to other people. Third World nations tend to see their need to experience real growth as the difference between what they have and what they see in the so-called developed countries.

Why Real Growth in the Numerical Sense Cannot Be Sustained. On the basis of what we know about ecosystems as described in the first several chapters of this book, it is clearly and definitely not within the natural order of things that the growth of anything can be sustained indefinitely. All living systems—and all economic activity has something to do with ecosystems either directly or indirectly—have finite capacities to yield products and to absorb by-products. Economic growth must be an exceptional state. Whatever economic growth there has been in the past, for however long, has been a reflection of exploitation of space and resources and expanding populations with expanding needs. We have now reached the point, or will at some time, at which either world population growth will level off—bringing the stimulus for sustained economic growth to the level with it—or there will be catastrophe that will rewrite the script entirely.

We seem to have developed two conclusions:

1. It is an absurdity, at least in terms of the limits of what the human brain can comprehend, to assume that economic growth or the growth of anything can be sustained indefinitely.
2. Growth is not necessarily an appropriate goal, and we would probably do well to develop a new breed of economics and economists whose instincts have not been nurtured during a period when economic health and growth occurred simultaneously for so long as to give the impression that they are related.

Growth or Change?

Change, development, and improvement all fall into a category that can be described very well by another connotation of the word *growth*. A person can grow in height (quantitatively) and can also grow in stature, maturity, and wisdom (qualitatively). If we can be permitted to use both senses of the word *growth* at the same time, we tend to agree with analysts who believe that some form of economic growth may be inevitable and perhaps even desirable. The time has come to discard the concept of growth embodied in most economic models of the world—models that permit planned obsolescence, inefficiency, and waste to be considered part of growth. It seems clear that the relevant ecological question is, How can economic activities be directed so as to improve the quality of life for all human beings?

To paraphrase William Ruckelshaus, former administrator of the EPA, the question is not whether there will be growth, because there will always be some kind of growth as long as humans have the inherent drive to achieve perfection. What we really must do is define growth and understand both what will happen as growth takes place and how to limit or avoid any deleterious consequences of growth. In other words, we must know where we are growing and go in that direction only if we wish to go in that direction.

Capitalist Versus Socialist Economic Systems

The issue of capitalism versus socialism has little to do with the relationship between economics and the environment. Both systems are based on productivity and are made even more similar by competition in the same world market. Environmental problems related to economics are therefore similar under both systems.

The World Commission on Environment and Development was established in 1983 by the U.N. General Assembly. In 1987 the commission issued its report and called for future economic growth "based on policies that sustain and expand the environmental resource base." Among other things, the commission recommended that ecological considerations be given equal status with other factors such as economics, agriculture, and trade in policymaking worldwide. Political systems should not make any difference in the need to develop ecologically sustainable economies.

Some Concluding Remarks

We have two points to make in conclusion. One is that in the past, many environmentalists have scorned economics, and this has not served the environmental protection movement well. Rather, it has tended to stereotype environmentalists as radical, out of tune with reality, and ignorant of the facts of life.

Environmental activists must learn as much as they can about the system by which things that people value are exchanged in society. Economics is too important to the environment to be left to economists. The other side of this is that the economists must also develop a more complete understanding of the environment and its relationship to economics. A better understanding of this relationship is needed by all. The world of the future may well look back and label this the "age of exploitation." A number of economists, notably Kenneth Boulding, William Nordhouse, and James Tobin, have begun to describe a new economics, one more in tune with ecology and natural laws and encompassing the notion that the ultimate derivation of value has to be based in part on the natural environment. Georgescú-Roegen (1977) has gone so far as to suggest that the new order might well begin with a bioeconomic program that takes into account not only contemporary fellow human beings but future generations of human beings as well.

Concepts to Remember

1. Our social institutions reflect our world view; as such, they also reflect the roots of our environmental difficulties. These institutions are being restructured as we adjust our world view in search of solutions.

2. Science and technology tend to offer us choices before the consequences of the choices are fully appreciated. We must continue to improve our ability to screen our technologies before they drag us unwittingly into problems in the future.

3. Medicine has historically focused on the sick and has left the well alone. Today's chronic diseases, some of which are connected to environmental quality, require that medicine adopt more of a preventive approach to keeping people well.

4. Nearly all of our cities were developed haphazardly, without regard to many important environmental realities. The cities of the future and changes in existing cities will have to be designed in harmony with the environment.

5. Ecological constraints suggest the need for more highly integrated, more energy-efficient and less polluting forms of transportation.

6. Education is a key to creating in all citizens a world view that is compatible with the laws of ecology. Education must give us environmental specialists with a good general view of the operation of the ecosphere.

7. The constitutional basis of the U.S. government's role in environmental protection is, among other places, found in the directive to provide for the general welfare and in the right to life, liberty (to breathe and drink water without getting sick), and the pursuit of happiness.

8. Government by the people—like the people themselves—has trouble dealing with the long term.

9. Ways must continue to be found to include the protection of the general environment in the cost of doing business.

10. Science can help us diminish our environmental problems by providing the basis for nonpolluting technologies. It can also help less directly but no less importantly by documenting the connections between environmental quality and the quality of life, thereby adding to our motivation to clean up the environment and keep it clean.

11. The laws of ecology say that quantitative economic growth—or any kind of quantitative growth for that matter—must be a temporary state.

Edith Felchle

Edith Felchle is an environmental education specialist with the City of Fort Collins (Colorado) Natural Resources Division. The Natural Resources Division has the job of planning for Fort Collins's environmental future and, in doing so, deals with air quality, recycling, wildlife habitat, and hazardous materials issues. Ms. Felchle's job is to facilitate communication, education, and motivation relative to environmental protection. Her background is in the areas of communication and environmental studies, and she says that her knowledge and skills in marketing, human learning processes, and communications are the most useful in the work she does. She organizes workshops for teachers and students, puts together environmental fairs, publishes a newsletter, designs exhibits, coordinates environmental communication within city government, writes articles and designs brochures, provides environmental consultation to individual citizens, and does some public speaking. She recommends the kind of work she does to all who enjoy meeting the public and who have a certain amount of tolerance for the frustrations normally associated with government bureaucracy and the political process.

References and Further Reading

References marked with an asterisk are cited in the chapter.

*Brown, L. R., and Jacobson, J. I., 1987. *The Future of Urbanization: Facing the Ecological and Economic Constraints.* Washington, D.C.: Worldwatch Institute.

*Carnow, B. W., 1971. "Air Pollution and Physician Responsibility," *Archives of Internal Medicine* **127**:91–95.

*Chandler, W. U., 1988. "Assessing SDI," in *State of the World, 1988.* New York: Norton.

French, H. 1992. *After the Earth Summit: The Future of Environmental Governance.* Washington, D.C.: Worldwatch Institute.

*Georgescú-Roegen, N., 1977. "The Steady State and Ecological Salvation: A Thermodynamic Analysis," *BioScience* **27**: 266–270.

*Greenwood, M. H., and Edwards, J. M. B., 1973. *Human Environments and Natural Systems: A Conflict of Domain.* North Scituate, Mass.: Duxbury Press.

*Hardin, G., 1969. "Not Peace but Ecology," In: Diversity and Stability in Ecological Systems, Brookhaven Symposium in Biology, No. 22, Upton, N.Y., pp. 151–161.

*Hueckel, G., 1975. "A Historical Approach to Future Economic Growth," *Science* **187**:925–931.

*Kupchella, C. E., and Levy, G. F., 1975. "Basic Principles in the Education of Environmentalists," *Journal of Environmental Education* **6**(3):2–6.

Landau, R., and Jorgenson, D., eds., 1986. *Technology and Economic Policy.* Cambridge, Mass.: Ballinger.

*Leopold, A., 1949. *The Sand County Almanac.* New York: Oxford University Press.

Lowe, M. D., 1990. *Alternatives to the Automobile: Transport for Livable Cities.* Washington, D.C.: Worldwatch Institute.

*Mishan, E. J., 1972. "Property Rights and Amenity Rights," in *Economics of the Environment: Selected Readings,* ed. R. Dorfman and N. S. Dorfman, pp. 187–193. New York: Norton.

*Murphy, E. F., 1967. *Governing Nature.* Chicago: Quadrangle Books.

Nash, S., 1991. "What Price Nature?" *BioScience* **41**(10):677–680.

Newcombe, K.; Kalma, J. D.; and Asta, A. R., 1978. "The Metabolism of a City: The Case of Hong Kong," *Ambio* **7**(1):3–15.

Outdoor News Bulletin, 1984. **38**(4):5.

Perkins, J. H., ed., 1986. *International Aspects of Environmental Education.* Troy, Ohio: North American Association for Environmental Education.

Postel, S., 1991. "Accounting for Nature," *World Watch,* 28–33.

*Repetto, R., 1989. "Wasting Assets," *Technology Review,* January, pp. 39–44.

*Rosenbaum, W. A., 1973. *The Politics of Environmental Concern.* New York: Praeger.

Rüdig, W. 1991. "Green Party Politics Around the World," *Environment,* **33**(8):6–9 ff.

Russell, M., 1987. "Environmental Protection for the 1990s and Beyond," *Environment* **29**:12–15ff.

Thompson, J., 1991. "East Europe's Dark Dawn," *National Geographic,* June, 35–68.

*Wallace, B., 1972. *People, Their Needs, Environment Ecology: Essays in Social Biology,* vol. 1. Englewood Cliffs, N.J.: Prentice Hall.

*Wandesforde-Smith, G., 1970. "National Policy for the Environment: Politics and the Concept of Stewardship," in *Congress and the Environment,* ed. R. A. Cooley and G. Wandesforde-Smith, pp. 205–226. Seattle: University of Washington Press.

20

From Problems to Promise: An Assessment

In 1972 a group of 18 internationally known ecologists met in Stockholm and identified what they considered to be the world's 10 most important environmental problems. Their list is as follows:

1. Too many people creating impossible demands on all natural resources
2. Pollution of the waters of the world
3. Pollution of the air
4. Lack of any significant worldwide research programs on food production
5. Lack of workable programs to preserve and protect the endangered wildlife of our planet
6. Worldwide inability to limit the indiscriminate use of persistent toxic substances
7. Failure to develop systems by which raw materials can be recycled and the effective loss of these resources to the people of the future
8. Failure to research and develop a plan for the use of various forms of energy in the future to improve living conditions for the people of the world
9. Inability to find ways to invest wisely, both public and private funds, in the improvement of the general environment

10. Inability of nations and their political subdivisions to develop workable systems of environmental control and cooperation

This is still basically a good list, but it is obviously only one of many ways to describe the most general problems of the environment. (See Enrichment Box 20.1) Our intention in this chapter is to carry this form of generalization a bit further, annotating and expanding the list with some of our opinions and the opinions of others concerning the general and philosophical nature of these problems and the prospects for diminishing them.

This chapter is intended as an editorialized summary. It has three sections. The first deals with the nature of the environmental problem and, in a general way, with some of the basic truths that will have to be taken into account in formulating solutions. The second section deals with more of the specifics of what must be done and why. In the final section we will peer into the future by guessing where present-day momentum is likely to carry us. Within each section we will present a series of individual statements with supporting discussion.

Forty-seven Issues

During the early 1980s there were several attempts to rank the major current and future environmental issues—by the Royal Swedish Academy of Sciences, the U.S. EPA, the government of France, the University of Michigan (with the EPA), the Congressional Clearinghouse on the Future, and the Conservation Foundation. In the volume *State of the Environment: An Assessment at Mid-Decade,* the Conservation Foundation consolidated the results of all of these surveys into a list of 47 "consensus" environmental issues as of 1985. Compared to the 1972 list given in the text, the 1985 list is more comprehensive and more specific, but it is basically the same. Would you add any new issues for the 1990s?

Wars, Accidents, and Natural Disasters

__ Wars
__ Nuclear accidents, terrorism
__ Chemical plant explosions
__ Failure of aging infrastructure (for example, dams, reservoirs, navigation channels, water supply systems, water treatment systems, sewers, highways, and bridges)
__ Intentional weather modification (unintentional effects)
__ Droughts
__ Floods
__ Earthquakes, volcanoes, and other natural disasters

Population Growth and Distribution

__ Population growth
__ Crowding and impacts of urbanization
__ Sprawl problems
__ Mass migration, immigration

Contaminants (Chemical, Physical, and Biological)

__ Radioactive waste disposal (including decommissioning of nuclear power plants)
__ Debris from space (especially, radioactive debris scattered by satellites reentering the atmosphere)
__ Microwave radiation
__ Electronic pollution
__ Solid waste disposal (including municipal waste management; landfills, incineration, and ocean disposal; sludge disposal or treatment; reduction at the source, resource recovery, and recycling; and littering and city cleanliness)
__ Noise
__ Pathogens from human wastes

__ Proliferation of biological organisms, bioengineering wastes, and mistakes
__ Genetic mutation
__ Carbon dioxide accumulation in the atmosphere (caused primarily by deforestation and the burning of fossil fuels—widely expected to absorb much of the solar heat escaping the earth's surface, thereby making the climate significantly warmer through the "greenhouse effect")
__ Acid deposition
__ Depletion of the ozone layer
__ Hazardous waste management
__ Conventional pollutants, ambient air
__ Toxic pollutants in air
__ Indoor air pollution (including carbon monoxide from stoves, heaters, and appliances; formaldehyde insulation; radon gas from building materials; and chemicals in household cleaners)
__ Conventional pollutants in water, from point sources
__ Nonpoint-source water pollution (including agricultural runoff of sediment, fertilizers, pesticides, and animal wastes; nonpoint municipal and industrial discharges; acid mine drainage; accelerated runoff from urban streets; storm water and sewer overflows)
__ Toxic pollutants in surface water
__ Groundwater, drinking-water contaminants
__ Pesticides
__ Chemical fertilizers
__ Chemicals in food chains

Natural Resource Depletion

__ Water scarcity
__ Loss of agricultural land because of salinization, desertification, or urbanization
__ Soil erosion and overexploitation of agricultural soils
__ Ocean fisheries depletion
__ Plant and animal species loss
__ Energy scarcity
__ Critical-materials scarcity
__ Damage to the marine environment (including damage caused by oil spills, ocean dumping, and ocean mining)
__ Loss of tropical forests
__ Degradation of coastal areas
__ Loss of wetlands
__ Degradation of wilderness areas, parks, and wild and scenic rivers

On the Nature of the Environmental Problem

We use energy in highly inefficient ways. Among the many problems that stem from our inefficient use of energy are the following:

1. Depletion of key energy resources
2. Acid rain
3. Acid mine drainage
4. CO_2 buildup in the atmosphere
5. Sulfur dioxide, ozone, particulate, and nitrogen oxide pollution of the air
6. Strip-mine-ravaged land

7. Destruction of forests
8. Oil pollution
9. Imbalances in international trade
10. Thermal pollution
11. Contamination of the environment by radioactive materials

We use mineral resources in a one-way pattern. This flies in the face of the principles of material cycling found in the rest of nature. Predictably, this pattern has led to resource depletion at one end and waste disposal problems at the other end of the line.

We not only pollute water, but we also waste it. We waste water by using it in a linear fashion and by using so much of it that water resources are being depleted in many places. Water pollution renders water less useful for humans directly and jeopardizes the aquatic ecosystems on which we depend indirectly. Water tables are falling in Africa, China, India, North America, and elsewhere. Water resource depletion will be a key environmental problem in our future.

People are getting sick and dying as a direct result of the contamination of air, water, land, and food by toxic substances. Cancer, birth defects, emphysema, and acute poisoning have all been linked to specific environmental contaminants. Some toxic materials reach us as a result of accidental leaks or spills, some as a consequence of war (dioxin in defoliants and fallout from weapons testing), and some as a result of deliberate illegal dumping. Still other toxic substances get to us because they were made, distributed, used, and scattered before we knew enough about their toxicities and ecological impacts (for example, PCBs, dioxin, certain pesticides). The United States has several thousand toxic waste sites that need to be cleaned up. The extent of toxic waste contamination worldwide is not known.

Our biological resources are in jeopardy. Tropical forests are shrinking by millions of hectares each year; several thousand species of plants and animals become extinct each year.

We use and abuse land in ways that irreversibly diminish this vital resource. Could there be a more fundamental problem of human ecology than rendering good land unproductive? Nothing can take the place of land. We should not tolerate any loss of good farmland. Worldwide, some 6 million hectares of new desert are formed annually, largely by mismanagement, and we lose billions of tons of topsoil each year.

We have altered and continue to alter global ecosystems in ways that destabilize them. Through habitat destruction, contamination of the environment, inadvertent manipulations (e.g., species introduction), and deliberate killing (e.g., pesticide overuse, predator poisoning, overfishing) we have changed our life support system, often without any thought of long-term consequences to that system. Thousands of lakes worldwide have fallen victim to acid rain. The mean temperature of the earth is projected to rise significantly by the middle of the twenty-first century as a consequence of the buildup of CO_2 and other gases. The protective shield of stratospheric ozone appears to be in jeopardy.

There are more people on earth than our ecological, social, economic, and agricultural systems can handle. Too many people? Too much per capita impact? Too little food? Inappropriate lifestyles? All of these are factors, but there is no doubt that we have a problem. Starvation, malnutrition, and related diseases constitute an enormous human tragedy.

Our surroundings are unnecessarily ugly, noisy, and cramped. This too is a very basic environmental problem. We do things in ways that unnecessarily diminish the quality of life. Perhaps if our surroundings were more pleasant, we would begin to see the environment more as something we are part of than as something to be dealt with, endured, or overcome.

The people of the world are largely ignorant of the laws of ecology and the workings of the system of nature. How can we expect the things people do to be ecologically sound when the average citizen is largely ignorant of ecological principles? All citizens must be educated about the system of nature.

National security and world peace are related to resource scarcity and vulnerability. The U.S. depends on large amounts of mineral and energy resources

---

For Further Thought . . . 20-1

Designing the Future

What do you think the world of 2030 will look like in terms of the environment? How will things operate—business, technology, government, education? What would you *like* the world to look like? What changes would be necessary for your preferred future to become a reality? What would be necessary to bring about these changes?

---

from other countries, many of which depend on the United States for foodstuffs and other products. Interdependence such as this among nations is both a stimulus for enhanced cooperation and a source of potentially explosive instability. Increasing global interdependence and awareness of this interdependence can work for us or against us; it cannot be ignored.

There is a growing gap between the rich and the poor at a time when it is being realized that there are not enough resources to go around. International environmental meetings point up the fact that the haves and the have-nots are separated by a wide breach. Developing countries want what the developed countries obtained at tremendous environmental expense and jeopardy. The wants and needs of developing countries cannot be ignored. A similar disparity exists within many countries in which some individuals have and others have not.

Certain kinds of environmental deterioration are, for all practical purposes, irreversible. Certain negative things that humans do to the environment can be cleaned up within reasonable periods of time, but many things that humans can do and have done to the environment can never really be restored, practically speaking. When farmland is given over to suburban development, that farmland is lost "forever." When especially rich ores are used and important minerals are dispersed rather than recycled, these materials are "lost," even though the absolute amounts remain unchanged. When areas of land are dedicated to hazardous or radioactive waste disposal, their use is restricted for generations to come. When major habitats, such as tropical rain forests, are disturbed, there is no way to restore them completely. When species are lost, they are gone forever.

Because we did not plan ahead, we have cramped, unattractive cities that cover what used to be productive farmland.

Problems of the environment are the result of the things people do and the numbers of people doing them. Most of our environmental problems are a result of technology more than it is a problem of overpopulation. The ratio of pollution generated to population size has increased sharply since the end of World War II. Although population and urbanization have increased in the past several decades, the rates of increase have been small in comparison to the rise in pollution levels.

Establishing the social cost of environmental deterioration is difficult. In Chapter 19 we considered this problem in some detail, the major point being that many of the social and environmental costs of various human activities are not easily brought into the equations used to determine the relationship between cost and benefit. It is easier to determine the value of ecosystem components like trees than it is to determine the value of the *function* of ecosystems. Economic assessment of such marketable products as fish, minerals, and forest products is straightforward and relatively easy. It is also relatively easy to assess the various standing elements of an ecosystem, for example, a forest used for recreation or other forms of enjoyment by people. But what ecosystems do for humankind in less obvious ways—for example, absorbing and breaking down pollutants, cycling nutrients, degrading organic wastes, maintaining the balance of gases in the atmosphere, and converting solar energy into chemical substances—are more difficult to evaluate in terms of dollars.

We often hear how much it will cost to clean our air, water, and land and to keep it clean. Not often enough do we hear the costs of *not* cleaning up these resources—costs to health, property, ecosystem stability, quality of life, and human survival. For too long air and water have been free commons for polluters. Polluters gained individually by increased production or consumption while everyone shared in the deterioration of the air, water, and land.

Inertia is part of the problem. Throughout this book we have talked about things like the power of population growth. If we think of human existence on earth as a process, what we focus on here is the fact that this process has momentum as well as an inertia—it is difficult to get new things started, and it is difficult to get old things stopped. More than a few prophets of doom have predicted that these features of humanity will be its downfall now that rapidly changing times require quick starts and quick stops.

An especially important corollary to this point is that not all of what now has momentum is very good or positive. A lot of it simply happened because of an infinite number of small decisions. Many of our problems are the results of individually insignificant decisions by large numbers of people. For example, even though a single automobile is not a pollution problem, if 60 million people each make a decision to buy and drive an automobile, society has a problem with traffic and pollution. The inertia established as millions of people adopt an automobile-centered lifestyle makes the associated pollution problem difficult to attack.

Homo sapiens is a natural agent. We are sometimes led to believe that the differences between our species and the other elements of the system of nature are more important than the similarities. In our opinion, this stands in the way of recognizing and dealing with our environmental problems for what they are. We do not cause environmental problems because we are different from other species; many, if not all, of our impacts are the result of the fact that we are similar. Many animal species do things to the environment that would ultimately render it inhospitable whenever and wherever certain restraints are temporarily removed. Habitat destruction by the species occupying a habitat is not unheard of in nature. As we try to address environmental problems, it will help to know that we may be up against the same forces of biological drive, aggressiveness, selfishness, and competitiveness that brought us out of the dark ages of the human past.

People cannot live on earth without having an impact. It should be realized that human beings cannot live on this planet without causing some form of environmental change. This stems from biological principles unique neither to the human species nor to modern women, men, and children. The fact that it would be a problem for Americans to learn to live within the beauty of nature without somehow destroying that beauty bothered some of the earliest settlers. It is useless to expect that we can live on this planet and leave the environment in a condition absolutely free of impact.

Nature may not always know best. In his book *The Closing Circle* (1971), Barry Commoner says that the third law of ecology is that nature knows best. He anticipated, however, that this principle was likely to encounter resistance.

First, the implied distinction between nature and humankind is inappropriate. Humankind is an integral part of nature. Nature without us cannot be assumed to be more all-knowing than nature with us. Beyond this, Commoner's analogy that the human impact on nature is equivalent to random poking into the works of a watch is worse than most analogies. The fact is that much human intervention is done with careful consideration of nature. The human intellect and human capacity to learn being what they are, human beings can improve on nature from a human point of view at times and in certain places in concert with nature's own laws.

In sharp contrast to the view that nature knows best, René Dubos (1976) presents a picture of nature as a great river of forces that can be directed or channeled in certain positive ways by human intervention. Dubos points out that nearly every part of the globe has humanized areas that have remained attractive, pleasing, productive, and ecologically sound for long periods of time. He cites as particularly notable examples China, Holland, Japan, Italy, and Sweden. These offer what René Dubos calls evidence for a symbiotic relationship between humankind and environment— evidence that we can have environmental impacts that, though profound, do not desecrate the environment. He goes on to argue that natural channels are not always best, particularly for the human species itself and not always even for other species. Certainly, learning to control population size is preferable to an otherwise inevitable population crash brought only by war, famine, or disease. Nature is really quite mindless.

We will probably always have difficulty dealing with the long term and the big picture. The history of humankind has been a long series of trial and tribulation, always seeming to have to do with the immediate problem of survival. Our species has never risen very far above crisis management. As long as there are large numbers of human beings having trouble ensuring their existence today, there will probably not be any concerted effort to deal with long-range and or even intermediate-range problems. Humans seem to have been bred to crisis management. There are many examples of our inability to deal with long-term effects; we have discussed some of these in other chapters.

Similarly, humankind has a problem dealing with the big picture. This is another harmful form of narrowness, although it may well be that nature's general pattern has been to favor the survival of the species by selecting for selfishness and narrowness in each individual member. In almost all aspects of our institutional approaches to science and education we seem to work from a reductionist viewpoint rather than a holistic one. We have far more lung specialists, kidney specialists, heart specialists, and cancer researchers than we have medical generalists or other types of generalists. Similarly, there are too few specialists in the relationship between humans and the environment who are paid to study, understand, and derive a better general understanding of the interaction between humans and the environment.

We need more facts. We need more and better data about the impacts of humans on the natural systems on which they depend and how some of these impacts actually affect us. Only with such data will we be able to continue to make it ever clearer that environmentalism is *not* an emotional movement based on exaggerated and unfounded claims. Research that documents specific cause-and-effect relationships and that can help to define thresholds for environmental impacts is crucial to the establishment of sound environmental policies.

Most environmental problems are complicated and have no perfect solutions. People who want to work toward an improved environment with a minimum of frustration should realize that it is sometimes best to work to diminish problems and to learn more about them rather than to solve them straightaway. Many problems of the environment are unsolvable in the absolute sense. They can be minimized, but they cannot be eliminated.

There will be no ultimate technical solution. In 1968, Garrett Hardin proposed that there is no technical solution to the population problem and acknowledged the difficulty in modern times of admitting the fact that there may not be a technical solution. There will likewise be no strictly technical solutions to *any* of our environmental problems. What we have now is a need for changes in values and attitudes. As long as we cling to the assumption that technology alone will provide the answer, we simply delay the inevitable point at which we will be forced into a new world view.

Whatever solutions are proposed for our environmental problems must take into consideration how the natural system operates. Unless our solutions operate in concert with the system of nature, they will inevitably lead to other problems and eventually to failure. Likewise, when new technologies are proposed, they must be weighed in terms of their compatibility with the operating laws of the universe.

Environmental improvement can be approached within the system. People who would work toward a better relationship between humankind and the environment should become familiar with how things work—the business system, the economic system, and the system of government—and proceed as any member of society would to bring about change.

People are not moved much by things that "experts" know. The public of the world has to be educated to the *need* to deal with environmental problems before anything significant can be done about them. It has never been effective for a group of informed individuals to say to the rest of the world, "Look, I know what needs to be done; let's get on with it." The ineffectiveness of such appeals is perhaps best illustrated by the small impact that warnings about smoking and health have had for so long on cigarette consumption in the United States. Imagine how much more difficult it will be to convince people to make changes in their behavior out of concern for the environment—when the connection to health is much less visible and less direct.

On the Nature of What Must Be Done

We must establish systems to review emergent and existing technologies. Our ability to alter the environment has grown much faster than our understanding of the environment. Many negative environmental impacts are the result of unanticipated side effects. New technologies that have emerged in recent decades have been particularly troublesome. We now need systematic controls on the emergence of new technologies, controls that insist that such technologies be ecologically sound. There is, after all, nothing inherently bad about technologies. The task is clearly one of taking stock of where we are and where we would like to go with our technologies and to consider new possibilities carefully—to assume control of our destiny, in other words.

We have to learn how to consider the relative importance of 100 people being killed in an explosion and 100 people being killed by a cancer initiated insidiously by exposure to a carcinogen. We have to develop equations to take into account the relative irreversibility of certain kinds of impacts like the accidental discharge of radioactive isotopes or the extinction of a species.

We must establish more environmentally cognizant approaches to technology transfer from developed to developing countries. We live in a diverse world—diverse in biomes, cultures, and development. All human beings obviously have similar needs for food, clothing, and shelter. There are also similar desires for a high standard of living that now constitute a conflict between what is fair and what is possible. Like it or not, the world cannot stand the spread of energy-intensive, material-intensive cultural practices to places where they do not yet exist.

We must develop technologies and economies that are more ecologically appropriate and sustainable. The current level of production and consumption in the United States cannot be maintained indefinitely. Our technologies are too environmentally costly (see Enrichment Box 20.2). It seems that we must now begin to move to smaller-scale and more environmentally appropriate technologies, as are advocated for developing nations. It is unfortunate that we have for a time put all of our eggs in the fossil fuel basket. And it would be unfortunate if we placed too heavy an emphasis on any one new energy source. We must begin to diversify again—to have technologies and ways of doing things that differ from place to place according to differences in local environments.

We need environmental multidisciplinarians. The task before us requires that some of us become trained

Sustainability and the Future of Humankind

Our ways of life and living must be ecologically sustainable. The concept that the word *sustainable* encapsulates is that the things we humans do must be compatible with the ecosphere such that our activities do not degrade the ecosphere. Nor must they be allowed to set us on a course that will ultimately degrade the ecosphere. The degree to which human activity can be sustained by the ecosphere, by definition, determines the duration of human civilization. When we say that agriculture must be sustainable, we mean that agricultural technology must not diminish the ability of the land to produce. Likewise our use of water and energy resources must be such that there will be sustained access to these resources. Sustainability is the basis for the emphasis on renewable energy resources.

The word *sustainable* is used most often today in connection with the word *development* in the phrase *sustainable development*. In a very fundamental way, the phrase *sustainable development* is an oxymoron: It sounds wise, but the words cannot really be juxtaposed because they are mutually exclusive. Most people use the word *development* to mean growth, and growth cannot be sustained in a finite world. *Sustainable* implies balances and limits; *development* implies the expectation of more. The phrase *sustainable development* makes sense only if *sustainable* is understood to apply to a new and different kind of development, one not based on growth. The phrase makes sense only if it serves to identify the shortcomings of the current idea of development—which would have all cultures of the world adopt a Western-style culture. Some commentators have gone so far as to suggest that the West must make it clear to other cultures that it has ''overshot its efforts to satisfy people's needs and now must take drastic steps to de-develop'' (Ehrlich, 1990).

Clearly, the world needs to put a sustainable spin and even sustainable limits on development efforts throughout the world. The prime minister of Norway, Gro Harlem Brundtland, put it this way:

> The time has come to break out of the negative development patterns of the past. . . . The overriding political concept must be ''sustainable development''—paths of human progress that meet the needs and aspirations of the present generation without compromising the ability of future generations to meet *their* needs and aspirations. . . . Governments, aid agencies, and others concerned with development must integrate environmental considerations into economic decision making and planning at all levels. (Gandhi and Brundtland, 1988)

Simply put, the things that we do must be compatible with the system of nature; otherwise, the system of nature will be unable to support us. We must come to see ourselves as living within the system of nature.

---

to deal with multidisciplinary problems. The optimal format for such training would probably consist of a solid grounding in one major discipline, but with training in a number of others as well. As simple as this might sound, it does go against the grain of the education and occupational trends of the "age of specialization."

Population growth for the world cannot continue. We are not sure where the population problem fits into the order of priorities of our environmental problems. We disagree with people who say that the population problem is *the* environmental problem. In industrialized nations like the United States, the problem is more one of how we do things than the number of people. But regional population problems will be with us for some time; population control programs must be encouraged.

We must protect critical natural areas now. We have much to learn about the many ecological communities of our world. But we are losing tropical rain forests, old-growth forests, coral reefs, and other habitats at alarming rates. This process must be slowed. We need time to know how large the tracts of habitat we preserve must be in order to maintain the ecological system. We need time to inventory and identify species. We need time to understand the value of biodiversity to the ecosystem.

We need environmentally sound energy policies. As we have seen throughout this text, without energy, nothing happens. Our world, especially in developed nations, is generally extremely vulnerable because of our dependence on fossil fuels. This dependence causes us to disturb and pollute pristine areas to find more fuels; it leads to problems of acid rain, global warming, and air pollution; it provides opportunities for ocean and shoreline pollution from spills; it even leads us to go to war. A concerted effort is needed to address the issue of sustainable energy supplies.

We must develop strategies to encourage source reduction of pollution. Many of our pollution problems are the result of systems that exploit resources and add nothing to our standard of living. Much of our present packaging falls into this category. An example is jars packaged in boxes that are promptly tossed away, adding to the solid waste problem. Disposable containers, razors, and cameras are other examples.

Likewise, many of our industrial processes that generate pollutants and hazardous wastes can be altered to produce less waste and less toxic by-products without altering the final product to any great extent.

The biggest obstacle to recycling is habit.

We must find ways to encourage these kinds of changes at the source.

Reuse and recycling must become a way of life. The problem with material resources is one of keeping materials in forms in which they are usable. There has been some movement toward recycling in industry, but what we have accomplished so far really amounts to just a tentative first step. If we can make recycling an important part of all of the things we do in society, we will have solved several problems:

1. We will not have to destroy large expanses of territory searching for less rich ores as fast as we would otherwise.
2. The reuse of materials would diminish the solid waste problem.
3. Reuse would obviate or lessen the problems of increasing shortages of various raw materials.

We must continue to internalize environmental costs in the cost of doing business. We must make it more expensive to pollute. If environmental costs are included in the price of the product, consumers are more likely to buy the less expensive, more environmentally compatible products. Ways must be found to internalize environmental costs along with other production costs.

Environmentalists must organize and collaborate. The institutionalization of environmental concerns since the 1960s has given environmentalism credibility and permanence. Politics in a democracy is based on compromise. Government responds to pressure—economic, social, and public—such that the final compromise is determined by the pressures.

The final compromise is also a function of where the points of the two extremes lie.

Environmental activist groups have long served as rallying points for public opinion, and environmental special interests have served the public well in competition with other special interests.

Environmentalists are a varied and a diverse lot. Some are interested in safe working conditions in factories, some in saving a single threatened or endangered species, some in preserving the wilderness, some in cleaning up air or water, and some in protecting us from radioactive materials, to name just a few. It is important that increasing solidarity be developed within this diversity of environmental concerns if we are to move ahead and win the important battles.

Where conflicts exist between individual freedom and the quality of the human environment, individual freedom must give way. In many places throughout this book, we have described our traditional reverence for individual freedom and how this sometimes obstructs intelligent stewardship of the environment. We have implied that (despite its not being a popular notion) government might actually have to assume a larger role in human affairs if the environment is to be looked after properly. If there are critics who would brand this attitude as an environmental form of socialism, so be it. We recognize that societies are made up of individuals and that, in general, as an individual goes, so goes the society. This is true only up to a point, however, and we believe that many of the environmental problems we face are direct consequences of our having carried this ideal past the point of optimization.

Ecological paradigms must become the basis for all ethical systems. In Chapter 19 we advocated the establishment of ecological principles as the core of general education. In other words, all training and education should emanate from the concepts having to do with human-environment relationships. Clearly, our most impressive hope for the future is that we will find new ways to think about our environment and our relationship with it. A new philosophical attitude must emerge. Van R. Potter (1971, 1977) has suggested that since ethical values cannot be separated from biological facts and since the survival of ecosystems really is the ultimate test for any system of values, bioethics or environmental ethics constitute an appropriate base for the ethical systems by which humans express humanity.

Our religions must place more emphasis on environmental stewardship. The basis of our system of laws must change from its emphasis on property ownership and possession to an emphasis on property stewardship and responsibility. Under such a system, destruction of "one's own" land so that future generations are deprived of it should be considered an act of theft.

Global Environmentalism

Protecting the environment must become global in its orientation and scope. The World Commission on Environment and Development (WCED) in 1987 published a report titled *Our Common Future* that stated that "a new development path is required, one that sustains human progress not just in a few places for a few years, but for the entire planet into the distant future." Issues such as global warming, acid rain, and the depletion of the ozone layer present all the peoples of the world with a common threat. It has been said that the nations of the world will probably not get along until confronted by some common enemy. We remember hearing this years ago in the context of the possibility of earth being visited by beings from another planet. Perhaps we don't have to wait for that visit. Perhaps *ecological* threats to the planet will be the glue that ultimately binds us together.

The need to address environmental problems on a worldwide scale was expressed by Rajiv Gandhi, then prime minister of India, this way: "There is no political boundary that delimits the spread of poisonous gases, no line on a map that radiation cannot cross, no national frontier at which effluents can be turned back" (Gandhi and Brundtland, 1988). Gro Harlem Brundtland, prime minister of Norway and chair of the World Commission on Environment and Development, expressed the same sentiment this way:

> We must break away from our sectoral ways of viewing economy and ecology. We must learn to accept the fact that environmental considerations and economic growth are parts of a unified management of countries and our planet. The one is dependent on the other. (Gandhi and Brundtland, 1988)

The United Nations Environment Programme and other such worldwide initiatives are beginning to give shape to a global approach to environmental problem solving. We have pointed out throughout this book that environmental problems are beginning to be addressed more vigorously throughout the world. This is being driven by a notable global rise in environmentalism—a trend that should be helped to coalesce into a single global trend.

In recent years, tree-planting brigades have been organized in Kenya and in China. In Italy, environmental concern is said to be the source of the sharpest ideological shifts and political realignments since World War II. A summit meeting held in Paris a few years ago came to be called the "green summit" because the environment took up such a large portion of the final communiqué. This provided some hope that nations will soon realize that long-range security depends more on environmental protection than it does on military strength.

The World Bank is now calling the environment its top priority. Media coverage of the environment has expanded vigorously in the United States in the past few years. According to a recent report, environmental stories were broadcast on U.S. television networks an average of one every third night in 1987, one every other night in 1988, and two each night in 1989. All the major U.S. newsmagazines have devoted whole issues to environmental affairs. Polls taken throughout the world have consistently shown that environmental values are drawing increasing support. Memberships in environmental organizations have been growing rapidly worldwide—membership in the World Wildlife Fund for instance, more than tripled between 1986 and 1989.

Bhutan

Bhutan has been compared to the mythical Shangri-La. Bhutan is a tiny mountainous country nestled between China and India in the eastern Himalayas. In May 1990, Bhutan's government officials adopted a national policy of ecologically sustainable development, steering clear of the disastrous course of development in India and Nepal that has left forests stripped and soils eroded (see Flavin, 1990). Bhutan's eco-compatible course will be built on an already well-established conservation ethic. Bhutan is a poor nation but one in which most of the basic human needs are fairly well met; the natural environment is still largely unspoiled.

The Republics of the Former Soviet Union

There is a rapidly growing interest in the protection of nature in the republics of the former USSR. Some of the most serious environmental problems in that part of the world are described in earlier chapters. The rapid industrialization of the Soviet Union following the Second World War came at the expense of air quality and many of the USSR's lakes, forests, and rivers. Environmental organizations are springing up in all of the republics; people are protesting nuclear power, water pollution, and other environmental problems. Nuclear power plants and chemical plants have been closed, and plans for others have been shelved in response to public opposition.

Eastern Europe

Eastern Europe's water, air, and soil have all been badly damaged by industrialization. Along with the appearance of *glasnost* ("openness") in the Soviet Union in the 1980s, overt concern for the environment began to gain momentum throughout the Eastern Bloc countries. The nuclear accident at Chernobyl catalyzed the movement to the extent that there were already some 2,000 environmental organizations in Poland alone by 1988 (French, 1988). As a result of all this, many eastern European countries established environmental protection agencies, and their governments increased funding for environmental protection. Eighty-three percent of the people of Czechoslovakia believe that solving environmental problems is the most essential task of the government (Miller, 1990). Energy conservation is the centerpiece of a strategy to reduce air pollution throughout this energy-inefficient region of the world.

Brazil

In a nation that appears to the rest of the world to be hell-bent on destroying every bit of its tropical rain forest, there is nevertheless considerable interest to protect the rain forest. Environmentalism has become a popular cause throughout Brazil; there were some 1,000 environmental organizations in the country by 1990 (Worcman, 1990). Although there has been little environmental progress in Brazil thus far, the second United Nations Environmental Conference was held there at the invitation of the government, and this is seen by some observers as cause for optimism.

We must all be concerned about the ecosphere because there is just one ecosphere, and we are in it together.

Some environmental progress has been made by people who keep the long term in mind. These Douglas fir trees were planted in the 1930s.

We need to develop a sound environmental ethic that runs through everything we do as a people of the world. We need a more ecologically compatible way of looking at the world.

We must proceed with environmental protection measures and cleaning up the environment even though there will be temporary negative impacts. We have touched on this idea in several places throughout the book. There are workers who will be displaced from jobs when plants are shut down because the operations are too dirty. Passing certain environmental costs along to consumers will undoubtedly have dis-

proportionate impacts on the poor. It is clear that no method of pollution control and cleanup will be easy, and it is also certain that the cost will not be borne equally by everyone. These problems cannot be cited as reasons why nothing should be done; rather they must be addressed via institutional mechanisms designed to assist those in need (see the section on the role of government in Chapter 19; see also Enrichment Box 20.3).

On the Future

It sometimes strikes us as every bit as ludicrous to discuss what life will be like in the future as it would be to talk about a special dish that no one has even seen or tasted. Yet we cannot resist the temptation—and it seems logical to do so—to end this text with a look into what may lie ahead. First, we would like to drive home the point that the future cannot really be known. Second, we would like to make the related point that even if the future could be known, it could not be appreciated. Third, we would like to make a few points about the general significance and usefulness and dangers of predictions. Finally, we will risk a few predictions.

The future cannot be known. The future cannot really be known because there is not, for example, any way to know what raw materials will be used in the future. We are all well aware that humankind has historically made use of resources that had in years earlier been unrecognized as resources. We see in print on occasion the concept of "proven reserves" being used as a basis for making future projections—even though proven reserves are, by definition, the amount of a resource that can be profitably recovered and used with *present* technology. The error here is that the future is being measured in terms of the present. Indeed, even though this is really the only way we can ever look into the future and even though we may qualify what we are doing when we do it, we should all be aware of the inherent shortcomings of this approach.

The future could not be appreciated even if it could be known. Buckminster Fuller said that within a short time we might see the disappearance of private property, the end of the use of hydrocarbons as fuel, and the end of sovereign nations and certain religions. He admitted that these are incomprehensible changes to those of us alive today. But the future has always been incomprehensible. We should expect that some of the most sacred tenets of our ways of life and our attitudes toward free enterprise, religion, government, and other relationships between human beings will all change in ways that would be judged today as terrible or even unspeakable. We are all prisoners of our own time, of the current climate of opinion. The things we read today, the people we talk to today, and the people among whom we live and from whom we derive our perspective and our points of view would keep us from understanding and appreciating the future even if we were somehow miraculously given a crystal-clear picture of what will come to pass. It seems to be a common error among people who make projections and predictions about the future that they inevitably introduce their own personal preferences, tacitly assuming that these preferences are those of all members of society and will continue to be held dear by the people of tomorrow. This is an inexcusable error and one that was illustrated very well in a classroom discussion not long ago. During a session in which a scenario was being developed of the future and what it might be like to live 100 years hence, one student said that he would rather die than live in the world being described. Another student pointed out that this wish would probably be granted. Still another student remarked that if Daniel Boone could somehow be brought back to life to catch a glimpse of the world of today, he would probably opt to remain dead.

On the Significance and Usefulness of Predictions

The foregoing notwithstanding, there might be some utility in trying to project the future. It could be argued that our present-day problems came about because our forebears did *not* consider the future—or that most present-day progress has been the result of people trying to see the future and to understand and anticipate it.

Now for some low-risk predictions.

There will be better ways. The imaginations of the people of the immediate future will be tested—but not very much—in trying to come up with improvements on our ways of doing things. A lot of things we have done in the past and continue to do are ugly and environmentally ignorant. Telephone poles offer one pedestrian example of the former; automobiles that get 8 miles per gallon exemplify the latter. The people of the future will come up with new views of the world and humankind's place within the scheme.

Health implications of environmental deterioration will grow in importance. Two aspects of our daily existence will increasingly force the individual to stop short and take notice of problems of the environment. One aspect relates to our wallet, the second to our health. To date, much time and energy have been devoted to bemoaning the costs relating to environmental protection. But as we have pointed out, more needs to be done about the costs associated with *not* cleaning up and protecting the environment. Gradually, research has brought the health ramifications of pollution into sharper focus in the public eye. As these health implications are brought more and more to the forefront, they will provide a favorable impetus to increased environmental protection.

There is room for optimism, but there is no room for blind faith. Despite the many contraindications scattered throughout this volume, we must say in summary that our position is one of optimism concerning the future. We have emphasized the frailty of the system of nature. But we have also stressed the fact that the system of nature is, above all else, highly resilient. Even when we speak of imminent disasters, we have the local scale in mind. *Homo sapiens* is not in any immediate danger of disappearing from the face of the earth. Secondarily, even though *Homo sapiens* may have gotten into a lot of environmental trouble following the dictates of biological nature, we foresee the emergence of the more rational side of this species, bringing intellect to bear on the solutions to some of these very same problems. We expect that, given the pattern of human history, the phases we have gone through with respect to our environment and that we will go through in the future reflect a common general pattern of (1) recognition, (2) overreaction, (3) backlash, and (4) serious accomplishment. In the late 1960s we went through a phase in which our awareness of the environment increased significantly. This was followed by a period of mixed productive reaction and overreaction—some progress and some reversals. This is now being succeeded by a period of quiet determination and steady progress. Concern for the environment has begun to become part of an innate way of thinking in many parts of the world, it has begun to form a baseline of care for the systems that support us, and it has begun to move our institutions in an appropriate direction in a subtle but significant way.

Our purpose in writing this book was to describe from a natural sciences perspective the environmental problems facing the human species. Our strategy was to expose and illuminate the underlying principles that have to be considered in order to recognize environmental problems for exactly what they are. You who stayed with us through these pages are now among the people who will be counted on to continue to make a difference.

Christanne J. Gallagher

Chris Gallagher is a park ranger and education coordinator with the U.S. Army Corps of Engineers. The Corps of Engineers is a branch of the U.S. Army that serves as its construction agency. The Corps is a diverse agency responsible for construction, operation and maintenance of both military and civil works. Some of its mission includes navigation, flood control, recreation, regulatory and emergency assistance. The Corps' recreation responsibility includes the building and maintenance of recreational and resource management areas.

Ms. Gallagher is stationed at the San Francisco Bay and Delta Model in Sausalito, California. A large-scale model of the San Francisco Bay and Delta was built in 1956–1958 to study the placement of solid barriers within the bay and the delta, dispersion of municipal and industrial waste, the effects of reclamation on tidal hydraulics, and the shoaling of dredged sediments. In 1980 the Corps dedicated a regional visitor center at the Bay Model, which is staffed by Ms. Gallagher and other Corps park rangers, aided by volunteers. Ms. Gallagher's typical day might include developing or conducting workshops for teachers, developing learning activities for students, giving a guided tour, meetings with environmental groups or with the center's advisory board, planning activities to mark such events as Earth Day, or doing work related to her membership in national organizations such as the North American Association for Environmental Education. Safety and security are another part of Ms. Gallagher's responsibility. She works with local safety agencies, conducts staff training, and prepares and presents programs for the public on such subjects as safe boating. She is also the computer hardware and software contact for the center; she prepares reports on visitation and

maintains the center's mailing list, among other things. What Ms. Gallagher likes most about her job is that no two days are alike; she loves the opportunity her job gives her to work with many interesting people. She also likes the fact that she feels that she is making a difference.

Ms. Gallagher has an associate degree in forestry and a bachelor's degree in environmental education. She worked as a volunteer for the National Park Service for four years and spent one year as Student Conservation Association intern before assuming her present position.

References and Further Reading

References marked with an asterisk are cited in the chapter.

Charles, D., 1990. "East German Environment Comes into the Light," *Science* **247**:274–275.

*Commoner, B., 1971. *The Closing Circle.* New York: Alfred A. Knopf.

*Conservation Foundation, 1984. *State of the Environment: An Assessment at Mid-decade.* Washington, D.C.: Conservation Foundation.

*Dubos, R., 1976. "Symbiosis between Earth and Humankind," *Science* **193**:459–462.

*Ehrlich, P., 1990. "Changing Our Minds," *Earth Ethics* **1**(2):6–7.

*Flavin, C., 1990. "Last Road to Shangri-La," *Worldwatch*, July-August, pp. 18–26.

*French, H. F., 1988. "Industrial Wasteland." *Worldwatch*, November-December, pp. 21–30.

*French, H. F., 1989. "The Greening of the Soviet Union," *Worldwatch*, May-June, pp. 21–29.

*Gandhi, R., and Brundtland, G. H., 1988. "Two Declarations of Interdependence," *Issues in Science and Technology* **V**(2):17–20.

*Hardin, G., 1968. "The Tragedy of the Commons," *Science* **162**:1243–1248.

*Miller, M. S., 1990. "A Green Wind Hits the East," *Technology Review*, October, pp. 53–63.

*Potter, V. R., 1971. *Bioethics.* Englewood Cliffs, N.J.: Prentice Hall.

*Potter, V. R., 1978. "Evolving Ethical Concepts," *BioScience* **27**:251–253.

Renner, M., 1991. *Jobs in a Sustainable Economy.* Washington, D.C.: Worldwatch Institute.

Rothkrug, P., 1991. *Mending the Earth: A World for Our Grandchildren.* Berkeley, CA: North Atlantic Books.

Rüdig, W., 1991. "Green Party Politics," *Environment* **33**(8):6–9ff.

*Worcman, N. B., 1990. "Brazil's Thriving Environmental Movement," *Technology Review*, October, pp. 43–51.

*World Commission on Environment and Development, 1987. *Our Common Future.* New York: Oxford University Press.

Appendix

Think Metric

As a nation, the U.S. uses both metric and English units of measurement. Depending on what you read, you may encounter either system. Scientists use the metric system, but many federal agencies report data in English units. We have used both in this book in the interests of clarity and familiarity. In most places we have provided conversion equivalents.

The conversion charts included here will be useful in putting comparable volumes, weights, distances, areas, or temperatures into perspective.

Temperature Conversion

Celsius, °C	Fahrenheit, °F
160	320
100	212 (water boils)
60	140°
37	98.6 (body temp)
32	90
20	68
10	50
0	32 (freezing)

(Celsius temperature × 1.8) + 32 = Fahrenheit temperature

(Fahrenheit temperature − 32) × 0.56 = Celsius temperature

From Metric (approximate to within 2%)			To Metric		
To Convert			To Convert		
From	To	Multiply by	From	To (Symbol)	Multiply by
millimeters	inches	0.04	inches	millimeters (mm)	25
centimeters	inches	0.4	inches	centimeters (cm)	2.5
meters	feet	3.3	feet	meters (m)	0.3
meters	yards	1.1	yards	meters (m)	0.9
kilometers	miles	0.62	miles	kilometers (km)	1.6
square centimeters	square inches	0.155	square inches	square centimeters (cm^2)	6.5
square meters	square yards	1.2	square yards	square meters (m^2)	0.84
hectares	acres	2.5	acres	hectares (ha)	0.4
square kilometers	square miles	0.39	square miles	square kilometers (km^2)	2.6
cubic centimeters	cubic inches	0.06	cubic inches	cubic centimeters (cm^3)	16.4
cubic meters	cubic feet	35	cubic feet	cubic meters (m^3)	0.028
cubic meters	cubic yards	1.3	cubic yards	cubic meters (m^3)	0.76
liters	pints	2.1	pints	liters (L)	0.47
liters	quarts	1.06	quarts	liters (L)	0.95
liters	gallons	0.26	gallons	liters (L)	3.8
grams	ounces	0.035	ounces	grams (g)	28
kilograms	pounds	2.2	pounds	kilograms (kg)	0.45
metric tons	short tons (2000 lb)	1.1	short tons (2000 lb)	metric tons (t)	0.91
metric tons	long tons (2240 lb)	1.0	long tons (2240 lb)	metric tons (t)	1.0

Credits

For figure and table sources, see reference list at the end of the relevant chapter.

Part One

Main photo: Philip Gendreau, Bettman Archives. Insert photo: Tom Stack/Tom Stack and Associates.

Chapter 1

Photographs
P. 4: National Oceanic and Atmospheric Administration. P. 7: U.S. Forest Service. P. 11: Grant Heilman Photography. P. 16: U.S. Department of Agriculture, Soil Conservation Service.

Chapter 2

Photographs
P. 23: U.S. Department of Agriculture, Office of Governmental and Public Affairs. P. 32: Grant Heilman Photography, Runk/Schoenberger. P. 32: Grant Heilman Photography, Hal Harrison. P. 39: U.S. Department of Agriculture, Office of Governmental and Public Affairs.

Figures
P. 25; Fig. 2.4: Data from Kormondy (1984), Odum (1983). P. 30: graph in middle from *Life: The Science of Biology* (1983) by W. K. Purves and G. H. Orians; on bottom: From Purves and Orians, *Life: The Science of Biology* (1983). P. 31; Fig. 2.8: Data from Odum (1983), Kormondy (1984), Woodwell (1970).

Tables
P. 27; Table 2.1: Adapted excerpts from Eugene P. Odum, Copyright © 1983 by Saunders College Publishing, a division of Holt, Rinehart and Winston, Inc., reprinted by permission.

Chapter 3

Photographs
P. 44: Courtesy of National Parks Service. P. 50: Grant Heilman Photography. P. 52: Grant Heilman Photography.

Figures
P. 57; Fig. 3.10: From *Limnology* by William H. Amos, 1969, published by LaMotte Chemical Co., Box 329, Chestertown, MD 21620. P. 57; Fig. 3.11: Mertz (1981). P. 58; Fig. 3.12: Brush (1982), p. 23.

Tables
P. 42; Table 3.1: Data from Kormondy (1984).

Chapter 4

Photographs
P. 63: U.S. Department of the Interior, Fish and Wildlife Service. P. 71: Moulin Photo. P. 75: Andrews Stepniewski. P. 77: UPI/Bettmann.

Figures
P. 74; Fig. 4.11: Data from Gause (1934).

Tables
P. 68; Table 4.1: Adapted from Deevey (1947), used with permission of the *Quarterly Review of Biology*, Vol. 22, p. 296, 1947.

Chapter 5

Photographs
P. 100: U.S. Department of the Interior. P. 104: Kenneth W. Fink, Bruce Coleman, Inc., New York. P. 105: Alaska Division of Tourism.

Figures
P. 88; Fig. 5.1: From E. Peter Volpe, *Understanding Evolution*, 5th ed. Copyright © 1967, 1970, 1977, 1981, 1985 Wm. C. Brown Publishers, Dubuque, Iowa. All Rights Reserved. Reprinted by permission. P. 97; Fig. 5.6: From Purves and Orians, *Life: The Science of Biology* (1983). P. 99; Fig. 5.7: Swanson (1973). P. 102: © The Washington Post.

Tables
Table 5.1: R. Dasmann, *Environmental Science*, © John Wiley and Sons, Inc.

Part Two

P. 108: Main photo: U.S. Department of the Interior, National Park Service. Insert photo: P.G. and E. News Bureau.

Chapter 6

Photographs
P. 127: U.S. Department of Energy. P. 132: Alyeska Pipeline Service Company. P. 134: U.S. Department of Energy. P. 136: U.S. Department of Energy. P. 142: U.S. Department of Energy. P. 145: Gary Millburn/Tom Stack and Associates. P. 152: Bureau of Reclamation. P. 155: Courtesy of Camp Dresser and McKee, Inc., Boston, MA. P. 158: U.S. Department of Energy; P. 159: U.S. Department of Energy, Lockheed Missiles and Space Company, Inc. P. 176: Courtesy of Solaron Corporation, Englewood, CO.

Cartoon
P. 149: Don Wright, *The Miami News.*

Figures
P. 111; Fig. 6.1: Adapted from M. K. Hubbert, *Energy Resources,* National Research Council Publication 1000-D, 1962, p. 91, National Academy of Sciences, Washington, D.C. P. 113; Fig. 6.2: Adapted from E. Cook, "Energy for Millenium Three," *Technology Review,* Dec. 1971, pp. 16–23. P. 113; Fig. 6.3: U.S. Department of Energy (1983b, 1987a); Conservation Foundation (1982). P. 115; Fig. 6.6: U.S. Bureau of Census, U.S. Bureau of Mines, Energy Information Administration, *Annual Energy Review 1986.* P. 116; Fig. 6.7: U.S. Department of Energy, Energy Information Administration, *Annual Energy Review 1982, Annual Energy Review 1986.* P. 117; Fig. 6.8: U.S. Department of Energy, Energy Information Administration, *Annual Energy Review 1986.* P. 118; Fig. 6.9: Dorf (1978), U.S. Department of Energy, Energy Information Administration, *Annual Energy Review 1983, Annual Review 1986.* P. 118; Fig. 6.10: U.S. Department of Energy, *Annual Energy Review 1986.* P. 120; Fig. 6.12: From "The Vulnerability of Oil-Based Farming" by L. Brown *Worldwatch* March/April 1988, p. 25. Used with permission. P. 123; Fig. 6.13: Federal Energy Administration, Office of Conservation, *Energy Conservation,* 1975. P. 124; Fig. 6.14: W. Dupree, E. Herman, S. Miller, and D. Hillier, *Energy Prospectives 2,* Washington, D.C.: U.S. Department of the Interior, 1976; U.S. Department of Energy, Energy Information Administration, *Annual Energy Review 1986.* P. 124; Fig. 6.15: U.S. Department of Energy, Energy Information Administration, *Annual Energy Review 1986.* P. 125; Fig. 6.16: U.S. Department of Energy, Energy Information Administration, *Coal Production Annual,* 1982, 1986. P. 127; Fig. 6.17: Adapted from Tennessee Valley Authority. Copyright © 1983. P. 128; Fig. 6.18: Office of Fossil Energy, Quarterly Reports, Coal Gasification, 1976 and 1977, Energy Research and Development Administration. P. 129; Fig. 6.19: From "Coal-Fired Power Plants for the Future," copyright © 1987 by Scientific American, Inc., all rights reserved. P. 130; Fig. 6.20: U.S. Department of Energy, Energy Information Administration, *Annual Energy Review 1983.* P. 131; Fig. 6.21: Crawford (1987). Copyright 1987 by the AAAS. P. 131; Fig. 6.22: Data from Dolton et al. (1982). P. 132; Fig. 6.23: Standard Oil, *Scene,* Spring 1987. P. 133; Fig. 6.24: Fossil Energy Research Program of the Energy Research and Development Administration, 1977; illustration adapted from *Exxon USA.* Courtesy Exxon Corporation, © 1982. P. 135; Fig. 6.25: U.S. Department of Energy, Energy Information Adminstration, *International Energy Annual 1986.* P. 150; Fig. 6.34: U.S. Department of Energy, Energy Information Administration, *Electric Power Annual 1985.* P. 150–51; Fig. 6.31: U.S. Department of Energy, Energy Information Administration, *Commercial Nu-* *clear Power.* P. 155; Fig. 6.34: Bjork and Granelli (1978). P. 156; Fig. 6.35: Reed and Lerner (1973). Copyright 1973 by AAAS. Courtesy of authors. P. 167; Fig. 6.40: U.S. Department of Energy, Energy Information Administration, *Annual Energy Review 1986.* P. 173; Fig. 6.42: U.S. Department of Energy, Energy Information Administration, *Annual Energy Outlook 1990.*

Tables
P. 112; Table 6.1: Adapted from "The American Energy Consumer," Copyright the Ford Foundation, 1975. P. 117; Table 6.4: 1920–1940, Dorf (1978); 1950–1986, U.S. Department of Energy, Energy Information Administration, *Annual Energy Review 1982, Annual Energy Review 1986;* 1995 projection, Energy Information Administration (1987d). P. 144; Table 6.10: National Academy of Sciences.

Quote
P. 149: Quote from Dr. Alfvén reprinted by permission of the *Bulletin of the Atomic Scientists,* a magazine of science and world affairs. Copyright © 1972 by the Educational Foundation for Nuclear Science, 1501 S. Kenwood Ave., Chicago, IL 60637.

Chapter 7

Photographs
P. 187: Fredrik D. Bodin. P. 201: Courtesy of Los Angeles Department of Water and Power.

Figures
P. 188; Fig. 7.4: National Commission on Materials Policy (1973), p. 4D-5. P. 190; Fig. 7.5: U.S. Bureau of Mines (1987). P. 192; Fig. 7.7: U.S. Geological Survey (1977). P. 193; Fig. 7.8: U.S. Water Resources Council (1980), p. 449. P. 193; Fig. 7.9: Adapted from U.S. Geological Survey (1984). P. 195; Fig. 7.10: Adapted from "Threats to the World's Water" by J. W. Maurits LaRiviere Sept. 1889. Copyright © 1989 Scientific American Inc. All rights reserved. Used with permission. P. 196; Fig. 7.12: Data from James et al. (1976).

Chapter 8

Photographs
P. 210: Courtesy of the United Nations. P. 219: Courtesy of the New York Convention and Visitors Bureau. P. 225: Richard J. Quataeri/Taurus Photos, Inc. P. 226: Eric Kroll/Joel Gordon Photography. P. 233: Courtesy of the Canadian Government Travel Bureau. P. 235: UNICEF/Massa Diabate.

Figures
P. 213; Fig. 8.2: Council on Environmental Quality (1980b), Vol. II, p. 17. T. Frejka and W. P. Mauldin in Mauldin (1980), p. 156. Copyright 1980 by the AAAS. P. 215; Fig. 8.5: U.S. Department of Commerce, Bureau of the Census (1987a). P. 217; Fig. 8.6: Coale (1983), p. 829. P. 218; Fig. 8.7: U.S. Department of Commerce, Bureau of the Census. P. 219; Fig. 8.8: U.S. Department of Commerce, Bureau of the Census (1987a). P. 226; Fig. 8.9: From "The Growing Grain Gap" by Lester R. Brown *Worldwatch* Washington, D.C. Sept/Oct 1988, p. 12. Used with permission. P. 229; Fig. 8.10: U.S. Department of Agriculture. P. 231; Fig. 8.11: Council on Environmental Quality (1980b), Vol. II, p. 87.

P. 233; Fig. 8.12: From "The Ocean Blues" by Nicholas Lenssen *Worldwatch* Washington, D. C. July/August 1989, p. 31. Used with permission.

Tables
P. 216; Table 8.1: *State of the World 1987*, Worldwatch Institute, © 1987; Population Reference Bureau, *1988 World Population Data Sheet*. P. 217; Table 8.2: *State of the World 1987*, Worldwatch Institute, © 1987; World Bank, *World Bank Development Report 1985*; Population Reference Bureau, *1988 World Population Data Sheet*. P. 222; Table 8.3: Based on A. L. Schirm et al., "Contraceptive Failure in the United States: The Impact of Social, Economic, and Demographic Factors," *Family Planning Perspectives*, Vol. 14, No. 2, p. 68, as found in *Basics of Birth Control*, Planned Parenthood Federation of America, Inc., New York, 1984. P. 224; Table 8.4: Berelson (1969).

Part Three

P. 242: Main photo: U.S. Department of the Interior, National Park Service. Insert photo: Courtesy of Camp Dresser and McKee, Inc., Boston, MA.

Chapter 9

Photographs
P. 245: Grant Heilman Photography. P. 246 (top and bottom): Clerk of the Works, Dean and Chapter of London. P. 252: Environmental Protection Agency. P. 303: Courtesy of Camp Dresser and McKee, Inc., Boston, MA.

Figures
P. 252; Fig. 9.2: U.S. Environmental Protection Agency, *Air Quality Criteria for Nitric Oxide*, April 1977, p. 4.4-2. P. 256; Fig. 9.4: U.S. Department of Health, Education and Welfare, National Air Pollution Control Administration, *Air Quality Criteria for Photochemical Oxidants*, March 1970, and U.S. Environmental Protection Agency, *Air Quality Criteria for Ozone and Other Photochemical Oxidants*, EPA 600/8-87-004, 1978. P. 259; Fig. 9.5: Peterson and Salvia, *Environment*, Vol. 19, pp. 66–79, 1968; a publication of the Helen Dwight Reid Educational Foundation. P. 267; Fig. 9.7: U.S. Environmental Protection Agency.

Tables
P. 250: Adapted from a compilation by Dr. Warren B. Crummett, Dow Chemical Co. P. 263; Table 9.2: From various sources as reported by Schlesinger (1979). P. 265; Table 9.4: Reproduced by permission from G. L. Waldbott, *Health Effects of Environmental Pollution*, 2nd ed., St. Louis, 1978, The C.V. Mosby Co.

Chapter 10

Photographs
P. 288: Grant Heilman Photography, Alan Pitcairn. P. 297: The Columbia Daily Spectator/Art Resource, NY.

Figures
P. 279; Fig. 10.6: Reproduced by permission from G. L. Waldbott, *Health Effects of Environmental Pollution*, 2nd ed., St. Louis, 1978, The C.V. Mosby Co. P. 289; Fig. 10.7(b), (c), (d), (e): Patrick et al., *Science*, Vol. 211, pp. 446–448, 1981. P. 290; Fig. 10.8: Adapted from Ontario Ministry of the Environment and Commission on Air Quality, 1981. P. 292; Fig.

10.10: Adapted from U.S. Environmental Protection Agency, 1981. P. 296; Fig. 10.11: (a) Herrmann and Johnson, "Acid Rain: A Water Resources Issue for the '80's," *Proceedings of the Water Resources Association International Symposium on Hydrometeorology*, 1983; (b) Brookhaven National Laboratory. Reprinted with permission from *Technology Review*, copyright 1982. P. 300; Fig. 10.13: Council on Environmental Quality (1983); from G. M. Woodwell et al., "The Carbon Dioxide Question." Copyright © 1978 by Scientific American, Inc. All rights reserved. Data for 1973–1985 from Houghton (1987).

Tables
P. 278; Table 10.4: From *Chemical Contamination in the Human Environment* by Morton Lippmann and Richard B. Schlesinger. Copyright © 1979 by Oxford University Press, Inc. Reprinted by permission. P. 284; Table 10.5: Mausner and Bahn (1974). P. 286; Table 10.6: From *Chemical Contamination in the Human Environment* by Morton Lippmann and Richard B. Schlesinger. Copyright © 1979 by Oxford University Press, Inc. Adapted by permission. P. 298; Table 10.8: From *Chemical Contamination in the Human Environment* by Morton Lippmann and Richard B. Schlesinger. Copyright © 1979 by Oxford University Press, Inc. Reprinted by permission.

Chapter 11

Photographs
P. 315: Courtesy of Carnegie Library, Pittsburgh. P. 317: Dave Baird/Tom Stack and Associates. P. 333: Courtesy of Bethlehem Steel.

Figures
P. 313; Fig. 11.2: U.S. Department of Commerce, Environmental Protection Agency, 230/3-79-001, p. 44. P. 324; Fig. 11.6: U.S. Environmental Protection Agency (1988a). P. 328; Fig. 11.12: Data from U.S. Environmental Protection Agency and *MMVA Motor Vehicle Facts and Figures '87*.

Tables
P. 314; Table 11.1: Barrett and Waddell (1973), p. 9. P. 315; Table 11.2: Barrett and Waddell (1973), p. 35. P. 320; Table 11.3: Annual reports of the Council on Environmental Quality and various publications of the U.S. Environmental Protection Agency. P. 325; Table 11.4: Council on Environmental Quality, Tenth Annual Report.

Chapter 12

Photographs
P. 346: Courtesy of the Bureau of Sport Fisheries and Wildlife. P. 350: U.S. Department of Agriculture. P. 353: U.S. Department of Energy. P. 354: Courtesy of the Bureau of Sport Fisheries and Wildlife.

Figures
P. 345; Fig. 12.6: R. H. Wagner, *Man and the Environment*, W. W. Norton & Company, Inc., 1971. P. 353; Fig. 12.8: Reproduced from "Petroleum in the Marine Environment," 1975, with permission of the National Academy of Sciences, Washington, D.C. P. 374; Fig. 12.16: Freeman (1979).

Tables
P. 341; Table 12.1: Lund, *Ambio*, Vol. 7, No. 2, 1978. P. 346; Table 12.3: Federal Water Pollution Control Administration, *Report of the Committee on Water Quality Criteria*, 1968. P. 350; Table 12.4: U.S. Department of Agriculture, *Wastes in Relation to Agriculture and Forestry*, Miscellaneous Pub-

lication No. 1065, March 1968. P. 351; Table 12.5: "Acid Mine Water: Its Control Reduces Stream Pollution," *Mechanization*, Parts I and II, Vol. 15. Data from natural waterways from M. Lippmann and R. B. Schlesinger, *Chemical Contamination in the Human Environment*, Oxford University Press, New York, 1979. P. 365; Table 12.6: Schlesinger (1979), pp. 97 and 99. P. 368; Table 12.7: Adapted from American Chemical Society, *Cleaning Our Environment: The Chemical Basis for Action*, 1969.

Mini-Glossary
P. 364: U.S. Environmental Protection Agency, *A Primer on Waste Water Treatment*, 1971.

Chapter 13

Photographs
P. 387: U.S. Department of Agriculture, Soil Conservation Service. P. 399: Grant Heilman Photography. P. 407: Eric Kroll/Taurus Photos, Inc.

Figures
P. 389; Fig. 13.4: U.S. Soil Conservation Service, *America's Soil and Water: Conditions and Trends*, December 1980. P. 390; Fig. 13.5: From "A First Look at the 1987 National Resources Inventory" by Tommy A. George and Jody Choate *Journal of Soil and Water Conservation* Nov/Dec 1989. Used with permission. P. 392; Fig. 13.6: Adapted from P. Barlow, *Land Resource Economics: The Political Economy of Rural and Urban Land Use*, © 1958, p. 14. Adapted by permission of Prentice-Hall, Inc., Englewood Cliffs, N.J. P. 406; Fig. 13.11: T. Lash, "Radioactive Waste," *Amicus*, Fall 1979, pp. 24–34.

Tables
P. 388; Table 13.1: Pimentel et al. (1987), p. 278. Copyright 1987 by The American Institute of Biological Sciences. P. 390; Table 13.2: U.S. Department of Agriculture and Council on Environmental Quality, *National Agricultural Land Study*, 1981. P. 402; Table 13.4: Martha K. Ritter, *National Resources Defense Council Newsletter*, March/April 1979, p. 25.

Chapter 14

Photographs
P. 416: Grant Heilman Photography. P. 422: (left) Barbara Rios/Photo Researchers; (right) Jen and Des Bartlett/Photo Researchers. P. 428: Robert Harbison. P. 433: Lola B. Graham/Photo Researchers.

Figures
P. 412; Fig. 14.1: Illustration by Patricia J. Wynne from "Threats to Biodiversity," by Edward O. Wilson, Sept. 1989, *Scientific American*. Copyright © 1989 Scientific American all rights reserved. Used with permission. P. 414; Fig. 14.2: From "Beyond Making Trades and Buying Time: Wetlands Conservation in the 1990s" by John Hodges-Copple Southern Growth Policies Board, 1989 p. 4. Used with permission.

Tables
P. 411; Table 14.1: Wolf (1987). Based on Edward O. Wilson, Museum of Comparative Zoology, Harvard University, Cambridge, Mass., private communications, 1987; Peter H. Raven, "The Significance of Biological Diversity" (unpublished), Missouri Botanical Garden, St. Louis, Mo., 1987; insect figures from Terry Erwin, National Museum of Natural History, Smithsonian Institution, Washington, D.C., private communication, 1987. P. 413; Table 14.2: Edward C. Wolf, *On the Brink of Extinction*, Worldwatch Institute, 1987;

based on William D. Newmark, "A Land-Bridge Island Perspective on Mammalian Extinctions in Western North American Parks," *Nature*, Jan. 29, 1987. P. 415; Table 14.3: Council on Environmental Quality (1987), p. 280; modified from John A. Kutler, *Our National Wetland Heritage—Wetlands: Their Use and Regulation*, Office of Technology Assessment, OTA-0-206, March 1984. P. 418; Table 14.4: From "Plight of the Other Rain Forest," by John C. Ryan *Worldwatch* 1989, Washington, D.C. p. 41. Used with permission.

Chapter 15

Photographs
P. 440: Eric Kroll/Taurus Photos, Inc. P. 466: Tom McHugh/Photo Researchers.

Figures
P. 440; Fig. 15.2: Adapted from Mollner (1975). P. 444; Fig. 15.4: Adapted from Moller (1975).

Chapter 16

Photographs
P. 452: Grant Heilman Photography. P. 453: U.S. Department of Agriculture.

Figures
P. 451; Fig. 16.1: Data from Cottam, "The Ecologist's Role in Problems of Pesticide Pollution," *BioScience*, Vol. 15, 1965, pp. 457–463, and Rudd, *Pesticides and the Living Landscape*, University of Wisconsin Press. P. 453; Fig. 16.2: U.S. Department of Health, Education and Welfare (1969). P. 457; Fig. 16.3: Grier (1982). Copyright 1982 by the AAAS. P. 528; Fig. 16.5: Westöö and Noren, *Ambio*, Vol. 7, No. 2, 1978. P. 460; Fig. 16.4: National Academy of Sciences as found in Conservation Foundation, *State of the Environment: An Assessment of Mid-Decade*, 1984. Reproduced with permission from *Toxicity Testing: Strategies to Determine Needs and Priorities* (1984), National Academy Press, Washington, D.C. P. 463; Fig. 16.5: *EPA Journal*, July/August 1979, Office of Public Affairs, Washington, D.C. P. 464; Fig. 16.6: P. L. Adkisson et al., "Controlling Cotton's Insect Pests: A New System," *Science*, Vol. 216, No. 4541, 1982, pp. 19–22. Copyright 1982 by the AAAS.

Tables
P. 456; Table 16.2: Courtesy: Scientists Institute for Public Information, New York, NY. Reprinted from *Pesticides*, 1970. P. 458; Table 16.3: A. D. Campbell et al., "Food Additives and Contaminants," pp. 167–169 in *Handbook of Physiology*, Section 9: *Reaction to Environmental Agents*, D. H. K. Lee, H. L. Falk, and S. D. Murphy (eds.), American Physiological Society, Bethesda, Md., 1977.

Chapter 17

Photographs
P. 474: Robert Harbison. P. 476: Dave Baird/Tom Stack and Associates. P. 487: Courtesy of Camp Dresser and McKee, Inc., Boston, MA. P. 494: Eric Kroll/Taurus Photos, Inc.

Figures
P. 477; Fig. 17.4: Lahey and Connor (1983). Reprinted with permission from *Technology Review*, copyright 1983. P. 486; Fig. 17.9: U.S. Environmental Protection Agency, *Review of Activities of Major Firms in the Commercial Hazardous Waste Management Industry*; 1981 update, May 7, 1982. P. 488; Fig. 17.10: U.S. Environmental Protection Agency, *Minimization of Hazardous Waste*, October 1986.

P. 493; Fig. 17.11: U.S. General Accounting Office (1977). P. 493; Fig. 17.12: From "Discarding the Throwaway Society" by John E. Young *Worldwatch* Washington, D.C. 1991, p. 24. Used with permission.

Tables
P. 477; Table 17.2: U.S. General Accounting Office (1977). P. 479; Table 17.4: Douglas R. Nichols, Jr., "Are We Ready to Convert Solid Waste to Energy—Profitably?" *Resource Recovery and Energy Review*, Vol. 4, No. 4, Fall 1977, p. 9. P. 485; Table 17.5: From "Recycling Study Guide" by Wisconsin Department for Natural Resources, Madison, WI, 1991, p. 25. Used with permission. P. 486; Table 17.6: U.S. Environmental Protection Agency, Office of Water and Waste Management, 1980. P. 487; Table 17.7: U.S. Office of Technology Assessment. P. 489; Table 17.8: U.S. Environmental Protection Agency, *Minimization of Hazardous Waste*, October 1987. P. 490; Table 17.9: U.S. Environmental Protection Agency, Office of Solid Waste and Emergency Response, *Using Compensation and Incentives When Siting Hazardous Waste Management Facilities*, July 1982.

Chapter 18

Photographs
P. 504: Taurus Photos, Inc. P. 513: Taurus Photos, Inc. P. 517: Grant Heilman Photography. P. 521: R. P. Kingston/ Stock Boston. P. 534: A. Smith. P. 536: P. N. Joseph/Massachusetts Bay Transportation Authority. P. 559: U.S. Environmental Protection Agency. P. 561: Courtesy of U.S. Forest Service.

Figures
P. 502; Fig. 18.1: From *Cancer Facts and Figures*, Courtesy of American Cancer Society; data from National Center for Health Statistics and Bureau of the Census. P. 503; Fig. 18.2: Adapted from R. Suss et al., *Cancer–Experiments and Concepts*, Springer Verlag, New York, 1973. P. 508; Fig. 18.4: Courtesy American Cancer Society, *Cancer News*, 1983. P. 509; Fig. 18.5 and P. 510; Fig. 18.6: Pickle et al (1987). Pp. 511 and 514; Fig. 18.7 and Fig. 18.8: From J. Cairns, "The Cancer Problem," Copyright © 1975 by Scientific American, Inc. All rights reserved. Pp. 519 and 520;

Fig. 18.11 and Fig. 18.12: From R. Suss et al., *Cancer–Experiments and Concepts*, Springer Verlag, New York, 1973. P. 520; Fig. 18.13: Lyon et al. (1976). Reprinted by permission of the *New England Journal of Medicine*, Vol. 294, No. 3, p. 131.

Chapter 19

Figures
P. 535; Fig. 19.5: After Greenwood and Edwards (1973).

Tables
P. 531; Table 19.2: L. R. Brown and J. L. Jacobson, *The Future of Urbanization*, Worldwatch Institute, © 1987; 1986 data from Population Reference Bureau, *1986 World Population Data Sheet*; 1950 and 2000 data from Carl Haub, Population Reference Bureau, personal communication, Aug. 28, 1986.

Chapter 20

Enrichment Box
P. 553; Box 20.1: Reprinted from STATE OF THE ENVIRONMENT: AN ASSESSMENT AT MID-DECADE, with permission of The Conservation Foundation, Washington.

Insert One

Savannah: Todd Hoffman; deciduous forest: Charles E. Kupchella; rain forest: Robert Harbison; prairie: Brian Parker/ Tom Stack and Associates; desert in bloom: Robert Harbison; Sonoran desert: John Shaw/Tom Stack and Associates; tundra: Andrew Stepniewski; coniferous forest: Charles E. Kupchella; Mount Rainier: Robert Harbison; salt marsh: Brian Parker/Tom Stack and Associates.

Insert Two

Opening page: NASA, Grant Heilman. Spread, left page, top: Image Workshop, Charles E. Zirkle; center: Grant Heilman Photography; bottom: Tom Stack/Tom Stack and Associates. Spread, right page, top: Peter Menzel/Stock Boston; bottom: Stewart M. Green/Tom Stack and Associates. Closing page: National Aeronautics and Space Administration.

Index

Note: Boldface indicates page on which definition is given;
f indicates coverage of item in a table or a figure.